Engineering Sciences and Fundamentals 2013

AA003023

Core Programming Area at the 2013 AIChE Annual Meeting:
Global Challenges for Engineering a Sustainable Future

San Francisco, California, USA
3 – 8 November 2013

Volume 1 of 2

ISBN: 978-1-63439-039-2

Printed from e-media with permission by:

Curran Associates, Inc.
57 Morehouse Lane
Red Hook, NY 12571

Printed by Curran Associates, Inc. (2014)

AIChE
3 Park Avenue
New York, NY 10016-5991

Phone: (203) 702-7660
Fax: (203) 775-5177

www.aiche.org

Additional copies of this publication are available from:

Curran Associates, Inc.
57 Morehouse Lane
Red Hook, NY 12571 USA
Phone: 845-758-0400
Fax: 845-758-2634
Email: curran@proceedings.com
Web: www.proceedings.com

TABLE OF CONTENTS

VOLUME 1

VOLUME 2

(49a) Identifying Optimal Active Sites for Heterogeneous Catalysis By Metal Alloys Based On Molecular Descriptors and Electronic Structure Engineering

Xin, H., University of Michigan
Holewinski, A., University of Michigan
Nikolla, E., Wayne State University
Linic, S., University of Michigan
Originally presented on: 11/4/2013 8:30:00 - 8:50:00

While one metal might be chemically inert for a particular catalytic transformation, its first neighbor in the periodic table might be overly reactive and equally inefficient. The main reason of this is that neighboring metals in the periodic table exhibit very large differences in binding energies of various adsorbates (as large as ~1 eV (~100 kJ/mol)). Simply put, the periodic table does not provide enough flexibility in identifying optimal catalysts. One avenue to address this problem is through fine-tuning the chemical reactivity of surface atoms in metals by changing their local coordination, for example by creating metal alloys. The phase space of alloys is immense, and it is almost impossible to screen through even a small fraction of potential alloys using experimental studies or even quantum-chemical calculations on model systems.

In this talk we will discuss our work on developing predictive models of chemical reactivity for alloys based on detailed understanding of the molecular transformations that govern catalytic processes. More specifically, we analyzed how the change in the local coordination of metal surface atoms due to alloying changes the local chemical reactivity of the metal site. The central finding of our studies is that it is possible to reliably predict the change in the local electronic structure of an active site induced by the formation of an alloy, and the change in the local chemical reactivity, based on very simple analytical models which employ only easily accessible physical characteristics of the elements that form the alloy (mainly their electronegativity, distance between neighboring atoms, and the spatial extent of metal orbitals). We show how these predictive structure-reactivity relationships can be employed to rapidly screen through large libraries of alloy compositions and structures to identify optimal alloy catalysts. We describe our findings by focusing on an example of the design of optimal Pt alloy electrocatalysts for the oxygen reduction reaction (ORR) in fuel cell cathodes. The model permits rapid screening through an enormous phase space of alloy structures and compositions using analytical expressions instead of expensive quantum-chemical calculations. Since the model is grounded on validated theories of chemisorption on metal surfaces, it can be used for the identification and targeted manipulation of the critical electronic structure descriptors of catalytic activity and propose how these features can be obtained based on electronic structure engineering.

1. **H. Xin**, N. Schweitzer, E. Nikolla, S. Linic, J. Chem. Phys., 132, 111101, (2010)
2. N. Schweitzer, **H. Xin**, E. Nikolla, J.T. Miller, S. Linic, Top Catal, 53, 348, (2010)
3. **H. Xin**, S. Linic, J. Chem. Phys., 132, 221101, (2010)
4. **H. Xin**, A. Holewinski, S. Linic, ACS Catal., 2, 12, (2012)
5. **H. Xin**, A. Holewinski, N. Schweitzer, E. Nikolla, S. Linic, Top Catal, 55, 376, (2012)
6. A. Holewinski, **H. Xin**, E. Nikolla and S. Linic, Curr Opin Chem Eng., accepted (2013)

(49b) A Computational Evaluation of Steady-State Bifunctional Catalysis

Doan, H. A.,
Grabow, L. C., University of Houston
Originally presented on: 11/4/2013 8:50:00 - 9:10:00

Significant improvements in computer hardware, increasingly efficient and accurate density functional theory (DFT) codes, and the use of scaling and Brønsted-Evans-Polanyi relations in heterogeneous catalysis, have enabled us to predict optimal materials for various catalytic processes computationally. However, as the best catalysts are not always practical due to their high cost, it is essential to continue the discovery new and affordable materials with high catalytic activity. In the literature, several examples of high performance catalysts with multiple site-specific functionalities at steady-state have been reported [1-4]. The most common systems consist of two distinct sites that catalyze different reaction steps independently and are often referred to as "bifunctional catalysts". Using DFT and microkinetic modeling, we aim to understand the mechanisms that are responsible for bifunctional activity and provide a computational framework for screening of bifunctional catalysts. Our results indicate that there are theoretical limits for the achievable activity improvement and bifunctional catalysts do not necessarily outperform single-site catalysts. More specifically, for CO oxidation on bimetallic systems we found that the overall activity is not significantly altered when bifunctional catalysts are considered, but equally active bifunctional catalysts may be tailored from less active and cheaper components.

1. Kotobuki M, Watanabe A, Uchida H, et al. (2005) Journal of Catalysis 236:262–269.
2. Thinon O, Diehl F, Avenier P, Schuurman Y (2008) Catalysis Today 137:29–35.
3. Neurock M (2010) Industrial & Engineering Chemistry Research 49:10183–10199.
4. Kyriakou G, Boucher MB, Jewell a. D, et al. (2012) Science 335:1209–1212.

(49c) Decoupled Scaling Relationships of Atomic and Molecular Adsorption Energies On Transition-Metal Carbide Catalysts

Michalsky, R., Brown University
Zhang, Y.,
Medford, A., Stanford University
Peterson, A. A., Stanford University
Originally presented on: 11/4/2013 9:10:00 - 9:30:00

Currently in the United States, natural gas is the primary feedstock for the production of many major chemical commodities such hydrogen and ethylene. The sustainable synthesis of these chemical energy carriers via the catalytic reduction of abundant resources such as carbon dioxide and water with renewable energy and without fossil fuel requires the development of efficient and inexpensive catalysts for large-scale production. The design of transition metal or metal alloy catalysts is often limited by the scaling of adsorption energies of reactants and reaction products. This work uses electronic-structure calculations to present how transition-metal carbide surfaces can decouple the scaling of hydrogen, carbon and oxygen adsorption energies - thereby increasing the degrees of freedom for the rational design of novel catalytic materials. Throughout the 12 studied transition metal carbide surfaces, carbides show an oxophilic departure from the scaling relations of adsorption energies on transition metal surfaces and a carbophobic shift of carbon adsorption energies by 0.30 to 1.8 eV relative to the (211) surface of the parent metal. Given a highly hybridized and delocalized electronic structure, the metal-projected d-band centers correlate with the reported adsorption energies. However, the preference of carbides for oxygen adsorption, site preferences of adsorbates, and the tendency of a surface to form C-C bonds between the lattice carbon and adsorbed carbon may be understood due to shifts in the sp-band center and an increasing degree of electron localization with decreasing electronegativity of the metal. Correlation of atomic adsorption energies with relevant molecular adsorbates such as CO, HCO, COOH or HCO2 as well as the activating or passivating influence of carbon vacancies or oxygen adatoms on adsorption energies respectively will be demonstrated.

(49d) Elucidating Adsorption Mechanisms On Oxides By Comparing Adsorption On FCC Metals and Reduced B1 Monoxide Surfaces

Xu, Z., Carnegie Mellon University
Kitchin, J. R., Carnegie Mellon University
Originally presented on: 11/4/2013 9:30:00 - 9:50:00

In the past decade, we have been able to use density functional theory to connect adsorption on metal surfaces with their electronic structures. This has allowed us to use calculated electronic structures to both explain adsorption phenomenon and create predictive models to accelerate materials design. However, our understanding of oxides pales in comparison. This is in large part due to both their diverse atomic and complex electronic structures. In an effort to understand the relationship between oxides' electronic structure and their adsorption properties, we compare side by side the bulk thermodynamics and surface adsorption properties of 3d transition metals in fcc structure with their corresponding B1 transition metal monoxides. Understanding that the B1 monoxide structure constitutes two interpenetrating face centered cubic lattices of cations (metals) and anions (oxygen), we construct a thermodynamic path from fcc metals to B1 monoxides through the expansion of the metal lattice and the insertion of an oxygen fcc lattice. We then studied how these structural transformations affected both the adsorption energies and electronic structure. We found that, in general, the expansion of the lattice corresponds to a weakening of the bond, which maximizes for metals with half filled d-orbitals. Interestingly, the introduction of oxygen into the expanded lattice strengthens the bond, which also maximizes for metals with half filled d-orbitals. The net effect is that the trends for the adsorption on the (111) surface of fcc metals mirrors that of their corresponding (111) B1 monoxide surfaces. We also concluded that irrespective of the material, the key descriptor of the oxygen adsorption energy is the center of the adsorbate oxygen p-band, which we then used to interpret the relationship between adsorption energies and bulk electronic structures. In the case of the metal, the oxygen p-band center is inherently tied to the d-band center, but for the oxide, the oxygen p-band center is tied to the properties of the d-band bonding, non-bonding, and bonding orbitals that are created when oxygen is inserted into the expanded metal lattice. We discuss possible structural and electronic structure descriptors of reactivities on oxide surfaces.

(49e) Reactivity of Acetaldehyde On CeO2(111) Surfaces and the Roles of Oxygen Vacancies

Xu, Y., Louisiana State University

Originally presented on: 11/4/2013 9:50:00 - 10:10:00

Ceria is a widely used catalytic and functional catalyst support material, well known for its ability to store oxygen and change oxidation state. There is a growing body of evidence that the surface reactivity of ceria toward organic molecules can vary significantly with the extent of reduction. In this study we use acetaldehyde as a probe molecule. First-principles density functional theory (DFT) calculations have been performed in close collaboration with surface characterization techniques to elucidate the roles of oxygen vacancies in redox reactions on the CeO_2(111) surface. A microkinetic model based on proposed surface intermediates and their calculated energetics has been constructed and yields close agreement with experimental temperature-program desorption results. Acetaldehyde desorbs without reaction from the stoichiometric CeO_2(111) surface near 200 K. When the surface is partially reduced, acetaldehyde loses its carbonyl bond character and forms C-O coupled dimers at very low temperatures. Increasing temperature to 360 K leads to the decomposition of this dimer species into acetaldehyde, which desorbs immediately, and another surface species, which has been conclusively identified to be the enolate form of acetaldehyde (CH_2CHO) and has not been captured on ceria surfaces previously. This enolate species can recombinatively desorb, again as acetaldehyde, at 550 K. Our findings demonstrate that surface oxygen vacancies are key to activating acetaldehyde and stabilizing it for further reactions, and that the dominant surface reaction pathway can change as a function of vacancy concentration. This work has relevance to the conversion of biomass-derived oxygenates because enolate species are key intermediates in C-C coupling reactions including aldol condensation.

(49f) Dopant Selection Guided By Sabatier's Principle: The Example of Doped Ceria for Promoting the Two-Step Water Dissociation

Botu, V., University of Connecticut
Mhadeshwar, A. B., University of Connecticut
Ramprasad, R., University of Connecticut
Originally presented on: 11/4/2013 10:10:00 - 10:30:00

Use of *ab-initio* methodologies, based on Density Functional Theory, in describing surface science phenomenon has allowed for an increasing predictive ability of materials. In the case of H_2 synthesis, cerium based oxides have garnered significant interest given their oxygen buffering capability that facilitates the creation of active surface defect sites (vacancies). Tailoring the activity and selectivity of these materials towards low temperatures is paramount to improving reactivity. We started investigating the fundamentals governing the surface phases of cerium (IV) oxide under a range of environments. In the case of a pure ceria system in an O_2 environment, it was found that T > 1500 K or $P_{O2} < 10^{-13}$ atm are required for any appreciable surface reduction. However, exposing the surface to a highly reducing environment such as CO or H_2 could help circumvent these harsh conditions. Given the aforementioned observations we sought to alter surface reducibility by doping with various alkali and transition metal elements. Doping ceria had an effect on the surface vacancy formation; low valence dopants favor defect formation, whilst high valence dopants suppress it. Armed with the insight dopants have on surface vacancy formation, their corresponding impact on the consecutive dissociation of water at these sites was studied. Using a carefully strategized multi-step screening approach abiding with the Sabatier's Principle, we were able to identify a subset of dopants that would promote surface reactivity towards H_2 synthesis. Dopants such as Sc, Au, Co, Pd, La, Y, Hg, Mn, Zr, Cr, and Fe possessed a critical balance in improving surface reducibility while sustaining the dissociation of water, hence are promising candidates in enhancing the two-step water dissociation process.

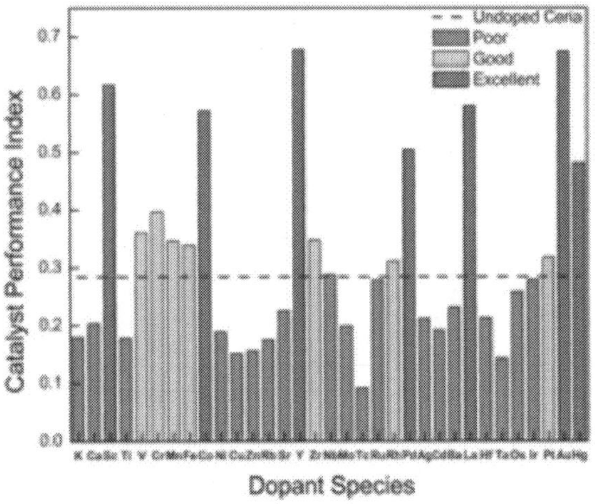

Figure 1: Measure of the Catalyst Performance Index for the class of dopants considered within ceria

(49g) Fundamental Insight Into the Unique Reactivity of NO Oxidation On the Mn-Ce Mixed Oxide

Wang, X., Tianjin University
Wang, J.,
Shen, M.,
Mei, D., Institute for Integrated Catalysis
Originally presented on: 11/4/2013 10:30:00 - 10:50:00

NO oxidation is the key step for NOx trapping, soot combustion under NOx/O_2, and NH_3-SCR. Recently, MnOx-CeO_2 catalyst has been found to be a promising catalyst to substitute the precious metals (e.g., Pt) for NO oxidation for its high activity. To understand the synergistic effect of MnO_2/CeO_2 catalyst, NO oxidation over CeO_2, β-MnO_2 and MnO_2/CeO_2 catalysts was systematically investigated using DFT+U method. For the perfect CeO_2 (111), we found that the NO adsorption energy is strong (-1.14 eV), which suggests that the extraction of lattice oxygen from CeO_2 (111) to form NO_2 is preferred. However, NO_2 desorption from the CeO_2 (111) is very difficult, which needs 2.51 eV. On the other hand, the adsorption of NO over the β-MnO_2 (110) surface is also strongly exothermic with an adsorption energy of -2.50 eV. Compared with CeO_2 (111), the NO_2 desorption on the β-MnO_2 (110) surface is much lower (0.48 eV). This is consistent with the experimental observations that MnOx sample exerts obviously higher NO oxidation activity than CeO_2. For MnO_2/CeO_2, the interaction between MnO_2 and CeO_2 (111) surface forms the interface Mn—O—Ce structure. The NO adsorption energy increases as the Ce site (-1.10 eV) ≈ the Mn site (-1.00 eV) < the interface site (-0.47 eV). While the NO_2 desorption energy increases as the interface site (0.48 eV) < the Mn site (1.02 eV). This implies that the NO_2 desorption on MnO_2/CeO_2 has been significantly facilitated by the interface sites, which only requires the lowest NO_2 desorption energy. To complete the catalytic cycle, the oxygen vacancies generated by the NO_2 desorption have to be replenished. Our calculations show that the adsorption energy of single oxygen atom over MnO_{2-x} is -0.33 eV while it is -1.88 eV at the interface site over MnO_{2-x}/CeO_2 catalyst. This indicates that the MnO_{2-x}/CeO_2 catalyst regenerates the active lattice oxygen much easier than the MnO_{2-x}. Our theoretical modeling results elucidate the experimental observation well.

(50a) Crosslinked Anion Exchange Membranes for Redox Flow Batteries

Jung, M. S., Illinois Institute of Technology
Parrondo, J., Illinois Institute of Technology
Arges, C., Illinois Institute of Technology
Ramani, V., Illinois Institute of Technology
Originally presented on: 11/4/2013 8:30:00 - 8:55:00

All-vanadium redox flow batteries (VRFB) have been very popular as they use the same electrochemical species in both electrode chambers and therefore they do not suffer from electrolyte cross contamination. A core component of any RFB is the electrolyte membrane (PEM or AEM) used to separate the positive and negative electrolytes. The key requirements for the separator are good chemical and electrochemical stability, good mechanical stability, high ionic conductivity, and low permeability for the vanadium ions. Nafion® membranes have been the most popular separators in RFBs because of their excellent oxidative stability and high proton conductivity, however, they exhibit high vanadium crossover. Anion exchange membranes (AEMs) are ideal candidates as RFB separators as they have much lower vanadium ion crossover than proton exchange membranes (10-100 times less) due to Donnan exclusion. Moreover, the reduced penetration of VO_2^+ into the AEM minimizes chemical degradation.

We selected poly (2,6-dimethyl 1,4-phenylene) oxide (PPO) based AEMs as battery separators as they possess good conductivity, low water uptake and good mechanical stability. PPO is commercially available, relatively inexpensive, and can be readily modified by attaching bromine to the backbone to induce a phase-separated morphology. PPO can be easily functionalized into an AEM by bromination followed by reaction with a tertiary amine. In this talk, we will discuss the effect of polymer crosslinking on the conductivity, water uptake, vanadium (IV) permeability, ultimate tensile stress, glass transition temperature and storage modulus of PPO- based AEMs. Polymer crosslinking has been performed by the reaction of brominated PPO (Br-PPO) with tetramethylhexanediamine (TMHDA) in liquid phase. Non crosslinked AEMs have been prepared by reaction of the Br-PPO with quinuclidine (PPO-$ABCO^+$) and trimethylamine (PPO-TMA^+). Nuclear Magnetic Resonance techniques [^1H-NMR, ^{13}C-NMR, correlation spectroscopy (COSY), and ^1H-^{13}C heteronuclear multiple quantum correlation (HMQC) spectroscopy] have been used to investigate the polymer structure before and after exposure to VO_2^+ solution (over time) and to identify any possible degradation products and mechanisms. Crosslinked membranes will be tested (and compared against Nafion® and non crosslinked AEMs) in a vanadium redox flow battery to evaluate its performance and durability. Postmortem analysis of the membranes via NMR, conductivity/IEC measurements, and mechanical testing will be performed to unearth key degradation modes.

(50b) Challenges to Implementation of High-Rate Energy Storage in Military Applications

Heinzel, J., Naval Sea Systems Command - Philadelphia
Ostanek, J.,
Hoffman, D., Office of Naval Research
Peek, I., Naval Sea Systems Command - Philadelphia
Originally presented on: 11/4/2013 8:55:00 - 9:20:00

Future military platforms will require more extensive use of electronic power systems to achieve required performance levels. Energy storage is becoming a key enabler for power systems operating with high levels of transient power requirements and supporting fuel-efficiency operations while maintaining capability. High density energy storage which operates continuously at high charge and discharge rates to support these loads under the combination of finite space, high power and energy requirements, stringent safety characteristics and wide upper temperature ranges drive the requirement for advancement in a multiphysical mix of electrical, chemical, thermal, safety and controls elements, which must be considered and have development performed to enable widespread implementation. These will be discussed from the perspective of requirements and considerations for system design and engineering for applications associated with high rate support of transient loads, power quality and optimization of generator and electrical architecture dynamics.

(50c) Novel Pt/TixM1-xO2 Nanocatalysts for Fuel Cells

Hwang, B. J., National Taiwan University of Science and Technology
Ho, V. T. T.,
Pan, C. J., National Taiwan University of Science and Technology
Su, W. N.,
Originally presented on: 11/4/2013 9:20:00 - 9:45:00

This work presents a new approach by exploring robust non-carbon $Ti_xM_{1-x}O_2$ as a novel functionalized co-catalytic support for Pt. This new approach is based on the novel nanostructure of the $Ti_xM_{1-x}O_2$ supports which holds strong metal-support interactions (SMSI) between $Ti_xM_{1-x}O_2$ and Pt that can not only modify surface electronic structure of Pt, evidenced from a shift in the d-band centre of the surface Pt atoms but also provide dual reaction sites for methanol oxidation. Furthermore, another benefit of $Ti_xM_{1-x}O_2$ is the extremely high stability of Pt/ v during potential cycling, which is attributable to the SMSI between Pt and $Ti_xM_{1-x}O_2$. Interestingly, $Ti_xM_{1-x}O_2$ can be fabricated as a much thinner catalyst layer, resulting in significantly improved the mass transport kinetics and performance of the resulting membrane-electrode-assembly (MEA).The new approach presented in this work opens a reliable path to the discovery of advanced concepts that may lead to the design of new catalyst materials that can replace the traditional catalytic structures and motivate further research in this field.

(50d) A Tunable Hierarchical Porous Carbon Structure for Studying Electrochemical Energy Conversion

Chan, K. Y., The University of Hong Kong

Yang, C.,

Originally presented on: 11/4/2013 9:45:00 - 10:10:00

A Tunable Hierarchical Porous Carbon Structure for Studying Electrochemical Energy Conversion

Chunzhen YANG and Kwong-Yu CHAN*, Department of Chemistry, The University of Hong Kong, Pokfulam Road, Hong Kong

Abstract

A well-defined porous carbon structure with tunable parameters provides a platform for systematic study of transport and confinement effects on electrochemical conversion. Synthesized via silica template techniques, the hierarchical structure has a hollow core within a mesoporous shell (HCMS) with uniform pore size, shell thickness, hollow core diameter, as shown in Fig. 1.

By varying the synthesis parameters, the structural parameters can be tuned to study

1) combined effects of surface area, mesopore size, macropore size, and hierarchy of porosity on ionic transport and adsorption in electrochemical capacitance behavior;

2) how distribution of platinum or other electrocatalyst nanoparticles in the porous structure affect catalytic faradaic reactions;

3) how distribution of mixed metal nanoparticles such as PtRu and varying of Pt and Ru content in the porous structure affects faradaic reactions such as methanol oxidation; and

4) the effectiveness of a mathematic model in analyzing electrochemical conversion in porous structures.

Electrochemical capacitance is studied with a series with meso-shell thickness stepwise increased from 0, 25, 50 to 100 nm while keeping an identical 330 nm hollow core and mesopore of 3.9 nm.[1] A thicker shell has a higher surface area with a proportional increase of electrochemical capacitance which however, can only be fully realized at low scan rates/currents. At high currents, ionic transport limits the electrochemical capacitance of a thick mesoporous shell. Electrochemical impedance spectra (EIS) obtained on the family of HCMS carbon structures were fitted to equivalence circuits described by models of network structures.[2] The results reveal the need to match the AC frequency with the characteristic time constant of a structure for optimum active and reactive power. For fuel-cell electrode conversion, the distribution of Pt and PtRu nanoparticles in a mesoporous carbon can vary in different synthetic protocol. Uniform Pt particle distribution can be achieved with a CPDP method [3] as shown in Fig. 2. Different electrocatalysts with varying Pt distribution can be compared for activity and utilization in methanol oxidation. The particles dispersed into a thicker shell are more stable and give better performance over time, as shown in voltage cycling and extended methanol oxidation.

SHAPE * MERGEFORMAT

Fig. 1. A hollow core mesoporous shell (HCMS) carbon structure with three varying geometrical parameters: pore diameter (d_p), shell thickness (S), and diameter of core (C).

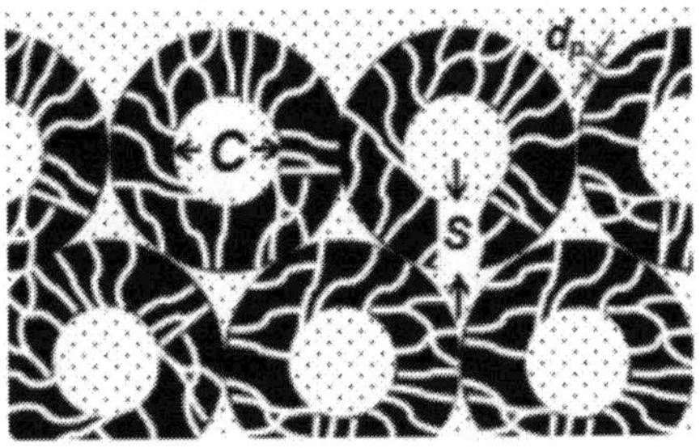

Fig. 1. A hollow core mesoporous shell (HCMS) carbon structure with three varying geometrical parameters: pore diameter (d_p), shell thickness (S), and diameter of core (C).

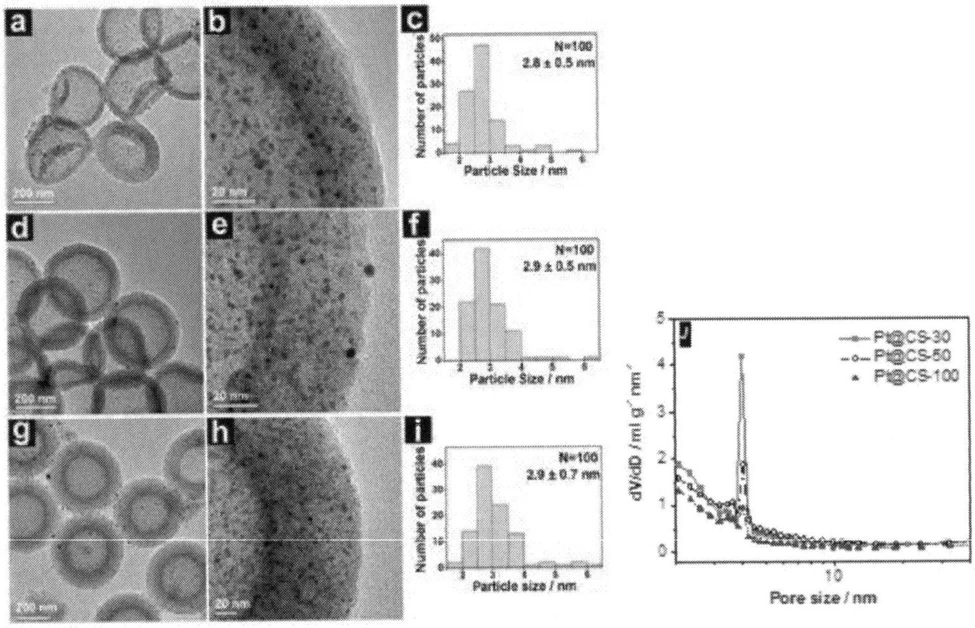

SHAPE * MERGEFORMAT

Fig. 2 TEM images of Pt nanoparticles loaded in mesoporous carbons of same hollow core diameter, $C = 330$ nm but different Shell thickness (a)-(c) Pt@CS-30, S=30 nm ; (d)-(f) Pt@CS-50, S=50 nm ; (g)-(i) Pt@CS-100, S=100 nm; (j) pore size distribution of the porous structures loaded with Pt.

Fig. 2 TEM images of Pt nanoparticles loaded in mesoporous carbons of same hollow core diameter, C=330 nm but different Shell thickness (a)-(c) Pt@CS-30, S=30 nm ; (d)-(f) Pt@CS-50, S=50 nm ; (g)-(i) Pt@CS-100, S=100 nm; (j) pore size distribution of the porous structures loaded with Pt.

References:

[1]. F. Li, M. Morris and K.Y. Chan*, ?Electrochemical Capacitance and Ionic Transport in the Mesoporous Shell of a Hierarchical Porous Core-Shell Carbon Structure?, *J. Mater. Chem.* 21 (2011) 8880-8886.

[2] Chunzhen Yang, Chi-Ying Vanessa Li, Fujun Li, Kwong-Yu Chan*, ?Complex Impedance with Transmission Line Model and Complex Capacitance Analysis of Ion Transport and Accumulation in Hierarchical Core-Shell Porous Carbons?, *J. Electrochem. Soc.* (2013) to appear.

[3] F. Li, K.Y. Chan*, and H. Yung, ?Carbonization over PFA-Protected Dispersed Platinum: An Effective Route to Synthesize High Performance Mesoporous-Carbon Supported Platinum?, *J. Mater. Chem.* 21 (2011) 12139-12144.

(50e) High-Performance Electrodes for Hydroxide Exchange Membrane Fuel Cells

Kaspar, R. B., University of Delaware
Letterio, M., University of Delaware
Jensen, K., University of Delaware
Gu, S., University of Delaware
Yan, Y., University of Delaware
Originally presented on: 11/4/2013 10:10:00 - 10:35:00

Hydroxide exchange membrane fuel cells (HEMFCs) are an attractive alternative to conventional proton exchange membrane fuel cells (PEMFCs) because they are compatible with non-precious-metal catalysts. However, HEMFC performance is lower than the PEMFC benchmark, in part because HEMFCs continue to borrow electrode architectures that were originally designed for PEMFCs. In PEMFCs, water is produced at the cathode, so hydrophobic gas diffusion layers (GDLs) are employed to control flooding. But hydrophobic GDLs may not be necessary for HEMFC cathodes, in which water is consumed instead of produced, reversing the direction of water transport in the cell. This work demonstrates that the performance and durability of membrane-electrode assemblies (MEAs) can be improved by tailoring electrode design to reflect the unique kinetics and water transport behavior of HEMFCs.

(50f) Capacitive Mixing in Porous Electrodes for Energy Production

Gabitto, J. F., PVAMU
Tsouris, C., Oak Ridge National Laboratory
Originally presented on: 11/4/2013 10:35:00 - 11:00:00

Energy obtained by salinity-gradients (blue energy) is a promising renewable energy source. Mixing of a high-concentration saline solution, such as seawater, with fresh water (river water) to produce a brackish solution dissipates more than 2.2 MJ of free energy per m^3 of fresh water. The charging-discharging cycle inside the pores of the solid electrodes is strongly influenced by the pore size distribution. Macroporous materials behave like standard plate capacitors, while mesoporous materials behave like cylindrical capacitors. The goal of this project was to optimize the recovery efficiency of free energy change by using tailored-design carbon electrodes. Morphology, pore size, and carbon electrode composition were chosen to optimize the charge-discharge steps of the reversible cycle. The physicochemical processes inside the electrode and the corresponding effects on the reversible energy cycle were investigated. A proposed energy cycle comprises a charging step, a discharging step and two switching steps that replace a saltwater solution inside the pores by a fresh water solution and vice versa. Equations derived using the volume averaging method have been employed to simulate the charging-discharging cycle in homogeneous porous materials of a macropore size distribution. Ionic transport by diffusion and electrical double layer expansion processes were considered to simulate the two alternating steps. Improved energy production was achieved by optimizing the duration of each step and variation of the operating variables.

(56a) Computational Approach of Thermodynamic Fragmentation Applied to Formulating Badly Soluble Actives

Fraaije, J. G. E. M., Culgi BV
Bulacu, M., Culgi
Gracia, R. S., Culgi
Nath, S., Culgi
Originally presented on: 11/4/2013 8:30:00 - 9:00:00

We address one of the major computational problems in the pharmaceutical and agrochemical business: the rational design of delivery vehicles for badly soluble actives. For example, in a typical scenario in a discovery/development cycle, one has only a few months, or even weeks, to find a proper formulation, driven by very costly time expenditure way down the pipeline. The chemical space for formulations is enormous, and even if one considers modern high-throughput experimentation, chances are the optimal formulation cannot be found by experimentation alone. For chemical informatics, such design also poses quite some challenges, since classical descriptor technology is almost always unitary (component specific), and therefore not suited for the non-linear binary interactions between active and delivery matrix. The more so, since in many cases one seeks or tests delivery systems that are based on some self-assembly structure, such as an emulsion, micelle or liposome. In contrast, in existing chemical informatics methods for the much simpler logP predictions one can rely on such unitary descriptors quite well, since the (iso-octanol/water) matrix in this case is known and constant over a known training set. Our strategy is completely different. We develop computational screening technologies that do take into account the structural resolution of both active and matrix, by employing a novel concept of thermodynamic fragmentation. Thereby the interactions of drug-matrix is described as a interaction between fragments, with the advantage that the fragmentation can be calibrated by (non-pharma) engineering databases. The presentation discusses in depth the motivation, the algorithm and some highlights from recent applications.

(56b) Computational Support of Solid Form Selection in Pharmaceutical Industry

Abramov, Y., Pfizer Global Research & Development
Originally presented on: 11/4/2013 9:00:00 - 9:20:00

In the pharmaceutical industry, drug polymorphism can be a critical problem, and is the subject of various regulatory considerations. One of the principal concerns is based on an effect that polymorphism may have on a drug's bioavailability due to change of its solubility and dissolution rate. Therefore, in Pharmaceutical industry, it is quite crucial to get comprehensive information on the available drug polymorphs and their relative stability and solubility.

State-of-the art computational approaches were proposed and developed to support the commercial solid state nomination. Rational solvent selection is adopted to guide polymorph screening experiments. In addition, computational methods enable an estimation of the likelihood of existence of an unknown, more stable form of the drug, as well as its potential impact on drug's solubility (and bioavailability). These computational methods involve crystal structure prediction, prediction of drug solubility and conformer distribution in organic solvents and solvent mixtures, *in silico* hydrogen bonding propensity estimation. A detailed application of these computational tools in the pharmaceutical industry is illustrated on a selective dual ALK and c-MET inhibitor crizotinib (trade name Xalkori) and on a VEGF inhibitor axitinib (trade name Inlyta).

(56c) Reaction Path and Coverage Effects Analyses of Benzene Hydrogenation On Pd(111)

Canduela, G., Gent University
Sabbe, M. K., Universiteit Gent
Joly, J. F., IFPEN
Reyniers, M. F., Ghent University
Marin, G. B., Ghent University
Originally presented on: 11/4/2013 9:20:00 - 9:40:00

1 Introduction

Benzene hydrogenation is a key reaction in various industrially important processes such as hydrotreatment of petroleum feedstocks, caprolactam and adipic acid production, and also plays an important role in environmental protective technologies. Palladium catalysts are often employed for hydrogenation reactions because of their high activity and selectivity. The (111) surface is normally chosen in fundamental studies to evaluate the catalytic performance in benzene hydrogenation [1-3] because of its abundance on supported industrial catalysts.

Micro-kinetic modeling allows to analyze catalytic surface reactions in detail, by incorporating the surface chemistry of all elementary reaction steps into a kinetic model [4]. The catalytic hydrogenation of benzene is a complex process. Even if only one single benzene adsorption site is considered, the hydrogenation of benzene can proceed via 13 different reaction paths. Performing an extended reaction path analysis based on the results of theoretical calculations allows to identify the most important reaction steps and surface intermediates. This may lead to simplification of the complex reaction network and can contribute optimizing the industrial catalytic process.

Periodic Density Functional Theory (DFT) calculations are usually performed at low to moderate coverage. However, these theoretical calculations can also be performed at higher surface coverages, enabling the modeling of coverage effects on the kinetics and thermodynamics of the adsorption and surface reaction steps. This feature is highly important for hydrogenation reactions, since in industry these are typically performed at high hydrogen pressures, and hence, at higher hydrogen coverages.

Three different techniques are employed to perform a reaction path analysis of the full reaction network of benzene hydrogenation from periodic DFT calculations at low coverage (θ_H=0.11). Additionally, the micro-kinetic model for benzene hydrogenation is also calculated at higher hydrogen coverages (θ_H = 0.44 and 0.67), in order to evaluate the coverage effects on the simulated activities by comparison with experimental data.

2 Methodology

The adsorption energies, geometries and vibrational frequencies at 0 K are calculated from Periodic DFT calculation using the Vienna Ab initio Simulation Package (VASP 4.6) [5]. This program solves the Kohn-Sham equations of DFT by using pseudopotentials with the projector augmented wave method (PAW) and periodic boundary conditions by expansions of the wave function in terms of plane-wave basis sets for which a converged cutoff energy of 400 eV has been used.

The *Nudged Elastic Band method* [6] and *Dimer* method [7] are used to locate transition states on the potential energy surface. In addition, the harmonic oscillator approach (HO) is used to numerically calculate the vibrational frequencies of all species. In combination with statistical thermodynamics, the frequencies are used to calculate the kinetic and thermodynamic parameters of every reaction, and therefore to construct the micro-kinetic model.

In the periodic DFT calculations, one hydrocarbon molecule and one hydrogen atom are adsorbed on a (3×3) Pd(111) unit cell to evaluate the kinetics at low coverage, leading to a fractional coverage of $\theta_H = 0.11$ and of $q_{HC} = 0.11$ for the hydrocarbon species. The effect of hydrogen coverage on the kinetics and thermodynamics is evaluated by increasing the number of hydrogen atoms to 4 and 6 atoms on the (3×3) unit cell, corresponding to hydrogen coverages of 0.44 and 0.67.

Three tools are implemented in an in-house FORTRAN code for the reaction path analysis: *Degree of Rate control* (DRC) [8, 9], *Sensitivity Analysis* (SA) and *Rate Of Production* (ROP). The equations for the two first are shown below. The rate of production analysis identifies the most important reactions and/or species within the reaction network by calculating the net production rates for all intermediate species occurring in the network

3 Results and discussion

The DFT results at 0 K for $\theta_H = 0.11$ indicate that the hydrogenation of benzene to cyclohexane follows the same minimum energy path for the two preferred *bridge30°* and *hollow-hcp0°* adsorption sites of benzene on Pd(111). These two networks are defined as *bridge-site* and *hollow-site* networks, respectively. The minimum energy path is the one that runs from benzene to monohydrobenzene (BH), proceeding over cyclohexa-1,3-diene (13CHD), trihydro-1,2,3-benzene (123THB), cyclohexene (CHE) and cyclohexyl (c-hexyl) to cyclohexane (CHA). For both adsorption sites, the calculated kinetics and thermodynamics confirm this path to be potentially dominant at 450 K and indicate that the first surface reaction step has the largest activation energy. Hence, the first hydrogen addition could potentially be the step with the largest control on the rate of cyclohexane formation along the dominant reaction path, as already previously suggested by Morin et al. based from BEP relationships [10]. Furthermore, simulations of the activity for both networks indicate that the *hollow-site* is more reactive than the *bridge-site*network.

For both *bridge-site* and *hollow-site* networks, ROP analysis at actual conditions and coverages (450 K, $P_{H2} = 0.1$ bar and $P_{benzene} = 0.02$ bar) agrees with the proposed dominant path from the calculated DFT rate coefficients, where the reactions along this dominant path contribute at least 90% to the net cyclohexane rate of formation. DRC and SA indicate that for the *bridge-site* network, the second hydrogen addition along the proposed dominant path is the kinetically most significant step, while for the *hollow-site*network it is the third hydrogen addition step. Clearly, inclusion of actual surface concentrations at reaction conditions is required to identify the step with the largest control along the dominant path, since the first step has the highest activation barrier.

Next, additional periodic DFT calculations were performed at higher hydrogen coverage ($\theta_H = 0.44$ and

0.67) for the *hollow-site* dominant path obtained at $\theta_H = 0.11$. For the 0.44 hydrogen coverage case, the calculated kinetics are much faster and lead to much higher simulated catalytic activities than for the $\theta_H = 0.11$ case. These activities are in the same order of magnitude as experimentally observed at similar reaction conditions, e.g. 0.064 $mmol_{Benzene}\ s^{-1}\ kg_{cat}^{-1}$ [11]. The reaction barriers calculated at a hydrogen coverage $\theta_H = 0.67$ are even lower than for $\theta_H = 0.44$. The very weak benzene adsorption calculated at $\theta_H = 0.67$ leads however to a lower activity than using the kinetics obtained at $\theta_H = 0.44$ coverage, since the simulated benzene coverages are too low. The activity using the kinetics obtained at $\theta_H = 0.67$ remains faster than using those at $\theta_H = 0.11$. 4 Conclusions

Benzene preferentially adsorbs at the *bridge30°* and *hollow-hcp0°* adsorption sites of Pd(111). At low hydrogen coverage, the dominant path of benzene to cyclohexane at these two sites proceeds through 1,3-cyclohexadiene and cyclohexene. The first hydrogen addition step is potentially the kinetically most significant step, due to its appreciable higher activation energy compared to the other elementary reaction steps of the dominant path. The reaction path analysis shows, however, that when surface concentrations are included for the simulation of activities, the second and third reaction steps along this dominant path are the ones controlling the rate of formation for the *bridge-site* and *hollow-site*network respectively. Therefore, it is required to include actual surface concentrations at reaction conditions to identify the step with the largest control on the net rate along the dominant path.

Inclusion of the effect of hydrogen coverage on the rate coefficients in the micro-kinetic model increases the rate of formation of cyclohexane, and the simulated activities using kinetics calculated at θ_H=0.44 are in the same order of magnitude as experimental observations.

Acknowledgements

The computational resources (Stevin Supercomputer Infrastructure) and services used in this work were provided by Ghent University, the Hercules Foundation and the Flemish Government – department EWI. Finally the authors thank IFPEN and the Fund for Scientific Research Flanders (FWO). References

[1] C. Morin, D. Simon, P. Sautet, J Phys Chem B, 108 (2004) 5653-5665.

[2] F. Mittendorfer, C. Thomazeau, P. Raybaud, H. Toulhoat, J Phys Chem B, 107 (2003) 12287-12295.

[3] M. Saeys, M.F. Reyniers, M. Neurock, G.B. Marin, J Phys Chem B, 109 (2005) 2064-2073.

[4] J.A. Dumesic, D.F. Rudd, L.M. Aparicio, J.E. Rekoske, A.A. Treviño, The Microkinetics of heterogeneous catalysis, American Chemical Society, Washington, D.C., 1993.

[5] G. Kresse, J. Furthmüller, Comput. Mat. Sci., 6 (1996) 15.

[6] G. Mills, H. Jonsson, G.K. Schenter, Surf Sci, 324 (1995) 305-337.

[7] G. Henkelman, H. Jonsson, Journal of Chemical Physics, 111 (1999) 7010-7022.

[8] C.T. Campbell, Journal of catalysis, 204 (2001) 520-524.

[9] C. Stegelmann, A. Andreasen, C.T. Campbell, J Am Chem Soc, 131 (2009) 8077-8082.

[10] C. Morin, D. Simon, P. Sautet, Surf Sci, 600 (2006) 1339.

[11] P. Chou, M.A. Vannice, Journal of catalysis, 107 (1987) 129-139.

(56d) Generalized Non-Linear QSPR Model for Liquid Viscosity

Yerramsetty, K., Oklahoma State University
Neely, B. J., Oklahoma State University
Gasem, K. A. M., Oklahoma State University
Originally presented on: 11/4/2013 9:40:00 - 10:00:00

Liquid viscosity (LV) is an important thermophysical property involved in many chemical processes and phenomena. Although accurate models have been developed for classes of chemicals, a need for a generalized model exists. Further, empirically developed models require experimental data, which are costly and laborious to obtain. Therefore, there exists a need for generalized models capable of providing *a priori* predictions for compounds that have no relevant experimental data or for compounds that are yet to be synthesized.

Previously, researchers at Oklahoma State University (OSU) have developed a unified framework for correlating saturation properties. This scaled-variable-reduced-coordinates (SVRC) model is based on the corresponding states theory (CST) and scaling-law behavior and in general is capable of representing saturation properties within their experimental uncertainties. The modeling approach involves the use of quantitative structure-property relationship (QSPR) methodology to generalize the model parameters of the developed SVRC model. Specifically, we used SVRC to develop the behavior model, and QSPR to generalize the SVRC model parameters. This approach, in the past, has proven to be more effective than the typical efforts to develop generalized models directly using QSPR techniques. In this work, we extended the SVRC model to correlate LV values and generalized the model parameters using structure-property modeling. A database of 5,800 data points involving 335 fluids was used in the development of this model. The predictions on the LV values were within 4% of the modeling set values. When applied to an external dataset containing 16,450 data points involving 1235 fluids, the LV model predictions were, on average, within 9% of the reported values.

(56e) New Insights On Mechanisms of Backbiting and β-Scission Reactions in Self-Initiated Polymerization of Methyl Acrylate

Liu, S., University of Pennsylvania
Srinivasan, S., Arkema
Grady, M., DuPont
Soroush, M., Drexel University
Rappe, A. M., University of Pennsylvania
Originally presented on: 11/4/2013 10:00:00 - 10:20:00

At high temperature, secondary reactions such as spontaneous thermal initiation, chain transfer reactions and β-scission reactions can significantly influence the polymerization process and the properties of the formed polymer [1-4]. The backbiting (intra-molecular chain transfer) and β-scission reactions have been studied extensively via various experimental techniques [5-8]. However, first-principles studies of these polymerization reactions have been hindered due to the large computational costs required for modeling polymeric systems of large size. Density functional theory (DFT) is generally regarded as a cost-effective method [9-14]. The accuracy of DFT depends on the approximation of the exchange-correlation function, and there is no obvious way to decide which density functional is the best for studying a given type of chemical reaction.

In this work, we firstly carry out a benchmark study using G4(MP2)-6X, a composite quantum chemistry method,[15] to predict the thermochemical properties of two representative 1:5 backbiting reactions. We then apply three different functionals (B3LYP, M06-2X and PBE0) and wavefunction-based method, MP2 to identify the mechanisms governing backbiting and β-scission reactions and the species generated by these reactions in high-temperature self-initiated homo-polymerization of methyl acrylate. Both hindered rotor approximations [16] and harmonic approximations are applied for the estimation of activation entropy. Four mechanisms of backbiting reactions and two types of β-scission reactions are studied. Our computational results show that the 1:5 backbiting mechanism with a six-membered ring transition state and 1:7 backbiting with an eight-membered ring transition state are energetically more favored than 1:3 backbiting and 1:9 backbiting. Two types of β-scission reactions have nearly equal activation energy and rate constants. The ^{13}C-NMR and ^{1}H-NMR spectrums are simulated and compared with experimental results in order to validate the proposed mechanisms.

References

[1] Grady, M. C.; Simonsick, W. J.; Hutchinson, R. A. *Macromol. Symp.* **2002**, *182*, 149.

[2] Quan, C. L.; Soroush, M.; Grady, M. C.; Hansen, J. E.; Simonsick, W. J. *Macromolecules* **2005**, *38*,

7619.

[3] Chiefari, J.; Jeffery, J.; Mayadunne, R. T. A.; Moad, G.; Rizzardo, E.; Thang, S. H. *Macromolecules* **1999**, *32*, 7700.

[4] Zorn, A. M.; Junkers, T.; Barner-Kowollik, C. *Macromol. Rapid Commun.* **2009**, *30*, 2028.

[5] Junkers, T.; Bennet, F.; Koo, S. P. S.; Barner-Kowollik, C. *Journal of Polymer Science Part a-Polymer Chemistry* **2008**, *46*, 3433.

[6] Yamada, B.; Oku, F.; Harada, T. *Journal of Polymer Science Part a-Polymer Chemistry* **2003**, *41*, 645.

[7] Asua, J. M.; Beuermann, S.; Buback, M.; Castignolles, P.; Charleux, B.; Gilbert, R. G.; Hutchinson, R. A.; Leiza, J. R.; Nikitin, A. N.; Vairon, J. P.; van Herk, A. M. *Macromol. Chem. Phys.* **2004**, *205*, 2151.

[8] Wang, W.; Hutchinson, R. A. *Chemical Engineering & Technology* **2010**, *33*, 1745.

[9] Degirmenci, I.; Aviyente, V.; Van Speybroeck, V.; Waroquier, M. *Macromolecules* **2009**, *42*, 3033.

[10] Van Speybroeck, V.; Van Cauter, K.; Coussens, B.; Waroquier, M. *ChemPhysChem* **2005**, *6*, 180.

[11] Furuncuoglu, T.; Ugur, I.; Degirmenci, I.; Aviyente, V. *Macromolecules* **2010**, *43*, 1823.

[12] Srinivasan, S.; Lee, M. W.; Grady, M. C.; Soroush, M.; Rappe, A. M. *J. Phys. Chem. A* **2010**, *114*, 7975.

[13] Liu, S.; Srinivasan, S.; Grady, M. C.; Soroush, M.; Rappe, A. M. *J. Phys. Chem. A* **2012**, *116*, 5337.

[14] Moghadam, N.; Liu, S.; Srinivasan, S.; Grady, M. C.; Soroush, M.; Rappe, A. M. *J. Phys. Chem. A* **2013**, *117*, 2605.

[15] B. Chan, J. Deng, L. Radom J. Chem. Theory Comput., 7, 112-120 (2011).

[16] Yu, X. R.; Pfaendtner, J.; Broadbelt, L. J. *J. Phys. Chem. A* **2008**, *112*, 6772.

(56f) A Theoretical Study of Mechanisms of Alkyl Acrylate Polymer Chain Transfer to Several Solvents

Moghadam, N., Drexel University
Liu, S., University of Pennsylvania
Srinivasan, S., Arkema
Grady, M. C., DuPont Experimental Station
Soroush, M., Drexel University
Rappe, A. M., University of Pennsylvania
Originally presented on: 11/4/2013 10:20:00 - 10:40:00

Acrylates are principal monomers in the production of coatings, adhesives, and polymers used in medical and also pharmaceutical applications, due to their transparency and resistance to breakage [1, 2]. The basic design of resins used in automobile coating has been changed because of the environmental limits on allowable volatile organic contents (VOCs) of resins [3, 4]. High-temperature (>100°C) polymerization of alkyl acrylates allows for reducing the solvent content of the resins, while keeping the resin viscosity within a desired range [5, 6]. However, at the high temperatures secondary reactions such as spontaneous initiation, backbiting, β-scission, and chain transfer to monomer, polymer and solvent reactions have stronger effects on the quality of the resins [7-10].

Previous studies [11-14] showed the successful use of computational quantum chemistry to better understand self-initiation, propagation, and co-initiation reactions in polymerization of alkyl acrylates. In our previous work [15, 16], we used different functionals (B3LYP, X3LYP, M06-2X) to study several mechanisms of chain transfer to monomer (CTM) reaction in thermal polymerization of methyl, ethyl, and *n*-butyl acrylate (MA, EA, *n*-BA). The kinetics of the most likely chain transfer to polymer (CTP) mechanisms was also studied for MA, EA, and *n*-BA [17]. To the best of our knowledge, no computational study of chain transfer to solvent (CTS) reactions in thermal polymerization of alkyl acrylates has been conducted as of yet.

This paper presents a computational study of CTS reaction mechanisms in self-initiated high-temperature polymerization of MA, EA, and *n*-BA in solvents such as butanol (polar, protic), methyl ethyl ketone (MEK) (polar, aprotic), and *p*-xylene (nonpolar). Three types of hybrid functionals (B3LYP, X3LYP, and M06-2X) and three different basis sets (6-31G(d,p), 6-311G(d), and 6-311G(d,p)) have been applied to predict energy barriers and molecular geometries of reactants, products, and transition states. The sensitivity of the levels of theory to barriers and rate constants has been evaluated. The effects of live polymer chain length and the type of mono-radical that initiated the live polymer chain on the kinetics of CTS reactions have also been investigated. Solvent effects have been explored for CTS reactions using polarizable continuum model (PCM). Among chain transfer to *n*-butanol, *sec*-butanol, and *tert*-butanol reactions, the *tert*-butanol reaction has the highest energy barrier and the lowest rate constant. We have identified that the type of mono-radicals generated via self-initiation had little or no effect on the ability of the MA, EA, and *n*-BA live polymer chains to undergo CTS reactions. The energy barriers do not change significantly with the length of the live polymer chains. Lower activation energy and higher rate constant of chain transfer to *n*-butanol reaction (than those estimated for chain transfer to MEK and *p*-xylene reactions) indicate the higher ability of *n*-butanol to facilitate chain transfer.

References

[1] Drin, A. P.; Efanova, V. V.; Shut, N. I. J. Eng. Phys. Thermophys. 1994, 66, 164-171.

[2] Okor, R. S. J. Controlled Release 1990, 12, 195-200.

[3] Schultz, K.R., in *Proc. Adv. Coat. Technol. Conf.* 1992: Chicago, IL.

[4] VOC's Directive, EU Committee of the American Chamber of Commerce in Belgium, ASBL/VZw, Brussels, July 8. 1996.

[5] Campbell, J. D., F. Teymour, and M. Morbidelli, High temperature free radical polymerization, 1. Investigation of continuous styrene polymerization, Macromolecules, 2003. 36(15): p.5491-5501.

[6] Wang, W. and R. A. Hutchinson, Recent advances in the study of high-temperature free radical acrylic solution copolymerization, Macromolecular Reaction Engineering, 2008. 2(3):p. 199-214.

[7] Grady, M. C.; Simonsick, W. J.; Hutchinson, R. A.; Studies of Higher Temperature Polymerization of *n*-Butyl Methacrylate and *n*-Butyl Acrylate, *Macromol. Symp.* 2002, 182, 149-168.

[8] Rantow, F. S.; Soroush, M.; Grady, M.; Kalfas, G.; Spontaneous Polymerization and Chain Microstructure Evolution in High Temperature Solution Polymerization of *n*-Butyl Acrylate, *Polymer* 2006, 47, 1423-1435.

[9] Nikitin, A. N.; Hutchinson, R. A.; Wang, W.; Kalfas, G. A.; Richards, J. R.; Bruni, C.; Effect of Intramolecular Transfer to Polymer on Stationary Free-Radical Polymerization of Alkyl Acrylates, 5 – Consideration of Solution Polymerization up to High Temperatures, *Macromol. React. Eng.* 2010, 4, 691-706.

[10] Quan, C.; Soroush, M.; Grady, M. C.; Hansen, J. E.; Simonsick, W. J.; High-Temperature Homopolymerization of Ethyl Acrylate and *n*-Butyl Acrylate: Polymer Characterization, *Macromolecules* 2005, 38, 7619-7628.

[11] Yu, X.; Pfaendtner, J.; Broadbelt, L.J.; Ab Initio Study of Acrylate Polymerization Reactions: Methyl Methacrylate and Methyl Acrylate Propagation, J. Phys. Chem. A. 2008, 112, 6772-82.

[12] Srinivasan, S.; Lee, M.W.; Grady, M.C.; Soroush, M.; Rappe, A.M.; Computational Study of the Self-Initiation Mechanism in Thermal Polymerization of Methyl Acrylate, J. Phys. Chem. A 2009, 113, 10787-94.

[13] Srinivasan, S.; Lee, M.W.; Grady, M.C.; Soroush, M.; Rappe, A.M.; Self-Initiation Mechanism in Spontaneous Thermal Polymerization of Ethyl and n-Butyl Acrylate: A Theoretical Study, J. Phys. Chem. A 2010, 114, 7975-83.

[14] Liu, S., S. Srinivasan, M.C. Grady, M. Soroush, and A.M. Rappe, Computational Study of Cyclohexanone-Monomer Co-initiation Mechanism in Thermal Homo-polymerization of Methyl Acrylate

and Methyl Methacrylate. *Journal of Physical Chemistry A*, 2012. 116(22): p. 5337-5348.

[15] Moghadam, N.; Soroush, M.; Liu, S.; Srinivasan, S.; Rappe, A. M.; Grady, M. C.; A Theoretical Study of Mechanisms for Chain Transfer to Monomer Reactions in Alkyl Acrylates, Paper 279e, AIChE Annual Meeting 2012.

[16] Moghadam, N.; Liu, S.; Srinivasan, S.; Grady, M. C.; Soroush, M.; Rappe, A. M.; Computational Study of Chain Transfer to Monomer Reactions in High-Temperature Polymerization of Alkyl Acrylates, J. Phys. Chem. A 2013, 117, 2605-2618.

[17] Moghadam, N.; Soroush, M.; Liu, S.; Srinivasan, S.; Rappe, A. M.; Grady, M. C.; Kinetics of Mechanisms of Chain Transfer to Polymer Reactions in Alkyl Acrylates: A Theoretical Study, Paper 508e, AIChE Annual Meeting 2012.

(56g) Computer-Aided Solvent and Process Design for Carbon Capture

Qadir, A., The University of Sydney
Dyson, A., The University of Sydney
Sharma, M., The University of Sydney
Arab, M., The University of Sydney
Khalilpour, R., The University of Sydney
Chiesa, M., Laboratory of Energy and Nano-Science (LENS)
Abbas, A., The University of Sydney
Originally presented on: 11/4/2013 10:40:00 - 11:00:00

The design of novel solvents for process improvements has stemmed much interest in recent years (Bardow et al., 2010, Folić et al., 2007, Oyarzun et al., 2011). Solvents are used in many industrial processes for separation of useful products from a feed. Consequently improvements in solvent performance are sought after for a more efficient and cost effective process. The properties desired in a solvent vary from process to process and include selectivity, volatility, flammability, toxicity, viscosity, heat capacity and reaction kinetics. Often a trade-off has to be reached between the desired properties as not all desired properties may be achievable within one solvent.

In order to optimize system performance, both process and solvent parameters must be optimized. The simultaneous optimization of a process and solvent is a challenging task. Although process parameters may be continuously optimized, solvent properties are discrete and therefore pose a major hurdle in the optimization process. Complex mixed-integer nonlinear programming (MILP) methods have to be implemented in order to optimize the nonlinear process problem and choose solvent parameters from a discrete solvent property space. This problem has traditionally been circumvented by decoupling the solvent and process optimization processes and by optimizing the solvent to specific desired property targets. The Computer Aided Molecular Design- Continuous Molecular Targeting (CAMD-CoMT) framework proposed by Bardow et al is an effective way of tackling the integrated process and solvent design problem and does not require decoupling the two optimization problems. Instead, the optimization problem assumes a hypothetical solvent with continuous solvent parameters based on a physically based thermodynamic model. The perturbed-chain-statistical associating fluid theory (PC-SAFT) equation of state is used to calculate the solvent properties based on five solvent parameters.

In the current paper, we calculate optimal process parameters and determine two optimal solvents (pure and binary mixture) for a carbon capture process in contrast to the pure solvent search in presented in (Bardow et al., 2010). The process is modelled in ASPEN Plus (Aspentech, USA), while an optimization is executed in Matlab (Mathworks, USA) to calculate the vapour liquid equilibrium properties of the solvent using the PC-SAFT thermodynamic model. This simultaneous process and solvent optimization enhances the performance of the system considerably, and has good potential for application across a myriad of other process systems.

BARDOW, A., STEUR, K. & GROSS, J. 2010. Continuous-Molecular Targeting for Integrated Solvent and Process Design. *Industrial & Engineering Chemistry Research,* 49,2834-2840.

FOLIĆ, M., ADJIMAN, C. S. & PISTIKOPOULOS, E. N. 2007. Design of solvents for optimal reaction rate constants. *AIChE Journal,* 53,1240-1256.

OYARZUN, B., DARDOW, A. & GROSS, H. 2011. Integration of process and solvent design towards a novel generation of CO2 absorption capture systems. *Energy Procedia,* 4, 282-290.

(57a) Gary Leal: Researcher, Teacher, Mentor
Fuller, G. G., Stanford University
Stone, H. A., Princeton University
Originally presented on: 11/4/2013 8:30:00 - 8:45:00

This session is a special session in honor of Gary Leal's 70th birthday. To begin the session, Professors Gerry Fuller and Howard Stone with present a retrospective of Gary Leal's many contributions. Specifically, they will discuss his key contributions as a researcher but also how he has impacted the fluid mechanics community as a teacher and mentor.

(57b) Simultaneous Measurements of Interfacial Forces and Electrochemical Reactions in the EC-Surface Forces Apparatus

Valtiner, M., Max-Planck-Institut f r Eisenforschung GmbH
Yu, J., University of California, Santa Barbara
Gebbie, M. A., University of California, Santa Barbara
Rapp, M., University of California, Santa Barbara
Kristiansen, K., University of California, Santa Barbara
Lee, D. W., University of California, Santa Barbara
Donaldson, S. H. Jr., University of California, Santa Barbara
Das, S., University of California, Santa Barbara
Banquy, X., University of California, Santa Barbara
Greene, G. W., University of California at Santa Barbara
Alcantar, N., University of South Florida
Israelachvili, J., University of California, Santa Barbara
Originally presented on: 11/4/2013 8:45:00 - 9:05:00

Recent developments in the SFA technique now allows for (electro)chemical reactions, including oxidation-reduction, adsorption and dissolution, to be quantitatively measured simultaneously with the long-range (e.g., DLVO) or short-range (e.g., adhesion) forces. The EC-attachment involves 3 electrodes at different locations within the SFA chamber, the solution and the surfaces (which can be conducting or non-conducting), and include reference, counter, working, Pt electrodes, etc. Examples will be given of how various electrochemical reactions and effects are being measured, including triboelectrical effects, the phenomenon of Pressure Solution, where one surface causes the dissolution of another nearby surface when there is a natural or applied potential difference between them (analogous to Chemical Mechanical Polishing (CMP), and corrosion), surface roughening or smoothing effects during reactions, flow patterns in thin films induced by electric fields, and oxidation-reduction reactions where the buildup of an oxide layer on one surface with time, coupled to the measured current flowing through the interface allows for the chemical reaction to be uniquely determined.

(57c) Viscoelastic Fluid Suspension Problems in Fracking and Drilling
Shaqfeh, E. S. G., Stanford University
Originally presented on: 11/4/2013 9:05:00 - 9:25:00

There are no comprehensive simulation-based tools for engineering the flows of viscoelastic fluid-particle suspensions in fully three-dimensional geometries. On the other hand, the need for such a tool in engineering applications, combined with the associated potential for understanding new physics related to elastic fluid/particulate suspensions is immense. Suspensions of rigid particles in viscoelastic fluids play key roles in many energy applications. For example, in oil drilling the "drilling mud" is a very viscous, viscoelastic fluid designed to shear-thin during drilling, but thicken at stoppage so that the "cuttings" can remain suspended. In a related application known as hydraulic fracturing suspensions of solids called "proppant" are used to prop open the fracture by pumping them into the well. It is well-known that particle settling in a viscoelastic fluid can be quite different from that which is observed in Newtonian fluids. An issue of importance in this context is that in Newtonian fluids, the presence of an imposed shear flow in the direction perpendicular to gravity (which we term *a cross or orthogonal shear flow*) has no effect on the settling of a spherical particle in Stokes flow (i.e. at vanishingly small Reynolds number or inertial forces). By contrast, in a non-Newtonian liquid, the complex rheological properties induce a nonlinear coupling between the sedimentation and shear flow. Recent experimental data have shown that this coupling is affected by both the shear thinning and the elasticity of the suspending polymeric solutions thus significantly affect the settling rate of the solids. In the present work, we use (a) simulations of viscoelastic flow past a single, torque-free sphere and (b) immersed boundary simulations of multiparticle suspensions including sedimentation with a cross shear flow to study the effect of carrier fluid elasticity on the drag experienced by a given sphere and thus on its settling rate. In the numerical simulations, we use different rheological models to represent the viscoelastic fluids with both highly shear thinning and non-shear thinning fluids considered. Fluid parameters are obtained by fitting rheological data. For weakly shear thinning fluids, we obtain an **increase** in drag, i.e. a decrease in settling rate, as shear Weissenberg number is increased in both the numerical simulations and the experiments. The simulations are in quantitative agreement with the experiments at small Weissenberg number (Wi <2) and qualitative agreement at higher Weissenberg number. For strongly shear thinning fluids we find a **decrease** in the drag for increasing Weissenberg numbers. We present the detailed physical mechanism for this remarkable behavior, and demonstrate that there is an optimum polymer concentration if reduction of particle sedimentation rate is the desired engineering criterion for these fluids.

(57d) Chemical Swarming

Brady, J. F., California Institute of Technology / Division of Chemistry and Chemical Engineering
Originally presented on: 11/4/2013 9:25:00 - 9:45:00

Nano- and micro-scale objects can undergo autonomous motion owing to chemical activity at their surfaces – so-called catalytic nanomotors. A surface chemical reaction creates local concentration gradients of the reactant (the fuel) and product species and, as these species diffuse in an attempt to re-establish equilibrium, they entrain the motor causing it to move. A single chemically active particle acts as a source (or sink) creating a reactant concentration disturbance that decays as 1/r, where r is the distance from the active particle. A second particle will, in general, undergo normal diffusiophoresis in the concentration gradient created by the first particle. For particles that move down the concentration gradient two sink particles will attract each other, while two source particles will repel each other. (A source and a sink particle will not move relative to one another.) Thus, a collection of source particles will tend repel each other and drift apart, while sink particles will attract each other and tend to 'collapse' into a dense clump – they 'swarm.' However, if fuel is supplied uniformly at the rate at which it is consumed by the active particles a 'chemically neutral' steady sate can be achieved in which the active particles exhibit Debye-like screening and phase behaviors analogous to those of a one-component plasma.

(57e) Particle Sedimentation in Structured Liquids

Hu, Y. T., Unilever R&D
Chung, J.,
Tonmukayakul, N., Halliburton
Originally presented on: 11/4/2013 9:45:00 - 10:05:00

Particle sedimentation in liquids is of both fundamental and practical importance. Most research on the subject has focused on the sedimentation in the absence of an external flow. However, there are many practical situations where particle sedimentation occurs under flow. In this study, sedimentation of a particle in various types of structured liquids is studied with and without flow. The structured liquids include entangled linear polymers, cross-linked polymers, and nanoparticle suspensions. The sedimentation rate of a single particle in these liquids is measured precisely using a CCD camera and an automatic particle tracking software while the particle is subjected to various flow conditions include steady and oscillatory shear as well as planar elongational flow. It is found that the static sedimentation rate depends on the low shear viscosity whereas the sedimentation rate under shear depends on the specific flow conditions and viscoelasticity of the sample. The interplay between the sedimentation, flow type, and viscoelasticity of the sample is discussed.

(57f) Computational Studies of Polymer Morphology Evolution in Drying Films

Fredrickson, G. H., University of California, Santa Barbara
Paradiso, S., University of California
Delaney, K.,
Ceniceros, H.,
Garcia-Cervera, C.,
Originally presented on: 11/4/2013 10:05:00 - 10:25:00

Film drying processes are important to achieving polymer films and membranes with micro and nano-structures tailored for applications ranging from advanced lithography to nanofiltration. We discuss a model and computational approach that enables the investigation of polymer morphology development by diffusion during the unidirectional drying of a polymer solution. The model is built on thermodynamic forces constructed from self-consistent field theory and can address the kinetics of phase transformations associated with the drying and ordering of complex polymer components such as block copolymers. As a first application of the model, we examine the drying of a solution of an asymmetric diblock copolymer, where the diblock assumes a hexagonal cylindrical mesophase in the dry state. We discuss the solution and solvent evaporation conditions for which cylinders formed at the drying front are oriented parallel or perpendicular to the substrate of the film and propagate in the same orientation through the film as the drying process continues. A particular focus is on identifying conditions where the technologically significant perpendicular orientation can be consistently achieved.

(57g) Polyelectrolyte Complex Formation: Structure and Properties

Tirrell, M., University of Chicago

Originally presented on: 11/4/2013 10:25:00 - 10:45:00

Our work is aimed at understanding complex coacervation in two model systems: branched poly(ethyleneimine) with linear poly(D,L-glutamic acid) or poly(D,L-aspartic acid), and identify differences and similarities with previously reported systems. Three different techniques (turbidity, isothermal titration microcalorimetry-ITC and rheology) were used in a comprehensive study of coacervation. Sample turbidity was used to show how various parameters (salt, stoichiometry, pH, temperature) affect complex coacervation. Complex coacervation decreases with increase in salt and coacervate formation is maximum when a 31:69 mol% acid:base ratio is used. Rare in literature phase diagrams revealed that coacervates are formed over a wide range of acid:base ratios (15–88 mol% NH3 groups), significantly broader compared to other systems. ITC was used for the thermodynamic characterization of the complexation between the polyelectrolytes, and showed that complex coacervation is entropy-driven (from the release of counter-ions) and enthalpically unfavored process. Composition and viscoelastic properties of the complex coacervates were examined gravimetrically and through rheology. Coacervate water content depends on the salt concentration and the stoichiometry. Coacervates exhibit a viscoelastic behavior that is dependent on the salt concentration. Master curves that can predict behavior at a wide range of time scales, not accessible by conventional rheological measurements, were created. (Work done in collaboration with Dimitris Priftis.)

(57h) The Elasto-Capillary Landau-Levich Coating Flow Problem

Homsy, G. M., Univ British Columbia
Dixit, H.,
Originally presented on: 11/4/2013 10:45:00 - 11:00:00

This paper concerns interfacial flows of complex fluids, a subject to which Gary Leal has made significant contributions. We consider the dip-coating flow problem when the interface has both an elastic bending stiffness and a constant surface tension. When the elasticity is absent, the classical analysis of Landau & Levich predicts a unique coating thickness as a function of the capillary number of withdrawal. We consider both a purely elastic problem when surface tension is absent, and an elasto-capillary problem when both elasticity and surface tension are present. In the case where interfacial tension is negligible, we assume the elasticity number *El* – the ratio of surface elasticity to viscous forces - is small and develop the solution for the free boundary as a matched asymptotic expansion in the small parameter $El^{1/7}$, thus determining the film thickness as a function of *El*. A remarkable aspect of the problem is the occurrence of multiple solutions, and five of these are found numerically. In any event, the film thickness varies as $El^{4/7}$, or equivalently, $U^{4/7}$, where U is the plate speed, in agreement with previous experiments. The solution for the elasto-capillary problem is formulated in a similar way, with an elasto-capillary number, *e*, (the ratio of elasticity to surface tension), as an additional parameter. It is shown that it is possible to connect the problems of pure elasticity and elasto-capillarity respectively through the parameter *e*, but that connecting one of the five elasto-capillary branches to the classical result remains an elusive goal.

(60a) Layered Biomimetic Nanocomposites: From Fundamental Advances to Scalable Nanotechnology

Kotov, N., University of Michigan

Originally presented on: 11/4/2013 8:35:00 - 9:20:00

Finding materials with combinations of several extreme properties is one of the key requirements for the successful engineering of adaptive systems. Many of such challenges are represented by the blank areas of Ashby plots, but others are less known. Successful realization of such materials requires new choices for materials components and new approaches for materials "assembly". Layer-by-layer assembly (LBL) is materials manufacturing technique from nanomaterials that affords engineering of nanocomposite materials based on sequential adsorption of nanometer scale layers of polymers and inorganic particle, nanowires, nanotubes, sheets, etc. In this presentation we demonstrate that LBL and related techniques can lead to the materials with seemingly "impossible" combinations of physical properties encompassing mechanical, electrical, optical, and biological properties. We will make particular emphasis in this presentation on nanoparticles composites which consistently demonstrate exceptional performance and display the potential for scalability. Hard-to-reach combinations of electrical and mechanical properties necessary for a number of technologies will be discussed. Finding composites with high stiffness properties + high damping and as well as high stiffness + transparency will be demonstrated. The pathway toward successful realization of two, three, and four different properties based on the structural designs and computer simulations will be shown. The last part of the talk will describe our latest exploits in the area of composites from plasmonic nanoparticles including those with superchiral properties.

(60b) Three Dimensional Self-Assembly Using Surface Forces

Gracias, D. H., Johns Hopkins University

Originally presented on: 11/4/2013 9:20:00 - 10:05:00

The magnitude of surface forces becomes increasingly large at small size scales often having deleterious effects as seen in the stiction of microelectromechanical devices. However, surface forces can be harnessed to overcome a critical challenge in human engineering by enabling parallel assembly of devices at small size scales and in three dimensions. In this talk, I will discuss the use of surface forces in directing the folding and aggregative assembly of curved and polyhedral structures. Of intellectual significance, polyhedra are the most fundamental solids that have been studied for over two thousand years since the days of Plato. Of technological importance, the surface tension driven assembly of lithographically-patterned nano to millimeter scale curved and polyhedral structures composed of metals, semiconductors and dielectrics opens up new opportunities in colloidal science, optics and electronics.

(60c) Polymer Hybrids Through Control Over Nanoparticle Interfaces

Kumar, S., Columbia University

Originally presented on: 11/4/2013 10:05:00 - 10:50:00

While much interest has been focused on the practical uses of polymer-nanoparticle hybrids, little is understood about the science that underpins this important class of materials. For example, the factors governing the equilibrium miscibility of polymers and particles are not understood. Here we demonstrate that control over the nanoparticle-polymer interface, through physical or chemical means, provides for a facile means of regulating the spatial dispersion state of the nanoparticles. Crucially, the availability of this control allows us to systematically explore and thus optimize the engineering properties of the resulting polymer nanocomposites. This talk will present both experimental and theoretical results from our group which focus on the underpinning role of nanoparticle interfaces on the properties of the composites that result.

(63a) Relative Free Enthalpies for Point Mutations in Two Proteins With Highly Similar Sequence But Different Folds

Hansen, N., Swiss Federal Institute of Technology
Allison, J. R.,
Hodel, F. H., Swiss Federal Institute of Technology
van Gunsteren, W. F., Swiss Federal Institute of Technology
Originally presented on: 11/4/2013 8:30:00 - 8:45:00

Understanding the capability of a protein to change to a completely different fold as a result of minor mutagenic perturbation is central to understanding how new protein structures and functions evolve [1]. Currently the structures of protein sequences at the interface between folds cannot be predicted reliably with both *ab initio* and knowledge based prediction algorithms [1]. A pair of proteins derived from Streptococcus protein G, GA95 and GB95, that have different folds despite being different by only three amino acids [2] is one striking example demonstrating that small local changes can cause large global changes and has implications for sequence-based structure-prediction methods. To understand why the amino acid changes bring about this large conformational change we use molecular dynamics simulations in explicit solvent and employ the enveloping distribution sampling methodology [3,4] to calculate the changes in free enthalpy associated with point mutations in the two different folds. The results show that point mutations at different positions in the amino acid sequence do not contribute equally to the free enthalpy difference between the native and non-native sequence-structure combinations. The individual free-enthalpy changes are almost sequence independent in the four-strand/one-helix structure, the stable form of GB95, while in the three-helix bundle structure, the stable form of GA95, an interplay between residues 20 and 45 is observed, supporting the experimental findings for this protein.

[1] P. N. Bryan and J. Orban, *Curr. Opin. Struct. Biol.* (2013) http://dx.doi.org/10.1016/j.sbi.2013.03.001

[2] P. A. Alexander, Y. He, Y. Chen, J. Orban and P. N. Bryan, *Proc. Natl. Acad. Sci. USA* **106** (2009) 21149

[3] C. D. Christ and W. F. van Gunsteren, *J. Chem. Phys.* **126** (2007) 184110

[4] N. Hansen, J. Dolenc, M. Knecht, S. Riniker and W. F. van Gunsteren, *J. Comput. Chem.* **33** (2012) 640

(63b) Activation Mechanism of c-Src Tyrosine Kinase

Shukla, D., Stanford University
Roux, B., University of Chicago
Pande, V., Stanford University
Originally presented on: 11/4/2013 8:45:00 - 9:00:00

Mechanistic understanding of the large scale conformational transformations coupled with activation of kinases is of enormous importance for treating aberrant kinase signaling, which leads to uncontrolled growth and differentiation of cancerous cells. Atomistic molecular dynamics simulations of kinases could capture these conformational transitions. However, due to the long timescales (100's of μs) associated with activation, such simulations are not feasible except on dedicated hardware. In this study, we employ an alternative computational paradigm that couples transition pathway generation techniques and Markov State Model (MSM) based massively distributed simulations for mapping the conformational landscape of c-src tyrosine kinase. MSMs of c-src tyrosine kinase capture the thermodynamics and kinetics of kinase activation for the first time, and help identify key intermediates along the activation pathway. We also predict the existence of a novel allosteric site in c-src intermediate state, which could be potentially utilized for drug design. Furthermore, kinetic model of kinase autophosphorylation built using the data obtained from atomistic simulations predict the experimentally observed induction time in the time evolution of active kinase population.

(63c) Dynamics of Torsionally Stressed Chromatin Fiber

Nam, G. M., University of California, San Diego
Arya, G., University of California, San Diego
Originally presented on: 11/4/2013 9:00:00 - 9:15:00

Torsional stresses serve critical functions for chromatin fiber to participate in biological activities such as transcription and replication of DNA. Despite the biological significance, most experimental and theoretical work so far has focused on understanding supercoiling in naked DNA, and little is known about how torsional forces are propagated, stored, and mitigated within chromatin fiber. In this work, we investigate the dynamic properties of nucleosome arrays subjected to torsional forces using the Brownian Dynamic simulations of a coarse-grained model of the chromatin fiber. By applying the twist to one end of array in stepwise manner, we monitor the conformational changes of array by measuring its vertical extension as a function of the number of turns. The obtained extension-rotations curve shows asymmetric shape with a shifted maximal extension toward negative rotations, which is in good qualitative agreement with the earlier experiments. These features imply that the nucleosome array exhibits different torsional behaviors in the two opposite directions of rotations: Negative rotations promote the linker DNA wrapping around nucleosome, promoting a one-start solenoidal structure while positive rotations cause the linker DNA to be bent in opposite direction to the handedness of nucleosome, leading to formation of a zig-zag structure. These higher-order structures are characterized by the two-angle model and their dynamical organization can be elucidated in terms of twist propagation. By means of rotational autocorrelation function for each segment along the array, we show that nonuniform twist profiles reflect the existence of geometrical barrier for twist propagation, which modulate the internal and global organization of array under torsional stresses.

(63d) Effects of DNA Methylation On DNA-Histone Interactions and Nucleosome Positions

Mathew, S., Columbia University
Wong, K. Y.,
Ortiz, V., Columbia University
Originally presented on: 11/4/2013 9:15:00 - 9:30:00

DNA occurs in CpG dinucleotides when a methyl group is added at the carbon 5 position of the C base ring, and it has an important role in regulating gene expression in higher eukaryotes. It is known that when cytosine bases are methylated, gene is silenced, but the precise role of methylated DNA in this process is still debated, and the mechanism by which methylation affects gene expression remains poorly understood. One way in which DNA methylation could regulate expression is by modifying the positions of nucleosomes, which in turn can regulate gene transcription by blocking the access of transcription factors to the gene. Here, we test this hypothesis by performing MD simulations of short DNA segments in solution and in the nucleosome, using an all atom force field. DNA conformations are characterized using the six rigid body base-pair step parameters, minor groove width, hybridization state and phosphate solvent exposure, and the changes in free-energy of the nucleosome are calculated upon cytosine methylation.

(63e) Scaling Laws and Essential Physics of DNA Oligomer Hybridization Rate Constants

Hinckley, D. M., University of Wisconsin-Madison
de Pablo, J. J., University of Wisconsin-Madison
Originally presented on: 11/4/2013 9:30:00 - 9:45:00

DNA hybridization is a fundamental process in biology and, increasingly so, in engineering. Theory and experiment have been developed to predict the scaling of hybridization rates constants for strands of 100 or more base pairs in length; however, the theory for hybridization of short oligomers remains undeveloped. In this work, we couple a recently developed version of the 3-Site-Per-Nucleotide (3SPN) DNA model with Forward Flux Sampling to examine the hybridization of short oligomers. We begin by studying the scaling of the rate constants on length for homogeneous and heterogenous sequences. We then determine the possible sites for nucleation and the mechanisms of hybridization. Lastly, we contrast the mechanisms of hybridization to those observed for DNA melting on the same sequences. These results are discussed in the context of polymer physics theory and previous simulations using simpler models.

(63f) Developing Predictive Models for Peptoids

Weiser, L., North Carolina State University
Santiso, E. E., Massachusetts Institute of Technology
Originally presented on: 11/4/2013 9:45:00 - 10:00:00

Peptoids (poly-n-substituted glycines) are peptide-like macromolecules composed of modified glycine units which have side chains attached to the nitrogen atoms [1]. Peptoids are biocompatible and resistant to protease degradation [2], and peptoid sequences can be synthesized using more than 300 commercially available amines [1], [3], making peptoids versatile and appealing compounds for combinatorial materials design. Additionally, peptide backbones don't have hydrogen bonding making developing peptoid models less challenging than developing models for their peptide counterparts.

The goal of this project is to develop predictive coarse-grained models for peptoids that can be used for *in silico* combinatorial design. This will be achieved in two stages: (1) Development of a model for the peptoid backbone that can reproduce its structural properties under different conditions, and (2) Development of models for a library of side chains that can later be used for combinatorial design. We will start by developing atomistic models using the OPLS force field with missing parameters fit to *ab initio* calculations. The models will then be fine-tuned to reproduce experimental structure data.

We will present models of peptoids with simple side chains. Atomistic simulations will be used to develop coarse-grained models. Our models will be tuned to accurately reproduce structural properties obtained by NMR measurements, and compared to available literature data. In future work, we will determine interaction potentials for other peptoid side-chains. These models will serve as starting points in developing predictive models for the combinatorial design of more complex peptoid sequences.

References:

[1] J. Seo, B. Lee, Z. R. N. P. Synthesis, N. I. P. Ducheyne, K. E. Healy, D. W. Hutmacher, D. W. Grainger, and C. J. Kirkpatrick, *Comp Biomaterials.* **2**, 53 (2011)

[2] K. T. Nam, S. a Shelby, P. H. Choi, A. B. Marciel, R. Chen, L. Tan, T. K. Chu, R. a Mesch, B.-C. Lee, M. D. Connolly, C. Kisielowski, and R. N. Zuckermann, *Nature materials.* **9**, 454 (2010)

[3] R. N. Zuckerman, S. B. H. Kent, and W. H. Moost. *J. Am. Chem. Soc.* **6,** 10646 (1992)

(63g) Structure and Dynamics of Mixtures of Ionic Liquids and Polyethers

Sharma, A., University of Notre Dame
Maginn, E. J., University of Notre Dame
McCready, M. J., University of Notre Dame

Originally presented on: 11/4/2013 10:00:00 - 10:15:00

Ionic liquids are salts that remain liquid near ambient temperatures. Characteristics such as extremely low vapor pressure and excellent thermal stability have led them to receive significant attention as potential solvents for separation processes and a plethora of other applications. One drawback of ionic liquids, however, is that they tend to be more viscous than conventional solvents. The large viscosity may be a limiting factor for the use of ionic liquids in industrial operations due to high mass transfer resistance and increased pumping and handling costs. One strategy being used to reduce the viscosity of ionic liquids is to mix them with small amounts of low viscosity additives such as polyethers or other low volatility fluids.

In order to understand the role additives have on the structure and dynamics of ionic liquids, we have carried out extensive molecular dynamics simulations on mixtures containing imidazolium-based ionic liquids and polyethers. Viscosities, self-diffusivities and rotational dynamics were computed as a function of polyether concentration and temperature. Structural properties such as radial and spatial distribution functions and end-to-end vectors were computed and used to help understand the trends in the dynamics. Comparison with available experimental data shows that the simulations are able to accurately capture both pure fluid and mixture dynamics.

(63h) Liquid-Vapor Phase Behavior of Asphaltene-Like Molecules

Kumar, V., University at Buffalo
Rane, K. S., University at Buffalo, The State University of New York
Wierzchowski, S., Shell Global Solutions
Errington, J. R., University at Buffalo, The State University of New York
Originally presented on: 11/4/2013 10:15:00 - 10:30:00

Asphaltenes have a significant effect on the thermodynamic and transport properties of crude oil. These molecules are known to clog pipes as well as clog the pores within rocks in the oil-field, thereby impeding oil extraction. The identification and design of solutions to mitigate the impact of asphaltenes would benefit from a better understanding of their thermophysical properties, including their phase behavior. In this presentation, we describe our recent efforts aimed at using molecular modeling to construct liquid-vapor phase diagrams for asphaltene-like molecules. A collection of histogram-based Monte Carlo (MC) techniques are used to deduce saturation properties. We first use direct grand canonical simulations to acquire coexistence properties at relatively high temperatures. These calculations are followed by temperature expanded ensemble simulations, which provide a means to trace saturation curves to relatively low temperature. Reservoir grand canonical MC, hybrid MC, and growth expanded ensemble strategies are employed to facilitate sampling the configuration space of these complex molecules. We first provide results for the phase behavior of a collection of polycyclic aromatic molecules. Such structures form the central unit of asphaltenes, and therefore represent a natural starting point for understanding the behavior of these molecules. Next, we probe the extent to which nitrogen- and sulfur-containing chemistries influence the phase behavior of polycyclic aromatics. Finally, we examine the phase behavior of realistic asphaltenes by decorating polycyclic aromatic molecules with tethered alkane chains.

(63i) Replica Exchange Dynamic Molecular Simulations and Studies of Novel Cyclic and Linear Amphiphilic Homopolymer

Liu, L., Tulane University
Originally presented on: 11/4/2013 10:30:00 - 10:45:00

Replica Exchange Dynamic Molecular simulations of Cyclic and Linear Amphiphilic Homopolymer

Lixin Liu, Henry S. Ashbaugh*, Steve W. Rick*, Scott Grayson*
Department of Chemical and Biomolecular Engineering, Tulane University, New Orleans, LA 70118

Abstract

Polymer-based micelles had been raised particular interests in application as drug delivery material because their spontaneously forming core-shell structure with hydrophobic core encapsulating solvent-compatible particles as long as hydrophilic corona imparting water-solubility upon the aggregation. In our research work, Molecular Dynamic (MD) simulations based on a novel Amphiphilic Homopolymer c-PHS-Amph first reported by Scott Grayson's lab were performed to investigate polymer's unique topologies as well as its resulting prominent solvent-dependent character, which would progressively shed a in-depth interpretation to the molecular driving force leading amphiphilic behavior. Since Macromolecule requiring relative large scale computational work, we have cooperated with Steve Rick's group to employ Replica Exchange Dynamic Simulation (REDs) to improve sampling space and determine the global polymer conformation. Through applying REDs technique, exchange among replicas split in large potential barriers occurred efficiently and successfully deliver linear and cyclic polymer's characteristic structure corresponding with different solvents' environment. Proximal radial distribution of solvent around hydrophobic arms and hydrophilic arms was performed to validate the thermodynamic signatures as well as used to reproduce the density profile of interacting solvents.

(73b) Preparation of Bioactive Insulin Microparticles for Pulmonary Delivery Using Supercritical Carbon Dioxide

Tang, C., Zhejiang University
Du, Z., Zhejiang University
Guan, Y. X., Zhejiang University
Yao, S. J., Zhejiang University
Zhu, Z. Q.,
Originally presented on: 11/4/2013 8:55:00 - 9:20:00

Insulin is an important pharmaceutical protein, which is delivered into the body exclusively via subcutaneous injection for diabetes therapy. Pulmonary delivery, which is a noninvasive route, has been considered as an alternative to injection. For pulmonary delivery, fine particles with narrow size distribution are essential. For example, the efficient inhalation therapy needs microparticles with size ranging from 1 to 5 μm to reach the deep lung. Moreover, the surface morphology of particles affects the dispersibility and flowability of the powders. Therefore, elaborate control of particle size and morphology plays a key role in the development of insulin formulations for inhalation.

Supercritical fluid assisted atomization (SAA) is a promising process in producing micrometric particles that may find much application in medical and pharmaceutical fields. An improved SAA process, i.e. supercritical fluid assisted atomization introduced by hydrodynamic cavitation mixer (SAA-HCM), was proposed by our group, which uses supercritical carbon dioxide to assist atomization and a cavitation mixer to intensify mass transfer. This novel technique can be operated not only on organic solvents but also aqueous solutions. In this paper, insulin was micronized using SAA-HCM process from aqueous solution without use of any organic solvents. Insulin microparticles produced under different conditions including solution type, solution concentration and precipitator temperature presented distinct morphologies such as highly folded, partly deflated, corrugated or smooth hollow spherical shape. Solution concentration had a striking influence on particle size, and insulin microparticles produced from acidic solution had volume median diameters increasing from 2.3 μm to 4.3 μm when protein concentration increased from 3 g/L to 50 g/L. HPLC chromatograms showed no degradation of insulin after SAA-HCM processing and FTIR, CD and fluorescence data further confirmed the structural stability. TGA analysis revealed that insulin microparticles remained moderate moisture content compared with raw material. In vivo study showed that insulin processed by SAA-HCM retained identical bioactivity.

SAA-HCM is demonstrated to be successful in producing insulin microparticles with different surface morphologies and well controlled size distribution. Therefore, it presents great potential in developing insulin pulmonary delivery formulations, especially in inhaled dry powders.

(73c) Supercritical Fluid Extraction and Chromatography of Bioactive Nutraceuticals

Samimi, R., Western University
Alsharari, Q., Western University
Salarian, M., Western University
Charpentier, P. A., Western University
Originally presented on: 11/4/2013 9:20:00 - 9:45:00

North American ginseng extracts are complex phytochemical mixtures with ginsenosides as the major biologically active components. There is tremendous interest in examining the phytochemistry and bioactivity of these extracts based on recent commercial products using North American ginseng extract for treating colds and flu (e.g. ColdFX™). This work examined supercritical fluid chromatography (SFC) coupled with an evaporative light scattering detector (ELSD) for separation and identification of ginseng extracts. Supercritical CO_2 extraction method was used for ginsenosides extraction from North American ginseng roots using DMSO as modifier. A cyanopropyl packed column stationary phase was used with a CO_2-based mobile phase containing methanol as modifier. SFC conditions were examined on ginseng extract profiles including pressure (100-200 bar), amount of modifier (5-50%), and the effect of water, acidic (trifluoroacetic acid), and basic (diisopropylamine) additives, integrated into methanol as modifier. The nature of additive have great effect on ginsenosides separation and retention time whereas not all the ginsenosides could elute without the presence of additives. The most effective elution of the six common ginsenosides with the best peak shape were obtained between 10-13 min in the case of using 0.1% acidic additive (trifluoroacetic acid) at a pressure of 150 bar, flow rate of 1.5 ml/min, and mobile phase gradient of 15-31% methanol in CO_2. These results show that SFC is a promising technique for separation and identification of novel extracts.

(73d) Supercritical Fluid Extraction of Polyphenolic Compounds From Pomegranate Peels

Wenzel, J. E., Kettering University
DeCrane, S., Kettering University
Wang, L., Kettering University
Zand, A., Kettering University
Originally presented on: 11/4/2013 9:45:00 - 10:10:00

Pomegranates contain a wide variety of antioxidant compounds that are thought to have numerous health benefits. Extracts of pomegranate are presently marketed for antioxidant support as a nutritional supplement. Due to unique solvent properties, including selectivity, supercritical fluids may be used to extract antioxidant compounds from plants, including pomegranates. The extraction of anti-oxidant compounds in pomegranate peels was evaluated using several supercritical fluids and supercritical mixtures. Fluids include supercritical carbon dioxide and mixtures of carbon dioxide with methanol, ethanol, and acetone. The extractions were performed batch-wise in a Parr Instrument 26 mL 316 stainless steel bolt-closure reactor. Extraction conditions were selected considering the critical point of the solvent. The extracts were collected and analyzed using HPLC, GC-MS, and tested for antioxidant properties. The chemical species extracted, selectivity of the extraction based upon supercritical fluid used, and antioxidant characteristics will be discussed.

(73e) Supercritical Solvent Impregnation for Drug-Loaded Lenses Using Carbon Dioxide With Water

Yokozaki, Y., Tokyo Institute of Technology
Sakabe, J., Tokyo Institute of Technology
Ng, B.,
Shimoyama, Y., Tokyo Institute of Technology
Originally presented on: 11/4/2013 10:10:00 - 10:35:00

Soft contact lens (SCL) has been expected as an alternative drug delivery system in order to solve the problem the low bioavailability in the case of eye drops. It is found that the impregnation of SCL with drug in aqueous solution results in the low amount of drug and burst-type drug release profile. The impregnation using supercritical carbon dioxid, supercritical solvent impregnation method (SSI) has been proposed as the alternative method in this work. The effect of the amount of water dissolving in supercritical carbon dioxide on the performance of the drug release profile from SCLs were studied. Salicylic acid was used as a drug impregnating into SCLs. The preparation of the SCLs with drug by SSI method were carried out at the constant temperature and pressure, 313 K and 11.0 MPa, and the depressurization rate of 0.1 MPa per minute.. The amount of water added into supercritical carbon dioxide was from 0 to 2.18 ml L^{-1}. (The saturated solubility of water in supercritical CO_2 is 1.19 ml L^{-1}). The Drug-loaded SCLs were soaked in the phosphate buffer solution at 310 K for the measurement of the released drug amount from SCLs. The concentrations of salicylic acid in the buffer solution were measured by an ultraviolet spectrometer. The performances of the drug-loaded SCLs were evaluated by the release profile and the amount of the drug. The SCLs processed in SSI without water present the burst-type drug release profile and the small amount of the drug. On the other hand, the SCLs processed with water present the sustained-type drug release profiles. In this case, the amount of water is larger than those in the case without water. The drug amount increases with increasing of water amount in supercritical carbon dioxide.

(83a) Molecular Simulation and SAFT Modeling of Solvation in Ionic Liquids

Maginn, E. J., University of Notre Dame
Vahid, A., University of Akron
Originally presented on: 11/4/2013 8:30:00 - 8:50:00

The solubility and other thermodynamic properties of ethanol, propanol, dimethyl formamide, and dimethyl sulfide in the ionic liquid trimethylbutylammonium bis(trifluoromethylsulfonyl)imide has been computed via Monte Carlo simulations. A united atom model (UAM) force field has been developed for both the organic solutes and the ionic liquid. The force field gives accurate vapor liquid coexistence curves and vapor pressures for the solutes and accurate liquid density and surface tension for the ionic liquid. Computed mixture properties including isotherms, activity coefficients, excess entropy, excess enthalpy, and mixing volume are all modeled accurately and agree with the experimental data of Massel et al. The experimental data are also modeled accurately using the Perturbed-Chain Induced-Polar Statistical Associating Fluid Theory (PCIP-SAFT) equation of state. Experimental results show a negative excess entropy trend for mixing of dimethyl formamide and dimethyl sulfoxide in the ionic liquid, whereas ethanol and propanol exhibit a positive enthalpy of mixing. The simulations and PCIP-SAFT capture this trend and analyses of the simulation trajectories show that this is driven by differences in hydrogen bonding tendencies of the solutes and that the entropy of solvation dominates for all four systems.

(83b) Using Monte Carlo Simulations to Compute Liquid-Vapor Saturation Properties of Realistic Models of Room Temperature Ionic Liquids

Rane, K. S., University at Buffalo, The State University of New York
Errington, J. R., University at Buffalo, The State University of New York
Originally presented on: 11/4/2013 8:50:00 - 9:10:00

Understanding the liquid-vapor saturation properties of room temperature ionic liquids (RTILs) is important for several industrial applications of these fluids. In this presentation, we discuss the use of Monte Carlo simulation to evaluate liquid-vapor saturation properties of the homologous series 1-alkyl-3-methylimidazolium bis(trifluoromethylsulfonyl)imide ($[C_n mim][NTf_2]$) at temperatures ranging from room temperature to those very near the predicted critical points. We also relate the trends observed in these properties to the microstructure of the saturated liquid and vapor phase. To compute these properties, we perform direct grand canonical Monte Carlo simulations at two temperatures near the critical point. These are followed by isothermal-isobaric temperature expanded simulations to trace the saturation curves to low temperatures. To address the challenges generally encountered in simulating complex molecules like those of RTILs, we use advanced simulation tools like Hybrid Monte Carlo, reservoir Monte Carlo and distance-biased moves. We present results for the above series with n ranging from 2 to 12. The calculated properties include saturated liquid and vapor densities, vapor pressures, enthalpies and entropies of vaporization. We also show how the liquid phase microstructure, especially the nano-domains observed for some of these fluids evolve over the temperature range investigated.

(83c) VLE of Lipidic Ionic Liquid Systems

West, K. N., University of South Alabama
Davis, J. H. Jr., University of South Alabama
O'Brien, R. A., University of South Alabama
Langham, J., University of South Alabama
Originally presented on: 11/4/2013 9:10:00 - 9:30:00

Ionic liquids (ILs) are organic salts, which are liquids at temperatures below 100°C; they are a unique class of compounds that are virtually non-volatile and have highly tunable properties. As one might expect, *ionic* liquids typically exhibit the same solvent characteristics as *polar*molecular solvents, readily dissolving moderately-polar and polar solutes, but often being poor solvents for non-polar compounds. This limits the applicability of ILs for chemical reactions and separations processes that involve non-polar compounds, including those of biological origin such as fatty acids and cholesterol.

For the past few years, our group has been studying a new class of ILs that contains long alkyl chains, incorporated to impart *non-polar-like solvent properties*, which *remain room temperature liquids*. Typically, ILs with long alkyl chains remain solids at room temperature due to the enhanced interactions created by the long chains; however, recent work has overcome this problem. This was accomplished by taking cues from biological systems, i.e. the manner in which certain organisms regulate membrane fluidity in colder temperatures by including unsaturation in the alkyl chains of phospholipids. In addition to unsaturations, replacement of a carbon atom in the chain with a sulfur atom, to produce an equivalent length thioether chain, has also proven to lower melting points. The effect of the position of the sulfur atom on the melting point is similar to that to in pure, molecular thioethers.

In this work, we focus on our latest binary VLE data for lipidic ionic liquid in mixtures with permanent gases and vapors of other molecular species. Specifically, we are interested in binary systems related to the separation of gases in biological systems (anesthetic gases) as well as selective separations for energy and environmental separations. Earlier work has demonstrated that N_2O has a higher solubility than CO_2 in ionic liquids containing a long chain with a *cis*-unsaturation, which is the same solubility trends as in olive oil, often used as a screening agent for potential anesthetic gases. Here, we examine the effect of the structure and composition of the alkyl chain on the relative solubility of these gases in lipidic ionic liquids.

(83d) Liquid-Liquid Equilibria in Systems Containing Ionic Liquids Using COSMO-RS

Mokrushina, L., University of Erlangen-Nuremberg
Sponsel, E., University of Erlangen-Nuremberg
Hausmann, P.,
Arlt, W., University of Erlangen-Nuremberg
Originally presented on: 11/4/2013 9:30:00 - 9:50:00

In the field of chemical engineering and life sciences, the improvement of processes is of continuous interest. In the case of separation, e.g. extraction, an improvement could be made by the use of ionic liquids (ILs). ILs are "designer solvents", thus the properties necessary for an application can be achieved by the systematic combination of anion and cation. Since there are many possible combinations, predictive methods are needed to replace/minimize measurements and to enable an a-priori screening of possible combinations.

One of the most popular and effective methods here is the COSMO-RS model. Based on the efficient combination of quantum mechanics and statistical thermodynamics, COSMO-RS allows for the phase equilibrium predictions based only on the chemical structure of the system components. In the frame of this model, IL can be treated as one compound (molecular form/meta-files) or as an equimolar mixture of the ions as two single compounds (dissociated form/single ions).

In this study, the influence of such treatment on the prediction results is studied for liquid-liquid equilibria (LLE). LLE in binary as well as in ternary systems (miscibility gaps and partition coefficients) containing ILs and water or alkanes, alcohols, ketones are analyzed.

In reality, both dissociation forms of IL can simultaneously be present in the solution. Therefore, an additional method is developed by introducing a degree of dissociation and considering the partial dissociation of IL. Based on the predictions for a series of ILs with different cations and anions, the effect of the solvent polarity and the strength of the ion-ion interaction between anion and cation on the IL's degree of dissociation are investigated systematically.

In general, the presented method leads to improved prediction quality and guarantees a reliable screening result. In addition, it gives theoretical insight into the IL's ionic interaction strength.

(83e) Selection of Ionic Liquids for Acetonitrile/ Water Separations and Determination of VLE Data Supported By COSMO-RS

Fang, J., Hebei University of Technology
Li, C., Hebei University of Technology
Liu, J., Hebei University of Technology
Originally presented on: 11/4/2013 9:50:00 - 10:10:00

The separation of acetonitrile and water is challenging, due to formation of azeotrope. Conventional solvents for those separations are volatile solvents. In recent years, Ionic liquids (ILs) as a new kind of solvent have many unique attributes, e.g. negligible vapor pressure, good thermal stability and environment-friendly, which make them attractive and potential being used as a suitable solvent in separation processes.

Ionic liquids are molten salts composed of large organic cations and various anions. The prediction of the ideal IL for each separation process is nowadays impossible, Since there are about 10^{18} possible ionic liquids by combination of cation and anion. Hence, it is possible to tailor a specific ionic liquid with the desired properties. In order to be able to make a selection out of all possible combinations COSMO-RS was found to be a suitable screening tool. By comparing the σ-profiles of acetonitrile - water compounds and an ionic liquid, it is possible to make an assumption whether the ionic liquid is suitable for the separation of the mixture.

In order to select a suitable IL entrainer for extractive distillation, the infinite dilution activity coefficients of acetonitrile and water in hundreds of ILs using the COAMO-RS (Conductor-like Screening Model For Real Solvents) model was calculated in this work. Then selectivity and solubility were determined according the infinite dilution activity coefficients. 1-ethyl-3-metnylimidazolium acetate ([EMIM]OAc), N-butyl-4-methylpyridine acetate ([4MePy]OAc), and 1-hexyl-2,3-dimethylimidazolium acetate ([HDMIM]OAc) has high selectivity and good solubility by screening. The isobaric VLE data for the ternary system of acetonitrile + water containing ILs were measured at 101.3 kPa, and the effect of the ILs on the separation performance of the acetonitrile and water system was discussed. The results show that the three ILs could increase the relative volatility of acetonitrile and even made the azeotropic phenomenon eliminate. The separation ability of ILs is in the order [HDMIM]OAc > [4MePy]OAc > [EMIM]OAc. COSMO-RS can thus be very useful in the scanning of the growing number of known ILs to find suitable candidates, or help designing new ILs, for specific applications before extensive experimental measurements.

(83f) Solubilities, Densities and Viscosities of Two Low-Density Ionic Liquids Containing Small Hydrocarbons

Liu, X., Xi'an Jiaotong University
Afzal, W., UC Berkeley
Yang, G., University of California, Berkeley
Prausnitz, J. M., UC Berkeley
Originally presented on: 11/4/2013 10:10:00 - 10:30:00

Experimentally, we find that low-density ionic liquid tetrabutylphosphonium bis(2,4,4-trimethylpentyl)phosphinate [P4444][TMPP] shows large solubilities for small paraffins and olefins (methane, ethane, ethylene, propane and propylene). Solubilities of paraffins in [P4444][TMPP] are larger than those of corresponding olefins. These results may make [P4444][TMPP] a useful solvent for separation of olefin and paraffin because the vapor pressure of an olefin is higher than that of its corresponding paraffin at the same temperature. However, the viscosity of [P4444][TMPP] is 1740 mPa·s at 25Â°C, too large for a separation process.

To reduce the viscosity, we use low-density ionic liquid 1-butyl-3-H-imidazolium acetate [BHMIM][AC] as diluent for [P4444]][TMPP]. We choose [BHMIM][AC] because it has low viscosity, about 7 mPa·s at 25Â°C. Low-density ionic liquids have higher solubility of gases. We present viscosities and densities of mixtures of [P4444]][TMPP] with 20 or 50 mass % [BHMIM][AC] from 20 to 70Â°C at atmospheric pressure. We present the solubilities of gaseous methane, ethane, ethylene, and propane in a mixture of [P4444]][TMPP] with 20 or 50 mass % [BHMIM][AC] from 25 to 50Â°C up to 5 MPa. We find that at 25Â°C, addition of 20 mass % [BHMIM][AC] to [P4444][TMPP] lowers the viscosity to 183 mPa·s, addition of 50 mass % [BHMIM][AC] to [P4444][TMPP] lowers the viscosity to 77 mPa·s. The solubilities of light olefins and paraffins are reduced by about 10 % when the solvent contains of 20 mass % [BHMIM][AC] and about 45 % when the solvent contains 50 mass %.

(83g) Thermal Conductivity of High Temperature Molten Salts for Concentrated Solar Power Applications

Kassaee, M. H., Georgia Institute of Technology
Hossain, M. Z.,
Jeter, S.,
Teja, A. S., Georgia Institute of Technology
Originally presented on: 11/4/2013 10:30:00 - 10:50:00

High temperature heat transfer fluids can play an important role in increasing the efficiency of solar thermal systems. Being able to operate at higher tempetaures than existing limit of traditional heat transfer fluids (synthetic oil at 400 °C, and eutectic nitrate salts at 550 °C), greately increases the efficiency of the system. However, identification and characterization of a suitable fluid with low melting point and high tempearature stability has been a challenge. It is imprative to have accurate measurement or correlation of the thermal conductivity (and other physical properties) of any newly developed fluid. The accurate assesment of these properties would enable the optimal design of other components of the solar plant such as piping system, and storage tanks. Several novel halide eutectic molten salt mixtures are proposed as future high temperature heat transfer fluids, to meet critical criteria that will advance the efficiency of concentrated solar power (CSP).

In this study, the thermal conductivity of low melting point (< 250 °C) eutectics in three tertiary chlorinated compounds (1)NaCl, KCl, $AlCl_3$, (2) NaCl, KCl, $ZnCl_2$, and (3) NaCl, KCl, $FeCl_3$, are predicted based on the modified rough hard sphere theory. We have also measured the thermal conductivities of these systems at low temperature, and compared them with the predicted results from the model. All of the studied eutectic systems have melting points of less than 250 °C, and low vapor pressure (less than 2 atm) at 1000 °C. These novel salt mixtures are evaluated for their sutability as high tempearature heat transfer fluids in CSP technology.

(83h) A New Perturbation Approach for Electrolyte Solutions

Drunsel, F., Institute of Thermodynamics and Thermal Process Engineering, University of Stuttgart
Gross, J., Delft University of Technology
Originally presented on: 11/4/2013 10:50:00 - 11:10:00

The treatment of electrolyte systems in fluid theories is demanding because of the long-range decay of the coulombic interactions. Integral equation approaches are the most prominent path to developing theoretical descriptions of electrolyte solutions. A promising perturbation approach, however, has been formulated by Henderson et al. [1]. The problem of diverging correlation integrals is alleviated by resumming the perturbation terms of different order and approximating a converging expression from the sum of diverging integral equations. The initial theory is of third order and the resulting equations are formulated in powers of ½ in inverse temperature. Due to inaccurate approximations and erroneous angle averaging, this approach does not predict the Helmholtz energy of the system correctly.

In this contribution we present a new approach of a perturbation theory for electrolyte solutions. Similar to the ewald sum the electrostatic potential is subdivided into a short ranged part and a long raged tail. The long-range contribution of the electrostatic potentials is accounted for in a powerful analytical theory (local molecular field theory, LMF) proposed by Rodgers and Weeks [2, 3]. We reformulate the correlation integrals up to third order for only the short-ranged part of the pair potentials. Diverging integrals and thus inaccurate approximations are avoided.
The underlying LMF theory is validated by computer simulations comparing the results with ewald summation calculations. Furthermore, the theoretical work is supplemented with molecular simulations applying thermodynamic integration for the estimation of the Helmholtz energy of charged and dipolar hard spheres. This allows us to test the theory without any further assumptions.

[1] D. Henderson, L. Blum, A. Tani: Equation of state of ionic fluids. *ACS Symp. Ser. No. 300* (1986), 281.

[2] J. M. Rodgers, J. D. Weeks: Local molecular field theory for the treatment of electrostatics. *J. Phys.: Condens. Matter 20.* (2008), 494206.

[3] J. M. Rodgers, J. D. Weeks: Accurate Thermodynamics for Short-Ranged Truncations of Coulomb Interactions in Site-Site Molecular Models. *J. Chem. Phys. 131.* (2009), 244108.

(98d) Investigation of Hydrodynamic Effects in Separation Columns Using X-Ray Computed Tomography

Schug, S., Chair of Separation Science and Technology
Schmidt, I.,
Minceva, M., University of Erlangen-Nuremberg
Arlt, W., University of Erlangen-Nuremberg

In the chemical industry packed columns are widely-used for different unit operations as for instance distillation, extraction or chromatography. To increase the separation efficiency of these processes it is necessary to determine relevant parameters like the packing structure or geometry and the wettability of the internals, without influencing the fluid flow through the measurement. Therefore we used a common medical X-ray computer tomograph as a non-invasive measuring technique. With this device a flow through horizontal separation columns can be investigated.

Due to the fact that in nearly every technical application the fluid flow proceeds in vertical direction we put a customized computed tomograph into operation at our institute. In this computer tomograph the detector and X-ray source are rotating around the column during the measurement and they can be moved in vertical direction which enables scanning of vertical objects. As at our institute the column is not moved the fluid flow is not influenced by centrifugal forces and based on a high spatial (83 µm) and temporal resolution (1000 projections per second), dynamic processes inside separation columns can be analyzed with a high precision.

We have shown that X-ray computed tomography is a well suited method for the measurement of local parameters, like the porosity and axial dispersion coefficient in chromatographic columns without influencing the fluid flow. With this technique it is possible to observe the fluid flow along the column in axial and radial direction. In a further step the measured parameters are used to establish and validate a CFD-model for the simulation of chromatographic separation processes and selection of a stationary phase for a specific separation problem.

Hence to the good applicability of computed tomography as measurement method we want to transfer this technique to three-phase rectification to determine the liquid distribution on column internals, the liquid hold-up and the entrainment of liquid droplets due to the gas flow.

(98q) Stability of Thin Liquid Films: Influence of Interfacial Viscoelasticity

Rosenfeld, L., Stanford University
Fuller, G. G., Stanford University

The phenomenon of dewetting is frequently observed in our everyday life. It is of central importance in many technological applications as well as in a variety of physical and biological systems. The presence of non-soluble surfactants at an air/liquid interface may affect the dewetting properties of the aqueous layer.

The tear film is comprised of an aqueous layer covered with a ~100nm thick blanket of lipids, known as the meibomian lipids. Interfacial rheological measurements of meibomian lipids reveal that these films are remarkably viscoelastic. Tear film dewetting is an extremely important parameter that influences tear film stability.

In this research, in order to better understand the role of the meibomian lipids in the tear film stability, we have developed a methodology to characterize dewetting of thin aqueous layers covered with insoluble surfactants. Meibum was used to form a viscoelastic monolayer. As comparison to meibum we have studied DPPC, a phospholipid which is also viscoelastic, and the Newtonian arachidyl alcohol (AA). The apparatus allows control of the surface pressure of the monolayer, which is a key feature since this variable controls the surface viscoelasticity.

We have found that monolayers of viscoelastic surfactants are able to stabilize thin films against spontaneous dewetting. As the surface pressure of these layers is increased, their effectiveness is enhanced. Moreover, these surfactants are able to reduce the critical thickness for dewetting. Meibum is particularly effective in stabilizing thin films. Our results suggest that the meibomian lipids play a vital role in maintaining tear film stability in addition to suppressing evaporation. Furthermore, changes in meibomian lipid composition that alter the lipid melt temperature and viscoelastic properties, may ultimately influence tear film stability, which in clinical cases may result in dry eye disease.

(98c) Transition and Turbulence in a Lid-Driven Cavity Flow At High Mach Number

Pradhan, S., Indian Institute of Science
Kumaran, V., Indian Institute of Science

Title : Transition and turbulence in a lid-driven cavity flow at high Mach number

Author(s) : S. Pradhan and V. Kumaran

Affiliation : Department of Chemical Engineering, Indian Institute of Science, Bangalore-560 012, India

Abstract :

The flow in a 2D and 3D lid-driven cavity, with two opposite walls moving in opposite directions with equal velocity, simulated using the direct simulation Monte Carlo (DSMC) method, has been used as a test bed for examining different aspects of transition and turbulence at high Mach $M = (U_w / \text{sqrt}(k_B T_w / m))$ and Reynolds numbers $Re = (\rho_{av} U_w L_y / \mu_w)$. Here, L_y is the smallest box dimension, U_w and T_w are the wall velocity and temperature, and ρ_{av} is the volume-averaged gas density and μ_w is the gas viscosity at the temperature corresponding to the wall temperature.

The transition is found to be highly sub-critical in 2D even at high Mach number, with well-separated lower and upper critical Reynolds numbers for the transition from turbulent-laminar and laminar-turbulent transitions. The transition Reynolds number increases faster than linearly with Mach number. This implies that the Knudsen number at transition (ratio of mean-free-path and system size, also proportional to the ratio of Mach and Reynolds numbers) passes through a maximum as the Mach number is increased. This maximum value is small, less than 0.025, indicating that transition is a continuum phenomenon even at high Mach numbers. The transition to turbulence is sub-critical in 3D as well. The transition Reynolds number does increase with Mach number, and the Knudsen number increases monotonically with Mach number over the parameter range studied here. There appear to be two linear regimes in the variation of the transition Reynolds number with Mach number, one with higher slope at lower Mach number, and the other with lower slope at higher Mach number.

In the present analysis of turbulent flows using DSMC method, wall slip in the temperature and the velocities are found to be significant. Slip occurs because the temperature/velocity of the molecules incident on the wall could be very different from that of the wall, even though the temperature/velocity of the reflected molecules is equal to that of the wall. There is slip even in the mean velocity as well as the intensity of the turbulent velocity fluctuations tangential to the wall. Due to the non-zero variation in the tangential velocities, it is found that the wall-normal fluctuating velocity increases proportional to the distance from the wall, in contrast to an incompressible flow with no-slip boundary conditions where the velocity increases as the square of the distance from the wall. We find that the amplitudes of the tangential fluctuating velocities increase close to the one-third power of the distance from the wall.

In a compressible turbulent flow, we examine the result that the ratio of the mean free path (lambda) and Kolmogorov scale (eta) increases proportional as ($M/Re^{1/4}$), and it increases asymptotically with Mach number in the high Mach number limit. The simulations show that the ratio does decrease as $Re^{-1/4}$, but it does not increase linearly with Mach number. The resolution suggested by our simulation is that even though the Mach number based on the wall velocity and temperature is large, the local Mach number based on the local dissipation velocity in regions of high shear decreases due to an increase in temperature. Due to this, the ratio of the mean free path and Kolmogorov scale appears to taper off in the high Mach number limit.

An important finding is that the ratio of the mean free path and Kolmogorov scale shows very little variation across the domain, even though the mean free path and Kolmogorov scale individually show larger variations. The ratio of the mean free path and Kolmogorov scale was also shown to be equal to the local Mach number based on the local dissipation velocity, and can also be interpreted as the square root of the ratio of the strain rate and collision frequency. All of these quantities are remarkably invariant across the domain, indicating a coupling between the local temperature and the dissipation rate in a high Mach number turbulent flow.

(98e) Asymmetric Cytoplasmic Streaming Across Septal Pore in Fungal Hyphae

Leong, F. Y., Institute of High Performance Computing
Pieuchot, L. P. A.,
Jedd, G.,
Chiam, K. H., A*STAR, Institute of High Performance Computing

We investigate the phenomenon of cytoplasmic streaming across a septal pore in the filamentous fungi, *Neurospora crassa*. Experimentally, we found that the flow profile across some septal pores in mature hyphae to be asymmetric, where the inlet flow resembles a long and narrow converging funnel in contrast with the broad diverging funnel at the outlet. The emergence of asymmetric flow behavior is somewhat surprising, since the Reynolds numbers are vanishingly small. To this, we propose and discuss two hypotheses. One considers the accumulation of organelles and other cytoplasmic deposits before the cross wall, which can be modeled as a porous region using Darcy-Brinkmann equations. With a perfectly impermeable deposition region, fluid is displaced towards the centerline so that the entrance bulk flow in the core approaches that of a converging pipe. In practice, however, it is difficult to assess the extent of deposition and its effect on the flow. Another approach considers the non-Newtonian behavior of the cytoplasmic fluid, where the stretching of semi-rigid polymers could lead to extensional flow at the accelerating inlet side, and relaxation of polymers on the decelerating outlet side. This proposition appears simpler in requirements, but such anisotropic flow behavior in cytoplasmic fluids is neither widely observed, nor well-understood. Since both hypotheses are not mutually exclusive, the actual phenomena may in fact be attributed to some combination of the two effects. This work has implications in cytoplasmic streaming processes in fungi, as well as the transport of suspensions through a microfluidic orifice.

(98a) Dimensional Analysis and Geometric Scaling Relations for the Viscous Flow Resistance of Exhaust Particulate Filters

Yin, D. W., Core R&D Engineering & Process Science, The Dow Chemical Company

Diesel and gasoline particulate filters, hereinafter exhaust particulate filters, are used to remove soot and ash from the exhaust gases of Diesel and gasoline direct-injection engines, thereby lowering particulate matter emissions and improving air quality. A large pressure drop across the filter during operation adversely affects fuel efficiency as it corresponds to increased engine power required to discharge the exhaust gas through the filter. Hence an objective in designing exhaust particulate filters is to minimize the pressure drop, subject to mechanical, thermal, filtration efficiency, particulates storage capacity, and other constraints. Experimental data, when plotted as the Euler number versus the Reynolds number, show that under on-engine conditions approximately 70% to 90% of the pressure drop arises from viscous effects and 10% to 30% is due to inertial effects. Our present work accordingly focuses on the analysis of the viscous flow resistance of exhaust particulate filters; for the inertial loss coefficient one may refer to prior literature [1–4].

Exhaust particulate filters are typically made of gas-permeable porous ceramic substrates, extruded as honeycombs consisting of hundreds to thousands of parallel square channels. Adjacent channels are closed at alternate ends with plugs to form pairs of inlet and outlet channels. Soot and ash are filtered onto the side of the partitioning ceramic wall facing the inlet channel as exhaust gas passes from the inlet channel through the wall to the outlet channel. The substrate permeability κ [dimensions L^2] and the four principal variables that describe the filter geometry—the diameter D [L], the channel length L [L], the channel density (the reciprocal honeycomb unit cell cross-sectional area) σ [L^{-2}], and the wall thickness w [L]—determine the viscous flow resistance of a clean filter. Adding to the above five variables are the mass density ρ [ML^{-3}] and the dynamic viscosity μ [$ML^{-1}t^{-1}$] of the exhaust gas, and the volumetric flow rate Q [L^3t^{-1}] and the pressure drop Δp [$ML^{-1}t^{-2}$]. Analysis of the pressure drop of a clean exhaust particulate filter thus involves nine variables [5,6].

Previous investigations on optimizing the design of exhaust particulate filters have relied on empirical rules such as fixing the aspect ratio and the volume of the filter to make the problem more tractable [7]. Such approaches, however, do not easily lend to a general understanding of the scaling relations between the filter geometry and the viscous flow resistance. We instead prefer to apply dimensional analysis in our study, both to reduce the number of degrees of freedom and to find the dimensionless groups that best describe the underlying physics.

By drawing inspiration from the analysis of the classical Z-type parallel-flow manifold [8–10], we construct a ladder-shaped resistor network to model gas flow through a pair of adjacent inlet and outlet channels in an exhaust particulate filter. This changes the flow basis of the problem from Q to the average volumetric flow rate per filter channel $q = 4Q/\pi D^2\sigma$ [L^3t^{-1}], effectively reducing the number of degrees of freedom by two. It follows from Buckingham's Π theorem [2,11] that seven variables spanning three basis dimensions can be reorganized into four dimensionless groups. We make the customary choice of selecting two of these to be the Reynolds number $\mathrm{Re} = \rho\kappa^{1/2}v/\mu$ and the Euler number $\mathrm{Eu} = 2\Delta p/\rho v^2$, where $v = 4Q/\pi D^2$ [Lt^{-1}] is the superficial velocity of the exhaust gas flow through the entire filter. The remaining two dimensionless groups will then define the dimensionless viscous flow resistance $R_v\kappa^{1/2}$. Several choices are possible for formulating the two remaining dimensionless groups:

- (1) Since the solution to the resistor network model, $R_v = R_{ch}\{1 + (2R_w/R_{ch})^{1/2}\coth[(R_{ch}/2R_w)^{1/2}]\}$, gives the overall viscous flow resistance R_v [L^{-1}] as a function of the channel flow resistance $R_{ch} = s_{ch}L/d^4\sigma$ [L^{-1}] and the wall flow resistance $R_w = s_w w/d\sigma L\kappa$ [L^{-1}], where $d = \sigma^{-1/2} - w$ [L] is the filter channel width, and s_{ch} and s_w are constants, a logical choice for the two groups is the respective pair of dimensionless resistances $R_{ch}\kappa^{1/2} = s_{ch}L\kappa^{1/2}/d^4\sigma$ and $R_w\kappa^{1/2} = s_w w/d\sigma L\kappa^{1/2}$. Two exhaust particulate filters are similar for the purpose of comparing or scaling their viscous flow resistances if they have the same value for $R_{ch}\kappa^{1/2}$ and the same value for $R_w\kappa^{1/2}$.

- (2) An alternative is to use the wall flow-to-channel flow resistance ratio $\alpha = (s_{ch}/s_w)(R_w/R_{ch}) = wd^3/L^2\kappa$ along with the honeycomb cross-sectional open area fraction $\beta = d^2\sigma = d^2/(d + w)^2$, where β is related to the product of the dimensionless channel flow and wall flow resistances by $[(1/\beta^{1/2}) - 1]/\beta^2 = R_{ch}R_w\kappa/s_{ch}s_w = w/d^5\sigma^2$. Hence two exhaust particulate filters that are said to be similar have the same resistance ratio α and the same cross-sectional open area fraction β.

- (3) A third possibility is to use the dimensionless channel density $1/\zeta^2 = \sigma L\kappa^{1/2}$ and the dimensionless wall thickness $\eta = w/L^{1/2}\kappa^{1/4}$. Two exhaust particulate filters have scale-similar viscous flow resistances if they have the same dimensionless channel density $1/\zeta^2$ and the same dimensionless wall thickness η. From this choice of dimensionless groups we also see that $L\kappa^{1/2}$ scales as $1/\sigma$, w^2, or d^2 for scale-similar exhaust particulate filters.

Having identified the above dimensionless groups that can be used to determine the overall dimensionless viscous flow resistance of an exhaust particulate filter, we can construct a universal viscous flow resistance diagram for exhaust particulate filters, where curves of constant $R_v\kappa^{1/2}$ are plotted in a plane with $\ln(R_{ch}\kappa^{1/2})$ as the abscissa and $\ln(R_w\kappa^{1/2})$ as the ordinate. On this diagram, lines of constant α have a slope of +1 and lines of constant β have a slope of −1, and curves of constant ζ and constant η form a curvilinear grid system. The aforementioned geometric scaling relations for the viscous flow resistance of the exhaust particulate filters can be demonstrated on this diagram, and comparisons between different filter designs are easily made.

Literature Cited

[1] Kays, W. M. *Trans. ASME* **72**:1067, 1950.
[2] Bird, R. B., Stewart, W. E., and Lightfoot, E. N. *Transport Phenomena*, Wiley, 2002.
[3] Konstandopoulos, A. G., Skaperdas, E., Masoudi, M. *SAE Technical Paper* 2001-01-0909, 2001.
[4] Payri, F., Broatch, A. Serrano, J. R., and Piqueras, P. *Energy* **36**(12):6731–6744, 2011.
[5] Konstandopoulos, A. G., Skaperdas, E., Warren, J., and Allansson, R. *SAE Technical Paper* 1999-01-

0468, 1999.
[6] Konstandopoulos, A. G. and Kladopoulou, E. *SAE Technical Paper* 2004-01-1133, 2004.
[7] Konstandopoulos, A. G. and Johnson, J. H. *SAE Technical Paper* 890405, 1989.
[8] Acrisvos, A., Babcock, B. D., and Pigford, R. L. *Chem. Eng. Sci.* **10**(1-2):112–124, 1959.
[9] Datta, A. B. and Majumdar, A, K. *Int. J. Heat Fluid Flow* **2**(4):253–262, 1980.
[10] Pigford, R. L., Ashraf, M. and Miron, Y. D. *Ind. Eng. Chem. Fundam.* **22**(4):463–471, 1983.
[11] Tropea, C., Yarin, A. L., and Foss, J. F., eds. *Handbook of Experimental Fluid Mechanics*, Springer-Verlag, 2007.

(98f) Controlled Production of Double Emulsions in Dual-Coaxial Capillaries Device for Millimeter-Sized Hollow Polymer Spheres

Shao, T., Tsinghua University
Jin, Y., Tsinghua University
Cheng, Y., Tsinghua University

Hollow polymer particles with diameters ranging from nanometers to millimeters have wide applications in various fields such as sensors, drug delivery, food industry, pigments, microcapsules and catalyst loading because of their unique properties. Such particles with different diameters usually lead to different applications. Thin-walled shells with diameters in millimeters have their special applications in the area of high energy density physics such as inertial confinement fusion (ICF) experiments. In the ICF experiment, controlling the size, sphericity and shell thickness uniformity of the targets is particularly important. Therefore, controlling the geometries of the hollow polymer spheres used as mandrels for the fabrication of the ICF targets is quite essential. One important aspect of this task is to control the size and shell thickness of the hollow polymer spheres.

The most commonly used method to produce millimeter-scale polymer shells is the emulsion-based microencapsulation technique which involve three steps: the emulsification of the polymer solution encapsulating other substance(s), the solidification of the middle layers of the double-emulsion droplets, and the removal of the encapsulated substance(s). The first step is quite essential to determine the final sizes and their uniformity of the hollow particles. Therefore, size control of the initial water-oil-water (W/O/W) double emulsions is quite important.

Microfluidic technique based on milli- or micro-channels has its special advantages on the controlled production of double emulsions. Based on the developed milli- or microfluidic devices, numerous studies, experimental or theoretical, have been conducted on the single-emulsion droplet size and double-emulsion formation mechanisms. However, few researches have been done systematically on the sizes of the double-emulsion droplets, especially of the millimeter-scale ones.

In present work, we first designed an easy assembling-disassembling dual-coaxial capillaries device for the production of double-emulsion droplets, where polystyrene dissolved in fluobenzene was employed as the oil phase. Two configurations of the co-flowing devices were employed in the experiments. In the so-called one-step device, the inner capillary tip locates at the same cross-section of the middle capillary tip. Thus, the core drop and the shell drop depart from their capillaries ends simultaneously, forming a double-emulsion drop in one step. While in the so-called two step device, the inner capillary tip locates upstream to the middle capillary tip. The core drop and the shell drop break off from their respective capillaries ends successively, forming a double-emulsion drop in two steps. The effect of the fluids flow rates on the double-emulsion droplets size as well as their formation mechanism was studied comprehensively in the proposed devices. Experimental results implied that slight difference in device configurations would lead to significant difference in droplet formation mechanisms and thus size-control laws. In the two-step device, the inner stream of fluid had little influence on the breakage and the outer diameter of the double-emulsion drops, while in the one-step device the effect was significant. With the produced double-emulsion droplets, millimeter-sized polymer capsules with single hollow cavity and ultra-thin shells were obtained by solidifying the oil layers of the compound drops using solvent evaporation method.

(98h) Numerical Simulation of Wet Gas Flow At Orifice Plate Meters

Almeida, L., ESSS
Teixeira, B. D. P. A.,
Moreira, R. P. M., ESSS
Costa, C. B.,
Quintella Coelho, E. F. A., Petrobras
Ropelato, K., ESSS

Orifice plate meters are one of the most widely used technologies in industry for gas flow metering, due to their simplicity. The measurement principle of an orifice plate meter is based on the pressure differential across the orifice plate due to the passage restriction. Their performances in single phase flow metering have been studied for many years, leading to extensive data sets and are well documented in the standards.

Wet gas metering is an important problem to many industries, in particular the oil and gas industry, and it is common for gas flow meters to be installed in applications where some liquid entrainment exist in a predominantly gas flow. The effect of wet gas flow on differential pressure meters is complicated and there are several on going research programs worldwide aiming to improve the understanding of the response of these meters to wet gas flow, as the accuracy of such meter systems is crucial in applications like production process control and fiscal metering.

Assuming this pressure differential to be caused only by the gaseous phase, orifice plate meters tend to over-read the true gas flow due the presence of the liquid. The uncorrected gas mass flow rate prediction is often called the apparent gas mass flow, and the over-reading is defined as the ratio of the apparent to the actual gas mass flow. The Lockhart-Martinelli parameter (X_{LM}) is commonly used to relate the liquid mass flow to the gas mass flow and several researchers concluded, from experimental data, that the over-reading is a function of the X_{LM} parameter and gas to liquid density ratio. Therefore, some correlations that have been proposed to correct the over-reading are based on it.

The use of computational fluid dynamics (CFD) as a design tool is increasing as the computational power and available CFD codes efficiency rises. It allows obtaining details of hydrodynamics within the device which would be impossible to be obtained by experimental methods, giving a deeper insight into underlying physical mechanisms.

This work aims to develop a methodology to investigate the flow of wet gas across a standard orifice plate by using CFD and compare the obtained results with correlations available in literature. The two-fluid model, which is based on an Eulerian-Eulerian approach, is used to model the two-phase flow inside the device. It solves a set momentum and continuity equations for each phase and coupling is achieved through the pressure and interphase exchange coefficients. The gas is considered the continuous phase, following the ideal gas law, whereas the liquid is treated as the secondary phase, Newtonian and incompressible fluid. To deal with the turbulence influence, the RANS modeling is used, in which the Navier-Stokes equations are time averaged and the arising terms are modeled using the turbulent viscosity concept, according to Boussinesq hypothesis. The realizable κ-ε model, which is based on the modeling of the turbulent kinetic energy and turbulent dissipation rate, is used, with the enhanced version of the wall function treatment.

The pressure is measured 1 in upstream and downstream the plate. The predicted pressure drop is then compared to experimental data from NEL (National Engineering Laboratory) for a 4 inch diameter pipe for different liquid amounts (X_{LM} ranging from 0 to 0.3) at 15 bar. The over-reading is then compared and several correlations, as the Chisholm equation, are evaluated.

(98j) Shear-Induced Ordering and Aggregation in Dense Soft Particle Suspensions

Chen, Y. L., Academia Sinica
Huang, W. Y.,
Wu, I. F.,

The flow of dense soft particulate suspensions is ubiquitous in food processing and blood flow. We investigate the effect of particle aggregation and shear-induced ordering on suspension properties with multi-scale computational models. Under shear, dense particles suspension (> 65% packing) with amorphous structure may be driven to crystalline packing. In addition, particle deformation under shear forces leads to anisotropic structure, which confirms previous experimental observations. In dense suspensions, non-Newtonian dynamics of a mono-disperse soft particle suspensions may be due to the change in the inter- and intra-particle structure under flow. We closely examined the coupling between flow and particle aggregation. Flow is found to both enhance and hinder particle aggregate and cluster-formation depending on the shear force, particle elasticity, particle shape, and inter-particle attraction. Depending on the average aggregate size and particle deformation under shear, the non-linear dependence of suspension viscosity on the shear rate is characterized and found to agree with prior measurements of blood rheology. The computational model is capable of predicting particulate suspension rheology given parameter-engineered particles.

(98k) Flexible Branched Polymers for Single Molecule Rheology

Mai, D. J., University of Illinois at Urbana-Champaign
Marciel, A. B., University of Illinois Urbana-Champaign
Brockman, C., University of Illinois at Urbana-Champaign
Schroeder, C. M., University of Illinois at Urbana-Champaign

In this work, we report the synthesis of well-characterized branched biopolymers for single molecule rheology. We extended the method of enzymatic polymerization of long linear polymers consisting of single-stranded DNA (ssDNA) to include bioorthogonal attachment chemistries, thereby enabling the synthesis of well-defined, branched ssDNA polymer molecules. The resulting polymers are suitable for single-molecule flow studies, which have potential to improve the development of molecular-based theories for topologically-complex polymer melts and solutions. Recently, our lab has pioneered the use of ssDNA as a new model system to study the dynamics of flexible polymer chains. ssDNA provides a superior alternative to double-stranded DNA (dsDNA), which has served as a model for single molecule polymer dynamics in flow for the past two decades. dsDNA is well known as a semi-flexible polymer with markedly different local molecular properties and a large persistence length ($l \approx 53$ nm) compared to flexible polymer chains, such as polystyrene ($l \approx 0.7$ nm) and ssDNA ($l \approx 0.6$ nm). In prior work, we have successfully synthesized long, fluorescently labeled ssDNA chains, which we have studied in extensional flow with epifluorescence microscopy.

In the current study, we synthesize and characterize branched polymers based on ssDNA for single molecule visualization. We rely on enzymatic template-directed polymerization to incorporate a chemically modified deoxyribonucleotide at pre-determined locations along the backbone. We utilize the modified nucleotides as "grafting sites" for facile integration of azide-terminated single stranded DNA branches via copper-mediated alkyne-azide cycloaddition or strain-promoted Cu(I)-free [2+3] cycloaddition "click" reactions. Copper-free click reactions are bio-orthogonal and nearly quantitative when carried out under mild conditions. Using this approach, we have synthesized branched polymers with three arm star, H-shaped, and comb architectures. In future work, we will utilize a microfluidic-based, automated, hydrodynamic trap to observe various timescales of branched polymer behavior by varying and oscillating strain rates. Overall, we seek a molecular-based understanding of the non-equilibrium dynamics of flexible polymer chains, which is crucial for control in processing and molecular self-assembly.

(98i) Forced Spreading and Evaporation of Films and Droplets of Colloidal Suspensions

Esp n, L., University of Minnesota
Kumar, S., University of Minnesota

When a thin film of a colloidal suspension flows over a substrate, uneven distribution of the suspended particles can lead to an uneven coating. To gain insight into this phenomenon, we analyze the flow of films and droplets of colloidal suspensions down an inclined plane. Lubrication theory and the rapid-vertical-diffusion approximation are used to derive a coupled pair of one-dimensional partial differential equations which describe the evolution of the liquid height and particle concentration. A precursor film is used at moving contact lines, the colloidal particles are assumed to be hard spheres, and particle and liquid dynamics are coupled through a concentration-dependent viscosity and diffusivity. We find that for sufficiently high Peclet numbers, even small initial concentration inhomogeneities produce viscosity gradients that cause the film front to evolve continuously in time instead of traveling without changing shape as happens in the absence of colloidal particles. At high enough particle concentrations, particle diffusion can lead to the formation of long-lived secondary flow fronts. The influence of evaporation is also examined, and it is found that evaporation can strongly suppress the effects of viscosity gradients. Our observations suggest a possible mechanism for the onset of patterns that are observed experimentally in flowing films of colloidal suspensions.

(98m) Pump Characterization for Integration With a Microfabricated CNF Array Sensor

Floyd-Smith, T., Tuskegee University
Amey, M.,
Douglas, D.,
Truss, W.,

Tuskegee University, in collaboration with NASA Ames, is integrating a **C**arbon **N**ano**F**iber (CNF) array sensor with microchannels, external pumps and detection capability. The focus of this study is the external pumps that were considered. Peristaltic, piezoelectric and electro-osmotic pumps were evaluated for use with the microfabricated CNF array biosensor. Specifically, applied voltage versus flow rate curves were obtained for all three pumps with water as the pumped fluid. Additionally, the source frequency was considered for the piezoelectric pump. For the peristaltic pump, the application of 5 V yielded a flow rate of 1,000 μL/min with linear behavior over the range of applied voltages. For the piezoelectric pump, at 50 Hz and 250 V, the flow rate was 4,200 μL/min. Lastly, for the electro-osmotic pump, an applied voltage of 12 V yielded a flow rate of 60 μL/min. Overall, the results are consistent with the manufacturer's specifications and suggest that each of the pumps is suitable for different flow rate and power requirements. The results of this study will be presented in poster format.

(98l) Research On the Dynamics of Liquid Film Flowing Down Different Corrugated Wall

Li, J.,
Tong, Z. Y., ZheJiang University
Hong, W. R., ZheJiang University

Abstract: The packing structure and surface geometry can change the flow pattern, and has significant influence on the mass transfer between liquid film and gas phase in packing column of chemical industry. It is necessary to investigate the flow pattern and factors that affect the tower efficiency. The Packing with different structure including triangular corrugation, trapezoid corrugation, rectangular corrugation and sinusoidal corrugation are investigated in this paper. Computational fluid dynamic simulations based on volume of fluid method by open source software OpenFOAM, are conducted to study the liquid film over these corrugations. It is found that the area of the free surface between liquid and gas reaches its maximum when the resonance phenomenon occurs, and the rectangular corrugation is of biggest free surface. It is also found that liquid film with bigger volume in the trough of corrugations encounters higher resonance Reynolds number, and the free surface has a bigger area in this condition. The obtained conclusions provide a theoretical basis for the study of mass transfer in packing column industry applications.

(98n) Dynamics and CFD Modeling of Viscoelastic Polymeric Jets Under Rayleigh Instability As a Result of a Controlled Vibration to Produce Size- and Span-Controlled Microdroplets

Rodriguez-Rivero, C., University of Salamanca
MartÃn del Valle, E. M., University of Salamanca
Galan, M. A., University of Salamanca

We explore the dynamics of laminar Rayleigh-type instabilities of viscoelastic jets in inviscid mediums (namely a liquid jet injection into quiescent air at atmospheric pressure). It has been long known that this type of surface-tension-driven instabilities evolve as free-surface waves whose amplitudes grow with time and leads to an eventual break of the jet, forming droplets.

These instabilities are of primary importance for certain processes such as ink-jet printing, drug processing or biomedical applications, where a proper control of the interfacial instability leads to a controlled breakup into mono-sized droplets.

Besides the understanding of the physical process, the primary goal of this study lies in optimizing a modified commercial method to produce solid microparticles (diameters ranging from 300 to 600 um) with applications in areas such as biomedicine.

The system that we are examining applies a mechanical vibration to a capillary laminar flow of high viscous polymer in order to destabilize the jet in a controlled way. The natural polymer used – sodium alginate – shows high biocompatibility and biodegradability. In order to produce microparticles with high mechanical stability, a medium- to high-viscosity sodium alginate solution must be used. In a previous work we obtained semiempirical expressions from experimental conditions in which the effect of surface tension, flow rate and zero-shear viscosity of the samples enabled us to roughly predict microparticle size but where the effect of the vibration conditions or the viscoelastic nature of the samples were not considered.

We found that the equations from a linear approximation considering Oldroyd-B as the viscoelastic model describe the process accurately, taking into account a larger number of parameters than those analyzed before. The dispersion equation derived from the system of equations relates perturbation growth rate and superficial wave characteristics and the obtained dispersion curves enable us to establish the best conditions for the perturbation source (theoretical dominant wavelength and the optimal frequency) and to control and minimize the droplet size and distribution span.

On the other hand, we have conducted CFD studies to validate the system behaviour with an Euler-Euler approach. We found good qualitative results as for the bulk liquid characteristics (breakup length and droplet size) by applying generalized-Newtonian models. We are carrying out further simulations to obtain a more accurate ligament-bead description.

(98p) Theoretico - Experimental Based Design of a Fan Jet Nozzle for Microparticle Production Using Non Newtonian Fluids

Rodriguez-Rivero, C., University of Salamanca
Graf, B.,
S nchez, C., University of Salamanca
Mart n, M., University of Salamanca
Mart n del Valle, E. M., University of Salamanca
Galan, M. A., University of Salamanca

Particle size is one of the key parameters for chemical and biomedical purposes when producing microparticles. Many current and new applications in this field are based on the breakup of liquid jets. In particular we deal with viscous polymeric solutions where the control of the particle size is carried out by the proper hydrodynamics of the jet generated. The hydrodynamics is mainly governed by the combination of the rheology of the polymer and the physical design of the nozzle. On the one hand, the microparticle properties must meet certain characteristics depending on the final use (i.e. biocompatibility, biodegradability, mono modal particle size distribution, particle strength when formed, etc.) defining the type of polymers used. On the other hand, the design of the nozzle determines the stresses, and consequently the breakup, on the jet.

The basic geometry of a fan-jet nozzle consists of a pipe that guides the liquid into a gas chamber just before the exit. This chamber is responsible for the air focusing effect on the liquid jet. The gas-liquid interaction generates instabilities into the jet that determine droplet (size) and particle dispersion. Typically, the nozzle designs, this is their geometrical features, are fixed and the operating parameters are reduced to the air and liquid pressures together with the liquid properties [1-4].

In this work we have used a fan jet nozzle with a new design done by the group. It has a variable height for the injection point for the liquid into the gas chamber in order to be able to vary the stream lines of the gas-liquid contact. We used high speed video techniques (Phanton v-310), laser particle measurement, and a proprietary pressure control design to investigate on the nozzle design that allows controlled particle size and jet break up. The study has two parts. First we evaluate jet stability as function of the gas – liquid stream lines and pressure patterns, in order to determine the break up characteristics of the jet as function of the liquid phase and the operating conditions for different positions of the liquid injection. Second, we evaluate how this translates into particle size by recovering the droplets in a Barium Chloride aqueous solution. We use a range of viscous Non-Newtonian polymeric solutions, whose rheological behavior is determined in the lab and are characterized using Carreau model.

From the experimental point of view, in terms of jet instabilization, we correlate the operating conditions with the break up period, length and the amplitude of the wave generated. For long oscillating periods generated on the jet, the elasticity of the liquid solution maintains the stability of the jet while for shorter periods; the jet breaks up into droplets. Furthermore, pulse regime was also identified. With respect to nozzle design, there is an injection location where the particle size obtained is minimum, but this minimum is more pronounced for larger pressure difference between the gas and the liquid phases.

We follow a hybrid experimental and simulated based approach with ANSYS –FLUENT® to validate the flow pattern in the chamber and the jet break up. CFD analysis is also being carried out to evaluate the process towards quicker nozzle design and optimization. We simulate the liquid flow through the pipes, the air flow inside the gas chamber, the contact between both phases and the instabilization of the liquid jet (primary break-up) based on an Euler-Euler approach. This part of the study allows for evaluating the effect of location of the injection point on the pressure and velocity patterns on the jet. On the other hand the second breakup is modeled using a Lagrange –Euler approach to account for the final particle size.

[1] .M. Gañán-Calvo, Generation of Steady Liquid Microthreads and Micron-Sized Monodisperse Sprays in Gas Streams Phys. Rev. Lett. 80 (1998) 285- 288

[2] X. Jiang, G.A. Siamas, K. Jagus, T.G. Karayiannis, Physical modelling and advanced simulations of gas–liquid two-phase jet flows in atomization and sprays Progress Energ. Comb. Sci. 36 (2010) 131-167

[3] J.M. Cervero, J. Nogareda, E.M.M. del Valle, M.A. Galan Development of a technology to produce monodispersed microparticles based on the formation of drops from viscous non-Newtonian liquids sprayed through a fan jet nozzle. Chem. Eng. J. 174(2-3) (2011) 699-708

[4] Nogareda, J, Rodríguez-Rivero, C., Martín, M, E.M.M. Del Valle, M.A. Galán, Non Newtonian CFD modeling of drops generated through fan jet nozzles for biomedical applications (2013) Rev. Submitted.

(98r) Ultrahigh-Salt-Tolerant Friction Reducer for Slickwater Fracturing in Shale Reservoirs

Zhou, M. J., Tulane University
Sun, H., Baker Hughes
Qu, Q., Baker Hughes
Guerin, M.,

Currently inverse-emulsion polymers are the most popular friction reducers used in fresh water slickwater fracturing in shale reservoirs. The fluid volumes in fracturing treatments have increased substantially, while water supply has become more of a public concern. Rather than paying to treat and dispose of produced and flowback water, operators would like to reuse it in subsequent stimulation treatments. Produced water, especially from shale plays such as Marcellus and Bakken, is known for its high total dissolved solids (TDS) and high divalent cation content. This poses extreme challenges for current friction reducers because cations hinder the inversion of friction reducers and cause loss of efficiency of friction reduction to below 30%. Treating produced water to the quality suitable for conventional fracturing fluids is time-consuming and often cost-prohibitive.

A salt-tolerant, water-based friction reducer was developed to address the challenges of high-TDS produced water. It was tested in a friction loop in a wide variety of produced water types from Bakken, Marcellus, Eagle Ford, Permian basin and other shale plays at different temperatures and was found to be highly effective. In a produced water sample with TDS over 300,000 ppm and total hardness as $CaCO_3$ over 90,000 ppm as representative water, the new polymer hydrates within 10 seconds and gives a friction reduction profile similar to that of current inverse-emulsion friction reducers in fresh water. The fluid is compatible with other common stimulation additives such as scale inhibitors, biocides, clay stabilizers, surfactants, and breakers, as substantiated with both friction loop and core flood tests.

This paper will discuss the evolution of the technology and present friction reduction performance in various high-TDS water samples under lab conditions. Significant performance improvement over the conventional friction reducers will be shown. All results illustrate the novel friction reducer is efficient in many different waters. It provides the oilfield industry a cost-effective solution of reducing produced water disposal and fresh water demands, thereby ultimately improving environmental impact of well operations.

(98s) Micro-Convective Mixing Influenced Size Selection of Liposomes Self-Assembled in Microfluidic Devices

Sunthar, P., Indian Institute of Technology Bombay
Phapal, S. M., Indian Institute of Technology Bombay
Has, C., Indian Institute of Technology Bombay

Microfluidic synthesis of liposomes encapsulating drug molecules has been suggested as a way to make in-situ formulations for drug delivery. The size and concentration of liposomes so produced are important parameters in a formulation design as these determine the dosage of the drug. It was thought that diffusion and self-assembly of lipids in laminar flow micro-fluidic devices, aided by length scales imposed by hydrodynamic flow-focussing could determine the size of the liposomes. Here we present experimental evidence and preliminary calculations that show that it is actually an instability that induces micro-convective mixing in the flow, which is otherwise in a high Peclet regime for diffusion to have any influence on the mixing and hence on the size of liposomes. The instability arises due to viscosity stratification between ethanol (used to dissolve the lipids) and water, that form the core and annular flow regions, respectively, in the microfluidic flow. Detailed predictions of size may now be possible when the instability is studied carefully by numerical methods.

(98t) Dynamic Rheology and Microstructure of Shear-Banding Wormlike Micellar Solutions Using 1-2 Plane Flow-SANS

Gurnon, A. K., University of Delaware
L pez-Barr n, C. R., University of Delaware
Porcar, L., Institut Laue-Langevin
Wagner, N. J., University of Delaware

Polymer-like micelles (PLMs, also worm-like or thread-like micelles) are models for understanding polymer physics as well as of technological interest across a broad range of applications ranging from enhanced oil recovery to many consumer products. A key phenomenon is the ability to formulate PLM solutions with very high zero shear shear viscosities (e.g. yield stress) that exhibit a "breaking viscosity", such that they flow rapidly upon a sufficient applied stress. Prior investigations have elucidated many aspects of the shear banding responsible for this viscosity behavior, yet despite its relevance, our understanding of the dynamic rheology is incomplete. To address this, large amplitude oscillatory shear (LAOS) measurements have been proposed as a characterization method and these have revealed intriguing, nonlinear dynamic rheological behavior. In this work we study this nonlinear dynamic rheology and understand its significance and source by measuring the microstructure of the PLMs during deformation using newly developed instrumentation implemented on SANS beamlines at the ILL Grenoble and the NCNR, NIST. A well-characterized formulation of 6 wt% cetylpyridinium chloride (CPCl) and sodium salicylate (NaSal) (2:1 molar ratio) with 0.5 M NaCl in D_2O is studied over a range of frequencies and amplitudes of LAOS. The microstructure on the segmental length scale is measured during LAOS via a newly developed method of time-resolved SANS in the velocity-gradient (1-2) plane and velocity-vorticity (1-3) planes of shear (tOR-SANS) [C. R. Lopez-Barron, L. Porcar, A. P. R. Eberle and N. J. Wagner, *Physical Review Letters*. **2012**, 108 (25), 897-908)]. This enables both time and spatial resolution of the microstructure during oscillatory flow. Measurements made in the velocity –velocity gradient (1-2) plane of shear are used to quantitatively define the microstucture by the micellar segmental alignment factor and orientation angle. The shear stress measured by rheology is successfully predicted using these microstructure parameters and the stress-SANS rule derived from complementary steady-shear data. We attribute the anomalous rheology observed in LAOS to an apparent shear-induced phase separation (SIPS) that occurs at longer length scales than those measured during SANS, but which can be observed as a "butterfly pattern" during a complementary rheo-small angle light scattering methods (SALS) [P. Thareja, I. H. Hoffmann, M. W. Liberatore, M. E. Helgeson, Y. T. Hu, M. Gradzielski and N. J. Wagner, *Journal of Rheology*. **2010**, 55 (6), 1375-1397]. The techniques used here to probe the material properties and microstructures of this WLM solution elucidate the microstructure responsible for the intriguing and useful nonlinear dynamic rheology. These measurements are invaluable for testing rheological constitutive models proposed for these solutions.

(98u) An Efficient Stochastic-Based Approach For Determination of Effective Diffusion Coefficient in Poiseuille Flow

Golbayani, P., Tennessee Technological University
Machida, M., Tennessee Technological University
Sanders, J. R., Tennessee Technological University
Arce, P. E., Tennessee Technological University

AN EFFICIENT STOCHASTIC-BASED APPROACH FOR DETERMINATION OF EFFECTIVE DIFFUSION COEFFICIENT IN POISEUILLE FLOW

PARVIN GOLBAYANI[1], MOTOYA MACHIDA[2], J.R.SANDERS[1], PEDRO E. ARCE[1]

[1]Department of Chemical Engineering, Tennessee Technological University, Cookeville, Tennessee

[2]Department of Mathematics, Tennessee Technological University, Cookeville, Tennessee

ABSTRACT

Taylor- Aris dispersion between two parallel plates with Poiseuille flow has been calculated, generally, by up-scaling the species-continuity equation using, for example, area averaging approach. In contrast, in this study, a stochastic-based method, to calculate Tylor-Aris dispersion, is explained where the process is modeled by stochastic differential equation (SDE). This is based on an exact solution of SDE with the hydrodynamic biased force (drag force) as the biased agent. The analysis involves the analytical solution of the equation of motion of a Brownian particle by using a Langevin equation as opposed to its counterpart, the Fokker-Plank equation. Results of this investigation are consistent with Taylor- Aris dispersion limit based on continuum mechanics. This, to the best of our knowledge, would be the first effort to determine biased diffusion coefficients directly from fundamental principles and without a scaling-based approach. The approach is the same that the one used in determine diffusion coefficients by using a Brownian motion-based methodology. Generalization to other regimes (beyond) the Taylor-Aris limit will be also illustrated. Numerous applications in bioseparation, mixing, and other similar ones are relevant to this effort.

(98v) Molecular Dynamics Simulations of Superspreaders

Isele-Holder, R. E., RWTH Aachen University
Ismail, A. E., RWTH Aachen University

Superspreading is the greatly enhanced spreading of aqueous droplets on hydrophobic surfaces due to the addition of trisiloxane surfactants, the so-called superspreaders. Because of the relatively small number of known superspreaders, and because of the wide range of industrial applications of superspreading—including coatings, cosmetics, paints, and textiles—numerous theoretical, experimental, and computational studies have been undertaken to uncover the secrets behind this process. However, the driving mechanisms of superspreading are still barely understood, largely because the relevant processes at the molecular scale cannot be resolved in experiments. The behavior of trisiloxane surfactants at the water-air and water-substrate interfaces and in solution, especially at the leading edge of the spreading droplet, is currently still subject to speculation.

We use advanced molecular dynamics simulations and optimized force fields to address open questions regarding the physical behavior of trisiloxane surfactants. We perform simulations of trisiloxane surfactants at the water-air interface as well as at the interfaces of hydrophobic substrate materials in contact with water. Moreover, we perform simulations of droplet spreading for aqueous solutions of trisiloxane surfactants and droplets of pure trisiloxane surfactants on various hydrophobic substrates to explore the mechanisms of superspreading and terraced wetting. In addition, we perform reference simulations with surfactants that do not lead to superspreading to gain additional insight into the relevant mechanisms of this fascinating phenomenon.

(98b) Departure From Stokes Drag Coefficients for Spherical and Non-Spherical Particles in Microducts

Hensley, Z. D.,
Papavassiliou, D. V., The University of Oklahoma

Drag has been studied in great detail over the last century providing many equations that predict drag at various Reynolds numbers. Creeping flow has become an increasingly more studied regime with the development of DPD (dissipative particle dynamics) and micro fluidics. Originally Stokes' Law was generated to describe drag in creeping flow, however it neglects the concepts of wall correction and non uniform velocity profiles. Fayon and Happel introduced a force correction for sphere sedimentation based on the data obtained by McNown in the early 1960's (Brenner and Happel, 1965). Fidleris has ran experiments that confirmed that the radius of the sphere divided by the radius of the pipe was the critical parameter when evaluating drag in non uniform flow (Fidleris and Whitmore, 1958). Levenspiel expanded these empirical equations to include a larger range of Reynolds numbers and different shaped particles (Haider and Levenspiel, 1988). Recently Ganser generated correction factors K_1 and K_2 along with a C_D function similar to that of Levenspiel to apply for all particles for a large range of Reynolds (Ganser, 1993). This paper uses computational fluid dynamics to establish the error associated with these equations at low Reynolds number. Simpler equations using the aspect ratios of (D/H diameter of particle/height of channel) and (L/D length/diameter) are used to describe the geometries of spheres, ellipses, and cylinders.

References

Haider, A. & Levenspiel, O. (1989). Drag coefficient and Terminal Velocity of Spherical and Nonspherical Particles. *Powder Technology*, 63-70.

Ganser, G. (1993). A rational approach to drag perdiction of spherical and nonspherical particles. *Powder Technology*, 143-152.

Happel, J. & Brenner, H. (1965). *Low Reynolds Number Hydrodynamics.* Englewood Cliffs: Prentice-Hall, Inc.

Fidleris, V. & Whitmore, R. L. (1958). The Physical Interaction of Spherical Particles in Suspension. *University of Nottingham*, 572-579.

(98x) Reduction of Turbophoresis By Polymer Additives

Nowbahar, A., University of California Santa Barbara
Sardina, G.,
Picano, F.,
Brandt, L.,

Turbophoresis is the migration and accumulation of particles towards the wall in wall-bounded turbulent flow. Its importance appears in many applications; for example, industrial processes related to coatings, paints, or inks, where metallic powders are added to polymeric solutions to obtain desired chromatic or thermo-chromatic features. In this study, we qualitatively and quantitatively observe the effects of drag reducing polymer additives on turbophoresis of inertial particles in turbulent channel flow. The analysis is based on data from a direct numerical simulation (DNS) of a polymeric turbulent channel flow modeled with FENE-P and laden with particles with different inertia, i.e. Stokes numbers. Data show that polymer additives decrease the turbophoretic drift. Reduction in turbophoresis is associated to an increase in particle transport through the channel. We establish that turbophoresis is reduced because of the smaller derivative of wall-normal fluid velocity fluctuations that are typical of drag reducing flows. Hence a reduction of turbophoresis should be a common feature of all drag reducing flows, e.g. with fibers or bubbles and MHD.

(98w) Origins of Periodic and Chaotic Dynamics in Microfluidic Loop Device

Maddala, J., Texas Tech University
Wang, W. S., Texas Tech University
Vanapalli, S. A., Texas Tech University
Rengasamy, R., Texas Tech University

The binary decision making of droplets in a microfluidic loop device produces complex spatio-temporal dynamics. It has been shown that a train of droplets entering the loop at fixed inlet spacing could exit the loop at periodic or chaotic time intervals. We observed that periodic behavior is an outcome of carrier phase mass conservation principle, which translates into a droplet spacing quantization rule. This rule implies that the summation of exit spacing is equal to an integral multiple of inlet spacing. This principle also enables identification of periodicity in experimental systems with input scatter. We find that the origin of chaotic behavior is through intermittency, which arises when drops enter and leave the junctions at the same time. We derive an analytical expression to estimate the occurrence of these chaotic regions as a function of system parameters. A network model is used to study the dynamic behavior of different inlet feeding frequencies. Using this model, a bifurcation map is generated. It is observed that the chaotic regions are observed in between islands of periodic behavior. We propose key insights regarding the behavior of droplets at bifurcations and show that these insights are not only important for the design of droplet based fluidic devices but more broadly, has relevance in understanding how decision making leads to multi-stability and memory storage in biological networks

(98ab) Producing Stable Nanoparticle Dispersions With Microfluidic Devices

Sun, J. Y.,
Liao, Y. C., National Taiwan University

In this study, we experimentally examined the effectiveness of applying conventional microfluidic devices to enhance dispersion stability of nanoparticles in liquid phase. Regularly, homogenizer and sonication bath are widely used to disperse nanoparticle powders in liquids by breaking large particles into small ones with vibrational energy. In the literature, similar breakage of particle-particle aggregates can also be accomplished by introducing large shear rate in microfluidic channels with obstructions. Following this idea, a microfluidic device able to provide large shear rates for particle disintegration was fabricated from plolydimethylsiloxane (PDMS) using soft lithography method. Rectangular and circular obstructions were placed inside the channel to introduce large transverse velocity components and shear rates to tear particle aggregates apart. Computational fluid mechanical models from commercial software were also used to calculate the shear stress along microchannels with optimum obstruction patterns. The inlet fluid stream containing nanoparticle aggregates (~ um) were driven into the channels by a high pressure syringe pump. The outlet stream fluids were collected for particle size analysis. After the microfluidic treatment, the particle sizes were found to be greatly reduced and thus the dispersion stability was enhanced.

(98y) The Nature of Branching During Hydrodynamic Instability of Miscible Displacement in Porous Media

Johns, L. E., University of Florida
Narayanan, R., University of Florida

Miscible displacement in a bounded fluid-saturated porous layer introduces instabilities due to the viscosity variation caused by temperature variation in the layer and due to a viscosity difference across the surfaces separating the layer from the surrounding fluid. We find that the problem in the layer itself is very similar in many ways to the B´enard problem with the neutral curve exhibiting the characteristic dip leading to the possibility of many flow cells at the critical speed of displacement.

The viscosity difference across the surface modifies the dip, slightly at large surface tensions, completely at low surface tensions. The effect of the surface deflection is present whether or not the surface is stable and can be omitted only if the viscosity of the driving fluid is very high or the surface tension is very low.

The nature of the branching at the critical speed depends on the cross section of the fluid layer and on how much of the dip in the neutral curve remains.

(98aa) A New Boundary Treatment for Complex Geometries in Smoothed Particle Hydrodynamics

Eitzlmayr, A., Graz University of Technology
Radl, S., Graz University of Technology
Kloss, C., Johannes Kepler University Linz
Koscher, G., Research Center Pharmaceutical Engineering GmbH
Khinast, J. G., Graz University of Technology

Smoothed Particle Hydrodynamics (SPH) is a Lagrangian particle method, which was originally developed in astrophysics to simulate flows in boundary-less domains. Nowadays, the method is mostly used to solve the governing equations of fluid flow, including free surface flows[1].

However, the treatment of wall boundaries in SPH is not unique and different approaches exists, mostly based on particles[2,3]. However, engineering applications often require complex shaped geometries, usually generated by CAD programs, e.g. in the STL format, where an appropriate use of boundary particles or ghost particles would be highly complicated or even lead to wrong results (e.g. at corners). Today it is not obvious how to handle such STL geometries properly in SPH.

In our work, we developed a new approach to model the interaction between SPH particles and wall triangles (i.e., the STL geometry), which can easily be implemented in a SPH code and lead to a proper behavior of fluid particles near walls.

We use the open-source particle simulator LIGGGHTS (www.liggghts.com), originally developed for the simulation of granular flow via the Discrete Element Method (DEM), also providing an SPH module. Our boundary treatment together with the viscosity model of Morris et al.[2] leads to a close agreement of the velocity profiles with the analytical solution for a channel flow between two plates, specifically an accurately fulfilled no-slip condition (see Figure 1).

The main focus of our research is the application of SPH to the simulation of mixing in hot melt extrusion processes (HME), which attracted increasing attention in pharmaceutical manufacturing in recent years. The typically used co-rotating twin-screw design consists of a complex shaped, rotating geometry, highly challenging for mesh-based methods due to the deformation of the free volume caused by the screw rotation. Our new method for boundary handling was used to simulate the mixing process within the geometry of a co-rotating twin-screw extruder (snapshots of the distribution of two different particle types in the cross section of a twin-screw are shown in Figure 2).

The presented approach provides a robust basis for efficient simulations of fluid flow in complex, moving geometries in the STL format.

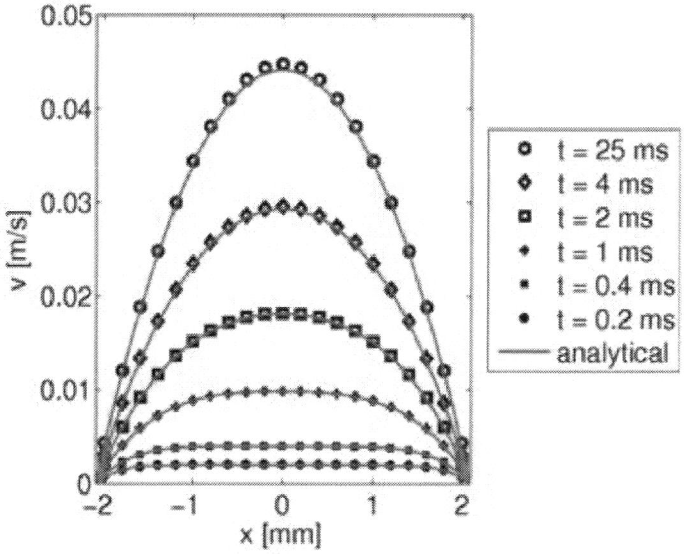

Figure 1: Velocity profiles in a pressure-driven channel flow: analytical solution, dots: SPH solution using the developed boundary conditions.

Figure 2: Snapshots of mixing in a cross-section of a co-rotating twin-screw extruder. Top: initial state, center after 0.05 revolutions, bottom after 1.05 revolutions.

References

1. Monaghan JJ. Smoothed particle hydrodynamics. *Reports on Progress in Physics*. 2005;68(8):1703-1759.

2. Morris JP, Fox PJ, Zhu Y. Modeling Low Reynolds Number Incompressible Flows Using SPH. *Journal of Computational Physics*. 1997;136(1);214-226.

3. Colagrossi A, Landrini M. Numerical simulation of interfacial flows by smoothed particle hydrodynamics. *Journal of Computational Physics*. 2003;191(2):448-475

(98ae) Kinetics of Colloidal Assembly By Direct Current Electric Fields

Ferrar, J. A., University of Michigan, Ann Arbor
Solomon, M. J., University of Michigan

We quantify and model the kinetics of the assembly of initially dilute colloidal spheres subjected to a DC electric field in a thin gap between parallel electrodes into crystalline structures, as well as the kinetics of the disassembly of these structures. We test our model's predictions with experimental measurements for a system of poly(12-hydroxystearic acid) (PHSA)-stabilized poly(methyl methacrylate) (PMMA) spheres dispersed in a mixture of cyclohexylbromide (CHB) and decalin containing 1 mM tetrabutylammonium chloride (TBAC). The transient evolution of the volume fraction of colloids in the electrode gap is quantified, under conditions of assembly and disassembly, by confocal laser scanning microscopy (CLSM) imaging. We model the kinetics by adapting a model for sedimentation (Davis and Russel, *Phys. Fluids A*, **1**, 82, 1989) to the DC electric field-assisted assembly and disassembly process. Parameters in the model are the electrophoretic velocity of the spheres in a DC field (which we characterize by direct measurement), the diffusivity of the spheres, and the volume fraction-dependent osmotic pressure of the suspension. The model's predictions agree well with the measured kinetics of the assembly and disassembly when the screened Coulombic interactions between the particles are accounted for. We also demonstrate that the assembly-disassembly process can be repeated multiple times by controlling the DC field. The results can inform the design of devices that rely on reversible assembly for potential applications in photonics, optics, and sensing.

(98ac) Biphasic Liquid Flows and Mass Transfer Characterization in Y Junction Microstructured Contactors

Sahu, A., Indian Institute of Technology, Madras, Chennai (T.N.) India
Pushpavanam, S., Indian Institute of Technology Madras

Microstructured contactor forms an important tool for process intensification in multiphase systems. Liquid-Liquid extraction is a key mass transfer operation in the chemical, food process and pharmaceutical industries. High mass transfer rates compared to conventional equipment and ease of handling of two-phase flows make microstructured contactors an ideal choice for extraction studies. Our work focuses on quantifying this improvement in mass transfer by carrying out extraction experiments in Y-Junction glass micro structured contactor with square cross-section of 150 x 150 µm. A mathematical treatment of these physical phenomena is done from basic flow equations and a model is developed for the same. Cyclohexane (organic) and water (aqueous) are chosen as solvents for flow regime characterization. Propionic acid is used as the solute to be extracted from the organic to the aqueous phase. The experimental results are used to determine the mass transfer coefficient.

Comparative analysis will be done for estimating mass transfer coefficients in batch, slug and stratified flows.

Effect of different solvents, slug sizes and the presence of thin film between slug and wall will also be highlighted. Conditions under which
microstructured contactors can be preferred to the conventional systems will be obtained.

(98ad) Intermittent Gas/Liquid Interfaces in a Micro-Channel for Liposome Production

Hidema, R., Kobe University
Suzuki, H., Kobe University
Mori, J., Kobe University
Komoda, Y., Kobe University

Liposomes produced from lipids have been studied as promising materials for drug delivery systems, since the liposomes are relatively non-toxic and biodegradable. Liposomes can keep drug insides and deliver the drug to tumor cells [1]. For using liposomes as a drug carrier, the size of liposomes and the homogeneity in their structure are the most important characteristics because they govern the stability and drug-release properties in the human body [2]. For example, liposomes whose diameter is less than 50 nm can be egested by kidney or caught by healthy cells, and liposomes whose diameter is larger than 200 nm can be eliminated by a liver. Thus, the sizes of 50 to 200 nm of liposome are suitable as a drug carrier to tumor cells. Therefore, in the production of liposomal formulations, the control of these characteristics is the main issue, and the reliable and cost-effective production method is common concern.

Bangham method is the most common technique to form liposomes [3]. Firstly, a liposomal lipid such as egg yolk lecithin is spread on the inner surface of a flask, and then an aqueous solution of a drug is added. This flask is shaken or churned to cause shear stresses, which peel off the thin layer of lipid and help to encapsulate the small droplet of aqueous solution with the lipid bilayer. Here, liposomes can be obtained. In this method, however, it is difficult to achieve a homogeneous and constant field of fluid stress, and therefore the resulting liposomes often have a variety of sizes and structures. This is disadvantageous because a screening process is necessary to obtain the liposomes of a desired size and structure, and consequently the yield becomes low. There are other some methods proposed to produce liposomes, but all of them are still required to be improved. For example, many methods use organic solvent that is harmful for human body, and many methods cannot control the size of the liposomes [4, 5].

In our previous studies [6], we proposed a micro-tube production system, which is a kind of micro-reactor [7]. In this method, a thin lipid film attached to the inner wall of a micro-tube as the first step. As the second step, the aqueous solution to be encapsulated in liposomes flows in the micro-tube. Indeed, this system achieved to obtain homogeneous liposomes compared with Bangham's method. The peak size of liposomes was determined by Reynolds number. The mechanism of liposome formation of this method is as follows. First, lipids on the wall of the micro-tube adsorb to the head of the flow, that is, the liquid/gas interface. Second, rod-like micelles were generated at the interface due to shear stresses caused by flow circulation in the flow close to the surface. After that, the rod-like micelles were separated to spherical liposomes. Thus, not the film of lipids but the gas/liquid interface catching lipids was confirmed only to be necessary for liposomes production [7].

In this study, a novel technique of liposome production using a gas/liquid interface in a micro-channel has been suggested. By this method, the liposomes can be obtained without harmful organic solvents and the miniaturization of liposome production can be realized.

As sample materials, egg yolk lecithin of 0.1 g as phospholipids is dispersed in 50 ml of deionized water by using ultrasonic. T-shape micro channel was used for experiments. A main channel has 400 mm width and 250 mm depth. A branch channel has 150 mm width and 250 mm depth. The main channel was used to flow dispersed phospholipids solution, and the branched channel was used to flow air. A syringe driver controlled the flow rate. The solution rate was changed from 0.0125 to 0.100 ml/min, and the air rate was changed from 0.025 to 0.200 ml/min. The flow condition of gas/liquid flow under each flow rate was observed. The effluent including liposomes has been recovered and analyzed. The effects of the flow velocity on the liposome size and on the yield of liposomes were investigated. The diameter of liposomes was analyzed with a dynamic light scattering system of Zetasizer NANO (Malvern Instruments Ltd.). The velocity of the gas/liquid interface was measured from recorded movies.

The apparent velocity determined from liquid and gas flow was the same as the real interface velocity measured by the movies, when the flow rate was low. However, the real flow rate calculated by the velocity of the gas/liquid interface in the recording movies became smaller than the apparent velocity. This implies the liquid slip through between the wall of the tube and gas column, or the gas was compressed. This mechanism was tried to be confirmed by observing the length of liquid column and the gas column under the real velocities.

The diameter of the liposomes determined as a peak diameter decreased with increase of the real flow velocities. The peak diameter divided by hydraulic diameter was plotted to Reynold's numbers defined by the hydraulic diameter of the micro-channel and water viscosity. The value of the peak diameter divided by the hydraulic diameter in the present study was compared to the previous data [8]. When the liquid flow rate was fixed at 0.0125 ml/min and when the gas flow rates were 0.025 and 0.5 ml/min, the value was fit to the previous data. On the other hand, when the gas flow rates were 0.0125 and 0.1 ml/min, the value did not agree with the previous data. This result indicates that the suitable length of the liquid column to form liposome exists. Lower gas flow rate leads to the formation of too short liquid column and higher gas flow rate makes too long liquid column for liposome production. The reasonable explanation we propose here is that too short liquid column cannot generate enough shear to make rod-like micelles and too long liquid column prohibit adsorption of lipids. It is concluded that there exists the optimum condition on the gas/liquid flow rate for producing suitable liposome in the present method.

[1] R. Schwendener, Bio-Applications of Nanoparticles Advances in Experimental Medicine and Biology, 620, 117 (2007)

[2] J. Zhu, F. Yan, Z. Guo, R. E. Marchant, Journal of Colloid and Interface Science, 289, 542 (2005)

[3] A. D. Bangham, Journal of Molecular Biology, 13, 238 (1965)

[4] A. Jahn, W. N. Vreeland, D. L. DeVoe, L. E. Locascio, M. Gaitan, Langmuir, 23, 6289 (2007)

[5] L. Lesoin, C. Crampon, O. Boutin, E. Badens, The Journal of Supercritical Fluids, 57, 162 (2011)

[6] H. Suzuki, J. Hamamura, T. Katsuda, Y. Komoda, S. Katoh, H. Usui, Journal of Chemical Engineering Japan, 41, 739 (2008)

[7] K. Mae, Chemical Engineering Science, 62, 4842 (2005)

[8] A. Fujiwara, H. Suzuki, T. Katsuda, Y. Komoda, The 13th Asia Pacific Confederation of Chemical Engineering Congress, 10388 (2010)

(98ag) CFD-Based Enhancement of Temperature Uniformity in Industrial Furnaces With Self-Recuperative Burners

Alan s, R. D., TecnolÃ³gico de Monterrey
Cant , A., Nutec Bickley
Montesinos, A.,

Natural gas used in industrial furnaces amounts for an important portion of the fuel global consumption. As much as 50% of the energy released by combustion inside hot enclosures is lost through the exhaust. A necessary and sustainable proposal is to seek the recovery of some of this energy, but centralized heat exchangers are severely limited in their effectiveness by sizing and temperature constraints. An alternative is to use a self-recuperative burner that merges a casing to create a premixed flame with a heat exchanger. In this process, up to 95% of the hot flue gases produced immediately preheat incoming combustion air with a recovery of up to 70% of the energy, that else would be lost. This approach ideally increases furnace efficiency and air preheat temperature, but also poses a design challenge on key specifications for product quality: Such as the pressure distribution, temperature uniformity and furnace fuel requirements. In this work, Computational Fluid Dynamics is employed to characterize turbulent flow and heat transfer profiles for two existing industrial furnaces: One with a self-recuperative setup and another with a traditional burner-exhaust configuration. In the former, profiles related to the burner configuration are investigated as well as the mixing efficiency; whereas in the latter, the conventional furnace is benchmarked against the recuperative setup thoroughly comparing impact of the burner design choice between the current. Relevant changes in design are then proposed for the conventional furnace to be successfully modified to a self-recuperative setup. Results indicate that a furnace with this new configuration, with a sufficiently controlled pressure in the recovery outlets, presents a sufficiently uniform temperature profile. In accordance to this result, the benefit obtained by using a self-recuperative configuration in fuel consumption and quality of heating should then be weighed against the related operative and capital costs of this implementation.

(98af) Impact of Nano-Drops On Solid Surfaces

Koplik, J., City College of the City University of New York
Zhang, R., CCNY
Farokhirad, S., CCNY
Lee, T., The City College of City University of New York

The impact of nanometer-sized drops on solid surfaces is studied by molecular dynamics computer simulations. Volatile and non-volatile
drops of pure liquid, surfactant solutions, and "liquid marble" drops consisting of particulate suspensions are considered. The simulation protocol is to equilibrate a drop made of short chains of Lennard-Jones atoms, with or without added surfactant (longer amphiphilic chains) or rigid solid (atomistic) particles, and direct them towards a solid substrate lattice. At high impact velocities we focus on the transition to splashing and comparison with the behavior of laboratory-sized drops. At lower velocities we address the effects of solid surface texture and wettability, and compare the results to lattice Boltzmann calculations for continuum drops.

(98ah) Capillary Instabilities of An Array of Parallel Polymer Stripes Confined in Another Immiscible Viscous Matrix

Zhang, Z., University of Colorado at Boulder
Ahn, D. U.,
Wang, L., University of Colorado at Boulder
Ding, Y., University of Colorado at Boulder

We study the simultaneous morphological evolution of parallel polymer stripes (PS in particular), which were non-axisymmetrically embedded in another immiscible polymer layer (PMMA), which was confined on a rigid substrate. Capillary instabilities of the confined PS stripes were observed when annealed at a temperature above the glass transition temperatures of both polymers. The influences of the stripe-to-matrix viscosity ratios and volume ratios on the morphological development were systematically examined. Both were found to modulate the instabilities and result in distinctive morphologies. When the matrix layer was thin, we observed the influences of the substrate confinement on the phase-correlation in the breakup of neighboring stripes, engendered by the non-deformable nature of the substrate. With the increase of substrate confinement, the simultaneous breakup of the parallel polymer stripes transitioned from non-correlation to in-phase, irrespective of the viscosity ratio between PS and PMMA.

(98g) Microstructural Processes During Yielding of Nanoemulsion Gels

Helgeson, M. E., Massachusetts Institute of Technology
Kim, J., University of California, Santa Barbara
Gao, Y., University of California, Santa Barbara

The transition to flow in colloidal gels is known to depend significantly on their network microstructure. However, little is understood about the differences in yielding between gels with the homogeneous structure typically found in dilute suspensions, and those with heterogeneity characteristic of more concentrated systems. In this work, we use nanoemulsion "organohydrogels" [Helgeson et al., Nature Materials, 2012, 11(4): 344-352] as a model system to explore these differences, whereby the interdroplet attractions can be dynamically tuned through a thermoresponsive polymer, allowing for the formation of gels with either homogeneous or heterogeneous cluster-scale microstructure. Large-amplitude oscillatory shear (LAOS) measurements reveal that the non-linear response shows an abrupt transition from elastoplastic strain softening to viscoplastic strain thinning at the yield point, whereas for heterogeneous gels a new behavior emerges where effective strain hardening appears and then disappears during a broad two-stage yielding transition . To probe the microstructural origins of this behavior, we present studies of rheo-small and ultra-small angle neutron scattering during yielding which provide detailed structural information on length scales ranging from the primary particle size to several microns. We find that, in general, yielding arises from a "top-down" cascade of microstructural processes, beginning with large-scale rupture of weak cluster bonds near the yield point, further cluster bond rupture and suspension beyond the yield point, and finally densification and breakage of clusters en route to terminal flow. The broad yielding transition in heterogeneous gels is related to the progressive melting of heterogeneous cluster-cluster correlations. These results suggest more attention should be given to the influence of large-scale microstructure on the yielding of attractive suspensions.

(98ai) Encapsulation of Anti-Traction Material to Maintain Water Content

Zukas, B., University of New Hampshire
Gupta, N. R., University of New Hampshire

Encapsulation of Anti-Traction Material to Maintain Water Content

Brian Zukas and Nivedita Gupta

Department of Chemical Engineering

University of New Hampshire, Durham, NH

An aqueous high concentration polymer solution has been proven effective as an anti-traction material (ATM). The anti-traction property however is highly dependent on the water content of the solution. Using emulsion polymerization, the ATM material will be encapsulated in a semi-impermeable encapsulant to increase the time over which the material remains effective regardless of the external environment. An effective encapsulant will prevent water loss over a wide range of temperatures, allow for the encapsulation of a large volume of material relative to the capsule shell thickness, and rupture easily under foot pressure. Different polymer shell materials, reaction times, and additives are tested to determine the optimum conditions for encapsulation. Water permeability of the capsule shell is measured by monitoring osmotic swelling of the capsule in an aqueous solution. The minimum force to rupture the capsule shell is measured as a function of reaction parameters and shell thickness. The capsules' carrying capacity of anti-traction material will be evaluated in conjunction with the force required to rupture the shell.

(98aj) Systematic Coarse-Graining of the Wormlike Chain Model for Dynamic Simulations

Koslover, E. F., Stanford University
Spakowitz, A. J., Stanford University

One of the key goals of macromolecular modeling is to elucidate how macroscale physical properties arise from the microscale behavior of the polymer constituents. For many biological and industrial applications, a direct simulation approach is impractical due to to the wide range of length and time scales that must be spanned by the model, necessitating physically sound and practically relevant procedures for coarse-graining polymer systems. We present a highly general systematic coarse-graining procedure that maps any detailed polymer model onto effective elastic-chain models at intermediate and large length scales, and we specifically focus on the wormlike chain model of semiflexible polymers. Our approach defines a continuous flow of coarse-grained models starting from the wormlike chain model, proceeding through an intermediate-scale stretchable, shearable wormlike chain, and finally resolving to a Gaussian chain at the longest lengths. Using Brownian dynamics simulations of our coarse-grained polymer, we show that this approach to coarse graining the wormlike chain model captures analytical predictions for stress relaxation in a semiflexible polymer. Since we can arbitrarily coarse grain the polymer in these dynamic simulations, our approach greatly accelerates simulations to capture time scales that were previously inaccessible. This efficient simulation methodology has the potential to dramatically broaden the range of materials and biophysical problems that can be captured by simulation and theory.

(98bd) A Simple Shear Cell for the Direct Visualization of Step-Stress Deformation Without Creep Ringing in Soft Materials

Chan, H. K., University of California, Irvine
Mohraz, A., University of California, Irvine

The application of external stresses to a viscoelastic material can induce complex rheological responses such as creep, shear-induced strengthening, and delayed yielding. Understanding the underlying physics of these phenomena requires the ability to characterize how the material's microstructure evolves during the deformation. Here, We introduce a custom-built stress-controlled shear cell coupled to a fast-scanning confocal microscope for direct visualization of constant-stress shear deformation in soft materials and complex fluids. The torque generator is a cylindrical Taylor-Couette system with a Newtonian fluid between a rotating inner bob and a free-to-move outer cup. A spindle / cone assembly is coaxially coupled to the cup, and transfers the torque exerted by the fluid to the sample of interest in a cone-and-plate geometry. We model and experimentally demonstrate the performance of our device in both steady state and transient experiments with different viscoelastic materials. Our apparatus can conduct unidirectional constant-stress experiments as accurately as most commercial rheometers, with the capability to directly visualize the flow field. Further, our step-stress experiments on viscoelastic materials are devoid of creep ringing, which is an advantageous aspect of the viscous coupling between the torque generator and the viscoelastic sample. We specifically demonstrate this aspect of our shear cell both theoretically and experimentally with a rubber-like crosslinked polymer whose viscoelasticity is well approximated by the Kelvin-Voigt model.

(98ak) Microfluidic Synthesis of Methane (sI) and Propane (sII) Hydrates

Flowers, B., The University of Alabama
Hartman, R. L., University of Alabama

Clathrate gas hydrates have long fascinated researchers and have been the focus of much in depth study over the past fifty years. Gas hydrates impact societal problems related to conventional and non-conventional energy production and storage in addition to global climate
change. Most laboratory studies on hydrate formation and agglomeration phenomena have been
carried out in high-pressure batch reactors. Thermal resistances and mass transport limitations
introduce challenges when hydrate formation and dissociation rates are of similar time scales.
However, the study of clathrate hydrates in microscale laminar flow offers that opportunity
to add new knowledge on hydrate science. Our high-pressure microfluidic synthesis and dissociation of methane (sI) and propane (sII) hydrates, on demand and in a matter of minutes,
shows the potential to bridge the knowledge gap between the laboratory study of hydrates and
the production systems that encounter them. In the present work, we will elucidate the rational
behind the high-pressure microfluidic synthesis of hydrates and the recent advances we have
made in the field.

(98al) Investigation of Multiphase Multicomponent Aerosol Spray From Pmdi Through Commercial Spacers

Sarkar, S., University of Connecticut
Chaudhuri, B., University of Connecticut
Kakhi, M., US Food & Drug Administration
Peri, P., US Food & Drug Administration

The purpose of the current study is to develop a numerical model which can characterize flow and the resulting particle metrics (surface deposition, PSD etc.) when a commercial suspension based pMDI is sprayed into a spacer. The effect of different variables like volumetric rate of coaxially flowing air, spacer geometry and delay are investigated in order to develop labeling guidelines for patient use for similar products. Flow behavior of the multiphase multicomponent aerosolic spray, in static air, and within the spacers (Aerochamber Plus and Optichamber Advantage) in the presence of coaxial air were simulated using phenomenological 2 equation RANS models . Computational geometries of the inhaler and spacers were realized and meshed (tet dominant) using ANSYS Workbench tools. The standard k- ω model with Lagrangian tracking available within ANSYS Fluent 14.0 was used to simulate spray emanating out of the Proventil inhaler in a 2 way coupling mode with concurrent coaxial air flow at specified volumetric flow rates of 28.3 and 11 LPM flowing through the entire domain. The cone angle of the spray used as a model input was experimentally measured by high speed video imaging. The predictions made by the model were validated against known experimental results. The effect of delay (0 and 5 seconds) was also investigated for both the spacers.

The CFD model predictions favorably compared with experimental data when velocity and size data were compared at two different locations when the inhaler was sprayed in static air. The results for inhaler-spacer systems show an increase of drug escaping the spacers on increasing the volumetric flow rate in accordance with experimental determinations. The recirculation observed near obstructions in axial path of the spray within the spacer is determined to be critical in enhancing spray retention within the spacer. Introduction of a delay is found to trap more drug in the spacer. The MMAD of the drug that escapes the spacer is smaller than what is injected for all cases, while the coarser particles are trapped within the spacer surfaces. In conclusion, a validated CFD model was developed for characterizing inhaler-spacer interactions which can be used for similar systems to predict desired particle metrics.

(98an) CFD-Aided Characterization Of A Spinning Disc Reactor (SDR)

Habla, F., Technische Universität München

Ren, F., Technische Universität München

Haseidl, F., Technische Universität München

Hinrichsen, O., Technische Universität München

Background

In the last years spinning disc reactors (SDR) are becoming more and more popular in the field of chemical reaction engineering. The radial acceleration of the liquid caused by the centrifugal forces generates a very thin film on the disc, typically in the range of 20 ? 300 µm. This thin film not only has superior micromixing properties when adding reagents to the film but also creates large heat transfer coefficients from the disc to the film and mass transfer coefficients from gas to liquid. These properties make the SDR technology attractive for many reactions such as polymerization, organic synthesis, photocatalysis and precipitation reactions. The film thickness and regime and as a result the performance of a SDR is strongly affected by the operating conditions, e.g. rotational speed, volumetric flow rate, feeding point locations, disc diameter and disc design. Although there is an increasing research in academia on the use of SDRs for various reactions, still very little is known about the film properties and most researchers rely on either experimental methods, which only provide film heights but do not disclose local film properties and factors affecting the film structure and regime, or very simplistic models such as the Nusselt film model [1]. Understanding and predicting the film properties is crucial for optimization of the operating conditions.

Aims

In this work we aim at characterizing and predicting film properties with CFD. In a recent publication of our group the micromixing efficiency was experimentally determined by use of the iodide-iodate reaction [2]. Focus in this work is also set on testing the predictability of CFD in terms of describing the micromixing efficiency by numerically modeling the reaction kinetics.

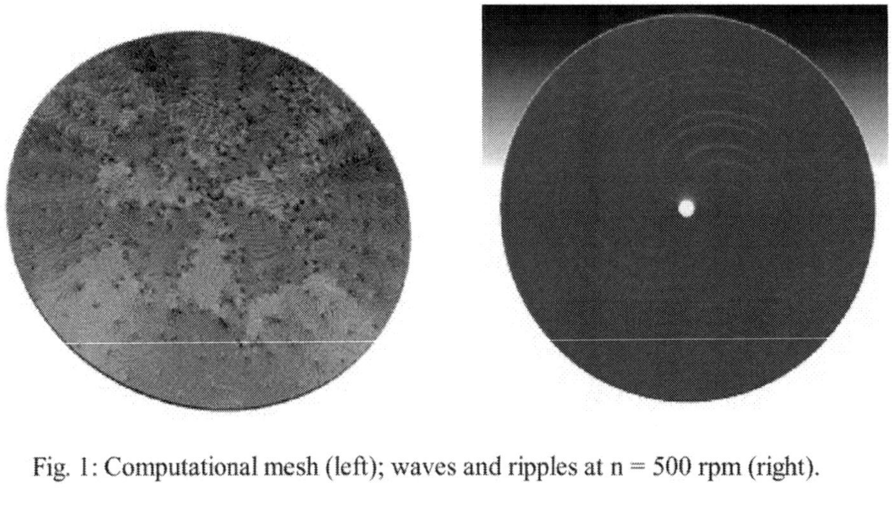

Fig. 1: Computational mesh (left); waves and ripples at n = 500 rpm (right).

Fig. 2: Nusselt model thickness (orange) [2] and simulated film thickness (marks) for different rotational speeds; error bars indicate radially minimum and maximum film height in the simulation.

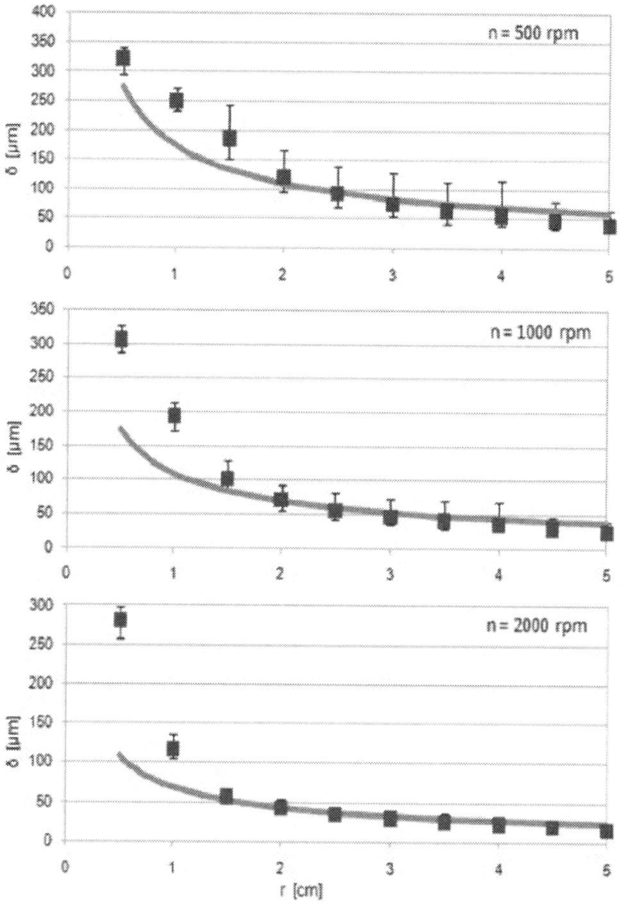

Results

Creating a computational mesh is very challenging due to the very large disc radius compared to the very small scales, which need to be resolved. One of the meshes used in this study is shown in Fig. 1 (left). In Fig. 1 (right) we present the top view of the liquid film at a rotational speed of $n = 500\ rpm$. The disc diameter is $d = 5\ cm$ and the volumetric flow rate is $Q = 3\ mL/s$. Close to the inlet spiral waves can be found, while further away from the inlet ripple structures prevail. The mean film height is generally in good agreement with the Nusselt film thickness at all rotational speeds for $r \gtrsim 2\ cm$, see Fig. 2. For $r \lesssim 2\ cm$ the influence of the inlet is still large. The height of the waves and ripples can be depicted from the error bars in Fig. 2 denoting the radial minimum and maximum film height. For higher rotational speeds the height of the waves and ripples decreases.

Conclusions

The hydrodynamics of the liquid film on a SDR were predicted with CFD. We point out that, to our knowledge, this is the first CFD study on SDRs to present a mesh convergence study to determine the influence of the numerical grid on the results. Current investigations focus on validation of the model with experimental film heights and on determining the micromixing efficiency by modeling the reaction kinetics of the iodite-iodate reaction.

References

[1] Wood, R.M., and Watts, B.E.: Trans. Inst. Chem. Eng. 51, p. 315 (1973).

[2] Jacobsen, N.C., and Hinrichsen, O.: Ind. Eng. Chem. Res. 51, p. 11643 (2012).

(98ao) Flow of Capsules Through Expansion and Contraction Geometries

Carroll, R., University of New Hampshire
Gupta, N., University of New Hampshire

We numerically simulate the flow of a membrane bound capsule as it moves through a straight cylindrical channel with a sudden expansion or constriction. A three-dimensional axisymmetric volume of fluid method with a front tracking scheme is used to simulate the flow. The capsule flow dynamics are characterized by calculating the deformation parameter as the capsule flows through the expansion / contraction and the excess pressure drop required to maintain a constant flow rate. We study the capsule dynamics at a low but finite Reynolds number for a variety of drop sizes, elastic capillary numbers, and channel expansion ratios. The dynamic deformation of the capsule is determined for different constitutive equations for the membrane tension such as Hooke's law, Mooney-Rivlin law and Evans and Skalak's law for biological membranes. For small deformations, the non-linear constitutive equations reduce to the linear Hooke's law. Geometries and elastic capillary numbers are chosen to ensure large deformations to highlight the effect of membrane tension laws on capsule behavior. In addition, the effect of membrane viscosity is also included in some of the dynamic studies. The viscosity of the encapsulated fluid also has a strong effect on the overall flow behavior of the capsule. At large capsule fluid viscosities, the motion and deformation of the capsule is damped and the effect of the different membrane constitutive laws on capsule dynamics diminishes.

(98ap) Experiments and Simulations of Droplet Flow Through Sharp Contractions

Blais, C.,
Carroll, R., University of New Hampshire
Gupta, N., University of New Hampshire

The stability of droplets flowing in channels is an important issue in the design of microfluidic devices. A sudden 90 degree contraction in a cylindrical channel geometry can create large extensional strains leading to droplet breakup. The critical capillary number for breakup depends strongly on the viscosity ratio as well as the geometry of the contraction. In this study, the dynamics of a droplet as it flows through a sudden contraction is studied both experimentally and numerically. Experimentally, we fabricate cylindrical PDMS channels with sharp contractions and varying contraction ratios. The deformation and break up of glycerol-water drops in silicon oil of varying viscosities is recorded at varying flow rates. Critical capillary numbers at which drop break up occurs is identified. The experimental results are validated by comparing them with numerical simulations using a front tracking volume of fluid method. In addition, the effect of surfactants on droplet breakup is also studied. Drop breakup pattern similar to tip streaming is observed in the presence of surfactants.

(98aq) Influence of Solvent Volatility On the Formation of Core-Shell Hollow Microfibers By Triaxial Electrospinning

Khalf, A., Oklahoma State University
Madihally, S. V., Oklahoma State University

Abstract

Electrospinning is a polymer processing technique that uses an electric field to produce nano and microfibers. Co-axial and triaxial electrospinning are modified methods for preparing fibers with multi layered walls. In triaxial electrospinning three different polymers solutions are supplied into a compound Taylor cone (three flow streams) through a spinneret and charged liquid at the nozzle orifice is drawn out by electric field to form a liquid jet. As the solvent of spinning solution evaporates, the jet solidifies and a polymer core-shell is formed. Reasons for these modifications include i) increasing the surface area to deliver important molecules, ii) improving the mechanical properties, and iii) enhancing the structural morphology. In this study, designing and fabrication of triaxial electrospinning spinneret capable of electrospun multi wall core-shell fibers. We investigated the effect of solvent system properties and molecular weight on the fabrication of core shell hollow nanofibers made of polycaprolactone (PCL), cellulose acetate (CA), and polyvinyl alcohol (PVA). The effects of the solvent mixture properties on the electrospinability of polymer solution were investigated. Switching the configuration of polymers in some systems led to a failure in obtaining fibers due to the rapid solvent evaporation (volatility) of the inner shell liquid which caused phase separation during electrospinning process. SEM images of the fabricated fibers indicated the formation of a multi layered hollow core-shell microfibers of hydrophobic hollow PCL/CA, PCL/PVA and hydrophilic CA/PCL. The obtained results showed that solvent properties include solvent ratio solvent boiling point and solvent miscibility are critical factors that are key to jet stability and fibers uniformity. Hollow microfibers of PCL, CA and PVA were successfully produced by triaxial electrospinning. While switching the configuration of polymers in these systems, rapid solvent evaporation (volatility) of the inner core caused phase separation, leading to failure in obtaining fibers. Selection of solvents with appropriate boiling points relative to the outer shell is critical: outer shell should have a boiling point less than that in the inner cores. Inner core MW should be smaller or comparable to that of the outer shell. Endothelial Cells attached and spread on these fibers suggesting no toxicity from the solvents.

(98ar) Structure Formation During Vibration of Viscoplastic Suspensions of Carbopol Microsponges

Wolf, J.,
Phillips, R. J., University of California, Davis
Dungan, S. R., University of California, Davis

Complex fluids, including particulate suspensions, emulsions, granular media and Bingham fluids, exhibit a variety of interesting surface structures when subjected to vertical vibration. Although several explanations for these phenomena have been proposed, there is currently no quantitative, accepted theory or model that explains them. We have studied the behavior of gelled solutions filled with swollen, polymeric Carbopol "microsponges" neutralized with triethanolamine (TEA). New, previously unreported features have been observed, including the spontaneous formation of holes at the corners of regular polygons. The number of holes, and therefore the type of polygon observed, depend on the mass of fluid studied, as well as the components in the Carbopol solution. Lubrication theory is used to obtain qualitative insights into the observed behavior.

(98be) Simulation Of Viscoelastic Flows In Complex Geometries At High Elasticities

Habla, F., Technische Universität München

Hasslberger, J., Catalysis Research Center and Chemistry Department

Dietsche, L. J., The Dow Chemical Company

Hinrichsen, O., Technische Universität München

Simulation of viscoelastic flows in complex geometries at high elasticities

Florian Habla[1], Johannes Hasslberger[1], Laura J. Dietsche[2] and Olaf Hinrichsen[1]

[1]Catalysis Research Center and Chemistry Department, Lichtenbergstraße 4, D-85748 Garching b. München, Germany

[2]The Dow Chemical Company, Midland, MI 48674, USA

Introduction

Viscoelastic flows can be found in a wide range of industrial applications such as food processing, injection molding or polymer blending. It is of major importance to understand and predict such type of flows.

Flows of viscoelastic fluids are very complex in nature and often result in complex secondary flows and transient flow patterns even in very simple geometries. Thus, a numerical method for viscoelastic flows must necessarily be transient and three-dimensional. Due to that, finite volume methods are becoming more and more popular in the field of numerical rheology due to their efficiency in time and memory savings.

Our research focuses on developing a comprehensive finite-volume CFD method for the open-source software OpenFOAM® to simulate complex viscoelastic flows including multi-phase flows[1-2] and non-isothermal flows[3]. In this work, we will provide an overview of our developments and show simulation results for various complex problems.

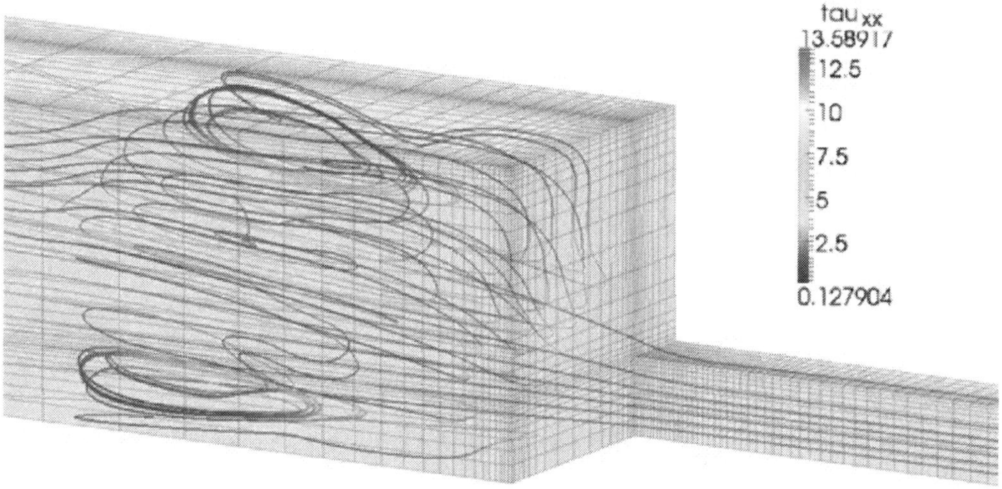

Fig. 1: Three-dimensional simulation of the flow of a simplified Phan-Thien-Tanner fluid through a 4:1 square-square contraction at Wi = 500. Streamlines are colored with the axial normal stress.

Simulation of single-phase flows

The high Weissenberg number problem (HWNP) is probably the central problem in the numerical simulation of viscoelastic flows for over three decades now. The Weissenberg number is a measure of elasticity of the flow. In critical regions of the domain stresses show exponential profiles. Due to the deficiency of approximating these profiles with polynomial-based discretization schemes, a recent contribution by Fattal and Kupferman[4] suggests to logarithmize the constitutive equation to overcome this issue. Use of this technique perceivably extenuates the HWNP, however, the problem still remains at hand.

Additionally to using that method, we developed a reformulation, which allows to handle the constitutive equation semi-implicitly, which strengthens the diagonal-dominance of the matrix and thereby strongly increases the stability of our numerical method[5]. In Fig. 1 the flow of a viscoelastic fluid through a three-dimensional contraction at a very large Weissenberg number is shown proofing the robustness of our method.

Simulation of multi-phase flows

In many applications multi-phase flows of viscoelastic fluids are present such as polymer blending or extrusion processing. In Fig. 2 a fully three-dimensional simulation of a droplet deforming in a viscoelastic matrix due to an external shear flow is shown using a highly accurate two-phase model derived from conditional volume averaging[2]. Validation of our method with experimental data shows generally good agreement.

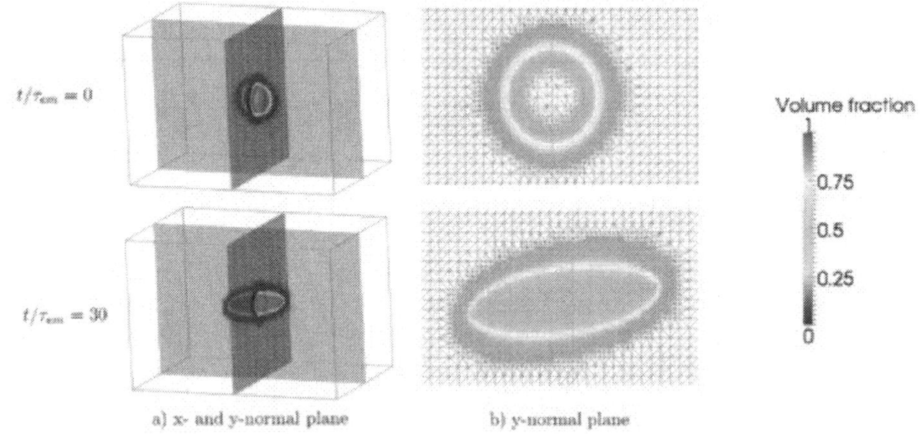

Fig. 2: Three-dimensional simulation of a droplet deforming due to a shear flow in a viscoelastic matrix with adaptive mesh refinement of the interfacial region.

Besides the conditionally volume-averaged two-phase model, another method to simulate multi-phase flows of viscoelastic fluids based on the well-known volume-of-fluid (VoF) approach was implemented. The method was validated with a number of different flows such as the die-swell effect, the Weissenberg effect and droplet deformations[1]. In Fig. 3 the VoF model is used in to simulate the flow in a partially filled twin-screw extruder. The complex movement of the screws is coped with a transient Immersed Boundary Method (IBM).

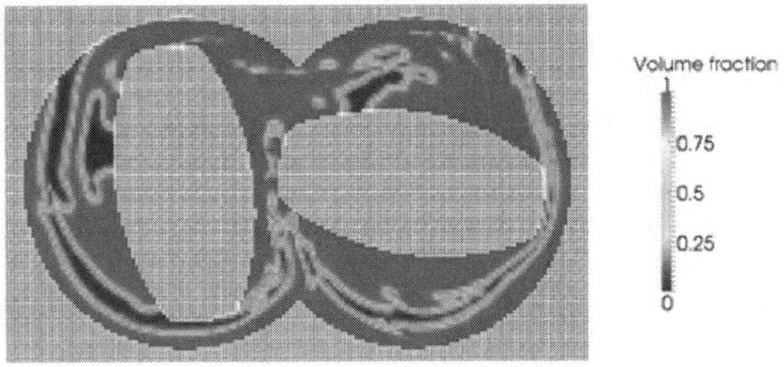

Fig. 3: Two-dimensional simulation of a partially filled twin-screw extruder using the Immersed Boundary Method (IBM). The blue phase is air, the red phase is the extruded viscoelastic fluid governed by the Oldroyd-B equation.

References

[1] Habla, F., Marschall, M., Hinrichsen, O., Dietsche, L.: Numerical simulation of viscoelastic two-phase flows using OpenFOAM®. Chem. Eng. Sci. 66 (2011) 5487-5496.

[2] Habla, F., Dietsche, L., Hinrichsen, O.: Modeling and simulation of conditionally volume averaged viscoelastic two-phase flows. AIChE Journal. DOI: 10.1002/aic.14095.

[3] Habla, F., Woitalka, A., Neuner, S., Hinrichsen, O.: Development of a methodology for numerical simulation of non-isothermal viscoelastic fluid flows with application to axisymetric 4:1 contraction flows.

Chem. Eng. J. 207-208 (2012) 772-784.

[4] Fattal, R., Kupferman, R.: Constitutive laws for the matrix-logarithm of the conformation tensor. J. Non-Newtonian Fluid Mech. 123 (2004) 281-285.

[5] Habla, F., Obermeier, A., Hinrichsen, O.: Semi-implicit formulation for viscoelastic models: Application to three-dimensional contraction flows. Submitted to Journal of Non-Newtonian Fluid Mechanics.

(98as) Transverse Migration of Polyelectrolytes in the Presence of Pressure and Electrical Fields

Arca, M.,
Butler, J., University of Florida
Ladd, A. J. C., University of Florida

Simultaneously applying an electric field and pressure gradient induces a net migration of DNA in a direction perpendicular to the fields. We have proposed that velocity disturbances generated by the electric field acting upon the polyelectrolyte (DNA), which is distorted from its equilibrium conformation by the local shear, drives the migration. The strength of the disturbance velocity scales quadratically with the Debye length, consequently the rate and extent of migration can be altered by changing the ionic strength of the suspending fluid according to the model. In experiments presented here, we systematically varied the ionic concentration and measured the effect upon both the rate and extent of migration as functions of the flow and electric field strengths. Comparisons of the measurements with Brownian dynamics simulations support the hypothesis that the electrophoretic mobility of DNA depends upon its conformation.

(98at) Understanding the Dynamics of An Organic Droplet Released in Quiescent Water Column

Rao, A., Louisiana State University
Reddy, R., Louisiana State University
Ehrenhauser, F. S., Louisiana State University
Nandakumar, K., Louisiana State University
Valsaraj, K. T., Louisiana State University

Through this work, we have made an effort to understand the mechanisms involved when oil containing light end is released from deep water ruptured well as in the case of BP Deep water horizon spill. We present results from our experimental studies that mimic the rise and fall of the oil droplet due to dissolution of the lighter hydrocarbons. A model organic droplet composed of chlorobenzene (insoluble in water) and acetonitrile (soluble in water) has been used to study this phenomenon. The experimental setup consists of a tank filled with water and a nozzle through which the dispersed phase (organic droplet) is injected. The mixture density of organic droplet is adjusted (0.89-0.96g/cc) such that it is initially less than that of water. The droplet initially rises and starts losing acetonitrile to surrounding water phase. As a consequence, the density of the droplet increases gradually. A stage is reached when density of droplet becomes same as that of surrounding fluid and droplet becomes stationary. Further loss of acetonitrile causes the droplet to sink. A numerical model has also been developed to match the trajectory of droplet exhibiting the above phenomenon.

As an extension to the above work, we have also analyzed the effect of surfactant addition on the dynamics of crude oil droplet. Experiments have been conducted in a tank with a capacity of about 100 liters with an oil droplet being released into the quiescent water column through a nozzle. The size of droplets is varied using nozzles of different diameters. The shape adopted by the emanating droplets varies from spherical to oblate. On addition of the surfactant (SDS) to water column the reduction of interfacial tension brings in variation in the internal circulation and causes the droplet to flatten. The change in droplet rise time, the time taken by droplet to reach the water- air interface serves as an observable parameter to support this variation in droplet dynamics. A numerical model based on Volume of Fluid method has been used to emulate the above experimental observations.

(98am) Three Dimensional Hydrodynamics of Liquid Falling Film Flow Over Inclined Micro Topography

Marek, A., Technische UniversitÃ¤t Bergakademie Freiberg
Repke, J. U., TU Bergakademie Freiberg

Structured packings used in a wide range of industrial engineering intensify the mass transfer between the gas and liquid phase up to 80% compared to the unstructured surface. Usually structured packings consist of microstructures in the range of less than one millimeter. It is known that these structures promote the stability and the spreading of the gravitational driven liquid film flow.
The phenomenon of mass transfer amplification induced by the micro structured topography is investigated by means of the three dimensional velocity field in the liquid film flow. Therefore a three dimensional stereoscopic micro particle image velocimetry system is adapted. The velocity field, film thickness profile and interfacial shape are measured for different industrial applied micro-structures.
The results for a wavy micro structured topography illustrate that the liquid interface follows the geometry but a phase shift appears, which is dominated by flow parameters. For a pyramidal micro structured topography a recession of solitary waves of the flow is documented through the development of a three dimensional flow structure. It can be concluded that the enhancement of the mass transfer is supported by the induction of stationary and non-stationary eddies caused by micro-structured topography. In the contribution results of the new adapted measurement system on micro structured topographies are presented.

(98av) Active Diffusiophoretic Motion of Submicron Swimming Colloids With Large Surface Reaction Rates and Solute Convection

Sharifi-Mood, N., Department of Chemical Engineering, Levich institute
Koplik, J., City College of the City University of New York
Maldarelli, C., The City College of New York

Harnessing forces to propel colloidal particles as mechanical engines are central to the design of locomoting objects for nanocolloid assembly, for transport of biomolecules in targeted drug delivery and as roving vehicles for sensing and detection. One of the potential mechanisms is diffusiophoresis, a chemo-mechanical transduction mechanism in which a concentration gradient of solutes in a thin interaction layer (L=O(1-10)nm) applies an unbalanced intermolecular interaction on the colloid and ultimately brings about its motion in a particular direction. We study active diffusiophoretic motion (or self-diffusiophoresis) in which a partially coated colloid with catalyst consumes reactant from the environment and releases product in an irreversible first order reaction, creating solutal fields in both reactant and product across the colloid which sustains diffusiophoretic motion. As technological interest progresses towards miniaturization of the motors, it's crucially important to understand the dependence of the swimming velocity, U, on the particle radius, a. Here we focus on two factors which affect this dependence, the reaction rate and the solute advection.

Numerical solutions are obtained for the case in which only the reactant solute interacts with the colloid, with a potential interaction which decreases exponentially with the distance between the particle and the colloid. The hydrodynamics at small Reynolds number but finite PÃ©clet number, the ratio of convective transport to diffusion (Pe=U'L/D, U' is the characteristic swimming velocity), is coupled with solute mass balance conservation. Our algorithm relies on an iterative method for obtaining the solute concentration field based on Galerkin finite elements in conjunction with an analytical solution we found for the hydrodynamics. The effect of the reaction rate is defined by the Damköhler number, Da, the ratio of the rate of reaction of solute to the diffusion (Da=kL/D, where k is the rate constant and D is the solute diffusion coefficient), and the effect of advection by the PÃ©clet number.

When convection is negligible (Pe equals zero), we find that for small Da, the swimming velocity decreases as the particle radius becomes smaller. For finite values of Da, the scaled velocity is non-monotonic: For particle radii much smaller than L, the velocity increases with decreasing radius. For particle radii of order L, the velocity decreases with increasing radius. This leads to the interesting and technologically promising conclusion that if the reaction rate is fast enough, swimmers will increase their velocity as their size decreases, as long as their size does not approach the length scale of the intermolecular interaction. At any particle radius, the effect of advection is shown to decrease the swimming velocity due to a reduction in the gradient across the particle. Lastly, we evaluate the swimming velocity for all various values of the nondimensional groups illuminating the swimming velocity dependency on colloid size which is a critical issue for engineering and designing the next generation of "nanoengines".

(98au) Suspension Displaying On Through-Hole Array Membranes

Lee, J., University of Michigan at Ann Arbor
Burns, M. A., University of Michigan

Microbeads are currently being used in a variety of biomedical assays due to inherent benefits such as the ease of separation, high surface-to-volume ratio, and flexible modification of targets. Although several platforms have been suggested for analyzing the results of bead-based assays, there has been few detection platforms suitable for portable use. The reason for this absence is the difficulty of incorporating portability into decoding systems for suspension arrays without the use of complicated and cumbersome optical set-ups. Here, we present a decoding unit for suspension processing that uses a micro-machined through-hole array membrane. Based on the bead sorting and capturing capability of the membranes, the displaying unit successfully decodes the information from mixed poly-disperse suspension arrays. Also, a quantitative guide for device design is provided based on the study of the hydrodynamic interaction between the beads and the through-hole membrane. As a proof-of-concept, the displaying unit is used as a detection platform for multiplexed suspension ligase detection reaction (LDR).

(98ay) Atomistic Simulation of Dynamics of Individual Molecules in Entangled Polymers Undergoing Homogenous Shear Flow

Nafar, H., University of Tennessee
Edwards, B., University of Tennessee
Khomami, B., University of Tennessee

Nonequilibrium molecular dynamics (NEMD) simulations of an entangled linear polyethylene ($C_{400}H_{802}$) were performed to study the chain dynamics over a wide range of Weissenberg numbers (*Wi*). The distribution of the end-to-end vector, $|\boldsymbol{R}_{ete}|$, exhibited Gaussian behavior at low *Wi*. At high *Wi* the distribution was bimodal with two peaks associated with rotation and stretching of the chains. To understand the underlying physics of the aforementioned distributions, the relevant system time scales were determined. Specifically, the autocorrelation function of the unit end-to-end vector showed a periodic behavior at high *Wi* which is associated with the onset of rotation of individual chains within a background of highly stretched chains. A functional form of $A\ \exp(-t/\tau_R)\cos(2\pi t/\tau_{rot})$ was fitted to the autocorrelation function data to extract the Rouse (τ_R) and rotational (τ_{rot}) relaxation times of the system. It is shown that the Rouse time calculated using this method is considerably different than the one calculated using a stretched exponential (as in previous works[1]) for all shear rates. Both the Rouse and rotational relaxation times exhibited a shear-thinning behavior that scaled as $Wi^{-0.75}$ at high shear rates. The number and lifetime of entanglements, which is related to the average time of creation and destruction of entanglements, was also computed as a function *Wi*. This characteristic timescale exhibited a shear-thinning behavior which scaled as $Wi^{-0.63}$ at high shear rates. The total number of entanglements also has a shear-thinning behavior at high *Wi*. The implications of these findings will be discussed in this presentation with regard to convective constraint release and contour fluctuations in reputation theory at high *Wi*.

---

[1] Baig, C.; Mavrantzas, V. G.; Kr ger, M. *Macromolecules* **2010**, *43*, 6886.

(98z) Anomalous Structure and Transport Properties of Soft Colloids

Srivastava, S., Cornell University
Archer, L. A., Cornell University

Structure, dynamics and rheology of model nanoparticle suspensions comprising of silica nanoparticles, densely grafted with oligomeric polyethylene glycol (PEG) chains and suspended in similar PEG oligomers are presented. The particle volume fraction corresponding to particle overlap is found to have unique signatures in particle arrangement and dynamics as well as the mechanical rheology of the suspensions. Small angle X-ray scattering measurements reveal that the particle-particle correlations, corresponding to the first maxima in particle structure factors, exhibit a maximum at the volume fraction corresponding to particle contact in conjunction with a transition from a liquid-like to a solid-like response in the linear viscoelastic regime and substantial increments in the viscosities. These anomalous structural trends are in stark contrast to hard sphere suspensions where the particle correlations increase monotonically. In addition, a maximum in particle relaxation time appears at a marginally higher volume fraction; the particles speed up upon further increase in volume fraction. Systems including complex molecular fluids like water and molten silica as well as particles interacting via soft repulsive potentials are known to exhibit similar "cascade of anomalies", and an analogous competition among configurational entropy and resistance to deformation is argued to lead to the origin of these anomalous trends in all the cases.

(98bf) A Fluid-Structure Interaction Study of Surface Acoustic Wave Interaction With a Sub-Micron Sized Droplet

Sankaranarayanan, S., Argonne National Laboratory
Suthar, K., Argonne National Laboratory
Richardson, M. B., University of South Florida
Bhethanabotla, V. R., University of South Florida

Liquid droplet based microfluidic systems find extensive applicability in biological applications, such as carriers to target drug delivery to the intended diseased tissue, encapsulation agents for various biological entities, micro-reactors for rapid mixing and reaction of reagents, liquid reaction vessels for protein crystallization, biosensing, DNA analysis, sampling glucose concentrations in bodily fluids, to name a few. In droplet-based microfluidics, discrete droplets are created via the use of two immiscible phases in micro-channels. All applications require a great control on droplet size, shape, and distribution; mono-dispersed droplets are typically required which can be used as micro-reactors and for achieving hands-free contact of reagents in biological applications such as in Lab-On-a-Chip systems. Recent studies suggest that surface acoustic wave (SAW) devices, with hydro-phobically modified surfaces, can be used for efficient microfluidic control and generation as well as actuation of droplets. In this work, we investigate using 2-D and 3-D Fluid-Structure Interaction Model, the phenomenon of droplet actuation in SAW devices as well as the acoustic streaming velocity fields generated in micro-liter size droplets as a result of interaction with SAW devices. Simulations indicate that the Rayleigh wave generated in the SAW device impinges on the fluid medium and imparts momentum generating fluid motion known as acoustic streaming. The simulated fluid velocity profiles inside the droplet exhibit recirculation, thereby indicating an enhancement in mixing and agitation in droplets. The influence of various parameters, namely fluid viscosity, applied voltage, and droplet size on the droplet actuation and acoustic streaming is studied using the developed fluid-structure finite element model.

(98aw) Nonequilibrium Work Relations for Dilute Polymer Solutions

Latinwo, F., University of Illinois at Urbana-Champaign
Schroeder, C. M., University of Illinois at Urbana-Champaign

In this work, we present a general framework based on the Jarzynski equality to study polymer solutions. This framework allows for the determination of equilibrium and nonequilibrium free energy landscapes of dilute polymer solutions in flow. The rheology of polymer solutions strongly depends on their equilibrium and nonequilibrium steady state properties. Using Brownian dynamics simulation in conjunction with experimental analyses, we apply our framework to individual polymer stretching trajectories in general flows. In addition, we show that equilibrium properties such as polymer relaxation time and chain elasticity can be determined from a wide array of fundamental steady-state properties. Overall, connecting equilibrium and nonequilibrium properties of dilute polymer solutions provides a new platform to study the dynamics of flowing soft matter systems.

(98ax) Inverse Approach in CFD Models Using Optimized Lattice Boltzmann Method

Vemuri, S. H., Carnegie Mellon University
Jhon, M. S., Carnegie Mellon University
Biegler, L. T., Carnegie Mellon University

Uncertainties in the output of a complicated phenomena such as turbulent flow simulation are not only caused by uncertainties in its input but may also result from heuristically motivated artificial model parameters. These additional equations in modeling this flow, constituting the so-called turbulence models with higher degrees of freedom (i.e., two equation models), typically involve certain parameters which can be justified only heuristically from empirical information and experimental data. The adjustment of these turbulence parameters is often not obvious, guided principally by "what worked before," even if the model is now used in a di erent application. Thus, an important question in the CFD community now is how the results of the overall simulation depend on the turbulence parameters, where if the results are strongly a ected by these parameters, the uncertainty of the simulation is considered to be high.

Recently, mesoscale methodologies based on kinetic theory approach via Boltzmann transport equation, namely lattice Boltzmann method (LBM) is increasingly competing with conventional computational fluid mechanics (CFD) methodologies in understanding complex phenomena of turbulence. LBM approaches to the asymptotically correct hydrodynamic [1,2] and local thermo-hydrodynamic [3,4] limits, and turbulence modeling methodologies can be efficiently incorporated into LBM via its advantages including highly efficient massively parallel computation ability, ease in handling complex geometries, and a "rule based" approach which offers significant flexibility.

Thus, in this work, to gain deeper understanding into the heuristic approach of turbulence modeling, using LBM we perform the parameter estimation of the closure parameters which predict the experimental data, and asses the sensitivity of the results of such a turbulence simulation, a derivative-based sensitivity analysis is carried out via automatic differentiation to investigate the closure parameters. Automatic differentiation (AD) [5] is a chain-rule-based technique for evaluating the sensitivity derivatives with respect to their input variables. In contrast to the approximation of derivatives by difference methods, AD does not incur any truncation error and is computationally efficient. For our benchmark example of turbulent flow simulation, the derivatives are computed via AD by specifying the closure parameters as the independent variables and the velocity components **u** and **v**of the velocity in directions x and y as the dependent variables.

In summary, this sensitivity analysis methodology will be used to adjust the approximation models within the system in order to improve its physical simulation or to obtain desired numerical solution properties such as stability or convergence. Furthermore, this analysis will help us to assess the robustness of the LBM, and can lead to deeper insight into the area of turbulence simulation that has never been done before and will lead to the best turbulence models by simply adopting an inverse approach.

1. H. Chen, S. Chen, and W. Matthaeus, "Recovery of the Navier-Stokes equations using a lattice-gas Boltzmann method," *Phys. Rev. E*, **45**, 5339 (1992).
2. L.S. Luo and S.S. Girimaji, "Theory of the lattice Boltzmann method: Two-fluid model for binary mixtures," *Phys. Rev. E*, **67**, 036302 (2003).H. Chen, C. Teixeira, and K. Molvig, "Realization of fluid boundary conditions via discrete Boltzmann dynamics," *Int. J. Mod. Phys.*, **9**(8), 1281 (1998).
3. H. Chen, S. Kandasamy, S. Orszag, R. Shock, S. Succi, and V. Yahkot, "Extended Boltzmann kinetic equation for turbulent flows," *Science*, **301**, 633 (2003).
4. H. Chen, C. Teixeira, and K. Molvig, "Realization of fluid boundary conditions via discrete Boltzmann dynamics," *Int. J. Mod. Phys.*, **9**(8), 1281 (1998).
5. Rall, Louis B. (1981). Automatic Differentiation: Techniques and Applications. Lecture Notes in Computer Science. 120. Springer, Heidelberg.

(98az) Numerical Simulation of Concentrated Emulsion Flow Through a Pore in Oil/Water Separation

Kagawa, Y., Kobe university, Center for Membrane and Film Technology
Ishigami, T., Kobe university, Center for Membrane and Film Technology
Fuse, H., Kobe university, Center for Membrane and Film Technology
Hayashi, K., Kobe university
Matsuyama, H., Kobe university, Center for Membrane and Film Technology

1. Introduction

A large amount of liquid waste, including oil-in-water (W/O) and water -in- oil (O/W) emulsions, are generated in chemical, metallurgical and food industries. Various technologies such as static separation, centrifuge demulsification, chemical demulsification, and freeze/thaw demulsification have been developed to separate disperse phase from continuous phase. Basically, these technologies make disperse phase much bigger. However, these technologies have disadvantages such as long operation time, big footprint, high cost and low efficiency. Membrane separation process attracts the attention due to the advantages of short operation time, small footprint, low cost and higher efficiency. The porous membranes act as a wetting and a coalescing medium. If feedwater is O/W emulsion, oil droplets attach on the pore walls and then coalesce in the pores. Consequently, oil droplets in permeate side become much bigger than those in feed side and oil phase is then easily separated from water phase by buoyancy force.

To improve the demulsification efficiency, a number of experimental studies have been performed using various kinds of porous membranes. They investigated the effects of the size of droplet, pore size and surface hydrophilicity of membrane[1-3]. However, the mechanism of membrane oil-water separation process is still not well understood and such influence factors have not been optimized completely. Though it is important for the clarification of the mechanism and optimization of parameters to understand concentrated droplet motion within the pore, it is difficult to experimentally observe fluid and droplet motion within the pore. In addition, it is also difficult to evaluate the effects of various parameters independently, because current porous membranes are usually prepared by phase separation methods. The phase separation methods can not control the membrane properties such as pore size, tortuosity, porosity and hydrophilicity, independently.

Numerical simulation is very useful to investigate the fluid and droplet motion within the membrane pores. In the numerical simulation, motions of oil droplets and fluid within the pore can be easily observed and various parameters can be independently evaluated. Darvishzadeh et al investigated the effect of the crossflow velocity and transmembrane pressure on the motion of a single droplet by numerical simulation [4]. However, they did not study the demulsification efficiency and dealt only with a single droplet, not dispersed droplets. To our knowledge, there have been no investigations about membrane demulsification by numerical simulation.

The aim of this study was to newly develop the numerical simulation model to investigate the behavior of concentrated O/W emulsion through a pore, and to clarify the effects of wettability of the pore walls, filtration flux and pore size on the motion of oil droplets and the demulsification efficiency by the developed simulation model.

2. Numerical method

In this study, we dealt with the large deformation of the oil-water interface, including coalescence and break up of dispersed droplets. In addition, it is necessary to estimate the volume of the droplet in permeate side to evaluate the demulsification efficiency. We used the coupled level set and volume of fluid (CLSVOF) method [5] for capturing the interface motion. CLSVOF method can satisfy volume conservation and evaluate the interfacial tension accurately. The VOF method can treat large deformation of the interface accurately and satisfies volume conservation, while this method cannot evaluate the interfacial tension. By contrast, the level set method can calculate the interfacial tension accurately, while this method cannot satisfy volume conservation. CLSVOF method consists of only advantages of both methods.

The governing equations of fluid flow are the equation of continuity and the Navier-Stokes equation. A Cartesian grid system was used for arrangement of the Eulerian variables. We used a fractional step method to solve the governing equations. Navier-Stokes equations was splitted into four terms consists of the convection term, the diffusive term, the surface tension force term and the pressure gradient term. The convection term was solved by the cubic interpolated pseudo-particle (CIP) method. The central finite difference was applied for diffusive term. The continuum surface tension force (CSF) model developed by Brackbill et al [6] was adopted to evaluate the normal component of the surface tension force. The projection method was applied for calculating the pressure gradient term. To impose the contact angle on the membrane surface, we used a method developed by Sussman [7]. The interface described by VOF function is extrapolated into the membrane wall as the specified contact angle is satisfied.

The size of the computational domain was L_x=50 μm, L_y=10 μm. The numbers of the main mesh were 502×102. A membrane with a straight pore was mounted in the computational domain. The membrane thickness and the diameter of a oil droplet set at 10 μm and 3.0 μm, respectively. N-dodecane was used as the dispersed oil phase. O/W emulsion with 51% of volume fraction, flowed into the left boundary and exited from right boundary. For boundary conditions, the inlet velocity on the feed side was constant and uniform. A non-slip boundary condition at walls of membrane was presented. The periodic boundary condition was applied for the upper and lower boundaries. In this study, the effects of the static contact angle walls of the membrane, the filtration flux and the pore size were investigated.

3. Results and discussion

To investigate the effect of wettability of membrane surface on the motions of oil droplets in a pore, the static contact angle between oil phase and membrane surface was changed in three steps; 45° , 90° and 135° and the filtration flux and the pore size were set at 1.0 m/s and 4.0 μm, respectively. In the case of 45° of static contact angle, oil droplets attach on the wall of a pore and the oil droplets move on a pore surface with wetting. After that, the droplet then stays near the membrane surface at the outlet of the pore due to the high wettability and the existence of wakes having low velocities. The droplet which flows from behind then coalesces with that at outlet of the pore one after another. It is found that this effect brings about enhances the coalescence between the droplets. In the case of 90° of static contact angle, oil droplets also attach on the wall of a pore and the oil droplets move on a pore surface with wetting. However, the droplets hardly stay the outlet of the pore because of the lower wettability. Therefore, the coalescence between oil droplets near the exit of the pore rarely occurs and the size of droplets do not increase so much. In the case of 135° of static contact angle, oil droplets hardly attach on the wall of a pore due to the much lower wettability. Consequently, oil droplets exit the pore without coalescence.

Based on these investigations, we can indicate that the coalescence of oil droplets is significantly affected by static contact angle of membrane surface. We will also discuss about the effects of the filtration flux and the pore size on the motion of oil droplets in a pore in the presentation.

[1] M. Hlavacek, *J. Membr. Sci.*, **102**, 1995, 1-7.

[2] U. Daiminger et al, J. Membr. Sci., **99**, 1995, 197-203

[3] T. Kawakatsu et al, AIChE J., **45**, 1999, 967-975

[4] T. Darvishzadeh et al, *J. Membr. Sci.*, **423-424**, 2012, 468–476.

[5] K. Yokoi, *J. Sci. Comput*, **35** (2008) 372-396.

[6] J. Brackbill et al., *J. Comput. Phys.*, **100** (1992) 335-354.

[7] M. Sussman, in: *Adaptive Method of Lines,* Chapman and Hall/CRC, (2001) 207-227.

(98bb) Extensional Flow of Entangled DNA Solution in Microfluidic Devices

Boukany, P. E., The Ohio State University
Berghout, P.,

Complex fluids from colloids, DNA and synthetic polymers, liquid crystals, micelles, foams, gels, emulsions to membranes are ubiquitous around. These materials respond elastically or solid-like with a deformation rate exceeding slowest relaxation rate (Wi >1.0). Extensional flow of complex fluids is important in many practical operations from spraying of paint, roll coating of adhesive to fiber spinning, extrusion, and film blowing of polymeric fluid. Typically, the extensional flow behavior of non-Newtonian fluids is terminated by the appearance of unstable necking or rupture-like failure. Although there have been theoretical attempts to interpret failure during extension, still molecular mechanism for necking and rupture-like failure is unknown. The aim of this work is to use entangled DNA solutions as model to unravel the molecular origin of necking during extension.

Here, we employ micro-droplet microfluidics and fluorescently labeled DNA to study the flow instability with respect of necking of a DNA solution undergoing extensional flow. We investigated different kinds of failure in entangled DNA solutions by capturing the molecular response of DNA during extension. These new experimental findings provide new insight into the mechanistic origins of different failure in the extensional flow of entangled fluids.

(98ba) Simulations of CSF Flow Dynamics in a Global CNS Model With Magnetically Targeted Intrathecal Drug Delivery

Linninger, A., University of Illinois at Chicago

Tangen, K., University of Illinois at Chicago

Venugopal, I.,

Lueshen, E., University of Illinois at Chicago

Introduction: Therapeutic intervention for central nervous system diseases has limited effectiveness due to drug transport restriction across the blood brain barrier. Intrathecal (IT) drug delivery aims to circumvent this obstruction, however interpatient variability impacts treatment outcomes. Drug distribution is affected by patient anatomy, cerebrospinal fluid (CSF) properties (volume, pressure, and density), heart rate, and drug infusion properties. A secondary problem with IT drug delivery is localization of the compound to a region of interest. External non-invasive magnetic fields have been shown to direct and concentrate magnetic nanoparticles for treatment of several diseases such as tumors. We propose to understand and explore the efficacy of treating CNS diseases with IT administration of magnetic nanocarriers functionalized with therapeutic molecules. Multi-phase simulations of a complete patient specific model of the CSF filled spaces can predict dispersion patterns of nanoparticle-therapeutic complexes both in a native state and under magnetic field influence. An *in vitro* bench top human spinal canal model with magnetic field guidance is used to corroborate computational simulation results.

Computational CSF Fluid Flow Methods: Phase contrast magnetic resonance imaging (PC-MRI) in combination with computational fluid mechanics is beginning to provide a complete picture of patient specific CSF flow fields. Patient specific MRI data allows for generation of an anatomically consistent mesh for computer modeling giving accurate predictions of drug dispersion patterns in the CNS.Modeling of spinal CSF flow is focusing on the role of the micro-anatomy including nerve roots and trabeculae. Using simulations, our lab showed that CSF pulsatility and nerve roots play a significant role in the bio-distribution of drug molecules during intrathecal drug delivery. Our work also illustrates the formation of complex micro-mixing patterns which are attributable to nerve roots and trabeculae.

We constructed a patient-specific model of the fluid filled spaces inside the CNS based on PC-MRI data. Pulsatile motion of the CSF for the full model is generated by explicitly moving the choroid plexus and parenchyma according to a Fourier series that captures the velocity wave of the basilar artery. The model includes regions of production and absorption at the choroid plexus and arachnoid villi respectively. Nerve roots were reconstructed from the image data and incorporated into the full SAS model. We added micro-anatomical detail below the image resolution threshold of MRI to quantify the effect of trabeculae (6 micron diameter) on CSF flow. We compared complex flow profiles to simulations in structures lacking the anatomical detail. The effect on spinal flow resistance and their impact on complex flow profiles as a function of varying trabeculae density. The full spinal SAS model visualizes flow profiles at 31 nerve pairs over a complete cardiac cycle predicting the effect of successive nerve roots.The computational fluid dynamic solver ANSYS Fluent 14.0 is used for multi-phase fluid flow simulations to predict dispersion of magnetic nanoparticle based drug infusate within the spinal canal. Magnetic fields of varying strength were also applied to the complete CNS model simulations to understand the effect on magnetic nanocarrier transport and accumulation in a global model.

Bench top Testing Methods of Magnetic Nanoparticle Guidance: Most intrathecally injected drugs are required only at certain locations within the spinal canal where it is essential for these drugs to diffuse into the spinal cord to produce a therapeutic effect. Currently, in order to reach the crucial drug concentration at the target site and aid diffusion into the spinal cord, high drug doses are used. With no method to deliver the large drug doses locally, current intrathecal drug delivery treatments are hindered with wide drug distributions throughout the CNS which cause harmful side effects. Our lab proposed a magnetic nanoparticle based drug delivery technique within the CNS by IT administration. An externally applied magnetic field localizes the magnetic drug coated nanoparticles at desired locations. Detailed simulations in COMSOL Multi-physics optimized the placement and strength of the external magnet. A physiologically consistent bench top model of the human spinal canal was constructed to determine the effect of magnetic fields on magnetic nanoparticle distribution, shown in Figure 1.

A

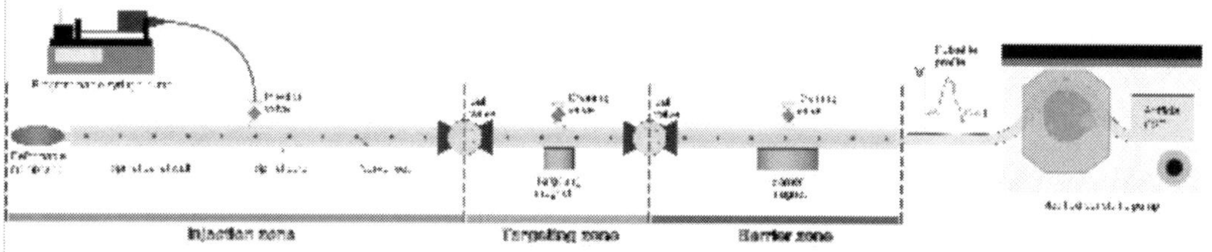

Figure 1: Schematic of entire experimental setup with the *in vitro* human spine model which clearly shows the three different zones (injection, targeting, and barrier zones).

co-precipitation technique using ferrous and ferric salts was used to form the 8-10 nm diameter superparamagnetic magnetite nanoparticle cores, which were then coated with a highly functionalizable gold layer of thickness between 8-15 nm. The nanoparticles were characterized by TEM, EDS, x-ray diffraction spectroscopy and SQUID magnetometry. The *in vitro* spinal canal model was physiologically consistent, and constructed out of a plastic tube (mimicking the spinal canal wall), hard wood cylinder (representing the spinal cord), and rubber projections (serving as spinal nerve roots). The proximal end of the model was connected to a peristaltic pump which produced pulsatile fluid motion. The model was filled with artificial cerebrospinal fluid, and a distensible rubber balloon was fixed at the distal end to serve as the filumterminale. A programmable syringe pump was used to inject the nanoparticles into the lumbar region, and NdFeB magnets of different size and field strength were placed below the targeting region. A strong 1.05 T magnet was used in the cervical/barrier region to prevent any remaining particles from leaving the system. Nanoparticle collection in different regions was achieved using ball valves to separate the model into three zones (infusion, targeting, and barrier zones) from which the particles were eluted, dried, and weighed. The *in vitro* experiments were also simulated in Comsol 4.2a to help analyze model parameters and to determine optimum magnetic field strength and orientation.

Global CSF Fluid Flow Simulation Results and Discussion: CSF flow is driven by a pulsatile waveform and Womersley (W_O) number is calculated to characterize the flow profile as in phase (<1) or lagging behind the pressure wave (>1). The Womersley number at the C4 region of the spinal SAS is W_O=4.63, T6 region W_O=4.38, and L4 region W_O=5.38. Flow at all spinal regions of interest has a phase lag behind the pressure wave. Velocity streamlines, shown in Figure 2, are

generated which visualizes recirculating flow patterns in the spinal SAS at multiple regions adjacent to nerve roots. Reynolds number is calculated at multiple regions of the spinal canal to determine if fluid is turbulent or laminar. Simulation results predict all regions of the spine to have laminar flow; C4 Re<81, T6 had Re<37, and L4 has Re<17. The presence of CSF recirculation patterns indicates micro-anatomy influence the fluid flow profiles. The pressure drop increased as the density of manually added trabeculae increased. Micro-anatomy has a significant role in the dispersion of fluid flow which will impact the distribution of drug nanoparticle complexes. Magnetic two-phasic flow predicts nanoparticle localization at regions of interest under increasing external magnetic field strength and gradient.It is plausible that the CSF pulsations and micro-anatomy are both important for preventing the formation of dead zones in the spinal SAS in which cell debris or metabolites might otherwise stagnate and even accumulate.

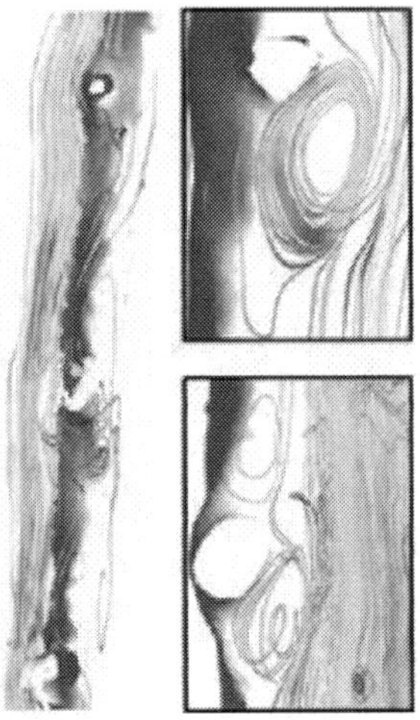

Figure 2: Fluid flow patterns in the spinal SAS at nerve roots. Micro-anatomy induced mixing profiles are visualized at the C4 and T7 nerves with eddies.

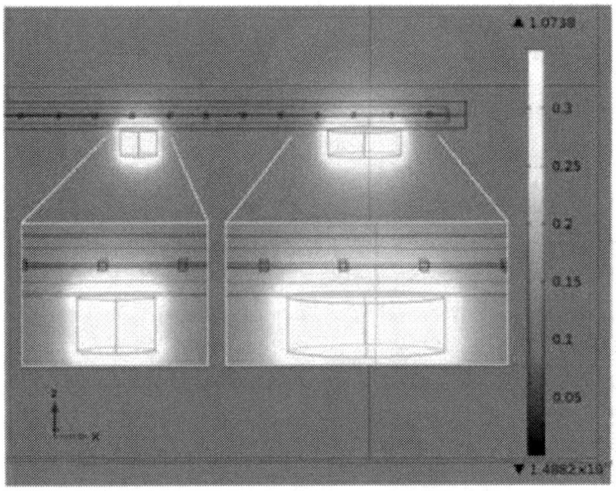

Figure 3: Magnetic field produced by Ne-Fe-B magnets at various locations along the CSF filled spaces in the bench top model. Magnetic field ranges from 0.051-0.098 T (left magnet), and 0.092-0.167 T (right magnet). Color scale visualizes the intensity of the magnetic field; white is highest, red and black lowest.

In vitro Model Results and Discussion: Multiple magentic field strengths were tested using COMSOL Multi-physics as shown in Figure 3. In the bench top model, collection efficiency at regions of interest, using a magnetic field of 0.528 T, increased with time and reached a steady collection level of 71% after 15 minutes (p-value = 0.341). No change in the magnetic retention value was observed beyond 60 minutes. The nanoparticle recovery at the region of interest increased with the rising field strength of the magnet. Percentage retentions of 48%, 61%, and 71% were observed for the surface field strengths of 0.396 T, 0.507 T, and 0.528 T respectively. Experiments utilizing the same 0.528 T magnet placed at one of the three regions of interest (C4, T6, or L4) for 15 minutes showed that location of the applied magnetic field always corresponded to the site of maximum particle retention around 76-81%.

Therapeutic interventions for diseases of the CNS can benefit from a patient specific understanding of the effect that micro-anatomy has on the transport phenomena in the CNS leading to improved applications of novel drug delivery techniques such as usingmagnetically targeted nanocarriers.

(98bc) High Deborah Number Elastic Instabilities Around Microfluidic Confined Cylinders

Christopher, G. F., Texas Tech University

Flow of viscoelastic fluids around a confined cylinder is a hallmark fluids problem because it resembles industrial processes and is easily simulated. This flow displays multiple regimes with elastic and inertio-elastic instabilities that occur at critical Deborah (De) or Reynolds (Re) numbers. Practical limitations have restricted the experimentally explored range of De; using microfluidic channels, these limitations have been overcome, and allowed the study of high De flows. We have systematically characterized the flow around symmetrically placed, highly confined\ microfluidic cylinders at $10^{-2} < De < 10^{3}$, using low viscosity Boger fluids. By changing fluids, flow rates and geometry, non-dimensional parameters are independently tuned, allowing a controlled and detailed study.

At low De, the nature of the flow around a confined cylinder displays inertio-elastic instabilities forming downstream of the cylinder at critical values of Re and De consistent with predictions in literature. At larger values of De attainable due to microfluidic channels, new elastic instabilities are observed that have not been previously reported. We first identify a downstream instability of temporally and spatially varying streamlines at De~50. This instability is a precursor to a new class of upstream instability at De > 100. During this instability, the upstream stagnation point separates from the cylinder face up to ~ 5 diameters upstream of the cylinder, and its position oscillates in a growth and collapse cycle. We characterize flow patterns and onset of both instabilities as functions of the Elasticity number and the viscoelastic Mach number, and conclude that the instability is purely elastic in nature but affected by inertia. Furthermore, we characterize the role of geometry on the onset of both instabilities. We have identified two new and unique elastic instabilities around confined cylinder flows and provide a basic description of their behavior and underlying physical mechanisms.

(98bg) Surface Wetting and Contamination On Alumina Ceramic

Grillet, A. M., Sandia National Laboratories
Barringer, D. A., Sandia National Laboratories
Brooks, C. F., Sandia National Laboratories
Archuleta, K., Sandia National Laboratories
Ohlhausen, J. A., Sandia National Laboratories
Brumbach, M. T., Sandia National Laboratories
Garcia, D.,
Roach, A.,

Surface energy quantifies the chemical interactions on a surface which are available for coating and adhesion. At a basic level, a liquid will coat a surface if the surface energy is greater than the liquid's surface tension. At a more complex level, the types of intermolecular interactions, such as dispersion forces or hydrogen bonding, at the surface are important for determining the degree of wetting. The surface energy of 94% alumina ceramic will be presented, including the relative influence of polar and non-polar interactions at the surface. These measurements will be used to identify factors which may influence the quality of an epoxy adhesive joint on the ceramic. The effect of cleaning method, heat treatment, aging and several contaminants at the surface will be examined. Finally we will demonstrate that for this 94% alumina ceramic, the contact angle of water correlates very well with the total surface energy and can be used as a quick method for analyzing surface cleanliness.

(104a) Use Vesicle Membrane As a Designable Platform for Molecular Recognition ~ Toward Creation of Bio-Inspired Chemical Engineering Based On Membranome ~

Umakoshi, H., Osaka University
Suga, K., Osaka University
Originally presented on: 11/4/2013 12:30:00 - 13:00:00

A "*Biomembrane*" is a highly-organized self-assembly of biomolecules (i.e. lipid, protein etc.) and a key interface for the survival of biological cell. The "*Membranome*" can be defined as the properties of vesicle (or liposome), which arise from the bilayer molecular assembly of amphiphiles, focusing on "*emergent properties*" which are not present in the individual components, and is gradually recognized as an important research methodology to investigate the potential functions of vesicles (or liposome) and to apply them for the bioprocess design. A liposome possesses several benefits in the recognition of (bio)molecules, where it can recognize them with (i) electrostatic, (ii) hydrophobic interaction, and (iii) stabilization effect of hydrogen bonds in the hydrophobic lipid environment. A molecular recognition with high selectivity can be achieved by the simple liposome membrane on its surface through the combination of (i)-(iii). We are now focusing on the creation of "*Bio-Inspired Chemical Engineering*" based on the "*Membranome*" by selecting "liposome or vesicle membrane" as a "*platform*" to perform the "*recognition*" and "*conversion*" of biomolecules, such as (a) *Artificial Chaperone* , (b) A*rtificial Membrane Protein* , (c) *Artificial Enzyme*, (d) *Artificial Virus,* (e) *Artificial Organ* (liposome-immobilized hollow fiber membrane or liposome-immobilized hydrogel matrix) and etc. The author will make a presentation on the basic concept of "*membranome*", together with some examples.

(104b) Morphological Transitions in Block Copolymers Directed By Superparamagnetic Nanoparticles

Raman, V., Massachusetts Institute of Technology
Olsen, B. D., Massachusetts Institute of Technology
Hatton, T. A., Massachusetts Institute of Technology
Originally presented on: 11/4/2013 13:00:00 - 13:20:00

This computational and theoretical study investigates the self-assembly of superparamagnetic nanoparticles and block copolymers under external magnetic fields. A variety of morphological transitions are observed based on the field orientation, nanoparticle loading, and selectivity of the nanoparticles for the blocks. For symmetric block copolymers, chaining of superparamagnetic nanoparticles under in-plane magnetic fields is shown to achieve long range orientational order of the block copolymer nanodomains and is found to be dependent on nanoparticle size, volume fraction and magnetization strength. A critical selectivity of the particles for one nanodomain is observed, above which strong alignment results and below which comparatively disordered structures are formed. Higher magnetization strengths are found to reduce equilibrium defect densities in the nematic-isotropic ordering of lamellar thin films, as corroborated by scaling arguments. For asymmetric coil fractions forming hexagonal block copolymer nanostructure, the in-plane field induced chaining of the nanoparticles selective for the minority block, leads to the formation of stripe phases oriented parallel to the magnetic field. Furthermore, in-plane field induced chaining of nanoparticles selective for the majority block leads to alignment of hexagonal morphology with <100> direction oriented parallel to the external magnetic field. Out of plane magnetic fields induce repulsive dipolar interactions between the nanoparticles that annihilate the defects in the hexagonal morphology of the block copolymer when the nanoparticle is selective for the minority block. Honeycomb lattices are obtained using nanoparticles selective for majority block under out of plane magnetic fields for certain specific nanoparticle loadings. Commensurability of nanoparticle size and loadings with the block copolymer structure is critical in optimizing the ordering of the final composite.

(104c) Design of Autonomously Moving Colloidal Objects - Systematic Motion of Vesicles, Particles and Oil/Water Interface in a Homogenous Chemical Field -

Shioi, A., Doshisha University
Originally presented on: 11/4/2013 13:20:00 - 13:50:00

The second law of thermodynamics states that all systematic motions are transformed into a random motion with molecular length scale. In biological system, however, spontaneous motions driven by chemical reaction with a molecular length scale generates a systematic motion with macroscopic scale. To study how and why the systematic motion can evolve over the hierarchy from an apparent random dynamics with smaller length scale is fascinating for designing a biomimetic chemical system. Design of autonomously moving colloidal objects is one of the methodologies that enable to understand the mechanism for the energy transduction. Moreover, the study of self-moving system may develop into various applications related to biomimetics and microfluidics. Autonomously moving colloids can demonstrate numerous significant characteristics such as transduction of chemical potential into work without heat, chemosensitive motion, and pulse-like motion with periodicities responding to the chemical environment. Here, three types of colloidal systems that exhibit autonomous motions will be discussed. Chemosensitive motion of oil/water interface that is caused by chemical Marangoni instability is the first example. The convection caused by the instability can be rectified to make one dimensional motion of a float at the interface. Second, a vesicle with rhythmic shape change is discussed. This vesicle shows a sustainable motion under a pH gradient. A vesicle that swims in a solution with specific chemical is also shown. Finally, autonomous motion of catalytic particles is discussed. The motion of homogeneous catalytic particle can be controlled by its geometry.

(104d) Development and Properties of Novel Photo-Cleavable Surfactants

Sakai, H., Tokyo University of Science
Aikawa, S., Tokyo University of Science
Fukuda, K., Tokyo University of Science
Shrestha, R.,
Endo, T., Tokyo University of Science
Sakai, K., Tokyo University of Science
Abe, M., Tokyo University of Science
Originally presented on: 11/4/2013 13:50:00 - 14:20:00

Recently, we have reported a new cinnamic acid-type photo-cleavable surfactant, C4-C-N-PEG9 that experiences a photo-cleavage through UV-induced cyclization in aqueous solution, yielding a coumarin derivative and an aminated polyoxyethylene compound[1)]. Here, we have studied the effects of C4-C-N-PEG9 on the photorheological behavior of viscoelastic wormlike micelles formed by aqueous mixture of nonionic surfactants, polyoxyethylene phytosterol ether ($PhyEO_{20}$) and tetraoxyethylene dodecyl ether ($C_{12}EO_4$). When C4-C-N-PEG9 is added to the viscous, non-Newtonian 4.9 wt% $PhyEO_{20}/H_2O$ + 2.4 wt% $C_{12}EO_4$ fluid, a decrease in viscosity is observed, while upon UV irradiation C4-C-N-PEG9 added solution regains the viscosity. It increases by 200 times, suggesting a modulation in micellar shape and size. The UV-Vis spectroscopy and HPLC measurements confirm that the *trans*-C4-C-N-PEG9 with large head group and its photodecomposition to form less surface-active coumarin derivative and an aminated polyoxyethylene compound are responsible for different rheological behavior[2)].

References 1) H. Sakai, et al., *Journal of Colloid and Interface Science,* 376, 160-164, (2012).

2) H. Sakai, et al., *Langmuir,* Article ASAP, DOI: 10.1021/la4001939.

(104e) Ordering Transitions Triggered By Specific Binding of Vesicles to Protein-Decorated Interfaces of Thermotropic Liquid Crystals

Abbott, N. L., University of Wisconsin-Madison
Tan, L. N., University of Wisconsin-Madison
Originally presented on: 11/4/2013 14:20:00 - 14:40:00

This presentation will address an investigation of the interactions of phospholipid vesicles with protein-decorated interfaces of thermotropic liquid crystals (LCs). We have observed that following the capture of vesicles at a protein-decorated LC interface via specific binding interactions, phospholipids are transferred from the vesicles onto the LC interface to form a monolayer, partially displacing proteins from the LC interface and thereby triggering a continuous ordering transition in the LC. The dynamics of this process are accelerated substantially by the specific binding event relative to a protein-decorated interface of a LC that does not bind the ligands presented by vesicles. The observation of the *continuous* change in the ordering of the LC is significant as it is consistent with the presence of sub-optical domains of proteins and phospholipids on the LC interface. An additional significant hypothesis that emerges from the work to be reported in this presentation is that the ordering transition of the LC is strongly impacted by the bound state of the protein adsorbed on the LC interface, as evidenced by the influence on the LC of (i) "crowding" of the protein within a monolayer formed at the LC interface and (ii) aging of the proteins on the LC interface. Overall, these results demonstrate that ordering transitions in LCs can be used to provide fundamental insights into the competitive adsorption of proteins and lipids at oil-water interfaces, and that LC ordering transitions have the potential to be useful for reporting specific binding events involving vesicles and proteins.

(104f) Directed Aggregation and Fusion of Vesicles Induced By a DNA-Surfactant

Goto, M., Kyushu University

Originally presented on: 11/4/2013 14:40:00 - 15:00:00

Conventional synthetic strategies have provided a number of attractive molecular devices, but the sizes of the devices produced using these strategies are limited to a few hundred nanometers at most. Molecular self-assembly is an alternative approach for constructing systems that are larger (submicrometers to micrometers) and more complex. In particular, lipid vesicles, which are spherical bilayer shells encapsulating an aqueous interior, have been widely studied as a biomimetic example of molecular self-assemblies. Such lipid vesicles can range in size from 50nm to ~1mm, and contain various compounds, including metal ions, macromolecules and biomolecules. Furthermore, modification of the lipid bilayer can confer diverse functions onto lipid vesicles. Higher order or secondary levels of self-assembly of lipid vesicles have been achieved through vesicle aggregation using electrostatic and ligand–receptor interactions. These higher ordered assemblies of lipid vesicles were also studied as a simple model for biological tissues.

Over the last decade, oligonucleotides have attracted much attention as intelligent materials due to their highly specific and designable binding properties toward their complementary sequences. Many researchers have employed oligonucleotides as building blocks to construct nanostructured systems with well-designed geometric shapes and sizes. Besides, oligonucleotides have been used as molecular glues for cross-linking, immobilizing and aggregating proteins, nanoparticles, colloids and cells. Oligonucleotides were also introduced to the vesicles and utilized for the immobilization of vesicles on an electrode or a plate. Patolsky et al. utilized the DNA-tethered vesicles for amplification of hybridization signal. Yoshina-Ishii and Boxer first reported the sequence-specific immobilization of lipid vesicles. These studies suggested the utility of DNA-tethered vesicles. Most of them mainly focused on surface modification and nanofabricated structures, rather than on the construction of macroscopic structured systems. Wu et al. reported that an extremely high concentration (10mg/ml) of long linear DNA-induced non-selective aggregation of a small portion of zwitter ionic vesicles, where the aggregation was supposed to be induced not by the DNA hybridization but simply by the electrostatic interaction. Graneli et al. first reported the interaction among DNA-tethered vesicles on a DNA-modified gold substrate and succeeded in the formation of three-dimensional vesicle matrices based on DNA hybridization. There are, however, considerable uncertainties on the DNA-directed aggregation of vesicles in solution. For example, are the DNA-tethered vesicles responsive to external stimuli such as temperature, ionic strength, etc.? Here, we investigate DNA directed aggregation of vesicles in solution using DNA-surfactants and clarify if the DNA-tethered vesicle possesses the DNA-derived characteristics to be responsive to external stimuli. Furthermore, we successfully accomplished DNA-directed fusion of vesicles based on this assembly procedure, leading to the production of a micro-reactor involving an enzyme-catalyzed reaction in the giant vesicle produced.

(112a) The Solubility of Carbon Dioxide, Hydrogen and Their Mixture in the Ionic Liquids 1-Alkyl-3-Methylimidazolium Bis(trifluoromethanesulfonyl)Amide ([CnMIM+][TF2N-])

Singh, R., University of Notre Dame
Marin-Rimoldi, E.,
Maginn, E., University of Notre Dame
Originally presented on: 11/4/2013 12:30:00 - 12:55:00

Ionic liquids are being investigated as potential solvents for pre- and post-combustion CO_2 capture. In pre-combustion applications, knowledge of the selectivity of CO_2 over H_2 in high pressure and high temperature gas mixtures is crucial. In the present work, the solubility of carbon dioxide, hydrogen and their mixture in the ionic liquids 1-alkyl-3-methylimidazolium bis(trifluoromethanesulfonyl)amide ([C_nMIM$^+$][TF2N$^-$]) is computed using the isothermal-isobaric Gibbs ensemble Monte Carlo (NPT-GEMC) method. In the case of 1-3-dimethylimidazolium bis(trifluoromethanesulfonyl)amide ([DMIM$^+$][TF_2N^-]) and 1-ethyl-3-methylimidazolium bis(trifluoromethanesulfonyl)amide ([EMIM$^+$][TF_2N^-]), the predicted absorption isotherms for CO_2 are compared with the experimental results of Ren *et al.* [J. Chem. Thermodyn., 42, 2010, 305]. The predicted solubility for CO_2 in 1-butyl-3-methylimidazolium bis(trifluoromethanesulfonyl)amide ([BMIM$^+$][TF_2N^-]) and 1-hexyl-3-methylimidazolium bis(trifluoromethanesulfonyl)amide ([HMIM$^+$][TF_2N^-]) are compared with the experimental results of Maurer *et al.* [*J. Chem. Thermodyn., 2006,38,1396*] and Brennecke *et al.* [*J. Phys. Chem. B, 2004, 108, 20355*] at 333 K. It is found that the predicted absorption isotherms for pure CO_2 in the studied ionic liquids are in good agreement with the experimental results.

The absorption isotherms for H_2 in the ionic liquid [HMIM$^+$][TF_2N^-] are computed at various temperatures ranging from 313 to 573 K. The computed absorption isotherms and Henry's law constant agree well with the experimental results of Maurer *et al. [J. Chem. Eng. Data, 51, 2006, 1364*], Peters *et al.* [*J. Chem. Eng. Data, 56, 2011, 1105*], and Costa Gomes [*J. Chem. Eng. Data, 52, 2007, 472* but there is a large discrepancy with the experimental results of Noble *et al.* [*Ind. Eng. Chem. Res., 47,2007,3453*]. The partial molar enthalpy was calculated for H_2 using the van't Hoff relationship and a value of 1.49 kJ/mole was determined. The value is in reasonable agreement with the experiment value of 4.09 kJ/mole reported by Maurer *et al.* [*J. Chem. Eng. Data, 51, 2006, 1364*]. Unlike CO_2 , which has a negative enthalpy of absorption (and thus the solubility increases with decreasing temperature), the simulations correctly predict that the solubility of H_2 increases with increasing temperature. The absorption isotherms for H_2 are also computed in [DMIM$^+$][TF_2N^-], [EMIM$^+$][TF_2N^-], and [BMIM$^+$][TF_2N^-] but unfortunately no experimental data are available for comparison. The studies have shown that hydrogen is sparingly soluble in ionic liquids, making experimental measurement difficult.

Mixed gas (CO_2/H_2) experiments are much harder to conduct that pure gas experiments and so it is usually assumed that absorption selectivity is ideal. To test this, mixed gas simulations were also carried out using NPT-GEMC method. The simulations indicate that the solubility selectivity of CO_2 over H_2 in the [HMIM$^+$][TF_2N^-] is greater than that predicted by assuming an ideal mixture. The selectivity decreases from about 45 at 333 K to about 3 at 573 K, consistent with the simulation study of Shi *et al.* [*J. Phys. Chem. B 2010,114,6531*]. This is due to the fact that CO_2 solubility decreases with increasing temperature whereas H_2 solubility increases. The solubility selectivity of CO_2 over H_2 is computed for the remaining ionic liquids and the effect of the cation alkyl-chain is determined. These systems show more ideal behavior than liquid [HMIM$^+$][TF_2N^-].

(112b) Effect of the Impurities On Thermophysical and Tranport Properties of CO_2 for Transportation

Llovell, F., MATGAS Research Center and ICMAB-CSIC
Vega, L. F., MATGAS Research Center and Carburos MetÃ¡licos, Air Products Group
Originally presented on: 11/4/2013 12:55:00 - 13:20:00

Carbon dioxide (CO_2) obtained from different carbon-capture processes usually contains small amounts of other gases, such as N_2, H_2, CO, CH_4, H_2S, NO_2 and SO_2, as well as traces of liquid water. The presence of these impurities may lead to challenging issues, as they can importantly modify the conditions of transport of supercritical CO_2. Usually, the gaseous CO_2-rich stream is generally compressed to be transported as liquid or dense-supercritical phase in order to avoid two-phase flow. However, the inclusion of substances other than CO_2 can severely change the phase diagram of carbon dioxide, producing changes on the density of the system, the bubble point pressure, the supercritical region limits or the viscosity, hence affecting the compression needs. Therefore, it is of great interest to have accurate equations of state (EoSs) for evaluating those effects. Empirical correlations and cubic equations of state have been commonly proposed to model those mixtures, achieving a good degree of success. However, these equations do not fully consider some important effects, such as the hydrogen-bonding interactions of water and other associating compounds, or the quadrupole moment of CO_2, leading to some inaccuracies and a lack of extrapolation capability. In this regard, molecular-based EoSs can become an excellent platform so as to account for these interactions and provide a complete thermodynamic description of CO_2 mixtures with a high degree of accuracy. The goal of this work is to identify the most important effects of CO_2-impurities in its thermodynamic and transport properties using the refined molecular-based soft-SAFT EoS [1]. Soft-SAFT considers the molecular effects due to repulsive and attractive interactions, chain formation and hydrogen bonding and builds molecular models for each molecule based on their physical properties and chemical structures, finding a characteristic set of molecular parameters for each compound. We will analyze the effect of the inclusion of other gases and water in the density, solubility and pressure conditions of the CO_2 system. A specific treatment for the critical region of CO_2 is used for an accurate calculation around this area [2]. Finally, the effect of these compounds on the stream viscosity is also studied by using the Free Volume Theory (FVT) coupled with the soft-SAFT equation [3].

[1] Blas, F. J.; Vega, L. F. Mol. Phys. 1997, 92, 135-150. [2] Llovell, F.; Pàmies, J. C.; Vega, L. F. J. Chem. Phys. 2004, 121, 10715-10724. [3] Llovell, F.; Marcos, R.M.; Vega, L. F. J. Phys. Chem. B 2013 (submitted).

(112c) Modeling the Temperature Dependence of Supercritical Gas Adsorption: A Comparative Study

Sayeed, A., Oklahoma State University
Charoensuppanimit, P., Oklahoma State University
Robinson, R. L. Jr., Oklahoma State University
Gasem, K. A. M., Oklahoma State University
Originally presented on: 11/4/2013 13:20:00 - 13:45:00

Simulations of enhanced coalbed methane (CBM) recovery and carbon dioxide sequestration in coalbeds require reliable adsorption models to describe the gas adsorption behavior under wide ranges of temperature and pressure. Due to the geothermal gradients in coalbed reservoirs, a significant variation in temperatures may exist in these reservoirs. Thus, a robust CBM adsorption model should be capable of describing the temperature dependence of supercritical gas adsorption expected in these systems.

In this work, we investigate the adsorption prediction capabilities of the simplified local-density (SLD) model. The SLD model is used to investigate systematically the predictions of supercritical gas adsorption behavior, examine the temperature dependence and extend the model's applicability to a wider temperature range.

To conduct this study, an expanded database was compiled that comprised of experimental adsorption measurements of pure- and mixed-gas adsorption. Emphasis was placed on data on gas species that are found typically in natural gas systems. The database comprised of measurements on activated carbons at several temperatures. The SLD adsorption model was then used to both correlate and predict the pure-gas adsorption over wide ranges of temperatures. The temperature dependence relations were incorporated in the model to improve the predictive capabilities. Using the parameters obtained from pure-gas adsorption, *a priori* predictions were obtained for mixture gas adsorption. The temperature dependence relations developed were found capable of reliable predictions of pure-gas adsorption. Further, the model was found capable of providing *a priori* predictions of mixed-gas adsorption for the systems studied.

(112d) Progress On the Hydrate Based Gas Separation (HBGS) Process for Carbon Dioxide Capture

Babu, P., National University of Singapore
Kumar, R., National Chemical Laboratory
Linga, P., National University of Singapore
Originally presented on: 11/4/2013 13:45:00 - 14:10:00

Hydrate based gas separation (HBGS) process is one of the promising new approaches for CO_2 separation from fuel gas through gas hydrate formation. The basis for separation is the selective partition of CO_2 component of a fuel gas mixture between the hydrate phase and the gaseous phase upon hydrate formation. The advantage of HBGS process are energy efficiency and clean technology, and it can simultaneously capture H_2S and other sulfur compounds. In this study, 2.5 mol% propane was added to the fuel gas mixture as additive to decrease the operating pressure in a novel fixed bed reactor employing silica sand and silica gel as media. Kinetic experiments were carried out at three different pressures (4.5, 5.5 and 6.0 MPa respectively) and at 274.15 K. Water conversion of up to 38% was achieved in silica sand bed where as in silica gel bed water conversion of 13% was achieved. Our results suggest that inter-particle pore space enhances the hydrate formation compared to the internal pore space in the particle. Hence, we replaced silica sand with inter connected HydraPor medium in which the inter-particle pore space is the dominant contributor to bed porosity. Water conversion of up to 65% was achieved at 6.0 MPa and 274.15 K. Our results on HydraPor media validates that the inter-particle pore space is the reason for enhanced kinetics and a fixed bed set up can be employed for potential scale up of the HBGS process. Our study also suggests that presence of propane not only decreases the operating pressure, it also enhances the kinetics of hydrate formation.

(112e) Design Consideration in Acid Gas Injection Wells

Matonis, D., Flucomp Enterprises, Inc.

Originally presented on: 11/4/2013 14:10:00 - 14:35:00

Acid gas injection (AGI) is becoming an increasingly popular choice for the disposal of H_2S and CO_2 sequestration in New Mexico. The reservoir engineering aspect will be covered periphally with the key focus on the process from the top of the amine regenerator column to the bottom of the injection well. As a past Senior Engineer of one such NM AGI well, three basic components of the actual AGI Process Schematic will be presented. These are: compression, pipeline and the injection well. Other basic components as the non-aqueous phase behavior, water content of the acid gas, and the density and viscosity of the gas will be presented from the compression and pipeline considerations at the well site last stage compression to the bottom of the well.

(112f) Operating Experience of BASF Amdea in a CO2 Removal System

Bin Khamis Butt, M.,

Originally presented on: 11/4/2013 14:35:00 - 15:00:00

Operating E

Date	April 3rd, 2013
Author	**Muhammad Bin Khamis Butt,**
	Bilal Mustafa, Muddassar Rathore
Company	Engro Fertilizers Ltd. Pakistan
Company address	ECPL, Daharki, Distt. Ghotki, Sindh. Postal Code 65020.
Business Tel.	+92-723-641001 Ext 4402
E-Mail	mkbutt@engro.com

Abstract:

This paper shares Engro's experience of BASF aMDEA in a CO2 removal system of this new Ammonia Plant commissioning and normal operation phase with aMDEA system. The commissioning & stabilization of aMDI planning and preparations, the new CO2 removal system's commissioning was not only accomplished well wit successful in realizing its excellent performance. The decision to rely on in house operational capabilities whil improvement paid heavy dividends.

No doubt, some problems were faced during the course of the project and the team got stuck at some poir operations have some unrevealed areas and the team is putting efforts to get hold on them for achieving excelle

---

To give you an idea regarding content of the paper, please find the following highlights.

Comparison of CO2 removal system in Engro:

Engro Fertilizers Limited operates two ammonia plants. Both plants have design capacity of 1550 MTPD and 2 fuel. The first plant was uses a hot potassium carbonate system (Catacarb) for CO2 removal. However for the s

Journey of aMDEA Operation at Engro:

1. Pre-commissioning of aMDEA system

a. Pre-cleaning of tower's packing (external cleaning)

b. Water flushing, purge and tightness test of all vessels

c. Flushing with potassium solution

d. Two consecutives flushing with demineralized water

e. Preparation of aMDEA solution

2. Commissioning / Operating experience

a. Antifoam dosing issues

b. Extensive foaming in columns

c. Amine smell

d. Vibration issues in piping & pumps

e. Recycle gas compressor problematic operation

Flooding Incident (aMDEA showering):

On 2nd April, 2011 sudden fluctuation in level of LP Flash Drum occurred followed by rapid level increase in c high level in V-19, CO2 Compressor suction KO Vessel tripped CO2 compressor and I-4 (trip of aMDEA circul sec after the fluctuation commenced. Significant amount of aMDEA discharged through the CO2 vent downstr events that pursued are mentioned below

Modifications in the System:

To ensure safety & integrity of the unit modifications have been made based on operating experiences & issues

(124a) Granular Materials and Flows: A Test Bed for Framing and Analysis

Ottino, J. M., Northwestern University

Originally presented on: 11/4/2013 12:30:00 - 12:50:00

Many significant problems in chemical engineering have resisted couching in elegant formalisms, the flow of granular materials being a particularly important example. There are valuable lessons that come sharply into focus by examining the nature of the difficulties. The objective here is to bring to the forefront questions of framing and analysis: decisions regarding the level of description and caveats constraining the range of solutions. We will present several examples from our own research and that of others to illustrate these two interrelated issues. The first issue is the challenge that lies in picking the appropriate level of description to address a specific question; the choice often being one from among a huge hierarchy of length scales and a range of approaches: continuum, discrete, purely geometrical, with much in between. The second issue, a characteristic of many if not most problems in granular matter, is that solving a problem comes with conditions, often associated with various time and length scales. When one digs deeper one finds more questions. But simple pictures often yield considerable insight.

(124b) Dissipation Potentials in the Mechanics of Fluid-Particle Suspensions

Goddard, J. D., University of California, San Diego

Originally presented on: 11/4/2013 12:50:00 - 13:10:00

As a generalization of the classical Rayleigh-Onsager dissipation function, D.G.B. Edelen (1972-73) provides a purely mathematical construct that represents any strictly dissipative force-dependent velocity or "thermodynamic flux" as the gradient of a dissipation potential plus an additive non-dissipative "gyroscopic" flux. The present work casts Edelen's formulae in a transparent form and provides the extension to (Legendre-Fenchel) duality. Among other things, this allows for a compact and general representation of constitutive equations for visco-plastic materials such as fluid-particle suspensions and granular media.

A connection is made to theories such as those of Ziegler (1963-83) and others that are based on the assumption of extremal dissipation. While such models give a correct form for the dissipative components of force or flux they generally fail to describe non-dissipative but mechanically relevant components or to provide valid variational principles.

The current theory allows for evolutionary microstructure and for particle diffusion with size segregation in inhomogeneous systems. The resulting general models are compared to certain contemporary models for suspension mechanics.

(124c) The Evolution of Bubble Strings in Taylor Vortices of a Couette Device

Prakash, J., Technion - Israel Institute of Technology
Lavrenteva, O. M., Technion
Nir, A., Technion
Originally presented on: 11/4/2013 13:10:00 - 13:30:00

Bubbles trapped inside a stable Taylor vortex in a Couette device, at the first shear flow instability, tend to repulse each other. The long time migration of equal-sized bubbles results in "bubble crystals", which have the form of ordered circular strings with equal separation distances between the neighbors. Hence, two bubbles assume opposite positions, three bubbles form a triangle, four a square and so on. The evolution of the initially shortest separation distance appears to be monotonic, except for the case when only two bubbles are present, where an overshoot above the final separation distance and some oscillations were experimentally evident.

We assume that inertia forces are the driving mechanism for this observed dynamics. The interaction of two bubbles embedded in a laminar simple shear flow has been solved at low and high Reynolds numbers. These solutions were used as a basis for the analysis of the experimental findings. For the case of low Re we adapted the method of Ho & Leal (1974) where the Lorenz reciprocal theorem is used to calculate the small inertia induced repulsive force, thereby bypassing the need to explicitly employ singular perturbations. For the case of high Re, when inertia effects are dominant, we solved the time-dependent inviscid problem, accounting for the added mass, and introduced viscous effects when calculating the stress in the fluid layer near the interface. The asymptotic solutions for the small and large inertia effects were obtained by employing conformal mapping techniques. These solutions were corroborated by the approximate method of reflections.

The two-bubble solutions were super imposed to describe the long time dynamic change of the separation distances between multiple bubbles in the Couette-Taylor vortex. For the low Re cases, when viscous forces are dominant over inertia, there appears to be a difference in the evolution pattern of the various separation distances within the string of bubbles. Distances that are initially shortest or longest remain such along the entire evolution process, and they approach their final magnitude monotonically. However, other separation distances may exhibit non monotonic evolution. When the inertia is dominant, oscillations are present and accompany the long term process. The shape, frequency, amplitude and rate of decay of the oscillations depend on the initial separations and on the relative level of viscous effects.

Ho, B. P. and Leal, L. G., 1974, "Inertial migration of rigid spheres in two-dimensional unidirectional flows", *J. Fluid Mech.* **65**(2), 365-400.

(124d) A Constitutive Equation for the Droplet Distribution in the Unidirectional Flow of a Dilute Emulsion of Deformable Drops Suspended in a Second Order Fluid

Ramachandran, A., University of Toronto

Originally presented on: 11/4/2013 13:30:00 - 13:45:00

This work elucidates, for the first time, the first effect of the non-Newtonian behaviors of the fluids in an emulsion on the distribution of droplets in a unidirectional flow, by employing (in part) the classic single-droplet-drift work of Gary from the late 70s. The limit considered in the work is $(Ca,\lambda) \ll \varphi \ll 1$, where Ca is the capillary number, λ is the Weissenberg number that compares the intrinsic relaxation time of the fluid to the convective time scale of fluid motion, and φ is the volume fraction of the drops. In this limit, the droplet drift normal to the unidirectional flow arises due to deformation of the drop, and the second-order, non-Newtonian behavior of the drop fluids. The particle flux can then be written as the sum of the drift fluxes due to the above two mechanisms, and the flux due to shear-induced migration. The drift of the droplet is modeled using the theoretical results of Chan and Leal [J. Fluid Mech., 92, 131-170, 1979]. The flux due to shear-induced migration is modeled using the suspension balance approach of Nott and Brady [J. Fluid Mech., 275, 157-199, 1994] , as explained in a prior publication [Ramachandran, Loewenberg and Leighton, Phys. Fluids, 2010]. As an example, the model is applied to the tube Poiseuille flow of a dilute emulsion. It is demonstrated that the unsteady concentration distribution of droplets resulting from arbitrary time-dependent average velocity follows a self-similar solution in certain limits.

(124e) Dilute Rigid Dumbbell Suspensions In Large-Amplitude Oscillatory Shear FLOW: Shear Stress Response

Bird, R. B., University of Wisconsin
Giacomin, A. J., University of Wisconsin
Aumnate, C., University of Wisconsin
Schmalzer, A. M., University of Wisconsin
Originally presented on: 11/4/2013 13:45:00 - 14:00:00

We evaluate the simplest relevant molecular model, a dilute suspension of rigid dumbbells, in large-amplitude oscillatory shear flow. We solve for the shear stress response, and specifically, we derive explicit expressions for the first and third harmonics. We thus connect the properties of the rigid dumbbells with the shear stress response, producing new structure-property relations for the interpretation of large-amplitude oscillatory shear flow measurements. We further explore the orientation distribution of the rigid dumbbells in large-amplitude oscillatory shear flow.

(124f) Measuring the Full Three-Dimensional, Nonequilibrium Microstructure of Complex Fluids Under Steady and Time-Dependent Shear, Including Laos, Using Spatial and Time-Resolved Small-Angle Neutron Scattering

Wagner, N. J., University of Delaware
Butler, P., National Institute of Standards and Technology
Eberle, A. P. R., NIST
Gurnon, A. K., University of Delaware
Helgeson, M. E., Massachusetts Institute of Technology
Liberatore, M. W., Colorado School of Mines
L pez-Barr n, C. R., University of Delaware
Porcar, L., Institut Laue-Langevin
Originally presented on: 11/4/2013 14:00:00 - 14:15:00

Understanding the nonlinear shear rheology of complex fluids - including polymers, self-assembling solutions, colloids and nanoparticle suspensions - is advanced by directly measuring the microstructure(s) responsible for the stress under deformation. Rheo-optics, direct visualization, rheo-NMR, and light and x-ray scattering under flow are among the various methods applied to a wide variety of model and industrial systems to aid in the development of structure-property relationships and often, to rationalize seemingly anomalous rheological behavior. These methods are often system specific, however, and a quantitative, robust, and broadly applicable method capable of measuring the full three-dimensional microstructure remains an experimental challenge. Indeed, some of the more interesting nonlinear effects, such as shear-banding and shear-induced phase transformations, where the flow-field and microstructure are intimately coupled, require the ability to resolve the microstructure both with spatial and time resolution.

In this presentation we review a robust and broadly applicable method using neutron scattering to measure the microstructure in all three planes of flow, and with spatial and time-resolution. Neutron scattering has a distinct advantage as most materials are amenable to study by neutron scattering, unlike x-Ray and light scattering. Further, truly unique experiments are possible by selective isotopic substitution, which allows probing components in a mixture by contrast matching. We have addressed this challenge by developing a Couette geometry to access the 1-2 (velocity-velocity gradient plane of flow) by small angle and ultra-small angle neutron scattering. Further, spatial resolution of order 100 microns is achieved by an aperture that scans across the flow. Combined with a now commercial, rheo-SANS Couette geometry that enables radial (velocity-vorticity) and tangential (velocity gradient-vorticity) measurements, these instruments provide the first measurements of microstructure in all three projections of the flow and thus, reconstructing the full, three dimensional flowing microstructure. Further, time dependent deformations that are often used to test physically-based constitutive models, such as flow start-up and large amplitude oscillatory shear (LAOS), are probed by stroboscopically synchronizing the deformation field and the scattered neutron collection through time stamping and binning methods. Finally, neutron scattering is an absolute scattering method such that local chemical composition can also be measured under flow. This enabled determining such effects as shear-induced concentration gradients. These instruments are now available for use by the community at the Institute Laue Langevin in Europe and at the NIST Center for Neutron Research in the U.S.

Recent results for the shear-crystallization of self-assembled block copolymer micelles in ionic liquids under LAOS, the shear banding of worm-like micelles under start-up flow, and the time-dependent microstructure of shear thickening colloidal dispersions and colloidal gels under LAOS are presented to demonstrate the breadth of the method and how these measurements provide fundamental microstructural data essential for understanding the rheology, as well as for critically testing rheological constitutive models of these complex fluids.

(124g) Assessment of Mesoscopic Particle-Based Methods for Simulation of Complex Fluids in Microfluidic Geometries

Larson, R. G., University of Michigan
Jiang, L.,
Zhao, T., Tsinghua Univ.
Originally presented on: 11/4/2013 14:15:00 - 14:30:00

We assess the accuracy and efficiency of mesoscopic simulation methods, namely Stochastic Rotation Dynamics (SRD) and Dissipative Particle Dynamics (DPD), for flows of complex fluids, such as polymer solutions, through microfluidic geometries. Since both SRD and DPD use soft or weakly interacting solvent "beads" to carry momentum, both methods contain unwanted particle inertial effects, and high fluid compressibility. To assess these effects, we compare the speed and accuracy of predictions of SRD and DPD for computing flow in a straight channel with periodic slip boundary conditions, representing a periodic electroosmotic flow, which is a complex flow with an analytical solution for low Reynolds number. We find that SRD is roughly ten-fold faster than DPD in predicting the flow field, with similar accuracy at low Reynolds number. However, SRD has more severe problems with compressibility effects than does DPD, which limits the Reynolds numbers attainable in SRD at reasonable computational cost to Re < 100, while DPD can achieve Re higher than this before compressibility effects become too large. The results help determine the range of conditions for which SRD or DPD is preferable for mesoscopic simulations.

(124h) Effect of Cholesterol Nanodomains On Monolayer Morphology and Dynamics

Kim, K., Univerisity of California, Santa Barbara
Choi, S. Q., University of Minnesota
Squires, T. M., University of California, Santa Barbara
Zasadzinski, J. A., University of Minnesota
Originally presented on: 11/4/2013 14:30:00 - 14:45:00

Cholesterol segregates into 10 – 100 nm diameter nanodomains in dipalmitoylphosphatidylcholine (DPPC) monolayers and reduces the surface viscosity of the monolayer by an order of magnitude for each wt% cholesterol in the film. The nanodomains consist of 6:1 lipid:cholesterol "complexes" that segregate to and lengthen lipid domain boundaries, consistent with a reduced line tension λ. The surface viscosity of the monolayer, η_s, decreases exponentially with the area fraction of the complex at fixed surface pressure, similar to simple polymer blends. However, η_s increases exponentially with surface pressure at fixed cholesterol content, as predicted by a free area model that relates η_s to monolayer compressibility and collapse pressure. The elastic modulus, G', initially decreases with cholesterol fraction, consistent with the decrease in λ expected from the line-active in analogy to 3-D emulsions. However, increasing cholesterol further causes a sharp increase in G' between 4 and 5 mol% cholesterol due to an evolution in the domain morphology, so that the monolayer is elastic rather than viscous. Understanding the effects of small mole fractions of cholesterol should help resolve the role of cholesterol in human lung surfactants and may give clues as to how cholesterol influences raft formation in cell membranes.

We find using a combination of fluorescence optical microscopy, atomic force microscopy, Langmuir isotherms and a novel micron scale monolayer rheometer that minute fractions of cholesterol lead to dramatic changes in lipid monolayer morphology and have equally dramatic effects on monolayer dynamic properties. One wt% cholesterol reduces the surface viscosity of DPPC monolayers by an order of magnitude, two wt% reduces surface viscosity by two orders of magnitude. We can relate these changes in surface viscous and elastic properties to characteristic features of the isotherms and monolayer morphology, and hence to the molecular organization of the films. Our findings suggest a possible role for cholesterol in lung surfactant (LS), a lipid-protein monolayer necessary to reduce the surface tension in the lung alveoli. At present, even the existence of cholesterol in native LS is questioned, as the lung lavage required to harvest LS inevitably causes blood and cell debris to be co-extracted, potentially contaminating LS with cholesterol. This lack of consensus over the role of cholesterol is reflected in the composition of replacement lung surfactants for neonatal respiratory distress syndrome (NRDS), which occurs in 20,000-30,000 premature births each year. Survanta and Curosurf, two clinically approved animal extract replacement surfactants for treatment of NRDS, have all cholesterol removed after harvesting. Infasurf, the third clinically approved surfactant, retains 4-5 wt% cholesterol. Resolving this controversy is difficult, as there is little information on the effects of small mole fractions of cholesterol on the organization and dynamics of phospholipid monolayers at the molecular, monolayer or cell membrane scale (mitochondrial membranes have ~ 5 mol% cholesterol, endoplasmic reticulum ~ 10 mol%). Cholesterol is also implicated in the formation of stable, 10 – 100 nm "rafts" within the plasma membrane of cells, which may serve as platforms for organizing proteins responsible for cell signaling or membrane trafficking. The local viscous, elastic and line tension properties of the rafts relative to their surrounding membrane likely have important implications for their formation, stability and function.

AFM images show that cholesterol segregates into line-active nanodomains of a 6:1 DPPC:cholesterol complex that lowers surface viscosity by orders of magnitude, by introducing sources of free area for the film. Rheology and isotherm measurements for films with ≥ 5 mol% cholesterol suggest a fundamentally different molecular organization that at lower cholesterol fractions; a reasonable hypothesis is that cholesterol begins to directly intercalates into the DPPC lattice at higher mole fractions, while being excluded at lower mole fractions. The small fractions of cholesterol significantly change the spreading and other dynamic properties of replacement lung surfactants, which consist of 50 – 70 mol% DPPC. Upon exhalation, i.e. at small alveolar area and high surface pressure, the exponentially increasing surface viscosity should minimize the Marangoni flow that would otherwise occur due to surface tension gradients between the alveoli and the bronchi. At the ~1s time scales common for respiration, the elastic resistance to flow that occurs with increasing cholesterol should also oppose any Marangoni flow. Upon inhalation, and large alveolar area, lower surface pressures are associated with exponentially lowered viscosity, as would be needed to quickly cover an expanding alveolar interface. DPPC films with >5 mol% cholesterol have a significant elastic component over the ~1s time scales common to respiration. If a film is primarily viscous over this time scale, microstructural elements in the film have time to relax and re-arrange. Films that are primarily elastic over this time scale, on the other hand, store deformational energy and work to "undo" the deformations imposed previously. On exhalation, the surfactant films may store deformation energy elastically, making it easier for the alveoli to re-expand on inhalation. This complex relationship between surface viscosity, surface pressure, and cholesterol content may make for an optimal rheological response to minimize the energy of breathing. At high surface pressures, a high surface viscosity coupled with an elastic response may slow the kinetics of three-dimensional buckling that occurs on monolayer collapse. Understanding the effects of cholesterol will enable these ideas to be tested by providing a simple way to alter monolayer mechanical properties by orders of magnitude with only subtle changes in lung surfactant composition and pressure-area isotherms.

(124i) Blood Flow in Cortical Microvessels

Santisakultarm, T. P., Cornell University
Nishimura, N., Cornell University
Farber, G.,
Olbricht, W. L., Cornell University
Schaffer, C., Cornell University
Originally presented on: 11/4/2013 14:45:00 - 15:00:00

Koh, Hookam, and Leal (JFM, 1994) adapted laser-Doppler anemometry to study velocity and concentration profiles in concentrated suspensions flowing through rectangular channels. To allow measurements inside the concentrated suspensions, indices of refraction of the particles and fluid were closely matched.

We report the results of measurements for a similar problem in physiology – the flow of blood, a highly concentrated suspension of red blood cells, in cortical vessels of live, anesthetized rodents. We are especially interested in measuring flow when the hematocrit – the volume fraction occupied by the red cells – exceeds normal physiological values, which are typically around 42%. Our interest stems from observations of flow deficits associated with various myeloproliferative diseases, which are characterized by an overproduction of blood cells.

Using a two-photon excited fluorescence (2PEF) microscope as a velocimeter, we have measured red cell velocity profiles inside arterioles, capillaries, and venules, both in healthy animals and in animals with elevated hematocrit (up to 60%) due to myeloproliferative disease. Results demonstrate that the spatial resolution of the method is sufficient to construct detailed velocity profiles, despite the high concentration of red cells that limits other measurement methods. Furthermore, the temporal resolution of 2PEF microscopy is sufficient to determine how local velocity profiles depend on heartbeat and respiration in individual blood vessels and at vessel bifurcations.

The data show that velocity profiles are blunted at all times during a cardiac cycle. The time-averaged blood flow speed decreases with vessel diameter in arterioles and in capillaries, and increases in venules. Furthermore, animals with elevated hematocrit show an abnormally large number of "stalled" capillaries that have little or no blood flow. A simple empirical model of capillary networks suggests these capillary stalls start as a result of hemodynamics and rheology. The experiments show that they can persist for long times, which can be a consequence of cell adhesion to the vessel wall. Diminished cerebral blood flow may be a cause of cognitive decline associated with myeoloproliferative disease.

(133a) Multiscale Modeling of Membrane Sculpting By the Protein exo70

Bradley, R. P., University of Pennsylvania
Ramakrishnan, N., University of Pennsylvania
Zhao, Y.,
Guo, W., University of Pennsylvania
Radhakrishnan, R., University of Pennsylvania
Originally presented on: 11/4/2013 12:30:00 - 12:45:00

A collection of membrane-binding proteins are responsible for sculpting cell membranes during cell processes, including endocytosis, exocytosis, and cell morphogenesis. The protein exo70 is a member of the exocyst complex, which mediates the secretion of matrix metalloproteinases (MMPs) at focal degradation sites during morphogenesis [1]. Overexpression of exo70 promotes tumor cell invadopodia formation, and in vitro experiments show that exo70 generates tubules on giant unilamellar vesicles (GUVs). We employ coarse-grained molecular dynamics simulations and continuum mechanics methods to quantify the membrane-bending action of exo70. Coarse-grained molecular dynamics simulations (CGMD) using the MARTINI force field [2] show that exo70 produces the strongest curvature when bound to a key lipid, phosphatidylinositol (4,5)bisphosphate (PIP2), and when bound to another monomer. To relate these simulations to experiments, we first quantify the deformation field generated by exo70 monomers and dimers using CGMD simulation, and then investigate the resulting cellular morphologies using a continuum mechanics model for membrane bending. In the first step, we characterize the deformation field by directly analyzing curved membrane conformations generated by CGMD simulations of mutant and wild-type exo70 on mixed lipid bilayers. Further analysis of the three-dimensional stress tensor [3] relates the observed membrane bending to changes in the balance of forces across the bilayer. In the second step, the calculated deformation field serves as the input for a continuum mechanics model for membrane bending which is based on the Helfrich Hamiltonian [4]. From the stand-point of methodological advance, we uncover new principles coupling the induced curvature field to membrane thermal undulations and use these results to perform system size scaling analysis. Monte Carlo simulation of this model yields membrane morphologies which we relate to cryo-transmission electron microscopy images of exo70 remodeling a GUV. This model agrees with experiments which show that PIP2 binding and dimerization are necessary for generating membrane tubules and protrusions. This multi-scale model allows us to relate the chemical details of exo70 membrane-binding and self-association to the induced deformation field and resulting cellular morphologies.

[1] Liu, J.; Yue, P.; Artym, V. V.; Mueller, S. C. & Guo, W. The Role of the Exocyst in Matrix Metalloproteinase Secretion and Actin Dynamics during Tumor Cell Invadopodia Formation, Molecular Biology of the Cell, 20,16,3763-3771.
[2] Marrink, S. J.; Risselada, H. J.; Yefimov, S.; Tieleman, D. P. & de Vries, A. H. The MARTINI Force Field: Coarse Grained Model for Biomolecular SimulationsJ. Phys. Chem. B, 2007, 111, 7812-7824.
[3] Ollila, O. H. S.; Risselada, H. J.; Louhivuori, M.; Lindahl, E.; Vattulainen, I. & Marrink, S. J., 3D Pressure Field in Lipid Membranes and Membrane-Protein Complexes, Phys. Rev. Lett., 2009, 102, 078101.
[4] Liu, J.; Tourdot, R.; Ramanan, V.; Agrawal, N. J. & Radhakrishanan, R. Mesoscale simulations of curvature-inducing protein partitioning on lipid bilayer membranes in the presence of mean curvature fields, Mol. Phys., 2012, 110, 11-12, 1127-1137.

(133b) Multiscale Modeling of the Organization of Receptor Transmembrane Domains in Lipid Membranes

Dai, W., University of Houston
Doxastakis, D. M., University of Houston
Originally presented on: 11/4/2013 12:45:00 - 13:00:00

Multiscale Modeling of the Organization of Receptor Transmembrane Domains in Lipid Membranes

Receptor tyrosine kinases (RTKs) are a large group of cell-surface receptors which are recognized as essential components of signal transduction pathways for inter and intracellular communication. Almost all RTKs have similar molecular structure, a large extracellular domain (ECD), an intracellular tyrosine kinase domain (ICD) and a single pass transmembrane domain (TMD) connecting the previous two. RTKs act as key regulators of critical cellular processes such as growth, metabolism, differentiation, migration and apoptosis. As a result, either loss-of-function or gain-of-function of individual RTKs can result in carcinogenesis.

Due to the important role in diseases of receptors, their activity should be tightly controlled and regulated. Precise control of the activity of receptors requires molecular engineering of specific components that interfere with the activation mechanism. Any signals must be transferred through TMDs into intracellular domains and the current hypothesis is that different arrangements of TMDs are most likely critically involved in signaling of receptors by transferring the ICD into active conformation. It has even been proposed that TMDs can act as potential therapeutic targets.

Current experiment methods face challenges in exploring the arrangements of TMDs within a membrane environment. State-of-the-art all atom (AA) simulations face severe temporal and spatial limitations. In this work efficient free energy Monte Carlo (MC) calculations with coarse-grained (CG) models provided us the favorable arrangements of TMDs of the growth hormone receptor (GHR) and the erythropoietin receptor (EPOR) in lipid membranes. By detailed comparison of GHR TMDs with new intriguing experimental data, two favorable dimer organizations of TMDs were linked to different activity states and a new activation mechanism model for GHR is proposed.

Furthermore, a multiscale method was designed and employed to determine the organization states of TMD with atomistic details. Using the exhaustive sampling offered by the MC coarse-grained simulations we rigorously identified states at distinct separations and assisted by modified rebuilding algorithms we compare free energy profiles calculated at two different resolutions. The multiscale methodology proposed combines the efficiency of the CG model and the accuracy of AA model providing unique insight into the arrangement of TMDs in lipid membranes.

Reference

L. Janosi, A. Prakash and M. Doxastakis, "Lipid-modulated sequence-specific association of Glycophorin A in membranes", *Biophys. J.*, **99**, 284-292., 2010.

A. Prakash, L. Janosi and M. Doxastakis, "GxxxG motifs, phenylalanine, and cholesterol guide the self-association of transmembrane domains of ErbB2 receptors", *Biophys. J.*, **101**, 1949-1958. , 2011.

C. Finger, C. Escher and D. Schneider, "The single transmembrane domains of human receptor tyrosine kinases encode self-interactions", *Science. Signaling.*, **2(89)**, ra56., 2009.

A. J. Rzepiela, L. V. Schäfer, N. Goga, H. J. Risselada, A. H. De Vries and S. J. Marrink, "Reconstruction of atomistic details from coarse-grained structures Journal of computational chemistry", *Wiley. Online. Library.*, **31**, 1333-1343., 2010.

A. J. Brooks and M. J. Waters, "The growth hormone receptor: mechanism of activation and clinical implications", *Nat. Rre. Endocrinol.,* **6**, 515-525., 2010.

(133c) Characterizing the Structure of Biomineralization Proteins Using Replica-Exchange Molecular Dynamics Simulations

Arora, P., Iowa State University
Lamm, M. H., Iowa State University
Originally presented on: 11/4/2013 13:00:00 - 13:15:00

Small biomineralization proteins like Mms6 found in *Magnetospirillum magneticum*strain AMB-1, exhibit iron-binding activity and are known to be responsible for the formation of highly ordered magnetite nanoparticles. It is crucial to determine the structure of the Mms6 protein to understand the function and mechanism involved in the biomineralization process. A multidisciplinary study is underway to characterize the structure and function of the Mms6 protein, using NMR and molecular simulation. The focus of this presentation is on the computational aspect of the project. All-atom replica exchange molecular dynamics simulations of Mms6 in explicit water are used to generate well-sampled conformations of the protein. A detailed conformational analysis of the protein is performed that includes dihedral distributions of the protein residues and principal component analysis. A comparison between the protein secondary structural regions identified from the replica exchange simulations and the solution phase NMR experiments is discussed. The data obtained shows that the positions of the helical regions predicted by the simulations are largely consistent with the NMR structure.

This work was supported by the U. S. Department of Energy, Office of Basic Energy Science, Division of Materials Sciences and Engineering. The research was performed at the Ames Laboratory. Ames Laboratory is operated for the U.S. Department of Energy by Iowa State University under Contract No. DE-AC02-07CH11358.

(133d) Absolute Ligand Binding Free Energy of Glycoside Hydrolases As a Metric for Processivity and Polysaccharide Morphology Dependence

Payne, C. M., National Renewable Energy Laboratory
Jiang, W.,
Shirts, M. R., University of Virginia
Himmel, M. E., National Renewable Energy Laboratory
Crowley, M. F., National Renewable Energy Laboratory
Beckham, G. T., National Renewable Energy Laboratory
Originally presented on: 11/4/2013 13:15:00 - 13:30:00

Multiple glycoside hydrolase (GHs) families have evolved to deconstruct crystalline polysaccharides, such as cellulose and chitin, to sugars. These GHs generally act via a processive mechanism wherein a single carbohydrate chain is decrystallized from the surface, and the glycosidic linkage is processively cleaved along the chain without dissociation. This processive mechanism has long been the focus of structure-function studies, and recent biochemical studies have shown dissociation of processive cellulases is likely rate limiting in cellulose conversion. Despite being a central theme of GH research, a general molecular-level theory of processivity linked to chemical and structural features of enzyme active sites does not yet exist. Here, we hypothesize that the degree of GH processivity is directly linked to an enzyme's ability to decrystallize a single chain from a crystal and can be quantified by the absolute ligand binding free energy. To understand the thermodynamics of processivity at the molecular level, we calculate the absolute ligand binding free energy of cellulose chains to processive members of the biologically and industrially important enzymes from GH Family 7 with Free Energy Perturbation/Hamiltonian Replica-Exchange Molecular Dynamics. Alongside available experimental observations, the computational results suggest that ligand binding free energies within a GH family correlate well with processivity. When compared to the thermodynamic work required to decrystallize cellulose chains, the calculated binding free energies suggest polymer morphologies susceptible to decrystallization by a given enzyme. Generally, the ligand binding free energy may offer a molecular-level basis for developing a generalized theory of carbohydrate processivity.

(133e) Comparison of Ionic Liquid-Tolerant Glycoside Hydrolases Using Molecular Dynamics

Jaeger, V., University of Washington
Burney, P. R., University of Washington
Pfaendtner, J., University of Washington
Originally presented on: 11/4/2013 13:30:00 - 13:45:00

Glycoside hydrolases are a class of enzyme that catalyzes the hydrolysis of bonds between glycosides such as cellulose and xylose. This reaction is particularly important for the conversion of biomass into value-added chemicals. One problem for the application of enzymes toward the degradation of cellulosic biomass is the recalcitrance of the highly structured cellulose structure and its entwinement with hemicellulose and lignin. Certain ionic liquids (IL), which are organic salts that are liquid at temperatures below 100° C, have the rare ability to solvate cellulosic biomass to help processes overcome this limitation. By discovering enzymes that can tolerate the presence of ILs, new processes can be developed that take advantage of the unique properties of ILs.

Datta et al. [1] studied the relative activity of three different cellulases in various concentrations of the IL 1-ethyl-3-methylimidazolium acetate [EMIM][OAC]. While the structures of the enzymes appear similar, their tolerance of IL varied greatly. The cellulase from *Pyrococcus horikoshii* was tolerant of up to 20% (v/v) IL, the cellulase from *Thermatoga maritima* was moderately tolerant of the same concentration, and the cellulase from *Trichoderma viride* was completely deactivated. It has been suggested that enzymes from halophilic and thermophilic organisms will tend to have adaptations that allow the enzyme to also tolerate organic salts such as ILs.

We have conducted molecular dynamics simulations to compare the structure and modes of motion for these enzymes in mixtures of IL and water. Preliminary results show the IL did not induce unfolding of the enzyme on the time scale of hundreds of nanoseconds. Therefore, some other factor is likely influencing the activity of these enzymes. Principle component analysis suggests that in each of the systems the presence of IL affects the slow modes of motion. The effects are varied, but they are more exaggerated for higher concentrations of IL and the more IL-intolerant enzymes. The dynamic fluctuations of the enzymes were measured by root mean square fluctuation. Higher concentrations of IL correspond to lower magnitude fluctuations. At higher temperatures with high concentrations of IL, we observe differences across the sequence in the relative magnitude of fluctuations as well. We tracked the evolution of solvent-protein interactions over the simulations. Areas of strong IL interaction tend to correspond with those that display changes in fluctuation. If these simulations are indicative of the behavior of other glycoside hydrolases in binary mixtures of IL and water, we can use molecular dynamics to predict retention or loss of activity of an enzyme in such solvents by applying tests such as those mentioned herein. We have gained molecular-level insight into the effects of ILs on these cellulases and which properties of the enzyme lead to its deactivation in IL.

[1] S. Datta, B. Holmes and J. I. Park, Green Chem. 2010, 12, 338.

(133f) Molecular Dynamics Insights Into Targeting CXCR4 By Branched Peptides Based On DV3

Tamamis, P., Princeton University
Smadbeck, J., Princeton University
Khoury, G. A., Pennsylvania State University-University Park
Szekely, Z., Rutgers University
Sinko, P. J., Rutgers University
Floudas, C. A., Princeton University
Originally presented on: 11/4/2013 13:45:00 - 14:00:00

Chemokine Receptor Type 4 (CXCR4) is an emerging target in the development of cancer and HIV-1 drugs due to its co-receptor involvement in the migration of cancer cells and its key role in HIV-1 entry. Extensive investigations led to the discovery of an inverso antagonist peptide, DV3 [1] derived from the sequence of a natural CXCR4 ligand, the viral macrophage-inflammatory protein - II. Recent-unpublished experiments investigated peptides incorporating 1, 2 and 4 DV3 sequences branched off a core structure consisting of 1, 2 and 4 D-lysine(s) respectively and showed that the largest ligand candidate bearing four peptides (4DV3) possesses remarkable avidity to the CXCR4 receptor, blocking HIV-1 entry in a fusion-specific antiretroviral assay. We computationally modeled the structures for the 2DV3 : CXCR4 and 4DV3 : CXCR4 complexes, which include two types of non-identical asymmetrical binding pockets to accommodate the dimer and tetramer DV3 peptide ligands, respectively, and performed extensive molecular dynamics (MD) simulations of the determined dimeric and tetrameric complexes verifying the complexes' structural integrity. Analysis of the MD trajectories reveals the key interacting residues within the complexes, as well as peptide residues with potential for further optimization. In addition, we provide evidence on the antagonistic, anti HIV-1, role of DV3-based compounds against V3 loop binding, by comparing the present results with recent-unpublished insights from a computationally derived V3 loop : CXCR4 complex structure.

Reference:

[1] N. Zhou *et al.* Exploring the Stereochemistry of CXCR4-Peptide Recognition and Inhibiting HIV-1 Entry with D-Peptides Derived from Chemokines. *Journal of Biological Chemistry,* **2002,** 277, 17476-17485.

(133g) A Molecular Dynamics Study of Single-Walled Carbon Nanotubes (SWCNTs) Dispersed in Bile Salt Surfactants

Phelan, F. R. Jr., NIST
Sun, H.,
Originally presented on: 11/4/2013 14:00:00 - 14:15:00

A Molecular Dynamics Study of Single-Walled Carbon Nanotubes (SWCNTs) Dispersed in Bile Salt Surfactants[*]

Frederick R. Phelan Jr.[?],[1] Â and Huai Sun[2]

[1]Materials Science and Engineering Division, NIST, Gaithersburg, MD 20899

[2]Aeon Technology Inc. and School of Chemistry and Chemical Engineering, Shanghai Jiao Tong University, Shanghai, China

Abstract

Separation of single-walled carbon nanotubes by chirality is an important aspect of nanotube metrology. Density gradient ultra-centrifugation (DGU) has emerged as one of the most important tools for chirality separations, and relies heavily on the use of bile salt surfactants to create buoyant density differences between tubes of different chirality, based on the size and packing structure of the hydration shell surrounding the dispersed tube [1-4]. However, even though DGU is extensively used, a greater understanding the surfactant self-assembly mechanism and packing structures for bile salt surfactants and co-surfactants of different functionality is greatly needed [3-4], as the efficacyÂ of the separations depends critically on the details of the SWCNT/surfactant complex that forms during the dispersion stage. In this work, we study the behavior of a family of bile salt surfactant and co-surfactant mixtures for a variety of metallic and semi-conducting SWCNT chiralities. The goal is to provide a greater understanding of how small changes in surfactant functionality, e.g., sodium deoxycholate (DOC), sodium taurodeoxycholate (TDOC), and sodium cholate (SC), promote different packing densities critical to DGU separations, and compare with recent experimental results [4]. The force field parameters used in the simulations are developed based on the Consistent Force-Field (CFF) potential using the Direct Force-Field 7.0 (DFF7) software [5], and will be stored in WebDFF, an open and extensible, smart force-field repository being developed at NIST in collaboration with Aeon Technology, as part of the Materials Genome Initiative (MGI).

References

[1]Â Â M. S. Arnold, S. I. Stupp, and M. C. Hersam, ?Enrichment of Single-Walled Carbon Nanotubes by Diameter in Density Gradients,? *Nano Lett.*, vol. 5, no. 4, pp. 713?718, Apr. 2005.

[2]Â Â M. S. Arnold, A. A. Green, J. F. Hulvat, S. I. Stupp, and M. C. Hersam, ?Sorting carbon nanotubes by electronic structure using density differentiation,? *Nat Nano*, vol. 1, no. 1, pp. 60?65, Oct. 2006.

[3]Â Â F. Bonaccorso, T. Hasan, P. H. Tan, C. Sciascia, G. Privitera, G. Di Marco, P. G. Gucciardi, and A. C. Ferrari, ?Density Gradient Ultracentrifugation of Nanotubes: Interplay of Bundling and Surfactants Encapsulation,? *J. Phys. Chem. C*, vol. 114, no. 41, pp. 17267?17285, Oct. 2010.

[4]Â Â J. A. Fagan, M. Zheng, V. Rastogi, J. R. Simpson, C. Y. Khripin, C. A. Silvera Batista, and A. R. Hight Walker, ?Analyzing Surfactant Structures on Length and Chirality Resolved (6,5) Single-Wall Carbon Nanotubes by Analytical Ultracentrifugation,? *ACS Nano*, Mar. 2013.

[5]Â Â ?DFF 7, AeonTechnology.? [Online]. Available: http://www.aeontechnology.com/Product_DFF.php. [Accessed: 13-May-2013].

(133h) Different Self-Associations of Carboxybetaine and Sulfobetaine

Shao, Q., University of Washington
Jiang, S., University of Washington
Originally presented on: 11/4/2013 14:15:00 - 14:30:00

Zwitterionic polycarboxybetaine and polysulfobetaine highly resist nonspecific protein adsorption in complex media. This common nonfouling property provides them an important role in applications including drug delivery, medical devices and diagnostics. One potential issue of zwitterionic materials is the possible self-association between zwitterionic moieties. This association will influence many relevant properties of zwitterionic materials including solubility, hydration and swelling, thus their nonfouling performance ultimately. Polycarboxybetaine contains carboxylic and trimethyl ammonium groups, while polysulfobetaine contains sulfonate and trimethyl ammonium groups. The different combinations of these charged groups may cause different degrees of self-associations. In this work, we studied carboxybetaine-carboxybetaine and sulfobetaine-sulfobetaine associations and their effects on polymeric properties using both molecular simulations and experiments. We studied carboxybetaine-carboxybetaine and sulfobetaine-sulfobetaine associations at high concentrations ranging from 1 to 4 M and carbon spacer lengths from one to three using molecular dynamics simulations. The analysis of radial distribution functions between the cationic and anionic groups showed significant self-associations of sulfobetaine moieties in all concentrations studied. Carboxybetaine moieties do not associate with themselves. The different self-associations are mainly attributed to the different combinations of charged groups. We studied the radii of gyration of poly(sulfobetaine methacrylate) (pSBMA) and poly(carboxybetaine methacrylate) (pCBMA) with metadynamics simulation in order to investigate the influence of different self-associations on polymeric properties. We also studied the antipolyelectrolyte effects of pSBMA and pCBMA experimentally to testify our simulation predictions.

(133i) Molecular Interactions Between Fc Fragment of IgG and Peptide Ligand for Affinity Chromatography

Wang, R. Z., Zhejiang University
Tong, H. F., Zhejiang University
Lin, D. Q., Zhejiang University
Yao, S. J., Zhejiang University
Originally presented on: 11/4/2013 14:30:00 - 14:45:00

Affinity chromatography with synthetic ligands has been focused as the alternative to protein A-based chromatography for antibody capture during the downstream process due to its comparable selectivity and efficiency. Better understanding the molecular interactions between synthetic ligand and antibody is crucial for designing novel ligands. In this work, the binding mechanism between Fc fragment of IgG and a synthetic ligand (DAAG) was studied with molecular docking and dynamics simulation. The consensus binding site (CBS) between C_H2 and C_H3 domains of IgG molecule was focused. The docking results showed DAAG could bind to the CBS like a tripods for the top-ranked ligand-Fc fragment complexes, then the complexes were tested by molecular dynamics simulation at neutral condition (pH 7.0). The van der Vaals interaction energy and electrostatic interaction energy between Fc fragment and DAAG were calculated and compared for 10 ns simulation. The results indicated that the binding of DAAG on the CBS of Fc fragment was achieved by the mixed-mode molecular interactions, combining hydrophobic interaction, electrostatic interaction, hydrogen bonding, etc. In addition, based on the analysis of molecular interactions between individual residue and DAAG, the key residues on Fc fragment were identified as GLN311, LEU314, ASP315, LYS338 and GLU430. It was also found that multiple secondary interactions were formed and dominated the molecular recognition between Fc fragment and DAAG and endowed DAAG with an excellent selectivity to Fc fragment. Finally, molecular dynamics simulation was conducted at acidic condition (pH 3.0), and the departure of DAAG ligand from the surface of Fc fragment was observed as the result of the reduced interaction energy, which was highly related to the dissociation states of ASP315 and GLU430. These results not only shed light on the molecular mechanisms for DAAG, but also provided useful information for the improvement of ligand design.

Acknowledgment: supported by National Natural Science Foundation of China and Zhejiang Provincial Natural Science Foundation of China.

(148a) Adapting SAFT-γ Perturbation Theory to Site-Based Molecular Dynamics Simulation: An Application to Homogeneous and Inhomogeneous Systems

Ghobadi, A. F., The University of Akron
Elliott, J. R., The University of Akron
Originally presented on: 11/4/2013 12:30:00 - 12:50:00

Molecular simulation has become an indispensable tool in developing theories to describe fluid properties. Ideally, atomistic detail applied in the simulation is sufficient to characterize the properties of interest and theories are refined until they are consistent with the details of the simulations. Currently, the leading theory in the prediction of fluid properties is Statistical Associating Fluid Theory (SAFT). Equations of states (EOSs) based on SAFT, which benefit from Wertheim's first order Thermodynamic Perturbation Theory (TPT1), reproduce molecular simulation results and provide an analytical equation that facilitates interpolation and extrapolation of fluid properties on much shorter time scales than full potential molecular simulation.

In its original form, and most subsequent forms, SAFT assumed that polyatomic species were composed of tangent sphere segments. Although this form was shown to agree reasonably well with simulations of tangent hard sphere chains, significant discrepancies arise when compared to simulations of more realistic molecular models such as fused soft-sphere chains. One particular implementation of SAFT that ostensibly includes atomistic detail in fused chain description of molecules is the SAFT-γ EOS (Lymperiadis, *J. Chem. Phys*. **127**, 234903 (2007)). The SAFT-γ model assigns an interaction site to each site-type in a united-atom characterization of the molecule. This level of description is consistent with molecular simulations based on transferable force fields such as the TraPPE-UA model. However, characteristic repulsive and attractive adjustable parameters of SAFT-γ EOS are fitted to experimental bulk fluid properties. Therefore, an agreement at the atomistic level with molecular simulation is not guaranteed. In fact, we demonstrate that there will be significant deviations between simulation and theory when reference and perturbation contributions of EOS are lumped together and fitted to experimental data.

In this context, we seek to develop a form of SAFT-γ EOS such that molecular simulations - based on transferable and united atom force fields - can be checked for consistency with SAFT-γ predictions. While our objective is a theory of interfacial properties, we find it necessary to first elucidate the details of the SAFT-γ approach in terms of site-based simulations for homogeneous fluids. Once these details are well-defined at the bulk level, we adopt classical density functional theory (DFT) to address the interfacial properties. DFT involves estimating the Helmholtz energy at intermediate densities between the vapor and the liquid, in other words, at thermodynamically unstable points inside the binodal. This step requires that the EOS be accurate in order for extrapolation inside the binodal to provide meaningful results. We aim to establish that accuracy based on homogeneous (bulk) fluids, then to test the extrapolations based on molecular simulations of inhomogeneous fluids.

We apply the rapidly convergent Weeks-Chandler-Andersen (WCA) perturbation theory to investigate the properties of softly repulsive reference fluids for realistic and fused models of polyatomic molecules. The SAFT-γ methodology plays an important role in developing a master curve for each reference fluid that can be applied to multiple temperatures and multiple values of the well-depth parameter (ε). Once the reference fluid is characterized, we turn our attention to the perturbation contributions. In particular, we compute perturbation terms for multiple chain lengths rather than simply extrapolating corresponding contributions of spherical molecules, finding that such extrapolations are not entirely reliable. Consistent with recent work by Jackson and coworkers (Avendano, *J. Phys. Chem. B* 2013, 117, 2717-2733), we find that inclusion of the third order term substantially improves accuracy in the critical region, noting that this accuracy is important for interfacial properties.

The resulting EOS takes force field parameters (σ and ε) as input variables and has no binary interaction parameter for united-atom site-site interactions. Finally, we validate our adaptation of SAFT-γ with extensive comparisons to site-based molecular simulations of the full potential force fields such as TraPPE-UA, OPLS and NERD. The results show satisfactory agreement between EOS and simulation for the entire phase diagram including the critical point region. For example, the deviation of predicted liquid density and vapor pressure of n-alkanes from methane to n-dodecane compared to those of TraPPE-UA force field is less than 1% and 6%, respectively. The same level of agreement is also obtained for mixtures and derivative properties such as heat capacity. Furthermore, the density profile of monomers and mixtures of fused soft chain molecules against a hard wall as well as in confined systems are predicted by SAFT-γ + DFT EOS and compared to atomistic molecular simulation of real molecules. Excellent agreement with simulation is achieved not only for center of mass density profiles but also for each united-atom site type comprising the molecular structure of studied compounds. Surface tension, adsorption isotherms, and atomistic density profiles of real substances and their mixtures at planar vapor-liquid interfaces are also predicted and compared to molecular simulation with reasonable agreement.

In summary, we can show that a robust comparison between theory and simulation at the development stage can lead to an EOS that is consistent with molecular simulation of transferable united-atom force fields for both homogeneous and inhomogeneous systems. The resulting EOS would inherit many advantages of molecular simulation while facilitates engineering applications with imposing very short time scales for calculation process. Also, the direct correspondence between EOS and simulation can be recognized as a valuable tool to further optimize force field parameters efficiently for the system of interest.

(148b) Net Adsorption Thermodynamic Framework for Microporous Solids

Talu, D. O., Cleveland State University
Originally presented on: 11/4/2013 12:50:00 - 13:10:00

The *net* adsorption thermodynamic framework was recently introduced as an alternative to *excess* and *absolute* adsorption (Gumma and Talu, *Langmuir* 2010, 26(22), 17013-17023). It is a fundamentally different thermodynamic description of adsorption since the placement of Gibbs dividing surface, necessary for any treatment of adsorption, is different than the known/used alternatives. Net adsorption is a natural choice for microporous solids. It removes any ambiguity in theoretical treatment of adsorption and circumvents all questionable measurements/quantities in experiments.

Initial development of net adsorption was prompted by the interest in adsorptive gas storage and by the need to unambiguously compare experiments to molecular simulation results. Previously, our research group and others have shown that: 1) non-adsorbing helium assumption commonly used to place the Gibbs dividing surface in experiments does not hold (e.g. Gumma and Talu, *Adsorption J.* 2003, 9, 17-28), 2) molecular simulation results can only be compared to excess data by performing more simulations with helium (e.g. Talu and Myers, *AIChE J.* 2001, 47(5), 1160-1168), and 3) volume occupied and thermal mass of solid adsorbent play a significant role in equilibrium and dynamics of adsorptive storage systems (e.g. Chang and Talu, *Applied Thermal Engineering* 1996, 16(5), 359-374). As such, our previous work has concentrated on pure component adsorption equilibrium and dynamics.

This contribution involves extension of net adsorption framework to gas mixtures. Theoretical developments on equilibrium formulations will be included as well as its implications on adsorption processes such as column dynamics in separations and in storage. The theory will be supplement with data examples and implementation methods. The advantages/disadvantages and limitations of net adsorption will be discussed.

(148c) Prediction of Multiphase Equilibrium of Water-Alcohol-Hydrocarbons Using Theory-Based Equation of State

Al-Rehaili, A. M., University of British Columbia
Al-Hamdan, A. M. M., King Fahd University of Petroleum & Minerals
Englezos, P., University of British Columbia
Alsaifi, N., King Fahd University of Petroleum & Minerals
Originally presented on: 11/4/2013 13:10:00 - 13:30:00

In this work, the role of long-ranged electrostatic interactions is evaluated in predicting multiphase equilibrium in water-alcohol-hydrocarbon-carbon dioxide systems using statistical association fluid theory. Various long-ranged interactions are added to the statistical association fluid theory including dipole and quadrupole interactions. The liquid-liquid and vapor-liquid-liquid behavior of complex mixtures are predicted without introducing binary interaction parameter. Three different SAFT versions are utilized; namely CK-SAFT, simplified SAFT and PC-SAFT. The effect of the number of association sites is explored in studying various systems such as water-hydrocarbons and water-alcohols. The liquid-liquid behavior of water-hydrocarbons is quantitatively obtained without adjusting mixture experimental data.

(148d) Description of Thermophysical Properties of Multi-Component Mixtures Employing Virial Based Mixing Rules (VBMR)

Baghdasarian, S., California State Polytechnic University-Pomona
Sen, S., California State Polytechnic University-Pomona
Originally presented on: 11/4/2013 13:30:00 - 13:50:00

The last several decades have witnessed the development of a number of high fidelity methods for describing thermophysical properties of pure substances in condensed phases. They include corresponding states techniques and component-specific correlations, as well as statistical thermodynamics approaches. However, the science for describing mixtures has consistently trailed behind, often being limited to ideal mixing or similar formulations. For example, the mixing rules (typically implemented in process simulators) for liquid viscosity and thermal conductivity, are the Kendall-Monroe and Li correlations, respectively.

While the attractive feature of such mixing rules is that they do not require interaction parameters (IP's), their downside is that the do not allow IP's, even when necessary. In many applications, they incorrectly predict only a small deviation from ideal mixing, often with incorrect sign. Thus, while these mixing rules perform reasonably for hydrocarbon systems, their efficacy for non-ideal systems is questionable.

The state-of-the-art is the Redlich-Kister correlation (RKC), which can incorporate additional higher-order terms to improve the fit of data with higher precision. It is often the correlation of choice for the applied thermodynamics research community, and has been successfully applied to precisely correlate a variety of thermophysical properties for binary mixtures. However, the RK correlation does not naturally extend to multi-component systems.

This investigation draws upon the exact mathematics of mixing rules established for virial coefficients of gases. Perturbations from ideal mixing are modeled akin to successive terms in the virial expansion, resulting in the name "Virial-Based Mixing Rules" (VBMR). Each perturbation reduces to zero for pure components, and adopts the composition order of the corresponding virial coefficient it represents. For example, the first perturbation represents the second virial coefficient and implements a quadratic mixing rule, while allowing one adjustable binary interaction parameter. The next term employs cubic mixing and allows multiple interaction parameters, including a ternary interaction parameter which may be fine-tuned for ternary and higher mixtures. All interaction parameters are initialized to zero.

Applying the VBMR methodology, Sen and co-workers have recently demonstrated excellent results for the correlation of several thermophysical properties for a number of binary liquid mixtures, including density, enthalpy, and viscosity. In this investigation, the VBMR methodology is extended to ternary (and multicomponent systems), yielding non-zero ternary (and higher-order) interaction parameters accounting for three-body interations, without compromising the goodness-of-fit for binary constituents. Comparisions are made with RKC, emphasizing how the VBMR naturally extends to multicomponent systems, while the RKC does not.

The VBMR model separates the temperature dependence of pure component and interaction parameters. This is postulated as a key reason for the smooth as well as weak temperature dependence of the interaction parameters. By contract, the temperature dependence of model parameters for the RKC is often siginifcantly jagged. This investigation implements and demonstrates the smotth temperature dependence for several binary and multicomponent mixtures.

(148e) Force-Field Based Quasi-Chemical Method for Rapid Evaluation of Binary Phase Diagrams

Sweere, T. J. M.,
Fraaije, J. G. E. M., Culgi BV
Originally presented on: 11/4/2013 13:50:00 - 14:10:00

For many years the cornerstone of chemical engineering thermodynamics has been the evaluation of segment based activity coefficients through some surface interaction site model; of which we address two well known variants: UNIFAC (group interactions) and COSMO-RS or COSMO-SAC (charge panel interactions). In the case of UNIFAC one relies heavily on (a) group definitions, that more than often are based on history or chemical intuition than a sound principle, and (b) parameterization through fitting on experimental phase diagrams. In COSMO variants, one does not use groups per se, but instead surface charge panels; with a parameter space that is only a fraction of that in UNIFAC. In quite a different corner one has a collection of force fields, that have been and will remain popular in molecular dynamics; in this case with full atomistic detail, and potentially a smaller parameter space (more 'ab initio') than either UNIFAC or COSMO, but unfortunately too slow and unreliable to be of practical use to the engineer. What now, if we could combine the power of the quasi-chemical derived activity coefficients directly with some force field, without the need for a condensed simulation? We developed such method: fast (no condensed simulation needed), also no ad hoc groups, and no surface charges needed, and *in principle* a relatively small parameter set. As we will demonstrate, by showing results against experimental data from the NIST databank, sometimes the predictions are of almost unbelievable accuracy. Not always though, sometimes the method fails badly, that we can attribute to the absence of correlations in molecular packing. We also compare the method with other popular approaches, for example Blanco's method for prediction of Flory-Huggins parameters, where we do much better. In the new method, from a practical point of view, we have removed the burden of group assignment, and introduced the difficulty of atom typing. Provided such atom types can be and will be defined by the biomolecular simulation community, this does open the door to application in a wide range of problems in the engineering of specialty chemicals, polymers, and pharmaceuticals.

(148f) Generalized Activity Coefficient Models for VLE Mixture Predictions

Gebreyohannes, S. B., Oklahoma State University
Dadmohammadi, Y., Oklahoma State University
Venugopal, V., Oklahoma State University
Yerramsetty, K. M., Oklahoma State University
Neely, B. J., Oklahoma State University
Gasem, K. A. M., Oklahoma State University
Originally presented on: 11/4/2013 14:10:00 - 14:30:00

Accurate prediction of the phase behavior properties of chemical species and their mixtures is essential for the design of chemical processes involving separation operations. Predicting phase equilibrium properties using reliable models offers an attractive alternative to costly and time consuming experimental measurements. In this work, we focus on liquid activity coefficient modeling.

Activity coefficient is a basic phase equilibria property that accounts for liquid mixture deviations from ideal behavior. Although a number of activity coefficient models exist in the literature, their use is limited by the availability of experimental data. Among these models, non-random two, liquid (NRTL), universal quasi-chemical (UNIQAUC) and WILSON are widely used to correlate fluid phase equilibrium data. These models require two or three adjustable interaction parameters that are determined through regression of experimental data for a specific system. Typically, to facilitate *a priori* predictions, group-contribution methods have been employed to generalize the UNIQAUC and Wilson models.

Our recent research has focused on developing an alternative method for generalizing the interaction parameters of the NRTL, UNIQUAC and WILSON models. Specifically, we applied theory-framed quantitative structure-property relationship (QSPR) modeling approach. In this modeling approach, theoretical frameworks, such as the NRTL, are used to develop the behavior models, and QSPR to generalize the substance-specific parameters of the models.

A database of 960 binary VLE data was employed to develop the QSPR models. Data regression analyses were performed to determine the interaction parameters of the NRTL, UNIQUAC and WILSON models. The structural descriptors of the molecules were generated and used in developing the QSPR models to predict the regressed interaction parameters. The predictive capabilities of the generalized models were assessed for phase equilibria properties including pressure, temperature, vapor mole fractions and equilibrium constants. The developed QSPR models provided phase equilibria property predictions within two times the errors obtained through the data regression analyses. The overall property predictions from the QSPR models are comparable to that of the results from the group-contribution method (UNIFAC-2006), which has over 4,000 model parameters. In contrast, the QSPR model has about 300 model parameters (neural network weights and biases). Thus, our methodology provides *a priori* and easy implementable models with comparable property predictions to those the UNIFAC model.

(148g) Calculation of the Vapor-Liquid Phase Coexistence of Polymer-Grafted Nanoparticles

Mukherjee, S., Vanderbilt University
Haley, J. D., Vanderbilt University
Iacovella, C. R., Vanderbilt University
McCabe, C., Vanderbilt University
Cummings, P. T., Oak Ridge National Laboratory
Originally presented on: 11/4/2013 14:30:00 - 14:50:00

It is well known that the aggregation/dispersion of nanoparticles can have a significant impact on the properties of the system and precisely controlling this behavior is crucial for the rational design of novel materials. One approach for controlling the aggregation is to graft polymers to the surface of the nanoparticles. The resulting behavior is therefore dictated by contributions from both the nanoparticles and polymers interactions and significant changes to the behaviorcan be achieved by tuning the relative strength of these interactions. The large number of interconnected parameters of grafted nanoparticles systems (e.g., nanoparticle size and interaction strength, polymer length and grafting density) makes it challenging to predict the behavior of a system *a priori.* Towards the goal of predictive assembly, we examine the vapor-liquid equilibrium (VLE) of polymer-grafted nanoparticles [1]. The main goal of this work is to explore the balance between nanoparticle-polymer interactions and provide general trends that can be applied to a wide range of systems. In this work, the VLE is calculated using both molecular simulation via quench dynamics [1,2] and the statistical associating fluid theory for potentials of variable attractive range (SAFT-VR) [3]. A coarse-grained model is used, generic enough to be applicable to a wide range of systems but with parameters informed by recent studies of the interactions between silica nanoparticles [1,4] and the interactions between alkane chains [5]. Quench dynamics simulations and SAFT-VR calculations exhibit similar trends demonstrating a strong connection between the VLE and the grafting architecture. It is observed from both the simulations and theoretical predictions that the critical temperature is reduced as the number of grafts and length of the grafts is increased. The VLE behavior is also investigated by increasing the nanoparticle-nanoparticle strength for both the low and high grafting density systems. We find that the critical temperature increases with the increase in nanoparticle-nanoparticle interaction strength for both these systems, but the effect of increasing interaction is seen to be much more prominent in the low grafting density system. Also the study of VLE of higher molecular weight chains via SAFT-VR reveals that for these systems the polymer grafts shields almost all the accessible surface area of the nanoparticle, thus shielding the strong nanoparticle interactions. As a result, the VLE behavior is dominated by the polymer-polymer interactions. The role of graft topology for systems with low surface coverage is also presented in this work.

References:

1. Iacovella CR, Varga G, Sallai J, Mukherjee S, Ledeczi A, Cummings PT, Theoretical

Chemistry Accounts, (2013), 132:1315

2. P.J. in' t Veld, M.A. Horsch, J.B. Lechman, G.S. Grest, J Chem Phys 129 (2008).

3. Y. Peng, C. McCabe, Mol Phys 105 (2007) 261-272.

4. C.K. Lee and C.C. Hua, J. Chem. Phys. 132, 224904 (2010).

5. M.G. Martin and J.I. Siepmann, J. Phys. Chem. B 102 (1998).

(148h) Hard-Disk Equation of State: First-Order Liquid-Hexatic Transition in Two Dimensions With Three Simulation Methods

Engel, M., University of Michigan
Anderson, J. A., University of Michigan
Glotzer, S. C., University of Michigan
Isobe, M., Nagoya Institute of Technology
Bernard, E. P.,
Krauth, W.,
Originally presented on: 11/4/2013 14:50:00 - 15:10:00

We report large-scale computer simulations of the hard-disk system at high densities in the region of the melting transition. Our simulations reproduce the equation of state, previously obtained using the event-chain Monte Carlo algorithm, with a massively parallel implementation of the local Monte Carlo method and with event-driven molecular dynamics. We analyze the relative performance of these simulation methods to sample configuration space and approach equilibrium. Our results confirm the first-order nature of the melting phase transition in hard disks. Phase coexistence is visualized for individual configurations via the orientational order parameter field. The analysis of positional order confirms the existence of the hexatic phase.

Reference: Physical Review E 87, 042134 (2013)

(116a) Propane Ammoxidation Pathway Over Mo-V-Te-Nb-O M1 Catalyst Investigated By Density Functional Theory

Yu, J., University of Cincinnati
Guliants, V. V., University of Cincinnati
Xu, Y., Oak Ridge National Laboratory
Originally presented on: 11/4/2013 12:30:00 - 12:50:00

The selective ammoxidation of propane into acrylonitrile catalyzed by the bulk Mo-V-Te-Nb-O system received considerable attention because it is more environmentally benign than the current process of propylene ammoxidation and relies on a more abundant feedstock. This process consists of a series of elementary steps including propane oxidative dehydrogenation (ODH), ammonia and O_2 activation, and N-insertion into C_3 surface intermediates. However, the limited fundamental understanding of the reaction mechanism and the roles of the different cations has hindered the progress in further improving its activity and selectivity required for the commercial application.

In this talk, we discuss the results of the density functional theory (DFT) calculations performed to investigate overall propane ammoxidation pathway on the cluster models of the proposed selective and active sites present in the surface *ab* plane of the M1 phase. The activation energies for the oxidative dehydrogenation (ODH) of propane and sequential intermediates (isopropyl, propene, and allyl) were calculated on different cation sites. Propane activation on V^{5+}=O was found to be the rate-limiting step (E_a = 1.2 eV), consistent with the current proposed reaction mechanism for propane activation on the Mo-V-Te-Nb oxides and the current understanding of V^{5+} as the active site for alkane activation present in V-based mixed oxides. The unselective pathway leading to C3 combustion starting from propene activation on Mo=O site was also explored.

Furthermore, a linear relationship was established between the H adsorption energy and the activation energy for H abstraction from various C3 intermediates, which is highly useful for predicting the energy barriers of H abstraction from C3 species based solely on H adsorption energy. The energy barriers of ammonia activation on different surface sites and NH insertion into the allyl species were investigated and compared with the hypothetical pathway reported in the literature. These elementary reaction steps were indicated to be energetically barrier-less. The distribution of NH_x and OH_x species over surface sites were then explored using micro-kinetic models to study the impact on selectivity of propane ammoxidation to acrylonitrile on the Mo-V-Te-Nb-O M1 phase.

(116b) Simulation of Au Nanoparticles Deposited On Metal Oxide Support

Garcia, J., Worcester Polytechnic Institute
Deskins, N. A., Worcester Polytechnic Institute

Originally presented on: 11/4/2013 12:50:00 - 13:10:00

Gold is a typically inert metal, but since the discovery that gold nanoparticles can be catalytically active, there has been intense research to understand the origin of this phenomena. The activity of gold nanoparticles can also change drastically depending on the support, usually a metal oxide, indicating that support effects are relevant. Au catalysts have application in oxidation of CO, the water-gas shift reaction, and several other oxidation reactions. There has been much debate about the nature of Au catalysts, and why certain systems (e.g. Au/TiO_2 or Au/CeO_2) are more reactive than others. Understanding the nature of nano-sized Au and support effects are crucial to developing better Au catalysts.

Au over metal oxide surfaces have been previously modeled, however the size of the Au particles have typically been limited to small clusters, perhaps up to 1 nm in size. Experimentally relevant systems often involve Au particles several nm in size. In this work we modeled Au nanoparticles deposited on metal oxide supports using density functional theory. We considered a range of Au particles, including relevant larger particles up to 219 atoms. Au particles deposited on TiO_2 and MgO were simulated at the DFT and DFT+U level. We probed several reaction sites through model reactions CO oxidation and the water-gas shift reaction. Reaction sites considered include nanoparticle corners, steps, terraces, and interfacial sites between Au and metal oxide. From this work we have provided meaningful insights on the the role of the substrate, surface defects, particle size and shape.

(116c) First Principle Studies of Cu and Au Growth Morphology On the Non-Polar ZnO(101-0) Surface

Nie, X., The Ohio State University
Asthagiri, A., University of Florida
Originally presented on: 11/4/2013 13:10:00 - 13:30:00

Catalysts consisting of Cu or Au supported on ZnO surface have been widely used in a variety of hydrogenation reactions, especially in methanol synthesis. Recent experimental studies from our collaborators at Louisiana State University show that Cu grows as metallic 3D clusters preferring to nucleate at [0001] step edges at low coverage, whereas Au initially grows as well-dispersed single-layer islands with nonmetallic character and no preferred nucleation sites below a certain critical coverage. To understand these experimental results, density functional theory (DFT) calculations were performed to investigate the single metal atom adsorption and diffusion, along with the morphology of small Cu and Au clusters on the non-polar ZnO(10-10) surface.

DFT results show that Au and Cu bond with Zn and O atoms of the ZnO(10-10) surface, respectively, with the Cu atom binding around ~40% more strongly than the Au atom. Nudged elastic band calculations show Cu adatom diffuses rapidly along the [1-210] direction, while the diffusion along [0001] is hindered by large barriers. For the Au adatom, no anisotropy is observed, with relatively smaller diffusion barriers than found for Cu. These DFT results suggest that the experimentally observed differences in dispersion of Au versus Cu clusters is due to differences in adatom diffusion, where Cu adatoms diffuse preferentially to the [0001] step edge and nucleate while the Au adatoms diffuse and nucleate in a random manner. The nucleation process and the behavior of the metal atoms at the step edges are currently being explored.

We also examined small clusters including the dimer, trimer, tetramer, and pentamer. Based on the structure of these small clusters, we generated larger clusters including 10 and 16 metal atoms in a single layer and a double layer configuration. The adsorption energy results for these large clusters verify a preference for Cu to grow in a stacking configuration, while the Au clusters prefer to spread out into 1D islands. These DFT results on the clusters again match well with experimental observations of metal clustering behavior. We will also present ongoing DFT and experimental studies correlating the chemistry of CO_2 and CO on the metal cluster structure and charge distribution. Such studies will be important in identifying potential active sites for the metal-supported ZnO catalysts used in methanol synthesis.

(116d) Modeling Pt/Sn/SnO2 Catalysts

Deskins, N. A., Worcester Polytechnic Institute
Teng, X., University of New Hampshire
Originally presented on: 11/4/2013 13:30:00 - 13:50:00

Pt-Sn alloys have been used in a variety of catalytic reactions, including organic molecule oxidation (such as in fuel cells) or in hydrocarbon hydrogenation/dehydrogenation. Under reaction conditions however a synthesized alloy may change drastically. For instance segregation of Sn may occur or oxidation of Sn is possible leading to formation of SnO_x. We have used density functional theory to model possible Pt-Sn catalyst surfaces, namely pure Pt , pure Sn, a Pt-Sn alloy, and the interface between Pt-SnO_2. This latter interface was formed by modeling Pt nanorods over SnO_2(110), similar to previous work that modeled Au nanorods over TiO_2[1]. We present results on the oxygen reduction reaction and ethanol oxidation and highlight the differences between the various possible reaction systems under study. Our goal was to identify potential reaction sites on the various surfaces as well as reaction pathways. The nature of the Pt-SnO_2 interface in particular is explained, including the degree of charge transfer, role of surface defects, and the reactivity of interfacial sites between the two materials.

[1] L M Molina, M D Rasmussen, and B Hammer. The Journal of Chemical Physics 120, no. 16, 2004, p. 7673–80.

(116e) Theoretical Studies of Propene Oligomerization in Ni-Na-X

Gomes, J. S., University of California Berkeley
Head-Gordon, M., University of California at Berkeley
Bell, A. T., University of California - Berkeley
Originally presented on: 11/4/2013 13:50:00 - 14:10:00

Oligomerization of C_2-C_5 olefins offers a route for the reincorporation of light olefins created during the production of transportation fuels from the cracking of petroleum or during Fischer-Tropsch synthesis back into the final gasoline, diesel, and jet fuel product. It has been previously shown that Ni-exchanged zeolites are active for light olefin oligomerization and exhibit greater than 97% selectivity to oligomers[1]. A combined DFT-D + QM/MM methodology is used to analyze the reaction mechanism of propene oligomerization over a Ni-exchanged Na-X zeolite material. The calculated apparent activation barrier of our proposed mechanism are in good agreement with experimental measurements. The impact of zeolite framework and co-cation on the selectivity to branched or linear oligomers is discussed.

[1]A.N. Mlinar, G.B. Baur, G.G. Bong, A.B. Getsoian, A.T. Bell, J. Catal. 296 (2012) 156-164.

(583ag) Theoretical Insights Into the Surface Reactivity of Au/Co3O4 Catalysts in the Oxidative Dehydrogenation of Cyclohexane

Qian, Q., University of Virginia
Neurock, M., University of Virginia
Originally presented on: 11/4/2013 14:10:00 - 14:30:00

The catalytic combustion of jet fuels shows promise in the application of hypersonic propulsion systems where catalysts are used to lower the ignition temperature and reduce soot formation [1]. In previous efforts [2] we have shown that Co_3O_4 nanoparticles are active for the oxidative dehydrogenation of cyclohexane (a jet fuel surrogate) at 300 °C. Recent experimental studies [3] have explored the influence on catalytic activity by doping Co_3O_4 with ~1 nm size Au nanoparticles. In this work, we carry out first-principles theoretical studies on the oxidative dehydrogenation of cyclohexane over model Au nanoparticles (approximated by a nanometer-sized Au rod) supported on Co_3O_4.

Density functional theory with on-site correction for Coulomb interactions of Co d-electrons (DFT+U method) calculations were carried out to examine the adsorption energies, reaction energies and activation barriers for a series of different elementary steps involved in the oxidative dehydrogenation of cyclohexane in order to compare the catalytic reactivity of Au/Co_3O_4 with that of pure Co_3O_4. The results show that the Au nanorod transfers electrons to the Co atoms at the interface between the Au and the Co_3O_4 support. The increased electron density at the interfacial Co^{octa} cations weakens the Co-O bond strength and increases the hydrogen affinity of the neighboring O site. The increase in hydrogen affinity at the interfacial O sites results in a lower activation barrier for the initial C-H bond activation of cyclohexane at Au-Co_3O_4 interface. The interfacial hydroxyl intermediate reacts further to form water which ultimately desorbs from the surface and creates an O vacancy at the interface. The partially reduced Co^{octa} cation at the vacancy allows for the adsorption and easy activation of O_2 from the gas phase to form reactive mono-oxygen surface species. The results provide a fundamental understanding of the experimentally observed reactivity increase in the oxidative dehydrogenation of cyclohexane on Au/Co_3O_4 compared to that on pure Co_3O_4 surfaces.

References

1. Wickham, D.T., et al. *Journal of Propulsion and Power*, 2001. **17**(6): p. 1253-1257.
2. Tyo, E.C., et al. *ACS Catalysis*, 2012. **2**(11): p. 2409-2423.
3. Tyo, E.C., et al. *Prepr. - Am. Chem. Soc., Div. Energy & Fuels*, 2013. **58**: p. 1059

(116g) Calculations of Catalytic Activity of Metal Nanoclusters

Garden, A. L., University of Iceland
Pedersen, A.,
Sk lason, E., University of Iceland
J nsson, H., University of Iceland
Originally presented on: 11/4/2013 14:30:00 - 14:50:00

Metal nanoclusters with up to a few thousand atoms are known to exhibit interesting and novel properties, which differ drastically from those of the bulk crystal. The catalytic activity of such clusters is often very high and strongly dependent on the cluster size. In addition to size effects, it is known that catalytic activity on metal surfaces can be dominated by steps and defects in the surface. Thus it follows that, to accurately predict the catalytic activity of a cluster, it is necessary to determine the structure of the cluster, including a full description of edges, corners and any defects.

The task of determining the lowest energy arrangement of atoms in nanoclusters is a serious challenge already for a cluster containing a hundred atoms. As the size of the system grows, the number of minima on the potential energy surface increases exponentially and a thorough sampling becomes difficult. In the current work, an optimization algorithm based on the adaptive kinetic Monte Carlo (AKMC) algorithm is employed to determine low energy configurations of metal nanoclusters [1,2]. In this algorithm, the system is advanced from one local minimum to another via first-order saddle points on the potential energy surface. An empirical EMT potential is used in the AKMC calculations but the low energy structures are then refined and catalytic activity predicted using density functional theory calculations.

The current optimization scheme has been utilized to determine and analyze the structure of gold nanoclusters near the magic sizes in the range from 55 to 561 atoms. These clusters exhibit distorted icosahedral or decahedral structure, the implications of which will be discussed within the context of catalysis, in particular CO oxidation.

We have also applied the same approach to copper nanoclusters. Copper is a unique catalyst in that it shows high efficiency for reduction of CO_2 to hydrocarbons. The catalytic activity can be expected to increase further when in nanocluster form. The density functional theory calculations are performed to predict the catalytic activity of the various facets, edges and corners present in the most stable structures of the nanoclusters.

References:

[1] G. Henkelman and H. Jónsson, 'Long time scale kinetic Monte Carlo simulations without lattice approximation and predefined event table', J. Chem. Phys.,**115**, 9657 (2001).

[2] A. Pedersen, J-C. Berthet and H. Jónsson, 'Simulated Annealing with Coarse Graining and Distributed Computing', Lecture Notes in Computer Science, **34**, 7134 (2012).

(162a) Squirming At Finite Reynolds Number

Khair, A. S., Carnegie Mellon University
Chisholm, N. G., Carnegie Mellon University
Originally presented on: 11/4/2013 15:15:00 - 15:30:00

The dynamics of swimming microorganisms at zero Reynolds number ($Re = 0$) has been the subject of extensive theoretical and experimental investigation over the past decade, and the study of locomotion at high Reynolds number ($Re \gg 1$), where inertial forces are dominant, has a venerable history. In this talk, we consider swimming between these limits, i.e. at finite Reynolds numbers, using the popular "squirmer" model of self-propulsion, wherein locomotion is achieved through surface distortions. We first utilize matched asymptotic expansions to derive an analytical expression for the swimming velocity of a squirmer through $O(Re^2)$, which highlights that inertia affects so-called "pusher"- and "puller-" swimmers in fundamentally different manners. (The equivalent of Whitehead's paradox for a self-propelled object is elucidated in the process.) Next, we employ numerical methods to compute the swimming velocity of pushers and pullers for higher Reynolds numbers. Finally, we demonstrate that inertia causes squirmers (which are non-chiral) to drift across the streamlines of an imposed shear flow.

(162b) Collective Behavior of Swimming Microorganisms in Viscoelastic Media

Karimi, A.,
Ardekani, A. M., University of Notre Dame
Originally presented on: 11/4/2013 15:30:00 - 15:45:00

Suspensions of motile microorganisms, such as bacteria, are often associated with the viscoelastic behavior due to secretion of the extracellular polymeric substances. This phenomenon has a profound impact on the hydrodynamic interactions of swimming microorganisms and the resulting complex dynamics. To elucidate this process, we used a kinetic model developed to study the behavior of self-propelled particles in conjunction with Oldroyd-B constitutive equation and the Stokes equations. Numerical simulations of the system have been conducted to scrutinize the effect of the viscoelasticity on the large-scale coordinated behavior of the microorganisms. In addition, different regimes of the flow have been explored by varying the corresponding parameters of the problem, such as Weissenberg number and viscosity ratio, in order to shed light on the characteristics of the flow patterns induced by the collective motion of microscopic swimmers.

(162c) Gait Adaptation Of Swimming C. Elegans: Hydrodynamic Efficiency and Neuromuscular Control

Blawzdziewicz, J., Texas Tech University
Bilbao, A.,
Khan, Z. S., Texas Tech University
Vanapalli, S. A., Texas Tech University
Bussel, F. V., Texas Tech University
Originally presented on: 11/4/2013 15:45:00 - 16:00:00

A millimeter-size nematode Caenorhabditis elegans is one of the most important model organisms used for investigations ranging from genetics to neurobiology. Applications such as mutant or drug screening require a thorough understanding of nematode locomotion in complex environments to quantify the interplay of sensing processes, control mechanisms, and mechanical interactions (including frictional, capillary, and hydrodynamic forces). Our analysis of nematodes swimming in fluids of different viscosities shows that the worms choose the optimal gait that minimizes the combined energy dissipation resulting from hydrodynamic and internal friction forces. Simple models of neuromuscular control of the nematode body will be presented to further elucidate the observed gradual change of the wavelength and frequency of nematode undulations when fluid viscosity is increased.

(162d) In Vitro Measurement of Particle Margination in Whole Blood

Fitzgibbon, S., Stanford University
Shaqfeh, E. S. G., Stanford University
Originally presented on: 11/4/2013 16:00:00 - 16:15:00

In pressure driven flow, the flexibility of red blood cells breaks axial symmetry and causes them to lift away from blood vessel walls, leaving behind a cell free layer near the wall. Platelets migrate into this region, positioning themselves to quickly respond to any injuries encountered in the blood vessel. A lack of platelets near the wall can cause long bleeding times post-injury. Recently, Zhao et. al. have run a large scale simulation of red blood cells and platelets in pressure driven flow. Based on shear induced collisions with red blood cells, they predict rates for platelet and particle margination. Furthermore, they show a qualitative agreement between platelet and sphere margination of comparable sizes. We have built an in vitro fluorescence microscopy experiment to measure their predictions in the lab. Using capillaries similar in size to the simulations, we report the margination of microparticles in pressure driven flow at several different red blood cell concentrations.

(162e) Red Blood Cell Suspensions With Polymer Additives: Orientation, Migration and Margination Dynamics

Sinha, K., University of Wisconsin Madison
Graham, M. D., University of Wisconsin-Madison
Originally presented on: 11/4/2013 16:15:00 - 16:30:00

In recent experimental studies, it has been shown that introducing drag reducing polymers (DRPs) to blood in nanomolar concentration has many beneficial physiological effects, including the prevention of lethality from hemorrhagic shock in animals. In micro-channel experiments with suspensions of red blood cells (RBCs), DRPs have been shown to reduce cell free layer thickness and affect the orientation of the cells. The underlying mechanisms for these phenomena are not understood. In the present study, we investigate the dynamics of RBCs in Newtonian fluids and in dilute solutions of DRPs in microcirculation. Cell membranes are modeled with the Yeoh strain energy function and an energy penalty for bending. Polymer molecules are modeled as bead-spring chains with finitely extensible nonlinearly elastic springs. Three dimensional simulations are performed with a Stokes flow formulation of the immersed boundary method for the RBCs, combined with Brownian dynamics for the polymer molecules. We find that the cell-free layer thickness is substantially reduced near wall with addition of polymer, while elongation of RBCs and their alignment with flow direction is significantly enhanced. The wall shear stress decreases with increase in capillary number and is reduced in presence of polymers. The wall normal migration velocity of single RBC is found to decrease by adding polymer. The combined effect of reduced migration velocity and increased alignment with flow can explain the paradoxical behaviour of reduced flow resistance even with reduction in cell free layer as observed in experimental work. These predictions are in qualitative agreement with experiments.

(162f) The Instability of a Single Vesicle Under Extensional Flow At Zero Reynolds Number

Narsimhan, V., Stanford University
Spann, A. P., Sanford University
Shaqfeh, E. S. G., Stanford University
Originally presented on: 11/4/2013 16:30:00 - 16:45:00

When floppy vesicles are placed in extensional flows (planar or uniaxial), they undergo two unique sets of shape transitions that to our knowledge have not been observed for droplets. At intermediate reduced volumes (i.e., intermediate particle aspect-ratio) and high shear rates, the vesicles stretch into an asymmetric dumbbell separated by a long, cylindrical thread. At low reduced volumes (i.e., high particle aspect-ratio), pearling occurs, where the cylindrical thread forms a series of bulges with a characteristic size on the order of the cylinder's radius. In this talk, we will describe our efforts to elucidate the physical mechanisms behind these two seemingly-unrelated sets of instabilities. We perform boundary integral simulations on single, fluid-filled particles whose interfacial dynamics are governed by a Helfrich energy (i.e., the membranes are inextensible with bending resistance). With this model, we perform a linear stability analysis to determine the conditions under which the instabilities occur. We also probe the modal structure of these shape transitions. Lastly, we develop relatively simple analytical theories to characterize the stretching dynamics in the limit of high shear rate (characterized by a large bending Capillary number), or high particle aspect ratio. We believe that this work will lend insight into the breakup and dynamics of many types of biological particles under strong tension (particularly those with soft membranes).

(162g) Mechanical Response of Cancer Cells in a Microfluidic Capillary

Khan, Z. S., Texas Tech University
Kamyabi, N.,
Vanapalli, S. A., Texas Tech University
Originally presented on: 11/4/2013 16:45:00 - 17:00:00

Circulating tumor cells (CTCs) are considered the primary cause of cancer metastasis. CTCs are disseminated from the primary tumor and transported to distant organs by blood flow, where new tumors are instigated. Despite the significance of CTC transport in microcirculation for cancer metastasis, quantitative understanding is lacking. *In vivo* studies directly examining CTCs in micro-capillaries lack adequate control of flow conditions and conduit geometry. Moreover, tracking individual cells at high spatial and temporal resolution becomes difficult due to limitations in deep tissue imaging and the out-of-plane motion of CTCs in microcirculatory networks.

To understand the hydrodynamic transport of CTCs through microcirculation, we have developed a minimal microfluidic capillary model that consists of a narrow channel to confine flowing brain and prostate cancer tumor cells. The driving pressures are controlled to subject the cells to hemodynamic range of shear stresses. By integrating an on-chip manometer and high-speed imaging, the device allows for the simultaneous characterization of five different parameters including the cell size, blockage pressure, velocity, elongation and entry time into the constriction. Our results show that the dependence of blockage pressure and entry times on cell confinement are the best metrics to identify the metastatic variations between cell lines, suggesting that these two parameters suffice to characterize CTC transport in capillaries.

By characterizing our device with simple model systems including viscous drops and soft elastic particles, we find that the blockage pressure and entry times show no apparent dependence on elastic modulus or drop interfacial tension, but depend significantly on drop internal viscosity. Thus, our results suggest that in the hemodynamic range of shear stresses, the metastatic potential of circulating cells can be characterized by an internal viscosity. We test this hypothesis with a variety of tumor cell lines (leukemia, prostate, brain cancer) and find strikingly that the blockage pressure (and entry times) for all the cancer cell types lies between two universal bounds determined by the viscous drop and elastic particle limit respectively. From a device perspective, our study opens new opportunities for single-cell microfluidic viscometry that has been difficult to realize experimentally.

(162h) Generalized Langevin Dynamics for Functionalized Nanocarrier Adhesion to Cell Surfaces in the Presence of Hydrodynamic Interactions

Yu, H. Y., University of Pennsylvania
Ramakrishnan, N., University of Pennsylvania
Eckmann, D. M., University of Pennsylvania
Ayyaswamy, P. S., University of Pennsylvania
Radhakrishnan, R., University of Pennsylvania
Originally presented on: 11/4/2013 17:00:00 - 17:15:00

Vascular delivery of antibody-functionalized nanocarriers (50 nm-1 μm in size) to selectively bind targeted receptors on the endothelial cells is a viable therapeutic strategy in systems pharmacology. However, the optimal design of such nanocarriers involves the appropriate selection of size, shape of nanocarriers and antibody density on the nanocarriers, and the choice of appropriate tethers/linkers bridging the antibodies to the nanocarrier surface depends on the physiological microenvironment including the hemodynamics and the rheological properties of the flow in the capillaries as well as the cellular state of the endothelial cells (which governs the expression of the targeted receptors such as the intracellular adhesion molecule-1 or ICAM-1). We have shown in recent work that a model-based optimization of this design can help guide the achievement of targeting selectivity and specificity in vivo [1,2]. Here we address the important aspects of nanocarrier adhesive dynamics in terms of how they impact binding selectivity and specificity.

The main premise of this work is the hypothesis that the efficiency of vascular targeted drug delivery using functionalized nanoparticles is affected simultaneously by several biophysical and biochemical factors such as the margination probability of the nanocarrier towards the red blood cell-free layer, hydrodynamic interactions due to the wall and the near-wall flow field, thermal fluctuations, and the specific ligand-receptor (or antibody-antigen) binding interactions. A complete description of hydrodynamic interactions for the system can be achieved by direct numerical simulations of the fluctuating hydrodynamic equations, which simultaneously resolve the velocities and the stresses of the fluid and the particles [3-5]. While such a model can be pursued for simple geometries for small systems such as short capillary segments not much longer than a few diameters of the blood cells, it is not realistic for a pharmacological model considering nanocarrier margination and binding in a vascular network tree. With the objective of developing a coarse-grained description of the hydrodynamics in complex geometries, we present a generalized Langevin dynamics aiming to capture the adhesive dynamics of the nanocarrier closed to the vessel wall in the presence of RBC-driven marginating potential, hydrodynamic, and Brownian forces. The functionalized nanoparticle is modeled as hard or soft sphere with surface-grafted ligands (antibodies) that interact with the receptors (such as ICAM-1) on the endothelial cells. To incorporate the memory effects of the nanocarirer due to coupling with the locally confined fluid and the relaxation of the ligand-receptor pair, a set of non-Markovian, generalized Langevin equations [6] for the nanocarrier motion (translational and rotational motion about the center of mass) and the stretching for each tether as well as antibody-antigen pair are solved simultaneously. The space-time dependent memory functions are obtained from independent experiments [7], microscopic simulations [8], or analytical theories [9]. We analyze the nanocarrier velocity autocorrelation function along suitable coordinates (center of mass translation/rotation, ligand-receptor bond, and tether coordinates) to characterize nanocarrier adhesive dynamics and the potential of mean force along a specified reaction coordinate to quantify the binding affinity [1] in the presence of various hydrodynamic couplings. We also discuss the generalization of the current theoretical framework to study the rolling motion and internalization of the nanocarrier.

References

[1] A computational model for nanocarrier binding to endothelium validated using in vivo, in vitro, and atomic force microscopy experiments, J. Liu, G. E. R. Weller, B. Zern, P. S. Ayyaswamy, D. M. Eckmann, V. Muzykantov, R. Radhakrishnan, Proc. Natl. Acad. Sci. U.S.A., 107(38), 16530-16535, 2010.

[2] Reduction of nanoparticle avidity enhances the selectivity of vascular targeting and PET detection of pulmonary inflammation, B. J. Zern, A.-M. Chacko, J. Liu, C. F. Greineder, E. R. Blankemeyer, R. Radhakrishnan, V. Muzykantov, ACS Nano, 7(3), 2461-2469, 2013.

[3] Nanoparticle Brownian motion and hydrodynamic interactions in the presence of flow fields, B. Uma, T. N. Swaminathan, R. Radhakrishnan, D. M. Eckmann, P. S. Ayyaswamy, Phys. Fluids, 23, 073602, 2011.

[4] Generalized Langevin dynamics of a nanoparticle using a finite element approach: Thermostating with correlated noise, B. Uma, T. N. Swaminathan, P. S. Ayyaswamy, D. M. Eckmann, R. Radhakrishnan, J. Chem. Phys., 135, 114104, 2011; Erratum, 136, 019901, 2012.

[5] A hybrid formalism combining fluctuating hydrodynamics and generalized Langevin dynamics for the simulation of nanoparticle thermal motion in an incompressible fluid medium, B. Uma, D. M. Eckmann, P. S.Ayyaswamy, R. Radhakrishnan, Mol. Phys., 110(11-12), 1057-1067, 2012.

[6] Molecular Hydrodynamics, J. P. Boon and S. Yip, McGraw-Hill, New York, 1980.

[7] Observation of a power-law memory kernel for fluctuations within a single protein molecule, W. Min, G. Luo, B. J. Cherayil, S. C. Kou, X. S. Xie, Phys. Rev. Lett., 94, 198302, 2005.

[8] Computer Simulation of Liquids, M. P. Allen and D. J. Tildesley, Oxford University Press, 1989.

[9] Low Reynolds Number Hydrodynamics, J. Happel and H. Brenner, Martinus Nijhoff Publishers, Hague, 1983.

(162i) When Is DNA Flexible and Non-Draining?

Tree, D. R., University of Minnesota
Muralidhar, A.,
Wang, Y.,
Doyle, P. S., Massachusetts Institute of Technology
Dorfman, K. D., University of Minnesota
Originally presented on: 11/4/2013 17:15:00 - 17:30:00

The details surrounding the cross-over from wormlike-specific to universal polymeric behavior has been the subject of debate and confusion in the literature in various geometries including slits, channels and even the dilute unconfined case. In order to elaborate on many of these details, we have directly computed the polymer size and diffusivity for unconfined, slit-confined and channel-confined wormlike chains near these limits. To do so, we use a chain-growth Monte Carlo algorithm, the pruned-enriched Rosenbluth Method (PERM), which allows us to estimate equilibrium and near-equilibrium dynamic properties (using a rigid-body approximation for the chain hydrodynamics) of wormlike chains over an extremely large dynamic range. For instance, we can accurately predict the diffusivity of an unconfined molecule of double-stranded DNA from about 100 bp up to several hundred kilobase pairs. From our calculations, we find that very large chains are required to reach the universal, flexible regimes including the de Gennes regime in slits and channels and the swollen coil regime for unconfined wormlike chains. Furthermore, our results indicate that the dynamic properties (i.e. the chain diffusivity) converge less quickly to universal, non-draining behavior than do the static properties. Accordingly, for the case of unconfined DNA in a high-ionic strength buffer, we predict that the chain becomes flexible and non-draining for contour lengths near 1 megabase pair. Qualitatively similar results are also obtained for DNA confined in slits and channels. This suggests that the flexible, non-draining limit is rarely (if ever) applicable to double-stranded DNA in situations commonly encountered in nanofluidic devices.

(162j) Single Molecule Studies of DNA Collapse in Slit-Like Confinement

Jones, J. J., Massachusetts Institute of Technology
Doyle, P. S., Massachusetts Institute of Technology
Originally presented on: 11/4/2013 17:30:00 - 17:45:00

Single molecule experiments of DNA in the presence of condensing agents have informed scientists about the high density packing of DNA in biological systems. Typically, these studies are performed in unconfined, *in vitro* environments with condensing agents such as multivalent cationic species (e.g. spermine^{3+}, $Co(NH_3)_6{}^{3+}$). However, relatively little attention has been given to single molecule DNA collapse in weaker condensing agents, such as ethanol. Moreover, studies on the effects of restricting DNA conformational space (e.g. via confinement) on kinetics of collapse are nearly non-existent. In this talk, we will discuss our experiments with single molecules of dsDNA in the presence of ethanol and confined to slit-like nanochannels. The compact states induced at our ethanol concentration (and in bulk) are shown to be much larger and amorphous than the highly condensed, orderly toroids seen in multivalent cation induced DNA collapse. We attribute this difference to the long lived, higher order intermediate racquet states that one sees when collapsing semiflexible chains in poor solvents. Furthermore, we study the kinetics of collapse of individual DNA molecules. We find that the collapse is a two step process much like a nucleation-propagation event. More strikingly, we observe that the time to collapse DNA is reduced by an order of magnitude, relative to bulk, when confining DNA to slit-like nanochannels. This behavior is further explained via scaling arguments. We believe these results will have future relevance in single molecule assays and biological physics.

(163a) Geological Labs On Chip for Studying Key Aspects of CO2 Geological Storage

Liu, N., Institute of Condensed Matter Chemistry of Bordeaux
Aymonier, C., Institute of condensed Matter Chemistry of Bordeaux
Lecoutre, C., Institute of Condensed Matter Chemistry of Bordeaux
Garrabos, Y.,
Hartman, R. L., University of Alabama
Marre, S., Institute of Condensed Matter Chemistry of Bordeaux
Originally presented on: 11/4/2013 15:15:00 - 15:40:00

The sustainable management of the CO2 industry is currently a key factor for preventing climate changes due to greenhouse effect. Among the proposed remediation solutions is the deep underground CO2 geological storage. Such storage required adequate scientific knowledge to prevent from unexpected events during injection or final storage. In this context, we develop and use high pressure / high temperature microfluidic tools to study and characterize typical problems related to CO2 trapping in deep saline aquifers. The experimental data could later feed the numerous numerical modeling works done in this field. The fabricated 2D porous networks are created in silicon/Pyrex microsystems to replicate the reservoir conditions and characteristics ($25 < T < 50°C$, $50 < p < 10$ MPa. These are able to mimic the actual storage conditions encountered in real storage reservoirs. Coupled to Raman spectroscopy and optical diagnoses techniques, such microsystems have been confirmed as a fast and accurate way to identify and quantify dissolved carbon dioxide in water and salted water, to study invasion mechanisms depending upon flow rates or wetting properties or to study carbonate precipitation. This technique is proved viable in the laboratory, and can be used to simulate and analyze various CO2 Underground Storage issues.

(163b) Ultrasupercritical CO2 Transport in Flexible Flowlines

Costa, J. B.,
Almeida, S.,
Lundsgaard, R., Danish Techn
Rubin, A., National Oilwell Varco
von Solms, N., Technical University of Denmark (DTU)
Originally presented on: 11/4/2013 15:40:00 - 16:05:00

In order to transport supercritical carbon dioxide safely in polymer pipelines, it is necessary to understand several thermodynamic and transport properties of the polymer/gas system. Measurements and modelling were carried out with supercritical CO_2 and a mixture of CO_2 and CH_4 in two polymers - PVDF and XLPE (a highly cross-linked polyethylene). Both of these polymers find service in the offshore industry and the samples were obtained from flexible pipes. The properties measured are solubility of the gas in the polymers, using a Magnetic Suspension Balance, as well as permeability through the polymers using a 2-D pressure permeation cell. Pressures up to 650 bar and temperatures up to 90 °C were studied.

An advanced equation of state (simplified PC-SAFT) which is suitable for modelling polymers was used to model the gas/polymer system. Since pure component parameters for this model are usually fitted to saturated properties remote from the critical region, a refit of the CO_2 pure-component parameters was required for application in the supercritical region. Pure component properties for polymers are also problematic, and here we fitted parameters to polymer density data combined with an extrapolation from the polymer monomer properties. Solubility was successfully modeled and it was found that correct modelling of the fluid phase (ultrasupercritical CO2) was most important for correlating binary phase data.

(163c) Transport and Storage of Captured Carbon Dioxide in South Korea

Zahid, U., Seoul National University
Kim, J., Seoul National University
An, J., Seoul National University
Han, C., Seoul National University
Originally presented on: 11/4/2013 16:05:00 - 16:30:00

The continuous increase of carbon dioxide (CO_2) emissions from power plants and other large industrial sources is major cause of global climate change. Carbon capture and storage (CCS) is widely seen as a practical technology for reducing CO_2 emissions. CCS mainly consists of capturing CO_2 from stationary sources and its transportation to a sequestration site where it can be stored safely for long period of time. Average CO_2 emission growth rate of Korea is 1.0% which is the second highest among Organization for Economic Co-operation and Development (OECD) countries. Korea has three potential storage basins namely Ulleung, Kunsan and Jeju located in the East, West and South of Korea mainland respectively. All of the three basins are located offshore; therefore, it becomes even more challenging when CO_2 is transported to an offshore storage since there is little experience with subsea pipelines for CO_2 transportation. In this study, a plausible transport and storage model scheme has been developed and then employed to study different offshore CO_2 transportation cases for South Korea as: CO_2 transport in liquid phase (Temperature= -20°C, Pressure= 6.50 MPa); CO_2 transport in liquid phase (Temperature= 5°C, Pressure= 9.30 MPa); CO_2 transport in supercritical phase (Temperature= 40°C, Pressure= 15.00 MPa). This study critically analyzes pipeline diameter calculation methods and optimum operational range for pipeline parameters. This study discusses economic aspects of various transport and storage scenarios in terms of thermodynamics conditions, and basin location and depth, respectively. This has been done by estimating the CO_2 storage capacity in Korea and then selecting the most promising storage site in terms of safety and cost 150 km away from the source location. CO_2 storage capacity for sedimentary basins of Korea was evaluated between 19- 27.2 Gt (giga-ton) of CO_2 using the Department of Energy (DOE) methodology. Transport cost varies between 10.9- 15.5 US$/t$CO_2$ while storage cost ranges between 20.8- 21.3US$/t$CO_2$ depending on specific scenario and depth at which CO_2 is stored. Sensitivity analysis showed a decrease in storage cost of 62.4% and 93.6% in 2030 and 2050 respectively for projected CO_2 volumes in Korea. This model is useful for performing an economic analysis to check feasibility of a project before a detailed design is made.

Acknowledgement: This research was supported by the second phase of the Brain Korea 21 Program in 2012, Institute of Chemical Processes in Seoul National University, Energy Efficiency & Resources Development of the Korea Institute of Energy Technology Evaluation and Planning (KETEP) grant funded by the Ministry of Knowledge Economy (MKE) and grant from the LNG Plant R&D Center funded by the Ministry of Land, Transportation and Maritime Affairs (MLTM) of the Korean government.

(163d) Model-Based Economic Optimisation of CO2 Compressor Train Design and Operation

Ramos, A., Process Systems Enterprise, Inc.
Originally presented on: 11/4/2013 16:30:00 - 16:55:00

Compressor trains are typically currently designed using heuristics, in an iterative fashion. This paper describes an approach for rapidly determining optimal design parameters of a CO_2 compression train for Carbon Capture and Storage (CCS) using rigorous model-based optimisation methods that can take into account time-varying flowrates or delivery pressures. Optimisation variables can include continuous decision variables such as inter-stage pressures, and integer or discrete decision variables, such as number of sections.

Rigorous models of a single-section compressor were constructed in the gPROMS advanced process modelling environment. These were used to construct a superstructure model that contained a maximum-allowable set of inter-connected compressor sections (arbitrarily set to 8 sections), including recycle loops and intercoolers. This was then subjected to a set of optimisations, with an objective function comprising total annualised capital and operating costs. Execution involved solution of a Mixed Integer Non-Linear Programming (MINLP) optimisation problem using the Outer Approximation / Equality Relaxation / Augmented Penalty (OAERAP) algorithm included in gPROMS platform.

The first optimisation determined the optimal number of sections, section diameters, required heat transfer area of the coolers and other key design attributes for a given service, based on identical pressure ratios across each section. The second performed the same optimisation, but with the inclusion of inter-section pressures as optimisation variables. The third case applied a multi-period optimisation approach to determine the optimal number of sections for a typical daily load profile that emulated the typical flow of CO_2 for a power station following grid demand.

The approach makes it possible accurately to determine trade-offs between higher capital expenditure (e.g. the use of more compressor sections) on the one hand, and increased flexibility of service and lower operating costs (through the application of optimised inter-section pressures), on the other hand, to minimise the overall annualised cost of operation under varying flowrate conditions.

(163e) Modeling and Optimization of CO2 Transport Infrastructure Systems With Time in South Korean Case

Kim, J., Seoul National University
Zahid, U., Seoul National University
An, J., Seoul National University
Han, C., Seoul National University
Originally presented on: 11/4/2013 16:55:00 - 17:20:00

Carbon Capture and Storage (CCS) is considered as one of the most practical approaches for the reduction of CO_2 emission. If CCS is applied to low carbon power generation, its deployment will require the construction of CO_2transport infrastructure on a massive scale. Because of its cost, estimated at many billions of dollars, decision makers who set the direction of the climate-change strategy must understand how to deploy the CCS infrastructure and when to extend it properly.

However, there is a dearth of data concerning CO_2 transport and storage. First, reservoir uncertainty has made it difficult to determine how many CO_2 sources can be connected to a reservoir. Second, it is hard to decide a transport method for locations and estimate its cost due to a lack of CO_2 transport data. Third, it is not clear whether to extend current transport infrastructure or change transport method when CO_2 emission rates and reduction objective increase with time. Because of those, it is necessary to provide analysis of CO_2transport network based on various transport scenarios.

In this presentation, authors present analysis of the CO_2 transport network deployment with time using South Korean case. To overcome a dearth of data, the cost model and configuration method of transport are estimated from similar cases, such as natural gas transport. After making the cost model per distance, authors match CO_2 source to reservoir and decide transport methods by the regular rules on reservoir scenarios. Lastly, CO_2transport network with time is optimized for cost savings.

Acknowledgement: This research was supported by the second phase of the Brain Korea 21 Program in 2013, Institute of Chemical Processes in Seoul National University, Energy Efficiency & Resources Development of the Korea Institute of Energy Technology Evaluation and Planning (KETEP) grant funded by the Ministry of Knowledge Economy (MKE) and grant from the LNG Plant R&D Center funded by the Ministry of Land, Transportation and Maritime Affairs (MLTM) of the Korean government.

(168a) Free-Energy Perturbation: Application to Chemical Reactions At Metal/Water Interfaces

Faheem, M., University of South Carolina
Heyden, A., University of South Carolina
Originally presented on: 11/4/2013 15:15:00 - 15:35:00

Computational investigations of structural and energetic changes in chemical reactions at metal/water interfaces pose a unique challenge of properly accounting for the effect of liquid-phase environment on free energies of reactants, products and transition states while keeping the computational model simple enough to be computationally affordable. A practical solution to this challenge is a hybrid QM/MM approach. By treating only the reaction zone and its immediate environment quantum mechanically (QM) for higher accuracy and the remainder of the metal surface and the bulk of the liquid water at the classical, molecular mechanical (MM) level of theory, a reasonably accurate energetic description of the complex metal/water system can be obtained while simultaneously realizing a computational speedup of multiple orders of magnitude.

In this talk, we present a novel solvation scheme that combines planewave density functional theory (DFT), periodic electrostatic embedded cluster (PEEC) calculations with Gaussian-type orbitals, and classical molecular dynamics (MD) calculations with the QM/MM free-energy perturbation (FEP) methodology to enable an accurate determination of free-energies of reaction and activation free-energy barriers at periodic metal/water interfaces. Optimization of local conformations (i.e., reactants, products, and transition states) are performed using an iterative approach with sequential MD sampling and QM optimization steps to reduce the number of QM calculations and the amount of MD sampling. Furthermore, we use a fixed-size, finite ensemble of MM conformations to allow a precise evaluation of the QM potential of mean force (PMF) and its gradient defined within this ensemble. As a result, we circumvent the challenges associated with statistical averaging during MD sampling and speed up the convergence of the optimization process. Application to C-C cleavage in double-dehydrogenated ethylene glycol on Pt (111) surface will be discussed as model reaction.

(168b) A Combined Density Functional Theory and Monte Carlo Approach for Quantifying Catalytic Energies in a Water Environment

Bollmann, L., Clemson University
Getman, R. B., Clemson University
Originally presented on: 11/4/2013 15:35:00 - 15:55:00

One of the biggest challenges in computational catalysis is modeling activity under realistic reaction environments. The main way that the reaction environment influences activity is by "dosing" the catalyst with atoms and molecules, which become chemically and/or physically adsorbed to the surface. These atoms and molecules alter binding and reaction energies through at least two effects: 1) molecules that form strong chemical bonds with the catalyst can accumulate on the catalyst surface, inducing coverage effects, and 2) molecules in condensed phase reaction environments can interact physically with adsorbates, inducing solvation effects. Several computational methods have been developed to treat these phenomena. For example, first-principles thermodynamics (FPT) is used to predict catalyst compositions under real reaction conditions, and a number of solvation models have been used to predict the energies of interaction between static fluid phases and solid catalyst environments. In this work, we extend these methods to more realistic, dynamic fluid environments. Specifically, we demonstrate a method to include entropic contributions. This method takes fluidity in the reaction environment and its influence on the free energies of relevant reactants, intermediates, and products into account. We use a combination of density functional theory (DFT) and Monte Carlo to quantify the different contributions to the overall free energies and make conclusions about how they influence catalytic activity. Our present focus is on reduction of nitrate over nanoparticulate Au in water. We chose Au because it is relatively non-reactive, so we can largely neglect coverage effects and instead focus on interactions between the reaction environment and the catalyst environment. Catalysts are modeled using 13-atom gold icosahedral clusters, which are computationally efficient yet comprise a number of typical adsorption sites for typical metal catalysts. We find that the water environment has a noticeable influence on reaction energetics. Ongoing work involves expanding these methods to more reactive transition metals and more realistic supported catalyst models.

(168c) Electric Field-Induced Reactions in Fuel Cells From First Principles

Che, F., Washington State University
McEwen, J. S., Washington State University
Originally presented on: 11/4/2013 15:55:00 - 16:15:00

Solid oxide fuel cells will no doubt find more widespread use in the new millennium in the conversion of chemical to electrical energy offering, as they do, very high efficiencies and a unique scalability in electricity-generation applications. To control the ongoing reactions in such a device one can modify the selectivity of the catalyst itself, which oxidizes the fuel at the anode and dissociates O_2 at the cathode. The selectivity of the catalyst in a solid oxide fuel cell will be altered by the presence of a huge electric field of the order of 1 Volts per Angstroms, which occurs near the electrode-electrolyte interface [1]. Given this background, it is surprising that the role of the electric field is often ignored in the optimization of such devices. *Ab initio* quantum chemistry methods coupled with detailed kinetic models allow the prediction of such field-induced effects. Here we report density-functional theory calculations of the dehydrogenation of methyl species on the Ni(111) surface with and without an electric field. The structures, adsorption energies, and reaction barriers for all methyl species on the Ni(111) surfaces are identified. Our results show that the presence of an electric field does not affect the structures nor does it affect the favorable adsorption sites of the adsorbed species, but causes the adsorption energies of the CH_x species and the dissociation energy barrier of a CH molecule to be significantly altered by about 0.2 eV. We expect that such field-induced effects to be even more important on a stepped surface, since the local electric field will be greatly enhanced at the step [2]. We compare these results to when a non-reactive metal (such Au) is added to the Ni catalyst with regard to the suppression of coking [3].

[1] E.M. Stuve, Ionization of water in interfacial electric fields: An electrochemical view, Chem Phys Lett, 519-520 (2012) 1-17.

[2] Y. Suchorski, N. Ernst, W.A. Schmidt, V.K. Medvedev, H.J. Kreuzer, R.L.C. Wang, Field desorption and field evaporation of metals, Prog. Surf. Sci., 53 (1996) 135.

[3] H.S. Bengaard, J.K. Nørskov, J. Sehested, B.S. Clausen, L.P. Nielsen, A.M. Molenbroek, J.R. Rostrup-Nielsen, Steam Reforming and Graphite Formation on Ni Catalysts, Journal of Catalysis, 209 (2002).

(168d) Parameterization of An Adsorbate-Adsorbate Interaction Model for Catalytic Trend Studies

Khan, T., Stanford University
Falsig, H., Technical University of Denmark
Guo, W., UNIVERSITY OF DELAWARE
Wang, S., Technical University of Denmark. Denmark
Dahl, S.,
Bligaard, T., SLAC National Accelerator Laboratory
Originally presented on: 11/4/2013 16:15:00 - 16:35:00

The lateral interaction between atoms and molecules adsorbed on the surface of a transition metal affects both the adsorption energies of the surface species and also the transition state energies for surface reactions. A simple model for obtaining the effect of such interactions on adsorption energies is proposed, where the adsorption energy varies linearly with the coverage of species at the surface. The model is parameterized using the DFT calculated adsorption energies at different coverages for a number of atoms and molecules on the most stable facet of different transition metals.

The number of parameters in a full description of the interactions varies as $N_{surface} \times N_{sp} (N_{sp} + 1)/2$, where N_{sp} is the number of adsorbed species and $N_{surface}$ is the number of surfaces we are interested in investigating. In order to reduce the number of independent parameters we discuss 3 submodels; i) Model Rh : based on the interaction parameters of Rh{111}, ii) Model GM : the cross-interaction parameters are obtained from the parameters of the interaction between the same adsorbates, iii) Model scaled : the interaction parameters are obtained from linear scaling to the adsorption energy at low coverage. We then discuss the limitations of the proposed models, and finally for the direct decomposition of NO, we show what effect the inclusion of a description of the interactions in the kinetic model, has on the catalytic activity.

(168e) Impact of Adsorbates On the Photochemistry of Metal Nanoclusters

Avanesian, T., University of California, Riverside
Kale, M., University of California, Riverside
Christopher, P., University of California, Riverside
Originally presented on: 11/4/2013 16:35:00 - 16:55:00

The possibility of photon-induced selective bond-activation of adsorbates on metal surfaces has sparked an interest in utilizing metal catalysts to drive photochemical reactions in excited states, which cannot be performed using thermal energy only. Ultrafast lasers have been used for many years to excite adsorbate covered single crystal metal surfaces and execute photo-catalytic reactions in excited states. The low efficiencies and high laser intensities required to drive these chemistries present significant barriers that must be overcome. Recent experimental work in our group has shown that spatial confinement of catalytic active sites in metal nanoparticles results in fundamentally different photo-catalytic behavior compared to extended metal surfaces. It has been demonstrated that Pt nanoparticles, diameter < 5 nm, and plasmonic Ag nanoparticles, diameter < 100 nm, are able to efficiently drive photo-catalytic reactions through resonant energy transfer processes. The goal of this study is to explain the mechanism of these nanoparticle mediated photochemical processes.

It is well known that the addition of adsorbates to metal surfaces induces electronic structure changes in the underlying metal and introduces adsorbate-derived states near the Fermi level. We have utilized density functional theory calculations to examine adsorbate induced changes in electronic structure of Pt surfaces upon addition of molecular (CO, NO) and atomic (O, C, N) adsorbates. The calculated densities of states were used as inputs into a model that describes the electron energy distribution induced by photon excitations of metal surfaces. In addition, an extended two-temperature model (ETTM) is used to describe the temporal evolution of non-equilibrium electron distributions in photo-excited metal nanoparticles. Our theoretical predictions accurately capture recent wavelength dependent photocatalytic measurements on Pt nanoparticles performed in our group and allow us to postulate general mechanisms regarding the effect of molecular and atomic adsorbates on metal nanocluster photocatalysis. The results of our study provide essential insights into the possibility of state-selected bond activation of adsorbates on metal nanoparticles under low-intensity visible light illumination.

(168f) First-Principles Modeling of Hot Electron Assisted Chemical Bond Activation On Metal Surfaces

Xin, H., University of Michigan
Linic, S., University of Michigan
Originally presented on: 11/4/2013 16:55:00 - 17:15:00

Chemical reactions on metal surfaces are typically driven by phonons, i.e., vibrational excitation of adsorbed molecules by thermal energies. Interestingly, it is well-established in the field of femtochemistry that energetic electrons, generated from optical excitation of metal single crystals with ultrashort laser pulses or other sources, can also be exploited to drive surface chemical reactions, often through novel reaction channels not accessible in thermal reactions. Recently, we showed that plasmonic nanostructures of coinage metals effectively couple thermal and low-intensity photonic stimuli to drive chemical transformations. Intensity- and wavelength-dependence of steady-state reaction rates together with kinetic isotope experiments suggest that this process is electron-driven in nature, attributed to the excitation of localized surface plasmon resonance (LSPR).

In this talk we will discuss our approach to study direct photocatalytic reactions on plasmonic nanostructures from first-principles. Photocatalytic ethylene epoxidation (catalyzed by silver (Ag) cubes) was used as the test reaction, where the O_2 dissociation is the rate-limiting step. We present a model that allows us to evaluate within a non-adiabatic dynamics the coupling of vibrational excitations of adsorbed O_2* and energetic electrons from the Landau damping of LSPR while the adsorbate-surface interaction is described by ab-initio potential energy surfaces (PESs). We used DFT calculations for the PES of O_2 dissociation on Ag(100) in its neutral state, while ΔSCF-DFT calculations were used to for the PES of the TNI state. The probability that an energetic electron with a particular energy scatters into the 2pi* anti-bonding orbital of O_2 gaining a given amount of vibrational energy is computed using the Green's function with the Newns-Anderson-type Hamiltonian. This Hamiltonian includes a non-adiabatic coupling of the energetic electron to adsorbate reaction coordinate. We substantiate that important aspects of the reaction dynamics are well captured by the model. In addition, we will also discuss electronic factors governing vibrational excitations of adsorbed species on metal surfaces and provides the rationale for manipulation of metal surfaces for electron-assisted surface reactions.

1. P. Christopher, **H. Xin**, S. Linic, Nature Chem., 3, 467, (2011)
2. P. Christopher, **H. Xin**, A. Marimuthu, S. Linic, Nature Mater. 11, 1044, (2012)
3. S. Linic, P. Christopher, **H. Xin**, A. Marimuthu, Acc. Chem. Res., accepted (2013)

(168g) Ab Initio Modeling of Direct Interaction of Surface Plasmons With Adsorbates On Metal Nanostructures

Morabito, M., University of Michigan
Xin, H., University of Michigan
Linic, S., University of Michigan
Originally presented on: 11/4/2013 17:15:00 - 17:35:00

We recently reported on a new family of photo-catalysts that can efficiently couple thermal energy and photonic stimuli to drive photo-chemical transformations with unprecedentedly high efficiencies.[1,2,3] These materials are plasmonic nanostructures of coinage metals which respond to low intensity visible light through a formation of a collective, oscillating electrons.

We will discuss our approach to study these reactions from first principles. We have employed Langevin dynamics simulations to investigate the activation of adsorbed diatomic molecules over plasmonic metal nanostructures.[4] Using these simulations we calculated molecular dissociation probabilities and reaction rates at different light intensities. These stochastic simulations combine Newton's equations of motion with electronic drag from the surface. The electronic friction is caused by partial charge transfer between substrate and adsorbed molecule due to changes in the electronic structure induced by direct interactions of adsorbates with low energy electron in plasmons. We calculated electronic friction from the adiabatic electronic wavefunctions from Density Functional Theory, using a quantum model to obtain an electron distribution in the surface plasmon, a collection of low-energy excited electrons.[5][6] We compare this approach to desorption induced by (multiple) electronic transitions (DIET/DIMET) model, another mechanism proposed for this type of plasmonic photocatalytic enhancement.

[1] P. Christopher, H. Xin, M. Andiappan and S. Linic. Nat. Mat. **11**, 1044 2012

[2] P. Christopher, H. Xin, S. Linic. Nat. Chem. **3**, 467 2011

[3] S. Linic, P. Christopher, D. B. Ingram. Nat. Mat. **10**, 911 2011

[4] T. Olsen and J. Schiøtz. J. Chem. Phys. **133**, 034115 2010

[5] V. Krishna and J. C. Tully. J. Chem. Phys. **125**, 054706 2006

[6] L. Genzel, T. P. Martin, U. Kreibig. Z. Physik B. **21**, 339 1975

(173a) Effect of Ionic Liquid Confinement On CO2 Solubility and Transport

Wang, D.,
Baltus, R. E., Clarkson University
Banu, L., Clarkson University
Originally presented on: 11/4/2013 15:15:00 - 15:35:00

Room temperature ionic liquids (RTIL) are salts consisting of a bulky cation and an inorganic anion with melting point below 100°C. Ongoing work in our laboratory has focused on examining the potential of these unique fluids for post combustion carbon capture processes. Experimental work has involved measurements of carbon dioxide solubility and diffusivity in different ionic liquids in bulk phase and when confined within a ceramic membrane support. Carbon dioxide and nitrogen permeation through supported ionic liquid membranes (SILMs) prepared using anodic alumina films as the support membrane has also been measured. More recently, properties of unconfined ionic liquids and ionic liquids near a solid interface have been simulated using Materials Studio.

Experimental results reveal that both carbon dioxide solubility and diffusivity are larger in ionic liquids confined within a porous support when compared to bulk phase properties. These results are consistent with our results from gas permeation measurements through SILM systems, which show measured carbon dioxide permeability to be larger than permeability predicted from bulk phase properties.

In this presentation, our results will be discussed in the context of other literature reports of SILM separations with different type of support membranes, with molecular simulations of RTIL characteristics near a solid interface and with reports of enhanced carbon dioxide solubility with confinement of non-ionic solvents.

(173b) Solvent Cavitation Under Solvophobic Confinement

Ashbaugh, H., Tulane University

Originally presented on: 11/4/2013 15:35:00 - 15:55:00

The stability of liquids under solvophobic confinement can tip in favor of the vapor phase, nucleating a liquid-to-vapor phase transition that induces attractive forces between confining surfaces. In the case of water adjacent to hydrophobic surfaces, experimental and theoretical evidence support confinement-mediated evaporation stabilization of biomolecular and colloidal assemblies. The macroscopic thermodynamic theory of cavitation under confinement establishes the connection between the size of the confining surfaces, interfacial free energies, and bulk solvent pressure with the critical evaporation separation and interfacial forces. While molecular simulations have confirmed the broad theoretical trends, a quantitative comparison based on independent measurements of the interfacial free energies and liquid-vapor coexistence properties has, to the best of our knowledge, not yet been performed. To overcome the challenges of simulating a large number of systems to validate scaling predictions for a three-dimensional fluid, we simulate both the forces and liquid-vapor coexistence properties of a two-dimensional Lennard-Jones fluid confined between solvophobic plates over a range of plate sizes and reservoir pressures. Our simulations quantitatively agree with theoretical predictions for solvent-mediated forces and critical evaporation separations once the length dependence of the solvation free energy of an individual confining plate is taken into account. The effective solid-liquid line tension length dependence results from molecular scale correlations for solvating microscopic plates and asymptotically decays to the macroscopic value for plates longer than 150 solvent diameters. The success of the macroscopic thermodynamic theory at describing two-dimensional liquids suggests application to surfactant monolayers to experimentally confirm confinement-mediated cavitation.

(173c) Water Flow in Nano Channels of Quartz, Titanium Dioxide and Carbon Nanotubes

Wei, M. J., Vanderbilt University
McCabe, C., Vanderbilt University
Cummings, P. T., Oak Ridge National Laboratory
Originally presented on: 11/4/2013 15:55:00 - 16:15:00

The flow of liquid through narrow pores is a key component in many natural and industrial processes. For example, the erosion of rock in the subsurface can be initiated as flow through nanopores, while the separation of mixtures in membranes by molecular sieving exploits the different flow rates of molecules through pores as a result of both attractive and steric interactions between the molecules and the pore material. Flow in such nanopores is driven by a pressure drop and/or chemical potential gradient. In order to understand such processes at the molecular level requires the use of non-equilibrium molecular dynamics (NEMD) simulation methods that allow the application of an external field [1,2]. In this work, we consider NEMD simulations of water flow in the mineral nanochannels. In order to model a pressure drop across the nanopore, we add a force to each water atom directly so as to induce the flow in the channels [3]. The application of this external force on the system requires that viscous heat generation must be removed in order to reach a steady state; this is achieved by applying a thermostat on the wall atoms [3].

Three types of channel wall were chosen for study: quartz slit nanopores, titanium dioxide slit nanopores and carbon nanotubes. The former two channels are hydrophilic while the last one is hydrophobic. We find that the pressure-drop-induced flow in nano channels does not alter in any appreciable way the structure and dynamic properties of the water inside the nanopore. The primary reason for this is that even a large pressure drop, such as 300 MPa, translates into an additional force per atom that is no more than 2% of the forces on the atoms in the absence of a pressure drop (i.,e., at eqiulibrium). Additionally, the consequent streaming velocity of flow is only 10% of the average speed of each water atoms. The flux in hydrophilic channels yields a velocity profile consistent with the classical continuum-theory Hagen-Poiseuille equation. In contrast, those in the hydrophobic nanotube pore do not, since we find that the no-slip boundary condition used to derive the Hagen-Poiseuille equation is not valid.

(1) Predota, M.; Cummings, P. T.; Wesolowski, D. J. The Journal of Physical Chemistry C 2007, 111, 3071.
(2) Joseph, S.; Aluru, N. R. Nano Letters 2008, 8, 452.
(3) Todd, B. D.; Evans, D. J.; Daivis, P. J. Physical Review E 1995, 52, 1627.

(173d) Anisotropic Diffusion in Zeolites and Metal-Organic Frameworks: Calculation of Diffusion Tensors

First, E. L., Princeton University
Gounaris, C. E., Princeton University
Floudas, C. A., Princeton University
Originally presented on: 11/4/2013 16:15:00 - 16:35:00

Microporous materials, such as zeolites and metal-organic frameworks, are composed of complex porous networks that give rise to intracrystalline diffusion anisotropy. Although the phenomenon is well-understood, experimental measurements of diffusion anisotropy are difficult to obtain. Computational techniques, which typically rely on molecular simulations or simplified lattice models, are also challenging. With recently developed pore characterization methods ZEOMICS [1] and MOFomics [2], it is now possible to exploit the exact geometry and connectivity of microporous networks and calculate energetic barriers to transport through various pathways [3-6]. We have developed a novel methodology [7] to estimate full diffusion tensors based on these detailed descriptions.

We have validated our approach through a number of computational studies that demonstrate good agreement with experimental and computational results reported in the literature. A comprehensive examination involving approximately 200 zeolite framework types elucidates classifications based on a number of anisotropy metrics.

The power of characterizing materials based on diffusion tensors is illustrated in the context of microporous membranes. In such membranes, the orientation of crystals is crucial because it controls transport along the desired direction. We have formulated mathematical optimization models that determine an orientation of a porous network to maximize the throughput of a species or the separation between two species of interest. These models are solved to global optimality using ANTIGONE [8-9], a branch-and-bound solver for general mixed-integer nonlinear models.

References:

1. First EL, Gounaris CE, Wei J, and Floudas CA. Computational characterization of zeolite porous networks: an automated approach. Phys. Chem. Chem. Phys. 13:17339-17358, 2011.
2. First EL and Floudas CA. MOFomics: Computational pore characterization of metal-organic frameworks. Micropor. Mesopor. Mater. 165:32-39, 2013.
3. Gounaris CE, Floudas CA, and Wei J. Rational design of shape selective separation and catalysis-I: Concepts and analysis. Chem. Eng. Sci. 61:7933-7948, 2006.
4. Gounaris CE, Wei, J, and Floudas CA. Rational design of shape selective separation and catalysis-II: Mathematical model and computational studies. Chem. Eng. Sci. 61:7949-7962, 2006.
5. Gounaris CE, Wei J, Floudas CA, Ranjan R, and Tsapatsis M. Rational design of shape selective separations and catalysis: Lattice relaxation and effective aperture size. AIChE J. 56:611-632, 2009.
6. First EL, Gounaris CE, and Floudas CA. Predictive framework for shape-selective separations in three-dimensional zeolites and metal-organic frameworks. Langmuir 29:5599-5608, 2013.
7. Gounaris CE, First, EL, and Floudas CA. Estimating diffusion anisotropy in microporous crystalline materials. Submitted for publication.
8. Misener R and Floudas CA. ANTIGONE: Algorithms for continuous / integer global optimization of nonlinear equations. Submitted for publication.
9. Misener R and Floudas CA. A framework for globally optimizing mixed-integer signomial programs. Submitted for publications.

(173e) Structural and Dynamical Properties of Atomic Thin Films

Haji-Akbari, A., Princeton University
Debenedetti, P. G., Princeton University
Originally presented on: 11/4/2013 16:35:00 - 16:55:00

Recent discovery of stable glasses by Ediger and coworkers [1] has spurred an increased interest in understanding the structural and dynamical properties of thin films. Here, we use molecular dynamics (MD) simulation to study thin films of the Kob-Anderson binary Lennard Jones system [2], and compute their structural and dynamical properties as a function of distance from the wall. The computed structural properties include profiles of densities, internal energies, normal and lateral stresses and lateral radial distribution functions, while the dynamical properties calculated on a layer-by-layer basis are the structural relaxation times and lateral diffusivities. A monotonic increase of relaxation times is observed along the free interface, which is consistent with earlier observations on free-standing films [3]. The dynamical heterogeneity in the vicinity of the wall is however more subtle and depends both on the strength of wall-liquid interactions and the local structure of the film in the vicinity of the wall. In addition, the film is strongly stratified in the liquid-wall interface, but this stratification is not usually accompanied with long-range lateral order, and the extent of such long-range order also depends on the strength of wall-liquid interaction.

1- Swallen SF, Kearns KL, Mapes MK, Kim YS, McMahon RJ, Ediger MJ, Wu T, Yu L, Satija S, *Science* **315(5810)**: 353-6 (**2007**).

2- Kob W, Andersen HC, *Phys. Rev. E* **51**: 4626-41 (**1995**).

3- Shi Z, Debenedetti PG, Stillinger FH, *J. Chem. Phys.* **134**: 114524 (**2011**).

(173f) Correlations Between Static Thermodynamic Properties and Dynamic Transport Properties of Confined Fluids: Application to Adsorption and Diffusion in Porous Solids

Krekelberg, W. P., National Institute of Standards and Technology
Siderius, D. W., National Institute of Standards and Technology
Shen, V. K., National Institute of Standards and Technology
Originally presented on: 11/4/2013 16:55:00 - 17:15:00

It is well appreciated that placing a fluid under confinement drastically alters both its static and dynamic properties relative to a bulk, unconfined fluid. However, predicting these changes (even semi-quantitatively) remains a scientific challenge, and thus, there exist fundamental knowledge gaps in a number of areas where confined fluids play a central role. Examples include lubrication, friction, and adsorption. In an effort to understand the impact of confinement on fluid properties, recent work has demonstrated that there exists a robust relationship between static fluid properties, such as excess entropy and insertion probabilities, and dynamic fluid properties, in particular, the self-diffusion coefficient, that is independent of the geometry of the system when the confinement is straightforward. However, such relationships have not been observed for fluids confined near highly attractive surfaces, which are prevalent in many applications. Using Monte Carlo and Molecular Dynamics simulations in tandem, we apply this combined static-dynamic analysis to models of light gas adsorption in simple confinements such as graphitic carbon materials to systematically investigate the effect of the fluid-material interaction strength on the correlation of static and dynamic properties. We observe that fluids confined in strongly attractive pores display three distinct diffusive regimes when cooled below the critical point. These regimes are directly related to the structural arrangement in the pore, and hence, to the thermodynamic properties of the fluid.

(173g) Spatially Distributed Fluid-Solid Friction in Confinement

Bhatia, S. K., The University of Queensland
Nicholson, D., University of Queensland
Originally presented on: 11/4/2013 17:15:00 - 17:35:00

Traditional hydrodynamics considers fluid-solid friction to reside at a sharp interface, incorporating this as a boundary condition via a slip length, or as a surface momentum accommodation coefficient, when modeling fluid flow in narrow pores and channels. Here we demonstrate, at the nanoscale, that the location of this slip boundary is ill-defined, since the range of fluid-solid interactions is of the same order of magnitude as the system size, and that instead, friction is distributed over the whole of the pore fluid. Using molecular dynamics simulations for the diffusion of several one-centre and multisite molecules in nanopores with an atomically detailed surface, we obtain position-dependent *dynamical* fluid-solid friction coefficients. These dynamical friction coefficient profiles are independent of the driving force, and only weakly dependent on fluid density. These are subsequently used in a modified hydrodynamic model, incorporating spatially distributed frictional momentum dissipation, to predict collective- and self-diffusion coefficients, which are found to be in very good agreement with those based on NEMD and EMD simulaton. We also determine theoretically a position-dependent *static* friction coefficient, based on equilibrium density distributions, which is shown to mimic the behaviour of the dynamical coefficient. These results suggest possibilities for tailoring nanomaterials and surfaces to engineer low friction pathways for fluid flow by appropriately tuning the fluid-solid potential energy landscape.

(175a) Synergistic Stabilization of Emulsions and Foams With Particles and Surfactants

Worthen, A. J., The University of Texas
Foster, L. M., The University of Texas at Austin
Dong, J., The University of Texas at Austin
Bollinger, J. A., The University of Texas at Austin
Truskett, T. M., The University of Texas at Austin
Bryant, S. L., The University of Texas at Austin
Bielawski, C. W., The University of Texas
Johnston, K. P., University of Texas at Austin
Originally presented on: 11/4/2013 15:15:00 - 15:30:00

The combination of zwitterionic surfactants and silica or clay nanoparticles and microparticles provides synergism in formation and stabilization of dodecane droplets in synthetic seawater. These combined amphiphiles are of interest for mitigation of oil leaks at the ocean floor. The surfactant lowers the oil/water interfacial tension and allows the formation of oil/water emulsions composed of small oil drops (<100 micron diameter), while the presence of nanoparticles at the oil/water interface enhances emulsion stability by providing a larger steric barrier to droplet coalescence than surfactants alone.

Droplet sizes were analyzed with optical microscopy and static light scattering, and the dispersion mechanisms for various formation pathways were explored via scanning electron microscopy. Droplet size and emulsion stability were characterized for a range of temperatures down to 4°C, to simulate deep-sea conditions, and over time-scales up to one month. The effect of particle hydrophilicity was also explored by using cation-exchange to tune surfactant adsorption on clay surface sites, which decreases hydrophilicity.

Similar synergistic benefits of combined surfactant/particle dispersant mixtures have been observed for emulsions and foams of carbon dioxide and water, which is of interest in enhanced oil recovery. Thus, the phenomena of particle/surfactant synergy may be of interest in many fields where stable emulsions or foams are desired.

(175b) Forces Between Micro-Bubbles in the Presence of Novel Responsive Peptide Surfactants

Balasuriya, T., THE UNIVERSITY OF MELBOURNE
Dagastine, R. R., The University of Melbourne
Originally presented on: 11/4/2013 15:30:00 - 15:45:00

Reversibly switchable surfactants are becoming increasingly important for controlling foam stability in many industrial applications because they can be recycled as foaming and antifoaming agents. Novel stimuli responsive peptide surfactants have been previously studied to classify the molecular confirmation at the air-water interface before and after switching. In this study, Atomic Force Microscopy (AFM) was used to correlate the peptide conformation to directly measured changes in colloidal interaction forces between immobilized air bubbles before and after switching(1). In addition the local surface tension of the micro-bubbles was measured from the AFM measurments as well. Results were compared to previous studies of macroscopic foam column stability using these peptides. Differences in foam stability and the AFM force measurements were apparent, highlighting variations in directly measured equilibrium colloidal interactions to macroscopic bulk behavior. These results help elucidate the connection between the switching mechanism of novel stimuli responsive peptide surfactants and their effect on colloidal scale interactions between bubbles.

1. Balasuriya, Dagastine, *Langmuir* **2012,** 28, 17230.

(175c) Particles At Liquid Interfaces: Effects of Particle Charge and Their Consequences for Particle-Stabilized Emulsions

Wang, H., Georgia Institute of Technology
Nguyen, T.,
Behrens, S. H., Georgia Institute of Technology
Originally presented on: 11/4/2013 15:45:00 - 16:00:00

It has been known for more than a century that colloidal particles with the "right" wetting properties can stabilize emulsions, much like surfactants or amphiphilic polymers do. Today, particle-stabilized emulsions (so-called Pickering emulsions) play an important role in many industrial applications, but our fundamental understanding of the connection between particle properties and emulsion characteristics is still surprisingly poor. We can, for example, prepare an emulsion by mixing equal amounts of an oil phase and an aqueous particle dispersion; but even with complete knowledge of both liquids and all particle properties, current theory fails to predict reliably the stability and even the type (w/o or o/w) of the emulsion formed.

This presentation will discuss electrostatic sources of such difficulties and propose several steps toward an improved theoretical description and better practical control of emulsion properties. In particular, we will address the role of electrostatic interactions between a particle and a liquid interface before and after particle adsorption, and show that widely neglected image charge effects can prevent particle adsorption and emulsion stabilization altogether. For the case of successful particle adsorption, the particle's contact angle with the interface is found to be influenced strongly by the particle charge in ways that are not usually accounted for, but have important consequences for the long term stability of Pickering emulsions.

(175d) Scale-Up and Influence of the Process Variables in the Preparation of Highly-Concentrated Emulsions

May-Masnou, A., Universitat de Barcelona

Ribo, J.,

Porras, M., Universitat de Barcelona

Maestro, A., Universitat de Barcelona

Gonzalez, C., Universitat de Barcelona

Gutierrez, J. M., Universitat de Barcelona

Originally presented on: 11/4/2013 16:00:00 - 16:15:00

Scale-up from experimental laboratory equipment to industrial plant size is one of the crucial issues in the field of process design. The processes working with low-viscosity Newtonian fluids are, usually, directly scalable. However, the scaling-up of processes involving high-viscosity non-Newtonian fluids is far more complicated, since fluid properties, like viscosity, and flow conditions can vary drastically during the process [1]. This is the case of emulsion manufacturing. Moreover, the final quality and properties of an emulsion are very dependent on the process variables, so a little change in the way of adding the components or in the vessel size can result in changes in the product quality, apart from a raw material, time and economical loss. The scale-up analysis of these processes and the study of the effect of process variables on the emulsion quality are necessary in order to predict what will happen at industrial scale and to optimize the process, saving money and time in unproductive tests.

In this study, preparation of highly-concentrated emulsions is used to investigate how the process variables and the scale-up influence emulsion properties, like droplet size, viscosity, yield stress, viscoelasticity and stability.

The w/o highly-concentrated emulsions (90 wt % dispersed phase) are prepared in batch mode following a two-step process: the first step consists in the addition of dispersed phase at a given flow rate (Q) while maintaining a constant stirring rate (N), the second step consists in the homogenization of the emulsion at the same stirring rate. Influence of composition variable surfactant/oil ratio (S/O) is also studied, using Span 80 as surfactant and dodecane as oil. Three different vessel sizes are used, with geometric similarity (1:2:4) and keeping the total emulsification time constant.

We use experimental design tools in order to minimize the number of experiments needed to obtain empirical models and response surfaces (Figure 1) that describe the influence of the variables on the emulsion properties and that can be used to predict further experiments. Additional experiments are performed to validate these empirical models obtained from few experiments. Similar response surfaces as function of the three input variables studied are obtained for each of the output variables. Empirical equations permit to predict corresponding emulsion property as a function of the input variables.

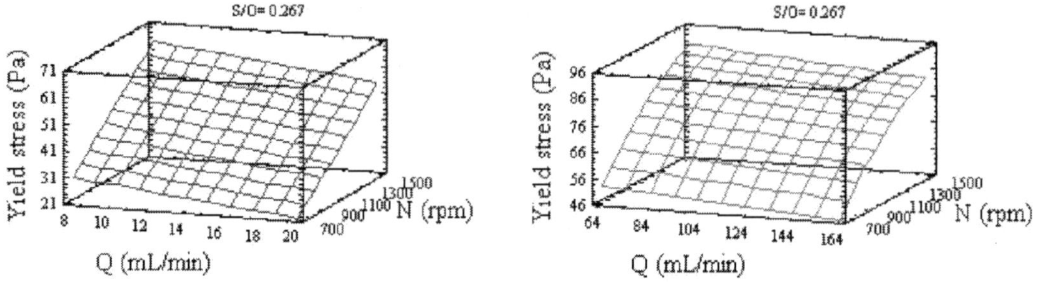

Figure 1. Response surface describing the change of yield stress with the addition flow rate (Q) and stirring rate (N) at small scale (left) and medium scale (right).

Experimentations at the three scales studied permit to obtain the scale invariants of the process, variables which have to remain invariants with scale in order to obtain emulsions with the same properties independently of the scale at which had been prepared.

From our results can be deduced that S/O is one of the invariants. Invariant related with flow rate is time of water addition, volume of reactor divided by flow rate. Finally, invariant related with stirring rate is not the Reynolds number, equivalent to ND^2, as occur for other processes. For preparation of highly concentrated emulsions, we obtained that the invariant have the form ND^a, where exponent a is between 0 and 1, and close to 0.5 for all the response variables chosen (including droplet size and rheological parameters).

For scaling up, we have empirical equations for each of the response variables as a function of the found invariants, and these empirical equations are valid for the three scales studied scales and, may be, it would be an extrapolation of empirical results, valid for any scale of magnufacturing that could be used.

[1] R. J. Wilkens, C. Henry, and L. E. Gates, ?How to scale-up mixing processes in non-Newtonian fluids,? *Chemical Engineering Progress*, vol. 99, no. 5, pp. 44?52, 2003.

(175e) Adsorption and Protein-Surfactant Interaction At the Air-Water Interface

Engelhardt, K., University of Erlangen-Nuremberg
Peukert, W., University of Erlangen-Nuremberg
Braunschweig, B., University of Erlangen-Nuremberg
Originally presented on: 11/4/2013 16:15:00 - 16:30:00

Adsorption and protein-surfactant interaction at the air-water interface

Kathrin Engelhardt, Wolfgang Peukert, Björn Braunschweig

Institute of Particle Technology

University Erlangen-Nuremberg,Germany

Taste, texture and appearance of foamy dairy products such as mousse or cream are influenced and partially controlled by the interfacial properties of milk proteins and their tendency to form foam stabilizing networks. The stability of aqueous foams is dominated by the molecular composition of the ubiquitous air-water interfaces and the hierarchical structure - from the interface via the complex lamella structure to the macroscopic foam. The relation of adsorption kinetics and competitive behaviour of proteins and surfactants are yet not fully understood, which is caused mainly by the lack of interface sensitive methods.

In this contribution we introduce vibrational sum-frequency generation (SFG) as an inherently interfacial specific probe that that is shown to be a powerful technique to address experimental challenges associated with liquid interfaces. In order to obtain molecular level information with SFG, tunable infrared pulses are combined at the air-water interface with visible pulses of a fixed frequency. The resulting SFG spectra provide a vibrational fingerprint with information on composition and conformation of the interfacial molecular layer[1].

A combination of SFG spectroscopy, surface dilational rheology and foam stability measurements is used to study the adsorption kinetics and interfacial properties of β-Lactoglobulin and its mixtures with ionic and nonionic surfactants such as SDS and hexadecyllactate. The adsorption kinetics of competitive adsorption and displacement reactions of the two species are studied with SFG. The vibrational fingerprint combined with kinetic information on the interaction between β-Lactoglobulin and surfactants provides us with new insights in the adsorption of mixed protein-surfactant systems. The behavior at the interface is correlated with the macroscopic foam stability which changes with increasing surfactant concentration.

[1] Engelhardt, K., Rumpel, A., Walter, J., Dombrowski, J., Kulozik, U., Braunschweig, B., Peukert, W., Protein adsorption at the electrified air–water interface: implications on foam stability, Langmuir 2012, 28, 7780–7787

(175f) Characterization of Concentrated Emulsions in the Energy Industry Using Acoustic Spectroscopy

Venkataramani, D.,
Clark, P., Oklahoma State University
Smay, J.,
Aichele, C. P., Oklahoma State University
Originally presented on: 11/4/2013 16:30:00 - 16:45:00

Concentrated emulsions have direct applications in the energy, food, agricultural, and pharmaceutical industries. In the energy industry, concentrated emulsions can be encountered during drilling, production, transporting, and processing of crude oils. These dispersions typically lead to high operating costs as they currently require costly treatment strategies. Fundamental characterization of these complex systems is required to develop strategies that decrease the associated operating costs of managing these systems. This paper highlights the efforts underway at Oklahoma State University in using optical and acoustic techniques for emulsion characterization, measurement of particle size distributions, and quantification of interfacial behavior. This paper will also evaluate the flow characteristics of dispersions using advanced rheological techniques. The acoustic technique utilizes sound waves and the attenuation coefficient to characterize particle size, zeta potential, and conductivity. The work explores the morphology and transient behavior of both conventional surfactant stabilized and solid stabilized emulsions. Both model oil and crude oil systems will be presented. Rheological characterization of select systems will be presented through the implementation of advanced rheological techniques. The rheological measurements explore the viscosity and elastic properties to measure complex rheological behavior of these dispersed systems. The particle size measurements coupled with rheology will lead to fundamental characterization of these complex systems that will lead to enhanced flow assurance strategies for the energy industry.

(175g) Engineering of Novel Classes of "Smart" Stimuli Responsive Foams

Lam, S., North Carolina State University
Fameau, A. L.,
Khan, S. A., North Carolina State University
Velev, O. D., North Carolina State University
Originally presented on: 11/4/2013 16:45:00 - 17:00:00

Soft materials, such as foam and emulsion systems, which respond to external stimuli, are on the leading edge of materials research and have recently been of interest to many scientists. We present a novel class of magnetically responsive Pickering foams, which remain stable at ambient conditions but can be quickly destroyed upon exposure to a gradient magnetic field (S. Lam, et al., *J. Am. Chem. Soc.* **2011**, 133, 13856). These foams are stabilized by particles made from hydrophobically modified cellulose (HP-55) with magnetic responsiveness imparted through the incorporation of carbonyl iron (CI) into the cellulose matrix. We studied the change in foam water fraction and collapse behavior over time, and correlated the evolution of these properties with foam viscoelasticity as a function of age. We observed differences in the mechanism of foam collapse based on the fraction of water in the foam as well as the age of the foam. This age-dependent response of the foams to a magnetic field was demonstrated using SQUID magnetometry as well as magnetorheology. By correlating the time-dependence of foam viscoelastic properties and magnetic response, we proposed two separate mechanisms of collapse - one for fresh foams and one for aged foams. In these mechanisms, the collapse of fresh foam is correlated with the expulsion of air bubbles from the foam matrix, whereas the collapse of aged foam results from the rupture of foam films. We will also discuss how these systems can be made to be photo- as well as thermally responsive; and, present in brief, a characterization of the collapse properties of photo-responsive foams as a function of system composition. Systems, such as those presented here, can find application in a wide range of industrial and environmental processes that require controlled, non-contact and on-demand defoaming.

(175h) Ellipsoidal Janus Nanoparticles Simulated At Water/Oil Interfaces

Luu, X. C., The University of Oklahoma
Yu, J.,
Striolo, A., University of Oklahoma
Originally presented on: 11/4/2013 17:00:00 - 17:15:00

Emulsions stabilized by nanoparticles, AKA Pickering emulsions, are finding many applications, including food processing and biofuel upgrading. Although spherical particles are often used for basic research, particles of different shapes are being produced and used, sometimes showing marked differences compared to results obtained when spherical particles are employed. We focus here on the behavior of ellipsoidal Janus nanoparticles. We employ dissipative particle dynamics simulations to uncover possible differences between the behaviors of these nanoparticles compared to that of spherical particles with similar characteristics. We report on the orientation of the nanoparticles with respect to the oil-water interface, the structure of the nanoparticles at the interface as a function of the surface coverage, with particular attention to the isotropic-to-nematic phase transition, and on the interfacial tension reduction caused by some of the nanoparticles considered. Implications of such observations on the interpretation of some experimental observations available in the literature will be discussed.

(175i) Stable Core-Shell Particles Produced By All-Aqueous Microfluidics

Ziemecka, I., Delft University of Technology
Olive, A.,
van Steijn, V., Delft University of Technology
Kreutzer, M. T., Delft University of Technology
van Esch, J. H., Delft University of Technology
Originally presented on: 11/4/2013 17:15:00 - 17:30:00

So far, the field of droplet microfluidics has relied on the use of organic solvents in combination with aqueous solutions to encapsulate materials and synthesize compartmentalized micro particles. A disadvantage of the use of organic solutions is that they harm biomaterial upon encapsulation.The obvious solution is to omit the use of organic solvents and produce compartmentalized micro particles from two or more immiscible aqueous solutions. Our group recently demonstrated the controlled and reliable formation of single and double emulsions from aqueous two phase systems (ATPS) [1-2]. Stabilization of these "water-in-water" and "water-in-water-in-water" droplets has turned out to be a challenge.

In this contribution, we teach how to stabilize these all-aqeuous compartmentalized micro droplets and report permeable microparticles composed of a hydrogel shell and a liquid core. Drying and rehydration experiments demonstrate the permeability of the shell, which enables the molecular-weight-dependent release and uptake of polar solutes.

References

[1] I. Ziemecka, V. van Steijn, G. J. M. Koper, M. T. Kreutzer and J. H. van Esch, Soft Matter 7, 9878-9880, 2011.
[2] I. Ziemecka, V. van Steijn, G. J. M. Koper, M. Rosso, A. M. Brizard, J. H. van Esch and M. T. Kreutzer, Lab on a Chip 11, 2011.

(175j) Modeling Polyurethane Foam Expansion and Cure

Mondy, L. A., Sandia National Laboratories
Rao, R. R., Sandia National Laboratories
Celina, M. C., Sandia National Laboratories
Wyatt, N. B., Sandia National Laboratories
Grillet, A., Sandia National Laboratories
O'Hern, T., Sandia National Laboratories
Soehnel, M., Sandia National Laboratories
Russick, E. M., Sandia National Laboratories
Originally presented on: 11/4/2013 17:30:00 - 17:45:00

We are developing computational models to elucidate the dynamic filling of a mold with a polyurethane foam. Polyurethane foam is created through two primary reactions that compete for isocyanate monomer. In the first, isocyanate reacts with polyol to produce polymer, eventually creating a rigid solid. In the second, isocyanate reacts with water to produce CO^2 gas that drives foam expansion. The kinetics of the resin polymerization was evaluated using infrared spectrophotometry on both wet and dry polyurethane precursor, in order to attempt to separate the curing and blowing reactions. The kinetics of the gas generating reaction was measured by tracking volume evolution with time in the curing system while simultaneously measuring the system temperature and pressure. As the material polymerizes, it becomes both more viscous and viscoelastic, hindering expansion; hence, gas bubbles become pressurized, since there is no change in volume though more gas has been created. Results show that the foaming reaction continues even after foam expansion has stopped. However, the curing reaction continues for some time after the gas generating reaction is complete. Because the isocyanate is in excess during the expansion stages of the foam formation, we treat the competing reactions as separable primary reactions. Both reactions show Arrhenius-type temperature dependence. The rheology of polyurethane, again both wet and dry, during reaction is examined through steady-shear and oscillatory tests. The rheological measurements are combined with the kinetic equations to estimate the effect of gas fraction and extent of polymerization on the viscosity. A finite element model was developed combining the equations of motion, an energy balance, and two rate equations for the polymerization and foaming reactions. This was joined with a level set method to track the location of the free surface as it evolves in time. The model predicts the evolution of the foam-gas interface and density of the foam. It is compared to experimental flow visualization data, temperature measurements, and x-ray images of density.

*This research is supported by the Laboratory Directed Research and Development program at Sandia National Laboratories. Sandia National Laboratories is a multi-program laboratory managed and operated by Sandia Corporation, a wholly owned subsidiary of Lockheed Martin Corporation, for the U.S. Department of Energy's National Nuclear Security Administration under contract DE-AC04-94AL85000.

(184a) Thermocapillary Instabilities During the Edge-Defined Film-Fed Growth of Scintillator Crystals

Yeckel, A., University of Minnesota
Feigelson, R.,
Derby, J. J., University of Minnesota
Originally presented on: 11/4/2013 15:15:00 - 15:30:00

A model of edge-defined film-fed crystal growth (EFG) is developed to study melt growth of cylinders of the scintillator crystal cesium iodide (CsI). The model, which includes fluid flow, heat transfer, and gravitational effects, is solved numerically by the finite element method to predict crystal diameter as a function of growth rate and die geometry. The model computes the location and shape of the crystal-melt interface and the meniscus to satisfy a self-consistent formulation of phase change and capillary dynamics. A sharp interface formulation solved on a deforming finite element mesh ensures accurate conservation of mass, energy, and momentum at the interfaces.

This system is characterized by strongly nonlinear interactions of heat transfer, capillarity, and die geometry that give rise to multiple stationary solution states under a single set of operating conditions. We use bifurcation theory and parameter continuation techniques to identify these solution states and characterize their stability.

Purely capillary instabilities are identified that result from the interaction of surface tension with gravity, and also the interaction of in-plane and out-of-plane (hoop stress) curvatures of the meniscus. These give rise to two solution families distinguished by a difference in gap width between die face and growth interface. The narrow gap solutions are shape stable, and the wide gap solutions are shape unstable, though these can be stabilized by heat transfer in a manner similar to Czochralski growth.

Additionally we identify several instabilities of convective heat transfer, which generally are of two types. One type is mostly convective in nature and largely independent of the interface or meniscus shapes. In this type the solution families are distinguished by different flow structures within the capillary bore. The other type is characterized by a strong interaction of convection with the geometry of the growth interface. Stronger convection moves the interface to a higher position, forming a larger gap. The larger gap allows a stronger flow to form, which enhances convection. This reinforcing behavior enables a family of solutions to exist in which the growth interface shape is concave in the central region, though it is convex elsewhere.

Supported in part by U.S. Department of Homeland Security, HSHQDC-11-C-00094, the content of which does not necessarily reflect the position or policy of the United States Government, and no official endorsement should be inferred.

(184b) Black Spots in the Human Tear Film: An Evaporation-Driven Instability

Peng, C. C., University of California, Berkeley
Cerretani, C., University of California, Berkeley
Braun, R. J.,
Radke, C. J., University of California Berkeley
Originally presented on: 11/4/2013 15:30:00 - 15:45:00

Tear-film instability is a major signature of eye health. When an interblink is prolonged, randomly distributed ruptures occur in the tear film. "black spots" and/or "black streaks" appear in 15 to 40 s for normal individuals. For people who suffer dry eye, tear-film breakup time can be less than a few seconds. In spite of decades of effort, there currently is no satisfactory quantitative explanation for the origin of tear rupture. Based on the experiments of Liu et al.,[1] we hypothesize that tear-film breakup is due to local tear-film lipid layer rupture and resulting locally elevated evaporation that drives tear-film rupture.

As observed in experiment, the 100-nm thick lipid layer covering the tear film initially ruptures giving a larger local evaporation rate than that in the surrounding lipid-covered tear film. Increased evaporative flux drives a hole in the tear film. As the hole deepens, local salinity increases. The growing hole is suppressed by curvature-driven healing flow and by osmotic-suction from the cornea due to the local salinity increase. Tear rupture occurs only when the locally high evaporative flux outweighs the two healing flows. The new proposed rupture mechanism leads to significant increased salinities at the bottom of the rupture spot (or streak) that we coin salinity "hot spots". Local salinity hot spots appear even when a growing instability does not rupture during a 5-s interblink.

Coupled nonlinear evolution equations for thin-film flow, evaporative heat transfer, and salt concentration are solved numerically. Tear-film instability and corresponding local increase of salinity is confirmed for higher, but yet reasonable, evaporation rates. Predicted roles of environmental conditions such as wind speed and relative humidity on tear-film stability agree with clinical observations. Most importantly, locally elevated evaporation leads to hyperosmolar spots in the tear film and, hence, vulnerability to epithelial irritation and dry-eye symptoms. Tear-film rupture is more likely with contact-lens wear because initial tear-film thickness is reduced.

This study provides the first (and only) physically consistent, quantitative explanation for observed black streaks and/or spots in the human tear film during interblink. Importantly, this study explains the formation of "hot spots" of locally high concentrated solutes in the tear film, such as salts, proteins, mucins. A tear film repeatedly peppered with salinity hot spots is vulnerable to dry eye even when tear-film breakup is not observed.

[1] Liu H, Begley CG, Chalmers R, Wilson G, Srinivas SP, Wilkinson JA. ***Optom Vis Sci***. 2006; 83:723-30.

(184c) Comparison of Different Gravure Cell Geometries Using a Potential Energy Framework

Boelens, A.,
de Pablo, J. J., University of Wisconsin-Madison
Originally presented on: 11/4/2013 15:45:00 - 16:00:00

Gravure cell printing is emerging as a viable strategy for mass production of flexible electronics, for applications like medical devices and solar cells. A key variable for design of efficient printing processes is the geometry of the wells used for ink transfer. In this work we examine various geometries and assess their effect on ink transfer. Previous work in this area includes investigation of the effect of different inclinations of the gravure cell walls, the scaling of ink transfer with gravure cell volume, and the effect of different aspect ratios of square gravure cells. The majority of past work has been limited to studies in two dimensions, and a general framework to investigate the effects of different gravure cell sizes and shapes in three dimensions is missing.

At the analytical level, a simple potential energy framework is introduced to analyze ink transfer. For a variety of geometries (disks, domes, and cones) we calculate the reversible work needed to transfer a droplet from a gravure cell onto a substrate, and the results of these calculations are compared to those of three dimensional continuum mechanics simulations. At the numerical level, our simulations rely on the Volume Of Fluid approach with a Brackbill surface tension model and a Hoffman dynamic contact angle model. The ink is described as a Bird-Carreau fluid, and automated remeshing is been implemented to account for topological changes during the printing process.

Our results indicate that the optimal geometry for a gravure cell consists of a dome. For lower contact angles, ink transfer increases dramatically in lower and wider shaped cells. There is good agreement between our model predictions and continuum mechanics simulations. For disks and cones, the model also predicts better ink transfer for wider and lower shapes, but the agreement between the model and simulations is not as good. We attribute the discrepancy to residual ink getting trapped in narrow spaces and in corners. The fluid time scales are too slow compared to the printing time scale. Overall, we believe that a potential energy description of gravure cell printing provides a general and simple framework to describe ink transfer for different geometries, contact angles, surface tensions, and sizes of gravure cells, and can be a useful tool for optimization of gravure cell printing processes.

(184d) Chiral Rheology of a Phospholipid Monolayer: Ductile Plasticity Vs. Brittle Fracture

Kim, K., Univerisity of California, Santa Barbara
Choi, S. Q., University of Minnesota
Zasadzinski, J. A., University of Minnesota
Squires, T. M., University of California, Santa Barbara
Originally presented on: 11/4/2013 16:00:00 - 16:15:00

The phospholipid DPPC contains two hydrocarbon tails attached to the headgroup in a chiral fashion, and biologically-synthesized DPPC is enantiomerically-pure (R). At moderate surface pressures, DPPC monolayers at the air-water interface exhibit phase coexistence between disordered liquid (LE) and condensed liquid (LC) phases, wherein crystals of LC-DPPC nucleate and grow, forming domains whose micron-scale shape belies the underlying molecular chirality. Using our microfabricated, ferromagnetic, amphiphilic "microbuttons" as rheological probes, we have previously shown LC-DPPC to behave as a concentrated 2D emulsion -- a viscoelastic solid with a finite, line-tension-mediated yield stress, and hour-long "healing" times following large-strain deformations.

Here, we reveal this single-component, uncrosslinked molecule to reveal a further surprise: its nonlinear rheological response shows a strong signature of its chirality. In particular, the nonlinear elastic response is stiffer when the microbutton is rotated counter-clockwise (CCW) than clockwise (CW), and the CCW yield stress is larger than the CW yield stress. This chiral rheology varies linearly with the enantiomeric excess of R vs. S. Direct visualization of the deforming domains shows qualitatively different nonlinear domain evolution for CCW and CW deformations. Furthermore, under applied CW stresses above the CW yield stress, domains exhibit ductile plasticity, deforming smoothly into long, thin domains. By contrast, domains exhibit brittle fracture under large CCW deformations. These properties can be understood in terms of simple mechanical models.

(184e) Directed Assembly of Carbon Nanotubes By Capillary Focusing

Wirth, C. L., Katholieke Universiteit Leuven
De Volder, M., Katholieke Universiteit Leuven
Vermant, J., Katholieke Universiteit
Originally presented on: 11/4/2013 16:15:00 - 16:30:00

Carbon nanotubes (CNTs) are promising candidates for components of advanced materials, microelectronics and electrochemical systems. Directed assembly of CNTs over multiple length scales is key to the performance of these technologies. Unfortunately, state of the art methods for the growth of assembled CNT structures typically require harsh conditions. In an effort to achieve controlled deposition and assembly of CNTs at mild conditions, we developed a technique called "capillary focusing" that uses the differences in capillarity experienced by fluid evaporating from a substrate with a chemically heterogeneous surface. A silicon wafer substrate modified to have hydrophilic and hydrophobic regions of specified geometry was coated with an aqueous suspension of electrostatically stabilized CNTs. The combined effect of convective and Marangoni flows and capillarity focuses particles suspended in solution to preferentially deposit onto hydrophilic regions during evaporation. The deposition process was studied by systematically changing the size, shape, and spacing of hydrophilic regions as well as the substrate coating method. These experiments demonstrated that capillary focusing is a robust method for the preferential deposition of CNTs at ambient conditions and could be used for CNT device fabrication.

(184f) Controlled Generation of Monodisperse Particle-Stabilized Droplets With Arbitrary Interfacial Coverage

Kotula, A. P., Carnegie Mellon University
Nelson, C. W., Carnegie Mellon University
Anna, S. L., Carnegie Mellon University
Originally presented on: 11/4/2013 16:30:00 - 16:45:00

The complex interfaces of particle-stabilized drops and bubbles have a significant effect on the flow properties of bulk emulsions and foams. Common high-shear methods of generating stable emulsions have no direct control over the rate of droplet generation, the final droplet size distribution, or the composition of the interface, all of which are important to the interfacial and bulk rheology. In this talk, we present a method that allows for independent control over the droplet size generated and the degree of particle loading on the interface. Droplets are formed on demand at a T-junction with a desired size via a pressure-controlled generation scheme. Subsequently, the droplets travel along the axis of a circular capillary containing a surface-active particle suspension. We use a particle transport model that we have previously developed to estimate the particle coverage on the bubble surface as a function of residence time. We use light scattering methods to assess the degree of depletion of particles from the bulk and thus to verify that surface coverage is indeed a function of residence time, for two different systems: air-in-water foams stabilized by silica nanoparticles, and water droplets stabilized by silica in a continuous phase of cyclohexane. Thus, we show that by controlling channel geometry, applied inlet pressures, and residence time, we can directly control droplet size, volume fraction, and particle loading on the bubble interface, all of which are critical parameters relevant to the stability and rheology of a particle-stabilized emulsion or foam.

(184g) Strong Surfactant-Dependent Lift Forces On Drops and Bubbles in Microchannels

Stan, C. A., Harvard University
Guglielmini, L., Stanford University
Ellerbee, A. K., Harvard University
Caviezel, D., ASCOMP GmbH
Whitesides, G. M., Harvard University
Stone, H. A., Princeton University
Originally presented on: 11/4/2013 16:45:00 - 17:00:00

The transverse motion of drops and bubbles within liquids flowing in pipes and channels is determined by the combination of several types of hydrodynamic lift forces with mechanical confinement, buoyancy and external-field forces. In centimeter-sized and larger pipes, lift forces are usually weak compared to the buoyant forces corresponding to the buoyant difference between air and water, but they can be observed in neutrally buoyant systems.

Hydrodynamic lift forces are more prominent in microfluidic applications, where they have been used to position and sort particles with high efficiency and high accuracy. We used experimental, analytical, and computational methods to investigate lift forces on drops and bubbles immersed in a liquid, flowing in microfluidic channels under conditions characterized by low particle capillary numbers ($0.0003 < Ca_p < 0.3$) and low particle Reynolds numbers ($0.0001 < Re_p < 0.1$). We measured the lift forces directly and we developed a method to compare their strength across more than 20 different systems that were chosen for their relevance to microfluidic applications.

The measured lift forces were often much larger (up to a factor of 1000) than the predictions of analytical models of inertial and deformation-induced lift. Numerical simulations that accounted for inertial and deformation-induced mechanisms revealed that the stronger forces can be explained in part by the confinement of the flow. The complete explanation, however, includes interfacial physicochemical effects: in one series of experiments with water drops in fluorocarbon carrier liquid, we observed a five-fold increase in the lift force as a fluorocarbon-soluble surfactant was gradually added to the system.

We will present measurements of lift forces and their dependence on hydrodynamic numbers. The dependencies separate the systems in (i) one group that exhibits a deformation-induced lift force enhanced by confinement effects, and (ii) another group that exhibits a lift force for which an analytical model does not exist yet, but to our best knowledge is based on physicochemical effects at the interfaces of drops and bubbles. The second group is most interesting from a practical point of view as well: these are the systems in which the lift forces were strongest—thus most useful to manipulate droplets in microchannels.

(184h) Effects of Drop-Fluid Interface and Hydrodynamic Interactions On Quincke Rotation

Ouriemi, M., Brown University
Vlahovska, P. M., Brown University
Originally presented on: 11/4/2013 17:00:00 - 17:15:00

The spontaneous rotation experienced by a weakly conductive sphere immersed in a slightly more conductive liquid when submitted to a strong DC electrical fields is known as the Quincke rotation. Similar phenomena have been observed for deformable particles, such as dielectric drops immersed in a more conductive liquid. Those effects have been widely explored due to the importance of deformable particle suspensions subjected to an external electric field in physics and biophysics. To understand the Quincke rotation of systems involved in nature or industry like biological cells, or electrorheological fluids, it is necessary to take into account more complex configurations, such as drops covered by a thin shell or the presence of hydrodynamic interaction between particles.

Here, we present an experimental study of two controlled configurations investigating these effects. Rotation thresholds and drop deformation were measured for a dielectric drop covered with particles and submitted to a strong electrical field. Our results show that the presence of highly conductive particles at the interface decreases the onset of rotation, indicating more complex interactions due to the presence of the shell.

In addition, we investigated the Quincke rotation of two similar dielectric spheres immersed in a slightly more conducting liquid. Our results show different rotation thresholds compared to a system composed of a single sphere. They are compared with the recent theoretical and numerical studies that focused on the interactions between two particles.

(184i) Rapidly Moving Contact-Lines and Inviscid Flows

Altieri, A. L., Cornell University
Cox, B. L., Cornell University
Steen, P. H., Cornell University
Originally presented on: 11/4/2013 17:15:00 - 17:30:00

Contact-lines that are driven far from equilibrium are found in a range of industrially important processes, including coating operations and immersion lithography for integrated chip fabrication. We study the way a liquid moves along a solid, advancing or receding against a gas under conditions where surface tension of the free interface and inertia of the underlying flow are in balance. The moving contact-line is defined as the common line between the free interface and the driving solid substrate. When the liquid flow adjacent to the interface is predominantly inviscid, the contact-line motion can be rapid. We report solutions to the inviscid equations of motion and interfacial boundary conditions that deliver meniscus dynamics and rapid contact-line motion. A perturbation analysis shows how the inviscid flow influences the meniscus dynamics and thereby the contact-line motion which can exhibit various stick-slip behaviors. Rapid contact-line motion (meters per second) can be captured and examined in experiments where the flow solidifies, as with quenched liquid metals frozen by the planar-flow process. We compare our predictions against experiment using the frozen-in flow patterns as a signature of contact-line motion.

(184j) On the Mechanism of Wetting Failure During Fluid Displacement Along a Moving Substrate

Vandre, E., University of Minnesota
Carvalho, M. S., Pontificia Universidade Catolica do Rio de Janeiro
Kumar, S., University of Minnesota
Originally presented on: 11/4/2013 17:30:00 - 17:45:00

This work investigates the onset of wetting failure for displacement of Newtonian fluids in parallel channels. A hydrodynamic model is developed for planar and axisymmetric geometries where an advancing fluid displaces a receding fluid along a moving substrate. The model is evaluated with three distinct approaches: (i) the low-speed asymptotic theory of Cox [J. Fluid Mech. 168 (1986) 169], (ii) a one-dimensional (1D) lubrication approach, and (iii) a two-dimensional (2D) flow model solved with the Galerkin finite element method (FEM). Approaches (ii) and (iii) predict the onset of wetting failure at a critical capillary number Ca_{crit}, which coincides with a turning point in the steady-state solution family for a given set of system parameters. The 1D model fails to accurately describe interface shapes near the three-phase contact line when air is the receding fluid, producing large errors in estimates of Ca_{crit} for these systems. Analysis of the 2D flow solution reveals that strong pressure gradients are needed to pump the receding fluid away from the contact line. A mechanism is proposed in which wetting failure results when capillary forces can no longer support the pressure gradients necessary to steadily displace the receding fluid. The effects of viscosity ratio, interface confinement, gravitational forces, substrate wettability, and fluid inertia are then investigated through comparisons of Ca_{crit} values and characteristics of the interface shape. Surprisingly, the low-speed asymptotic theory (i) matches trends computed from (iii) throughout the entire investigated parameter space. Furthermore, predictions of Ca_{crit} from the 2D flow model compare favorably to values measured in experimental air-entrainment studies, supporting the proposed wetting-failure mechanism.

(208b) Free Energy Analysis of the Liquid-Liquid Transition in Supercooled ST2 Water

Palmer, J. C., Princeton University
Martelli, F.,
Car, R., Princeton University
Debenedetti, P. G., Princeton University
Originally presented on: 11/4/2013 15:55:00 - 16:13:00

Many of the well-known thermodynamic anomalies of water, such as its negative thermal expansion and increased compressibility upon cooling, become more pronounced when it is cooled below its freezing point into a metastable liquid. One thermodynamically consistent interpretation of water's anomalies posits that water becomes highly compressible when cooled below its freezing point due to the presence of a second critical point associated with a first-order phase transition between two metastable liquid phases[1]. Since the region of the phase diagram where this hypothetical critical point would occur is below the homogenous nucleation temperature of bulk water, obtaining direct experimental evidence to falsify the second critical point hypothesis has so far proved to be a significant challenge.

We examine the phase behavior of the ST2 water model[2] under deeply supercooled conditions, using state-of-the-art molecular simulation techniques (umbrella sampling and well-tempered metadynamics) to compute the reversible free energy surface parameterized by density and bond-orientational order. Our analysis demonstrates that ST2 water phase-separates at sufficiently low temperatures, forming two liquid polymorphs that are metastable with respect to cubic ice. Such findings are consistent with our earlier umbrella sampling[3] and grand canonical Monte Carlo[4] calculations. The nature of the liquid-liquid phase transition is also investigated by examining the finite-size scaling behavior of the interfacial free energy between the two liquids. We find that the interfacial free energy obeys the $N^{2/3}$ scaling law expected for a first-order transition, suggesting the existence of an associated second critical point in ST2 water. Finally, we examine the mechanism for ice nucleation from the liquid state and discuss order parameters for describing this transition.

[1]P. H. Poole, F. Sciortino, U. Essmann, and H. E. Stanley, *Nature*, 360, 324 (1992)

[2]F.H. Stillinger and A. Rahman, *J Chem Phys*, 60, 1545 (1974)

[3]Y. Liu, J. C. Palmer, A. Z. Panagiotopoulous and P. G. Debenedetti, *J Chem Phys*, 137, 214595 (2012)

[4]Y. Liu, A. Z. Panagiotopoulous and P. G. Debenedetti, *J Chem Phys*, 131, 104508 (2009).

(208c) Critical Properties Extracted From Virial Series Using Approximants Consistent With Universal Scaling-Laws

Barlow, N. S., University at Buffalo, The State University of New York
Schultz, A. J., University at Buffalo, The State University of New York
Kofke, D. A., University at Buffalo, The State University of New York
Weinstein, S. J., Rochester Institute of Technology
Originally presented on: 11/4/2013 16:13:00 - 16:31:00

We report on our work towards establishing the necessary and su cient conditions for an asymptotically consistent approximant used to predict critical fluid properties based solely on the virial equation of state (virial series) and knowledge of the established universal critical behavior at the critical point. Given enough coe cients, accurate estimates of the critical temperature Tc and pressure Pc of a fluid can be extracted from the virial series. However, using this approach, the critical density ρc is consistently predicted as being too low. We provide evidence that this is a result of the non-singular nature of the virial series. In an e ort to explore techniques that would allow one to capture a more accurate estimate of ρc, we draw from known critical scaling-laws and cast the critical density as a branch-point singularity, using an approximant method to analytically continue the series. As a result, critical isotherms are constructed that capture the non-classical (singular) behavior at the critical point, while still retaining the low density behavior of the virial series. The approach to the critical region (from above, T >Tc) is also explored, using approximants that are asymptotically consistent with universal scaling laws.

(208d) Thermodynamic Properties of Water Under Tension: Measuring the Pull of H2O

Santiago, M., Cornell University
Pagay, V.,
Sessoms, D. A., Cornell University
Stroock, A. D., Cornell University
Originally presented on: 11/4/2013 16:31:00 - 16:49:00

Given the right conditions, liquids can exist at negative pressure, a metastable state where attractive intermolecular forces prevent spontaneous cavitation. The fluid then is said to be under tension, analogous to a stretched rubber band. This tensile state occurs under many situations of interest, such as liquids in nanoporous media and soils, flow in plants under drought conditions, cavitation near propeller blades, and the suction in octopi suckers. Yet, measurements of thermodynamic properties of liquids at negative pressure are limited due to experimental difficulties in creating, maintaining, and measuring this stretched state. Particularly, the properties of arguably the most important liquid in science—water—under tension are as yet scarcely explored. Experimental measurements of these properties are vital to predict the behavior of water under tension, understand the origin of the many anomalies in water, and ascertain which theoretical models and simulations best describe this liquid in metastable conditions.

In this paper, we explore the behavior of water under tension, presenting measurements of its thermodynamic properties in this state, and comparing our measurements with existing theoretical models. For these measurements, we developed a novel microfabricated sensor to manipulate and study a macroscopic volume of metastable water. Our sensor has three main components: (1) a millimeter-sized water reservoir, (2) a nanoporous membrane connecting the reservoir to ambient vapor, and (3) a piezoresistive pressure sensor in mechanical contact with the water reservoir. With this system, we equilibrate the water reservoir with a sub-saturated vapor phase of known activity, and measure the resulting (negative) pressure of the water reservoir with the piezoresistive sensor. All measurements are performed isothermally. Because we have 3 state variables (temperature, pressure and activity) we are thus able to fully specify the thermodynamic state of water. We finish by discussing the implications of these measurements for theoretical models of water, and our plans for future studies using this technology.

(208e) A Molecular-Level Model for Predicting Properties of Electrolytes At High Pressure and Temperature

Djamali, E., Rice University
Venkataraman, P., Tulane University
Fouad, W. A., Rice University
Marshall, B. D. II, Rice University
Cox, K. R., Rice University
Chapman, W. G., Rice University
Tomson, M. B., Rice University
Originally presented on: 11/4/2013 16:49:00 - 17:07:00

Predicting scaling and corrosion at conditions of high temperature, pressure, and TDS is essential for safe and successful operation of ultra-deepwater oil and gas production processes. However, experimental data and thermodynamic models for predicting scaling, corrosion, and stress cracking at these extreme conditions are scarce. The ability of a theoretical model to describe the fluid-phase properties of ionic species such as Na^+, Ca^{2+}, Ba^{2+}, Fe^{2+}, Fe^{3+}, Cl^-, SO_4^{2-}, CO_3^{2-}, and S^{2-} at pressures and temperatures as high as 24,000 psi and 250°C is key to the development of an accurate predictive tool. The current statistical associating fluid theory (SAFT) based equation of state models accounts for ionic interactions by incorporating Debye-Hückel theory or mean spherical approximation (MSA). These models treat ions as completely dissociated species and account for ion hydration by fitting association parameters from SAFT to experimental data at ambient conditions. At conditions of interest, the above assumptions do not hold well and hence there is a need for a model capable of accounting for ion hydration and ion-pair association at elevated temperature and pressure. A review of the current SAFT based models for electrolytes for prediction of equilibrium properties at ultra-deepwater conditions is presented. The authors also propose development of a new equation of state to account for ion hydration and ion pair association through modification to Wertheim's thermodynamic perturbation theory.

(208f) Liquid-Liquid Equilibria of Binary Water/1-Butanol and Water/THF Mixtures Studied By Molecular Simulation

Bai, P., University of Minnesota
Keasler, S. J., University of Minnesota
Siepmann, J. I., University of Minnesota
Originally presented on: 11/4/2013 17:07:00 - 17:25:00

Aqueous mixtures can exhibit a large variety of phase behavior. The water/1-butanol mixture exhibits only an upper critical solution temperature (UCST) below the critical points of both components and can be categorized as a type II phase diagram following van Konynenburg and Scott [1]. In contrast, the water/tetrahydrofuran (THF) mixture shows a type VI phase diagram, characterized by an UCST and a lower critical solution temperature (LCST) and, hence, a closed-loop miscibility gap. Although thermodynamic descriptions of both types of mixtures are well known, the microscopic origins of such behavior remain unclear. In this work, Gibbs ensemble Monte Carlo simulations using the TraPPE force field and the TIP4P water model were performed to directly model the coexisting fluid phases at atmospheric and elevated pressures. The calculated liquid-liquid coexistence curves are in good agreement with experimental data. The unfavorable entropy of mixing is explored in terms of hydrogen-bond speciation, aggregate distribution, and number integrals.

[1] van Konynenburg, P. H. and Scott, R. L., *Phil. Trans.* **A298**, 495 (1980)

(208g) Maxima and Minima of Thermodynamic Properties in Supercritical Fluids

Yigzawe, T., Swinburne University of Technology
Sadus, R. J., Swinburne University of Technology
Originally presented on: 11/4/2013 17:25:00 - 17:43:00

The thermodynamic properties of supercritical fluids are of considerable interest for both practical applications such as supercritical fluid extraction and from a scientific perspective. It is of particular interset that some properties such as heat capacities exhibit a locus of maxima in the supercritical phase. In this work, molecular dynamics simulation results are reported for the energy, pressure, isothermal pressure coefficient, thermal expansion coefficient, isothermal and adiabatic compressibilities, isobaric and isochoric heat capacities, Joule-Thomson coefficient and speed of sound of fluids interacting via both the Lennard-Jones and Weeks-Chandler-Andersen potentials. These properties were obtained for a wide range of temperatures, pressures and densities. For each thermodynamic property an excess value is determined to distinguish between attraction and repulsion. It is found that the contributions of intermolecular interactions have varying effects depending on the thermodynamic property. The maxima exhibited by the isochoric and isobaric heat capacities, isothermal compressibilities and thermal expansion coefficient are attributed to interactions in the Lennard-Jones well. Repulsion is required to obtain physically realistic speeds of sound and both repulsion and attraction are necessary to observe a Joule-Thomson inversion curve. Significantly, both maxima and minima are observed for the isobaric and isochoric heat capacities of the supercritical Lennard-Jones fluid. It is postulated that the loci of these maxima and minima converge to a common point via the same power law relationship as the phase coexistence curve with an exponent of $\beta = 0.32$. This provides an alternative explanation for the terminal isobaric heat capacity maximum in supercritical fluids.

(216a) Force Based Dynamic Contact Angles and Wetting Kinetics On a Wilhelmy Plate

Al-Shareef, A., Missouri S&T

FORCE BASED DYNAMIC CONTACT ANGLES AND WETTING KINETICS ON A WILHELMY PLATE A Wilhelmy plate method has been used here to measure the dynamic advancing and receding contact angles using a force balance technique. The liquid phase here was silicone oils of different viscosities and the substrate was a cover glass coated with a perfluorinated polymer. A model for such a system for comparison against the data was based on the de Gennes' theory that the rate of viscous dissipation is equal to the rate of surface work. de Gennes' original model has been extended here to cover larger contact angles. The comparison with the experimental data shows good agreement for both advancing and receding contact angles. Data over an extensive range have been provided for the first time that such detailed agreement can be established. In addition, when the speed of the plate is increased, the receding contact angle decreases with increasing capillary numbers but does not show any receding contact angle less than 30°. This sudden stop conforms to de Gennes' result that there is a minimum receding contact angle below which entrainment takes place. Photographs were taken of the dynamic meniscus to illustrate the onset of entrainment. The experimental results are supportive of de Gennes-Huh-Scriven model, and disagree with many conjectures on where the receding contact lines are entrained, or whether the dynamic contact angle is due to the failure of the equilibrium, etc.

(216o) Environmental Stress Cracking of Glassy Polymers

Downs, T.,
Griscom, M., Missouri University of Science and Technology
Neogi, P., Missouri University of Science and Technology
Zahedi, G.,

When a glassy polymer is exposed to an organic solvent, it can crack spontaneously or fail more readily under extension. This phenomenon is called environmental stress cracking and a model is proposed here. When the solvent diffuses into the polymer, stresses are developed due to the swelling of the polymer. In addition, infinitesimal fluctuations give rise to stresses due to the variation in surface tension from changes in the surface concentration of the solute, which is known as the Marangoni effect. A linear stability analysis is performed to show that such a system where the above effects are included, is unstable to infinitesimal disturbances. The analysis is confined to very short times following the initial contact between the solid and the fluid phases. The main cause is seen to be the Marangoni effect. As a result, large tensile stresses arise locally leading to failures which will be observed in the interfacial region of the solid. In particular, the unstable disturbances of smaller wavelengths grow faster. This is in keeping with the observation that the microscopic cracks that emanate from the surface are more profuse in cracks that are more closely spaced.

(216b) Thermal Rectification and Thermal Logic Devices Based On Manipulation of Solid-Liquid Interfaces

Murad, S., University of Illinois at Chicago
Puri, I. K., Virginia Tech

Thermal rectification, the origin of which lies in modifying the thermal resistance in a nonlinear manner, could significantly improve the thermal management of a wide range of nano-devices (both electronic and thermoelectric), thereby improving their efficiencies. Since rectification requires a material to be inhomogeneous, it has been typically associated with solids. However, the structure of solids is relatively difficult to manipulate, which makes the tuning of thermal rectification devices challenging. Since liquids are more amenable to tuning, this could open up new applications for thermal rectification. We use molecular dynamics simulations to demonstrate thermal rectification using liquid water. This is accomplished by creating an inhomogeneous water phase, either by changing the morphology of the surface in contact with the liquid. We then propose a thermal logic device based on asymmetric solid-fluid resistances that couples two fluid reservoirs separated by solid-fluid interfaces. It is the thermal analog of a three terminal transistor, the hot reservoir being the emitter, the cold reservoir the output, and smaller input reservoirs as the base. Changing the input temperature alters the transport factor and the flux gain as does the base current in a transistor.

(216p) Surface Hydrophobicity and Liquid Crystal Pretilt Angle in Polymer-Stabilized Liquid Crystal Alignment Systems

Liu, B. Y., National Taiwan University
Chen, L. J., National Taiwan University

Photocurable monomers lauryl acrylate and biphenyl diacrylate were added into the liquid crystal (LC) E7 in certain concentration range. Under a fixed voltage applied, LC cells were treated with UV light in a constant dosage. LC pretilt angle is stabilized during polymerization and varies continuously over the whole range from 0.8° to 90° by simply increasing the monomer concentration. Polymer aggregations were found on the inner surface of LC cells in a rather uniform size around 250 nm in diameter. The size of polymer aggregations was independent of monomer concentration, yet the number of aggregations increased with monomer concentration and polymer aggregations tended to be in chain-like cluster form. Advancing contact angle of water was utilized as index of surface hydrophobicity. Advancing contact angle of inner cell surface increased with increasing monomer concentration due to the formation of polymer layer. LC pretilt angle showed a great dependence on advancing contact angle. That is, LC pretilt angle increased monotonically with increasing advancing contact angle (or surface hydrophobicity). In other words, surface hydrophobicity plays an essential role in pretilt angle control in polymer-stabilized liquid crystal alignment system.

(216q) Role of Metal Layer Thickness and Surface Chemistry in the Self-Assembly of Metallodielectric Janus Spheres

Shemi, O.,
Solomon, M. J., University of Michigan

We study the clustering of colloidal polystyrene spheres with a hemispherical coating of gold and quantify the effects of gold thickness and self-assembled monolayer chemistry on their self-assembly. A gold layer varying from 10nm to 40nm in thickness was deposited on 1-micron diameter carboxylate-stabilized polystyrene spheres. The metallic surface was functionalized with a self-assembled monolayer of varying chemistry. The kinetics of self-assembly was studied in aqueous suspensions as a function of the concentration of added sodium chloride. The ordered aggregates of these Janus microspheres, formed via gold-gold binding, were studied using confocal laser scanning microscopy. From acquired images we quantify the abundance of each observed cluster size and show how the cluster size distribution changes with increasing gold thickness at a fixed salt concentration. We also investigate cluster formation and equilibration of cluster growth as a function of salt concentration varied from 1mM – 5mM and as a function of self-assembled monolayer chemistry. We discuss the kinetics and equilibrium structures observed in light of the variation in metal layer thickness, self-assembled monolayer chemistry, and Janus particle stability.

(216r) Vesicle Generating An Oscillatory Shape Change Under a pH Gradient

Nawa, E.,
Yamamoto, D., Doshisha University
Shioi, A., Doshisha University

Living matter exhibits autonomous and oscillatory motion in response to the surrounding environment. Some biological functions are maintained by proteins operating under pH gradient that is formed by an active transport of proton. The pH is also an important quantity of chemical reactions in an aqueous medium. Thus, the design of autonomously moving molecular assembly with sustainable motions similar to biological systems is interesting for biomimetic applications of a molecular assembly. In this presentation, a vesicle generating an oscillatory shape change is reported. The vesicle is oleic–acid based and exhibits a sustainable oscillatory shape change under pH gradient that is formed by a diffusion of basic aqueous solution. The vesicle was apparently double spherical, and its size was almost constant during the structural change. This vesicle continuously generates mechanical work from a pH gradient. In this presentation, we reported experimental results of this vesicle dynamics in detail and proposed a mechanism of vesicle dynamics that converts pH gradient to work.

(216c) Stability and Flocculation Behavior of Silica Microparticles in Ionic Micellar Solutions

James, G. K., Virginia Tech
Walz, J. Y., Virginia Tech

Using the technique of colloid probe atomic force microscopy (CP-AFM), measurements of the equilibrium depletion force between a silica particle and a flat silica plate in solutions of tetradecyltrimethylammonium bromide (C_{14}TAB) and sodium dodecyl sulfate (SDS) at varying concentrations were performed. With increasing surfactant concentration, the onset of a depletion well and structural forces was observed implying the possibility of depletion induced flocculation or stabilization in dispersions of colloidal silica. The force versus separation curves obtained through CP-AFM were converted into energy versus separation curves and the stability ratio analysis described by Marmur (*J. Colloid Interface Sci.* **1979**, *72*, 41-48.) was used to predict the stability behavior of colloidal silica dispersed in ionic micellar solutions. Comparisons are drawn between the stability ratio analysis and photographic analysis and ultraviolet-visible (UV-VIS) light spectroscopy results for the stability of dispersions of 0.5 μm colloidal silica particles in solutions of C_{14}TAB and SDS at varying concentrations.

(216s) Lattice Boltzmann Simulation of the Meniscus Shape Evolution in Vertical-Deposition Convective Self-Assembly

Inada, K., Kyoto University
Watanabe, S., Kyoto University
Miyahara, M., Kyoto University

Patterned structures of colloidal particles exhibit physically and chemically peculiar characteristics for various applications, including plasmonic materials, photonic materials, and sensors. Establishment of patterning techniques is thus of increasing importance for engineering materials functions. Among the fabrication techniques, bottom-up self-assembly methods are promising because of their inexpensive and energy saving approach. In the present study, we focus on the vertical-deposition convective self-assembly technique. In this technique, particles are carried into the contact line by a solvent convective flow induced by the evaporation, and arranged into an ordered array by attractive capillary forces acting between particles. So far, we have succeeded in producing colloidal patterns, such as stripes, grids and cluster arrays, and investigated the formation mechanism of the patterned structures by directly observing the processes. Through the investigation, we have found out that the deformation of the meniscus shape plays a crucial role in the pattern formation process. However, detailed mechanism has not yet been clarified, which is firstly because the self-organization process intrinsically contains intricate nano- and microscale phenomena, such as evaporation, convective flow, heat transfer, meniscus shape evolution, and particles motion, and secondary because experimental approaches are limited in giving one a deep understanding of the complex phenomenon due to finite time and space resolutions. In this regard, numerical simulation is a possible alternative and quite suitable for revealing the key physics behind the assembly process. In the present study, we conduct a lattice Boltzmann simulation of the meniscus shape evolution in order to capture the complex phenomena involved in the vertical-deposition convective self-assembly process. We clarify the relationship between self-assembled structure and dynamics of the meniscus and particles under different conditions.

(216d) Single-Molecule Observations of Protein-Protein Interactions At the Solid-Liquid Interface

Langdon, B. B., University of Colorado at Boulder
Walder, R., National Institute of Standards and Technology and University of Colorado at Boulder
Kastantin, M. J., University of Colorado
Schwartz, D. K., University of Colorado Boulder

Protein adsorption at the solid-aqueous interface is a widely studied and very complex phenomenon. Single-molecule techniques uniquely allow us to separate protein surface dynamics (e.g. protein adsorption, desorption, diffusion) and identify and characterize multiple protein populations and dynamic modes (e.g. oligomeric states, multiple diffusive modes). Recent work using single-molecule total internal reflection fluorescence microscopy (TIRFM) at low protein concentrations, such that protein-protein interactions are insignificant, has shown protein desorption and diffusion to be highly heterogeneous. For example, multiple residence time populations were found to correspond to protein oligomeric state where 99.9% of protein monomers reside on the surface for <1s while larger oligomers (trimers and tetramers) reside for up to hundreds of seconds. These findings suggest that protein-protein interactions, not just protein-surface interactions, must greatly influence protein layer formations and protein surface dynamics observed at high concentrations where protein unfolding and aggregation is thought to occur. By combining TIRFM and intermolecular resonance energy transfer (RET), the dynamics of protein-protein associations were directly observed and quantified at the solid-liquid interface. Using this technique, dynamic associations of two or more bovine serum albumin (BSA) molecules were found to be heterogeneous and fast with the majority (80-90%) of protein-protein associations lasting for <1s. Based on a distribution of RET efficiency (related to the extent of protein association), protein molecules were identified as unassociated, partially associated, or completely associated. The RET efficiency, or extent of association, are interpreted as clusters (complete RET) and free monomers (no RET). For increasing bulk protein solution concentrations the fraction of time a molecule on average spent completely associated increased while the fraction of time spent unassociated decreased. Surprisingly, the distributions of times spent completely associated remained constant for all concentrations. These observations suggest a pre-layer formation model of heterogeneous protein surface coverage, consisting clusters and free monomers, where local surface dynamics (near clusters or in free area) are independent of surface coverage.

(216u) Electric Field Induced Self-Assembly of Colloidal Particles

Shah, A. A., University of Michigan
Solomon, M. J., University of Michigan
Ganesan, M., University of Michigan, Ann Arbor
Jocz, J.,
Ferrar, J. A., University of Michigan, Ann Arbor

Electric Field Induced Self-Assembly of Colloidal Particles

Aayush A Shah[1], Mahesh Ganesan[2], Jennifer Jocz[2], Joseph A. Ferrar[2] and Michael J. Solomon[1, 2]

[1] Macromolecular Science and Engineering, University of Michigan, Ann Arbor, MI

[2] Chemical Engineering, University of Michigan, Ann Arbor, Ann Arbor, MI

e-mail: aayushs@umich.edu

We report the use of an applied electric field to produce crystalline order in micron and sub-micron sized Brownian colloidal particles. Colloidal suspensions, because of their slow Brownian dynamics, undergo equilibrium self-assembly only on long time scales. Application of an electric field significantly accelerates the kinetics of assembly. We analyze our results in terms of the underlying electrokinetics of the system, and connect our results to equilibrium self-assembly theory. We subsequently assemble colloidal rod shaped particles using the electric field assembly method, and report the formation of a body centered tetragonal (BCT) structure. We use this newly developed method to reversibly assemble sub-micron sized colloidal particles into crystals that show reflective color. We use light scattering, light microscopy and well as spectrophotometry to analyze the reflective color generated using this method.

(216v) Effect of Molecular Weight of Polyelectrolyte On the Interaction Between Surfactant and Polyelectrolyte

Tseng, H. W., National Taiwan University
Chen, P. C., National Taiwan University
Yen, H. Y., National Taiwan University
Chen, L. J., National Taiwan University

The interactions between oppositely charged tetradecyltrimethylammonium bromide ($C_{14}TAB$) and anionic polymers poly(acrylic acid) (PAA) with the molecular weight ranging from 1800 to 450000 g/mol were investigated by using isothermal titration calorimetry (ITC) at 25 °C. The ITC was performed by titrating $C_{14}TAB$ aqueous solution into 0.05wt% PAA aqueous solution Thermo-histogram exhibits an exothermic peak due to the formation of PAA/$C_{14}TAB$ complex. The driving force is simply the electrostatic attraction between opposite charged PAA and $C_{14}TAB$. When the $C_{14}TAB$ concentration reaches the saturated concentration (C_s), PAA is saturated with bound $C_{14}TAB$ molecules. And when the $C_{14}TAB$ concentration further increased up to the second critical micelle concentration (C_m), the $C_{14}TAB$ free micelle started to form. The PAA molecular weights would also affect the result. Based on the titration of $C_{14}TAB$ into PAA with six different molecular weights ranging from 1800 to 450000, it was found that when the PAA molecular weight larger than 25000 g/mol, they show the tendency as the above-mentioned, but there is a different mechanism in low molecular weights PAA/$C_{14}TAB$ system (1800 and 5000 g/mol), $C_{14}TAB$ started to form free micelles before the saturated concentration. The interaction of $C_{14}TAB$ and PAA is independent of PAA molecular weight when the PAA molecular weight is larger than 250000 g/mol.

(216w) Biopolymer Particles for Stabilization of Pickering Foams

Lam, S., North Carolina State University
von Klitzing, R., Technische Universit t Berlin
Velev, O. D., North Carolina State University

At the beginning of the 20th century, Spencer Pickering and Walter Ramsden determined that particles with appropriate wettability can impart greater stability onto foams and emulsions than molecular surfactant stabilizers.[1,2]Our goal is to synthesize particles from biologically derived materials – lignin and modified cellulose - and to tune synthesis parameters for the generation of particles with improved foam stabilizing attributes. Materials such as cellulose and lignin are sustainable materials which can find application in foods, personal hygiene products, as well as porous materials.

The increased stability imparted onto aqueous foams by the biopolymer particles studied is largely due to their anisotropic shape, which results in jammed particle layers in the lamella of the foam. A hydrophobically modified cellulose, HP-55, is used as a model biopolymer system since it had previously been shown to stabilize foams very well and has also been used as the stabilizing matrix for the creation of magnetically responsive foams.[3,4]We elucidate the mechanism behind foam stabilization from anisotropic biopolymer particles by studying the morphology of the thin films formed at varying particle concentrations for systems containing HP-55. We are currently evaluating the effect of particle concentration as well the bulk environment on the morphology of HP-55 foam films. When comparing results from the thin film studies to the bulk foam system, we found that a critical concentration of particles is necessary for the formation of a thick particle shell around all the bubbles in the foam to create superstable foams. Another biopolymer of interest is lignin, which can form particles through the same water-based antisolvent method used for the generation of HP-55 particles. Lignin is the second most abundant biopolymer found in nature with many potential uses as a stabilizer, encapsulation agent, and as a component in composite materials. We used Kraft lignin to generate particles for the stabilization of biodegradable, environmentally-friendly foams. We characterized the effect of a water-based antisolvent precipitation on the surface charge and structure of KL particles. This process was found to increase the effective volume of the KL particles by ~30 times, suggesting that reforming enhances the ability of the particles to sterically stabilize foams and emulsions.

References: [1] Pickering, S. U., *J. Chem. Soc.* **1907**, 91, 2001-2021; [2] Ramsden, W., *Proc. R. Soc. London* **1903**, 72, 156-164; [3] Wege, H. A. et al., *Langmuir* **2008**, 24, 9245-9253; [4] Lam, S. et al., *J. Am. Chem. Soc.* **2011**, 133, 13856–13859.

(216e) Synergestic Effects of Polymers and Nanoparticles On Colloidal Stability

Ji, S., Virginia Tech
Walz, J. Y., Virginia Tech

The focus of this study was on understanding the synergistic effects of nanoparticles and polyelectrolytes on the depletion and structural forces between colloidal particles (order microns in size). The experiments consisted of measuring the force-vs.-distance profile between a large silica particle and a silica plate in aqueous solutions using colloidal probe atomic force microscopy (CP-AFM). The depletants used were silica nanoparticles (22 nm in diameter) and polyacrylic acid (Mw of 100k). With low concentrations of either only nanoparticles or only polymer in solution, little, if any, depletion forces were detected between the microparticle and plate. When both polymer and nanoparticles were mixed, however, significant depletion and structural forces were observed. Calculation of the potential energy profile that would result from these forces indicated that the forces would be more than sufficient to induce flocculation of an otherwise stable dispersion of micron-sized particles. Our hypothesis is that polymer chains adsorb to the surface of the nanoparticles, creating much larger, charged complexes that are more easily excluded from the gap region between the microparticle and plate, thus increasing the depletion force.

(216x) Assembly of Two-Directionally Percolated Particle Networks Using Orthogonal Electric and Magnetic Fields

Bharti, B., North Carolina State University
Zhu, S., North Carolina State University
Velev, O. D., North Carolina State University

External field assisted assembly of colloidal particles is the most widely used method to obtain highly ordered one- two and three dimensional structures. However, only one type of external trigger viz., electric or magnetic or light is typically implemented to organize the particles into ordered arrays. We present a novel approach towards the formation of highly cross-linked, low density, two dimensional orthogonally percolated networks of microparticles under the action of conjugated electric and magnetic fields. A mixture of magnetic and non-magnetic spherical latex particles when subjected to orthogonal electric and magnetic fields assemble in a range of two-dimensional interconnected structures. The properties of such multidirectionally-gelled fluids are dependent on the extent of percolation in a given direction which in turn can be tuned by modulating the directionally oriented electric or magnetic field intensity. In addition, it was established that the order of application of fields governs the equilibrium morphology of 2D structure. The role of the number density of the biparticle dispersion in controlling the interconnectivity of the assembled structure was investigated and correlated to the field intensity and duration. This method for obtaining low density gelation is rapid, robust and reversible and the cross-linked structures exist as long as the external combination of electric and magnetic fields is applied. This experimental technique for the biparticle assembly offers a potential way for fabricating smart gels and biresponsive functional materials that can find electronics and biomedical applications.

(216t) Difference in Adsorption Behaviors of SDS and Gemini-Surfactants Onto Clathrate Hydrates

Lee, J. W., The City College of New York
Salako, O., CCNY
Lo, C., The City College of New York
Somasundaran, P., Columbia University
Couzis, A., The City College of New York

Clathrate hydrates are non-stoichiometric whitish, ice-like crystalline compound formed when water molecules form cages that trap low molecular weight molecules. It has been estimated that the amount of natural gas trapped in discovered natural gas hydrate deposits is 1.5 x 10^{16} m^3 . This makes natural gas hydrates a potentially viable energy resource. Natural gas can be stored in hydrate form; up to 170 volumes of natural gas per volume of hydrates. Other potential usages of clathrate hydrates are for separation of flue gases and desalination of seawater. There are negative aspects to clathrate hydrates as they can form in oil and gas pipelines which can result into the disruption of oil production. Understanding the kinetics of clathrate hydrate formation is important for the utilization of clathrate hydrate technologies in gas storage and separation or for the prevention of hydrate plugs. Sodium dodecyl sulfate is one of the most effective hydrate formation promoters. This work proposes other alternative surfactants, Dow Gemini surfactants that have comparable performance to SDS. We report the difference adsorption behaviors of SDS and two Gemini surfactants of C6L and 2AI at Cyclopentane (CP) hydrate-water interface. The adsorption of these surfactants was quantified using adsorption isotherms and the adsorption isotherms were generated using liquid-liquid titration. Zeta potentials of CP hydrate particles in surfactant solution of various concentrations of surfactants were measured to further explain the different adsorption behaviors at the CP hydrate-water interface. We will also propose the adsorption mechanisms.

(216y) Asymmetric Recognition of L-/-Amino Acid By Liposome Membrane

Ishigami, T.,
Sugita, K.,
Umakoshi, H., Osaka University

Liposome membrane is a self-assembly of the phospholipid molecules prepared in water and can be regarded as a model biomembrane. Recent research shows the possibility of the chiral recognition function of the liposome membrane through the binding of the dipeptide on the liposome membrane. The purpose of this study is development of new optical resolution method by using liposome membrane. We first investigated the chiral recognition function of liposome membrane by observing the adsorption of amino acids that have different chirality. We also investigated the ability of optical resolution by using equilibrium dialysis method with hydrogel entrapping liposome membranes. The adsorption of L-Trp on the DPPC liposome membrane was increased with the incubation time and the value was reached to approximately 100% in more than 48 hours, while that of D-Trp was retained at almost zero. The adsorption selectivity of chirality in liposome membrane was further studied by investigating the adsorption of other amino acids. Similar tendency was observed in the case of other amino acids. In general, amino acids at D-form were not adsorbed on the DPPC liposome membrane, while those at L-form were adsorbed, depending on their type. Polar or charged amino acids (His, Asp, etc.) show the higher adsorption behavior. Based on the above findings, the optical resolution was performed through the equibrillium dialysis, equipped with the liposome immobilized hydrogel, by selecting the L-/D-Trp as target materials, resulting in the higher chiral selectivity.

(216f) Property Investigation of Pseudo Polyelectrolytes Self-Assembled From PVP/SDS By Viscometry

Wu, Y., Jiangnan University
Chen, J., Jiangnan University
Fang, Y., Jiangnan University

Relative viscosity of the system self-assembled by polyvinylpyrrolidone(PVP) and sodium dodecyl sulfate(SDS) was determined by Ubbelohde viscosity meter. Flow time of PVP aqueous solutions, SDS aqueous solutions and PVP/SDS complex solutions at different concentrations were seperately measured. The results showed that there existed two inflection points, critical aggregation concentration (CAC) and polymer saturation point (PSP), on the reduced viscosity - SDS concentration curve for the PVP / SDS complex system. It was also determined that CAC < CMC < PSP. The reduced viscosity of PVP / SDS solution increased dramatically with increasing SDS concentration which had reached CAC, and the higher the PVP / SDS concentration, the lower the viscosity of the system, presenting the same characteristics with polyelectrolytes. The potential of PVP / SDS system was negative determined by Zeta-potential. All of the results showed that the PVP / SDS complex system presented a supermolecular structure and was an anionic pseudo polyelectrolyte.

(216z) Evaporation of a Liquid Drop Deposited On a Chemically Heterogeneous Striped Surface

Lin, S. Y.,
Chuang, Y. C.,
Yang, Y. J., National Taiwan University
Chen, L. J., National Taiwan University

Chemically heterogeneous striped surfaces consisting of alternating and parallel hydrophobic/hydrophilic self-assembled monolayer were prepared on silicon substrate by micro-contact printing. The evaporation of a liquid droplet on the heterogeneous surface was studied by using an enhanced video microscopy system to record the evolution of contact angle along two different viewing angles (normal or parallel to the stripe). At the very beginning of the evaporation process in these two viewing angles, the contact angle decreases dramatically while the radius of three phase contact line remains almost constant. Then, the angle decreases slowly while the radius decreases. It is interesting to note that the angle in the direction parallel to the stripe exhibits the stick-slip-jump phenomenon and the radius decreases with a step jump while the contact angle and radius in the direction normal to stripe decrease disorderly. The preliminary model of evaporation of liquid drop on the chemically heterogeneous striped surface by the Surface Evolver has been simulated and compared to the experimental results.

(216aa) Antifouling Biointerfaces Based On Photodefinable Reactive Coatings

Su, C. T., National Taiwan University
Yuan, R. H.,
Chen, Y. C.,
Lin, T. J.,
Hsieh, C. C., National Taiwan University
Chen, H. Y., National Taiwan University

Poly(ethylene glycol) (PEG) or PEG-like molecules have long been reported as promising substances to suppress non-specific protein adsorption, or to resist cell adhesion. However, sufficient knowledge, synthesis skill, and cost are required to induce terminal groups on PEG. In this present study, we investigate the usefulness of exploiting a photodefinable polymer for the immobilization of a variety of PEG-like molecules that do not require functional groups. The photodefinable polymer, poly(4-benzoyl-*p*-xylylene-*co*-*p*-xylylene), can be prepared by chemical vapor deposition (CVD) polymerization on a wide range of different substrates/materials. The photoactivated carbonyl groups of the polymer has the potential to enable light-induced cross-linking of molecules and can rapidly react via insertion into CH- or NH-bonds upon photo-illumination at 340 nm. Molecules ranging from poly(ethylene glycol) (average Mn 400), poly(ethylene glycol) methyl ether methacrylate, dextran, to ethanolamine, are used as model anti-fouling materials in the study. Their antifouling properties towards fribrinogen and streptavidin adsorptions were studied using a combination of fluorescence microscopy and quartz crystal microbalancing (QCM). In addition, controlled anti-fouling presentation was also demonstrated on a non-conventional substrate, such as a stent, by using this coating technology. The photoimmobilization procedure based on the photodefinable CVD coating described in this study indeed introduce a simple and cost effective pathway for non-fouling modification. This technique may find applications in microfluidics, cell culture study, diagnostic devices, and implant devices.

(216ac) Flow Properties and Morphology of Multilamellar Structures in Surfactant Solutions

Pommella, A., University of Naples Federico II
Donnarumma, D., University of Naples Federico II
Caserta, S., University of Naples Federico II
Guido, S., University of Naples Federico II

Surfactant systems morphology is a topic of high scientific interest that has various industrial applications. However, while many studies have been addressed to the phase behaviour of water-surfactants solutions, only limited information is available in literature on flow-induced morphology of such systems. The recent advances in rheo-optical techniques help now to elucidate the changes in microstructural conformation taking place in the lamellar phase during flow.

The focus of this work is to develop a quantitative investigation of the flow behaviour of single surfactant multilamellar vesicles (SMLVs). The fluid investigated is a solution of HLAS in water, a surfactant widely used in the detergent industry. Depending on concentration, different phases can be observed, such as planar lamellae, unilamerllar and multilamellar vesicles. We focused on a range of concentration where lamellar phase inclusions are surrounded by an isotropic matrix, and eventually form SMLVs. Vesicle shape has been quantified by image analysis techniques. Different setups allow the direct observation of the sample, either during the flow or in static conditions, by several microscopy imaging techniques including high speed, bright field, phase contrast and confocal. A simple shear flow cell, where the fluid is placed among two parallel transparent glass plates one of which is translating respect to the other, allow the sample observation both along the vorticity and velocity gradient directions, using two different setups, hence allowing the three-dimensional reconstruction of the system microstructure. A different flow cell can be used to investigate the Poiseuille flow using a cylindrical microcapillary.

The main result is that SMLVs are deformed and oriented by the action of shear flow while keeping constant volume, and exhibit complex dynamic modes (i.e., tumbling, breathing and tank-treading). SMLV deformation shows some analogies with the behaviour of a single droplet immersed in an immiscible fluid, thus suggesting possible non-dimensional scaling parameters such as the Capillary number, that is the ratio between the shear and the interfacial stress. Furthermore, 3D shape reconstruction of confocal images of SMLVs, was used to identify the presence of both superficial and internal defect, whose role in the deformation dynamics is discussed.

A possible application of the physical insight provided by this work is in the rationale design of processing methods of surfactant-based systems.

S. Caserta, S. Guido, "Vorticity banding in biphasic polymer blends", Langmuir 28, 16254−16262 2012.

A. Pommella, S. Caserta, V. Guida, S. Guido, "Shear-induced deformation of surfactant multilamellar vesicles", Physical Review Letters 2012 108, 138301.

A. Pommella, S. Caserta, S. Guido, "Dynamic flow behaviour of surfactant multilamellar vesicles under simple shear flow", Soft Matter, submitted.

(216g) Numerical Simulation On Unsteady Dispersion Characteristics of Particles in a Flow Between Coaxial Cylinders

Suzuki, H., Kobe University
Masuda, K.,
Hidema, R., Kobe University
Komoda, Y., Kobe University

Polymer composites including fine particles in polymer have been paid attention to because various functions are expected to appear in such materials. The composites are made from suspensions, which disperse particles in molten polymer. However, the dispersion of fine particles in polymer is not so easy. The fine particle has high specific surface area and then the inter-particle energy due to the surface energy becomes relatively large. Thus, fine particles easily aggregate with each other and form larger clusters. The functions of polymer composites severely depend on the particle dispersion. Thus, it is of importance to control the dispersion characteristics of particles in molten polymer.

In order to disperse particles in polymer, a kneading extruder with rotors is often used. In such an extruder, high shear stresses are added to and break up the clusters with rotors. However, stresses added vary in space and in time in the extruder. Then, the dispersion of particles in polymer composites occurs non-uniformly in space and in time. Due to this complexity, it is very difficult to know the dispersion characteristics in such an extruder with experimental studies.

In order to calculate the dispersion characteristics in suspensions, Usui (1999) suggested a thixotropy model. In the model, the average cluster size can simulate at a uniform shear rate by a balance of Brownian and shear coagulations and shear breakup and the viscosity characteristics can be also calculated with the model and a suitable viscosity model as Simha's cell model. In the previous studies (Hasegawa et al.,2009, Masuda et al., 2013), a new thixotropy model was developed for estimating the cluster size in an unsteady non-uniform flow. With this model, the time-variation cluster size distributions can be locally obtained by a local balance of Brownian and shear coagulations and shear break-up of clusters. In this study, this model has been applied to simulate the unsteady dispersion characteristics of fine particles in a flow between coaxial cylinders, when the inner cylinder rotates as a first step in order to investigate the dispersion characteristics in an extruder with rotors. The unsteady start-up flow when the inner cylinder moves in one way with a constant angular velocity was focused on before the flow was fully developed as a steady Couette flow. Under such a situation, fine particle clusters break up non-uniformly in space and in time. From the results, the mechanism of particle dispersion under an unsteady flow can be basically clarified.

In this study, the effects of inter-particle bonding energy on dispersion characteristics have been investigated with the above situation. The inter-particle bonding energy was changed in three steps; $5x10^{-15}$, $5x10^{-16}$ and $1x10^{-16}$ J. The computational domain was set in the region between coaxial cylinders, where the inner cylinder starts to rotate at the beginning of the computation. The inner and outer cylinder diameters and the rotating angular velocity of the inner cylinder were set at 22.8 mm, 38 mm and ???. Between coaxial cylinders, a Newtonian molten polymer with the viscosity of 195 Pas was filled and the fine particle with the diameter of 2.5 mm was included in the molten polymer with the solid volume fraction of 0.2. As the initial condition, the cluster size was assumed to distribute uniformly at any position in the domain from 2 to 50 of particle number in a cluster. Thus, the initial average particle number in a cluster was 25.5.

The governing equations are two-dimensional continuity and momentum equations for the simulation on the fluid flow. For the dispersion characteristics, the thixotropy model determining the time-variation of cluster size distribution affected by the flow shear rate was used. The suspension viscosity was calculated by use of Simha's model from the apparent solid fraction estimated with particle cluster size distributions. This was used for the fluid viscosity in the momentum equations for fluid flow. From the fluid flow, the local shear rate was obtained and leaded to local breakup of the clusters. Thus, two-way coupling method was used for the flow and for the cluster distributions.

From the start-up of the inner cylinder rotation, the velocity and momentum fields were developed from the region near the inner cylinder to the region near the outer cylinder. It was found from the results that the flow developing characteristics between coaxial cylinders was not severely affected by the inter-particle bonding energy under the present conditions. On the other hand, the inter-particle bonding energy affects the cluster breakup behaviors. The low inter-particle bonding energy accelerates breakup of clusters and high value decelerates the breakup rate. The cluster breakup should obey the characteristics time determined by the inter-particle bonding energy divided by the viscosity, the square value of the shear rate and the cube value of the particle diameter. It was found that the breakup behavior of clusters at each radius position can be rather well expressed with this characteristics time, though the difference due to the inter-particle bonding energy was not small in the outer region.

On the other hand, the breakup rate decreases in the radius direction even on the time normalized by the characteristic time. The shear rate is very high near the inner cylinder at the start-up while the shear rate is zero near the outer cylinder. Thus, the time-dependency of the breakup was also affected by the shear history. In order to take the shear history at each position into consideration, a normalized time was defined with the time multiplied by the integrated shear strain and divided by the characteristic time. This is found to express rather well the time dependency of the cluster breakup at any flow positions between coaxial cylinders.

However, the breakup of clusters was faster near the inner cylinder even on that normalized time. From the detailed investigation, the solid volume fraction was found to decrease near the inner cylinder. This occurs due to the fluid movement for the continuity at the start-up. The particle moves with the fluid in the radius direction at the start-up and this causes the lower shear coagulation rate near the inner wall due to the low solid volume fraction. Thus, it was concluded that the breakup of clusters quickly occurs in the region near the inner cylinder due to the fluid movement in the radius direction under the startup situation.

References

Hasegawa, E., H. Suzuki, K. Kameyama, Y. Komoda and H. Usui, 2009, "Model analysis on dispersion characteristics of fine particles in Newtonian molten polymer," Advanced Powder Technology, 20-2, pp.139-144.

Masuda, K., H. Suzuki , Y. Komoda and R. Hidema, 2013, Particle dispersion/aggregation model in a non-uniform shear flow," Nihon Reoroji Gakkaishi, 41-2, pp.69-75.

Usui, H., 1999, Rheology model of aggregable slurry on monodisperse silica particles," Kagaku Kogaku Ronbunshu, 25-3, pp.459-465.

(216ad) Quartz Crystal Microbalancing Based On Vapor-Deposited Reactive Coatings

Chen, Y. C., National Taiwan university
Lin, T. J., National Taiwan University
Chen, H. Y., National Taiwan University
Tsai, W. B.,

Quartz Crystal Microbalancing Based on Vapor-Deposited Reactive Coatings

Yung-Chih Chen[a], Ting-Ju Lin[a], Wei-Bor Tsai[a], Hsien-Yeh Chen[a]

Department of Chemical Engineering, National Taiwan University, Taipei 10617, Taiwan

E-mail: hsychen@ntu.edu.tw

Abstract

Quartz crystal microbalancing (QCM) is a simple, cost-effective, and high-resolution mass sensing technique that the analysis is based upon the piezoelectric effect for measure mass densities down to a level of below 1 $\mu g/cm^2$. QCM evolved a solution measurement capability in largely analytical chemistry and electrochemistry applications due to its sensitive solution-surface interface measurement capability, and has widely be used for investigating the behavior of biomolecule adsorption or desorption in many biological applications. In order to perform sophisticated biological sensing , controlled surface chemistry of introducing sensing molecules on surfaces of quartz crystals usually requires substantial knowledge to conduct surface modification on such crystals, and an optimal modification approach has to be identified in a case-by-case fashion with selected methods. Functionalized poly(*p*-xylylene) coatings are used to provide versatile anchoring sites for the immobilization of sensing molecules on quartz crystals, as they decouple surface design from bulk properties and require minimum knowledge on surface chemistry of the underlying materials.

A multicomponent poly(*p*-xylylene) coating that contains two distinctly-addressable reactive moieties of electron-deficient alkynes and unsaturated maleimides is chosen as the model system to modify on crystal surface in the current study. The platform has provided i) maleimide groups for immobilization of thiol-PEG via thiol-maleimide coupling to render a surface that eliminates undesired perturbance from fouling substances, and ii) alkyne sites for immobilization of azide-biotin via azide-alkyne click reaction to activate biological functions. The modified quartz crystal is allowed for the quantitative measurement of subsequently attached streptavidin and biotinylated antibody. The herein reported coating technology may be broadly applicable to a wide range of different functional groups of poly-*p*-xylylenes and are extendable to attach other sensing molecules of detection. We envision the technique can provide a solution to the development of a robust QCM sensing tool for diagnostic applications.

(216ab) Effect of Contact Angle On the Ink Transfer Rate in the Gravure Offset Printing

Huang, Y. L.,
Liu, B. Y., National Taiwan University
Wang, Y. M.,
Chen, L. J., National Taiwan University

The gravure offset printing technique has been attracted tremendous attention due to its high throughput and cost-effectiveness for printing process of electronic circuits. A plate-to-plate offset printing equipment was assembled to simulate the gravure offset printing. In addition, the Surface Evolver was used to verify the experiment results. The process of the ink transfer between two plates was deliberately examined by monitoring the dynamic contact angles. For simplicity, water and glycerol were used to emulate the ink. It is found that the correlation between ink transfer rate and the receding contact angle is significant. The transfer rate is large when the liquid droplet is transferred from a substrate (placed on the lower plate) with a higher receding contact angle to a substrate (placed on the upper plate) of a lower receding contact angle. On the other hand, the liquid droplet is hardly transferred from a substrate (placed on the lower plate) with a lower receding contact angle to a substrate (placed on the upper plate) of a higher receding contact angle. The simulation results of the Surface Evolver also confirm this observation qualitatively.

(216ae) Synthesis of Amphiphilic Copolymers and Their Effects On Rheology of Oil-in-Water Emulsion

Huang, J.,
Xu, J., East China University of Science and Technology
Li, L., East China University of Science and Technology
Guo, X., East China University of Science and Technology

In order to acquire stable oil-in-water (O/W) emulsion, two kinds of amphiphilic copolymers were synthesized by grafting poly(styrene-*co*-maleic anhydride) with 1-naphthylamine-6-sulfonic acid alone (SMANS) or together with methoxypolyethlene glycol (SMANP), and their effects on the emulsification and rheological behaviors of Shengli heavy oil emulsion were investigated. It was found that without emulsifier, it tend to form water-in-oil (W/O) emulsion, rather than O/W emulsion, with even much higher viscosity than that of the crude oil. After addition of amphiphilic copolymers, W/O emulsion converted into O/W emulsion, which remarkably reduced the viscosity and yield stress of the formed emulsion. With SMANP at the concentration of 0.2 %, stable O/W emulsion with the oil volume ratio at 70% was achieved. The electrostatic repulsion and steric hindrance generated by the copolymers are believed to play an important role on the emulsification of oil. With the presence of sulfonate groups, both kinds of copolymers can provide electrostatic repulsion among oil droplets after being adsorped on the surface of oil droplets. While for SMANP, the polyethlene glycol (PEG) pendants offer additional steric hindrance, which further stabilizes the O/W emulsion. Based on those results, the formation mechanism of O/W emulsion by amphiphilic copolymer was proposed.

(216af) Synthesis of Mussel Inspired Biocompatible Polyelectrolyte Hydrogel Adhesives

Wang, W.,
Wang, M., East China University of Science and Technology
Li, L., East China University of Science and Technology
Guo, X., East China University of Science and Technology

Mussel adhesive proteins (MAPs) are the natural products from marine organisms which are friendly to the environment with high curing rate, corrosion resistance, biocompatibility, and outstanding mechanical property. To mimic MAPs, DOPA- the major functional unit of MAP is selected and grafted onto poly(acrylic acid) (PAA) backbone to form polyelectrolyte hydrogel adhesives. The structure of the obtained polymer was confirmed by IR and ^{1}H NMR. The grafting ratio is calculated from elemental analysis. Concentration and reactivity of dopamine side chains are detected by UV-vis spectroscopy. The 180° peel strengths in the air and under water were determined. It was found that the adhesion under water is higher than that in the air for the obtained polyelectrolyte hydrogels which implied the successful synthesis of mussel inspired biocompatible polyelectrolyte hydrogel adhesives. The correlation between the flowing ability and adhesion was discussed after the rheological measurements. The obtained polyelectrolyte hydrogels should be ideal candidates for the clinical adhesives and biomaterials.

(216ag) Adhesion of Poly(acrylic acid) Hydrogel

Shen, M., East China University of Science and Technology
Heckner, R.,
Wang, M., East China University of Science and Technology
Li, L., East China University of Science and Technology
Guo, X., East China University of Science and Technology

A series of poly(acrylic acid) (PAA) hydrogels were synthesized by in-situ free radical polymerization of acrylic acid (AA) in the presence of cross-linker N, N ′-methylenebisacrylamide (BIS) in aqueous media. The rheological test was performed to investigate the effects of cross-linker BIS content on the modulus and viscosity of the resulting hydrogels, where a critical gel point was found during the transition from liquid to solid. As the cross-linker concentration increased, the effective elastic modulus were steady-state during a large range of strain and increased dramatically at low cross-linking densities. The influence of pH value and crosslinking density on the swelling ratio was studied. 180°peel strength test was used to study the adhesion of the synthesized hydrogels both in air and water. Furthermore, JKR indention tests were also performed to investigate effects of the cross-linking density on the adhesion and mechanical properties in terms of adhesive pull-off and Young modulus. PAA based hydrogels should be ideal candidate for bioadhesives.

(216h) Predicting the Nature of the Protein Corona Surrounding Nanoparticles

Ridge, L., North Carolina State University
Hall, C. K., North Carolina State University

Our knowledge of the risks of exposure to nanoparticles is essentially in its infancy since the current *in vivo* and *in vitro* studies of "nanotoxicology" span only a minute subset of the many different types of nanoparticles being made today. *In vivo* and *in vitro* risk assessment studies could be bolstered, however, by the addition of computational (*in silico*) tools to help pinpoint those nanoparticles that are most likely to pose a risk to the public health. A key step in analyzing the risks posed by nanoparticles is to determine the composition of the corona of plasma proteins adsorbed onto the surface of nanoparticles. The nature of the protein corona is expected to have a marked influence on the fate of nanoparticles in the body.

Our goal is to develop a computational toy model for the formation of a protein corona around a hydrophobic nanoparticle. Our simulations allow for illustrating and quantifying the protein corona phenomenon at a resolution that is currently difficult to access experimentally, and also provide the capability to examine molecular-level structural changes that may occur after protein adsorption. Preliminary discontinuous molecular dynamics simulations with the coarse-grained force field PRIME20 were used to calculate the adsorption of model small peptides (polyalanine, A_{16}) and model large peptides (polytryptophan, W_{16}) onto a hydrophobic nanoparticle with a radius of 4 nanometers. Systems containing 48 chains of A_{16} and one hydrophobic nanoparticle were simulated to investigate the degree of adsorption onto the nanoparticle as a function of both time and temperature. We also examined the effect that adsorption had on the folding behavior of A_{16}, namely whether the native folded α-helical structure was disrupted upon adsorption. Competitive adsorption in mixed systems of A_{16} and W_{16} where A_{16} was in excess of W_{16} has also been simulated and compared with the theoretical adsorption models proposed by Fang and Szleifer.

(216i) Stabilization of Weakly Charged Microparticles Using Highly Charged Nanoparticles

Herman, D., Virginia Tech
Walz, J. Y., Virginia Tech

An experimental study was performed to understand the ability of highly-charged nanoparticles to stabilize weakly-charged microsphere dispersions. The experiments involved adding either anionic (sulfate) or cationic (amidine) latex nanoparticles to dispersions of micron-sized silica spheres near the silica IEP. Although both types of nanoparticle increased the zeta potential of the silica microspheres above the value at which dispersions containing only silica spheres remained stable, only with the amidine nanoparticles was stability obtained. Adsorption tests on flat silica slides indicated that the amidine nanoparticles deposited more densely than sulfate nanoparticles, producing multi-layer surface coverage, roughly three times that of the sulfate nanoparticles. Calculated DLVO energy profiles between nanoparticle-coated microspheres predicted stability for both systems. It is hypothesized that the relatively low coverage of the sulfate nanoparticles ($\leq$25%) led to bare silica patches on the microspheres that could align during interaction due to Brownian motion. These results indicate that highly-charged nanoparticles can be effective stabilizers, provided sufficient levels of adsorption, and that zeta potential alone is insufficient for predicting coated microsphere stability.

(216k) Diffusion of n-Alkanols With in Self-Assembled Monolayer On Mica Surface: A Molecular Dynamics Study

Bhandary, D., Indian Institute of Technology Kanpur
Khan, S., Indian Institute of Technology Kanpur
Singh, J. K., Indian Institute of Technology, Kanpur

Hydrophobicity or hydrophilicity of surface greatly induces the orientation of amphiphilic molecules and give rise to different types of pattern on the surface which is also called as self-assembled layer. In this study, self-assembled mono layer of n-alkanols on mica surface has been investigated with the help of molecular dynamics simulations. In particular, we have investigated the effect of chain length of n-alkanols on the self-assembly and layering of n-alkanols on the surface as well as the surface diffusion of the n-alkanols with in the monolayer. This work provides some insight and clarifies the assumption of tilted n-alkanols monolayer on mica surface irrespective of the chain length and the ordering of the n-alkanols with in the layer. Under varying density of n-alkanols with in monolayer, the orientation profile of the molecules change, at the same time, transition from solid-like structure to liquid-like structure is also observed from the radial distribution function. This transition also ensures strong attractive force between surface and alkanols and hence strong hydrogen bonding. The isotherm typically follows Langmuir adsorption isotherm. Study of the nature of the surface diffusion of alkanols is found to be most interesting. We have tried to correlate the surface diffusion of alkanols with hydrogen bonding, both intermolecular as well as with surface.

(216aj) Protein Adsorption At An Environmentally Responsive Surface: Towards Improved Adsorptive Membrane Materials

Wood, V., South Dakota School of Mines and Technology
Menkhaus, T., South Dakota School of Mines and Technology

We examined molecular level protein adsorption and desorption behavior on poly(acrylic) acid brushes (PAA), an environmentally responsive surface. PAA brushes are of particular interest because they have been proposed as a promising adsorptive material and can be used as an ion exchange media, controlled in large part by environmental conditions such as pH and conductivity. This study aimed to characterize and then optimize adsorption kinetics for, egg white chicken lysozyme (PI~ 11) and bovine serum albumin (BSA) (PI~4.8), to PAA brushes.

A new approach using Si-Atom Transfer Radical Polymerization (Si-ATRP) was successfully used for the controlled polymerization of sodium acrylate that with additional sample modification resulted in homogenous PAA brushes. The effect of varying polymerization conditions, such as molar ratios of polymerization components and initiator concentrations, along with rinsing techniques, were examined. The surface chemistry of these PAA brushes on modified gold-coated substrates was verified and characterized by FTIR-hATR, ellipsometry, SEM images, AFM, and water contact angle measurements.

The PAA brushes were 'grafted from' a gold substrate/chip and used with Surface Plasmon Resonance (SPR) spectroscopy to examine the fundamental interactions and adsorption/desorption kinetics of model proteins to the PAA surface. SPR spectroscopy enabled the detection of molecular level transport phenomena, fundamental adsorption behavior, and intrinsic physicochemical properties at the solid/fluid interface for various microfluidic system conditions. SPR sensors are frequently used to examine molecular level binding events and are advantageous over many other detection techniques due to their high resolution and real-time, label-free detection abilities. Seven process variables (including temperature, protein concentration, and buffer conditions), were examined and correlated to performance metrics used for determining optimal adsorption properties.

(216j) Relating Single-Molecule Measurements of Adsorption and Diffusion: A Path Towards Enhanced Surface Capture

Mabry, J., University of Colorado Boulder
Schwartz, D. K., University of Colorado Boulder

Many processes, including heterogeneous catalysis and receptor-ligand binding, require adsorption of a molecule from bulk liquid to a specific site on a solid surface. Previous work in our group has demonstrated that many different types of molecules (including proteins and polymers) adsorb from bulk liquid to a solid surface and then diffuse via bulk-mediated transport. Using total internal reflection fluorescence microscopy, we have tracked single molecules adsorbing and diffusing at a solid-liquid interface. By modeling the surface diffusion as bulk-mediated transport, we can use our adsorption measurements to predict the nature of the surface diffusion and subsequently validate these predictions with independent measurements of surface diffusion. Understanding the interplay between adsorption and diffusion provides a new framework for measuring the affinity of a molecule to a surface and will enable optimization of technologies based on surface capture.

(216ah) Phospholipid:Cholesterol Bilayer Permeability Is Predictable

Nitsche, J. M., University at Buffalo, The State University at New York
Kasting, G. B., University of Cincinnati Academic Health Center

One of the outstanding problems related to biomembranes is the reliable prediction of the permeability of lipid bilayers.[1] This property is a strong indicator of the ability of molecules to penetrate the cells of any tissue of the body. As such, its prediction as a function of the molecular structure and properties of the penetrant (molecular weight MW, octanol/water partition coefficient $K^{oct/w}$, etc.) is directly relevant to the mechanistic understanding of molecular transport and bioavailability[2] in such crucial applications as absorption–distribution–metabolism–excretion (ADME) screening of drug candidates,[3,4] and permeability profiling in particular,[5] as well as risk assessment of chemical exposures through the skin[6-8] and other pathways.[8] A focus on phospholipids is logical because viable cell membranes are bilayers of such polar lipids (with intercalated cholesterol, membrane proteins and other constituents).[9-12] Although active transport should not be discounted, a focus on passive diffusion is also appropriate "[i]n light of recent conclusions that 80–95% of commercial drugs are absorbed primarily by passive diffusion."[5]

Despite a wealth of permeability data covering numerous phospholipids, wide variations in temperature and cholesterol content, and many permeant species, as well as considerable theoretical understanding, no correlations exist that effectively unify the entire collective database, and permit the direct prediction of passive permeability coefficients P^{lip} from readily available permeant properties such as MW and $K^{oct/w}$. Part of the difficulty lies in the fact that values of P^{lip} vary over some ten orders of magnitude in a manner that bears no obvious relation to permeant size or structure. Adding to the puzzle is an apparent complete disconnect between two types of data sets, namely:

(1) those quantifying variations of P^{lip} with varying permeant size for a bilayer under given conditions[13,14] (often described using a power law in permeant volume); and

(2) those quantifying variations of P^{lip} with varying temperature and cholesterol content of the bilayer for a given permeant[15-17] (often described using an exponential function involving permeant projected area).

This poster develops a convenient set of explicit formulas that unify the picture, and describe roughly 100 very diverse data points on fluid-phase bilayer permeability to within an rms error for $\log_{10} P^{lip}$ of only ~0.2. It is based on a judicious assessment of data on various underlying bilayer properties,[18-21] in addition to permeability per se,[13-17] informed by theoretical arguments. Permeability data for liquid crystalline phospholipid bilayers composed of egg lecithin, DLPC, DMPC, DPPC and DSPC are analyzed in terms of a mathematical model that accounts for free surface area and chain-ordering effects in the bilayer as well as size and lipophilicity of the permeating species. Free surface area and chain ordering are largely determined by temperature and cholesterol content of the membrane, molecular size is represented by molecular weight, and lipophilicity of the barrier region is represented by the 1,9-decadiene/water partition coefficient, following earlier work by Xiang, Anderson and coworkers.[13,14,16,22] A correlating variable $\chi = MW^n\sigma/(1-\sigma)$ is used to link the results from different membrane systems, where different values of n are tried, and σ denotes a reduced phospholipid density. The group $(1-\sigma)/\sigma$ is a measure of free surface area, but can also be interpreted in terms of free volume. A single exponential function of χ is developed that is able to correlate all the observations. The best fit found for $n \sim 0.9$ ultimately makes χ much closer to the ratio of molecular to free volumes than surface areas.

A significant advance over previous work[23] is a precise quantification of the effects of bilayer thickness as determined by fatty acid chain length, temperature, and cholesterol content of the bilayer. The formulas developed serve as a strong starting point for estimating passive permeability of cell membranes to non-ionized solutes as a function of temperature and cholesterol content of the membrane.

1. Xiang T-X 1999. Translational diffusion in lipid bilayers: Dynamic free-volume theory and molecular dynamics simulation. *J Phys Chem B* **103**(2):385-394.

2. Semple KT, Doick KJ, Jones KC, Burauel P, Craven A, Harms H 2004. Defining bioavailability and bioaccessibility of contaminated soil and sediment is complicated. *Environ Sci Technol* **38**(12):228A-231A.

3. Balani SK, Miwa GT, Gan LS, Wu JT, Lee FW 2005. Strategy of utilizing in vitro and in vivo ADME tools for lead optimization and drug candidate selection. *Curr Top Med Chem* **5**(11):1033-1038.

4. Mandagere AK, Thompson TN, Hwang KK 2002. Graphical model for estimating oral bioavailability of drugs in humans and other species from their Caco-2 permeability and in vitro liver enzyme metabolic stability rates. *J Med Chem* **45**(2):304-311.

5. Kerns EH, Di L, Petusky S, Farris M, Ley R, Jupp P 2004. Combined application of parallel artificial membrane permeability assay and Caco-2 permeability assays in drug discovery. *J Pharm Sci* **93**(6):1440-1453.

6. Poet TS, McDougal JN 2002. Skin absorption and human risk assessment. *Chemico-Biol Interact* **140**(1):19-34.

7. Fitzpatrick D, Corish J, Hayes B 2004. Modelling skin permeability in risk assessment - the future. *Chemosphere* **55**(10):1309-1314.

8. CDC. 2009. Fourth National Report on Human Exposure to Environmental Chemicals. Centers for Disease Control and Prevention, Atlanta, GA. URL: http://www.cdc.gov/exposurereport.

9. Singer SJ, Nicolson GL 1972. The fluid mosaic model of the structure of cell membranes. *Science* **175**(4023):720-731.

10. Xiang T-X, Anderson DB 2000. Influence of a transmembrane protein on the permeability of small molecules across lipid membranes. *J Membr Biol* **173**(3):187-201.

11. Xiang T-X, Chen J, Anderson BD 2000. A quantitative model for the dependence of solute permeability on peptide and cholesterol content in biomembranes. *J Membr Biol* **177**(2):137-148.

12. McIntosh TJ, Simon SA 2007. Bilayers as protein solvents: Role of bilayer structure and elastic properties. *J Gen Physiol* **130**(2):225-227.

13. Xiang T-X, Anderson BD 1994. The relationship between permeant size and permeability in lipid bilayer membranes. *J Membr Biol* **140**(2):111-122.

14. Cao Y, Xiang TX, Anderson BD 2008. Development of structure–lipid bilayer permeability relationships for peptide-like small organic molecules. *Mol Pharm* **5**(3):371-388.

15. Xiang T-X, Anderson BD 1995. Phospholipid surface density determines the partitioning and permeability of acetic acid in DMPC:cholesterol bilayers. *J Membr Biol* **148**(2):157-167.

16. Xiang T-X, Anderson BD 1997. Permeability of acetic acid across gel and liquid-crystalline lipid bilayers conforms to free-surface-area theory. *Biophys J* **72**(1):223-237.

17. Xiang T-X, Anderson BD 1998. Influence of chain ordering on the selectivity of dipalmitoylphosphatidylcholine bilayer membranes for permeant size and shape. *Biophys J* **75**(6):2658-2671.

18. Greenwood AI, Tristram-Nagle S, Nagle JF 2006. Partial molecular volumes of lipids and cholesterol. *Chem Phys Lipids* **143**(1-2):1-10.

19. Kucerka N, Pencer J, Nieh MP, Katsaras J 2007. Influence of cholesterol on the bilayer properties of monounsaturated phosphatidylcholine unilamellar vesicles. *Eur Phys J E* **23**(3):247-254.

20. Pan J, Tristram-Nagle S, Nagle JF 2009. Effect of cholesterol on structural and mechanical properties of membranes depends on lipid chain saturation. *Phys Rev E* **80**(2):12.

21. Kucerka N, Nieh M-P, Katsaras J 2011. Fluid phase lipid areas and bilayer thicknesses of commonly used phosphatidylcholines as a function of temperature. *Biochim Biophys Acta* **1808**(11):2761-2771.

22. Mayer PT, Anderson BD 2002. Transport across 1,9-decadiene precisely mimics the chemical selectivity of the barrier domain in egg lecithin bilayers. *J Pharm Sci* **91**(3):640-646.

23. Nitsche JM, Kasting GB 2013. Permeability of fluid-phase phospholipid bilayers: assessment and useful correlations for permeability screening and other applications. *J Pharm Sci* http://dxdoiorg/101002/jps23471.

(216ai) Characterization of Microbial Growth On Processing Equipment By Electrochemical Impedance Spectroscopy

Holser, R. A., USDA-ARS-RRC

Microbial activity that leads to the formation of biofilms on process equipment can accelerate corrosion, reduce heat transfer rates, and generally decrease process efficiencies. Additional concerns arise in the food and pharma industries where product quality and safety are a high priority. Pharmaceutical production with sterilized water lowers the risk of microbial growth and contamination. Within the food industry the cost would be prohibitive. In poultry processing, for example, bird carcasses are directly contacted by process water which becomes contaminated and remains on the wetted surfaces of the equipment. Immersion chilling is practiced in the United States to rapidly cool bird carcasses to inhibit bacterial growth. Biofilm formation was investigated in the presence of the disinfectant trisodium phosphate and hard water ions. Microbial growth was followed by optical density (OD) measurements and biofilm formation was characterized with electrochemical impedance spectroscopy (EIS).

(216am) Evidence of a Hydrophobic Region in Pluronic Brush Layers Formed On Silica Surfaces

Snider, J.,
Ryder, M., Oregon State University
McGuire, J., Oregon State University
Schilke, K., Oregon State University

The action of pendant, polyethylene oxide (PEO) brush layers as nonfouling coatings for biomedical materials is well understood. However progress toward clinical application of stable, high density PEO coatings has historically been impeded by the lack of cost effective, non-invasive methods for their preparation. Triblock polymers featuring a hydrophobic, polypropylene oxide (PPO) or polybutadiene (PB) centerblock flanked by two PEO side chains can be used to coat the surfaces of hydrophobic biomedical materials, and in this way render them less prone to protein adsorption and bacterial adhesion. Optimal coating requires awarenesss of triblock aggregation properties, and for this purpose pyrene fluorescence quenching was used to determine the critical aggregation concentration of the triblocks used in this work. Coatings produced by radiolytic grafting of PEO-PB-PEO triblocks to polyacrylonitrile membranes used in hemodialysis were evaluated in relation to impact on urea permeability through the membrane. Neither the presence of Â PEO-PBD-PEO triblocks nor the irradiation process was observed to have any effect on membrane permeability to urea. Beyond issues surrounding biocompatibility, PEO coatings can potentially be used to entrap and later release bioactive peptides for short-term medical device applications. In this context, pyrene fluorescence quenching (Figure 1) was used to demonstrate the existence of a hydrophobic inner region of PEO layers based on PEO-PPO-PEO triblocks, substantially explaining the high affinity for entrapment previously recorded for amphiphilic peptides.

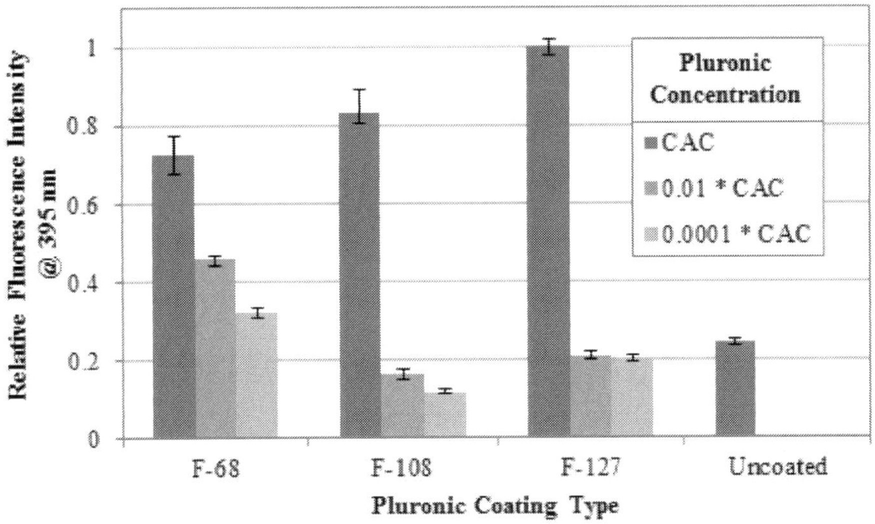

Figure 1: Plot of relative fluorescence intensity of pyrene at 395 nm for various Pluronic coatings on R816 nanoparticles. Coatings were made at different concentrations, denoted by the dilution factors. The dilution scheme is based on the CAC value for each Pluronic. The error bars show the minimum and maximum values for each data set of three samples.

(216ak) Synthesis and Applications of Anisotropic Colloidal Dimers

Wang, S., Colorado School of Mines
Ma, F., Colorado School of Mines
Wu, N., Colorado School of Mines

Anisotropic particles have attracted lots of attention due to their wide applications in optical display, emulsion stabilization, biomedical imaging, and drug delivery. We have successfully synthesized colloidal dimers with properties targeting for three different applications: directed-assembly under electric fields, colloidal surfactants, and self-propelling micro-motors. Colloidal dimers with geometric anisotropy showed interesting assembly behavior, such as chiral structures and Ising-like crystals, under electric fields. Dimers with interfacial or compositional anisotropy could also be used as colloidal surfactants. Stimuli-responsiveness was also incorporated into colloidal surfactants by adding functional polymers such as poly(N-isopropylacrylamide) and poly(acrylic acid). Self-propelling motors can also be made from dimers in which one lobe is coated by platinum. These dimers exhibited both linear and circular movement in a mixture of hydrogen peroxide and water.

(216al) Adsorption, Structural Alteration and Elution of Peptides At Pendant PEO Layers

Wu, X., New Jersey Institute of Technology
Ryder, M., Oregon State University
McGuire, J., Oregon State University
Schilke, K., Oregon State University

Adsorption, structural alteration and elution of poly-peptides at pendant PEO layers

Xiangming Wu, Matthew P Ryder, Joseph McGuire, Karl F Schilke*

A more quantitative understanding of peptide entrapment and elution from otherwise protein-repellent polyethylene oxide (PEO) brush layers will provide direction for development of new strategies for drug storage and delivery. In this work we recorded selected effects of peptide structure and amphiphilicity on peptide adsorption into PEO brush layers based on covalently stabilized Pluronic$^{Â®}$ F108. In particular we evaluated the adsorption of poly-*L*-arginine (PLR), and the cationic amphiphilic peptide WLBU2 to PEO brush layers. Each peptide is comprised of 24-30 amino acids with structure readily controllable between disordered and more ordered forms by varying solution conditions. Further, while WLBU2 is highly amphiphilic, PLR is not amphiphilic.

Self-assembled PEO brush layers were formed by suspension of hydrophobic silica nanoparticles in F108 solution. Uncoated and F108-coated nanoparticles were then incubated with PLR or WLBU2 under different solvent conditions for a desired period of time. Peptide structure in the presence or absence of nanoparticles was evaluated by circular dichroism (CD). In addition peptide-nanoparticle suspensions were washed by centrifugation, resuspended in different solvents, and CD spectra recorded again to detect changes in the amount of peptide remaining on the nanoparticles, and determine conformational changes associated with the washing step. Optical waveguide lightmode spectroscopy (OWLS) was used for direct determination of peptide adsorption and desorption kinetics on F108-coated sensors under selected solution conditions. In addition, pyrene fluorescence quenching was used to determine the existence of a hydrophobic inner region (favors peptide adsorption) of self-assembled PEO brush layers.

For both PLR and WLBU2 in their disordered states the CD spectra were similar, independent of the presence of F108-coated nanoparticles. Moreover, CD spectra indicated complete removal of the peptide from each sample after washing the F108-coated nanoparticle suspensions with water. On the other hand, when peptide is in a more ordered form (i.e., possessing α-helix content), CD spectra showed that considerable α-helical PLR remained on the F108-coated nanoparticle surface after washing the suspension with the helix-promoting solvent, while a completely removal of PLR was observed upon washing suspensions with the solvent favored the disordered form of the peptide, as shown in Figure 1. Interestingly, peptide α-helix content increased after location within the brush. Retention of the amphiphilic peptide in the brush was independent of whether the eluting solvent favored the structured or disordered form of the peptide. Complementary OWLS evaluation of adsorbed mass changes within the F108 layer upon washing was fully consistent with the CD results.

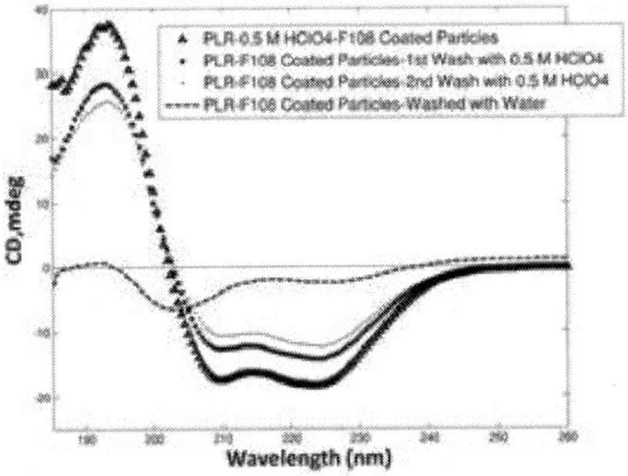

Figure 1. CD spectra of: Poly-*L*-arginine (PLR) in $HClO_4$, and in suspension with F108-coated nanoparticles before and after washing with $HClO_4$ or water. The non-amphiphilic peptide PLR entrapped in the PEO brush layer was partially elutable when the helix-promoting solvent (HClO4) was used for elution. However, in contact with water (which favors PLR's disordered, non-adsorbable conformation), the adsorbed PLR was entirely elutable.

In summary, an initially more ordered (*α*-helical) structure promotes peptide adsorption into the PEO layer. Further, a partially helical peptide undergoes a cooperative increase in helicity after entry, likely due to concomitant loss of capacity for peptide-solvent hydrogen bonding.Â Peptide interaction with the hydrophobic inner region of PEO chains resulted in entrapment and conformational change that was irreversible to elution with changing solution conditions in the case of the amphiphilic peptide.Â In contrast, the adsorption and conformational change of the non-amphiphilic peptide was reversible. These results indicate that responsive drug delivery systems based on peptide-loaded PEO layers can be controlled by modulation of solution conditions and peptide amphiphilicity.

(216an) A Microscopic View of the Effect of Chemical Demulsifiers At the Water-Brazilian Model Oil Interface From Langmuir and Langmuir-Blodgett Films

Carneiro, J. S. A., Universidade Tiradentes
Fortuny, M., Universidade Tiradentes
Santos, A. F., Universidade Tiradentes
Dariva, C., Universidade Tiradentes
Franceschi, E., Universidade Tiradentes
da Rocha, S. R. P., Wayne State University

Heavy oil fields represent a challenge for the oil industry due to their undesirable tendency to form stable water-in-oil emulsions during crude oil exploration, which increases transportation costs and negatively affects processing. Despite the recent findings of light medium crude oil reserves in Brazil, the heavier crude oil fields still account for most of the overall oil production. The abundance of asphaltenes in heavy oils is known to enhance the emulsion stability owing to the formation of a viscoelastic film preventing the coalescence of water droplets. In the presence of such highly rigid films, chemical agents are used to speed up the process of dehydration of the crude oil. It is known that the demulsifiers reduce the interfacial tension between oil and water and also influence the viscoelastic properties of the interfacial film. However, the disruption mechanism of the continuous interfacial film is still not well understood, and is expected to be a strong function of asphaltene chemistry and type of crude.

In the present work, we investigate the effect of chemical additives on the molecular properties of asphaltene films through Langmuir and Langmuir-Blodgett films. The elasticity of Langmuir films in the presence of the additives and as a function of additive concentration was correlated to the microstructure of the transferred films obtained by Atomic Force Microscopy (AFM). The interfacial properties and diffusivity of asphaltenes and demulsifiers were investigated by pedant drop tensiometry. These results are of relevance as they can serve as a guide for the design of demulsifiers for the petroleum industry.

(216ao) High-Pressure Interfacial Tension of the Brazilian Crude Oil - CO2 System

Borges, G., Universidade Tiradentes
Fortuny, M., Universidade Tiradentes
Santos, A. F., Universidade Tiradentes
Dariva, C., Universidade Tiradentes
Franceschi, E., Universidade Tiradentes
da Rocha, S. R. P., Wayne State University

Enhanced oil recovery (EOR) processes are once again becoming increasingly important to the petroleum industry. After the primary and secondary recovery, a typical residual oil saturation in a light or medium oil reservoir is in the range of 50–60% of the original oil-in-place (OOIP). In the EOR, gases such carbon dioxide (CO_2) and natural gas are injected in the petroleum reservoirs for crude oil recovering. It is well known that the interfacial tension between the injected gas and the crude oil has a major influence on the efficiency of displacement of oil by gas. When the injected gas becomes miscible with the crude oil, which means that there is no interface between the injected and displaced phases, the oil is displaced with maximum efficiency, resulting in high recoveries.

In this study, an experimental setup is established to measure the dynamic and equilibrium interfacial tensions of the Brazilian crude oil + CO_2 system under various conditions using the pendant drop technique. This experimental technique is used to determine the interfacial tension of the above system at different pressures and various isotherms. The measured interfacial tension data will provide a better understanding of the interfacial interactions between the crude oil and CO_2 and, consequently, to improve the economics of the gas injection process.

(216l) Effect of Surfactant Drag-Reduction Agent On Fluid Flow and Heat Transfer for DI Water Flowing in a Mock Aviation Coolant Loop

Narvaez, J. A., University of Dayton
Wilkens, R. J., University of Dayton

Cooling systems in military aircraft are insufficient to handle the additional thermal loads of advanced avionics and weaponry. As such, drop-in replacements are being sought. Surfactant Drag Reducing Agents (SDRA) have been proposed as one possible solution in that for equal pressure drop, fluids can circulate faster without undergoing permanent shear degradation.

However, these surfactants form rod-like micelles that - it is thought – assemble in larger structures that force the flow to become laminar even at high Reynolds numbers. Hence, significant drag reduction is obtained by the addition of SDRAs to DI water. However, this approach has two pitfalls. The first one is the effect of abrupt changes in geometry on pressure drop. When the surfactant solution passes though regions with abrupt change in geometry (for example, sharp elbows, valves, or pumps), the induced laminar flow is disrupted, causing a temporary increase in pressure drop until the induced laminar flow is recovered. The second one is the minimization of radial motion that reduces film heat transfer coefficient.

To avoid the first problem, changes in geometry have been minimized so that the savings in pressure drop with SDRAs is large enough to increase volumetric flow of the liquid solution. For the second problem, temporary disruption of the micellar structures before the surfactant solution enters the heat exchanger promotes turbulence, enhancing the heat transfer coefficient. It is proposed that at the system level, the net reduction in pressure drop in the system will be sufficient to increase the flow through the heat exchanger enough to enhance the heat transfer coefficient.

In this study, the thermal performance of water is compared to that of a water/SDRA solution as they flow through both laminar and turbulent style heat exchangers. The comparison is carried out at the same total pressure drop in the coolant loop. The effect of surfactants on heat transfer coefficient for each of the coolant and on each of the two types of heat exchangers is discussed.

(216aq) Fabrication of Functionalized Parylene Particles Based On Electrospinning

Ho, H. Y.,
Chen, H. Y., National Taiwan University

Fabrication of Functionalized Parylene particles Based on Electrospinning

Hsin-Ying Ho[a] and Hsien-Yeh Chen[a]*

Department of Chemical Engineering, National Taiwan University, Taipei 10617, Taiwan

E-mail: hsychen@ntu.edu.tw

Abstract

Functionalized poly-p-xylylenes are synthesized via CVD polymerization from substituted [2.2]paracyclophanes, and these functional coatings have recently been introduced as powerful interlayers that can flexibly decouple surface properties from bulk materials to achieve biointerface engineering on a variety of substrate materials.

Those coatings are belong to the parylene family, or may be referred to as "reactive coatings," have been reported to possess all of the advantages of their commercial siblings, and in addition, provide versatile anchor groups for sophisticated surface modifications to mimic biological functions. More attractively, a library of functional groups including amines, alcohols, aldehydes, anhydrides, activated carboxylic acids, activated esters, maleimides, and alkynes, have successfully been equipped on these polymers and used for performing manipulations on biotins, proteins, saccharides, DNA, hydrogels, polymers, enzymes, and cells. In this study, functionalized poly-p-xylylenes synthesized via CVD are verified by infrared reflection absorption spectroscopy (IRRAS) technique. Subsequently, electrospinning is performed at relatively low polymer concentrations to produces functionalized poly-p-xylylene particles. Two characterization method, scanning electronic microscope (SEM) and photon correlation spectroscopy, are then used to measure particle size distribution. We foresee the reported functionalized poly-p-xylylene particles may find important applications in developing sophisticated biomaterial.

(216ar) Phase Transition in HPMC Hydrogel

Dutt, M., Rutgers University
Vidyalakshmi, C.,
Lee, S. Y.,
Huynh, P., Rutgers University
Takhistov, P., Rutgers, The State University of New Jersey

The rearrangement of the polymer molecules within the film matrix during film drying, which defines its internal and surface structure, has an important role in the control release, optical and mechanical properties of the therapeutical oral films. Via a coarse-grained Molecular Dynamics-based simulation technique called Dissipative Particle Dynamics we investigate changes in the molecular conformational dynamics and morphology of polymer chains in films during the drying process. Dissipative Particle Dynamics is a multiscale mesoscopic simulation method that simultaneously captures the molecular details of the components through soft-sphere coarse-grained models and reproduces the hydrodynamic behavior of the system over extended time scales. We have developed a bead-spring model of individual polymer molecules where each bead or soft sphere is a coarse-grained representation of a group of atoms with two consecutively bonded beads along a polymer chain linked by a spring. The solvent is modeled by soft spheres which are coarse-grained representation of multiple solvent molecules. To capture the drying process, we simulate the polymer chain dynamics for different concentrations of the polymer, for a constant number of solvent and polymer soft spheres in the system. We characterize the polymer chain dynamics and conformation at the various stages of the drying process, or solvent concentration, via the correlation between measurements of the radius of gyration and end-to-end distance with the inter- and intra-polymer chain interactions. Obtained computational data are in good agreement with experimentally observed conformation changes in the polymeric matrix determined by means of Raman spectroscopy and diffusing wave spectroscopy. It is also confirmed that DPD methodology can provide a good prediction for the critical points of drying process.

(216ap) Designing Ionic Liquids for the Removal of Phenolic Compounds From Bio-Oils: A Microscopic Approach

De Conto, J. F., Universidade Tiradentes
Egues, S., Universidade Tiradentes
Santana, C. C., Universidade Tiradentes
da Rocha, S. R. P., Wayne State University

The production of biofuels from residual biomass has great potential as it not only considers the conversion of carbon rich waste sources into energy, but also provides for an opportunity to use agricultural waste instead of edible biomass. In the thermochemical process, which is a widely employed process, the biomass is submitted to high temperatures in the absence of oxygen, thus producing a biofuel called bio-oil. Bio-oil is a brown viscous liquid, chemically unstable and with large amounts of phenolic compounds. They are the ones responsible for the high viscosity, instability and aging (oxidation) of the bio-oil, which prevents its direct use as a fuel source. It is of great relevance, therefore, to develop new strategies to improve the quality of the bio-oil by removing such undesirable products.

In this work we determine the affinity of various phenols (representative of those undesirable compounds present in the bio-oil) with ionic liquid candidates using atomic force microscopy (AFM). The strategy consists in modifying a glass substrate with the ionic liquid and the AFM tip with a moiety possessing strong interaction to the ionic liquid on the substrate. A reduction in the force between tip and substrate in a solution containing the phenolic compounds of interest indicates strong affinity between phenol-ionic-liquid, and a relative scale of interaction is thus determined. Ionic-liquids investigated in this work include 1-propyl-3-methylimidazolium chloride and 1-propyl-3-methylimidazolium bis(trifluoro-methylsulfonyl) imide, and the phenolic compounds include phenol, 4-cyanophenol and 2-tert-butyl-6-methylphenol. We characterized the ionic-liquid modified substrate by contact angle goniometry and XPS, and report the force curves for the different phenolic compounds and ionic-liquids. This work is relevant as it will allow us to synthesize hybrid organic-inorganic materials such as silica modified with ionic liquids that can be used to remove phenolic compounds from bio-oils.

(216m) Block Copolymer Self-Assembly in Ionic Liquids

Tsoutsoura, A., University at Buffalo - The State University of New York (SUNY)
He, Z., University at Buffalo - The State University of New York (SUNY)
Alexandridis, P., University at Buffalo - The State University of New York (SUNY)

In the context of developing self-assembly methodologies for the synthesis and manufacturing of complex formulations and materials, our group has a long-standing interest on the utilization of solvents for the modulation of the organization of amphiphiles such as surfactants and block copolymers. Having low vapor pressure, high thermal stability, and novel solvation properties, ionic liquids can expand the operating range and applications of self-assembled amphiphiles. We discuss here the structuring of poly(ethylene oxide)-poly(propylene oxide)-poly(ethylene oxide) (Pluronic) amphiphilic block copolymers in the ionic liquids ethylammonium nitrate (EAN), 1-butyl-3-methylimidazolium tetrafluoroborate (bmimBF4), and 1-butyl-3-methylimidazolium hexafluorophosphate (bmimPF6), and their mixtures with water or polar organic solvents.

(439g) Effect of pH On the Luminescence Properties of CMC/Tb Nanocomposites

Li, Q. Y.,
Ye, J., South China University of Technology
Xiong, J., Sichuan University

the carboxymethyl cellulose (CMC) and rare earth metal Tb(III) Nanocomposites (CMC/Tb) were successfully synthesized over a pH range of 7-12 under water bath. The structure,surface topography and luminance were characterized by FT-IR,SEM and fluorescence spectrum. Emission and excitation spectra were obtained for the CMC/Tb Nanocomposites. And the fluorescence intensities at 544 nm of these Nanocomposites were measured under an excitation wavelength of 285 nm. It was learned that the sizes of CMC/Tb nanocomposites were from 100 to 500nm and the fluorescence intensities of Tb(III) $^5D_4 \rightarrow ^7F_5$ at 544 nm selective emission spectra showed that the fluorescence intensities were much stronger at pH7 and pH9 but weaker at pH10. Therefore, the pH had a great effect on the luminescence properties of CMC/Tb Nanocomposites and the results indicated that Tb(III) forming complexes with carboxymethyl cellulose were also influenced by pH.

Acknowledgments: We gratefully acknowledge the financial support provided by the National 973 project (2010CB732201), the National Natural Foundation of China (project 31270617) and the Opening Project of State Key Laboratory of Polymer Materials Engineering (Sichuan University, KF201301).

(216n) Effect of Hydrogen-Bonding Surfaces On Ice

McDougall, N.,
Hadley, K.,
Smith, J.,
Wohl, C.,
Kreeger, E.,

Ice formation on surfaces like wind turine blades and airplane wings is both hazardous and costly. By chemically modifying these surfaces, they will exhbiit passive anti-icing behavior to eliminate the need for ice removal. We have taken a novel approach to this problem by introducing hydrogen-bonding to the surface modifier in order to disrupt ice formation at the interface between crystalline water and a surface through experimental and molecular modeling techniques. In this presentation, we plan on showing how hydrogen-bonding on the surface will affect the ice structure in the bulk and at the adhesive interface as compared to more hydrophobic surfaces. In addition, the effects of surface roughness and mixing modifier types will be discussed.

(222a) Phase Equilibrium Properties of Ternary Mixtures Containing 1-Butanol (or 2-Butanol), Cyclohexane and Methylbenzene At 313.15 K

Negadi, L., Laboratoire de Thermodynamique Appliqu e et Mod lisation Mol culaire, University of Tlemcen.
Belabbaci, A., Laboratoire de Thermodynamique Appliqu e et Mod lisation Mol culaire, University of Tlemcen, Po Box 119, Tlemcen 13000, Algeria
VillamaÃ±an, R. M., Universidad de Valladolid
Ghezouali, C., Laboratoire de Thermodynamique Appliqu e et Mod lisation Mol culaire, University of Tlemcen, Po Box 119, Tlemcen 13000, Algeria
VillamaÃ±an, M. A., Universidad de Valladolid

Oxygenated additives, as ethers and alcohols have been traditionally used as a blending agent in the formulation of the new gasoline to enhance octane number in substitution of the traditional leaded products. To better understand and model the new formulated gasoline, we have started many years ago a research program on the thermodynamic characterization of ternary mixtures, as the simplest multicomponent system, containing oxygenated additives (ethers and alcohols) and different type of hydrocarbons (paraffins, cycloparaffins, aromatics, olefins).

In this work, new accurate vapor liquid equilibrium data for the two ternary systems: 1-butanol (1) + cyclohexane (2) + methylbenzene (3) or 2-butanol (1) + cyclohexane (2) + methylbenzene (3) at 313.15 K have been measured.

A static method using an isothermal total pressure cell (Van Ness' technique) has been employed.

The investigated systems present a positive deviation from ideal behavior. Reduction of the experimental data has been done by Barker's method using Wohl expansion, NRTL, Wilson and UNIQUAC models. Good results are obtained for the correlation by all the models.

(222b) Isothermal Vapor-Liquid Equilibria of 1-Ethyl-3-Methylimidazolium Thiocyanate + Water, + Ethanol, + 1-Propanol, Or + Acetonitrile Binary Mixtures At Several Temperatures

Negadi, L., Laboratoire de Thermodynamique Appliqu e et Mod lisation Mol culaire, University of Tlemcen.
Khelassi, A., University of Tlemcen
Mokbel, I., UMR 5280, Institut des Sciences Analytiques.

The field of ionic liquids is rapidly expanding with application being sought in an almost unlimited scope. Nevertheless, for many, even popular ionic liquids there is still a great lack of data on their physical properties and thermodynamic behavior. Such data is needed in order to verify the many models that are being developed to predict this behavior, models which are essential in an area with so many varieties and possibilities.

In our group, we have started a research program on a selected group of ionic liquids for which much basic data is unavailable in order to investigate the thermodynamic (VLE, LLE, ...) and thermophysical properties.

The present paper reports the vapor pressures of 1-ethyl-3-methylimidazolium thiocyanate + water, + ethanol, + 1-propanol, or + acetonitrile binary mixtures measured using a static device at temperatures between 273 and 363 K.

The data were correlated with the Antoine equation. From these data excess Gibbs functions (G^E) were calculated for several constant temperatures and fitted to a fourth-order Redlich-Kister equation using the Barker's method.

Acknowledgements

This work has been done in the framework of the bilateral project TASSILI (Ref. 09MDU761).

(222c) Densities, Speed of Sound and Derived Properties of the Binary Systems (Furfuryl Alcohol + Ethanol), (Furfuryl Alcohol + Toluene), Or (Ethanol + 1, 3-Propanediol) Binary Mixtures At Several Temperatures and Atmospheric Pressure

Bendiaf, L., Université de Tlemcen
Zaoui-Djelloul Daouadji, M., Université de Tlemcen
Negadi, A., Laboratoire de Thermodynamique Appliquée et Modélisation Moléculaire, University of Tlemcen, Po Box 119, Tlemcen 13000, Algeria
Negadi, L., Laboratoire de Thermodynamique Appliquée et Modélisation Moléculaire, University of Tlemcen.

With the steadily increasing energy consumption contributing to the depletion of fossil resources, the insecurity of energy supply and global warming, renewable energy resources emitting less CO_2 become popular alternatives to substitute fossil fuels, especially in the transportation sector which is responsible for a large part of the global CO_2 emissions.

Among many energy alternatives, biofuels, hydrogen, natural gas and syngas may likely emerge as the four strategically important sustainable fuel sources in the foreseeable future. Within these four, biofuels are the most environment friendly energy source. Hence, biofuels are being explored to replace fossil fuels. They are favorable choice of fuel consumption due to their renewability, biodegradability and generating acceptable quality exhaust gases. Biofuels are referred to liquid, gas and solid fuels predominantly produced from biomass. A variety of fuels can be produced from biomass such as ethanol, methanol, biodiesel, Fischer-Tropsch diesel, hydrogen and methane.

The present work is a continuation of a research program concerning the investigation of the thermodynamic properties of binary mixtures containing solvents derived from biomass.

In this paper, we report the experimental densities and speed of sound for three binary mixtures: (furfuryl alcohol + ethanol), (furfuryl alcohol + toluene), or (ethanol + 1,3-propanediol) measured with an Anton Paar DSA 5000 M vibrating U-tube densimeter and sound analyzer with a certified precision of $\pm 1.10^{-6}$ $g.cm^{-3}$ and ± 0.1 $m.s^{-1}$, respectively. The measurements have been performed for different compositions including the pure compounds at seven temperatures from (283.15 K to 343.15 K) and atmospheric pressure. Excess molar volumes and deviations in isentropic compressibility for the binary systems at the above-mentioned temperatures were calculated. These results were fitted to the Redlich–Kister equation to determine the fitting parameters and the root-mean-square deviations.

(222d) Uncertainty in Sour Gas Viscosities Estimation

Elsharkawy, A., Kuwait University

Viscosity of natural gases is necessary in most petroleum engineering calculations. Some of these calculations are design of pipeline and calculation of gas flow rate through reservoir rock. Usually the gas properties are measured in laboratory. Occasionally, experimental data become unavailable and it is estimated from equations of state or empirical correlations.

Natural gases containing CO_2 or H_2S the so-called acid/sour gases are often encountered in the petroleum industry. However, reliable estimate of their thermo-physical properties in reservoir conditions, on viscosity in particular, are available. From a modeling point of view, H_2S and carbon dioxide (CO_2) are polar compounds and as such are often considered rather difficult to model accurately.

In this project viscosity data of natural gases containing large amount of H_2S and/or CO_2 are collected. These data are used to study the reliability and range of application of the industry standard methods and the newly published ones. Finally the data bank is used to develop a new correlation to predict viscosity of the sour/acid gases containing considerable amounts of H2S and/or CO2.

(222e) Prediction of VOCs Adsorption Equilibria Under Supercritical Carbon Dioxide Conditions By Using Pure Component Data

Ushiki, I., Tohoku University
Ota, M., Tohoku University
Sato, Y., Tohoku University
Inomata, H., Tohoku University

Recycle of the supercritical carbon dioxide ($scCO_2$) is essential in terms of preventing CO_2 emissions and reducing running cost for industrial uses of the solvents. We have focused on adsorption processes du e to its removal capability of dilute solutes such as VOCs (Volatile organic compounds) in $scCO_2$ solution. In this paper, measurements and prediction of VOCs adsorption equilibria on activated carbon in $scCO_2$ were studied for designing the processes.

Experimental data of VOCs (Acetone, n-Hexane, Toluene, n-Decane, Methanol) adsorption equilibria in $scCO_2$ were measured by a fixed bed method in the pressure range of 4.2 - 15.0 MPa at 313 - 353 K [1, 2]. The Dubinin-Astakhov (DA) equation [3] and the ideal adsorbed solutions theory (IAST) [4] were applied to predict the adsorption equilibria by using pure component adsorption isotherms, that were also measured with a gravimetric method.

The DA-IAST model provided a good prediction of experimental results in the cases of non-polar substance like n-hexane and at lower pressure conditions. On the other hands, relatively large deviations occurred in the cases of polar substances like methanol and also at higher pressure conditions. These results could be caused by large non-ideality in the adsorbed phase in polar and high pressure conditions. Therefore, more suitable model for $scCO_2$ systems must be developed with consideration of the strong non-ideality.

References

[1] I. Ushiki *et al.*, *Fluid Phase Equilib.*, **344**, 101 (2013).

[2] I. Ushiki *et al.*, *J. Chem. Eng. Jpn.*, **45**, 931 (2012).

[3] M. M. Dubinin, V. A. Astakhov, *Adv. Chem. Ser.*, **102**, 69 (1971).

[4] A. L. Myers, J. M. Prausnitz, *AIChE J.*, **11**, 121 (1965).

(222ad) Prediction of Phase Behaviors of Ionic Liquids Over a Wide Range of Conditions

Lee, B. S., National Taiwan University
Lin, S. T., National Taiwan University

Here we compare several predictive approaches for the properties and phase behavior of mixtures containing ionic liquids (ILs). These models combine the Pitzer-Debye-Hückel (PDH) model for the long-range electrostatic interactions and the COSMO-SAC model for the short-range interactions between contacting molecules. While the COSMO-SAC model alone has been shown to be accurate for vapor-liquid equilibrium (VLE) and infinite dilution activity coefficient (IDAC) of solvents in IL, it is inaccurate for dilute IL solution properties such as osmotic coefficient (OC) and mean ionic activity coefficient (MIAC), where long-range interactions are important. The PDH model alone is accurate for dilute IL solutions; however, it is not applicable at high IL concentrations. Here we propose a novel method for attenuating the long-range effect from the PDH model at high IL concentrations. The resultant model is thus accurate over the whole concentration ranges of the IL solutions. We have examined this method for three versions of the COSMO-SAC model, denoted as COSMO-SAC(2007)+PDH(α), COSMO-SAC(2010)+PDH(α), and COSMO-SAC(ion,α), for mixtures involving 73 ILs formed from the combination of 44 cations and 31 anions. The value of absolute relative deviation (ARD) for IDAC (3552 points) is 107.89% (COSMO-SAC(2007)+PDH(α)), 165.37% (COSMO-SAC(2010)+PDH(α)) and 116.94% (COSMO-SAC(ion,α)), respectively. The errors for VLE (1626 points) are 23.98%, 28.88% and 22.02%. The prediction results for osmotic coefficient (1348 points) of solutions showed 40.86, 48.10% and 24.09% and the predicted MIAC (484 points) of ionic liquid showed a similar result with about 48%. The proposed COSMO-SAC models provide prediction of properties of IL fluids with consistent accuracy over the full range of concentrations.

(222f) An Estimation Method for the Boyle Temperature

M ller, K., University of Erlangen-Nuremberg
Lingeswaran, K., Friedrich-Alexander-University Erlangen-Nuremberg
Arlt, W., University of Erlangen-Nuremberg

The Boyle temperature, the temperature at which a substance at low pressures behaves like an ideal gas, could be estimated using the Hayden-O'Connell model by finding the zero of the second virial coefficient. However, this model was not designed for estimating the Boyle temperature, but rather to determine the second virial coefficient at far lower temperatures. Furthermore the critical temperature and pressure as well as the dipole moment and the radius of gyration are needed as input parameters, but are not available for all substances.

Analogously to the well-known Guldberg rule that correlates the critical temperature with the normal boiling point, as a rule of thumb the Boyle temperature can be estimated by multiplying the critical temperature with a factor of about 2 to 2.5. However, prediction quality is far lower than it is the case for the Guldberg rule. To overcome this poor prediction a group contribution method was developed, that evaluates the specific factor for every compound. This allows for an improved estimation of the Boyle temperature solely based on the critical temperature and the molecular structure.

A simplified form of the equation used by Joback and Reid for the determination of the critical temperature proved to be suited best. The square term used by Joback and Reid was skipped in our equation, since it is a correction term without physical background and resulted in no improvement of the prediction quality for the Boyle temperature.

Parameter fitting has been done with a training set consisting of 231 compounds and the prediction quality was evaluated using a test set consisting of 49 compounds. The experimental values could be reproduced with an relative error of 4.2 % for the training set and 6.9 % for the test set (the Hayden O'Connell model could only reproduce the experimental data with a relative error of 12 %).

(222g) Vapor-Liquid Equilibrium of the 3-Methyl-1-Butanol + 3-Methyl-1-Butyl Propionate and Propionic Acid + 3-Methyl-1-Butyl Propionate Systems At 50.00, 101.33 and 150.00 Kpa

Leyva, F., Universidad Nacional de Colombia
Orjuela, A., Universidad Nacional de Colombia
Gil, I. D., Universidad Nacional de Colombia
Vargas, J. C., Universidad Nacional de Colombia
Rodriguez, G., Universidad Nacional de Colombia

This work helps to construct a complete phase-equilibrium representation of the isoamyl alcohol - propionic acid esterification system, with the presentation of data not reported in the open literature. The data were correlated with phase equilibrium models that can be used for further studies on distillation-based separation processes for the reactive mixtures of isoamyl alcohol with propionic acid. Specifically, the isobaric vapor-liquid equilibrium for the binary systems 3-methyl-1-butanol + 3-methyl-1-butyl propionate, and propionic acid + 3-methyl-1-butyl propionate at three different pressures is reported. Measurements were done in a dynamic equilibrium cell (Fischer Labodest VLE 602) at 50.00, 101.33 and 150.00 kPa. The experimental data were tested for thermodynamic consistency (TC) with point, total area and infinite dilution tests, using the Hayden-O'Connell equation-of-state for the calculation of vapor fugacity coefficients and the Dortmund UNIFAC-modified group contribution method for the prediction of excess enthalpies. The approving criteria for the use of the point, total area and infinite dilution TC tests presented by Kurihara et al (Fluid Phase Equilib. 219, 2004, 75-85.), together with the evaluation and regression utilities available in the software Aspen Propierties ® 7.3, were used. VLE data were correlated using the NRTL activity model for each pressure and for all pressure range.

(222ae) Estimation of Normal Boiling Point, Critical Properties, and Lennard-Jones Parameters for Polycyclic Aromatic Hydrocarbons and Fullerenes

Pope, C., N/A

SUMMARY

Group contribution methods for the estimation of normal boiling points and critical properties are reviewed in depth. Several of these methods are used to predict the normal boiling point, critical temperature, and critical pressure for nearly 100 unsubstituted polycyclic aromatic hydrocarbons (PAH) and the fullerenes C_{60} and C_{70}. The methods vary greatly in their level of complexity, the sensitivity of their predictions to molecular structure, and their ability to predict realistic trends when extrapolated to large aromatic molecules. Application of some of the methods to fullerenes and PAH containing both 5- and 6-membered rings (PAH5/6) presents unusual challenges in group assignments. The predicted critical properties are not only of importance in themselves but also in deriving Lennard-Jones (LJ) parameters for use in calculating transport properties for these molecules.

BACKGROUND

PAH are present in crude oil, including shale oil and its derivatives. They can also be formed in the treatment of shale oil. Properties related to the phase behavior of PAH are needed for design of processes involving the treatment of feedstocks with large aromatic content. The vapor pressures of PAH are also pertinent to thermal remediation of contaminated soils.

PAH and fullerenes can form under fuel-rich or incomplete combustion conditions. Many PAH are known carcinogens. They are also key intermediates in the formation of soot. Fullerenes, spheroidal closed carbon cages, are a special class of polycyclic aromatic compounds; curved PAH5/6 are likely intermediates in their formation in flames [1].

In detailed kinetic modeling of chemical processes in combustion, elementary-step mechanisms for the formation of PAH, fullerenes, and soot can contain $O(10^2)$ species and $O(10^3\text{-}10^4)$ reactions, so there is a need for a computationally efficient way to calculate up to $O(10^4)$ binary diffusion coefficients, which subsequently are used to calculate multi-component diffusion coefficients. LJ parameters are used as inputs for transport properties by modeling software packages like CHEMKIN [2,3].

Correlations for LJ parameters are often based on the critical temperature (T_c) and critical pressure (P_c) [4,5], or these properties and the acentric factor (ω) [6]. There are correlations based on other properties such as liquid density and critical volume (V_c), but these properties are usually less available and less accurately estimated than T_c and P_c. Many of the estimation methods for T_c require the normal boiling point (T_b), so estimations of T_b also need to be considered.

Previous studies deriving LJ parameters for PAH [7,8] were limited by the estimation methods available then, and by lack of data for normal boiling points of most PAH. Pope [7] compared predictions from the method of Joback and Reid [9-11] and of Forman and Thodos [12-14] for a series of the most peri-condensed (having the most tightly packed ring structure) benzenoid (containing only 6-membered rings) PAH. This set of PAH, called the tau-upsilon ($\tau\upsilon$) series, extends from benzene (C_6H_6 -- 1 aromatic ring, 78 amu) to circumcircumcoronene ($C_{96}H_{24}$ -- 37 aromatic rings, 1177 amu), and includes the first ten compounds in the one-isomer series of Dias [15]. The $\tau\upsilon$ series is representative of the PAH found under combustion conditions. The method of Forman and Thodos [12-14] was chosen, since it does not rely upon T_b to estimate T_c. Also, extrapolation of the Joback and Reid method to larger species yielded questionable values for $\theta=T_b/T_c$. Correlations for the LJ parameters were chosen which only require T_c and P_c, and not ω [5], since θ is a key input into equations for ω. Wang and Frenklach [8] considered individual PAH up to coronene ($C_{24}H_{12}$ -- 7 aromatic rings, 300 amu). They used experimental values for T_b and the method of Somayajulu [16] to find T_c and P_c; these values were used to obtain ω from the Lee-Kesler vapor-pressure relation [11,17]. The correlation of Tee, Gotoh, and Stewart [6] was used to derive the LJ parameters for the individual species.

APPROACH AND RESULTS

Estimations of T_b, T_c, and P_c from group contribution methods are compared for three different series of PAH: (1) acenes -- kata-condensed benzenoid PAH with 1 to 23 aromatic rings ($C_{2+4n}H_{4+2n}$); (2) the $\tau\upsilon$ series described above; (3) PAH5/6, starting with acenaphthalene ($C_{12}H_8$), which are proposed as intermediates in the formation of fullerenes C_{60} and C_{70} in flames [1], called the FFM (fullerene formation mechanism) series. The majority of the PAH in the FFM series have the most condensed structures possible for PAH5/6 [18]; all of the structures obey the "isolated pentagon rule" [19]. All of the PAH considered have the maximum number of conjugated non-adjacent double bonds possible, with all the carbon atoms being sp^2 hybridized. In that sense, all the carbon atoms are involved in an extended aromatic π-bond structure.

The only estimation methods which are considered are those containing groups sufficient to predict properties for the PAH in all three series. This does not mean that only the most complex methods, containing the largest sets of groups, are chosen. For many of the methods considered only two groups are needed for all the PAH studied. The more complex estimation methods are better at distinguishing between isomers having the same empirical formula, but their implementation is less straightforward, especially for the species in the FFM series.

The estimation methods used include those of Forman and Thodos [12-14], Joback and Reid [9-11], Constantinou and Gani [20,21], Wilson and Jasperson [21], Marrero and Pardillo [21,22], AvaullÃ©e et al. [23,24], and Nannoolal, Rarey, and Ramjugernath [25,26]. Predicted values for T_b, T_c, and P_c vary greatly for the different methods, sometimes yielding nonsensical results. The predictions are compared to the very limited data available (which include computational studies). Given the sparse data for testing the estimations, the ability of the methods to predict plausible and consistent trends with respect to structure and size becomes a crucial factor in assessing their relative merits. Results for the LJ parameters are given for the three series of PAH, and are correlated with molecular weight when feasible.

REFERENCES

[1] Pope, C.J.; Marr, J.A.; Howard, J.B. (1993) "Chemistry of Fullerenes C_{60} and C_{70} Formation in Flames", *J.Phys.Chem.*, **97**:11001-11013.

[2] Kee, R.J.; Rupley, F.M.; Miller, J.A. (1989) "CHEMKIN-II: A Fortran Chemical Kinetics Package for the Analysis of Gas-Phase Chemical Kinetics", Rep. SAND89-8009, Sandia National Laboratories, Livermore, CA.

[3] CHEMKIN is currently licensed by Reaction Design, San Diego, California [http://www.reactiondesign.com/products/chemkin/].

[4] Hirschfelder, J.O.; Curtiss, C.F.; Bird, R.B. (1964) "Molecular Theory of Gases and Liquids", Wiley, New York.

[5] Bird, R.B.; Stewart, W.E.; Lightfoot, E.N. (1960) "Transport Phenomena", Wiley, New York.

[6] Tee, L.S.; Gotoh, S.; Stewart, W.E. (1966) "Molecular Parameters for Normal Fluids: The Lennard-Jones 12-6 Potential", *Ind.Eng.Chem.Fundam.*, **5**:356-363.

[7] Pope, C.J. (1988) "Fluxes and Net Reaction Rates of High Molecular Weight Material in a Near-Sooting Benzene-Oxygen Flame", S.M. Thesis, Massachusetts Institute of Technology.

[8] Wang, H.; Frenklach, M. (1994) "Transport Properties of Polycyclic Aromatic Hydrocarbons for Flame Modeling", *Combust.Flame*, **96**:163-170.

[9] Joback, K.G. (1984) "A unified approach to physical property estimation using multivariate statistical techniques", S.M. Thesis, Massachusetts Institute of Technology.

[10] Joback, K.G.; Reid, R.C. (1987) "Estimation of Pure-Component Properties from Group-Contributions", *Chem.Eng.Comm.*, **57**:233-243.

[11] Reid, R.C.; Prausnitz, J.M.; Poling, B.E. (1987) "The Properties of Gases and Liquids", Fourth Edition, McGraw-Hill, New York.

[12] Thodos, G. (1957) "Critical Constants of the Aromatic Hydrocarbons", *AIChE J.*, **3**:428-431.

[13] Forman, J.C.; Thodos, G. "Critical Temperatures and Pressures of Hydrocarbons", (1958) *AIChE J.*, **4**:356-361.

[14] Reid, R.C.; Sherwood, T.K. (1966) "The Properties of Gases and Liquids", Second Edition, McGraw-Hill, New York.

[15] Dias, J.R. (1984) "Isomer enumeration of nonradical strictly peri-condensed polycyclic aromatic hydrocarbons", *Can.J.Chem.*, **62**:2914-2922.

[16] Somayajulu, G.R. (1989) "Estimation Procedures for Critical Constants" *J.Chem.Eng.Data*, **34**:106-120.

[17] Lee, B.I.; Kesler, M.G. (1975) "A generalized thermodynamic correlation based on three-parameter corresponding states", *AIChE J.*, **21**:510-527.

[18] Smalley, R.E. (1992) "Self-Assembly of the Fullerenes", *Acc.Chem.Res.*, **25**:98-105.

[19] Schmalz, T.G.; Seitz, W.A.; Klein, D.J.; Hite, G.E. (1988) "Elemental Carbon Cages", *J.Am.Chem.Soc.*, **110**:1113-1127.

[20] Constantinou, L.; Gani, R. (1994) "New Group Contribution Method for Estimating Properties of Pure Compounds", *AIChE J.* **40**:1697–1710.

[21] Poling, B.E.; Prausnitz, J.M.; O'Connell, J.P. (2001) "The Properties of Gases and Liquids", Fifth Edition, McGraw-Hill, New York.

[22] Marrero-Marejon, J., and E. Pardillo-Fontdevila (1999) "Estimation of Pure Compound Properties Using Group-Interaction Contributions", *AIChE J.*, **45**:615-621.

[23] AvaullÃ©e, L.; Neau, E.; Zaborowski, G. (1997) "A Group Contribution Method for the Estimation of the Normal Boiling Point of Heavy Hydrocarbons", *Entropie*, **202/203**:36-40.

[24] AvaullÃ©e, L.; Trassy, L; Neau, E.; Jaubert, J.N. (1997) "Thermodynamic modeling for petroleum fluids I. Equation of state and group contribution for the estimation of thermodynamic parameters of heavy hydrocarbons", *Fluid Phase Equil.*, **139**:155-170.

[25] Nannoolal, Y.; Rarey, J.; Ramjugernath, D.; Cordes, W. (2004) "Estimation of pure component properties: Part 1. Estimation of the normal boiling point of non-electrolyte organic compounds via group contributions and group interactions" *Fluid Phase Equil.*, **226**:45-63.

[26] Nannoolal, Y.; Rarey, J.; Ramjugernath, D. (2007) "Estimation of pure component properties: Part 2. Estimation of critical property data by group contribution", *Fluid Phase Equil.*, **252**:1–27.

(222h) Literature-Reference, Correlation and Prediction of Physicochemical Properties for Ionic Liquid Systems Using Pcpce-Asogdb

Tochigi, K., Nihon University
Matsuda, H., Nihon University
Kurihara, K., Nihon University

Physicochemical properties are useful for designing a lot of chemical processes. Ionic liquids are discussed for some chemical process as environmental kindly solvent. Since 1990 age, many research works have been made for measurement, correlation and prediction of physicochemical properties for the systems with ionic liquid. PCPCE-ASOGDB produced by the authors is the databases for literature-reference, correlation and prediction of physicochemical Properties.

This article introduces the application of ionic liquid systems using PCPCE-ASOGDB as follows.

Introduction

1. Ionic liquid

2. PCPCE-ASOGDB

3. Application of Ionic Liquid Using PCPCE-ASOGDB

3.1 Literature-Reference of physicochemical properties using PCPCE-ASOGDB

3.2 Viscosities

3.2.1 Experimental data

3.2.2 Prediction of viscosities using group contribution method, and reverse design of ionic liquids

3.3 Infinite-Dilution Coefficients

3.3.1 Experimental Data

3.3.2 Prediction of infinite-dilution coefficients using group contribution method

3.4 Vapor-Liquid Equilibria

3.4.1 Experimental data

3.4.2 Prediction of VLE using group contribution method

3.5 Liquid-Liquid Equilibria

3.5.1 Experimental data

3.5.2 Prediction of LLE using group contribution method

Conclusions

(222ac) Vapor-Liquid-Liquid Equilibrium Measurements and Modeling of Ethanethiol + Methane + Water, 1-Propanethiol + Methane + Water and 1-Butanethiol + Methane + Water Ternary Systems At 303, 335 and 365 K and Pressure Up to 9 Mpa

Awan, J. A., Department of Chemical and Biochemical Engineering

New vapor-liquid-liquid equilibrium (VLLE) data for ethanethiol + methane + water, 1-propanethiol + methane + water and 1-butanethiol + methane + water ternary systems have been measured at three temperatures (303, 335, and 365 K) and pressures up to 9 MPa. A "static-analytic" method was used for performing the measurements; the total system pressure was maintained by CH_4. The objective of this work is to provide experimental VLLE data for mixtures of mercaptans (thiols) with other natural gas contents at its crude form for which no data are available in the open literature. Such data will help the industrial modeling of processes relevant to reduction of sulfur emissions. The Cubic-Plus-Association (CPA) equation of state was applied to describe the phase behavior of the investigated systems. It is shown that the CPA EoS satisfactorily describes the solubilities of mercaptans (thiols) in all phases,. On the other hand, the modeloverpredicts the water content of the vapor phase, especially at low pressures and at the highest investigated temperature, *i.e* at 365 K. Furthermore no cross association was found to be significant in 1-propanethiol + methane + water and 1-butanethiol + methane + water ternary systems.

(222af) Experimental and Modeling Study of the Phase Behavior of CO2-Hydrocabon-Water Mixtures

Al Ghafri, S., Imperial College London
Maitland, G., Imperial College London
Trusler, J. P. M., Imperial College London
Galindo, A., Qatar Carbonates and Carbon Storage Research Center, Imperial College
Forte, E.,

Experimental and Modeling Study of the Phase Behavior of CO_2-Hydrocabon-Water Mixtures

*Saif Zahir Al Ghafri, Esther Forte, Amparo Galindo, Geoffrey C. Maitland, and J. P. Martin Trusler**

Qatar Carbonates and Carbon Storage Research Centre, Department of Chemical Engineering, Imperial College London, South Kensington Campus, London SW7 2AZ

* Corresponding author e-mail: m.trusler@imperial.ac.uk.

1. Introduction

Experimental phase-behavior data for CO_2-hydrocarbon-water mixtures play a fundamental role in reservoir engineering and reservoir management, wherein phase equilibrium data at reservoir conditions are required for determination of the in-place volume and recovery factor. These data are also key inputs to reservoir simulations, used for evaluating reservoir development plans, and are required for the interpretation of well-test data and in the design of the surface facilities and processing plants. Knowledge of phase behavior for these mixtures is also crucial in carbon capture and storage (CCS) and enhanced oil recovery (EOR) processes where the miscibility of carbon dioxide with hydrocarbons plays an essential role. Â Furthermore, measuring the mutual solubility of water and hydrocarbon fluids and the water content of hydrocarbon gases and non-hydrocarbon gases such as CO_2 is necessary to assess corrosion problems caused by moisture condensation in pipelines or process equipment, and to determine the dosage of chemical inhibitor required to avoid hydrate crystal formation which may block pipelines, equipment and instruments.

Due to the complexity of oil mixtures, simpler systems that may represent certain characteristics of the real ones are often studied. Despite their importance as a simple model for reservoir fluids, experimental data for CO_2-hydrocarbon-water ternary mixtures are scarce and limited to VLE measurements in most cases. Although much effort has been dedicated to the development of integrated modeling approaches for the prediction of phase behavior and other thermophysical properties of mixtures, experimental data are still required to tune the interaction parameters of the models or to assess the predictive capability of models. Therefore, any new measurements on ternary mixtures of the type CO_2 + hydrocarbon + water is a valuable addition to the literature especially in the three phase, VLLE, region.

<>2. Experimental

In this work the phase equilibria of (n-heptane + carbon dioxide + water) and (methane + carbon dioxide + water) mixtures have been measured. The n-heptane system was chosen as being a representative model for (light oil fraction + carbon dioxide + water, while the methane system is considered representative of (natural gas + carbon dioxide + water). The phase equilibrium measurements were carried out using a high-pressure quasi-static-analytical apparatus that was previously described previously [1]. The apparatus relies on recirculation of two coexisting phases using a two-channel magnetically operated micro-pump with sampling and on-line compositional analysis by gas chromatography. The maximum operating temperature and pressure of this apparatus are 423 K and 45 MPa, respectively. The apparatus was calibrated and validated by means of comparison with the published available literature data of the systems (n-heptane + carbon dioxide) and (water + carbon dioxide). Â

In this work, new experimental data have been measured for the system (n-heptane + carbon dioxide + water) under conditions of three-phase equilibria. Compositions of the three coexisting phases have been obtained along five isotherms at temperatures from (323.15 to 413.15) K and at pressures up to an upper critical end point at which the heptane-rich and the carbon dioxide-rich phases become critical. In addition, new experimental data have been measured for the system (methane + carbon dioxide + water) under conditions of two-phase vapor-liquid equilibrium, three-phase vapor-liquid-liquid equilibrium, and four-phase vapor-liquid-liquid-hydrate equilibrium. Compositions of three coexisting fluid phases have been obtained along eight isotherms at temperatures from (285.15 to 303.5) K and at pressures up to either the upper critical end point (at which the carbon dioxide liquid-rich and the gas-rich phases become critical) or up to the hydrate formation locus. Compositions of coexisting vapor and liquid phases have been also obtained along three isotherms at temperatures from (323.15 to 423.15) K and pressures up to 20 MPa. The quadruple curve along which hydrates coexists with the three fluid phases was also measured.

<>3. Modeling

The experimental data obtained for these ternary mixtures have been compared with the predictions of the statistical associating fluid theory for potentials of variable range (SAFT-VR) in which pure-component model parameters are usually obtained by fitting the theory to the experimental values of the vapor pressure and saturated liquid densities [2] and unlike binary interaction parameters are usually determined by comparison to mixture data. In this work, we used the SAFT-VR parameters reported previously in the work of Míguez et. al [3] for the system (methane + carbon dioxide + water), while the parameters used for the system (n-heptane + carbon dioxide + water) were obtained in our own analysis. Furthermore, a detailed study of the ternary mixtures was carried out based on comparison with available data for the constituent binary subsystems. In this way, we analyzed the observed effects on the solubility when the third component was added.

<>References

1. Forte, E.; Galindo, A.; Trusler, J. P. M., J. Phys. Chem. 2011, 115, 14591.

2. Gil-Villegas, A.; Galindo, A.; Whitehead, P. J.; Mills, S. J.; Jackson, G.; Burgess, A. N., J. Chem. Phys. 1997, 106, 4168?4186.

3. Míguez, J. M.; dos Ramos, M. C.; Piñeiro, M. M.; Blas, F. J., J. Phys. Chem. 2011, 115, 9604.

Acknowledgment

We gratefully acknowledge the funding of QCCSRC provided jointly by Qatar Petroleum, Shell, and the Qatar Science and Technology Park, and their permission to publish this research.

Keywords: Phase behavior; modeling; SAFT, hydrocarbon; water, methane, heptane, carbon dioxide.

(222ag) Modelling Electrolyte Solutions Using the SAFT-VRE Mie Equation of State

Dufal, S., Imperial College London
Schreckenberg, J. M. A., Imperial College London
Adjiman, C. S., Imperial College London, Centre for Process Systems Engineering
Haslam, A. J., Qatar Carbonates and Carbon Storage Research Center, Imperial College
Jackson, G., Imperial College London
Galindo, A., Imperial College London, Centre for Process Systems Engineering

The use of an intermolecular potential of variable form, such as the Mie (generalized Lennard-Jones), within SAFT type equations of state [1] leads to an excellent description not only of thermodynamic fluid phase behaviour, as has become expected of the SAFT family of equations, with a much-improved description in the region of the critical point, but also of a remarkably good representation of second-derivative properties, such as heat capacities and Joule-Thomson coefficients [2,3]. It is highly desirable to be able to treat all fluids that one is likely to encounter in modern chemical engineering practice using a single platform. However, to this point, no version of the SAFT approach for Mie potentials has been available for the treatment of electrolytes. In this paper we report on the extension of SAFT-VR Mie to treat electrolyte systems. In the resulting SAFT-VRE Mie approach, as in its parent EOS, the Mie pair potential [3] is used to represent the short-range repulsive and dispersive interactions. As usual in SAFT implementations, hydrogen bond formation is treated using the first-order thermodynamic perturbation theory (TPT1) of Wertheim [4–7]. In our implementation, however, the association integral, Δ, is evaluated using an essentially exact radial distribution function of the reference fluid, obtained from integral-equation theory; this leads to much more accurate models for hydrogen bonding systems than can be obtained using previously used approximations of Δ, and provides for an excellent thermodynamic description of water. Coulombic interactions between ions are incorporated using an unrestricted primitive model, such as augmented Debye-Hückel theory [8,9] or the mean spherical approximation (MSA) [10,11]; the dielectric constant is represented using a new empirical temperature-dependent correlation with the density of the solvent (water) and a classic Born term is included in order to calculate the solvation energies of the ions. The ionic models proposed will be discussed in some detail, especially in terms of their relationships with measurable physical quantities. Results will be presented for single and mixed salt solutions, and for single and mixed solvent solutions including mixtures of brines with CO_2, which are of great importance in the topical area of sequestration of CO_2 in saline aquifers.

References

[1] T. Lafitte, A. Apostolakou, C. Avendaño, A. Galindo, C.S. Adjiman, E.A. Müller, G. Jackson (2013), in preparation.

[2] T. Lafitte, D. Bessieres, M. M. Piñeiro, J.-L. Daridon, *J. Chem. Phys.*, 124, 024509 (2006).

[3] G. Mie., *Annalen der Physik*, 316, 8, 657–697 (1903).

[4] M. S. Wertheim, *J. Stat. Phys.*, 35, 1–2, 19–34 (1984).

(222i) Density and Viscosity Study of the Binary System 2,5-Dimethylfuran and Decane

Zuniga-Moreno, A., INSTITUTO POLITECNICO NACIONAL-ESIQIE
Elizalde-Solis, O., INSTITUTO POLITECNICO NACIONAL-ESIQIE
Garcia-Cisneros, E.,
Varela-Rico, G., INSTITUTO POLITECNICO NACIONAL
Rosas-Ortega, A.,

Substitution of fossil fuels is a concern nowadays because two main reasons: effect of fossil fuels in the environment and depletion of oil reservoirs and the subsequent rise of petroleum products. Alternative fuels such as ethanol, and biofuels are under investigation for their use as fuels [1,2]. Recently, 2,5-dimethylfuran (DMF) has been proposed as an alternative fuel or as additive [3,4] for automotive engines [2]. As many alternative fuels under investigation, its use is proposed first in alternative-fossil fuels formulations. One advantage of using DMF blended with gasoline is that lubricity properties are good [2]. Attention to DMF has been redirected now because of new advances in its synthesis [3], however must of the physicochemical properties of DMF and DMF mixed with other chemicals are scarce. Hexane + 2,5-Dimethylfuran mixtures have been studied recently [5]. In this contribution, density and viscosity of blends of DMF and decane were study. Experimental measurements were performed form 293.15 to 333.15 K at atmospheric pressure. Density measurements were made with a vibrating tube densimeter (Anton Paar, model 4500 M) and viscosities were made using capillary viscometers (Cannon-Fenske).

[1] Hua, E.; Hua, X.; Wang, X.; Xu, Y. Dearn, K.D.; Xu, H. Tribology Tribology International 55 (2012) 119–125.

[2] Zhang, Q.; Chen, G.; Zheng, Z.; Liu, H.; Xu, J.; Yao, M. Fuel 103 (2013) 730–735.

[3] Daniel, R.; Xu, H.; Wang, C.; Richardson, D.; Shuai, S. Applied Energy 98 (2012) 59–68.

[4] Christensen, E.; Yanowitz, J.; Ratcliff, M.; McCormick, R.L. Energy Fuels 2011, 25, 4723–4733

[5] Mejía, A.; Segura, H.; Cartes, M.; Coutinho, J.AP. J. Chem. Eng. Data 2012, 57, 2681–2688

(222ai) Numerical Simulation for Coating of Titanium Nitride Thin Film in a Tubular Reactor By a Thermal CVD Method

Hatori, Y., Yamaguchi University
Yamamoto, H.,
Tanoue, K. I., Graduate school of science and enginnering, Yamaguchi University
Nishimura, T., Graduate school of science and enginnering, Yamaguchi University

In this study, numerical simulation for coating of titanium nitride (TiN) thin film in a tubular reactor by thermal CVD method has been conducted. From previous our experimental results, it was suggested that the coating progressed successively from gas phase reaction to surface reaction. Growth rate of the TiN film along the axis of the reactor was predicted by solving heat and mass transfer under fully developed Poiseuille flow. Governing equations and boundary conditions were discretized using the finite volume method. The values of physical property in each place were computed using mixed mean temperature. Moreover, the wall temperature conditions of a tubular reactor used the values of experimental data. Experimental results for dependences of gas flow rate and setting temperature on the growth rate could be reproduced partly by the calculated ones. Furthermore, the predictive accuracy of growth rate profile was raised by the nonlinear least-squares method using the initial reaction rate constants due to the experimental ones.

(222ah) Comprehensive Thermophysical Model Development for CO2 Pipeline Transport

Economou, I. G., National Center for Scientific Research Demokritos
Diamantonis, N. I., National Center for Scientific Research Demokritos
Boulougouris, G. C., National Center for Scientific Research "Demokritos"
Tsangaris, D. M., National Center for Scientific Research "Demokrtios"

Carbon Capture and Sequestration (CCS) is one of the most promising technologies for the reduction of CO_2 accumulation in the atmosphere. Flue gas sources such as power plants and other manufacturing processes that depend heavily on fossil fuels, can be equipped with systems that capture the CO_2 from the flue gas stream, and then transport the CO_2-rich stream via pipelines to places where oil reservoirs near depletion, saline aquifers, or other underground cavities, can receive and store the CO_2.

The part of CO_2 pipeline transport is often overlooked and simulated using models developed for natural gas transport. However, its importance raises the need for multi-disciplinary research on the details involved, that are substantially different than natural gas. In particular, since the CO_2 pipeline networks may run close to populated areas, thorough hazard assessment studies are required both for the regulatory frameworks and the social acceptance campaigns. A hazard assessment study is based on fluid calculations in and out of the pipeline, such as normal flow, and dispersion in the event of a rupture. All these calculations rely heavily on the models used for the prediction of thermophysical properties of the fluids involved, which are mainly CO_2 mixtures with other gases.

A large number of properties are necessary for the aforementioned calculations, ranging from density and compressibility, to derivative thermodynamic properties such as speed of sound and the Joule-Thomson inversion curve, and even further to transport properties that include viscosity and self-diffusion coefficient. By employing an accurate, robust, and reliable thermodynamic model that covers the entire range of properties and conditions, improved quality of hazard assessment studies can be ensured.

In this work, several equations of state (EoS) have been assessed for their capabilities of predicting accurately the thermodynamic properties of CO_2 mixtures with other gases. Extensive comparisons with literature experimental data have been performed, showing the similarities of the approaches in relatively simple properties such as density and vapor-liquid equilibria, while pointing out the superiority of higher order EoS (i.e. PC-SAFT) when it comes to more complex properties such as derivative thermodynamic and transport properties. More specifically, the derivative thermodynamic properties are calculated by analytically derived expressions, which means that the computational cost is kept low, while the physical background of each approach is tested. On the other hand, transport properties include the notion of time, thus it is conceptually impossible to be calculated by equilibrium thermodynamic models that do not take into account time. To overcome this obstacle, several established models of the literature have been combined with the EoS, and re-tuned, in order to extend the properties calculation framework to those properties.

Phase equilibria and saturated density calculations for pure components were the starting point, while moving toward mixtures, experimental data were used to optimize the binary interaction parameters. Analytical expressions for the thermodynamic derivative properties were derived for every EoS, and checked against the numerical derivatives, ending with the comparison of all EoS against experimental data available in the literature. Viscosity models both for pure components and mixtures were linked to the EoS, and the use of the meta-heuristic optimization method of Particle Swarm Optimization aided the production of the parameters' tables. Calculations for mixtures of CO_2 were compared with a few experimental data available in the literature and the agreement was excellent in all cases. Furthermore, gaps of experimental data were identified, in order to act as a suggestion for future experimental work. The combined approaches and the new optimized parameters constitute integral parts of a broader thermodynamic simulator that was developed in this project.

Several useful conclusions are drawn from this work, that can be used further to simulators dedicated to the pipeline transport part of the CCS process. Higher order EoS, such as PC-SAFT and the like, can predict more accurately the thermodynamic properties of the systems of interest, while the overhead computational cost is marginally higher than the cost for a cubic EoS. Especially the derivative thermodynamic properties point out the power of higher order EoS over cubic EoS since the latter usually exhibit high discrepancies compared to experimental data. Transport properties can be very efficiently calculated via the combination of an EoS with an appropriate model, given that the parameters have to be re-tuned in order to achieve a good fit for the respective reference systems and states. In this work, several optimization exercises have been performed in order to come up with the parameter values for each EoS and every model.

(222j) Mathematical Model and Validation of Parallel Plate Water Electrolyzers

Estejab, A., Ohio University

Daramola, D. A., Ohio University

Botte, G. G., Ohio University

Mathematical Model of Parallel Plate Alkaline Water Electrolyzers

Ali Estejab, Damilola A. Daramola, and Gerardine G. Botte
Center for Electrochemical Engineering Research
Chemical and Biomolecular Engineering
Ohio University
Athens, OH 45701

Objective

Electrolysis of chemical compounds, like H2O and NH3, can be a way to produce energy. These compounds can be dissociated by electrochemical reactions to generate hydrogen which is an energy carrier. The required energy input for this reaction can be obtained through intermittent sources like solar or wind energy[1]. Subsequently, the produced hydrogen can be an energy source when those intermittent sources are unavailable.

While water is abundant, safe and can be used in small scale units like homes, a less energy intensive process would use ammonia as the energy source [2]. Ammonia electrolysis would also be a method to remove ammonia from waste water [3,4]. The costs and safety issues of experimental methods to design and optimize the process are issues that motivate development of mathematical models to reduce experimental efforts. Thus, this work presents a mathematical model of water electrolysis as a foundation for modeling ammonia electrolysis.

Methodology

Mathematical modeling of electrolyzers has been previously performed for other electrochemical systems [5]. In this model, a parallel plate electrolyzer with numerous simultaneous electrochemical reactions at the electrodes has been simulated. Parallel plate electrolyzers are widely used in industrial processes and provide a closer representation to the overall goal of industrial-scale application of ammonia electrolysis. The model is an attempt to investigate the effect of conditions such as inlet concentration, velocity, temperature, cell potential, dimensions and so on. One of the parameters in the model which is very important in the kinetic of electrochemical reaction is the exchange current density. This parameter depends on temperature, concentration as well as catalyst surface and its morphology.

The following equations can be used to simulate the electrolyzer:

$$N_i = -z_i u_i F c_i \nabla\Phi - D_i \nabla c_i + v c_i$$

$$\frac{\partial c_i}{\partial t} = -\nabla \cdot N_i + R_i$$

$$i = F\sum_i z_i N_i$$

$$\sum_i z_i c_i = 0$$

The combination of these different equations yields a set of non-linear differential equations, with the following boundary conditions based on Butler-Volmer kinetics:

$$i_{nj} = i_{0j,ref}\left\{\prod_i \left(\frac{c_{i,0}}{c_{i,ref}}\right)^{p_{ij}} exp\left[\frac{\alpha_{aj}F}{RT}\left(V_a - \Phi_{0a} - U_{j,ref}\right)\right] - \prod_i \left(\frac{c_{i,0}}{c_{i,ref}}\right)^{q_{ij}} exp\left[\frac{-\alpha_{cj}F}{RT}\left(V_a - \Phi_{0a} - U_{j,ref}\right)\right]\right\}$$

Results

Initial experiments and calculations show the measured and predicted performance of the electrolyzer (Figure 1). Although, there are similar trends observed in the calculated and measured values, there is quite a deviation between these values especially as the current increased. These reasons will be discussed at the meeting. Additional results due to the effects of flow-rate, temperature and inlet concentration will also be presented.

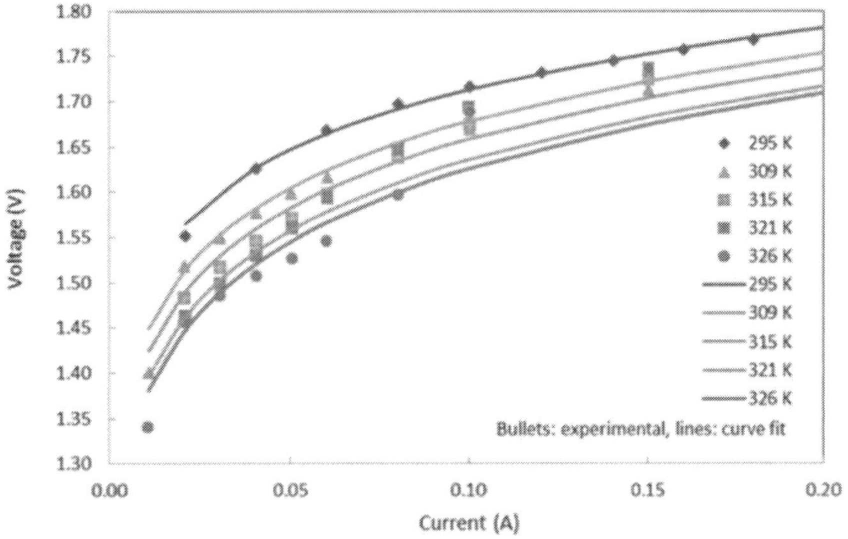

Figure 1. The voltage current (polarization) curve of water electrolysis

References

[1] O. Ulleberg, "Modeling of advanced alkaline electrolyzers: a system simulation approach," International Journal of Hydrogen Energy, vol. 28, pp. 21-33, 2003.

[2] F. d. r. Vitse, M. Cooper, and G. G. Botte, "On the use of ammonia electrolysis for hydrogen production," Journal of Power Sources, vol. 142, pp. 18-26, 2005.

[3] E. P. Bonnin, E. J. Biddinger, and G. G. Botte, "Effect of catalyst on electrolysis of ammonia effluents," Journal of Power Sources, vol. 182, pp. 284-290, 2008.

[4] L. A. Diaz and G. G. Botte, "Electrochemical Deammonification of Synthetic Swine Wastewater," Industrial & Engineering Chemistry Research, vol. 51, pp. 12167-12172.

[5] R. E. White, M. Bain, and M. Raible, "Parallel Plate Electrochemical Reactor Model," Journal of The Electrochemical Society, vol. 130, pp. 1037-1042, 1983.

(222k) Experiment On Superadiabatic Radiant Porous Burner With Augmented Preheating

Wu, H., Sungkyunkwan University
KIM, Y., Sungkyunkwan University
Vandadi, V.,
Park, C.,
Kaviany, M., University of Michigan
Kwon, O. C., Sungkyunkwan University

Most radiant porous burners use only a small portion of combustion heat (less than 30%), though they adopt the internal heat recirculation concept. A new radiant porous burner (i.e., superadiabatic radiant porous burner) configuration in which combustion heat is recovered internally as well as externally is suggested, computationally investigated and designed. A preheater that consists of several preheating pipes is used to externally recuperate heat from the exiting flue gas and preheat the inlet air for the burner. Also, radiation rods with high thermal conductivity are used to extract heat from flame and transfer it to radiating disk surfaces. Results show that the thermal efficiency over 40% can be obtained for the high-efficiency radiant porous burner due to the internal and external heat recirculation, indicating the superadiabatic radiation temperature higher than the subadiabatic flue gas exit temperature, and that the burner can be operated even at very low fuel-equivalence ratios. Using the computational results, the superadiabatic radiant porous burner (alumina, Al_2O_3, and silicon carbide, SiC) has been fabricated.

(222aj) Molecular Simulation Study of Triangle Well Fluids Confined in Slit Pores

Sengupta, A., Indian Institute of Technology, Bombay.

Behera, P.,

Adhikari, J., Indian Institute of Technology, Bombay.

Molecular simulation study of triangle well fluids confined in slit pores.

Angan Sengupta*, Pratik Behera# and Jhumpa Adhikari+

Department of Chemical Engineering, Indian Institute of Technology, Bombay.

Mumbai ? 400076. India.

Fluids confined inside geometries of molecular scale dimensions, do not obey macroscopic laws due to the pronounced wall-fluid interactions in addition to the fluid-fluid interactions. Adsorption and phase behaviour of such confined fluids are the result of the collective effect of the pore geometry, the interactions between the fluid particles, and the interactions between the fluid particles and the confining surfaces. Grand canonical Monte Carlo (GCMC) simulations have been performed in this study, to distinguish the effect of each of the above factors on adsorption and phase behaviour of confined fluids. To achieve this, we have selected a simple triangle well (TW) (equation 1)[1] fluid confined inside the simplest pore model available, the slit pore.

$$\varphi^*_{ff}(r^*_{ij}) = \begin{cases} \infty & r^*_{ij} \le 1 \\ \left[\frac{\lambda}{\lambda-1}\right]\left[\frac{r^*_{ij}}{\lambda}-1\right] & 1 < r^*_{ij} \le \lambda \\ 0 & r^*_{ij} > \lambda \end{cases}$$

[1]

Parameters of the TW model are: core diameter, σ_{ff}; well width, λ; and well depth, ε_{ff}.

The wall fluid interactions in the slit pore are given by equation 2[2].

$$\varphi^*_{wf}(z) = \begin{cases} -\varepsilon^*_{wf} & \left(-\frac{H^*}{2}+\frac{1}{2}\right) < z^* < \left(\frac{H^*}{2}-\frac{1}{2}\right) \\ \infty & \text{otherwise} \end{cases}$$

[2]

In literature, studies using hard sphere[3], square-well[4] and Lennard-Jones[5] (LJ) fluids have been reported which isolate the effect of the pore geometry, fluid-fluid interactions and wall-fluid interactions on adsorption, phase behaviour and surface tension. The TW potential has been used here as it is more realistic than the square-well model; and being segmental in nature this potential also enables us to study the effect of well-width (λ); which is not a parameter in the LJ model. The effect of slit height (pore geometry), TW potential width λ (fluid-fluid interactions) and ε_{wf} (wall-fluid interactions) on the adsorption and the phase behaviour of confined fluids is determined in this study. Verification of the in-house codes has been done by performing bulk-phase molecular simulations to obtain densities and pressures which show an excellent agreement with data reported by Betancourt-Cardenas and co-workers[6].

The adsorption isotherms for the confined TW fluids are determined using the GCMC simulations inside slit pores of reduced heights, H^* = 5, 7 and 9 at reduced temperature, T^* = 1, 1.5 and 2 and within a range of reduced pressure, P^* = 0.01 to 0.8. The adsorption isotherms follow the same trends as proposed by Evans *et al.*[7] and also as reported by Balbuena and Gubbins[5] for LJ fluids confined inside narrow slit pores. As is known from experiments, we also observe that the density inside a given slit pore decreases with increase in temperature. For a given pore volume, the change in the densities of the TW fluids (with different values of λ); indicate a vapour ? liquid phase transition at higher values of ε_{wf}.

The density changes continuously at relatively lower values of ε_{wf}, including for $\varepsilon_{wf} = 0$ (hard wall pores). The value of ε_{wf} above which this sudden change from vapour-like to liquid-like densities is observed, depends on the TW potential width (λ) and the reduced pore height (H^*). The density profiles perpendicular to the pore walls show that layering occurs at progressively lower pressures as the slit height increases at a given temperature in attractive pores. Layering is also observed inside the hard wall pores, though at a relatively higher pressure as compared to that in attractive pores. At lower pressures the fluid particles are uniformly distributed inside the hard wall pores. Also noted is that as λ increases; the number of layers formed inside the attractive pores remain unaffected, while the height of the peaks in the density profiles increase. With slit height increasing, the number of layers is found to be more because of the enhanced adsorption inside the pore. We have, thus, been able to observe the effect of each factor separately on the complex behaviour of confined fluids via molecular simulations.

References:

1. Feinberg, M. J. and Rocco, A. G. De. Intermolecular forces: the triangle well and some comparisons with the square well and Lennard-Jones. *Journal of Chemical Physics*, 41, 11, 3439 ? 3450. (1964).

2. Hamada, Y.; Koga, K. and Tanaka, H. Phase equilibria and interfacial tension of fluids confined in narrow pores. *Journal of Chemical Physics*, 127, 8, 084908, 1 ? 9. (2007).

3. Mittal, J; Errington, J. R. and Truskett, T. M. Does confining the hard-sphere fluid between hard walls change its average properties? *Journal of Chemical Physics*, 126, 24, 244708, 1 ? 8. (2007).

4. Jana, S; Singh, J. K. and Kwak, S. K. Vapour-liquid critical and interfacial properties of square-well fluids in slit pores. *Journal of Chemical Physics*, 130, 21, 214707, 1 ? 7. (2009).

5. Bulbuena, P. B. and Gubbins, K. E. Classification of adsorption behavior: simple fluids in pores of slit-shaped geometry. *Journal of Fluid Phase Equilibria*, 76, 21 ? 35. (1992).

6. Betancourt-Cardenas, F. F.; Galicia-Luna, L. A.; Benavides, A. L.; Ramirez, J. A. and Scholl-Paschinger, E. Thermodynamics of a long-range triangle-well fluid. *Journal of Molecular Physics*, 106, 1, 113 ? 126. (2008).

7. Evans, R.; Marconi, U. M. B. and Tarazona, P. Fluids in narrow pores: adsorption, capillary condensation, and critical points. *Journal of Chemical Physics*, 84, 5, 2376 ? 2399. (1985).

[+] *Corresponding author.* ***E-mail address:*** adhikari@iitb.ac.in

[*] *Presenting author.* ***E-mail address:*** *angan.sengupta@gmail.com*

[#] ***E-mail address:*** *pratik2005@gmail.com*

(222l) Choline Chloride-Based Eutectic Mixtures: An Experimental and Molecular Simulation Study

Perkins, S. L., Pennsylvania State University
Painter, P., Pennsylvania State University
Colina, C. M., Pennsylvania State University

Deep eutectic solvents (DES), considered readily available and relatively inexpensive ILs analogues, show promise for many engineering applications compared to typical ionic liquids (IL)[1]. A large portion of the eutectic mixtures that have been thus far studied experimentally as novel solvents use choline chloride as their ammonium salt component while varying the hydrogen bond donor (HBD) of the system at different compositions. The eutectic point of these mixtures can vary depending on the HBDs used and can thus be difficult to predict.

In this work, molecular simulations and experimental infrared spectroscopy are used to characterize hydrogen bonding and ionic interactions within each system to determine how the HDB affects choline chloride-based eutectic solvents. This combined experimental and computational approach is used to obtain atomistic detail of the hydrogen bonding interactions in the system at its eutectic point. Since only the HBD type is changed, conclusions are based on the ability of these HBD's ability to interact with choline chloride.

Experimental IR spectra are reported for choline chloride-malonic acid mixture and its pure components since this technique is very sensitive to hydrogen bonding. Our previous method[2]for modeling choline chloride-urea was implemented on this and two other systems, one having ethylene glycol and the other glycerol, to determine force field applicability for other DES. Through atomistic molecular dynamics, force field validation is conducted for each system by comparing experimental and simulated data, when possible. Detailed structure and hydrogen bond analyses between moieties show the greatest interactions for each system. Through this work, it is suggested that the anion-HBD and HBD-HBD are crucial interactions (based on RDFs and a hydrogen bond analysis) responsible for the behavior of these eutectic mixture.

1. Abbott, A.P., Capper, G., Davies, D.L., Rasheed, R.K. & Tambyrajah, V. Chem Commun. 2003, 70-71

2. Perkins, S.P., Painter, P., Colina, C.M. J. Phys. Chem. B. 2013, *submitted for publication.*

(222ak) Relevant Phase Equilibria for the Hydroformylation of Long Chain Olefins

Schrader, P.,
Enders, S., TU Berlin

Since the 1940's the Rh- or Co-catalyzed hydroformylation is the most important pathway to produce aldehydes in the chemical industry. Annually, more than 10 mio. metric tons were produced with help of this process (Marcus, et al., 2013). Today's state of the art process for hydroformylation of short chain alkenes is the heterogeneous Rh-catalyzed hydroformylation in aqueous media under mild reaction conditions named the Ruhrchemie/Rhône-Poulenc oxo process. This process has several advantages: The application of Rh as a catalyst provides an excellent selectivity to the desired product, the linear aldehyde. Furthermore, the required extreme low catalyst loss is affected by catalyst immobilization in the aqueous phase. This is absolutely essential due to Rh's extreme high price. After the reaction took place product and catalyst can be separated easily by a phase separation step. Based on the mild reaction conditions (low temperatures and moderate pressures) the energy consumption is very low. Though, if the alkene's chain length is increasing this process is no longer applicable having in mind that the water solubility of the long chain alkenes is decreasing dramatically.

Long chain alkenes are hydroformylated until today under harsh conditions with help of homogenous Co-catalyzed processes like the EXXON or BASF process (Robert, et al., 2012). However, this process has several disadvantages: The energy consumption is very high because of high pressures (up to 30 MPa) and high temperatures (up to 180°C) (Falbe & Bahrmann, 1984). The Co-catalyst shows in contrast to Rh a poor selectivity to the desired product, the linear aldehyde. Therefore often complicated product purification is needed. Also the catalyst recovery is complex and the loss of catalyst cannot be prevented, what makes a catalyst make up stream necessary.

One alternative to avoid the harsh conditions beside others (addition of co-solvent or supercritical fluid or ionic liquids) is the addition of a surfactant in order to solubilize the polar catalyst in the apolar environment (Haumann, et al., 2002), (Ünveren & Schomäcker, 2006). The hydroformylation should take place in the Winsor III region were the contact between aqueous regions and oil rich regions is maximal. Therefore the product achievement is maximal as well. With lowering the temperature catalyst and product can be separated easily with a phase separation process analogue to the Ruhrchemie/Rhône-Poulenc oxo process. In any case the knowledge of the phase behavior is necessary for application of these systems.

Therefore for basic understanding the phase behavior of the system water + otaethyleneglycolmonododecylether ($C_{12}E_8$) + 1-dodecene was studied (Schrader, et al., 2013), here the Winsor III phase region occurs in a for the reaction optimal temperature range. However, the application of a highly purified surfactant is not applicable in a technical process due to the enormous high price (about 250$

per g). Therefore, in industry technical grade surfactants have to be used. These technical grade surfactants are always a mixture of different surfactant molecules of different ethoxylate chain lengths and/or different carbonyl chain length. Also a large amount of not converted fatty alcohol can be enclosed in the technical grade surfactant. The influence of the ethoxylate chain length was studied by means of the so called Kahlweit's fish at a constant oil/water ratio α=0.5 at different surfactant weight fractions γ. Beside this, the fish was measured for a system containing the technical grade surfactant Genapol X080® and technical grade 1-dodecene. The major component of this technical grade surfactant (beside not converted fatty alcohols and surfactants with different alkyl and ethoxylate chain length) is $C_{12}E_8$. The three phase area is shifted to much higher temperatures (temperatures differ up to 30°C) thus additional cooling could be necessary in a technical process.

Furthermore, the phase prism in a range between 30°C and 90°C was detected for the system containing $C_{12}E_8$ + pure 1-dodecene + water and Genapol X080® + technical grade 1-dodecene + water. When technical grade surfactant is used at high temperatures and high surfactant concentrations a four phase equilibrium was found where the micro emulsion splits into two phases.

The last part of the contribution is highlighting the influence of the product and the catalyst on the phase equilibrium. It is obvious that this data is crucial for optimized reaction control. The influence of the product n-tridecyl aldehyde on the phase equilibrium is investigated at three different portions of n-tridecyl aldehyde in the oil rich phase with help of the Kahlweit's fish. Aside, this catalyst influence on the phase behavior was investigated, too. Literature

Falbe, J. & Bahrmann, H., 1984. Homogeneous catalysis - industrial applications. *J. Chem. Ed.*, Volume 61, pp. 961-967.

Haumann, M., Koch, H., Hugo, P. & Schomäcker, R., 2002. Hydroformylation of 1-dodecene using Rh-TPPTS in a microemulsion. *App. Catalys. A*, Volume 225, pp. 239-249.

Marcus, U., Stephan, D. & Armin, B., 2013. Rhodium catalyzed hydroformylation with formaldehyde and an external H_2-source. *Tetrahedron Lett.*, Volume 54, pp. 2209-2211.

Robert, F., Detlef, S. & Armin, B., 2012. Applied Hydroformylation. *Chem. Rev.*, Volume 112, pp. 5675-5732.

Schrader, P., Culaguin – Chicaroux, A. & Enders, S., 2013. Phase Behavior of the water + nonionic surfactant ($C_{12}E_8$) + 1-dodecene ternary system across a wide temperature range. *Chem. Eng. Sci.*, Volume 93, pp. 131-139.

Ünveren, H. & Schomäcker, R., 2006. Rhodium catalyzed hydroformylation of 1-octene in microemulsion: comparison with various catalytic systems. *Catalys. Lett.*, Volume 110, pp. 195-201.

(222m) Quantitative Measurements of CH4 Production From CH4 Hydrates With the Injection of CO2

Lee, B. R., Colorado School of Mines
Sloan, E. D., Colorado School of Mines
Koh, C. A., Colorado School Of Mines
Sum, A. K., Colorado School of Mines

Naturally occurring gas hydrates found in oceanic sediments and under the permafrost, contain mostly methane, and present a potential future energy resource. Carbon dioxide, a global climate change gas, has the potential to be used in enhanced gas recovery techniques, where carbon dioxide is injected into the hydrate reservoir with simultaneous release of methane from within the gas hydrate bearing sediments. However, most previous studies in this area have provided only qualitative information on the methane production rates during the methane-carbon dioxide hydrate exchange process. Consequently, the rate of methane production is not well quantified, and thus important information towards quantifying methane release from various methane hydrate saturations observed in natural settings is required

In this study, quantitative experiments have been performed to investigate the possible methane- carbon dioxide exchange kinetics from injection of carbon dioxide and carbon dioxide -nitrogen gas mixtures through methane hydrate. Methane hydrate was formed from ice particles (75~90 micron in diameter) at 1500 psig and 263 K. In order to reduce unexpected errors, nearly full conversion (> 99%) of ice particles to hydrates was achieved. Liquid carbon dioxide or carbon dioxide-nitrogen gas mixtures were injected (123 ml/hr and 286 ml/hr respectively) into the pressure cell to sweep the residual methane atmosphere, ensuring no free methane was left in the gas phase. After soaking the hydrate for several hours (e.g., 12, 24 and 48 hours), methane was produced from the hydrates and was transferred into a sample cylinder by injecting liquid carbon dioxide or a carbon dioxide-nitrogen mixture. The final composition and analysis of the produced methane was measured using an in-line gas chromatograph. The methane moles after hydrate dissociation were also measured to confirm the closure of the total mass balance of the experiment. The exchange between carbon dioxide and methane during the experiments was demonstrated through the pressure trace in terms of two possible systems, an exchange and no exchange system, and the penetration depth of hydrate particles was calculated to estimate specific kinetic trends. These experiments are essential in the quantification of the production of methane from methane hydrates with the injection of carbon dioxide or a carbon dioxide /nitrogen mixture.

(222n) Phase Behaviour of Mixtures Containing Imidazolium-Based Bis(trifluoromethylsulfonyl)Imide Ionic Liquids - Experiment and Modelling Using the Soft-SAFT Eqation of State

Wagner, Z., Institute of Chemical Process Fundamentals AS CR, v. v. i.
Andresova, A., Institute of Chemical Process Fundamentals AS CR, v. v. i.
Bendova, M.,

Reliable data on phase equilibria in mixtures of ionic liquids is of particular importance for their application in separation processes as well as for their use as reaction media in homogeneous and heterogeneous catalysis. For all these processes, knowledge of liquid-liquid equilibria in systems of ionic liquids with molecular solvents is essential [1, 2]. Ionic liquids may also be used in mixtures with carbon dioxide as advantageous reaction media – both in monophasic catalysis with supercritical CO_2 and in biphasic CO_2 as CO_2-expanded ionic liquid. To optimize reaction conditions in such applications the solubility in sub- and supercritical CO_2 should be first investigated [3].

In this contribution, comparison of previously published experimental data with thermodynamic modeling using the soft-SAFT equation of state [4] will be presented. Examples of liquid-liquid equilibria in binary [5, 6] and ternary systems [7] of imidazolium-based bis(trifluoromethylsulfonyl)imide ionic liquids and their solubility in CO_2 [8] will be used to compare the accuracy of prediction of the thermodynamic properties of mixtures of both chosen methods.

References

1. Müller, A.; Górak, A. *Sep. Purif. Technol.* **2012**, *97*, 130-136.
2. González, B.; Corderí, S.; Santamaría, A. G. *J. Chem. Thermodyn.* **2013**, *60*, 9-14.
3. Andanson, J.-M.; Marx, S.; Baiker, A. *Catal. Sci. Technol.* **2012**, *2*, 1403-1409.
4. Llovell, F.; Valente, E.; Vilaseca, O.; Vega, L. F. *J. Phys. Chem. B* **2011**, *115*, 4387-4398.
5. Jacquemin, J.; Bendová, M.; Sedláková, Z.; Blesic, M.; Holbrey, J. D.; Mullan, C. L.; Youngs, T. G. A.; Pison, L.; Wagner, Z.; Aim, K.; Gomes, M. F.; Hardacre, C. *ChemPhysChem* **2012**, *13*, 1825-1835.
6. Bendová, M.; Wagner, Z.; Moučka, M. *Int. J. Thermodyn.* **2008**, *11*, 109-114.
7. Bendová, M.; Sedláková, Z.; Andresová, A.; Wagner, Z. *J. Sol. Chem.* **2012**, *41*, 2164-2172.
8. Sedlakova, Z.; Wagner, Z. *Chem. Biochem. Eng. Q.* **2012**, *26*, 55-60.

Acknowledgments

The authors wish to acknowledge the financial support of the Ministry of Education, Youth, and Sports in the frames of the Kontakt programme under grant No. LH12020.

(222al) Creating Free-Standing Lipid Bilayers On Fused Silica Substrates With Nanograting Structure

Peng, P. Y., National Taiwan University
Chiang, P. C., National Taiwan University
Chao, L., National Taiwan University

Supported lipid bilayer membranes have been widely used to study protein-lipid membrane interactions due to the fixed position and planar geometry suitable for many surface analytical tools. However, the membrane physical properties are often influenced by the solid support, which may affect the interested interactions. Some previous studies have created free-standing bilayers by depositing unilamellar vesicles on the pores with sizes smaller than the vesicles or by using shear-driven bilayers to span over nanopores at high pH. However, these methods may not form continuous bilayers in a large scale or may not apply to various types of lipids. Here, we developed a theory, based on extended-Derjaguin-Landau-Verwey-Overbeek (DLVO) theory and Helfrich bending theory, to estimate the appropriate conditions for creating free-standing lipid bilayers on solid supports with nanograting structure. We calculated how the buffer conditions, lipid types, and the solid support geometry can influence the affinity between the lipid membrane and the solid support, and the bending energy of the lipid membrane. When the bending energy dominates the entire energy of the system, we predicted that the membrane would form free standing bilayers above the trenched area of the nanograting solid support. On the other hand, if the support-membrane affinity dominates, the lipid bilayer can easily bend and follow the contour of the solid support. In order to tune the relative size of the support-membrane affinity and the bending energy, we prepared buffers with a series of ionic strength (from 5 mM to 100 mM NaCl) and fused silica substrates with periodic 100 nm and 200 nm nanograting structure. The characterization of membrane states by fluorescence recovery after photobleaching measurement shows that our experiment results well match the theoretical predictions and that the developed theory could be used to predict the suitable conditions for the free-standing bilayers to form on solid supports with various kinds of geometry.

(222am) Optimal Heating and Energy Management for the Reheating Furnace Using Oxygen Enhanced Combustion

Wang, N.,
Wen, Z.,
Su, F.,

Oxygen enhanced combustion techniques can be applied to reheating furnaces by reducing nitrogen of combustion air, either partially or completely, for reducing gas emissions and improving the thermal efficiency and productivity. However, with the increase of the oxygen concentration, the furnace temperature and the heat transfer characteristic are different from conventional air combustion, and the heating process of billets in the furnace will be affected, which is a challenge to the optimized control of the reheating furnace.

The optimization of the furnace temperature system is the basis of optimized control□ In this paper□the dynamic model and furnace temperature optimization model were established□ According to the mathematical model steady quantitatively analyze the heating process influence that change of each billet temperature□the heuristic algorithm rule sets was formed□and the furnace temperature of the reheating furnace using oxygen enhanced combustion was optimal analyzed by the object function under various operation conditions□

Optimal heating and energy management are identified via a constrained, multi-parametric optimization approach. The objective function is the minimization of total process energy consumption (both oxygen-making device and reheating furnace) per ton of steel. The optimal variables are the oxygen concentration and production rate. Key performance indicators such as temperature uniformity, the discharging target temperature as well as the minimization of steel mass loss and also the minimization of CO_2 emissions to the atmosphere are included in the optimality criteria.

Energy consumption per ton of steel, production rate and thermal efficiency of furnace in various levels of oxygen enrichment were investigated by comparing with baseline furnace (21 % of O_2 in air). The results showed that the best range of oxygen enrichments was between 21% and 45% by volume, as the higher slope of energy consumption decrease and production increase occurs in this range.

(222o) Effect of Chain Length of Ionic Liquid On the Formation of Methane Hydrate

Chu, C. K.,
Chin, H. Y.,
Chen, P. C., Central Geological Survey
Chen, Y. P., National Taiwan University
Lin, S. T., National Taiwan University
Chen, L. J., National Taiwan University

The dissociation conditions for methane hydrates in the presence of ionic liquid 1–carbonyl–3–methylimidazolium chloride were experimentally measured within the pressure range from 10 to 35 MPa. A homologous series of ionic liquid 1-carbonyl-3-methylimidazolium chloride: 1-ethyl-3-methylimidazolium chloride, 1-hexyl-3-methylimidazolium chloride and 1-decyl-3-methylimidazolium chloride, with different chain lengths were used as methane hydrate inhibitors. The hydrate phase dissociation conditions were measured by using a differential scanning calorimetry operated at constant high pressure. It was found that the addition of 1-carbonyl-3-methylimidazolium chloride has an inhibition effect on the formation of methane hydrate. In addition, the longer the chain is, the weaker the inhibition is. In other word, the inhibition effect of three ionic liquids is in the order of 1-ethyl-3-methylimidazolium chloride > 1-hexyl-3-methylimidazolium chloride > 1-decyl-3-methylimidazolium chloride. In addition, a predictive method was applied to predict the vapor-liquid-hydrate three-phase equilibrium condition of methane hydrate in the presence of ionic liquids. The Peng-Robinson-Stryjek-Vera equation of state incorporated with COSMO-SAC activity coefficient model and the first order modified Huron-Vidal mixing rule were chosen to evaluate the fugacity of vapor and liquid phase. A modified van der Waals and Platteeuw model was applied to describe the hydrate phase. This predictive method can successfully describe the inhibition effect of these ionic liquids on the methane hydrate formation.

(222an) Measurements of the Dissociation Conditions of Gas Hydrates in the Presence of Hydrocarbon Additives

Chen, Y. P., National Taiwan University
Chen, L. J., National Taiwan University
Lin, S. T., National Taiwan University
Tang, M., Chinese Culture University

Measurements of the dissociation conditions of gas hydrates in the presence of hydrocarbon additives

Yan-Ping Chen*, (1), Li-Jen Chen(1), Shiang-Tai Lin(1), , Muoi Tang(2)

Department of Chemical Engineering

(1) National Taiwan University, (2) Chinese Culture University

Taipei, Taiwan

Abstract

This study measured the dissociation temperatures and pressures of gas hydrates in the presence of hydrocarbon additives. The gases include methane and carbon dioxide. The hydrocarbon additives include alcohol compound like tetrt-butanol, ring compound like 1,3,5-trioxane and furan compounds. The gas hydrate dissociation conditions were measured with a static and well-mixed apparatus which was operated at high pressure and low temperature conditions. The liquid water-hydrate-vapor (L_w-H-V) three-phase dissociation temperatures and pressures for methane and carbon dioxide hydrates in the presence of additives were determined by employing the isochoric method. New experimental data were reported within the pressure range from 4 to 15 MPa. It is demonstrated that the addition of some additives in gas hydrate systems shifted the original hydrate phase boundaries to lower pressure and higher temperature (the promotion effect). Some other additives showed the opposite inhibition effect. In addition, the liquid-hydrate-vapor (L-H-V) three-phase dissociation conditions for brine systems with 0.035 mass fraction of NaCl were also measured and reported in this study.

*Corresponding author, email address: ypchen@ntu.edu.tw

(222ap) Molecular Interactions and the Response to Nanoscale Broken Symmetries

Remsing, R. C., University of Maryland
Weeks, J. D., University of Maryland

The solvation of ions in water presents many conceptual and computational challenges to current models of ion and water interactions. The solvation free energy is usually described using a two step process: the formation of a cavity in water that accommodates the neutral ion core along with its associated ion-water dispersion interactions and the additional free energy resulting from charging the core to the full charge of the ion. The simplest Born model treats water outside the cavity as a continuum dielectric that responds linearly to the inserted charge, and predicts that the resulting electrostatic free energy contribution is independent of the sign of the charge. However, it has long been recognized that nonlinear electrostatic effects induced by the initial insertion of the ion core must be taken into account, in addition to other possible nonlinearities associated with the charging process. In order to understand where the various asymmetries and nonlinearities arise in the ion solvation process, we examine the structural and electrostatic consequences of various nanoscale broken symmetries that arise in ion hydration. We pay particular attention to donor-acceptor asymmetries of the hydrogen bond network of water, and the dependence of these asymmetries on the form of the molecular interaction potential by studying solvation in the SPC/E and TIP5P models of water. The results obtained from empirical, point-charge models are then compared with data from *ab initio* density functional theory-based simulations.

(222aq) Segmental Chemical-Potential Increments for Polymeric and Polyatomic Fluids: A Potential Distribution Based Investigation

Lee, L. L., The California State University Pomona

Segmental chemical-potential increments for polymeric and polyatomic fluids: A potential distribution based investigation.

We apply the *three* potential distribution theorems derive earlier [1] [2] to the characterization of the segmental increase of the chemical potentials for chain molecules. Both tangent hard-sphere chains and Lennard-Jones chains are investigated. We define the conditional singlet direct correlation functions as the incremental works of insertion of successive segments of the molecule. By employing Monte Carlo data, we arrive at a quantitative evaluation of the chemical-potential increments of the initial mers, and a clear demonstration of the non-congruent nature of first segments of a polymer. We suggest a purely geometrical obstruction theory that accounts for 90% of the non-congruence.

[1].. L. L. Lee, *The correlation functions of classical fluids III. The method of partition function variations applied to the chemical potential: cases of PY and HNC2*, Journal of Chemical Physics **60**, 1197 (1974).

[2] L.L. Lee, *A potential distribution approach to fused heterochain molecules. I. Mixtures of hard dumbbells and spheres* Journal of Chemical Physics, **103**, 4221-4233 (1995).

(222ar) Solid-Liquid Phase Equilibrium and Determination of Excess Enthalpy, Entropy and Gibbs Free Energy of Androstenedione in Lower Alcohols

Tang, W.,

Solid-liquid phase equilibrium is the basis of crystallization. To choose the proper solvent and to design an optimized crystallization processs, the solubilities of androstenedione (AD) in methanol, ethanol, isopropanol, and n-butanol were measured at the temperature range from (278.15 to 323.15) K using a multiple reactor set up. The temperature (clear point) at which the suspension turns into a clear solution was taken as the saturation temperature of AD. The experimental data were modeled by van't Hoff equation and activity coefficient models (Wilson, NRTL and UNIQUAC models), and the NRTL equation was able to estimate the solubility with the best accuracy. Based on the NRTL model and experimental data, the excess enthalpies, entropies, and Gibbs free energies of solutions and activity coefficients of AD were determinded in different alochols. Futhermore, other thermodynamic properties (infinite-dilution activity coefficients and excess enthalpies) were derived as well. The intermolecular interactions and structural effects were analyzed on the basis of these obtained properties.

(222as) SAFT-γ Mie Group Contribution Approach for Branched and Cyclic Molecules

Chremos, A., Imperial College
Papaioannou, V., Imperial College London
Galindo, A., Imperial College London, Centre for Process Systems Engineering
Jackson, G., Imperial College London, Centre for Process Systems Engineering
Adjiman, C. S., Imperial College London,Center for Process Systems Engineering

Branched hydrocarbons provide an excellent example of how molecular architecture affects macroscopic physical properties, since they are better lubricants and fuels as well as having lower critical temperatures and boiling points than the corresponding linear isomers. Within the framework of the Statistical Associating Fluid Theory (SAFT)[1] the modelling of chain molecules is undertaken by application of Wertheim's first-order thermodynamic perturbation theory (TPT1)[2] where isomers of the same chain length cannot be distinguished. Reformulations of the underlying theory for the explicit accounting of branching have been recently presented.[3] In this work, the modelling of branched alkanes is tackled within a group-contribution approach based on the SAFT-γ Mie framework.[4] Though this essentially involves a TPT1 treatment, the introduction of a functional group within the heteronuclear molecular backbone of the chain molecule will be shown to allow for an effective treatment of the different molecular architecture of branched alkanes, and how this impacts their thermodynamic properties. Parameters for the CH (tertiary) and C (quaternary) functional groups are presented, developed by examining the vapour-liquid equilibria (VLE) of pure branched alkanes as well as mixture data. The C chemical group parameters are particularly challenging to obtain due to their small and yet significant contribution to the thermodynamic properties. We have investigated and developed three approaches to describe the thermodynamic behavior of branched alkanes containing the C group. The differences between the three approaches lay the foundations on how the group contribution may be used within SAFT using the TPT1 approximation to tackle complex molecular architectures.

[1] W.G. Chapman, K.E. Gubbins, G. Jackson and M. Radosz, *Fluid Phase Equilib.* 52 (1989), pp. 31-38.

[2] M. S. Wertheim, *J. Stat. Phys.*, 35 (1984), pp. 19-34; *J. Stat. Phys.*, 35 (1984), pp. 35-47; *J. Stat. Phys.*, 42 (1986), pp. 459-476; *J. Chem. Phys.*, 87 (1987), p 7323.

[3] B. D. Marshall and W. G. Chapman, *J. Chem. Phys.*, 138 (2013), p. 174109.

[4] C. Avendano, T. Lafitte, C. S. Adjiman, A. Galindo, E. A. Muller and G. Jackson, *J. Phys. Chem. B*, 117 (2013), p. 2717.

(222p) Vapor-Liquid Equilibria and Excess Gibbs Energies of the Binary Systems (3,3'-diaminodipropylamine + water), (1,3-diaminopentane + water), Or (1,5-diamino-2-methylpentane + water) At Several Temperatures

Negadi, A., Laboratoire de Thermodynamique Appliqu e et Mod lisation Mol culaire, University of Tlemcen, Po Box 119, Tlemcen 13000, Algeria
Bouzina, Z., University of Tlemcen
Negadi, L., Laboratoire de Thermodynamique Appliqu e et Mod lisation Mol culaire, University of Tlemcen.
Mokbel, I., UMR 5280, Institut des Sciences Analytiques.
Jose, J., UCB-Lyon1

Amines are an important class of compounds used in different fields of industry, they represent a particularly interesting family of molecules for the purpose of testing group-contribution models, and analyze of the intramolecular effects, especially the proximity effect.

The present paper is part of a research program on thermodynamic properties of aqueous solutions of amines. The purpose of this work is to investigate vapor-liquid equilibria (VLE) of (3,3'-diaminodipropylamine + water), (1,3-diaminopentane + water),or (1,5-diamino-2-methylpentane + water) binary mixtures with a view to use the results to determine the interaction parameters for predictive group contribution methods. a survey of the literature shows that there is no data available in the literature.

The vapor pressures of the three binary mixtures and of all the pure components were measured by means of two static devices at temperatures between 263 and 363 K. The data were correlated with the Antoine equation. From these data, excess Gibbs functions were calculated for several constant temperatures and fitted to a fourth-order Redlich-Kister equation using Barker's metho.

The NRTL and UNIQUAC models have also been used and good results were obtained in the prediction of the total pressure using.

(222ao) Vapor-Liquid Equilibrium for Binary System of 2-Methyl-1,3-Propanediol(MPO)+Water, and MPO +1,4-Butandiol(BDO)

Li, H., Zhengzhou University
Guo, Y., HQC (Guangdong) Company
Li, H., Zhengzhou University
Shen, L., Zhengzhou University
Chen, H.,

2-methyl-1,3-propanediol (MPO), as a new glycol synthesis of block, has been widely applied to saturated polyester, unsaturated polyester, coating, personal care products, plasticizer and so on due to its colorless, odorless, water-soluble, low toxic and environmentally benign. Vapor-liquid equilibrium (VLE) data are absolutely necessary to modeling separation process, analyzing the efficiency of chemical plant, and optimizing operation conditions of vapor-liquid mass transfer. However, to the best of our knowledge, there are few reports about thermodynamic properties of MPO, and the VLE data of MPO+H_2O and MPO+BDO systems so far. So, Vapor pressures of MPO and VLE data of the two systems were measured under atmospheric pressure with the static method and improved ROSE vapor-liquid equilibrium still respectively in this work, which intend to provide the preliminary data and fundamental information for various applications of MPO. The reliability of static device and improved ROSE still were calibrated before measuring. And the consistency of thermodynamics was also verified.

Materials: 2-methyl-1,3-propanediol (MPO, Darian Co. H_2O<0.08%) was purified by distillation under reduced pressure after dried with 3A molecular sieve and analyzed by GC900 with a purity>99.6%. Chromaticity < 15 (APHA), carbonyl group (by -CHO)<500ppm, Fe< 0.5ppm; 1,4-butanediol (BDO) , double-distilled water.

Experimental devices: Analytical Balance (Mettler Toledo AB204-N with 0.1mg precision), isobaric gauge, improved ROSE equilibrium still, WZS-1 Abbe refractometer, GC900□micro-thermometer (1/10 precision, 150°C-200°C, 200°C-250°C); ZXZ-1 rotary-vane vacuum pump; potential transformer.

First, Vapor pressures of MPO at 100°C and BDO at 140°C and 180°C under atmospheric pressure were measured to calibrate the purity of the reagents and the reliability of experimental methods. Compared with literature data, the average relative error was less than 0.04%. Then, vapor pressures of MPO were measured under normal pressure at 298.15K-484.95K with a static apparatus, which there is a linear relationship between $\ln p^s$ and $1/T$ (square of correlation coefficient R^2=0.9984). And the average enthalpy of vaporization of MPO is 37.68KJ/mol at above temperature range.

Then, experimental VLE data of ethanol (1) + water (2) binary system were measured, and compared with the literature data to verify the reliability of the improved ROSE equilibrium still. A series of different concentrations of ethanol(1) + water(2) mixture were prepared as standard solutions. Refractive index in a corresponding concentration was measured by Abbe refractometer at 306.15K and the standard curve was obtained. Then, the solution to be tested was put into the improved ROSE equilibrium still. When the temperature of system was almost constant within 20 minutes, an equilibrium state was reached. Gas and liquid phase sample were taken from corresponding sample port, and equilibrium temperatures were recorded at the same time. The molar concentrations of gas and liquid phase were obtained in according to the standard curve.

Third, VLE data of MPO(1) + H_2O(2) were measured with the improved ROSE equilibrium still, which the measuring procedure was followed as stated above. The *x-y-t* diagram of MPO+H_2O system was done according to the experimental VLE data. The results showed that there was no azeotropic point of MPO+H_2O binary system and the larger area between gas and liquid phase curves meant the big difference of boiling point between pure 2-methly-1,3-propandiol and H_2O, which MPO+H_2O mixture can be completely separated by distillation easily.

VLE data of MPO(1) + BDO(2) were measured with the same ROSE still. Accurate concentration of prepared solutions was obtained by gas chromatography. The analysis of each sample was repeated three times. The *x-y-t* data of MPO+BDO system shown that an azeotropic point was found (x_1=0.9507; T=487.0K) at low concentration of BDO.

At last, the isobaric VLE data of MPO (1) + H_2O (2) and MPO(1) + BDO(2) binary systems were verified with the Herrington method of area test. The results showed that the experimental VLE data of two binary systems were in according with the consistency of thermodynamics.

Acknowledgement: Thanks for the financial support from Henan Natural Science Foundation (511021100).

(222at) Phase Behavior of Quaternary System of TAME Synthesis Under Reactive Distillation Conditions

Li, Y., Tianjin University

Etherification of light gasoline is a kind of promising method for raising octane number. This reaction is reversible and exothermic so that it is beneficial to conduct the etherification in a reactive distillation column. There is complex phase behavior in the reactive distillation column as the vapor-liquid phase equilibria and the reactions affect each other.

Etherification of methanol with 2-methyl-1-butylene and 2-methyl-2-butylene to Tert-Amyl Methyl Ether (TAME) is the major chemical reaction among the etherification of light gasoline. In this paper, experiments of vapor-liquid phase equilibria of the quaternary system, including methanol, TAME and the two isoamylenes, were performed in a circulative vapor-liquid equilibrium cell containing sulfonic resin catalysts. The vapor-liquid phase behaviors coupled with the chemical reaction were investigated at the temperature of 338.15K and 348.15K. The reactive equilibrium constants at the bubble-point temperature and equilibrium pressure were calculated. The activity coefficient models, including Wilson, NRTL and UNIQUAC, as well as the SRK equation of state were used to correlate the experimental data respectively. The results have shown very good thermodynamic consistency and better agreement of Wilson predictions with the experimental data. This study provides important thermodynamic information for the process design of etherification reactive distillation of light gasoline.

(222au) Development and Evaluation of the Simple Models of Aqueous Electrolytes

Vlcek, L., Oak Ridge National Laboratory
Chialvo, A. A., Oak Ridge National Laboratory
Simonson, J. M., ORNL

Simple non-polarizable models of aqueous electrolytes underlie the majority of current geochemical and biochemical molecular simulations. To evaluate the reliability and predictive qualities of such studies, it is necessary to understand the limitations of the molecular force fields. Here we present the results of our effort to identify and overcome some of the main limiting factors of the existing simple models.

First, we analyze relationships between experimental structural, dynamic, and thermodynamic properties and individual force field parameters. Our main focus is on the proper choice of reference experimental data for absolute single ion thermodynamics and utilizing known correlations between solvation entropies and tracer diffusivities. Subsequently, we use this information to construct empirical simple (Lennard-Jones potential and point charges) models of common mono- and divalent ions. These include alkali metal and alkaline earth cations paired with halide and oxo-anions. Finally, the limits of the resulting force field are evaluated and compared to other models.

Acknowledgements. This work was supported by the Division of Chemical Sciences, Geosciences, and Biosciences, Office of Basic Energy Sciences, U.S. Department of Energy.

(222av) The Potential of Improving the Thermal Efficiency of Steam Rankine Cycles By Reheating

Fu, C., Norwegian University of Science and Technology
Anantharaman, R.,
Jordal, K.,
Gundersen, T., Norwegian University of Science and Technology

The steam Rankine cycle has been in development for more than 100 years. Its basic principle has been described in many thermodynamic textbooks. The common measures for improving the thermal efficiency of the cycle are also presented in these books, e.g. increasing the temperature and pressure of the main steam and using superheating, reheating and regenerative preheating of the boiler feedwater. The primary reason for using reheating is to reduce the moisture content of the exhaust steam from the last stage of the steam turbines. Reheating may increase or decrease the thermal efficiency depending on the comparison between the mean temperatures of the heat input with and without reheating [1]. A key parameter to be optimized is the reheating pressure. One-stage reheating is commonly applied in modern power plants, while two-stage reheating is less commonly used due to the increased complexity (thus increased investment cost) and diminished increment of the thermal efficiency. More than two-stage reheating has not been applied in practice and less discussed in literature. Infinite-stage reheating represents an ideal isothermal expansion process [2].

With increasing concerns on CO_2 emissions and fossil fuel depletion, the improvement in the thermal efficiency of steam Rankine cycles is important. An interesting question is how the thermal efficiency can be improved by various measures in the long-term future. This paper investigates the improvement potential of using reheating from the thermodynamic point of view. The thermal efficiencies for one-, two-, three- and infinite-stage reheating are presented. The near optimal reheating pressures are obtained by sensitivity analysis for various efficiencies of the steam turbines. The mean temperature of heat input is compared. The results show that a higher thermal efficiency can be obtained if the exhaust steam from the last stage of the steam turbines is superheated when two or more reheating stages are used.

References

[1] H. Spliethoff. Power generation from solid fuels. Berlin, Heidelberg, Germany: Springer-Verlag Berlin Heidelberg; 2010.

[2] C. Fu, R. Anantharaman, K. Jordal, T. Gundersen. Thermal efficiency for coal to power: from theoretical to practical assessments. Accepted to be presented at the 26th International Conference on Efficiency, Cost, Optimization, Simulation and Enviromental Impact of Energy Systems, 16-19 July, 2013, Guilin, China.

(222q) Effect of pH On Surface Charge of Zwitterionic Surfactant and Its Hardness Tolerance

Maneedaeng, A., Suranaree University of Technology
Kingpho, P., Suranaree University of Technology

Precipitation of surfactant by hardness ions effects negatively on the cleaning result in the detergency industry since the unsightly soap scum precipitated removes the surfactant from the cleaning action. Zwitterionic surfactant has been synergistically mixed with anionic surfactant in order to serve as a foam booster and generate mildness at certain pH levels. However, the anionic form of this inner salt in hard water is often limited by its tendency to precipitate. Precipitation of zwitterionic surfactant in the anionic form with hardness ions depends on temperature, ionic strength and pH. Therefore, this work aims to study the effect of pH on the solubility of their precipitate and precipitation behavior of 3-(decyldimethylammonio) propanesulfonate with calcium ions at 30°C. A series of zwitterionic surfactant used with varying concentration was prepared in order to measure the zeta potential and to determine its CMC as a function of pH at a certain amount of swamping solution of 0.01 M NaCl. The isoelectric point (pKa) was determined from the curve of zeta potential vs. pH at various surfactant concentrations. The solubility of calcium salt (soap scum) of zwitterionic surfactant was measured as a function of pH at a certain amount of swamping solution of 0.01 M NaCl. The experiments above provide informative data to construct the precipitation phase boundaries of 3-(decyldimethylammonio) propanesulfonate precipitated with calcium ions as a function of pH at a certain amount of swamping solution of 0.01 M NaCl. These studies show that pH level is a key factor for zwitterionic surfactants when dealing with hardness ions. Knowledge gained from this work will be valuable in the understanding of hardness tolerance ability of zwitterionic surfactant in hard water.

(222aw) Vapor Pressures of Supercooled Liquid Water in No-Man's Land

Choi, E., Cornell University
Nahm, R., Cornell University
Engstrom, J. R., Cornell University
Stroock, A. D., Cornell University

Since Angell and co-workers [1] reported the rapid variation of thermodynamic properties of supercooled water upon approaching the homogeneous nucleation limit of ice ($T_h \sim 231$ K), finding the origins to the anomalous properties of supercooled water has been a field of great research interest [2]. One of the competing hypotheses suggests that the anomalous variations near the homogeneous nucleation limit may be due to a second critical point that terminates at the end of a liquid-liquid coexistence region at elevated pressures and more deeply supercooled temperatures [3]. In this scenario, the properties such as the density and the heat capacity should pass through extrema as the liquid passes near this buried critical point. However, the difficulty of experimentally studying liquid water at deeply supercooled temperatures – in so called no man's land ($T_g \sim 150\ \mathrm{K} < T < 231\ \mathrm{K} \sim T_h$) [2] – has limited tests of these predictions.

In this presentation, we discuss a new experimental approach with which to extract information about the liquid state in no man's land. We will introduce a method based on the kinetics of heterogeneous freezing from the vapor phase that allows us to extract values for the saturation vapor pressures of both the liquid and solid states at supercooled temperatures. We will show how these measurements constrain the form of the heat capacity of supercooled liquid water in no man's land and discuss the implications of these constraints on the hypothetical, buried critical point. We will also point to extensions of our approach aimed at measurements of structure and dynamics and at understanding the role of specific surface chemistries on the kinetics of freezing. We will conclude with a discussion of the potential importance of these measurements in the context of high altitude clouds.

[1] Angell C.A. *et al. Journal of Physical Chemistry*, **77**, 3092 (1973)

[2] Debenedetti P.G. *Journal of Physics: Condensed Matter*, **15**, 1669 (2003)

[3] Poole P.H. *et al. Nature*, **360**, 324 (1992)

(222ax) Isolating the Non-Polar Contributions to the Intermolecular Potential for Water

Ballal, D., Rice University
Chapman, W. G., Rice University

An accurate water intermolecular potential model is required to predict the structure and properties of water. Water is generally modeled as a non-polarizable rigid molecule with an effective pair potential consisting of a center-to-center Lennard-Jones (LJ) potential and coulombic interactions between charges representing hydrogen and oxygen sites. Due to the large number of parameters, a unique solution for the parameters cannot be obtained for even this simplified model of water. Several potential models have been developed depending on the data used to fit the parameters. Here we show that with the Lorentz-Berthelot rule, the SPC/E[1] model under predicts the solubility of water in alkanes by an order of magnitude compared with experimental data. Since the solubility of water in alkanes is very small, there are negligible water-water interactions and water interacts only with the alkanes through non-polar LJ interactions. This is a good test for the LJ parameters of the water model. We find that a much higher value of energy (1.829KJ/mol) for water is required to match the experimental solubility. Since in this limit charges do not affect the properties, LJ parameters obtained here are independent of the charges.

References

1. Berendsen, H., J. Grigera, and T. Straatsma, *The missing term in effective pair potentials.* Journal of Physical Chemistry, 1987. **91**(24): p. 6269-6271.

(222ay) Prediction of Incipient Conditions of Gas Mixtures Using Statistical Association Fluid Theory

Al-Hamdan, A. M. M., King Fahd University of Petroleum & Minerals
Al-Rehaili, A. M., University of British Columbia
Alsaifi, N., King Fahd University of Petroleum & Minerals

The present work predicts the hydrate dissociation conditions for gas mixtures using the statistical association fluid theory. The incipient hydrate formation calculations are conducted in the absence and the presence of inhibitors like methanol and ethylene glycol. The effect of polar interactions is taken into consideration. The van der Waals-Plauteeuw model is utilized for hydrate phase fugacities while the statistical association fluid theory is applied for the calculation of vapor and liquid phase fugacities. The study is truly predictive in the way that no mixture experimental data are adjusted and the binary interaction parameter is set to zero. The equilibrium phase compositions are also computed for multicomponent gas-inhibitor-water system in the hydrate region.

(222az) Liquid-Liquid Equilibria of Three Ternary Systems (Water + Propargyl Alcohol + n-Propyl Acetate, or + iso-Propyl Acetate, or + iso-Butyl Acetate)

Hu, X.,
Sun, C., Zhejiang University
Cao, W.,
Wu, J.,

The liquid-liquid equilibria of three ternary systems (water + propargyl alcohol + *n*-propyl acetate, water + propargyl alcohol + *iso*-propyl acetate, and water + propargyl alcohol + *iso*-butyl acetate) were experimentally investigated under the temperature of (279.15 - 321.15) K and the atmospheric pressure in this work. 15 profiles of isothermal liquid-liquid equilibria data measured by gas chromatograph were provided. The quality of experimental tie-line data was attested using the Othmer-Tobias equation and the Bachman equation. The consistency of the experimental data indicates that the detected data are confident. The experimental results shows that propargyl alcohol (PA) concentration in *n*-propyl acetate (NPA) phase and in *iso*-propyl acetate (IPA) phase raise a bit more rapidly than that in *iso*-butyl acetate (IBA) phase with PA concentration in water phase increasing. Each system behaves type of LLE. NRTL model was used to correlate the experimental data and the binary interaction parameters of three investigated systems were obtained. But the predicted values of PA concentration in water phase are some large than the measured data. The total root-mean-square deviation (RMSD) calculated for three systems were 0.0104 (NPA), 0.0157 (IPA) and 0.0117 (IBA) respectively.

(222ba) New Volume-Explicit Equations of State Based On Statistical-Mechanical Theory

Alsaifi, N., King Fahd University of Petroleum & Minerals

Almost all proposed theory-based equations of state are developed in canonical ensemble (NVT). In chemical engineering applications, however, pressure and temperature are the main specified variables. Therefore, it is more convenient to determine thermodynamic properties from volume-explicit equation of state. In this work, two new volume-explicit equations of state are proposed based on statistical-mechanical theory. The equations of state are developed by transferring a pressure-explicit equation of state in NVT ensemble to NPT ensemble. The transformation produced two-pair of volume-explicit equations; one for the gas phase and the second for the liquid phase. The new equations of state take into consideration both repulsive and dispersive interactions. The equations are tested to noble gases and short-chain hydrocarbons and show good agreement with experimental data.

(222bb) Understanding the Role of Dispersive Interactions in the COSMO-SAC Model

Hsieh, C. M., University of Paderborn
Lin, S. T., National Taiwan University
Vrabec, J., University of Paderborn

A reliable predictive thermodynamic model is highly desirable to reduce the amount of required experimental data, which are crucial for the design of separation or purification processes in chemical and pharmaceutical industries. In the past two decades, a new class of predictive methods, i.e. COSMO-based approaches (COSMO-SAC or COSMO-RS), has emerged. These methods utilize the results from modern computational chemistry and do not contain any species-dependent parameters. Therefore, the problem of missing parameters does not arise, and crucial thermophysical information can be predicted with the molecular structure as the only input. Recently, the COSMO-SAC model was revised to provide acceptable predictions for both vapor-liquid and liquid-liquid equilibria of organic mixtures. However, a well-known issue, which is the neglect of the dispersive molecular interactions in mixtures, remained unsolved in the COSMO-SAC model. In this study, classical force field based molecular modeling and simulation is used to supply data on a sound physical basis and a new dispersion term to consider the contribution of the dispersive interaction to the non-ideality of mixtures is introduced into COSMO-SAC model. With this modification, the accuracy in vapor-liquid equilibrium predictions can be improved by around 10%. For 391 different binary mixtures, the average deviation in terms of total pressure at a fixed liquid composition is 4.7% and the overall average deviation in terms of vapor phase composition is 1.7%, compared to 5.3% and 1.8% from the original model.

(222s) Predicting the Phase Behavior of Fluorinated Complex Systems Using the GC-SAFT-VR Equation of State

Haley, J. D., Vanderbilt University
dos Ramos, C., Vanderbilt University
McCabe, C., Vanderbilt University

Predicting the Phase Behavior of Fluorinated Complex Systems Using the GC-SAFT-VR Equation of State

Jessica D. Haley[1], M. Carolina dos Ramos[1] and Clare McCabe[1,2]

[1]Department of Chemical and Biomolecular Engineering, Vanderbilt University, Nashville, TN

[2]Department of Chemistry, Vanderbilt University, Nashville, TN

Fluorinated molecules such as fluoroalkanes, perfluoroalkanes (PFA), perfluroalkylalkanes (PFAA), and fluorinated ethers possess interesting physical properties that lead to their use in a wide range of applications, for example as catalysts, solvents, and pharmaceuticals. However, while there is an abundance of thermophysical data for alkanes, only limited studies have been performed that report the thermodynamic properties of the corresponding fluorinated species. With only limited data available, predictive approaches are needed to accurately and reliably determine the physical properties of these molecules. The statistical associating fluid theory (SAFT) [1] is a commonly used molecular-based equation of state that has been successfully applied to study a wide range of fluid systems. In recent work, the GC-SAFT-VR equation was developed, which combines the SAFT equation for potentials of variable range (VR) [2] with a group contribution (GC) approach [3] that allows for the description of hetero-segmented chains. The equation has been shown to provide an excellent description of the phase behavior of pure associating and non-associating fluids and their mixtures, with a minimal reliance on fitting the model parameters to experimental data [3-5]. Parameters for key functional groups (such as CH_3, CH_2, CH, CH_2=CH, C=O, C_6H_5, ether and ester, OH, NH_2, CH=O, COOH) were obtained by fitting to experimental vapor pressure and saturated liquid density data for selected low molecular weight fluids and then used to predict the phase behavior of pure non-associating fluids and their mixtures without adjusting the group parameters to binary mixture data. Here we report parameters for the CF_3,CF_2, CF, CH_2F, CHF_2, and CHF functional groups and their cross interactions. Comparing the theoretical predictions with experimental data for pure perfluoroalkanes and PFAAs as well as mixtures of alkanes, perfluoroalkanes, fluorinated ethers and PFAAs test the transferability of the fluorinated parameters. The GC-SAFT-VR approach is found to accurately predict the phase behavior of the systems studied without adjusting the group parameters to binary mixture data.

[1] W.G. Chapman, K. E. Gubbins, G. Jackson, M. Radosz. Fluid Phase Equilibria 52, 31-38 (1989).

[2] Gil-Villegas, A.; Galindo, A.; Whitehead, P. J.; Mills, S. J.; Jackson, G. and Burgess, A. N. Journal Of Chemical Physics, 106, 4168-4186. (1997).

[3] Y. Peng, K. D. Goff, M. C. dos Ramos, C. McCabe, Fluid Phase Equilibria 277(2), 131- 144 (2009).

[4] Y. Peng, K. D. Goff, M. C. dos Ramos and C. McCabe, Industrial & Engineering Chemistry Research, 49 (3), 1378-1394 (2010).

[5] M. C. dos Ramos, J. D. Haley, J. R. Westwood, and C. McCabe, Fluid Phase Equilibria, 306, 97-111 (2011).

(222bd) Composition Dependency of the Flory-Huggins χ Parameter in Isotopic Polymer Blends

Russell, T.,

Edwards, B., University of Tennessee

Khomami, B., University of Tennessee

Flory-Huggins Theory has been the basis for understanding polymer/solvent and blended polymer thermodynamics for much of the last 50 years. Within this theory, a parameter (χ) was developed to account for the energy of dispersion between distinct components. Thin film self-assembly of diblock copolymers and polymer melts depends critically on this parameter, and in application, χ has generally been assumed to be independent of the concentrations of individual components in the system. However, Small Angle Neutron Scattering (SANS) data on isotopic polymer blends, such as polyethylene (PE) and deuterated polyethylene (dPE), have shown a parabolic concentration dependency for χ. In order to understand better the nature of χ and develop more accurate morphological data for polymer systems, an investigation of this concentration dependency was undertaken.

The focus of this research was to use Molecular Dynamics and Configurational Bias Monte Carlo simulations to test the Flory-Huggins interaction parameter in the PE-dPE system via both its structural (χ_S) and thermodynamic (χ_T) definitions. Structural calculations for χ_S were based on the Random Phase Approximation developed by de Gennes, which assumes that the structure factor (S(Q)) is linked to the Flory-Huggins parameter as shown in equation 1.

$$S^{-1}(Q)=[N_D\phi_D g_D(R_{g,D},Q)]^{-1}+[N_H\phi_H g_D(R_{g,H},Q)]^{-1}-2\chi_S$$

1.

Here, g_D is the familiar Debye function,

$$g_D(R_g,Q)=2[R_g^{\,2}Q^2+e^{-R_g^{\,2}Q^2}-1]/R_g^{\,4}Q^4$$

2.

and R_g is the radius of gyration of the polymer. Q is the wavenumber, and Φ and N are the volume fraction and chain length, respectively.

Thermodynamic information was obtained through thermodynamic integration of the free energy with χ_T defined using the original Flory-Huggins Theory, as shown in equation 3.

$$\frac{\Delta G}{k_B T}=\frac{\phi_H\ln\phi_H}{N_H}+\frac{\phi_D\ln\phi_D}{N_D}+\chi_T\phi_H\phi_D$$

3.

Here, k_B is the Boltzmann constant; T is the absolute temperature of the system, and DG is the change in Gibbs free energy.

Calculation of the structural and thermodynamic interaction parameters allowed X_S and X_T to be compared for internal consistency and then examined with respect to the experimental SANS data for the PE/dPE system.

(222be) Are Ionic Liquids An Alternative for Gas Sweetening?

Mota Martinez, M. T., Petroleum Institute
Berrouk, A. S., Petroleum Institute
Kroon, M. C., Eindhoven University of Technology
Peters, C. J., Petroleum Institute

Natural Gas (NG) undergoes several treatment steps aiming to remove undesirable components such as carbon dioxide (CO_2) and hydrogen sulfide (H_2S), which are known as acid gases. A gas sweetening process is commonly performed by chemical or physical absorption. The latter is usually preferred when large amounts of acid gas should be removed because of the less energy demanding regeneration step.

Ionic liquids (ILs) have been proposed as alternative solvents for numerous chemical processes due to properties such as chemical stability and negligible vapour pressure. Extensive research has been executed on the solubility of CO_2 in ILs. It was found that CO_2 is highly soluble in certain ILs. This fact encouraged us to propose ILs as an alternative solvent for CO_2 capture.

The aim of this work is to study the thermodynamic aspects of the feasibility of using ILs for CO_2 separation from natural gas. The solubility of CO_2, CH_4 and other light hydrocarbons in the low-viscosity 1-hexyl-3-methylimidazolium tetracyanoborate ionic liquid ([hmim][TCB]) has been experimentally determined using a synthetic method (a Cailletet equipment) at temperatures up to 400 K and pressures up to 15 MPa.

The solubility of CO_2 in [hmim][TCB] has been found to be up to 20 times higher than the solubility of CH_4. Interestingly, while the solubility of CO_2 is highly dependent on the temperature, the solubility of CH_4 in the IL is almost not influenced by temperature. Also it will be discussed why Henry's law underestimates selectivities at elevated pressures.

The binary P,T phase diagram of ethane and propane in [hmim][TCB] shows maximum solubility near the respective critical points of the two gases. It was observed that both hydrocarbons have a lower solubility than CO_2 in [hmim][TCB]. However, there is a clear tendency that the higher the alkyl chain of the hydrocarbon, the higher is the solubility in [hmim][TCB]. Therefore, it is possible to find the optimum conditions to maximize the absorption of CO_2 and minimizing the absorption of CH_4 and other hydrocarbons.

(222bf) Volume-Translated Peng-Robinson Equation of State for Liquid Densities of Diverse Binary Mixtures

Abudour, A. M., Oklahoma State University
Mohammad, S. A., Oklahoma State University
Robinson, R. L. Jr., Oklahoma State University
Gasem, K. A. M., Oklahoma State University

Two-parameter cubic equations of state (CEOS) are used widely in process engineering calculations; however, inaccurate liquid density predictions remain a significant deficiency in these equations. To remedy this problem, a volume translation of the CEOS is employed frequently. In a recent work, we presented a volume-translated Peng-Robinson equation of state (VTPR EOS) that is capable of providing accurate density predictions for both saturated- and single-phase regions of pure fluids at high pressures. In the current work, we present an extension of that approach, employing conventional mixing rules, to predict densities of liquid mixtures over wide ranges of pressure and temperature. For this purpose, two databases were compiled for vapor-liquid equilibrium and liquid density measurements of 73 binary systems composed of diverse chemical species. The molecular species in the databases ranged widely in terms of molecular size, shape, asymmetry and polarity and, thus, were well suited to test the efficacy of our approach. Overall, the databases contained more than 5,000 data points for vapor-liquid equilibrium measurements and over 13,000 data points for liquid density measurements of mixtures.

The VTPR EOS predictions of mixture liquid densities yield errors that are three to five times lower than the corresponding predictions from the untranslated Peng-Robinson equation of state (PR EOS). Specifically, the overall percentage average absolute deviations (%AAD) from the VTPR EOS varied from 1.5 to 3 for binary mixtures. This represents a substantial improvement relative to the untranslated PR EOS, for which errors ranged from 2 to 15 %AAD. These results indicate that extension of the VTPR EOS to liquid mixtures is capable of providing reliable density predictions for diverse binary mixtures over a wide range of pressure and temperature.

(222r) Experimental Characterization of the Solubility of High Molecular Weight Liquids in Polydimethylsiloxane (PDMS) in Presence of a Low Molecular Weight Solvent

Cocchi, G., University of Bologna

De Angelis, M. G., University of Bologna

Doghieri, F., University of Bologna

Experimental characterization of the solubility of high molecular weight liquids in polydimethylsiloxane (PDMS) in presence of a low molecular weight solvent

Giovanni Cocchi, Maria Grazia De Angelis*, Ferruccio Doghieri

Department of Civil, Chemical, Environmental and Material Engineering - University of Bologna, Via Terracini 28, 40131 Bologna, Italy

* Corresponding Author Email Address: grazia.deangelis@unibo.it

In order to characterize the separation performances of polymeric membranes that could be applied in separation processes relevant to food and pharmaceutical industry, experimental values of the solubility of low molecular weight species in the polymer that forms the membrane are required [1]. In this work, the experimental characterization of the solubility of liquids of interest to the food and pharmaceutical industry in a crosslinked PDMS film is discussed, considering the effect of the presence of a low molecular weight solvent that swells the matrix, affecting the transport [2,3]. Crosslinked PDMS was selected for its wide use in applications, especially in membrane separation processes and for its resistance to solvents [4]. The measurement of liquid solubility in PDMS was performed gravimetrically, according to the blot and weight method: samples were immersed in vials filled with liquid, placed in a thermostatic bath and weighted at regular intervals with an analytical balance. In this way the mass uptake as a function of time was obtained. In the case of mixed liquids solubility, we considered solutions formed by a volatile component and a second one with negligible vapor pressure. Firstly the measurement of the total mass uptake of the liquid mixture was performed, according to the previously described procedure. Then, the samples are left under hood to let the volatile component evaporate, until constant weight was obtained. Mass balance calculations provided the separate mass uptake of the volatile and of the non volatile component. It was found that the solubility of the less soluble component is deeply enhanced by the presence of the other, more soluble component, that usually is the low molecular weight solvent. For example, as shown in the figure below, the swelling induced by n-pentane promotes the dissolution of eicosane in PDMS, which shows a maximum at an intermediate mass fraction of n-C_5 in the mixture. This behavior has been observed in many other mixtures such as the one formed by: acetone and squalene, pentane and oleic acid, pentane and olive oil, acetone and oleic acid, acetone and olive oil and tertbutanol and PEG 400. The Figure 1 depicts the solubility of pentane and eicosane in PDMS, as a function of pentane concentration. For the case of pentane and eicosane, an approximate thermodynamic analysis has been conducted in order to compare the infinity dilution activity coefficient of pentane in eicosane in this ternary mixture, to the value available in the literature.

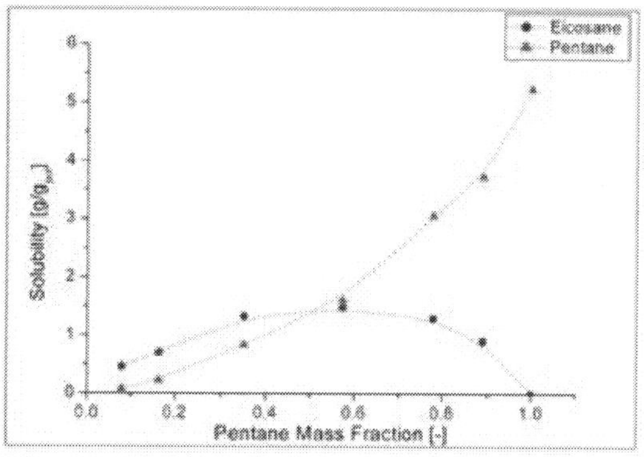

Figure 1: Solubility of Pentane and Eicosane in Crosslinked PDMS at 308 K.

[1] S.P. Nunes, K.V. Peinemann, Membrane Technology in the Chemical Industry, 2nd Edition, 2006, WILEY-VCH Verlag GmbH & Co. KGaA, Weinheim, Germany.

[2] R. Subramanian, M. Nakajima, T. Kawakatsu, *Processing of Vegetable Oils Using Polymeric Composite Membranes,* Journal of Food Engineering, 38, 1998, 41-56.

[3] R. Subramanian, K. Raghavarao, M. Nakajima, H. Nabetani, T, Yamaghuchi, T. Kimura, *Application of Dense Membrane Theory for Differential Permeation of Vegetable Oil Constituents,* Journal of Food Engineering, 60, 2003, 249-256.

[4] M.G. DeAngelis, T. C. Merkel, V.I. Bondar, B.D. Freeman, F. Doghieri, G.C. Sarti, *Hydrocarbon and Fluorocarbon Solubility and Dilation in Poly(DimethylSiloxane): Comparison of Experimental Data with Predictions of the Sanchez ? Lacombe Equation of State,* Journal of Polymer Science Part B: Polymer Physics, 37, 1999, 3011-3026.

(222bc) Modeling Speciation in Nitric Acid Solutions Using the Binding Mean Spherical Approximation

Hlushak, S., Institute for Condensed Matter Physics
Simonin, J. P., Universit P.M. Curie
De Sio, S.,
Bernard, O., Universit Pierre et Marie Curie
Ruas, A.,
Pochon, P.,
Jan, S.,
Moisy, P., CEA Marcoule, France

A model was developed[1] to describe the thermodynamic properties (osmotic and activity coefficients) and degree of association of electrolytes. It is assumed that the cation and the anion can form two different kinds of pairs: a solvent separated ion pair and a chemical pair (undissociated molecule). The model is an extension of the binding or associative mean spherical approximation (BiMSA or AMSA), in which the size and charge of the ions in the chemical pair may differ from those of the free ions. Corrections arising from the description of the chemical equilibrium conditions at Lewis-Randall (constants pressure) state, instead of McMillan-Mayer (constant solvent chemical potential) state, are taken into account in the model. The model is then applied to the solutions of nitric acid for which the experiment suggests the formation of two ion pairs. A fit of the osmotic coefficient and the proportion of free ions (obtained from Raman spectroscopy experiments), allowed us to determine[1] the speciation of nitric acid, i.e the concentrations of chemical and solvent separated ion pairs, which is in good agreement with experimental results.

[1] S. Hlushak, J. P. Simonin, S. De Sio, O. Bernard, A. Ruas, P. Pochon, S. Jan and P. Moisy, "Speciation in aqueous solutions of nitric acid" *Dalton Transactions*, *42*(8), 2853-2860 (2013)

(222bg) Modeling of Thermodynamic, Volumetric, and Electrical Properties With the Electrolyte CPA Equation of State

Maribo-Mogensen, B., Technical University of Denmark
Kontogeorgis, G., Technical University of Denmark
Thomsen, K., Technical University of Denmark

Complex mixtures of associating/polar components and electrolytes are often encountered in the oil- and gas and chemical industry. It is well-known in the oil- and gas industry that electrolytes have a substantial effect on e.g. solubilities of gases in water-hydrocarbon mixtures (salting-out effect), and furthermore, the presence of electrolytes may enhance the inhibitory effect of methanol and glycol on the formation of gas hydrates in natural gas pipelines, thus allowing for problem-free flow [1].

The osmotic coefficients and mean ionic activity coefficients of aqueous and mixed solvent electrolytes have been successfully modeled through activity coefficient models such as electrolyte NRTL [2] or Extended UNIQUAC [3]. These properties have also been represented by electrolyte equations of state (EoS) [1], but the e-EoS are still in their infancy as they require a more accurate physical representation to simultaneously account for volumetric and thermodynamic properties.

Molecular simulations have been used to assess our physical understanding of the complex interactions arising from the strong and long-range Columbic forces [4]. While the basic ion-ion interactions may be modeled through the Debye-Hückel or mean spherical approximation (MSA) models [4,5], these terms become less valid at high concentrations or in low-permittivity solvents where the electrostatic interactions lead to the formation solvation shells and short-lived ion pair complexes that may either be charged or neutral with dipole or multipole moments. While the thermodynamic properties represent a combination of all of the physical effects, the individual processes have also been studied indirectly through dielectric spectroscopy [6], density [7] and conductivity measurements [8].

CPA has successfully been applied in modeling of the phase equilibrium of natural gas components in mixtures of water, glycols or alcohols. CPA was combined with a simple geometrical model and the Onsager / Kirkwood / Fröhlich framework for calculation of the static permittivity [9] . We have studied how the effect of ion association can be used to extend the Cubic Plus Association (CPA) EoS to predict and correlate the thermodynamic, volumetric, dielectric, and electric properties of electrolytes in mixed solvents for a selection of ions; (Li^+, Na^+, K^+, Cs^+, Mg^{2+}, Ca^{2+}, Cl^-, Br^-, I^-, NO_3^-, SO_4^{2-}). Finally, we show how the electrolyte CPA can be used to predict important flow assurance properties including gas solubility and gas hydrate inhibition in mixed solvents.

[1] G. M. Kontogeorgis; G. Folas *Thermodynamic Models for Industrial Applications*, Wiley, Chichester, **2010**, ISBN: 978-0-470-69726-9

[2] Chen, C. C., & Evans, L. B. (1986). *A local composition model for the excess Gibbs energy of aqueous electrolyte systems*. AIChE Journal, *32*(3), 444-454.

[3] Thomsen, K., & Rasmussen, P. *Modeling of vapor–liquid–solid equilibrium in gas–aqueous electrolyte systems*. Chemical Engineering Science (**1999**), *54*(12), 1787-1802.

[4] Fisher, M. E., & Levin, Y.. *Criticality in ionic fluids: Debye-Hückel theory, Bjerrum, and beyond*. Physical review letters (**1993**), *71*(23), 3826-3829.

[5] B. Maribo-Mogensen, G. M. Kontogeorgis, K. Thomsen, *Comparison of the Debye–Hückel and the Mean Spherical Approximation Theories for Electrolyte Solutions*, Ind. Eng. Chem. Res. (2012), 51, 5353

[6] Buchner, R., & Barthel, J. (2001). *Dielectric relaxation in solutions*. Annual Reports Section" C"(Physical Chemistry), *97*, 349-382.

[7] Marcus, Y. (2010). On the intrinsic volumes of ions in aqueous solutions.*Journal of solution chemistry*, *39*(7), 1031-1038.

[8] Safonova, L. P., & Kolker, A. M.. *Conductometry of electrolyte solutions*. Russian Chemical Reviews, (**1992**), *61*(9), 959-973.

[9] B. Maribo-Mogensen, G. M. Kontogeorgis, K. Thomsen, *Modeling of Dielectric Properties of Complex Fluids with an Equation of State*, J. Phys. Chem. B (2013), 117, 3389

(222bi) A Statistical Mechanically-Based Cubic Equation of State for Athermal Hard-Sphere Chains

Gow, A. S., University of New Haven
Demir, S., University of New Haven
Alkhaldi, S., University of New Haven
Suss-Riter, D., Western Michigan University

Traditional engineering equations of state such as the Soave-Redlich-Kwong or Peng-Robinson EOS are desirable owing to their cubic in molar volume form. On the other hand these equations of state generally give poor results when applied to systems at extreme conditions or which contain polar and/or associating components. Conversely, molecularly-based equations of state (i.e., statistical associating fluid theory EOS and its many variants) are much more accurate when successfully applied to such systems (Muller and Gubbins, 2001); but they are sufficiently computationally complex on account of several molar volume roots, which make physically meaningful solutions often difficult to obtain. One approach to overcome this trade-off between accuracy and complexity is to incorporate molecular structure and behavior into a form of equation of state which has the attractive cubic in molar volume form.

We present here the athermal backbone for constructing a new class of statistical mechanically-based cubic equations of state. Specifically, we incorporate molecular simulation data for hard-sphere interaction, covalent bond formation and hard-dimer interaction into a simple expression which retains the cubic in molar volume form of the EOS while accurately describing the compressibility factor versus packing fraction behavior for homonuclear hard-sphere chains up to 201 segments. Firstly, molecular simulation data for hard-sphere fluids (Wu and Sadus, 2005) were nearly exactly fit to a simple two-parameter second-order polynomial in packing fraction which ensures the ideal gas limits of the compressibility factor Z=1 and for the radial distribution function at contact $g^{HS}(s)=1$ as packing fraction h=1. Next, Werthiem's first-order perturbation theory (TPT1) (Wertheim, 1987) is applied to obtain a simple, highly accurate and consistent expression for the compressibility factor of chain formation which retains the overall cubic in molar volume form of the EOS while introducing no new parameters. Finally, as shown in earlier published work (Ghonasgi and Chapman, 1994; Chang and Sandler, 1995), that the performance of equations of state for athermal hard-sphere chains is improved by including a contribution for the interaction between athermal hard-dimers. We refit $g^{HD}(s)$ versus h data produced from a simple theoretical expression based on the Percus-Yevick approximation (Chiew, 1990) a form which was derived from same simple second-order polynomial in packing fraction used for Z^{HS} which yielded two additional parameters. The combined equation of state including all three contributions yielded excellent results in representing molecular simulation data for Z versus h for athermal homonuclear hard-sphere chains up to 201 segments (Yu et. al., 1994).. Furthermore, decent results comparable with those obtained from considerably more complex equations of state (Chang and Sandler, 1995) were obtained for the second-virial coefficient B versus segment number m up to 128 segments. Finally, since the four *temperature-independent* parameters are *universal*(the same for all real fluids), the EOS is easily extended to mixtures. Results for Z versus h were also generally very good when compared with available molecular simulation data from the literature (Yu et. al., 1994).

In sum, a simple and accurate molecularly-based EOS was developed for athermal homonuclear hard-sphere chains using molecular simulation data for three key effects. The equation contains four *universal*parameters, is easily applied to mixtures and serves as a basis for developing fundamentally sound cubic equations of state including dispersion, multipolar interactions and association.

REFERENCES

Chang, J. and Sandler, S. I., Correlation Functions of Hard-Sphere Chain Fluids: Comparison of the Wertheim Integral Equation Theory with the Monte-Carlo Simulation, J. Chem. Phys., 102, 437-451 (1995).

Chiew, Y. C., Intermolecular Site-Site Correlation Function of Athermal Hard-Sphere Chains: Analytic Integral Equation Theory, J. Chem. Phys., 93, 5067-5074 (1990).

Ghonasgi D. and Chapman, W. G., A New Equation of State for Hard Chain Molecules, J. Chem. Phys., 100, 6633-6639 (1994).

Muller, E. A. and Gubbins, K. E., Molecular-Based Equations of State for Associating Fluids: A Review of SAFT and Related Approaches, Ind. Eng. Chem. Res., 40, 2193-2211 (2001).

Wertheim, M. S., Thermodynamic Perturbation Theory of Polymerization, J. Chem. Phys., 87, 7223-7231 (1987).

Wu, G.-W. and Sadus, R. J., Hard Sphere Compressibility Factors for Equation of State Development, AIChE J., 51, 309-313 (2005).

Yu, Y.- X, Lu, J.-F., Tong, J.- S. and Li, Y.- G., Equation of State for Hard-Sphere Chain Molecules, Fluid Phase Equilibria, 102, 159-172 (1994).

(222t) Experimental Vapor-Liquid Equilibria for Carbon Dioxide + 1-Propanol + Heptane

Ver nico-S nchez, F. J.,
Elizalde-Solis, O., INSTITUTO POLITECNICO NACIONAL-ESIQIE
Galicia-Luna, L. A., INSTITUTO POLITECNICO NACIONAL-ESIQIE

The aim of this work is to report the vapor-liquid equilibrium (VLE) for the ternary system carbon dioxide + 1-propanol + heptane at temperatures of (315, 344 and 374) K. Experiments were performed in an apparatus based on the static – analytic method. The main advantages of the apparatus are the on-line sampling via a ROLSI sampler-injector and the volume sample of about 1 uL on each analysis. VLE is reported for temperature, pressure and compositions with an uncertainty less than 0.02 K, 0.008 MPa, 0.0035 for liquid and 0.0015 for vapor composition in mole fraction for the last two values. Separation factors obtained from experimental x_i, y_i mole fractions for heptane over 1-propanol indicate that heptane is feasible to be separated from 2-propanol in the presence of carbon dioxide. The higher values for this variable are obtained at 315 K, and as temperature rises, the separation factors decrease. In consequence, the lower values correspond to the temperature of 374 K. Finally, the experimental data of VLE were represented using the Peng-Robinson Equation of State with the classical and Wong-Sandler mixing rules. The binary interaction parameters were taken from data reported in the literature about VLE of the concerning binary systems.

(222bh) Oxy-Fuel Combustion Optical Flame Analysis and Characterization

Carney, C. S., National Energy Technology Laboratory, U.S. Department of Energy
Woodside, R., National Energy Technology Laboratory, U.S. Department of Energy
Ochs, T., National Energy Technology Laboratory, U.S. Department of Energy
Gerdemann, S. J., National Energy Technology Laboratory, U.S. Department of Energy
Oryshchyn, D., National Energy Technology Laboratory, U.S. Department of Energy

Title: Oxy-Fuel Combustion Optical Flame Analysis and Characterization

Lead Presenter's Email: casey.carney@contr.netl.doe.gov
KeywordsCombustion Flame Analysis, Fossil Fuel, Heat Transfer, Oxy-Fuel

Oxy-Fuel Combustion Optical Flame Analysis and Characterization

C. S. Carney[1,2], R. Woodside[1], S. J. Gerdemann[1], D. Oryshchyn[1], T. Ochs[1]

[1]National Energy Technology Laboratory, 1450 Queen Ave SW, Albany, OR 97321

[2]URS Corporation, P.O. Box 1959, Albany, OR 97321-0521

Oxy-fuel combustion uses conventional technology for pulverized coal combustion, generating a flue gas composed primarily of carbon dioxide and water. While conventional air-fired coal combustion flame behavior is fairly well understood and optimized, the absence of nitrogen and consequent increase of carbon dioxide fraction and mass flow differences in oxy-fuel combustion results in differing flame behavior. Two notable examples are flame temperature and heat transfer within the boiler radiant zone. A unique suite of optical instrumentation devices has been designed for the purpose of oxy-fuel combustion flame study: two monochromators, four total radiometers, two fiber optic CCD spectrometers, and a video camera. These instruments view the flame via optical ports mounted along the flame axis – some along the burner centerline, and some 16" below the centerline. This system is capable of analyzing the following flame components: temperature, radiance, stability, morphology, and radical species. The radiometric components of the system will tend to observe emitted radiation from the soot particles, which will be used as a proxy for the flame's temperature characteristic.

The monochromator records irradiance vs. wavelength over the range of 800-5000 nm. Average radiant particle temperature can be calculated by fitting the Planck blackbody radiation curve to the shape of the spectrum over an optimized wavelength range for scan time minimization. The radiometers record total radiant flame intensity over the range of 200-7000 nm, can indicate flame oscillatory behavior, and are located along the burner centerline axis, allowing for a flame intensity profile along the flame length. The CCD spectrometers provide instantaneous and continuous radiance vs. wavelength spectra in the range of 230-1100 nm. CCD spectrometer data are applied to calculate flame temperatures in their viewing area using the three-color method (evaluating the radiance ratio at three wavelengths). Flame radical species can be identified by these spectrometers, which are of use for combustion diagnostics. Rapid scan times enable the spectrometers to also indicate oscillations in the flame. Flame stability and morphology are monitored with a high speed video camera with a 90° field of view that has been customized for high temperature flame use.

This optical flame measurement system has been installed at the Jupiter Oxygen Burner Test Facility (15 MWth) in Hammond, IN. The test facility also contains various non-optical devices that further aid flame characterization such as thermocouples installed in inter-tube webbing and "chordal" thermocouples which can be used to determine boiler tube heat flux. Online flue gas compositional analysis records combustion-product volume-fraction composition at the boiler exit. The optical analysis system was operational for both conventional air firing and oxy firing testing. These results will be presented and are used both for flame characterization and comparison to existing computational fluid dynamics boiler simulations, including radiant zone temperature and heat transfer mapping.

(222u) Modeling of Liquid-Liquid Equilibrium for Extraction of Free Fatty Acids (FFAs) in Corn Oil

Mills, P. L., Texas A&M University-Kingsville
Mehrbod, M., Texas A&M University-Kingsville
Moran, C., Texas A&M University-Kingsville

Modeling of Liquid-Liquid Equilibrium for Extraction of Free Fatty Acids (FFAs) in Corn Oil.

Removal of free-fatty acids (FFA's), whose content in crude corn oil and other sustainable vegetable oils varies from 3% to 40% [Srivastava and Prasad, 2000; Ma and Hanna, 1999], is an essential process for creating refined products that can be subsequently used in food and other important applications. The method chosen for oil extraction depends on the nature of the raw material, as well as on the capacity of the industrial plant. Pressing followed by solvent extraction is the method most widely employed for handling a wide variety of oilseeds [Young et al., 1994], which contribute nearly 50% of the total vegetable oil produced in the world. The crude oil that is extracted from the oilseeds is a mixture of FFA, mono-, di-, and triglycerides, phosphatides, pigments, sterols and tocopherols.

Several approaches for removal of the FFA's can be used that are based on non-reactive and reactive solvent extraction. In non-reactive extraction, for example, either pure methanol (MeOH) or MeOH-water mixtures are commonly used to selectively extract the FFA's from the crude oil. Reactive extraction involves contacting the crude oil with a solution of MeOH with dissolved sodium hydroxide (NaOH) to form the sodium salts of the corresponding acids. Design of either non-reactive or reactive extraction contactors requires accurate predictions of thermodynamic phase equilibrium properties.

In this poster, experimental liquid-liquid equilibrium data for ethanol-water-corn oil and pure ethanol-corn oil are used as the basis for discrimination between two solution–phase thermodynamic models. The resulting parameters are then used as the basis of an Aspen simulation for extraction of FFA's from corn oil using ethanol and ethanol/water mixtures as extraction solvents.

The liquid-liquid extraction is modeled using the Non-Random Two Liquids (NRTL) and the UNIversal QUAsi Chemical models in which the interaction parameters for both models are obtained by regression analysis of experimental liquid-liquid extraction data. It is shown that the NRTL model provides the best agreement between experimental and model-predictions. Simulation of the extraction using this models shows that realistic predictions for the concentration profiles for the various species. It also provides information needed to evaluate the performance of simultaneous reaction with extraction for the case where NaOH is added to the solvent.

References

1. Srivastava, A., Prasad, R., 2000. Triglycerides-based diesel fuels. Renew.Sustain. Energy Rev. 4, 111–133.

2. Ma, F., Hanna, M.A., 1999. Biodiesel production: a review. Bioresour. Technol. 70, 1–15.

3. Young, F. V. K., Poot, C., Biernoth, E., Krog, N., Davidson, N. G. J.,& Gunstone, F. D. (1994). Processing of fats and oils. In F. D. Gunstone, J. L. Harwood, & F. B. Padley (Eds.), The lipid handbook (2nd ed., pp. 249–276). London, UK: Chapman & Hall.

(222v) Simulating Nanoscale Thermal Transport in Nanoscale Systems

Vemuri, S. H., Carnegie Mellon University
Chung, P. S., Carnegie Mellon University
Park, S., Carnegie Mellon University
Jhon, M. S., Carnegie Mellon University

The ever increasing demand for higher performances in NEMS/MEMS and hard disk drives (HDDs) has resulted in decreasing the dimensions further into nanoscale, where thermal management via heat transfer is increasingly important due to significantly higher heat dissipation and operating temperatures. For example, state-of-the-art recording technology in the benchmark example of HDDs called heat assisted magnetic recording (HAMR) results in swift degradation of nanoscale lubricant, carbon overcoat (COC), and magnetic layers, leading to magnetic media corrosion which is detrimental to HDD operation. In addition, the lack of thorough understanding of the temperature profiles introduced by the hotspot and energy management throughout these materials also exacerbates the problem.

To address this issue, in this study we will investigate the transient heat transfer in nano-scale thin films in HDI when a hotspot is created. Since traditional conduction models like Fourier law are not suitable due to dominant sub-continuum effects such as ballistic thermal transport and temperature slip at the boundaries, we utilize a lattice Boltzmann method (LBM) for phonons and electrons (which are primary carriers of energy) in the nano-scale thermal modeling. LBM originates from the non-equilibrium methodology of Boltzmann transport equations (BTEs) and is computationally efficient due to easy parallelization with capabilities of easily handling complex geometries. Another advantage of LBM formulation in this system arises from the capability to accurately account for the temperature slip at the boundaries and interfaces, which is critical for energy management in HDDs. This is done by modeling phonon scattering (a) at the system boundaries, where a specularity factor P is introduced, which represents the fraction of carriers undergoing diffusive scattering at the boundary, accounting for the surface roughness, and (b) at the interface of two layers, where the transmission coefficient obtained from the diffusive mismatch model is applied, representing total fraction of energy transmitted across the interface. Our results of in-plane and out of plane thermal transport mechanism and temperature profiles in realistic thin film structures with complex grain boundary interfaces as well as heat transfer between each adjacent layers will provide a clear physical understanding of the dynamics at nanoscale systems. In addition, our study will also lead to providing novel design criteria for efficient thermal management in these systems.

(222bj) A Comprehensive Energy Analysis of Photovoltaic Thermal Water Electrolyzer for Hydrogen Production

Oruc, M. E., University of Illinois at Urbana-Champaign
Nuzzo, R. G., University of Illinois at Urbana-Champaign

Solar energy is often harnessed by using a photovoltaic/thermal (PVT) collector, where a photovoltaic (PV) module and a thermal collector convert solar energy into electricity and heat, respectively [1]. The heat energy has been conventionally used for domestic hot water and/or room heating; it is, however, highly desirable to convert the released energy into alternative storable forms. Hydrogen is environmentally friendly and a sustainable energy carrier that can be stored. Currently the PV-assisted water electrolyzers available on the market have PV conversion efficiencies of only 15%. The energy that is not transformed into electricity is just dissipated. To this end, we have developed a new integrated system – so-called "photovoltaic thermal water electrolyzer" – which consists of PV cells positioned on top of a planar micro water electrolyzer. Using this system, we utilize the released heat to raise the temperature of the electrolyte solution, thereby increasing the efficiency of hydrogen production. Herein, we provide a comprehensive analysis of the overall energy output efficiency of a PVT water electrolyzer.

The hourly radiation and ambient temperature data of Phoenix are used in our model. The temperature of PV module and the electrolyte are computed using COMSOL Multiphysics software. The 3D design of the PV system and water electrolyzer, which includes a 400 um thick electrolyzer chamber, is presented. The electrical output of the PV was calculated based on the work of presented in reference [2]. The efficiency of the water electrolyzer is calculated from the data reported in the literature at constant current flux as 10mA/cm^2 [3]. The analysis for the day June 21st 2010 indicates that the highest efficiency difference between the PV modules with and without water electrolyzer is 2%. At the same time the improvement in the efficiency of the water electrolyzer is around 6%, which means an overall increase of 4% in the efficiency of the integrated system. This new integrated approach is advantageous over PV module and water electrolyzer systems as it utilizes heat into efficient hydrogen production, less space occupation with architectural uniformity, and reduces installation costs.

(222x) Nitrogen Clathrate Hydrate Phase Equilibrium: Measurements of Hydrate Phase Equilibria for Nitrogen + Water + Ethanol

Galicia-Luna, L. A., INSTITUTO POLITECNICO NACIONAL-ESIQIE
P rez-L pez, H. I., Instituto Polit cnico Nacional
Garcia-Fuentes, D.,

Gas hydrates, or clathrate hydrates are, ice-like solid crystalline compounds, which are former through combination of water and small guest molecules, like nitrogen, under low temperatures and elevated pressures. Gas hydrates formation phenomenon has been used for developing separation processes. In this work we report experimental dissociation points for the systems N_2+H_2O+C_2H_5OH and N_2+H_2O+C_3H_7OH at temperatures between 262 and 282 K and pressures up to 33 MPa.

* To whom correspondence should be addressed

E-mail: lgalicial@ipn.mx

Tel: (52) 55 5729-6000, Ext. 55133 Fax: (52) 55 5586-2728

(222w) Densities for Binary Systems Alkane + Thiophene At Temperatures From 313.15 to 363.15 K in the Range of 1-20 Mpa

Arenas-Quevedo, M. G.,
Galicia-Luna, L. A., INSTITUTO POLITECNICO NACIONAL-ESIQIE
Elizalde-Solis, O., INSTITUTO POLITECNICO NACIONAL-ESIQIE

In this paper, we present experimental results for the volumetric properties of two binary systems heptane + thiophene and nonane + thiophene at temperatures from 313.15 to 363.15 and pressures of 1 to 20 MPa. These experiments were performed in a static method coupled to the vibrating tube densimeter technique. It uses a synthetic-cell made of titanium; it withstands high pressures and temperatures. Vibrating periods of the fluid are obtained in the experiment; then, density of the studied fluid is obtained from a calibration procedure. For this case, we use water and nitrogen as reference fluids. The obtained uncertainty in density for these binary systems is less than 0.2 kg/m^3. Afterwards, density was represented using an empirical equation of 6-parameters. Deviations between experimental and calculated values showed that these are within the experimental uncertainty. Thermodynamic derived properties are computed from the concerning expressions that involve the derivative of the empirical equation.

(222bk) Validation of An Empirical Interatomic Potential for the Si-Ge-O System: Application to Selective Epitaxial Growth of Ge On Si/SiO2

Chuang, C. Y., University of Pennsylvania
Sinno, T., University of Pennsylvania
Han, S. M.,

The fidelity of a Tersoff-based empirical potential model for the Ge-Si-O ternary atomic system [1,2] is studied in detail, with the ultimate aim of validating the potential for later use in large-scale simulations of Ge selective epitaxial growth (SEG) on crystalline silicon. The ability to grow thin Ge films on bulk Si wafers is of interest as a cost-effective means of producing substrates for high-efficiency III-V multijunction photovoltaic devices. In this context, SEG is a powerful approach for reducing the lattice and thermal expansion mismatch between two films of different materials that are epitaxially adhered. In the process of force-field validation, several comparisons are presented between the predictions of atomistic simulations based on the empirical potential studied here and prior experimental measurements and electronic structure calculations. The points of comparison include the structure and thermodynamics of bulk amorphous silica (a-SiO_2), the a-SiO_2 free surface, the c-Si/a-SiO_2 interface, the c-Ge/a-SiO_2 interface, and the desorption, wetting, and diffusion behavior of Ge atoms on a-SiO_2 surfaces. A single fitting parameter, which describes the strength of the Ge-O interaction, is used to establish good agreement between the empirical potential predictions and experimental measurements across all points of comparison. We conclude that a Tersoff-based empirical potential, while it neglects explicit Coulombic interactions and is highly simplified, is a reasonable basis for probing the Ge/SiO_2/Si material system. The resulting potential is used to generate some predictions for the behavior of Ge island coalescence, which is known to lead to defect formation but has not yet been understood.

[1] J. Tersoff, *Phys. Rev. B* **39**, 5566 (1989).

[2] S. Munetoh, T. Motooka, K. Moriguchi and A. Shintani, *Comput. Mater. Sci* **39**, 334 (2007).

(222y) Compressed Liquid Densities for the Ethanol + Heptane + Decane System At High Pressures and Temperatures

Zamora-L pez, H. S., Instituto Polit cnico Nacional - E.S.I.Q.I.E.
Galicia-Luna, L. A., INSTITUTO POLITECNICO NACIONAL-ESIQIE
Elizalde-Solis, O., INSTITUTO POLITECNICO NACIONAL-ESIQIE

Thermodynamic behavior of ternary systems has been studied in the last decades. In this work, volumetric properties for the ternary system x_1 ethanol + x_2 heptane + $(1-x_1-x_2)$ decane are reported in the range of temperature of (313 to 363) K and pressures up to 20 MPa. The experimental apparatus is based on the static – synthetic method coupled to a commercial vibrating tube densimeter from Anton Paar. Experimental densities are reported within an uncertainty of 0.2 $kg \cdot m^{-3}$. Deviations between experimental and calculated values agree and are within the experimental uncertainty. Compositions for the ternary system were prepared keeping constant the heptane/decane relation and by varying of ethanol. Density results were correlated using an empirical model with 5 parameters.

(222aa) Nanotribological Responses of Nanoparticle Solutions

Wang, J. C., Missouri University of Science and Technology
Wang, C., Missouri University of Science and Technology

Molecular dynamics modeling and simulation studies have been perform to investigate the nanotribilogical responses of nanoconfined nanoparticle solutions under time-continuous shearing. The results show that when nanoparticles are added as nanolubricant additive, the tribological responses of the resultant nanoparticle solution thin films depend on a number of factors, including surface separation (film thickness), shear rate (sliding speed), nanoparticle size, nanoparticle volume fraction, and surface microstructure, and exhibit complex interplay among these factors. At sufficiently large surface separations, nanoparticles tend to be well solvated but unable to break through the dense adjacent layers of solvent molecules to establish direct contact with the surfaces, thereby having only minimal effects on the apparent viscosity. However, when the surface separation becomes comparable to the size of nanoparticles, nanoparticles can significantly alter the layered configurations and in-plane ordering of the solvent molecules. It was found that the magnitude and trend of the apparent viscosity of a thin nanoparticle solution film are determined to a high extent by the interplay between the in-plane ordering of the solvent molecules in direct contact with the surfaces and the in-plane microscopic structure of the surface atoms.

(222ab) Modeling Self-Assembly of Objective Structures

Van Lengerich, H., Cornell University

Virus capsids have evolved for millions of years to efficiently self-assembled out of many protein sub-units. We wish to understand and model the fundamentals of this assembly to further synthetic self-assembly and hinder undesired virus assembly. All viruses have a symmetric structure, in the sense that each sub-unit sees the exact same environment in the assembled state. We call the class of all such structures "objective structures" and these can be characterized by translations and rotations of the sub-unit. To model the equilibrium and dynamics of assembly, the three-dimensional rotation and translation of each sub-unit must be included. Simulations using the overdamped Langevin equation with rotation show the equilibrium distribution as a function of temperature. The temperature at which an objective structure undergoes a phase transformation between disordered and assembled compare favourably to predictions from a Smoluchowski equation with rotation, which we derive. This equation could also be used to model the dynamics of assembly.

(222z) Experimental Determination of Hydrate Formation Systems CO2 + H2O and CO2 + H2O + n-Heptane

Mendo-S nchez, R. P., Instituto Polit cnico Nacional
Galicia-Luna, L. A., INSTITUTO POLITECNICO NACIONAL-ESIQIE
Mohammadi, A. H., MINES ParisTech

This paper presents experimental results for hydrate formation systems $CO_2 + H_2O$ (from 273.17 to 281.67 K and 12.51 to 37.41 bar) and $CO_2 + H_2O$ + n-heptane (270. 40 to 279.07 K and 9.58 to 24.40 bar) which were determined by the isochoric method using a blind cell supporting pressures up to 300 bar and temperatures from 258.15 to 343.15 K. The temperature control equipment is a PolyScience bath in which employed a water-ethanol mixture as cooling fluid and a ISCO pump for supplying the CO_2 and pressurize the cell. Experimental results for $CO_2 + H_2O$ system were compared with data reported in literature with deviations ≤0.5 bar.

* To whom correspondence should be addressed

E-mail: lgalicial@ipn.mx

Tel: (52) 55 5729-6000, Ext. 55133 Fax: (52) 55 5586-2728

(239a) Theoretical Investigation of Ethylene Glycol Reforming Over Pt(111) in Vapor and Aqueous Environments

Faheem, M., University of South Carolina
Heyden, A., University of South Carolina
Originally presented on: 11/5/2013 8:30:00 - 8:45:00

Catalytic conversion of biomass-derived oxygenates for production of hydrogen and transportation fuels has gained significant attention in recent years. Ethylene glycol has been used as a model molecule to gain mechanistic insights into the reaction mechanism. The main reaction pathway for the production of hydrogen involves cleavage of C-C, C-H, and O-H bonds with subsequent water-gas shift reaction. A parallel pathway leads to the production of alkanes through C-O bond scissions. While the reaction mechanism is to some degree understood at the metal-gas interface, lack of a well-established methodology for describing the influence of a complex liquid on a reaction across a solid-liquid interface with quantum mechanical accuracy has resulted in very little knowledge on the effect of a liquid-phase environment on the reaction mechanism.

In this talk, we present an application of our recently developed iSMS[1] methodology to study the mechanism of ethylene glycol reforming over Pt(111) in water. Periodic planewave density functional theory calculations have been performed to obtain optimized structures for all intermediates and to calculate free energies of reaction and activation free energy barriers in the vapor phase. The effect of water has then been described through a combination of Gaussian-type orbital calculations and COSMO-RS calculations as required by iSMS. Finally, detailed microkinetic models have been developed for both vapor and aqueous phases to provide insights into the effect of a liquid phase environment on the reaction mechanism and activity descriptors.

1. Faheem, M.; Suthirakun, S.; Heyden, A. *J. Phys. Chem. C* **2012,** *116*, 22458-22462.

(239b) Designing Catalysts Using Predictions of Hydrocarbon and Oxygenate Adsorption Energies
Montemore, M. M.,
Medlin, J. W., University of Colorado Boulder
Originally presented on: 11/5/2013 8:45:00 - 9:00:00

Oxygenates and hydrocarbons are important in most catalytic processes, such as alcohol oxidation, biomass conversion, petrochemical production, fuel cells, Fischer-Tropsch synthesis, and CO_2 electroreduction. The heterogeneous catalysts used in these processes are generally found through a trial-and-error process. Given the immense combinatorial space for multimetallic surfaces, more efficient design paradigms are needed. Adsorption energies have been shown to be good predictors of catalytic performance; hence, efficient predictions of adsorption energies could greatly improve catalytic design.

We have developed an accurate model for predicting adsorption energies of hydroxyl and methyl in the top site from electronic structure parameters. This model uses the d-band center, p-band center, matrix coupling element, d-band filling, and electronegativity of the surface. These quantities are readily available from density functional theory (DFT) calculations, and can also be estimated directly from alloy structures. Simple relations based on metallic parameters can also be used to predict the energy of translating these species to more highly coordinated sites. Hence, these relations allow very efficient screening of surfaces for their adsorption properties.

We have also developed accurate scaling relations that allow prediction of adsorption energies of other oxygenates and hydrocarbons from the adsorption energies of hydroxyl and methyl. For example, the adsorption energy of longer alkyls can be predicted using methyl's adsorption energy as well as gas-phase bond energies. Previous scaling relations among C_1 hydrocarbons have been made more accurate by scaling between these adsorbates in the sites where the tetravalency principle is fulfilled. These relations allow efficient prediction of potential energy surfaces for a variety of reaction pathways based on adsorption energies calculated from electronic structure parameters or calculated using DFT.

Together, these relations allow the rational design of a catalyst based on electronic structure parameters, as well as efficient screening from DFT calculations. Hence, we have carried out a comprehensive search for surfaces that catalyze alcohol oxidation, the oxygen reduction reaction, and alkane reforming.

(239c) Effect of Palladium Surface Structure On the Hydrodeoxygenation of Organic Acids

Behtash, S., University of South Carolina
Lu, J., University of South Carolina
Heyden, A., University of South Carolina
Originally presented on: 11/5/2013 9:00:00 - 9:15:00

Lipid feedstock can be utilized for the production of second-generation biofuels via a catalytic hydrodeoxygenation (HDO) process. The conversion of fatty acids plays an important role in the activity and selectivity of this process. Understanding the HDO reaction mechanism of organic acids on metal surfaces is a prerequisite for the rational design of new HDO catalysts specifically designed for lipid feedstock. In our previous study of the HDO of propionic acid over Pd (111), we found that the dehydrogenation of α-carbon and the C-OH bond dissociation steps are activity descriptors in the HDO of organic acids. However, it is well known that the higher-lying d-states in under-coordinated stepped surfaces lower the C-O and C-C bond dissociation barriers and consequently enhance the reactivity in comparison to flat surfaces. To investigate the sensitivity of the HDO of organic acids on the Pd surface structure, the reaction mechanism of the HDO of propionic acid has been studied on the Pd(211) model surface and the results are compared to our previous Pd(111) investigation. Periodic density functional theory (DFT) calculations have been carried out to obtain reaction rate parameters which are subsequently used in a microkinetic modeling study. Finally, we performed a sensitivity analysis and computed Campbell's degree of rate control to identify rate-determining steps.

(239d) Computational Design of Copper Containing Bimetallic Syn-Gas to Ethanol Catalysts

Bruce, D. A., Clemson University
He, M., Clemson University
Originally presented on: 11/5/2013 9:15:00 - 9:30:00

A major challenge associated with the synthesis of ethanol from syn-gas is an inability to find a low-cost catalyst that promotes the proper combination of CO dissociation and CO insertion steps, so as to yield ethanol as the primary reaction product and inhibit the formation of methane, methanol, longer chain alkanes, and other coking reaction products. Given the complexity of this reaction system, a simple trial-and-error approach to catalyst design is fraught with difficulties, which could severely limit efforts to identify an ideal catalyst material. Therefore, we used quantum mechanical simulations for the rational design of bimetallic catalysts that are optimally suited for the production of ethanol from syn-gas. Specifically, Density Functional Theory (DFT) simulations and Bronsted-Evans-Polanyi (BEP) relations were used to map out the full reaction mechanism from syn-gas to ethanol for 4 different copper containing bimetallic clusters (CuCo, CuNi, CuFe, and CuRu) that range in size from 13 to 38 metal atoms. For these bimetallic catalysts, microkinetic models were built, considering all necessary adsorption and reaction steps as well as the diffusion of intermediate species between varying metal surface sites. Results from additional computational studies examining surface coverage effects were also included in the microkinetic models. These simulations indicate the nature and stability of the various bimetallic nanocatalysts and more importantly identify specific metal combinations that are ideally suited for ethanol production. Select catalyst formulations (e.g., CuCO and CuNi) were prepared and tested for catalytic activity and selectivity. These experimental results were used to validate the computational results.

(239e) Improved Catalytic Reactivity of PdFe Bimetallic Catalysts Towards the Deoxygenation of Phenolic Compounds: A Density Functional Theory Study

Hensley, A., Washington State University
Zhang, R., Washington State University
Wang, Y., Pacific Northwest National Lab
McEwen, J. S., Washington State University
Originally presented on: 11/5/2013 9:30:00 - 9:45:00

With an increasing world population along with the emergence of newly industrialized nations, the availability of fossil fuels will significantly decrease as current oil wells and fields are depleted. One of the most promising renewable energy alternatives to fossil fuels are biofuels because of their compositional similarity to fossil fuels and their ability to use the current transportation fuel infrastructure[1]. The major problem facing biofuels is the high level of oxygen, up to 40 wt%, present in the crude bio-oil after pyrolysis which cause the resulting fuel to be corrosive and thermally unstable[2]. To address this problem, a novel PdFe catalyst was recently developed for the purpose of removing the oxygen functional groups from the phenolic compounds found in crude bio-oil[3]. Experimentally, this bimetallic catalyst showed a significant catalytic improvement for the hydrodeoxygenation (HDO) of guaiacol to benzene over its monometallic components[4]. To gain a deeper understanding of the interactions occurring within the PdFe bimetallic catalyst and their effect on the HDO of phenolic compounds, the adsorption of benzene, phenol, and guaiacol on model Fe (110), Pd (111), and mixed PdFe surfaces along with the HDO reaction path for the conversion of phenol to benzene on the monometallic surfaces were examined using density functional theory (DFT). Our results show that while the adsorption strength for these aromatic species on the pure surfaces was nearly identical, the Fe surface distorted the C-O bonds to a greater degree than the Pd surface. When these compounds were adsorbed onto mixed PdFe surfaces, their adsorption was found to be weakened by the interaction with the surface Pd. In particular, we find that benzene is physisorbed when one monolayer of Pd is deposited on Fe(110) while it is chemisorbed on a pure Fe(110) surface.[5] In addition to the adsorption studies, the reaction pathway and energy barriers for the HDO of phenol to benzene on the monometallic model surfaces were calculated. Preliminary results show that the energy barrier for the HDO reaction on the pure Fe surface is significantly reduced as compared to the pure Pd surface. Overall, these results strongly support the hypothesis that the Fe surface acts as the active catalytic surface.

References

1. Lovins, A. B.; Datta, E. K.; Bustnes, O.-E.; Koomey, J. G.; Glasgow, N. J., *Winning the Oil Endgame: Innovation for Profits, Jobs, and Security*. Rocky Mountain Institute: 2005.

2. Mohan, D.; Pittman, C. U.; Steele, P. H., Pyrolysis of Wood/Biomass for Bio-oil: A Critical Review.

Energy & Fuels 2006, *20*, 848-889.

3. Sun, J.; Karim, A. M.; Zhang, H.; Kovarik, L. S.; Wei, Z.; Li, X.; Wang, Y. In *Hydrodeoxygenation of Biomass-Derived Compounds to Biofuels*, AIChE Annual Meeting, Pittsburgh, PA, Pittsburgh, PA, 2012.

4. Sun, J.; Karim, A. M.; Zhang, H.; Kovarik, L.; Li, X.; Hensley, A. J.; McEwen, J.-S.; Wang, Y., Carbon Supported Bimetallic Pd-Fe Catalysts for Vapor-Phase Hydrodeoxygenation of Guaiacol. Washington State University: Journal of Catalysis (submitted), 2013.

5. Hensley, A. J.; Zhang, R.; Wang, Y.; McEwen, J.-S., Energetic and Electronic Interactions between Benzene and PdFe Surfaces: A Density Functional Theory Study. Washington State University: (in preparation), 2013.

(239f) Multiscale Kmc-DFT Simulation On Bimetallic Catalysts

Guo, W., UNIVERSITY OF DELAWARE

Originally presented on: 11/5/2013 9:45:00 - 10:00:00

Multiscale KMC-DFT Simulation on Bimetallic Catalysts

Wei Guo[1], Michail Stamatakis[1,2] and Dionisios G. Vlachos[1]*

[1]Center for Catalytic Science and Technology, Department of Chemical and Biomolecular Engineering, University of Delaware, Newark, Delaware 19716, USA

[2]Department of Chemical Engineering, University College London, Torrington Place, London WC1E 7JE, UK

* vlachos@udel.edu

Bimetallic catalysts have provided an opportunity for designing more active, stable and less expensive materials than the individual, pure metals catalysts. However, the reaction mechanism and the dynamics of the catalyst during catalytic reaction are not yet fully understood. The effect of fine microstructure of Ni clusters on catalyst reactivity remains elusive. Here we have developed a multiscale approach to simulate the nitrogen desorption from Ni clusters of different sizes and shapes on Pt(111) (Ni/Pt). We employed density functional theory (DFT) calculations to model the coverage dependent N binding energies on a variety of Ni/Pt configurations, and presented a model based on local coordination environment of Ni neighbors around a nitrogen atom [1]. Then the energetics of adsorption, surface diffusion, association and desorption of atomic N on Ni/Pt catalysts are incorporated into a recently developed kinetic Monte Carlo (KMC) framework to simulate the temperature-programed desorption (TPD) spectra of nitrogen on Ni /Pt [2]. This is the first study of first-principles KMC simulation on bimetallic catalysts that provides a conceptual platform for accounting for complex microstructure on kinetics. Our simulation framework can be employed to understand the structure sensitivity in catalysis and eventually tune bimetallic catalytic materials for complex chemistries.

1. W. Guo and D. G. Vlachos, J Chem Phys **138**(17), 174702 (2013).

2. W. Guo, M. Stamatakis and D. G. Vlachos, Submitted (2013).

(239g) Investigation of Hydrogen Spillover Pathways On a Bimetallic Hydrogenation Catalyst

Ruscillo, F., University College London
Sykes, C., Tufts University
Stamatakis, M., University of Delaware
Originally presented on: 11/5/2013 10:00:00 - 10:15:00

Hydrogenation processes are of paramount importance in the chemical industry with a vast range of applications in the food, petrochemical and pharmaceutical sectors. Heterogeneous catalysts are widely used for hydrogenation; however, controlling the spillover of hydrogen on catalytic surfaces is challenging. It has recently been shown experimentally that it is possible to achieve this type of control on a Cu/Pd catalyst by reversible adsorption of a spectator molecule (CO).[1] This phenomenon was termed the "molecular cork" effect and can have profound implications in catalysis. In particular, it is possible to trap hydrogen on the catalytic surface at temperatures higher than its normal desorption temperature, thereby increasing the rate of hydrogenation reactions due to the higher reactivity of hydrogen under these conditions. Yet, questions remain with regard to the origins of this phenomenon and how it can be leveraged for the design of improved hydrogenation catalysts.

Motivated by these questions we employ a multiscale computational approach to study the "molecular cork" system. At the molecular level we use density functional theory (DFT) to calculate the energetics and kinetics of CO adsorption onto and desorption from the Pd sites, as well as hydrogen dissociative adsorption and associative desorption in the absence and in the presence of CO. At the catalyst level, we use kinetic Monte Carlo simulation to reproduce the experimental observations and investigate in detail the H_2 dissociation and spillover pathways. Our simulations provide valuable information that cannot be directly accessed by experiments and offer a fundamental understanding of the molecular-level processes underlying the "molecular cork" effect.

References

Marcinkowski, M. D., A. D. Jewell, M. Stamatakis, M. B. Boucher, E. A. Lewis, C. J. Murphy, G. Kyriakou and E. C. H. Sykes, *Controlling a Spillover Pathway with the Molecular Cork Effect.* Nature Materials, 2013: Accepted for publication.

(239h) Understanding the Alcohol Dehydration On Metal Oxides

Mpourmpakis, G., University of Delaware
Originally presented on: 11/5/2013 10:15:00 - 10:30:00

A fundamental understanding on the dehydration of simple alcohols can help us elucidate and eventually control the selective dehydration of polyols and complicated biomass molecules to value-added chemicals. Herein, we use Density Functional Theory (DFT) calculations to study the alcohol dehydration on group XIII oxides, including γ-Al_2O_3 which is a commonly used support in heterogeneous catalysis. Both E1 and E2 type of dehydration mechanisms for primary (ethanol, 1-propanol), secondary (2-propanol), and tertiary (2-methyl-2-propanol) alcohols were investigated. In addition, we considered possible coupling reactions between reactants, products and dehydration intermediates. A key finding is that the dehydration barriers of the alcohols correlate linearly with the corresponding carbenium ion stability (alcohol property) [1]. This is because the transition state of the alcohol dehydration has carbenium characteristics. By accounting for the acid strength of the active site, the basic strength of the surface oxygens and the substitution degree of the alcohol, structure-property relationships were developed. Our results explain several experimental observations.

[1] Roy S., Mpourmpakis G., Hong D.-Y., Vlachos D.G., Bhan A. and Gorte R.J., *ACS Catal.* **2**, 1846–1853 (2012)

(239i) CeO2-Supported Ni Catalysts For CO2 Methanation

Hahn, K., University of Zurich
Hutter, J., University of Zurich
Originally presented on: 11/5/2013 10:30:00 - 10:45:00

Combined metal-metal oxide systems are important materials for numerous technological applications such as microelectronics, photovoltaics, gas sensors and in particular in oxide- supported metal catalysts [1]. Deposition of small metal particles on metal oxides such as Al_2O_3, CeO_2 and TiO_2 has led to enhanced catalytic activity, for example, for the water gas shift reaction or CO oxidation [2,3]. Metal oxide supported materials are also promising catalysts for hydrocarbon production from CO_2, a reaction process interesting for alternative energy production since it provides high energy density fuel from CO_2 which is omnipresent in our environment. In fact, CeO_2-supported Ni particles have proven to be promising materials for the activation of methanation of CO2 [4].

Here, density functional theory within the Gaussian and plane waves (GPW) implemented in the Quickstep [5] module of the CP2K [6] program package has been used for the simulation of Ni cluster formation on CeO_2(111) surfaces. Stabilization of Ni particles has been observed with increasing cluster size up to ten atoms. It has been found that O diffusion from the subsurface layer of CeO_2(111) to the surface layer is facilitated in the presence of Ni particles on the surface. Furthermore, the adsorption and dissociation of CO_2 has been investigated on clean CeO_2(111), on non-supported Ni clusters in the gas phase and on the Ni/CeO_2(111) system. This study gives fundamental insight into the metal-metal oxide interactions and its function for CO_2 catalysis.

[1] C.T. Campbell, Surf. Sci. Rep. 27 (1997), 1.
[2] Y. Zhai, D. Pierre, R. Si, W. Deng, P. Ferrin, A. U. Nilekar, G. Peng, J. A. Herron, D. C. Bell, H. Saltsburg, M. Mavrikakis and M. Flytzani-Stephanopoulos, Science 329 (2010), 1633.
[3] G. N. Vayssilov, Y. Lykhach, A. Migani, T. Staudt, G. P. Petrova, N. Tsud, T. Skala, A. Bruix, F. Illas, K. C. Prince, V. Matolin, K. M. Neyman and J. Libuda, Nat. Mater. 10 (2011), 310.
[4] W. Wang and J. Gong, Front. Chem. Sci. Eng. 5 (2011), 2.
[5] G. Lippert, J. Hutter and M. Parrinello, Mol. Phys. 92 (1997), 477.
[6] J. VandeVondele, M. Krack, F. Mohamed, M. Parrinello, T. Chassaing and J. Hutter, Comput. Phys. Commun., 167 (2005), 103.

(239j) Effects of Doped N On the Reaction of Alkyne Hydrochlorination Catalyzed By Graphene Supported Au Based Catalyst

Du, Q., Tianjin University
He, Z.,
Zhang, J., Tianjin University
Li, W., Tianjin University
Han, Y., Tianjin University
Originally presented on: 11/5/2013 10:45:00 - 11:00:00

Many studies have shown that the support can greatly affect the distribution and stability of the catalyst through the interfacial interaction. It is a trend to modify the support to improve the catalytic performance. The aim of our work is to reveal the effect mechanism of the support modification by N-doping on the reaction of alkyne hydrochlorination catalyzed by graphene supported Au based catalyst. This investigation will provide a theoretical basis of how to select optimum N-doped location and valence state to interact with the catalyst and improve the catalytic performance. And it is instructive to the experimental design of novel non-mercury catalysts with high performance.

The graphene edge has two types: zigzag-edge and armchair-edge. Herein, four different N-doped sites are considered in this study: graphitic N, pyridine N on zigzag edge(zigzag 6-N), pyrrolic(armchair 5-N) and pyridine N(armchair 6-N) on armchair. Our results are divided into three sections:**(a) The effects of doped N and its location on the stability of Au based catalyst**. **(b) The effects of doped N and its location on the adsorption of reactants and product. (c) The effects of doped N and its location on the reaction paths and barriers.**

(243a) Hydrophobic Forces Between Deformable Oil Drops

Tabor, R., The University of Melbourne
Wu, C., The University of Melbourne
Grieser, F., The University of Melbourne
Chan, D. Y. C., The University of Melbourne
Dagastine, R. R., The University of Melbourne
Originally presented on: 11/5/2013 8:30:00 - 8:45:00

The hydrophobic force is a fundamental interaction in nature, and is central to many processes ranging widely from dispersing oil spills to forming cell membranes. Yet, many studies by different groups over the past 30 years to directly measure the hydrophobic force have led to the observations of extra attractions that have varied in range from tens to hundreds of nanometers. Perhaps most surprisingly, the hydrophobic effect is often invoked to explain results using solid (mainly mica) surfaces in contact with water, but rarely in the study of drops and bubbles, arguable some of the most hydrophobic interfaces in nature. For solid surfaces many secondary phenomena, surface chemical heterogeneity and roughness, and the presence of nano-bubbles, have been uncovered without a clear picture of how to describe the hydrophobic effect on a microscopic scale.

We employ molecular smooth and inert oil droplets in aqueous solutions to isolate the intrinsic hydrophobic attraction, showing it to be far a shorter range and stronger attraction than the conflicting results in the literature suggest. This is also more consistent with an expected molecular origin. We have measured this force directly by suppressing any extraneous forces leaving only a "hydrophobic force" between two drops in aqueous solution. We also introduce separately different mild repulsive forces of quantum electrodynamic, electrical and hydrodynamic origin between the drops, and in each case the same attraction is observed.

(243b) Development of a General Interaction Potential for Hydrophobic and Hydrophilic Interactions

Donaldson, S. H. Jr., University of California, Santa Barbara
Valtiner, M., Max-Planck-Institut f r Eisenforschung GmbH
Kristiansen, K., University of California, Santa Barbara
Rapp, M., University of California, Santa Barbara
RÃ¸yne, A.,
Das, S., University of California, Santa Barbara
Gebbie, M. A., University of California, Santa Barbara
Chmelka, B. F., University of California
Israelachvili, J., University of California, Santa Barbara
Originally presented on: 11/5/2013 8:45:00 - 9:00:00

By performing surface force measurements, we have derived an interaction potential for the contribution of the attractive hydrophobic interaction as two surfactant bilayers approach, stress (causing the bilayers to thin and expose hydrophobic area), rupture, fuse, and are then separated from adhesive contact. The model quantitatively describes the complex interaction potential during bilayer hemi-fusion, including the long range forces, short range forces, hydrophobic breakthrough point, and final hydrophobic adhesion at contact. Between the bilayers, the hydrophobic interaction potential depends on the amount of exposed hydrophobic area, which progressively increases as the bilayers are further stressed. Generalization of the interaction potential results in a non-dimensional number, the "Hydra parameter," that can account for the amount of hydrophobicity at a given interface. The Hydra parameter, H_y, varies from $H_y = 1$ for a fully hydrophobic interaction (corresponding to an interfacial energy of 50 mJ/m^2 for a hydrocarbon surface in water) to $H_y = 0$ for no hydrophobic contribution. For $H_y < 0$, the equation also naturally accounts for hydrophilic interactions, i.e., the additional repulsive steric-hydration force between hydrophilic surfaces, where now the magnitude of the repulsion depends on the degree of hydration. An exponential decay length of ~1 nm is proposed for both the attractive (hydrophobic) and repulsive (hydration) forces. The interaction potential is tested for a variety of surfaces and parameters: surfactant bilayers and monolayers and self-assembled hydrophobic polymer surfaces will be discussed, as will the effects on the hydrophobicity of these surfaces by adjusting salt concentration, pH, specific ions, and solvent quality. A new picture is emerging that suggests that both hydrophobic and hydrophilic interactions can be described by a single unifying equation, suggesting a similar mechanism for both interactions, for example, a depleted water density (and/or decrease of the hydrogen bonding order parameter) near hydrophobic surfaces, and an increased water density (and/or hydrogen bonding order parameter) near hydrophilic surfaces.

(243c) A Closure-Based Density Functional Theory for the Depletive Adsorption of Fluids With Attraction

Lee, L. L., The California State University Pomona
Originally presented on: 11/5/2013 9:00:00 - 9:15:00

A new closure-based density functional theory is proposed for fluids with attractive and/or repulsive interactions. We pay special attention to the cases where depletion of the gas density occurs at the wall. Applications are made to the Lennard-Jones fluid and the Yukawa potential with a hard core. The closure is based on the exact third-order Ornstein-Zernike relation [1]. We discover that the bridge clusters is of the *functional* type dependent on the indirect correlation function, $B = B[\gamma]$. Upon invoking reasonable approximations to the triplet correlation functions, we obtain accurate prediction for the adsorption where fluid molecules are rarified compared to the bulk fluid when verified with machine data [2].

[1].. L. L. Lee, *The correlation functions of classical fluids III. The method of partition function variations applied to the chemical potential: cases of PY and HNC2*, Journal of Chemical Physics **60**, 1197 (1974).

[2].. L. L. Lee, "Constructing a new closure theory based on the third-order Ornstein-Zernike equation and a study of the adsorption of simple fluids", Journal of Chemical Physics **135**, 204706 (2011).

(243d) Calculation of Thermodynamic Properties of Pure Components and Mixtures At Vapor-Liquid Interfaces Using Classical Density Functional Theory

Klink, C.,
Gross, J., Delft University of Technology
Originally presented on: 11/5/2013 9:15:00 - 9:30:00

The accurate prediction of thermodynamic properties such as surface tension or transport resistivities across an interface is a challenging task in molecular thermodynamics especially for pure dipolar components or mixtures.

As a versatile and successful tool to describe interfaces we use the classical density functional theory in combination with the PC-SAFT equation of state. With this approach we are able to calculate a continuous density profile across the interface. All other thermodynamic properties can be calculated as derivatives of the density profile.

Based on a previous study [1] for pure non-polar or weakly dipolar inhomogeneous fluids the approach is here extended to pure strongly dipolar fluids and mixtures of non-associating components. Polar fluids show a non-isotropic orientation across interfaces. The orientation of molecules at the interface is a degree of freedom that is used to minimize the free energy of the system. It is therefore important to account for the orientational distribution for calculating thermodynamic properties.

We formulate and analyze a Helmholtz energy functional that properly accounts for the orientational distribution of molecules across vapor-liquid interfaces. The theory reduces to the PCP-SAFT equation of state for bulk phases. Our approach goes beyond previous work [2] that was based on a mean-field approximation in that we do not rely on a low-density limit and do not introduce approximations in solving the orientational distribution function.

For calculating mixtures it is necessary to reformulate the Helmholtz energy functional for both, the repulsive and attractive interactions. The repulsive contribution to the Helmholtz energy functional is formulated according to the fundamental measure theory [3,4] and the SAFT-chain term [5]. The functional for dispersive attractions is based on a non-mean field first order perturbation approach. This approach requires no adjustable parameters beyond the binary parameter k_{ij} that was independently adjusted to bulk-phase vapor-liquid data.

The mass- and heat-transport across interfaces requires the interfacial resistivities. Recent studies [6,7] show results of the resistivities for heat transfer, for mass transfer and for the coupling of heat and mass transfer for Lenard-Jones-fluids with different pair potentials. Our approach is not only capable of calculating model fluids as proposed but also of calculating real components and mixtures.

[1] Gross J., *J. Chem. Phys.* 131, 204705 (2009)

[2] Frodl P., Dietrich S., *Phys. Rev. A* 45, 7330 (1992)

[3] Roth R. et al., *J. Phy.: Condens. Matter* 14, 12063 (2002)

[4] Yu Y., Wu J., *J. Chem. Phys.* 117, 10156 (2002)

[5] Tripathi S., Chapman W. G., *J. Chem. Phys.* 122, 094506 (2005).

[6] Johannessen E. et al., *J. Chem. Phys.* 129, 184703 (2008)

[7] Glavatskiy K. S., Bedeaux D., *Phys. Rev. E* 77, 061101 (2008)

(243e) Application of Classical Density Functional Theory to the n-Alkane / Water Interface: Accurate Interfacial Properties Using Bulk Phase Parameters

Marshall, B. D. II, Rice University
Chapman, W. G., Rice University
Cox, K. R., Rice University
Originally presented on: 11/5/2013 9:30:00 - 9:45:00

The liquid / liquid interface between oil rich and water rich phases is a fundamental component of many physical and biological systems. Important properties of these interfaces include interfacial tension, width, density profiles and molecular orientation. The interfacial tension is relatively straight forward to measure using pendant drop techniques; however the isolated nature of the liquid / liquid interface makes measurement of structural quantities such as interfacial width and molecular orientations at the interface difficult. The structure of the n – alkane / water interface has been probed experimentally by X – ray reflectivity, neutron scattering and total internal reflection second harmonic generation spectroscopy; however these data are not always in agreement. Also, a number of molecular dynamics (MD) simulation studies have been conducted on the n – alkane / water interface. Calculations of interfacial tension using MD are difficult and fraught with uncertainty. Also, MD simulations have difficulty calculating properties of trace components (small mole fraction); for instance, is there a preferred orientation of alkane molecules in the water phase as the interface is approached? A powerful alternative to these approaches is classical density functional theory (DFT). DFT involves minimizing the grand free energy functional and allows for the calculation of the structure and properties of inhomogeneous fluids. Unlike molecular simulation, DFT is accurate for systems with trace components and is accurate for the calculation of interfacial tension. Computationally, DFT is typically much faster than molecular simulation allowing for a rapid exploration of parameter space. In this work we apply the interfacial statistical associating fluid theory density functional theory (i-SAFT) to the alkane – water interface. i-SAFT is the extension of the SAFT equation of state for associating chain fluids to interfacial systems. For homogeneous systems the i-SAFT DFT reduces to the SAFT equation of state. We obtain the parameters (alkane chain length, water association energy, molecular sizes and Lennard – Jones well depths) by fitting the SAFT equation of state to bulk phase equilibrium data, and then show how these same parameters can be used to accurately predict the interfacial properties of the oil / water interface (interfacial width, density profiles, interfacial tension etc…). Using the DFT, we are the first to show that alkane molecules in the water phase but close to the interface orient normal to the interface.

(243f) Computer-Simulation-Assisted Molecular-Thermodynamic Framework to Predict Interfacial Tensions of Non-Ionic Surfactants

Sresht, V., Massachusetts Institute of Technology
Blankschtein, D., Massachusetts Institute of Technology
Originally presented on: 11/5/2013 9:45:00 - 10:00:00

The reduction in interfacial tension by surfactants underlies several natural phenomena in multi-phase systems, including emulsions and foams. This effect is also responsible for many industrial processes such as spray painting, emulsion polymerization, distillation in packed bed columns, and froth flotation. The mechanisms underlying these phenomena have been described by several mathematical models in the colloid science literature. Since the interfacial tension is an essential input for the implementation of all these models, their accuracy depends on the accuracy of the estimate(s) of the interfacial tension(s) used. Here, we propose a method to evaluate the surface tensions of non-ionic surfactants using a combination of molecular dynamics (MD) simulations and molecular-thermodynamic (MT) theory.

To this end, we have developed an automated, computational workflow to carry out a series of simulations of surfactant molecules at a fluid-fluid interface. One set of simulations studies the pair-wise interactions between surfactant molecules, and attempts to characterize the intermolecular interactions in a surfactant monolayer in terms of two molecular parameters – a hard-disk radius (representative of strong intermolecular repulsions) and a second-virial coefficient (representative of weak, intermolecular, van der Waals attractions). These parameters are obtained by first calculating the potential of mean force between the two surfactant molecules at the interface, and then analyzing this potential using principles of statistical mechanics. A second set of simulations studies the change in the free energy of a surfactant molecule as it is brought from bulk solution to an interface. This free energy of adsorption is a measure of the affinity of the surfactant of interest for the interface of a given system, and provides a metric capable of ranking families of surfactants. Together with the intermolecular interaction parameters, the free energy of adsorption enables prediction of the interfacial tension as a function of the surfactant bulk concentration through the use of a suitable equation of state for the interface.

In this talk, we will present the details of our proposed computational methodology, including the results obtained by its application to a series of CiEj nonionic surfactants at fluid-fluid interfaces.

(243g) Probing Collapse Mechanisms of Two-Dimensional Lipid Monolayers

Choi, S. Q., University of Minnesota
Zasadzinski, J. A., University of Minnesota
Originally presented on: 11/5/2013 10:00:00 - 10:15:00

Lung surfactant is a lipid monoalyer and protein mixture that lines the alveolar air-liquid interface to lower surface tension. When lung surfactant is subjected to a large compression, such two-dimensional thin film collapses. To date, several collapse mechanisms are known for lipid monolayers: fracture, solubilization, and folds. We present a novel method that enables us to distinguish "folds" from other mechanisms and, furthermore, quantify folds. We simply measure fluctuations of the surface tension when the collapse occurs, and compute the mean squared fluctuation, which provides various properties of folds (e.g. a size and time that folds remain). We validate this technique using mixed lipid monolayers of DPPC and POPG, a good model system of lung surfactant, which exhibits the different mechanisms of collapses by varying POPG fractions. Moreover, visualization of such monolayers with the measurements reveals that formation of folds occurs randomly, and depends strongly on compression rates and a composition of POPG.

(243h) High-Resolution Simulations of Turbulently Flowing Liquid-Liquid Dispersion

Komrakova, A. E., University of Alberta
Eskin, D., Schlumberger
Derksen, J., University of Alberta
Originally presented on: 11/5/2013 10:15:00 - 10:30:00

The present study demonstrates feasibility of multi-scale direct numerical simulations of turbulently agitated liquid-liquid dispersion using a free energy lattice Boltzmann equation model revealing a spectrum of scales in the range from the dispersed phase size to integral eddy scale. The utilized method refers to the class of diffuse interface methods which require an explicit specification of the interface thickness. Two dimensionless parameters determine this additional degree of freedom: the interface Peclet number Pe that describes the diffusivity of the interface, and the Cahn number Ch that is the ratio of the interface thickness and the drop size. Based on simulations of a single drop deformation and breakup in simple shear flow, an influence of Pe, Ch and mesh resolution on accuracy and stability was investigated. A wide range of physical conditions (Reynolds number Re=0.0625-50) was examined. It was demonstrated that the accuracy of the results is primarily determined by the mesh resolution. The guidelines for a selection of Pe and Ch leading to stable simulations were outlined.

Based on the results obtained in the sheared drop simulations, the numerical parameters related to the interface were determined for turbulently flowing two-phase systems. Three-dimensional simulations were carried out in a fully-periodic cubic domain of 500^3. The liquids were of equal density and varying viscosities to provide the viscosity ratio from 0.33 to 3.0. Three levels of resolution at Kolmogorov length scale η_k=1, 5 and 10 [lu] were considered. High-resolution simulations with η_k=5 and 10 [lu] allowed to capture the drop/eddy interactions on scales smaller than Kolmogorov one where energy dissipation takes place. Thus, the numerical tool can be used to determine the drop size distributions of dispersed systems in the viscous sub-range of the turbulent energy spectrum. A velocity magnitude field over a drop cross-section (black curve is the interface) is shown in Fig.1.

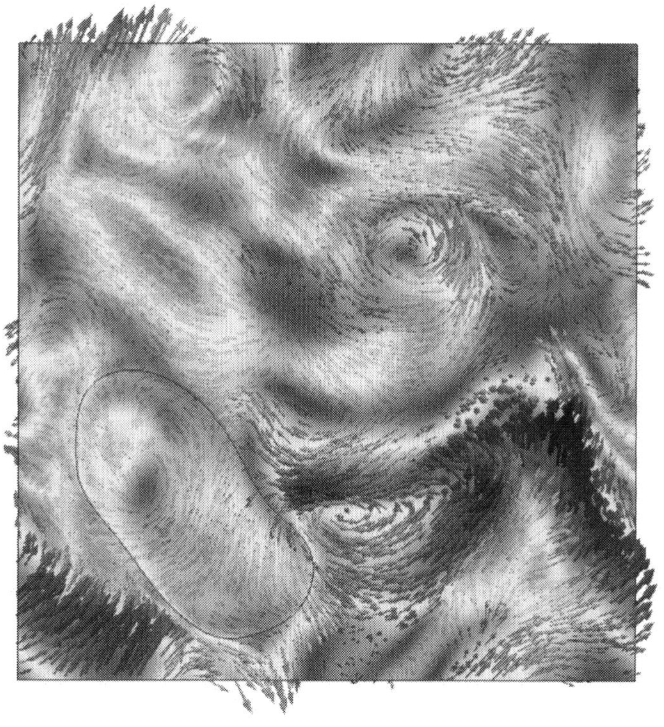

Fig.1. A velocity magnitude field in a section crossing the droplet at η_k=5 [lu].

(243i) Micro and Macrorheological Methods At Fluid-Fluid Interfaces

Samaniuk, J. R., Katholieke Universiteit Leuven
Vermant, J., Katholieke Universiteit
Originally presented on: 11/5/2013 10:30:00 - 10:45:00

Microrheological methods have been developed to probe the rheological properties of fluid-fluid interfaces, but orders-of-magnitude differences in interfacial properties measured with micro and macrorheological techniques have been reported in the literature. In bulk fluids, microrheological techniques have successfully been developed around the generalized Stokes-Einstein relation (GSER) and can be expected to agree with macroscopic measurements under conditions where the GSER is valid. For microrheological measurements at interfaces, the assumptions of GSER are not generally met due to the multi-component, multi-length scale environment of fluid-fluid interfaces. Instead, hydrodynamic models that relate the motion of a particle at an interface to the viscosity of the interface and the surrounding bulk fluids have been used to extract interfacial shear viscosities from particle-tracking experiments. Currently, models account for particle intrusion into the bulk phase, the effects of particle shape and the influence of Marangoni stresses. Notable omissions from these models are the effects of interface heterogeneity and electrocapillarity. We have investigated the orders-of-magnitude discrepancy between microrheological and macrorheological results at fluid-fluid interfaces and will discuss experimental results using both techniques on various interfaces.

(253a) Molecular Simulations Of Ionic Liquids

Zhang, S., Institute of Process Engineering, Chinese Academy of Sciences
Dong, K.,

Originally presented on: 11/5/2013 8:30:00 - 8:49:00

Molecular Simulations of Ionic Liquids

Suojiang Zhang ,Kun Dong

Beijing Key Laboratory of Ionic Liquids Clean Process, State Key Laboratory of Multiphase Complex System, Institute of Process Engineering, Chinese Academy of Sciences, Beijing 100190, China.

The complexity of interactions and diversity of combinations between cations and anions of have resulted in the further difficulty in applications of ionic liquids. The exploration of new functional ionic liquids is prompted to understand the structures and interactions in ionic liquids to correlate their properties.

In our group, a CUDA-based GPU program was developed to accelerate the computations of nano- and micrometer levels of ionic liquids. The volatile of ionic liquids, the interactions of ionic liquids and cellulose, and water, and CO_2, and the nano-aggragate in ionic liquids have been investigated by molecular simulations. We found that the ions are pairware in gas phase when aprotic ionic liquids were diluted into the gas. [1] The hydrogen bonds are important between anion and cellulose, but the role of cations can not been ignored in dissolving cellulose. [2] We also found the special stable structure of the hydrophobic [Bmim][PF_6] at [Bmim][PF_6]-water interface, and the water is difficult to hydrate the ions. In addition, it is found that the distribution of CO_2 molecules and $[Ppg]^+$ cations around [BF_4] anions are almost complementary when CO_2 molecules enter into [Ppg][BF_4] ionic liquids. In many neat ionic liquids and binary systems, ion-aggregate to form the nano-clusters is very common, in particular in the imidazolium-based ionic liquids with length of alkyl chain more than 4. [3]

[1] Kun Dong, Suojiang Zhang, et al. Phys. Chem. Chem. Phys., 2013, 15, 6034-6040.

[2] Yuling Zhao, Suojiang Zhang, et al. ChemPhysChem., 2012, 13, 3126– 3133

[3] Xiaomin Liu, Suojiang Zhang et al. J. Phys. Chem. B, 2012, 116, 4934-4942.

(253b) Monitoring Inter- and Intra-Molecular Interactions in Imidazolium Protic Ionic Liquids HCnImNTf2 (n=0-12)

Nikolakis, V., University of Delaware
Moschovi, A. M., FORTH/ ICEHT
Dracopoulos, V., FORTH/ ICEHT
Originally presented on: 11/5/2013 8:49:00 - 9:08:00

In this work we synthesized and studied the effect of the alkyl chain size in the structure of the imidazolium based Protic Ionic Liquids (PILs) 1-H-3-alkyl-imidazolium bis(trifluoromethanesulfonyl)imide HC_nImNTf_2 (n=0-12) , using vibrational spectroscopy (FTIR and Raman). The main characteristic of PILs is the presence of a proton on site 1 of the imidazolium ring. As a result PILs can form an extensive network of hydrogen bonds (HB).

The observed N-H vibrational modes in the high frequency spectral region at 3270-3150cm^{-1} indicate that the liquid phase of ionic liquids consists of ion pairs. The observed redshifts, of the N-H and $C_{(i)}$-H bands frequencies (in the same region) with increasing alkyl chain length, suggest the strengthening of the HB between the anion and the cation. However, the $C_{(i)}$-H vibrational modes are primarily affected when n is varied between 0 and 3 (short alkyl chains), whereas N-H modes are affected for when n is varied from 3 to 12.Thus, the shift of the $C_{(i)}$-H mode can be attributed to the induction *+I* effect due to the presence of the alkyl chain in the ring, whereas the shift of the N-H modes is attributed to the intercationic interactions between the imidazolium rings that can be attributed to the parallel packing in the liquid phase structure.

The parallel stacking of the imidazole rings was also suggested from FT-Raman high frequency region. It was shown that when n>3, alkyl groups tend to develop structures similar to those of n-alkane melts, consisting of parallel imidazolium rings with oriented chains, defining polar and non polar regions. Moreover, the conformation of the anion NTf_2^- was studied as a function of temperature and the alkyl chain length. It was shown that the elevation of the temperature favors the cis conformation of the anion, whereas trans conformation is dominant for bigger cations with long alkyl chains.

Finally, the enthalpy ΔH_{eq} of the conformational equilibrium of the anion NTf_2^- for the PILs HC_nImNTf_2 (n=0-12) was experimentally calculated from spectroscopic data. The calculated enthalpy values reduce as the alkyl chain increases. The change in the slope of the calculated values from –ethyl to –propyl may reflect the change from intramolecular to intermolecular interactions.

(253c) Beyond 1,3-Difunctionalized Imidazolium Cations: Properties of Ionic Liquids With Tri-, Tetra- and Pentafunctionalized Cations

Bara, J. E., University of Alabama
Shannon, M. S., University of Alabama
Originally presented on: 11/5/2013 9:08:00 - 9:27:00

Although perhaps thousands of ILs have been reported in the literature, many are essentially built upon the same structure: 1,3-difunctionalized imidazolium cations. While these difunctionalized cations offer immense possibilities for tuning structure-property relationships, it is well-known that imidazolium-based materials can be further tailored via the introduction of various functionalities at the "2," "4," and/or "5" positions within the imidazole ring. However, strategies which employ trifunctionalized (or greater) imidazolium cations are seldom employed in materials design. Our group has synthesized a library of imidazole-based precursors via a scalable synthetic technique so as to further explore imidazolium cations with greater degrees of functionality, with focus on new solvents, polymers, composites, and liquid crystalline materials. This presentation will address the challenges and opportunities of "branching out" the imidazole ring so as to gain new dimensions for materials design and the potential implications in applications such as polymer membrane separations for energy applications (e.g. CO_2 capture) and self-assembly.

(253d) Volumetric Properties of Mixtures of Ionic Liquids With CO2 Or CO2 + Ethanol

Grandelli, H., Virginia Tech
Kiran, E., Virginia Tech
Originally presented on: 11/5/2013 9:27:00 - 9:46:00

We are exploring new approaches for potential use of ionic liquids in processing of natural polymers and specifically those that are important in processing of lignocellulosic materials. Our approach is to try to combine the desirable attributes of ionic liquids in terms of their solvent power with supercritical fluids to explore the extent to which dissolution levels can be tuned with pressure or fluid compositions. As an initial phase of this activity we have been evaluating the miscibility and volumetric properties (in particular the density and compressibility) of selected ionic liquids with CO_2 and CO_2 + ethanol. In this talk, we will present new data on the volumetric properties of two ionic liquids, namely [EMIM]Ac and [EMIM]Cl and their mixtures with carbon dioxide or carbon dioxide + ethanol, and comment also on the reactive interaction between [EMIM]Ac with CO_2. We will also report on the volumetric properties of mixtures containing Avicel as a model cellulose polymer at high pressures. The data are being generated in a special a high pressure variable-volume view-cell in which position of a movable piston is continually recorded in real-time while its position is altered at controlled rates with the aid of a motorized pressure generator. This position information is then converted to real time changes in the internal volume and hence the density as a function of pressure and/or temperature [1, 2]. The densities at a given temperature are then correlated with a polynomial function of the form $\rho = A + BP^{-1} + CP + DP^{-2} + EP^{m}$ were A, B, C, D, E and m are adjustable parameters. These functions were then used to determine the isothermal compressibilities for each mixture.

Acknowledgment: This research is supported by the National Science Foundation, CBET - Energy for Sustainability Program (CBET Award # 1250663).

[1] H. Grandelli and E. Kiran, Volumetric properties of mixtures of ionic liquid [EMIM]Ac with CO_2 as potential processing fluids for lignocellulosic materials at high pressures, Proceedings of the 3rd Iberoamerican Conference on Supercritical Fluids, *Prosciba 2013*, Cartagena, Colombia, April 1-5, 2013.

[2] J. M. Milanesio, J. C. Hassler, and E. Kiran, Volumetric properties of propane, n-octane an their binary mixtures at high pressures, Industrial and Engineering Chemistry Research, dx.doi.org/10.1021/ie4007084

(253e) Molar Volume Dependence of N2 Solubility in Ionic Liquids

Bennett, J.,
DeSilva, A., University of Notre Dame
Brennecke, J. F., University of Notre Dame
Originally presented on: 11/5/2013 9:46:00 - 10:05:00

Ionic liquids (ILs) have received increasing attention in recent years for their potential use in many different processes, particularly carbon capture. Many gas streams from which one might want to remove CO_2 also contain nitrogen. ILs that could be used for these separations include a diverse range of possible cation/anion combinations. It is therefore necessary to know the solubility of nitrogen in a broad range of ionic liquids. Unfortunately, there is a fair amount of disagreement amount the relatively limited N_2 solubility data available in the literature, most likely due to the challenges in measuring low solubility gases. Here we present nitrogen solubility data in a set of ionic liquids composed primarily of imidazolium cations with BF_4 and Tf_2N anions, as well as several phosphonium and ammonium cations with Tf_2N anions. The different ion combinations have molar volumes ranging from 150 cm^3/mol to 650 cm^3/mol. We have used a gravimetric microbalance produced by Rubotherm GmbH for solubility measurements at 60°C and pressures up to 150 bar. The resulting isotherms can be used to determine Henry's law constants for nitrogen in each IL. We will present results showing that the primary effect on the solubility of nitrogen in ILs is the molar volume of the liquid, with some differences between imidazolium and ammonium cations. Because of the low solubility of nitrogen in ILs and the time intensive nature of solubility experiments, the molar volume can be used as a quick estimate of the Henry's law constant for nitrogen in ionic liquids.

(253f) Structure and Properties of Ionic Liquid/Organic Co-Solvent Mixtures

Bandlamudi, S., South Dakota School of Mines & Technology
Cooney, M., University of Hawaii
Chupas, P.,
Benjamin, K. M., South Dakota School of Mines and Technology
Originally presented on: 11/5/2013 10:05:00 - 10:24:00

The extensive use of ionic liquids (ILs) is a significant achievement in extraction processes, especially in extracting lipids and proteins from biomass for bio-fuel production from algae and other bio-based feedstocks. The particular effectiveness of ionic liquids combined with polar-covalent molecules (PCM), such as methanol, as an extraction solvent mixture in extracting bio-oils from micro-algae has been demonstrated successfully by Young et al., 2010, the main highlight being auto-partitioning of extracted lipids into a separate phase. While the efficacy of these IL/PCM mixtures has been demonstrated experimentally, there is no current molecular-level understanding of the structural and thermodynamic properties of these mixtures.

The thermodynamic and structural properties of IL/PCM mixtures are explored using both molecular simulation and computational quantum chemical methods. Molecular dynamics (MD) simulations are conducted to predict mixture volumetric, thermal, and structural properties. MD simulations of thermodynamic mixing properties for IL/PCM binary mixtures show virtually ideal mixing behavior, while excess heats of mixing are predicted. In addition, MD predictions of structural properties like radial distribution functions and structure factors, supported by simulation visualization, provide information on local fluid structure and interactions between ionic liquid and polar-covalent molecules, including the change in ion aggregation with mixture composition. These simulation predictions are compared to small angle X-ray scattering (SAXS) and polarization data for model validation. Lastly, density functional theory (DFT) quantum chemical calculations are used to probe the shift in frequency of O-H stretch in methanol at various concentrations of IL/methanol mixtures, and compared to FTIR measurements.

These efforts provide an improved understanding between the molecular-level behavior and macroscopic thermodynamic properties of IL/PCM mixtures, to characterize these new solvents mixtures for separation applications.

(253g) The Interactions of the Corn Protein Zein and Ionic Liquids

Tomlinson, S., Kansas State University
Schlup, J. R., Kansas State University
Anthony, J. L., Kansas State University
Originally presented on: 11/5/2013 10:24:00 - 10:43:00

The solubility and secondary structure of the hydrophobic corn protein zein in ionic liquids was studied as a function of temperature and solvent properties. Â The solubility of zein was measured gravimetrically from 30Â°C to 60Â°C. Â The secondary structure of zein as a function of temperature and solvent was examined via curve fitting of infrared spectroscopic data. Â Two-dimensional correlation infrared spectroscopy was employed to monitor changes in the secondary structure of the zein as a function of temperature.Â These data clearly illustrate changes in the temperature dependent behavior between different solvents.

Most of the ionic liquids investigated change zein's secondary structure. Â Large solvent molecules increase the amount of β-turn secondary structure through non-polar interactions between the aromatic imidazolium cations (with alkyl substitutions) and the non-polar portions of the zein.Â Strong hydrogen bond accepting molecules increase the amount of β-turn secondary structure.

Solubility - solvent property relationships were explored to better understand the solvent properties of importance and how to design an ionic liquid to better dissolve a hydrophobic protein. It is found that a good solvent for zein has a small molecular volume, has a low polarity, and is a poor hydrogen bond acceptor. This combination of properties will enhance zein's solubility and limit secondary structure changes that can harm protein properties.

(253h) Synthesis and Characterization of Novel Ternary Deep Eutectic Solvents

Liu, Y., Donghua University
Xing, Y., Donghua University
Originally presented on: 11/5/2013 10:43:00 - 11:02:00

A series of novel, moisture-stable, ternary deep eutectic solvents (TDESs) has been prepared and characterized by mixing appropriate molar ratios of 1-butyl-3-methylimidazolium halide ([BMIM]X, X=Cl, Br), zinc halide (ZnX_2, X=Cl, Br) and various amides (acetamide, urea and N,N'-dimethyl urea). The synthesized ternary deep eutectic solvents are liquid at ambient temperatures and have interesting solvent properties. It shows that the most synthesized TDESs have much lower melting points than the conventional binary system according to the data of differential scanning calorimetry (DSC). Data on the melting point, viscosity and conductivity are presented and these are shown to be significantly different to other reported binary system. The [BMIM]Cl-$ZnCl_2$-$(Acetamide)_2$ system showed the lowest melting points (<-60°C), lowest viscosity and highest conductivity in all synthesized TDESs. The results showed that both viscosity and conductivity of all synthesized TDESs decrease with increasing temperatures and can be fitted by *Arrhenius equation* versus temperature. The data of conductivity and viscosity also fits well to the Walden's Rule. The influence of amides, halides and molar ratio of amides in ternary mixtures on the physical properties of the synthesized TDESs has been investigated.

(257a) Kinetic and Diffusive Mass Transfer in the Binding of Targets to Probes Displayed On the Bead Surface in a Microfluidic Cell

Chen, X., The City College of New York, The School of Engineering
Maldarelli, C., The City College of New York
Originally presented on: 11/5/2013 8:30:00 - 8:45:00

Platforms which can display biomolecules in a microarray format as probes for the multiplexed screening of the binding interactions against a target are important tools in biomedical research (drug discovery), clinical diagnostics (biomarker identification) and sensor development. Biomolecular microarrays which integrate the array and the binding detection in a single, microfluidic format have the advantage of reduced volumes of reagents and targets, high sensitivity, and increased throughput. Microbead-based microarrays are biomolecular displays in which the probe molecules are first linked to micron-sized beads which are arranged and fixed to a surface. These arrays can easily be constructed in a microfluidic platform by using a concourse of traps designed to hydrodynamically capture the beads as they stream through the course. The arrangement of the traps serves as the template for the microbead array. By trapping the beads in a sequence of stages, in which at each stage, beads displaying only one probe are trapped, with their entrapment locations mapped, a fully indexed array can be formed. The array can then be screened against targets which are passed through the concourse.

The dynamic response of the array is determined by the rate of mass transfer of the target to the probe on the bead surface. The binding can be divided into two steps. After fluorescently labeled target is streamed into the cell, the target diffuses to the surface of the bead, and then binds to the receptor or probe on the surface. Because the beads are localized in the traps, most of the streamlines, instead of passing through the gap between the traps, tend to go around the trap. This reduced convection increases the size of diffusion boundary layers. For this reason, diffusion can play an important role on the dynamic response of the microbeads array. Numerical solutions are constructed for the surface density of bound targets as a function of time for different ratios of the surface reaction rate to the bulk diffusion (Damkohler number) and the convection of target to diffusion (Peclet number). Values for these numbers are determined for which the dynamic response is either diffusion or kinetically limited. The mass transfer limitation, which arises for small Peclet number, is especially important when there is a limited amount of target, and flow rates through the device are small. The theory is compared to experiments in which the dynamic response of the microbead array is measured for the binding of the target Neutravidin to the probe biotin on the bead surface, and for this prototype and the flow rates used the binding is shown to be kinetically limited.

(257b) Sorting Droplets Based On the Number of Encapsulated Particles

Cao, Z., Virginia Tech
Chen, F.,
Bao, N.,
He, H.,
Xu, P.,
Jana, S.,
Jung, S.,
Lian, H.,
Lu, C., Virginia Tech
Originally presented on: 11/5/2013 8:45:00 - 9:00:00

Droplet microfluidics provides a high-throughput platform for screening subjects and conditions involved in biology. Droplets with encapsulated beads and cells have been increasingly used for studying molecular and cellular biology. Droplet sorting is needed to isolate and analyze the subject of interest during such screening. The vast majority of current sorting techniques use fluorescence intensity emitted by each droplet as the only criterion. However, due to the randomness and imperfections in the encapsulation process, typically a mixed population of droplets with an uneven number of encapsulated particles results and is used for screening. Thus droplet sorting based on the number of encapsulated particles becomes necessary for isolating or enriching droplets with a specific occupancy. In this work, we developed a fluorescence-activated microfluidic droplet sorter that integrated a simple deflection mechanism based on the use of a solenoid valve and a sophisticated signal processing system with a microcontroller as the core. By passing droplets through a narrow interrogation channel, the encapsulated particles were detected individually. The microcontroller conducted the computation to determine the number of encapsulated particles in each droplet and made the sorting decision accordingly that led to actuation of the solenoid valve. We tested both fluorescent beads and stained cells and our results showed high efficiency and accuracy for sorting and enrichment.

(257c) Asynchronous Magnetic Bead Rotation (AMBR) Viscometer for Label-Free DNA Detection in Digestion and Quantitative PCR

Li, Y., University of Michigan
Albertson, T., University of Michigan
Kopelman, R., University of Michigan
Burns, M. A., University of Michigan
Originally presented on: 11/5/2013 9:00:00 - 9:15:00

Sensitive and low-cost DNA detection methods have a wide range of applications from clinical diagnostics and drug development to the food industry and forensic sciences. Here we report a cost-effective label-free DNA detection technology using asynchronous magnetic bead rotation (AMBR) viscometer. In our work, we have demonstrated theoretically and experimentally that the rotation period of the asynchronous motion is linearly proportional to the viscosity of a DNA solution surrounding a magnetic bead in a low Reynolds number regime. The viscosity of a DNA solution at a fixed temperature, which depends on the molecular weight of the DNA strands and the concentration of the solution, can be used to calculate DNA concentration. By measuring the rotation periods of magnetic beads, we have measured the viscosity of DNA aqueous solutions, digestion reaction solutions and quantitative PCR (qPCR) solutions with different initial template concentrations. The results prove the feasibility of viscosity-based DNA detection using AMBR method.

(257d) Motion of Carbon Nanotubes in Water Flowing Through a Silica Nano-Slit With a Dissipative Particle Dynamics Model

Papavassiliou, D. V., The University of Oklahoma
Vo, M., The University of Oklahoma
Originally presented on: 11/5/2013 9:15:00 - 9:30:00

Carbon nanotubes (CNTs) are an interesting class of materials with a wide range of practical applications, because of their unique mechanical, thermal and electrical properties [1]. Depending on the interaction of the surrounding fluid with CNTs, the mobility of a CNT in the flow can be affected significantly. The main focus of this work is on the hydrodynamic interactions of a CNT with the fluid flow, the interactions of a solid surface with a CNT, and their effect on the propagation of CNTs in microflows.

This study is a model of water flow between two parallel silica plates using dissipative particle dynamics (DPD) with no slip boundary conditions. This is representative of the case of a CNT traveling in a pore in the subsurface, for example in a sandstone pore. DPD is a coarse graining method that is quite promising for studying hydrodynamic behavior at mesoscopic scales [2]. A DPD model for each simulated case, however, requires the determination of physicochemical parameters that characterize the interaction between the fluids and the solids involved. In our case, the parameters for CNT-water and CNT-silica need to be determined. We use prior results from molecular dynamics simulations of flow around CNTs to validate the DPD parameters in simple cases, before applying them to the final simulations. We adopt these parameters in order to simulate clusters of molecules that allow the simulation of not only nanoflow but also microflows. Then, a cylindrical CNT is released at the center of the flow flied, after the flow reaches equilibrium. The criteria used to validate the DPD model parameters include the satisfaction of the no-slip boundary condition at the water-silica interface, the viscosity of the water, the drag force on a cylinder and the slip length on the water-CNT interface. Different sizes of CNTs are investigated in our simulation in order to examine the effect of the aspect ratio on the hydrodynamic forces and on the rotation of a CNT as it is moving in the flow. As the CNT moves close to the solid surface, its trajectory can provide a set of data to determine a model for the potential interactions between the CNT and silica. Such a model can be applied in macroscopic calculations to account for CNT adsorption on the solid surface.

ACKNOWLEDGEMENTS

The financial support of the Advanced Energy Consortium (AEC BEG08-022) and the computational support of XSEDE (CTS090017) are acknowledged.

REFERENCES

1. Ray H. Baughman, Anvar A. Zakhidov, Walt A. de Heer, "Carbon nanotubes – the route toward applications," *Science,* **297,**787-792, 2002
2. Robert D. Groot, Patrick B. Warren, "Dissipative particle dynamics: Bridging the gap between atomistic and mesoscopic simulation," *J. Chem. Phys.,* **107**(11), 4423-4435, 1997.

(257e) Horizontal and Vertical Gas-Liquid-Solid Slurry-Taylor Flow: Liquid-Solid Mass Transfer Measurements Via Ion Exchange Resin

Liedtke, A. K., Laboratoire de Génie des Procédés Catalytiques, UMR 5285 CNRS, CPE, UCB Lyon

Bornette, F., Laboratoire de Génie des Procédés Catalytiques, UMR 5285 CNRS, CPE, UCB Lyon

Zanota, M. L., Laboratoire de Génie des Procédés Catalytiques, UMR 5285 CNRS, CPE, UCB Lyon

Philippe, R., Laboratoire de Génie des Procédés Catalytiques, UMR 5285 CNRS, CPE, UCB Lyon

Bellefon, C. D., LGPC - CNRS/CPE Lyon

Originally presented on: 11/5/2013 9:30:00 - 9:45:00

In multiphase micro-reaction technology the solid catalyst is usually immobilized on the reactor walls, which not only leads to rather small catalyst amounts per unit volume but more seriously impinges on its flexibility. The catalyst removal in case of deactivation or change of operation is difficult, even impossible without damage to the reactor wall. Furthermore the coating process itself is specific for each catalyst and requires therefore additional development time.

A new approach to join beneficial properties of Taylor-flow with the operational flexibility of conventional slurry reactors is the slurry-Taylor flow where catalyst particles are suspended and kept in motion by the internal circulations in the liquid slugs.

The concept of a three phase slurry micro-reactor was first applied by Enache et al. [1] for gas-liquid-solid vertical flow. In contrast to other recent studies [2, 3, 4] on liquid-liquid horizontal flow where solid particles are placed in the dispersed liquid phase, we investigate gas-liquid horizontal and vertical flow with the solid localised (in general) in the continuous phase.

We were able to show that the performance of this new contact mode is comparable to a laboratory stirred tank vessel under semi-batch conditions [5] and we concentrate now on hydrodynamics and mass-transfer properties.

We identified different flow regimes by varying the fluids flow rates, solid charge and flow direction (figure 1).

Here we discuss our first results regarding the liquid-solid mass transfer capacity of slurry-Taylor flow. We chose to work with ion exchange resins as suspended particles [6] and apply and compare two different measurement strategies to follow the neutralization of caustic soda and thereby the ion exchange rate: we use a pH color indicator (figure 2) which allows not only to estimate the time required to reach a specific pH-value but also to detect the region of transfer positioned in the liquid slug which is especially interesting for heterogeneous flow regimes. Also we measure the change of conductivity with electrodes consisting of two platinum wires, connected to an alternating current source. Several of these electrodes are placed along the tube length which allows us to follow the sodium concentration online. Both methods are used to calculate liquid-solid mass transfer coefficients for different fluid flow rates, particle diameters and charges with the aim to propose a correlation law.

SHAPE * MERGEFORMAT

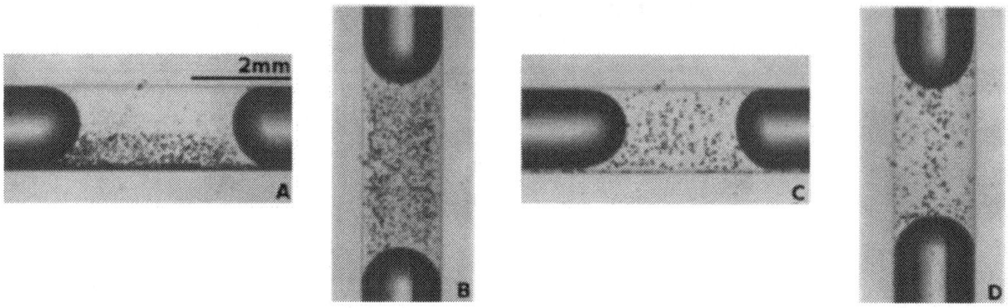

Figure 1) Examples for some typical flow patterns for horizontal (A, C) and vertical (B, D) flow. Materials : gas phase: N_2, liquid phase : EtOH, solid phase: SiO_2, 40-76 µm, impregnated with NiO_2, solid charge 5g/L (C, D) and 10 g/L (A,B). PFA tube, d_{tube}=1.65mm. Total velocity 47 mm/s (A, B) and 150 mm/s (C, D).

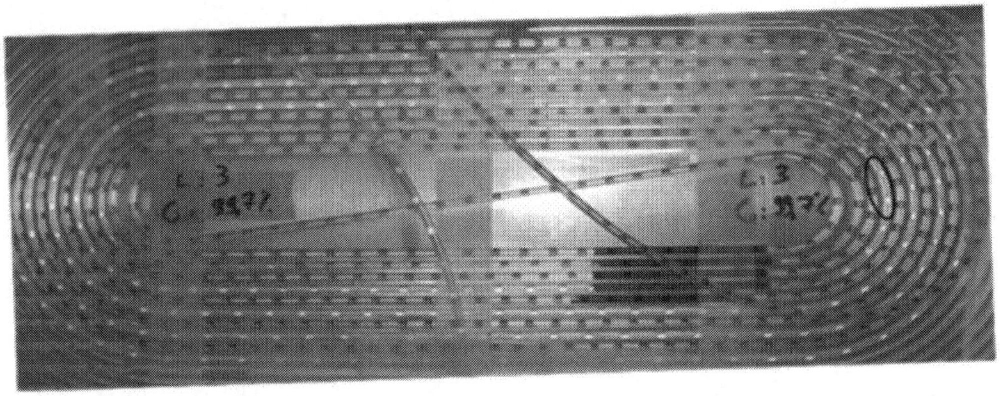

Figure 2) Example of liquid-solid mass transfer measurement using a pH color indicator: gas phase: N2, liquid phase: H_2O, initially 0.01 mol/l NaOH, 0.1 g/L cresol red, solid phase: Dowex 50WX8, 200-400 mesh; PFA tube, d_{tube}=1.65mm; total velocity 120 mm/s; complete color change and thus due to the usage of cresol red neutralization can be detected after a residence time of ≈40 seconds.

REFERENCES

[1] D.I. Enache, G.J. Hutchings, S.H. Taylor, R.Natividad, S.Raymahasay, J.M. Winterbottom, E.H. Stitt, Experimental evaluation of a three-phase downflow capillary reactor, Ind. Eng. Chem. Res. 44 (2005) 6295-6303.

[2] A. Ufer, D. Sudhoff, A. Mescher, D. W. Agar, Suspension catalysts in a liquid-liquid capillary microreactor, Chem. Eng. J. 167 (2011) 468-474.

[3] K. Olivon, F. Sarrazin, Heterogeneous reaction with solid catalyst in droplet-flow millifluidic device, Chem. Eng. J., in press, corrected proof.

[4] G. K. Kurup, A. S. Basu, Field-free particle focusing in microfluidic plugs, Biomicrofluidics 6 (2012) 22008-2200810.

[5] A.-K. Liedtke, F. Bornette, R. Philippe, C. de Bellefon, Gas?liquid?solid ??slurry Taylor" flow: Experimental evaluation through the catalytic hydrogenation of 3-methyl-1-pentyn-3-ol, Chem. Eng. J. in press, corrected proof.

[6] V. G. Pangarkar, A. A. Yawalkar, M. M. Sharma, A. A. C. M. Beenackers, Particle-liquid mass transfer coefficient in two-/ three-phase stirred tank reactors, Ind. Eng. Chem. Res. 41 (2002) 4141-4167.

(257f) Electrophoretic and Dielectrophoretic Assembly of Nanocolloids in Conic Nanopipettes

Liu, S., University of Notre Dame
Yan, Y., University of Notre Dame
Wang, Y., University of Notre Dame
Senapati, S., University of Notre Dame
Chang, H. C., University of Notre Dame
Originally presented on: 11/5/2013 9:45:00 - 10:00:00

We report and analyze a new electrokinetic technique for assembling fewer than one hundred 20-nm metal and dielectric nanoparticles at the tip of a nanopipette that is fabricated by laser-assisted drawing techniques. The nanocolloid assembly mechanism is driven by two fundamental phenomena at nanoscales: the high ionic strength at the tip due to sub-Debye length Boltzmann equilibrium and singular field focusing towards the tip. The enhanced Debye screening due to the former and the enhanced (reactive) induced particle dipole due to the latter conspire to produce a submicron slug of nanoparticle crystal at the tip under certain conditions. Using an extended DLVO theory, we are able to collapse the measured phase boundaries for such confined-domain nanoparticle assembly at different electric field intensity/frequency, particle size/property and bulk ionic strength. Interestingly, by monitoring the slow ion current dynamics, as well as short current jump events due to single nanoparticle or single aggregate translocation, we are able to decipher the packing dynamics and crystal geometry using a statistical scaling theory. Such sub-micron nanoparticle crystals are used for generating nm-sized plasmonic hot spots that allow single-molecule detection and even single-molecule spectroscopy. We report a single-stranded microRNA detection experiment with functionalized hairpin oligos and FRET (Forster Resonance Energy Transfer) fluorescent reporters on gold nanoparticles. Sub-femtomolar limit of detection is achieved with fewer than 100 target molecules. We also show single-mismatch selectivity because of the high shear rate during the nanoparticle assembly process.

(257g) Application of Electrochemical Impedance Spectroscopy (EIS) to Electrokinetic Characterization of Planar Microfluidic Substrates

Saini, R., Queen's University
Barz, D. P. J., Queen's University
Originally presented on: 11/5/2013 10:00:00 - 10:15:00

Electrokinetic phenomena play an important role in characterization of microfluidic and colloidal substrates. In terms of consolidated substrates, the measurement of streaming potential or streaming current can be used to infer the influence of liquid parameters on the zeta potential of the substrates. The classic Helmholtz-Smoluchowski equation is generally used to correlate the zeta potential to the streaming potential/current. In the current work, different sets of experiments are performed in order to investigate zeta potential's dependency on ion concentration and pH value for poly (methyl methacrylate) (PMMA) and borosilicate glass. Non-linear (electrochemical impedance spectroscopy) and linear (DC voltage/current) resistance measurements of a streaming potential setup are carried out under experimental conditions. We find that the resistance of much diluted electrolytes show considerable deviations from theory which cannot be explained by the phenomenon of surface conductivity. The results of the impedance measurements and the streaming current can be used to derive correlations for the correction of the classic streaming potential theory.

(257h) Concentration Profile of Rod-Like Particles in a New Hyperlayer Mode of the Field Flow Fractionation: Considering Hydrodynamic Interaction With a Channel Wall

Alfi, M.,
Park, J., The City College of City University of New York
Originally presented on: 11/5/2013 10:15:00 - 10:30:00

We present theoretical studies on the feasibility of the shape separation of rod-like micro/nanoparticles in a new hyperlayer mode of the field-flow fractionation (FFF). Park and coworkers analyzed a mechanism for the rod migration in microchannel flows [Park et al. (2007) Phys. Rev. E, 76, 040801; Park & Butler (2009) JFM, 630, p267-298]. They explained that the coupling of Brownian rotation and hydrodynamic interaction with a channel wall induces the rod migration. Among many rod migration mechanisms, it has never been utilized in any of the FFF applications. Therefore, we incorporated this rod migration mechanism into the dynamics of the rod-like particles in FFF to find a potential design for a separation device that sorts particles by shape. We show both the analytically derived concentration profiles of rods in the FFF channel and the corresponding retention ratios as a function of the aspect ratio of the rod with varying the Peclet number.

(257i) Chiral Chromatography of Nanomotors

Nourhani, A., The Pennsylvania State University
Lammert, P. E., The Pennsylvania State University
Borhan, A., The Pennsylvania State University
Crespi, V. H., The Pennsylvania State University
Originally presented on: 11/5/2013 10:30:00 - 10:45:00

Rotary nanomotors (nanorotors) with sizes in the range of 0.1-1 micron have been shown experimentally to execute circular motion on scales as small as the motor dimensions. Orientational diffusion of these nanorotors is unavoidable. The combination of a chiral object and a powered motion superposed by stochastic orientational dynamics leads to long-term translational diffusion. These nanorotors engage in an unusual chiral diffusion: after one period of deterministic rotation, the nanorotor is not equally likely to displace to its left as right. When these chiral nanorotors experience underdamped motion in a chiral potential, orientational diffusion can lead to a much more remarkable effect. An induced drift velocity, which depends on the sense of rotation of the nanorotor, is generic, and can even be comparable to the motor's speed in some parameter ranges. Depending on the form and strength of the chiral potential, the nanorotor demonstrates a range of dynamics from hopping between local minima to linear-like motion with sudden changes in the direction of motion. We will give a tour of the various regimes of motion in this problem, and discuss the possibilities for experimental realization of "chiral chromatography".

(257j) Numerical Simulation Study of the Deterministic Vector Chromatography of Rigid and Flexible Particles Over Slanted Open Cavities

Bernate, J. A., Johns Hopkins University
Yang, M.,
Zhao, H., Stanford University
Risbud, S. R., Johns Hopkins University
Paul, C. D., University of Arkansas
Dallas, M., Johns Hopkins University
Konstantopoulos, K., Johns Hopkins University
Drazer, G., Rutgers University
Shaqfeh, E. S. G., Stanford University
Originally presented on: 11/5/2013 10:45:00 - 11:00:00

Planar microfluidic platforms for vector chromatography, in which different species fan out in different directions and can be continuously sorted, are particularly promising for the high throughput separation of multicomponent mixtures. We carry out a computational study of the vector separation of dilute suspensions of rigid and flexible particles transported by a pressure-driven flow over an array of slanted open cavities. The numerical scheme is based on a Stokes flow boundary integral equation method. The simulations are performed in a periodic system without lateral confinement, relevant to microfluidic devices with much larger width than the dimensions of the open cavities and the ridges that create them. We study the deflection of rigid spherical particles, of flexible capsules as a model of white and red blood cells, and of rigid discoidal particles as a model of platelets. We characterize the deflection of different particles as a function of their size, shape, shear elasticity, their release position, and the geometric parameters of the channel. The simulations provide insight into the separation mechanism and allow the optimization of specific devices depending on the application. Good agreement is observed with microfluidic experiments measuring the deflection of polystyrene particles, white and red blood cells, and MCF-7 breast cancer cells. Promising applications of this platform include the fractionation of blood and the capture of circulating tumor cells. This platform also has the potential to yield samples with controlled polydispersity at high throughputs, which would allow the study of the rheological behavior of complex fluids with desired particle size distributions.

(258a) High Resolution Simulations of Droplet Collisions and Coalescence in Shear

Shardt, O., University of Alberta

Mitra, S. K., University of Alberta

Derksen, J., University of Alberta

Originally presented on: 11/5/2013 8:30:00 - 8:45:00

The outcomes of collisions between droplets determine the droplet size distribution and rheology of emulsions i turbulence, the flow patterns of emulsions in porous media, and the behaviour of microfluidic devices that man: to the large separation of length scales between typical droplet radii (1-100 μm) and the minimum thickness of separate drops (~10 nm), simulations that predict the outcome of a collision are challenging because both lengtl resolved. We recently reported [Shardt et al. Langmuir (2013) DOI: 10.1021/la304919p] large-scale three-dime droplet collisions in simple shear flow using a diffuse interface model. In these simulations, the film between th was resolved and the critical capillary number for coalescence was determined as a function of several geometr interface model parameters. For example, if the interface thickness is kept constant, the critical capillary numb droplet size increases.

While the previous results were obtained with a large initial vertical (see figures below) separation between the present simulations with half the previous vertical offset. By simulating droplets with double the previous size i twice as large (eight times the volume), the critical capillary number for this geometry is sufficiently low that it simulations. Due to the larger size of the droplets, we study coalescence at a nondimensional vertical offset of (between centres divided by droplet diameter), near the value in the experimental work of Chen et al. (Langmui 12885-12893). We analyze the results of the simulations to determine the critical capillary number for coalesce critical film thickness, and the droplet trajectories. These results can be used to provide insight into the conditio and eventually perform predictive simulations of coalescence phenomena.

Two sample collisions between droplets are shown below. At a capillary number Ca of 0.1 (upper sequence), th Ca=0.2 (lower sequence), they first coalesce with an internal droplet being formed and then break apart. In the: radius of the droplet is 200 lattice nodes, and the simulation domain is 2048×512×512 lattice nodes. Due to the boundary conditions, this represents a domain of 2048×1024×1024. These are the largest simulations of coalesc far, and they were made possible by the use of 66 Tesla M2070 GPUs, which reached typical processing speeds lattice updates per second.

Collision at Ca = 0.1.

Collision at Ca = 0.2.

(258b) Collision and Coalescence of Liquid Drops in a Dynamically Active Ambient Fluid

Sambath, K., Purdue University
Thete, S. S., Purdue University
Subramani, H. J., Chevron Energy Technology Company
Basaran, O. A., Purdue University
Originally presented on: 11/5/2013 8:45:00 - 9:00:00

The fluid dynamics of the collision and coalescence of liquid drops has intrigued scientists and engineers for more than a century owing to its ubiquitousness in nature, e.g. raindrop coalescence, and industry, e.g. breaking of emulsions in the oil and gas industry. The complexity of the underlying dynamics, e.g. occurrence of hydrodynamic singularities, has required study of the problem at different scales---macroscopic, mesoscopic and molecular---using stochastic and deterministic methods. In this work, we adopt a multiscale, deterministic method to simulate the approach, collision, and eventual coalescence of two drops where the drops as well as the ambient fluid are incompressible, Newtonian fluids. The free boundary problem governing the dynamics consists of the Navier-Stokes system and associated initial and boundary conditions that have been augmented to account for the effects of disjoining pressure as the separation between the drops becomes of the order of a few hundred nanometers. This free boundary problem is solved by a Galerkin finite element-based algorithm. The approach and results to be reported build on earlier work by Leal and coworkers, and are used to identify conditions conducive for coalescence in terms of flow and fluid properties.

(258c) Turbulent Confined Swirling Flow in a Microscale Multi-Inlet Vortex Reactor

Shi, Y., Iowa State University
Olsen, M., Iowa State University
Fox, R. O., Iowa State University

Originally presented on: 11/5/2013 9:00:00 - 9:15:00

The turbulent confined swirling flow in a microscale multi-inlet vortex reactor (MIVR) is investigated by means of microscopic particle image velocimetry (μ-PIV) as well as computational fluid dynamics (CFD). The multi-inlet vortex is designed in hope of providing fast and homogeneous mixing that is essential in production of nanoparticles via Flash Nanoprecipitation (FNP). Both Reynolds-averaged Navier-Stokes (RANS) and large-eddy simulations (LES) are carried out to simulate the velocity fields for three different jet Reynolds numbers, i.e., $Re_j = 53$, 93 and 240. The first two Reynolds numbers represent laminar and transition flow regime, whereas the last one is turbulent. For $Re_j = 240$, numerical predictions are validated against μ-PIV measurements. Comparisons show the RANS method with the widely-used standard k–ε turbulence model underpredicted the mean velocities. LES with the Smagorinsky model, on the other hand, compares well with experimental results. Features of the confined swirling flow are examined from the LES by looking at the axial velocities and streamlines. By computing the λ_2 values, the vortex cores are extracted. A great amount of vortices are observed at the contraction where the fluid exits the mixing chamber. In addition to the turbulent motions in the mixing chamber, the vortex breakdown in the outlet channel is also key to promoting the mixing performance of the reactor.

(258d) Interacting Length Scales in a Reactive-Infiltration Instability

Ladd, A. J. C., University of Florida
Szymczak, P., University of Warsaw
Originally presented on: 11/5/2013 9:15:00 - 9:30:00

Reactive-infiltration instabilities occur in a wide range of geophysical and geotechnical systems where an incoming flow of reactant dissolves the surrounding solid matrix, leading to local increases in permeability. The enhanced permeability increases the local flow in that region leading to further dissolution. The reactive-infiltration instability is related to, but distinct from, viscous fingering, where the permeability is constant but the viscosity of the fluid (and possibly its density) respond to variations in concentration.

The reactive-infiltration instability in porous materials contains two important length scales, unlike viscous fingering which only has one. In this presentation I will outline a linear stability analysis that simultaneously incorporates both scales. It shows that the commonly used ``thin-front" model is a limiting case of a more general theory, which also includes convection-dominated dissolution as another special case. In contrast to the thin-front theory, the wavelength of the instability is bounded from below, and lies In the range 1mm to 1km for physically reasonable flow rates and reaction rates. We have obtained an explicit expression for the growth rate when the change in porosity is small.

Theoretical investigations of the reactive-infiltration instability have typically focused on a steadily propagating dissolution front that separates regions of high and low porosity. However, this is not the only possible dissolutional instability in porous rocks; there is another instability that operates instantaneously on any initial porosity field, including an entirely uniform one. The relative importance of the two mechanisms depends on the ratio of the porosity increase to the initial porosity. We have shown that the "inlet" instability is likely to be important in limestone formations where the initial porosity is small and there is the possibility of a large increase in permeability. In quartz-rich sandstones, where the proportion of easily soluble material (e.g. carbonate cements) is small, the instability in the steady-state equations is dominant.

(258e) Effect of Flow Structure At the Onset of Instability On Barium Sulfate Precipitation in Taylor-Couette Crystallizers

Aljishi, M. F., Saudi Aramco
Ruo, A. C., Cornell University
Park, J. H., Cornell University
Nasser, B., Cornell University
Kim, W. S., Kyung Hee University
Joo, Y. L., Cornell University
Originally presented on: 11/5/2013 9:30:00 - 9:45:00

The effect of the vortex flow by primary instability on the precipitation of barium sulfate crystals was examined in Taylor–Couette crystallizers. Experiments involving precipitation of barium sulfate under laminar flow (before the onset of instability) and vortex flow (after the onset of primary instability) at three different axial flow rates revealed that the vortex motion plays a critical role in fine-tuning crystal properties such as internal crystal structure, particle microstructure and morphology. While no clear trends in crystalline size were observed for laminar flow, all flowrates at the onset of instability resulted in smaller crystalline sizes that ranged between 40 nm and 50 nm. A decrease in the unit cell dimensions of the orthorhombic lattice structure was also observed after the onset of instability. For both laminar and vortex flow, shortening the residence time by increasing axial flow rate led to a substantial decrease in crystal size. It is also observed that the crystal size distribution (CSD) became much narrower after the emergence of vortices. Under vortex flow, an increase in axial flow resulted in even narrower crystal size distributions. Finally, the observed phenomena of smaller crystal size and narrow crystal size distribution under vortex flow are qualitatively explained by the analysis of particle trajetory.

(258f) Mixing and Axial Dispersion in Taylor-Couette Flows: A Multi-Scale Study

Charton, S., CEA, DEN
Nemri, M., IMFT
Climent, E., IMFT
Originally presented on: 11/5/2013 9:45:00 - 10:00:00

Liquid-liquid extraction in an industrial apparatus is a complex process involving chemistry, interface dynamics, mass transfer and fluid mechanics. Among these phenomena, transport processes, which are size-dependent, deserve a particular attention in the scope of R&D studies (where size-reduction is encouraged, as in the nuclear industry) and for scale-up purpose. Hence, flow patterns and properties in extraction devices are the subject of increasing interest, involving numerical as well as experimental studies. In this aim, we choose to take advantage of Taylor-Couette flows, already used to perform small-scale solvent extraction studies [1], in order to investigate the specific hydrodynamic issues: mainly mixing and axial dispersion.

Taylor-Couette flows take place in the annular gap between two concentric cylinders with the inner one rotating and the outer one fixed. This flow is known to evolve towards turbulence through a sequence of successive hydrodynamic instabilities [2,3]. These different flow-patterns are likely to influence dramatically the mixing properties. Moreover, the simple and periodic geometric configuration of Taylor-Couette flows allows Direct Numerical Simulations to be performed.

Prior to the study of the mixing properties, the sequence of flow instabilities has been determined using a visualization technique for both the small-gap and large-gap columns we used. Thanks to a spectral analysis, the transition Reynolds and hydrodynamic characteristics of the achieved flow states were identified. Start-up procedures were also established in order to confidently reproduce these flow states [4].

At the reactor scale, it is commonly admitted that all the mechanisms responsible for flow non-ideality and mixing are represented by a single and assumed linear process, and are quantified by a lumped parameter: the axial dispersion coefficient Dx [4]. The Dx evolution along the successive flow states encountered in the small-gap R&D apparatus (g= 1.5 mm), was investigated thanks to dye's Residence Time Distribution measurements [5]. In agreement with literature results related to different aspect ratio configurations, axial dispersion was observed to increase with cylinder rotation [4,6,7]. Moreover, although the results reproducibility was checked, different Dx values have been measured, for a given Reynolds number, depending on the initial condition and start-up procedures. This flow hysteresis effect is well-known [8], although its effect on axial dispersion had not yet been demonstrated experimentally. The measured Dx evolutions were confirmed by Direct Numeric Simulations results, which highlighted furthermore that the larger the cylinder's gap, the higher the influence of the flow regimes.

The local mechanisms involved in mixing have been characterized by means of simultaneous PIV (particle image velocimetry) and PLIF (planar laser induced fluorescence) measurements. A specific apparatus, with a large-enough gap to perform accurate optical measurements (g = 11 mm) was specifically designed for this purpose. Simultaneous PIV and PLIF measurements were performed in different flow patterns: Taylor vortex flow, wavy and modulated wavy vortex flow.

PLIF visualizations showed clear evidences of different transport mechanisms going from diffusion-like intravortex mixing to fluid bundle transport between neighboring vortices. Moreover, the PLIF results have confirmed the occurrence of intervortex mixing in the steady Taylor Vortex Flow regime, especially in the near-wall regions and at inflow boundaries. The relative importance of each mechanism, as well as their evolution, depending on the flow structure and wave state, have been studied and elucidated, thus supplementing the conclusion of the "macroscopic" studies [5]. The results confirmed that the occurrence of different wavy regimes, depending on flow history (hysteresis), may have a dramatic effect on mixing for a given Reynolds.

This study demonstrates that the commonly used 1-parameter chemical engineering models (e.g. the "well-mixed stirred tanks in series" model) are not valid to model Taylor-Couette reactors: two parameters are at least required for an efficient description of mixing in Taylor-Couette flows.

References:

[1] M.W. Davis and E.J. Weber. Liquid-liquid extraction between rotating concentric cylinders, *Industrial and Engineering Chemistry*, 52, 1960.

[2] D. Coles. Transitions in circular Couette Flows, *Philosophical Transactions of the Royal Society of London*, 223, 1923.

[3] D. Andereck, S.S Liu and H.L. Swinney. Flow regimes in a circular Couette system with independently rotating cylinders, *Journal of Fluid Mechanics*, 164, 1985.

[4] C. Moore and C. Cooney. Axial dispersion in Taylor-Couette flow, *AIChE Journal*, 41, 1995.

[5] M. Nemri, S. Charton, E. Climent and J.Y. Lanoë. Experimental and numerical investigation on mixing and axial dispersion in Taylor-Couette flow patterns, *Chemical Engineering Research and Design*, 2012, *in press*.

[6] W.Y. Tam and H.L. Swinney. Mass transport in turbulent Taylor-Couette flow, *Physical Review A*, 36, 1987.

[7] G. Desmet, H. Verelst and G.V. Baron. Local and global dispersion effect in Couette-Taylor flow – II. Quantitative measurments ans discussion of the reactor performances, *Chemical Engineering Journal*, 51, 1996.

[8] M Rudman. Mixing and particle dispersion in the wavy vortex regime of Taylor-Couette flow, *AIChE Journal*, 44, 1998.

(258g) Lattice Boltzmann Simulation of Mixing Process Inside Micro-Droplets in a Gas - Liquid Taylor Flow

Zhao, S., Tsinghua University
Wang, W., Tsinghua University
Shao, T., Tsinghua University
Jin, Y., Tsinghua University
Cheng, Y., Tsinghua University
Originally presented on: 11/5/2013 10:00:00 - 10:15:00

Mixing of two fluids in microfluidic system is of essential interest in nowadays chemical engineering. Since the flows in simple microchannels under typical operating conditions are predominantly laminar and uniaxial, molecular diffusion becomes the major driving force for the mixing of fluid streams. Considering the low diffusion capability, mixing process is still not fast enough even at the micro-scale of ~100 mm. As an alternative, droplet-based microfluidic system has become one of the promising solutions, which takes advantage of the internal recirculation inside droplets to increase the convective mass transfer between streams. Micro-droplets in a continuous gas-liquid Taylor flow behave like micro-liter even nano-liter batch reactors to precisely control the underlying hydrodynamics and mixing processes inside droplets so that the product processes possess well-defined residence time distribution and intensified micro-scale mixing.

LBM has experienced rapid development and attracted increasing interests during the past two decades in simulating multiphase/multicomponent flows owing to its excellent numerical stability and constitutive versatility, especially on solving the challenge of moving and deformable interfaces. Three LB models for multiphase flows have emerged, i.e. free-energy approach, Shan & Chen scheme and color-gradient method, where the color-field LBM is quite appropriate to simulate the gas-liquid dispersed microfluidic system for its capability to independently tune the viscosity, interfacial tension and interface thickness under the modeling scheme. On the basis of the simple color-field two-phase method (Latva-Kokko and Rothman, 2005), we further introduced a third component, which perfectly acted as a standard tracer to track the behavior of one of fluid steams without any impact on bulk flow field. This newly established method was then used to study the hydrodynamics and liquid mixing performance inside micro-slugs/droplets in a microchannel with the gas-liquid Taylor flow. The micro-LIF measurements of ethanol mixing process in the ethanol-air system were carried out to provide experimental foundations to validate the simulation analysis. Also, the comparison of the liquid mixing behaviors in a simple laminar flow and a droplet-based flow in microchannels has been made to verify the intensification effect of the internal recirculation inside independent micro-droplets.

M. Latva-Kokko, Daniel H. Rothman, Diffusion properties of gradient-based lattice Boltzmann models of immiscible fluids. Physical Review E 71,056702 (2005)

---

* Corresponding author, Dr. Yi Cheng, Email: *yicheng@tsinghua.edu.cn*

(258h) Simulating Phenol Oxidation in An Annular Photoreactor Using Computational Fluid Dynamics

Peres, J. C. G., University of SÃ£o Paulo
Vianna, A. D. S. Jr.,
Originally presented on: 11/5/2013 10:15:00 - 10:30:00

Advanced oxidation processes are a promising technology for degradation of organic compounds resistant to conventional treatments such as phenol. Computational fluid dynamics (CFD) has recently emerged as a powerful tool that allows a deeper understanding of photochemical processes in reactor engineering by solving the coupled momentum, mass and radiation balances. Little attention has been given to the homogeneous UV/hydrogen peroxide (H_2O_2) systems, which are simpler to operate and design than photocatalytic equipments. The current state-of-the-art of UV/H_2O_2process simulation in CFD commercial codes considers simplifications to its kinetic modeling, mainly the pseudo-steady-state hypothesis (PSSH) and equilibrium conditions for radical species.

This work aimed to investigate the UV/H_2O_2 process in an annular photoreactor using CFD and a more realistic kinetic model. The fluid dynamics was studied in a previous work. Next step was introducing the mechanism of phenol degradation proposed by Edalatmanesh, Dhib and Mehrvar (2008) into the CFD model. It considers a total of 15 reactions and 6 key components: phenol, hydrogen peroxide, hydroxyl radical (OH•), hydroperoxyl anion (HO_2^-), hydroperoxyl radical (HO2•) and superoxide anion radical ($O2^{\bullet -}$); this kinetic model is based on dynamic equations for all species. The fluence rate field was calculated by the radial model. Simulations reproduced experimental data spanning a wide range of initial phenol concentrations (50-500 mg/L), H_2O_2/phenol molar ratios and three values for lamp power (36, 75 and 115 W, corresponding to 11.19, 14.89 and 17.66 W of 253.7 nm radiation emission). The volumetric inlet flow rate was fixed at 72 L/h. Convergence of the numerical solution was reached when the maximum of the scaled residuals of the equations reached less than 10^{-4}.

It was found that, for the 36 and 75 W lamps (diameter of 25 mm), 911,751 elements were necessary to reach grid independence, and 909,903 for the 115 W lamp (diameter of 37 mm). The velocity field depends on the volumetric flow rate: either it maintains a swirling motion through the whole reactor or might develop like a plug flow. The k-ε model did not represent the RTD data accurately, and the velocity field therefore, since it is not appropriate for swirling flows. The other turbulence models properly matched RTD data, especially the k-ω model.

The average conversion of phenol for a single pass is 14.3%, with an average deviation of 7.7% from experimental data. Fluence rate was symmetrical around the reactor axis. Due to the exponential relationship between distance and the fluid absorption coefficient, the radial model predicted a sharp decay of radiation intensity away from the lamp surface. It was possible to verify the concentration maps of the species through the reactor: due to the swirling inlet effects, reactants got concentrated close to the walls and migrated on the lamp direction along the reactor path. High radiation intensities close to the lamp surface created a layer around it where photolysis of H_2O_2 took place with higher rates. OH• radicals were generated in that layer and transported towards the outer wall by convection. This caused most of phenol to be consumed in the second half of the reactor and accumulation of the radical near the lamp and the reactor outlet, since the pollutant in this area was already oxidized. Satisfactory results indicated that CFD was an appropriate tool for analyzing this reactive flow.

(258i) Computationally Efficient CFD Simulations With Detailed Free-Radical Mechanisms

Schietekat, C. M., Ghent University
Van Geem, K. M., Ghent University
Marin, G. B., Ghent University
Originally presented on: 11/5/2013 10:30:00 - 10:45:00

Chemical engineers are continuously trying to increase process efficiency to increase process throughput and profitability while lowering environmental cost. One such example is the replacement of conventional tubular reactors by more advanced three-dimensional reactor configurations in steam cracking units such as the patented Swirl Flow Tube of Technip or the MERT technology of Kubota. By means of improved heat transfer, the reactor metal temperatures and the corresponding coke formation on the reactor inner wall decreases, which enables longer process operation and/or higher throughput [1]. To assess the effect of the geometry of these three-dimensional reactor configurations on product yields and coke formation, a three-dimensional reactor model is necessary. Furthermore, to accurately describe the chemical kinetics over a wide range of process conditions and feedstocks, a detailed microkinetic model is required. Only these microkinetic models capture the essential chemistry and allow prediction of the effect of reactor geometry on product yields.

The incorporation of these detailed kinetic models within complex computational fluid dynamics (CFD) simulations requires extremely long CPU times [2]. The reason is twofold. First, every species in the reaction network adds an extra continuity equation that needs to be integrated. Second, the large differences in the species time scales due to the distinct reactivity of molecules and radicals makes the resulting system inherently stiff. To alleviate the computational burden while retaining the accuracy of detailed kinetic models, different methodologies have been developed and evaluated [3-7]. In chemical reaction engineering the quasi-steady-state approximation (QSSA) is often applied [8, 9]. The applicability of QSSA on steam cracking kinetics has been validated previously using a 1D plug flow reactor model [9]. It allows to obtain the radical species' concentrations from a set of algebraic equations, greatly reducing stiffness and computational time.

In the present contribution, QSSA is applied in three-dimensional simulations of steam cracking reactors. The implemented reaction mechanisms contain over 50 species and a few hundred reactions. The Navier-Stokes equations, turbulence equations, energy equation and molecular species' continuity equations are solved using the commercial FLUENT/13.0 software. The radical concentrations are determined from algebraic equations by application of QSSA outside the main FLUENT solver. This reduces both the number of continuity equations and stiffness, yielding a speed-up of up to a factor 5. First, the approach is validated by comparison to reference simulations, i.e. without QSSA, for reaction networks of different sizes and complexity. Next industrial steam cracking reactors with a three-dimensional configuration are simulated and compared with conventional tubular reactors in terms of product yields and coking tendency. The advantages and disadvantages of using three-dimensional reactor configurations instead of conventional tubular reactors are discussed.

ACKNOWLEDGMENTS

CMS acknowledges financial support from a doctoral fellowship from the Fund for Scientific Research Flanders (FWO). The authors also acknowledge the financial support from the Long Term Structural Methusalem Funding by the Flemish Government – grant number BOF09/01M00409. The computational work was carried out using the STEVIN Supercomputer Infrastructure at Ghent University, funded by Ghent University, the Flemish Supercomputer Center (VSC), the Hercules Foundation and the Flemish Government – department EWI.

REFERENCES

[1] J.J. De Saegher, T. Detemmerman, G.J. Froment, Three dimensional simulation of high severity internally finned cracking coils for olefins production, Revue de l'Institut Francais du Petrole, 51 (1996) 245-260.

[2] W.J. Pitz, C.J. Mueller, Recent progress in the development of diesel surrogate fuels, Progress in Energy and Combustion Science, 37 (2011) 330-350.

[3] K.Y. He, I.P. Androulakis, M.G. Ierapetritou, Multi-element Flux Analysis for the Incorporation of Detailed Kinetic Mechanisms in Reactive Simulations, Energy & Fuels, 24 (2010) 309-317.

[4] K.Y. He, M.G. Ierapetritou, I.P. Androulakis, Integration of On-The-Fly Kinetic Reduction with Multidimensional CFD, Aiche J., 56 (2010) 1305-1314.

[5] L. Liang, J.G. Stevens, J.T. Farrell, A dynamic adaptive chemistry scheme for reactive flow computations, Proceedings of the Combustion Institute, 32 (2009) 527-534.

[6] Y. Shi, L. Liang, H.W. Ge, R.D. Reitz, Acceleration of the chemistry solver for modeling DI engine combustion using dynamic adaptive chemistry (DAC) schemes, Combustion Theory and Modelling, 14 (2010) 69-89.

[7] S.B. Pope, Computationally efficient implementation of combustion chemistry using in situ adaptive tabulation, Combustion Theory and Modelling, 1 (1997) 41-63.

[8] G.M. Come, Radical reaction-mechanisms - Mathematical-theory, Journal of Physical Chemistry, 81 (1977) 2560-2563.

[9] K.M. Van Geem, M.F. Reyniers, G.B. Marin, J. Song, W.H. Green, D.M. Matheu, Automatic reaction network generation using RMG for steam cracking of n-hexane, Aiche J., 52 (2006) 718-730.

(261a) Molecular Understanding of the Hydration of Zwitterionic Moieties: Effect of Charged Groups

Shao, Q., University of Washington
Jiang, S., University of Washington
Originally presented on: 11/5/2013 8:30:00 - 8:44:00

Strong hydration of a zwitterionic moiety largely determines the ultralow fouling property of zwitterionic materials important for applications including drug delivery, protein detection and marine coating. One of the fundamental questions is how the hydration depends on the types of charged groups within the zwitterionic moiety. In order to shed light on this issue, we studied the hydration of zwitterionic moieties containing different charged groups systematically using quantum mechanical calculations and molecular simulations. We selected three anionic groups (carboxylic group, sulfonate group and sulfate group) and four cationic groups (quaternary amine, tertiary amine, secondary amine and primary amine). The carbon spacer length between the cationic and anionic groups was fixed to one. Combining these seven charged groups generates twelve zwitterionic moieties including glycine betaine, a naturally occurring protein protectant. We analyzed the charge distributions of these zwitterionic molecules in order to investigate the effect of diffeent charged group combinations on the interactions between charged groups. We developed force field parameters for these zwitterionic moieties and studied their hydration using molecular dynamics simulation and metadynamics simulation. The analyzed hydration properties involve the hydration number, residence time of water molecules in the first hydration shell, hydration free energy and dehydration free energies of anionic and cationic groups. We found that varying the cationic group causes variations of the hydration in the order of carboxylic >>sulfonate and sulfate groups. We also investigated how the hydration of cationic group varies with the type of anionic group and found that the tertiary amine, secondary amine and primary amine groups show the same tendency. The simulation results provide a fundamental understanding of how the hydration of a zwitterionic moiety depends on its charged groups.

(261b) Biomolecular Simulations With the Transferable Potentials for Phase Equilibria: Extension to Phosphlipids

Bhatnagar, N., Wayne State University
Kamath, G., University of Missouri-Columbia
Potoff, J. J., Wayne State University
Originally presented on: 11/5/2013 8:44:00 - 8:58:00

The Transferable Potentials for Phase Equilibria (TraPPE) is extended to zwitterionic and charged lipids including phosphatidylcholine (PC), phosphatidylethanolamine (PE), phosphatidylserine (PS) and phosphatidylglycerol (PG). The performance of the force field is validated through isothermal-isobaric ensemble (NPT) molecular dynamics simulations of hydrated lipid bilayers performed with the aforementioned head groups combined with saturated and unsaturated alkyl tails containing 12 – 18 carbon atoms. To highlight the transferable nature of the TraPPE force field, Lennard-Jones parameters developed for the prediction of vapor-liquid equilibria for low molecular weight organic molecules are used without modification. Partial charges are determined from a CHELPG analysis of HF/6-31+G(d,p) *ab initio* calculations, while torsional parameters are refit to match the rotational barriers predicted by the CHARMM C36 lipid force field and MP2/6-31G(d) *ab initio* calculations. The effects of water model (TIP3P [1] vs. SPC/E [2]) and sodium ion parameter (Roux [3] vs. Aqvist [4]) on the performance of the lipid force field are determined. The predictions of the TraPPE force field for the area per lipid, bilayer thickness and volume per lipid are within 1-5% of experimental values. Key structural properties of the bilayer, such as order parameter splitting in the sn-2 chain and x-ray form factors are found to be in close agreement with experimental data. Overall, these results highlight the broad transferability of the Transferable Potentials for Phase Equilibria, which was originally developed for the prediction of vapor-liquid coexistence, to the simulation of complex biological systems.

[1] Jorgensen, W. L.; Chandrasekhar, J.; Madura, J. D.; Impey, R. W.; Klein, M. L. *J. Chem. Phys* **1983**, *79*, 926-935.

[2] Berendsen, H. J. C.; Grigera, J. R.; Straatsma, T. P., *J. Phys. Chem.* 1987, *91* (24), 6269-6271.

[3] Beglov, D.; Roux, B., *J. Chem. Phys.* 1994, *100* (12), 9050-9063.

[4] Aqvist, J., *J. Phys. Chem.* 1990, *94* (21), 8021-8024.

(261d) Curvature and Tension Effects On the Sorting of Proteins in Phase-Separated Model Lipid Bilayers

Uline, M. J., U. South Carolina
Originally presented on: 11/5/2013 9:12:00 - 9:26:00

The composition of lipid membranes greatly influences both their organization and properties. Highly organized assemblies of proteins and lipids in the membrane, sometimes called "lipid rafts", have been studied significantly for their evident medical importance due to the promotion of lateral segregation of proteins on the cell surface [K. Simons and R. Ehehalt, *J. Clin. Invest.*, **110**, p.597 (2002)]. Experimental model membranes containing cholesterol, dipalmitoyl-phophatidylcholine (DPPC), and dioleoylphosphatidylcholine (DOPC) are known to phase separate into liquid-ordered (lo) and liquid-disordered (ld) phases [S.L. Veatch and S.L. Keller, *Biochimica et Biophysica Acta*, **1746**, p. 172 (2006)]. It has long been hypothesized that certain proteins and lipid chain anchors would be enriched, either in the lo or in the ld phase, thereby increasing their efficacy. We use this model system to calculate the partition coefficient (mole fraction in the lo phase to the mole fraction in the ld phase) of proteins within the phase separated model system.

Using a theoretical model of a bilayer membrane containing cholesterol, DPPC, and DOPC that qualitatively reproduces phase diagrams of giant unilaminar vesicles (GUVs) of the same three components [R. Elliott, I. Szleifer, and M. Schick, *Phys. Rev. Lett.*, **96**, p.098101 (2006)], we calculate how the curvature of lipid vesicles determines the amount of binding of molecules with lipid tail anchors. By explicitly determining the chemical potential difference of species across a curved bilayer under different modes of deformation in both lo and ld phases, we are able to calculate the equilibrium binding concentrations of lipid tail anchors as a function of membrane curvature, concentration of lipids, and solution environment. Our results are in excellent agreement with recent experiments [N. S. Hatzakis *et. al.*, *Nature Chem. Bio.*, **5**, 835, 2009]. We also calculate the partition coefficients of protein chain anchors into lo and ld phases as a function of curvature, surface tension, temperature, and degree of saturation of the chain anchors. We are able to take into account the exact molecular architecture of protein chain anchors and calculate the orientations and positions of proteins as they arrange themselves at the order-disorder phase boundary.

(261e) Mechanism of a Hotdog-Fold Acyl-CoA Thioesterase By QM/MM Metadynamics Simulation

Cantu, D. C., Iowa State University
Ardevol, A., UniversitÃ della Svizzera italiana
Rovira, C., Parc CientÃfic Barcelona
Reilly, P. J., Iowa State University
Originally presented on: 11/5/2013 9:26:00 - 9:40:00

Many thioesterases have a structural HotDog fold, while others have alpha/beta folds. Based on crystal structures, putative mechanisms have been proposed for some HotDog-fold thioesterases. The reaction of a human thioesterase enzyme (hTHEM2), part of thioesterase family 8 in the ThYme database and one of 25 current families, was explored by first-principles methods (Car-Parrinello molecular dynamics) to elucidate atomic and electronic details of the mechanism, its transition-state conformation, and the energy of the process. Mixed quantum mechanics/molecular mechanics simulations using the metadynamics technique were used, and an acid-base-like mechanism was found. The activation of nucleophilic water by an aspartate residue acting as a base, as well as a tetrahedral-like intermediate, were confirmed as previously proposed. Additionally, new evidence pointing toward thioester substrate protonation from a histidine residue via a serine residue is presented. To our knowledge, this is the first time a thioesterase with any fold has been studied by first-principles methods.

(261f) DPD Simulation of Protein Conformations: From Alpha-Helices to Beta-Structures

Vishnyakov, A., Rutgers, The State University of New Jersey
Talaga, D. S.,
Neimark, A. V., Rutgers, The State University of New Jersey
Originally presented on: 11/5/2013 9:40:00 - 9:54:00

We suggest a coarse-grained model for DPD simulations of polypeptides in solutions. The model mimics hydrogen bonding that stabilizes alpha-helical and beta-structures using dissociable Morse bonds between quasiparticles representing the peptide groups amenable to hydrogen bonding. We demonstrate the capabilities of the model by simulating transitions between coil-like, globular, alpha-helical, and beta-hairpin configurations of model peptides, varying Morse potential parameters, the hydrophobicities of residue side chains, and pH, which determines the charges of residue side chains. We construct a model triblock polypeptide mimicking the sequence of residues alpha-synuclein at two different pH values. The conformations of this model polypeptide depend on pH similarly to the behavior observed experimentally. The suggested approach to accounting for hydrogen bond formation within the general DPD framework may make the DPD method a competitive alternative to CGMD for modeling equilibrium and dynamic properties of proteins and polypeptides, especially during their transport in confined environments.

(261g) Theoretical Model for HP1-Induced Heterochromatin Formation

Mulligan, P. J., Stanford University
Koslover, E. F., Stanford University
Spakowitz, A. J., Stanford University
Originally presented on: 11/5/2013 9:54:00 - 10:08:00

Gene regulation in eukaryotes requires the segregation of silenced genomic regions into densely packed heterochromatin, leaving the active genes in euchromatin regions more accessible. The stable maintenance of this process is critical for ensuring proper cell differentiation, and is present in organisms from yeast to humans. We introduce a model that connects the presence of epigenetically inherited histone marks, tri-methylation at histone 3 lysine-9 (H3K9), to the physical compaction of chromatin fibers via the binding of heterochromatin protein 1 (HP1). Using a model based on the mechanical properties of the DNA and the steric interactions of the DNA and nucleosomes, we generate optimized chromatin-fiber structures, which lay out the spatial positions of the H3K9 sites (1). Our model demonstrates some of the key physical features that are necessary to explain several experimental observations, including the phase segregation of heterochromatin regions and the increased HP1 concentration in these regions (2). Since each site within these arrays can be either methylated or unmethylated, the binding of HP1 will be affected not only by the total methylation level but also by the specific methylation profile for a given array. Calculation of the quenched average over all these profiles reveals increased fluctuations and a rising of the minimum interaction strength necessary to form a stable heterochromatin phase. We also explore under what circumstances there is a threshold of methylation over which the fibers will compact; this threshold provides a buffer against small losses of methylation such as occur during cell division (3). Finally, by extracting our model parameters from recent *in vitro* experiments, we find the biologically relevant HP1 concentrations that are needed to maintain the segregation of HP1 into highly methylated regions (4). Many of the observations that we make about the HP1 system are guided by general thermodynamics principles and thus could play a role in other DNA organizational processes such as the binding of linker histones.

[1]Koslover, Elena et al. (2010) *Biophys J*, 99.

[2]Muller, Katharina et al. (2009) *Biophys J*, 97.

[3]Xu, Mo et al. (2012) *EMBO Reports,* 13.

[4]Canzio, Daniele et al. (2011) *Molec Cell*, 41.

(261h) A Predictive Model to the Drug Solubility in Supercritical Carbon Dioxide

Wang, L. H., National Taiwan University
Lin, S. T., National Taiwan University
Originally presented on: 11/5/2013 10:08:00 - 10:22:00

In this research, we examined the prediction of drug solubility in supercritical carbon dioxide from a predictive Peng-Robison plus COSMOSAC equation of state (PR+COSMOSAC EOS). The melting temperature, T_m and enthalpy, H_m, of the solid drug and the critical properties (Tc,Pc) and acentric factor for fluid are needed for the solubility predictions. The average logrithmetic deviations (ALD-x) in predicted solubility of 46 drugs in subcritical and supercritical carbon dioxide (T= 293.15 K~473K, P= 8.5MPa~50MPa, and 1150 solubility data ranging from 10^{-7}~10^{-2}) is found to be 1.14(1280%). The prediction inaccuracy can be significantly reduced to 0.75(462%) when introducing an additional correction to the dispersion energy for aromatic and ring structures. The stability of predicted solubility is further examined by the calculation of solid-liquid-vapor three-phase coexisting curve. To test the generality of the predictive method, the solid solubility in a variety of solvents (60 solids including 34 different solvents (with different polarities) and 190 drug-solvent pairs) at ambient pressure was also examined. The ALD-x was found to be 0.59(2890%) which is comparable to other prediction model, ex: COSMO-SAC (0.68,379%). The solid sublimation pressure was also examined. The ALD-x for the 32 solids (10 aromatic rings, 16 rings with polar functional group and 6 linear compounds) is 0.70(401%) which shows significantly improvement from the original PR+COSMOSAC EOS (1.30,1895%). The modified PR+COSMOSAC EOS is thus a useful tool for a priori prediction of solid solubility in scCO2, as well as for other solvents and other solid physical properties.

(261i) A Computational Investigation Of Attrition-Enhanced Chiral Symmetry Breaking In Conglomerate Crystals

Ricci, F., Princeton University
Stillinger, F. H., Princeton University
Debenedetti, P. G., Princeton University
Originally presented on: 11/5/2013 10:22:00 - 10:36:00

Attrition-enhanced chiral symmetry breaking in crystals, also known as Viedma ripening, provides a direct route to solid-phase homochirality in a controllable manner. Hence, it is of inherent interest to those who study chiral symmetry-breaking/amplification mechanisms. When applied to intrinsically chiral molecules, Viedma ripening may have implications for the origin of biological homochirality, as well as applications in chiral drug resolution. To date, the mechanistic details underlying this phenomenon is still a matter of debate. We employ a Monte Carlo algorithm to study this driven system, in order to gain further insights into the mechanisms capable of reproducing key experimental signatures. We provide a comprehensive numerical investigation of how the model parameters (attrition rate, liquid-phase racemization kinetics, and the relative rates of growth and dissolution kinetics) impact the system's overall behavior. It is shown that size-dependent crystal solubility alone is insufficient to reproduce most of the experimental signatures of Viedma ripening, and that some form of a solid-phase chiral feedback mechanism is required in order to reproduce experimentally-observed behavior. In this work, such feedback mechanisms can take the form of agglomeration, or of artificial modification of the size dependent growth kinetics.

(265a) Nonlinear Elasticity: From Single Chain to Networks and Gels

Carrillo, J. M., Oak Ridge National Laboratory
MacKintosh, F. C., Vrije Universiteit
Dobrynin, A., University of Connecticut
Originally presented on: 11/5/2013 8:30:00 - 8:50:00

Biological and polymeric networks show highly nonlinear stress–strain behavior manifested in materials that stiffen with increasing deformation. Using a combination of the theoretical analysis and molecular dynamics simulations, we develop a model of network deformation that describes nonlinear mechanical properties of networks and gels by relating their macroscopic strain-hardening behavior to molecular parameters of the network strands. The starting point of our approach is a nonlinear force/elongation relation for discrete chains with varying bending rigidity. The derived expression for the network free energy is a universal function of the first deformation invariant and chain elongation ratio that depends on a ratio of the unperturbed chain size to chain dimension in a fully extended conformation. The model predictions for the nonlinear shear modulus and differential shear modulus for uniaxial and shear deformations are in very good agreement with both the results of molecular dynamics simulations of networks and with experimental data for biopolymer networks of actin, collagen, fibrin, vimentin, neurofilaments, and pectin.

(265b) Formation and Microstructure of Self-Assembled Molecular Gels

Dudukovic, N., University of Illinois at Urbana-Champaign
Zukoski, C. F., University of Illinois at Urbana Champaign
Originally presented on: 11/5/2013 8:50:00 - 9:10:00

Molecular gels are associated with the formation of strongly anisotropic space-spanning structures produced by self-assembly of low molecular weight gelator molecules that associate through hydrogen bonding, π-π stacking, acid-base or van der Waals interactions. One specific class of gel forming molecules is based on aromatic short peptide derivatives. Gelation in these systems is associated with π-π stacking of antiparallel peptide ß-sheets, resulting in the formation of fibrils that branch out to fill space. We explore structural changes taking place in solution during and after gelation of a hydrophobic dipeptide molecule, N-fluorenylmethoxycarbonyl diphenylalanine (Fmoc-FF). When water is added to a solution of Fmoc-FF in dimethyl sulfoxide (DMSO), the molecules self-assemble into rigid ($G' > 10^5$ Pa) gels at Fmoc-FF concentrations as low as 0.1 wt%, creating space-spanning fibrous networks (as confirmed by transmission electron microscopy). We investigate the changes in the system's microstructure at compositions below, at and above the gel point, along lines of increasing Fmoc-FF volume fractions and increasing water concentrations, using small-angle and wide angle X-ray scattering. We find that at compositions right below the critical gel point concentration, precursors with distinctive spacing are being formed. As the concentration is increased to and above the gel point, the scattering curves exhibit branched fibrous structure features and an absence of the pre-gel aggregates. Evidence of characteristic spacing of ß-sheet π-π stacking can be seen as well. Furthermore, scattering studies of the temporal changes in the microstructure during the gel transition, as well as the effects of temperature on the gel structure, are presented.

(265c) Cavitation Or Fracture? Measuring Material Properties With a Mechanical Instability

Hutchens, S. B., California Institute of Technology
Fakhouri, S., University of Massachusetts - Amherst
Crosby, A. J., University of Massachusetts-Amherst
Originally presented on: 11/5/2013 9:10:00 - 9:30:00

The quantitative characterization of mechanical properties for soft materials, such as gels and biological tissues, remains a significant challenge for the field of materials science. The difficulties lie in the following: large strain and low force responses; nonlinear constitutive behavior; and the impact of surface energy on deformation responses. This talk describes a simple, though not yet broadly adopted, procedure referred to as Cavitation Rheology, which has been shown to provide data that quantitatively corresponds to Young's modulus and strain energy release rate in gels.[1,2] Most importantly, this technique provides direct measurement of mechanical properties on local length scales, ranging from sub-micron to millimeter, allowing for characterization of material property gradients[1,3]. Although quantitative correlation has been demonstrated previously, many open questions remain with regard to the underlying deformation mechanisms related to Cavitation Rheology. The principal concern of this talk is the observed deformation response, which is comprised of a triggered instability whose partially understood physical mechanisms and limitations are explained in detail and verified by experiment.

Cavitation rheology consists of the insertion, pressurization, and monitoring of a fluid filled needle. Pressurization is accomplished through displacement of the plunger of an attached syringe. At the tip of the embedded needle, optical microscopy reveals that a cavity first grows in direct response to plunger displacement (corresponding to increased pressure). At a critical point, the cavity expands rapidly, independently of plunger movement, often resulting in a marked pressure drop within the needle/syringe system, which is monitored using a pressure sensor. It is this pressure peak value, or critical pressure, that has been shown to correspond to both fracture toughness and elastic moduli in several hydrogel materials depending on the size of needle used.

The rapidity of the bubble expansion at the critical pressure indicates that the deformation is associated with the onset of instability. The nature of this instability, either elastic (reversible) or corresponding to the onset of fracture, has been predicted through estimates of the deformation using a spherically symmetric cavity [4] and a membrane approximation for the embedded needle geometry[2]. Both approximations lack key elements, preventing their utilization in fully explaining experimental observations. This talk eliminates these shortcomings. It describes the limited parameter space for which the instability may correlate to an elastic instability while comparing the needle geometry with analytical spherical results. It also leverages existing mechanisms for fracture via a spherical void to describe a new formulation for fracture in the embedded needle geometry using a combination of analytical theory and finite element analysis. These deformation mechanisms are then translated into predicted critical pressures dependent on needle size, void surface energy, and ability of the pressurizing fluid to store energy. Using this new understanding of the relation between critical pressure and material properties, a transition between cavitation and fracture is illustrated for a triblock co-polymer gel system. These results not only provide a basis of understanding for a promising new experimental technique, but also hold implications for void formation governed failure mechanisms, such as crazing, in soft materials.

1. Zimberlin, J. A., Sanabria-DeLong, N., Tew, G. N. & Crosby, A. J. Cavitation rheology for soft materials. *Soft Matter* **3**, 763 (2007).

2. Kundu, S. & Crosby, A. J. Cavitation and fracture behavior of polyacrylamide hydrogels. *Soft Matter* **5**, 3963 (2009).

3. Cui, J., Lee, C. H., Delbos, A., McManus, J. J. & Crosby, A. J. Cavitation rheology of the eye lens. *Soft Matter* **7**, 7827 (2011).

4. Zhu, J., Li, T., Cai, S. & Suo, Z. Snap-through Expansion of a Gas Bubble in an Elastomer. *The Journal of Adhesion* **87**, 466–481 (2011).

(265d) Swelling of Crosslinked Elastomers: Frenkel, Flory, Rehner Revisited

Xu, B., Texas Tech University
Di, X., Texas Tech University
Wu, J., Sichuang University
McKenna, G. B., Texas Tech University
Originally presented on: 11/5/2013 9:30:00 - 9:50:00

The swelling and mechanical properties of networks are intimately tied together through the free energy function as it is affected either by chain stretching (elastic free energy) or through the mixing of the polymer with the solvent (usually through a Flory-Huggins estimate). Frenkel, then Flory and Rehner (FFR) developed a simple way of looking at this problem by simply balancing the elastic and mixing contributions to the chemical potential at swelling equilibrium. Much work has been done to examine this model, often in a way that suggests that something is missing or that the model itself is incorrect. In this presentation we examine the FFR model and show that it is correct for polymer networks swollen in organic liquids. We demonstrate that anomalous peaks in the swelling activity parameter S (or dilational modulus) are artifacts of experimental methods and we further show that, once the peak in S is removed, quantitative agreement is obtained between mixing and elastic contributions to the free energy with the simple addition of a cross-link dependent interaction parameter.

(265e) Invited Talk: Patterned Sheets and Multilayers of Photo-Crosslinkable Polymers

Hayward, R. C., University of Massachusetts, Amherst

Originally presented on: 11/5/2013 9:50:00 - 10:20:00

Our group is studying the use of photo-crosslinkable copolymers containing pendent benzophenone units as a platform for the fabrication of actuating micro-scale devices and sensors. We have studied the design considerations, in terms of copolymer chemistry, necessary to obtain efficient UV crosslinking through this route, and found a sensitive dependence on how hydrogens susceptible to radical abstraction by triplet-state benzophenone are located on the polymer chain. With this knowledge, it is straightforward to fabricate patterned polymer sheets and multilayers with a wide variety of geometries, material properties, and responses to stimuli. Here, we will focus on the opportunities presented by multi-layer polymer films. In one case, we focus on developing strategies to control the swelling-induced shape of thin polymer sheets through tailoring of the target mean and Gaussian curvatures prescribed by contrasts in degree of swelling and stiffness of the constituent materials. In another case, we design photonic multi-layer films that yield tailored optical changes in response to specific environmental changes or analytes.

(265f) Covalently Adaptable Networks As Biophysical-ECM Mimics for Cell Culture

McKinnon, D., University of Colorado-Boulder
Domaille, D., University of Colorado-Boulder
Cha, J., University of Colorado-Boulder
Anseth, K. S., University of Colorado-Boulder
Originally presented on: 11/5/2013 10:20:00 - 10:40:00

Covalently crosslinked synthetic hydrogels are especially suitable as tissue engineering scaffolds due to their well-defined and easily tunable biochemical and biophysical properties. In order to enable complex cell functions like ECM deposition, motility, and spreading, a mechanism for crosslink degradation must be engineered into the material; however, the presence of a degradation trigger can complicate the cellular biophysical microenvironment. Furthermore, covalently crosslinked polymers typically produce an elastic material, while native tissues are complex viscoelastic structures. Here, we present a step-growth poly(ethylene glycol) (PEG) hydrogel crosslinked by reversible hydrazone bonds. The macromer components are readily synthesized from commercially available precursors, and the resulting gels form rapidly under physiological conditions and provide a non-toxic matrix that is suitable for cell culture. This material is capable of mimicking aspects of the viscoelastic properties of native tissues, and the dynamic stress relaxing crosslinks permit complex cellular functions to occur while retaining the benefits of traditional covalently crosslinked hydrogels. Taken together, these attributes make hydrazone crosslinked hydrogels a unique tool for designing viscoelastic scaffolds and studying cellular responses to scaffold elasticity.

(265g) Hindered Diffusion in Contact-Lens-Material Hydrogels

Liu, D., UC Berkeley
Taylor, N., UC Berkeley
Dursch, T. J. Jr.,
Kotsmar, C.,
Radke, C. J., University of California Berkeley
Originally presented on: 11/5/2013 10:40:00 - 11:00:00

Solute diffusion in soft contact lenses (SCLs) is critical in designing optimal care-packaging solutions for wetting and disinfection, controlling ocular drug delivery, and mitigating lens fouling. In all applications, the extent of solute partitioning and rate of diffusion is regulated by solute and hydrogel charge, solute size, and mesh size. Solute uptake and release in hydrogels is governed by hindered diffusion and occurs primarily through the water-filled meshes. Diffusivities of aqueous solutes in hydrogels (D) are reduced relative to their bulk values (D_0) due to interaction with the polymer chains including hydrodynamic drag and physical obstruction.[1-3] We investigate relative diffusivities in SCL-material hydrogels experimentally and theoretically, and develop an *a priori* extended Brinkman-Effective-Medium model utilizing the distribution of meshes available for solute transport.

Characteristic of SCL material hydrogels is their relatively high polymer volume fraction (>15%) leading to smaller meshes and more pronounced effects on gel diffusivities. In comparison, previous studies have focused almost exclusively on low polymer content gels (<10%).[2-7] The polymer network contains a distribution of polymer-strand molecular weights between cross-links, and correspondingly a distribution of meshes.[8, 9] Since solutes are excluded from meshes smaller than their size, only a subset of the mesh-size-distribution is available to the solute as it diffuses. Thus, in addition to solute-chain interactions, significant size-exclusion can affect reduced diffusivities when solute size and gel mesh size are comparable.

We study diffusion of macromolecular solutes in representative SCL- material hydrogels (hydroxethyl methacrylate-co-methacrylic acid) of varying composition. The Stokes Diameters of the macromolecules tested (3.1-6.7nm) are of comparable size to the average gel mesh size (3.6-8.4nm) determined by oscillatory shear rheometry and elastic theory. Uptake and release experiments under perfect sink conditions are performed using fluorescence confocal microscopy and diffusivities are fit to resultant transient concentration profiles. Independent measurement of the average gel mesh size and equilibrium water volume content allows for computation of an approximate polymer fiber radius assuming an Ogston mesh-size-distribution.[9]

We interpret our data with an extended Brinkman-effective-medium theory using *no adjustable parameters*. The relative diffusivity is expressed as a product of a hydrodynamic factor, F, and a steric factor, S, related inversely to tortuosity or $D/D_0 = FS$.[10] Each factor is modified for the phenomena that solutes cannot access meshes smaller than their size and, thus, only experience drag and obstruction within accessible pores. In this way, we construct a Large-Pore-Effective-Medium (LPEM) only taking into account the distribution of meshes available for transport. Predicted diffusivities from this effective-medium model are in excellent agreement with those measured.

1. Pluen, A.; Netti, P. A.; Jain, R. K.; Berk, D. A., Diffusion of Macromolecules in Agarose Gels: Comparison of Linear and Globular Configurations. *Biophysical Journal* **1999,**77, (1), 542-552.

2. Phillips, R. J., A hydrodynamic model for hindered diffusion of proteins and micelles in hydrogels. *Biophysical journal* **2000,**79, (6), 3350.

3. Johansson, L.; Elvingson, C.; Loefroth, J. E., Diffusion and interaction in gels and solutions. 3. Theoretical results on the obstruction effect. *Macromolecules* **1991,**24, (22), 6024-6029.

4. Amsden, B., Solute diffusion within hydrogels. Mechanisms and models. *Macromolecules* **1998,**31, (23), 8382-8395.

5. Johnson, E. M.; Berk, D. A.; Jain, R. K.; Deen, W. M., Hindered diffusion in agarose gels: test of effective medium model. *Biophysical Journal* **1996,**70, (2), 1017-1023.

6. Tong, J.; Anderson, J. L., Partitioning and diffusion of proteins and linear polymers in polyacrylamide gels. *Biophysical Journal* **1996,**70, (3), 1505-1513.

7. Kong, D. D.; Kosar, T. F.; Dungan, S. R.; Phillips, R. J., Diffusion of proteins and nonionic micelles in agarose gels by holographic interferometry. *AIChE Journal* **1997,**43, (1), 25-32.

8. Amsden, B., Solute diffusion in hydrogels.: An examination of the retardation effect. *Polymer Gels and Networks* **1998,**6, (1), 13-43.

9. Ogston, A. G., The spaces in a uniform random suspension of fibres. *Transactions of the Faraday Society* **1958,**54, (0), 1754-1757.

10. Brady, J. *Hindered diffusion*, Extended Abstracts, AIChE Annual Meeting, San Fransisco, CA; 1994; p 320.

(266a) Invited Talk: Polymer Thin Film Surface Area and Stiffness Influences Bacteria Inactivation and Biofilm Formation

Schiffman, J. D., University of Massachusetts Amherst
Rieger, K. A., University of Massachusetts Amherst
Kolewe, K. W., University of Massachusetts Amherst
Originally presented on: 11/5/2013 8:30:00 - 9:00:00

The development of environmentally benign polymer coatings that effectively prevent surface contamination by microorganisms is a pressing necessity. As both microbial resistance and infectious diseases remain, new approaches to delay the onset of biofilm formation with less evolutionary pressure on pathogens are needed. In this work, we explore the effect that materials properties have on inactivating planktonic bacteria and altering the structure of young microbial biofilms. In the first study, we focus on the effect that the available surface-to-volume ratio has on *Escherichia coli*, *Pseudomonas aeruginosa*, and *Bacillus anthracis* inactivation when the bacteria is challenged with chitosan, a copolymer of 2-acetamido-2-deoxy-D-glucopyranose and 2-amino-2-deoxy-D-glucopyranose units joined by $\beta(1\rightarrow4)$ bonds. Due to the availability of primary amine groups, chitosan exhibits inherent antimicrobial activity; the proposed mechanism is the interaction between the positive charge on the C-2 of the glucosamine monomer and negatively charged microbial cell membrane. In a second system, we look towards the field of mechanobiology – where it is theorized that fundamental cell processes are modulated by the substrate matrix stiffness. Here, we characterize the formation and growth of *P. aeruginosa* biofilms grown on mechanically variable polymer films. Characterization of the extrapolymeric substances (EPS) matrix of the biofilm as a function of the mechanical properties of a model polymer film were conducted. Results on the interaction of the nano- to macro- structured thin films with bacteria viability and EPS formation will be presented.

(266b) Polymer Coatings for Biomedical Applications Using Low Temperature, Atmospheric Pressure Plasma

Farhat, S., Kettering University
Gilliam, M., University of Missouri - Columbia
Zand, A., Kettering University
Rabago-Smith, M., Kettering University
Originally presented on: 11/5/2013 9:00:00 - 9:20:00

Polyethylene is used as a biocompatible surface for ball and socket joint replacements, which are under constant wear. The interface could be improved by the addition of a hydrophilic coating to provide a lubricating effect and possibly reduce wear. Presently, these coatings can be synthesized using wet chemistry techniques, which can often be time consuming and result in poor adhesion or coating quality. Current research has been dedicated to investigating the viability of atmospheric pressure plasmas for use in coating technology. In addition to being more cost effective and efficient, atmospheric pressure plasma offers a more streamlined process, as it can be employed directly into the production line. Atmospheric pressure plasma has been used in applications including biocompatibility, hydrophilicity/hydrophobicity, and coating with antibacterial films.

Atmospheric pressure plasma was used to change the surface chemistry of polyethylene by grafting various biocompatible polymers to the substrate surface, as methods of providing wear resistance as well as providing a self-lubricating surface. The organic coatings included biocompatible polymers, such as poly(2-hydroxyethylmethacrylate), polyethylenimine, and polyethylene glycol. Low temperature, atmospheric pressure plasma was used, along with an in-house constructed spray delivery system, to coat high density polyethylene substrates. Coatings were characterized with Fourier transform infrared spectroscopy (FTIR), contact angle analysis, and adhesion testing. A significant decrease in contact angle was noted for various coatings produced with this method, indicating an increased wettability. Plasma processing conditions, specifically the pretreatment of the substrate and the input power, greatly affected the adhesion and uniformity of the polymerized layer. Polymer concentrations were varied to understand the effect on coating adhesion and hydrophilicity.

(266c) Vapor-Phase Deposition of Polymers Onto Liquid Substrates

Haller, P., University of Southern California
Bradley, L., University of Southern California
Gupta, M., University of Southern California
Originally presented on: 11/5/2013 9:20:00 - 9:40:00

The vapor-phase deposition of polymers is typically used to coat solid substrates, but we recently introduced liquid substrates into the deposition process.[1] The liquid surface adds complexity to the system by offering different initial conditions. For example, we have observed the formation of both continuous polymer films and polymer nanoparticles on the liquid/vapor interface. We have recently demonstrated that these different polymer morphologies can be used to fabricate free-standing polymer films[2] and to encapsulate liquids within polymer shells.[3] In this talk, we will discuss our current work in determining the mechanism that governs film and nanoparticle formation. Specifically, we will discuss the effect of the surface tension interactions between the liquid and polymer, which determine whether it is energetically favorable for the polymer to spread over the liquid surface and form a film or aggregate into particles. We will also discuss special cases in which the polymer deposition rate can influence polymer morphology and we will also discuss the effect of liquid viscosity on polymer morphology.

1. Haller, P. D.; Frank-Finney, R. J.; Gupta, M. "Vapor-Phase Free Radical Polymerization in the Presence of an Ionic Liquid." *Macromolecules,* 2011, 44, 2653-2659.

2. Frank-Finney, R. J.; Haller, P. D.; Gupta, M. "Ultrathin Free-Standing Polymer Films Deposited onto Patterned Ionic Liquids and Silicone Oil." *Macromolecules*, 2012, 45, 165-170.

3. Bradley, L. C.; Gutpa, M. "Encapsulation of Ionic Liquids within Polymer Shells via Vapor Phase Deposition." *Langmuir*, 2012, 28, 10276-10280.

(266d) Photodirected Wrinkle Formation On a Thiol-Ene Elastomer

Ma, S.,
Mannino, S., University of Delaware
Wagner, N. J., University of Delaware
Kloxin, C. J., University of Delaware
Originally presented on: 11/5/2013 9:40:00 - 10:00:00

Wrinkles are a cost-efficient method to produced surface topography, and have a wide range of potential applications including antifouling coatings[1-3], adhesives[4], and microarray lenses with tunable focal lengths.[5] However, the precise control over wrinkle confinement and orientation necessary to engineering textured surfaces remains a challenge.

Photopolymerization is an attractive method to spatiotemporally direct wrinkle formation. Using acrylate-rich thiol-ene elastomers, embedded with a photoinitiator and photoabsorber, we can irradiate strained elastomers with UV light to generate wrinkles in as little as five seconds. While the photointiator aids in radical polymerization of the acrylates, the photoabsorber confines the UV light to a thin skin layer, creating the necessary conditions for wrinkle formation upon release of the strain. In conjunction with photomasked UV light, these wrinkles are confined and oriented, leading limitless complex patterns of multiple distinct wavelengths and the formation of gradients containing a continuum of length scales. Moreover, with a photoorthogonal photoinitator, these wrinkles can subsequently be locked in through additional bulk radical polymerization. With the wide array of thiol and ene monomers, photoiniators, photomasks and stoichiometric control, we present an alternative approach to wrinkle formation that is rapid, facile, and highly versatile, further enabling the *a priori* design and engineering of textured surfaces

References:

(1) Efimenko, K., Finlay, J., Callow, M. E., Callow, J. A. and Genzer, J. *ACS Appl. Mater. Interfaces* **2009**, 1, 1031-1040

(2) Schumacher, J. F., Aldred, N., Callow, M. E., Finlay, J. A., Callow, J. A., Clare, A. S. and Brennan, A. B. *Biofouling* **2007**, 23, 307-317

(3) Schumacher, J. F., Carman, M. L., Estes, T. G., Feinberg, A. W., Wilson, L. H., Callow, M. E., Callow, J. A., Finlay, J. A. and Brennan, A. B. *Biofouling* **2007**, 23, 55-62

(4) Davis, C. S., Martina, D., Creton, C., Lindner, A. and Crosby, A. J. *Langmuir* **2012**, 28, 14899-14908

(5) Chandra, D., Yang, S. and Lin, P.-C. *Appl. Phys. Lett.* **2007**, 91,

(266e) Locomotion Based On Hygromorphic Bilayer Actuator: Humidity-Induced Swelling/Deswelling of Layer-By-Layer Films

Lee, D., University of Pennsylvania
Lee, S. W.,
Prosser, J.,
Originally presented on: 11/5/2013 10:00:00 - 10:20:00

Several organisms in nature exhibit complex actuation in response to changes in environmental factors. Humidity-induced motions and actuations are utilized for seed dispersal by pine cones and wild wheat seeds. In general, humidity-induced actuations observed in these biological systems rely on the differential swelling and deswelling of bilayer structures under cyclically changing humidity conditions. Inspired by these natural actuators, we develop a hygromophic bilayer actuator (HBA) that shows locomotion on a ratchet track in response to changes in relative humidity (RH). The basic principle of the HBA lies in the rapid swelling and deswelling of highly hygroscopic layer-by-layer (LbL) films of poly(allylamine hydrochloride) (PAH) and poly(acrylic acid) (PAA) deposited on a moisture-resistant Teflon ribbon. We demonstrate that the swelling of the PAH/PAA LbL films is a governing parameter that controls the step length of the HBA on the ratchet track. We also present the humidity-induced swelling/deswelling behavior of (PAH/PAA) films assembled under different conditions, which provides fundamental insights into the interactions between water and these LbL films.

(266f) Photochemical Transfer Printing of Block Copolymer Patterns

Janes, D. W., Columbia University
McCoy, B. D., The University of Texas at Austin
Inoue, T., University of Chicago
Madan, I., The University of Texas at Austin
Thode, C. J., University of Wisconsin-Madison
Willson, C. G., The University of Texas at Austin
Nealey, P. F., University of Chicago
Ellison, C. J., The University of Texas at Austin
Originally presented on: 11/5/2013 10:20:00 - 10:40:00

Thin films of block copolymers are promising patterning templates for next generation microprocessors, hard drives, bio-arrays, and separation membranes because they rapidly self-assemble into periodic structures with characteristic feature sizes of 1-100 nm. However, guiding patterns must be employed in order to direct the self-assembly of BCP thin films into prescribed, device-oriented patterns or regular arrays with long-range order. Fabrication of these guiding patterns requires performing lithographic patterning with resolution near that of the BCP periodicity, which can be a relatively lengthy and expensive process step. Therefore, strategies to replicate the patterns formed with BCP thin films onto other substrates while recovering the original for more replication cycles could be practically useful because the guiding pattern would only have to be fabricated once. Here a procedure is described that uses light as an externally applied stimuli to facilitate the transfer printing of BCP top surface patterns from one substrate to another. Because photo-polymerizable liquid conformal layers are placed between the samples, the pattern is replicated continuously and over large areas. The strategy has been used to replicate patterns with periodicities as small as 20 nm, as well as films in which the self-assembled structures have been directed by an underlying guiding pattern. Additionally, efforts to improve process throughput and conformal layer uniformity will be described.

(266g) Zwitter-Wettability and Antifogging Coatings With Frost-Resisting Capabilities

Lee, H., MIT
Alcaraz, M. L., MIT
Rubner, M. F., Massachusetts Institute of Technology
Cohen, R. E., Massachusetts Institute of Technology
Originally presented on: 11/5/2013 10:40:00 - 11:00:00

Antifogging coatings with hydrophilic or even superhydrophilic wetting behavior have received significant attention due to their ability to reduce light scattering by film-like condensation. However, under aggressive fogging conditions, these surfaces may exhibit frost formation or excess and non-uniform water condensation, which results in poor optical performance of the coating. In this work, we show that a zwitter-wettable surface, a surface that has the ability to rapidly absorb molecular water from the environment while simultaneously appearing hydrophobic when probed with water droplets, can be prepared by using hydrogen-bonding assisted Layer-by-Layer (LbL) assembly of poly(vinyl alcohol)(PVA) and poly(acrylic acid)(PAA). An additional step of functionalizing the nano-blended PVA/PAA multilayer with poly(ethylene glycol methyl ether)(PEG) segments produced a significantly enhanced antifog and frost resistant behavior. The addition of the PEG segments was needed to further increase the non-freezing water capacity of the multilayer film. The desirable high optical quality of these thin films arises from the nanoscale control of the macromolecular complexation process that is afforded by the LbL processing scheme. An experimental protocol that not only allows for the exploration of a variety of aggressive antifogging challenges but also enables quantitative analysis of the antifogging performance *via* real-time monitoring of transmission levels as well as image distortion is also described.

(278a) Miscibility and Dynamics of Pressure-Induced Phase Separation in Polymers Solutions: Poly(epsilon-caprolactone) + Acetone + CO2

Takahashi, S., Virginia Tech
Hassler, J. C., Virginia Tech
Kiran, E., Virginia Tech
Originally presented on: 11/5/2013 8:30:00 - 8:51:00

We have recently developed a new experimental system which permits determination of the volumetric properties, phase conditions, and also the dynamics of phase separation in terms of phase settlement times and mechanism being Nucleation and Growth of Spinodal Decomposition [1]. The system is a unique high pressure view-cell (Figure 1) equipped with a dual set of pistons and dual set of sapphire windows. One set of the windows separated by 25.4 mm allows the assessment of the phase state and is used to monitor the transmitted light intensities. The second set of windows separated by 50 μm is used to monitor the scattered light intensities over a wide range of scattering vector q (from 0- 4 μm^{-1}) which allows the assessment of the mechanism of phase separation. The dual set of pistons that are employed which are synchronized and actuated by motorized pressure generators creates a churn-like action in the cell insuring effective mixing, even at high polymer concentrations by translating the cell content across a magnetically-coupled rotating mixer impeller. The piston actions assure also that the solution is effectively introduced into the narrow gap between the scattering windows, and refreshed. Piston positions are determined with real-time positions sensors that help determine the internal volume at any given temperature and pressure, hence permit assessment of density and the volumetric (P, ρ, T) properties of the solutions.

Using this new system, we have investigated the miscibility and the kinetics of pressure-induced phase separation in solutions of poly(e-caprolactone) (PCL) in acetone + CO_2 binary fluid mixtures has been studied Investigations have been carried out for a wide range of polymer concentrations, from 2.0 to 35 wt %, while holding the acetone-to-CO_2 (wt:wt) ratio in each solution at a constant value of 2:1. The results show that these solutions all display LCST type behavior. In the concentration range from 9-15 wt %, phase separation after crossing the phase boundary, proceeds by spinodal decomposition which is characterized by the formation and evolution of the spinodal ring patterns corresponding to a maximum in the angular variation of the scattered light intensities. At concentrations below 9 % and above 15 %, the solutions were observed to undergo phase separation via nucleation and growth mechanism which is characterized by circular symmetric patterns in their light scattering patterns. In the nucleation and growth regime, Debye-Bueche type scattering function was used to analyze the domain size of the new phase that forms and develops after a pressure quench. The early stage of the spinodal decomposition was captured with the wide angular range that the present system permits, and its time evolutions was found to be describable by the linearized Cahn theory. The variation of the scattered light intensity maximum I_m and its location in scattering vectors q_m with time in the later stage of the spinodal decomposition was found to obey power-law scaling according to $I_m \sim t^b$ and $q_m \sim t^{-a}$. The results for the 9.0 and 12.0 wt % solutions show that b/a changes its value from $b/a > 3$ to $b/a \approx 3$ with time, indicating the progression of the spinodal decomposition from intermediate to late stage.

[1] E. Kiran, J. C. Hassler and R. Srivastava, Miscibility, phase separation and phase settlement dynamics in solutions of EPDM in propane + n-octane binary fluid mixtures at high pressures, *Industrial & Engineering Chemistry-Research,* **2013**, 52, 1806-1818.

(278b) Phase Behavior of a Five-Component Model Gas Condensate

Shariati, A., Shiraz University
Straver, E., Delft University of Technology
Florusse, L. J., Delft University of Technology, Faculty of Applied Sciences, DelftChemTech (PCMT)
Peters, C. J., Petroleum Institute
Originally presented on: 11/5/2013 8:51:00 - 9:12:00

In this work, the bubble points and dew points of a multicomponent mixture of methane, butane, heptane, decane and tetradecane as a model mixture representative of a gas condensate, have been measured experimentally. Ten samples with approximately the same composition were prepared and their bubble-point and/or their dew-point pressures were measured at several temperatures. The phase-behavior measurements were carried out in the temperature range of 240 to 410 K, and at pressures up to 21.6 MPa. The critical point for this system was measured and the cricondenbar and cricondentherm of the phase envelope were determined. The Peng-Robinson equation of state was used for modeling of the phase behavior of this system. The Peng-Robinson equation of state was able to successfully determine the bubble- and dew-points of this model gas condensate. Using the PR EOS, the retrograde region and liquid drop out characteristic of this model gas condensate were determined.

(278c) Application of Computational Fluid Dynamics (CFD) to Boiling Liquid Expanding Vapor Explosions (BLEVEs)

Hamaoui, A.,

Originally presented on: 11/5/2013 9:12:00 - 9:33:00

A boiling-liquid expanding-vapor explosion (BLEVE) occurs when a vessel containing pressurized liquefied gas ruptures, resulting in the instantaneous release of the boiling liquid, vessel fragments, and blast wave. For structural design loading, the overpressure and impulse from a postulated BLEVE uses the method provided in *Guidelines for Vapor Cloud Explosion, Pressure Vessel Burst, BLEVE and Flash Fire Hazard.* The method's conservative assumptions provide a bounding estimate for the overpressure and impulse that would most likely occur during any BLEVE. However, using CFD to simulate a pressure vessel burst can provide a more realistic estimate of the blast wave where complex geometries exist. Using the modeling approach of approximating a BLEVE as a sudden gas expansion, CFD may be used to predict the initial shock wave produced by the expansion to determine a detailed pressure-time histories on structures from BLEVE generated blast waves and an understanding of the far field directionality of the blast waves on the surroundings. The results may be used to optimize vessel orientation and mitigation methods (i.e. blast walls, berms). CFD can also be used to predict the expansion of the gas cloud behind the shock wave to determine the concentrations of the evaporated BLEVE fluid in the environment following the BLEVE event to assess the potential toxic impact on humans. Because CFD is providing the case-specific pressure-time history using the exact structures, it is not encumbered with the necessary conservatism that a generic guideline must include. Hence, the CFD results will likely produce weaker overpressures than those derived from generic guidelines, which yield a less conservative design of the structure.

(278d) Evaluation of Predictive Models for High-Pressure Natural Gas Viscosity

Elms, R. D., Texas A&M University
Atilhan, M., Qatar University
Aparicio, S., University of Burgos
Fitz, C.,
El-Halwagi, M., Texas A&M University
Originally presented on: 11/5/2013 9:33:00 - 9:54:00

Accurate knowledge and prediction of natural gas thermo-physical properties, such as density and viscosity, is requisite to appropriate, reliable, and economical design of natural gas related production, transmission, and processing. Of increasing importance are accurate predictions at higher pressures and temperatures due to, among other things, deeper drilling, higher wellhead pressures, and discovery of new reservoirs with such conditions. Historically, viscosity data in this range of pressures and temperatures has been limited. Towards addressing this need, viscosity data for 5 synthetic natural gas mixtures over a temperature range of 250 to 450 K and pressure range of 10 to 65 MPa was recently obtained and reported. These previously reported results are utilized to investigate and compare the predictive ability of several viscosity models (PFCT, LBC, CLS, and NBS) and the PR and SRK Equations of State. ProMax simulations are utilized to facilitate evaluation of PR and SRK. The effect of sample composition is also discussed.

(278e) Experimental Measurements and Modeling of Viscosity for Liquid Hydrocarbons At Pressures Up to 243 Mpa and Temperatures Up to 534 K

Baled, H., University of Pittsburgh
Xing, D., University of Pittsburgh
Enick, R. M., University of Pittsburgh
Gamwo, I. K., National Energy Technology Laboratory (NETL), Office of Research and Development, Department of Energy
Tapriyal, D., U.S. Department of Energy, National Energy Technology Laboratory
Bamgbade, B., Virginia Commonwealth University
Wu, Y., Virginia Commonwealth University
McHugh, M., Virginia Commonwealth University
Originally presented on: 11/5/2013 9:54:00 - 10:15:00

Viscosity data are reported for n-octane, isooctane, cyclooctane, n-hexadecane, n-octadecane, and n-eicosane at pressures up to ~243 MPa and temperatures up to ~534 K. These extreme conditions are representative of those encountered in ultra-deep petroleum formations beneath the deepwaters of the Gulf of Mexico. The measurements are taken with a novel windowed rolling ball viscometer designed by our team. The reported viscosity values are in good agreement with the available literature data that cover limited pressure and temperature ranges; but our results extend the database of viscosity to the high temperature-high pressure region where most gaps occur in the literature data. The viscosity results are modeled with the free volume theory model in conjunction with the Tait, Peng-Robinson, and PC-SAFT equations of state. The mean absolute percent deviation (MAPD) values obtained with this model range from 2.0 to 3.5%. The data are also correlated by a nonlinear surface fit as a simultaneous function of temperature and pressure that yields MAPD values less than 0.5% for all hydrocarbons considered in this study.

(278f) Accurate Viscosity Modeling of CO2-Hydrocarbon Mixtures

Macias-Salinas, R., Instituto Polit cnico Nacional
Aquino-Olivos, M., stituto Mexicano del Petr leo
Garc a-S nchez, F., Instituto Mexicano del Petr leo
Originally presented on: 11/5/2013 10:15:00 - 10:36:00

A knowledge of the fluid property behavior of CO_2-hydrocarbon mixtures is paramount in enhanced oil recovery. In particular, the viscosity of such a mixtures must be known with accuracy for a suitable simulation of CO_2 displacement processes. Using the Significant Structure Theory (SST) of Eyring, we present in this work a formal modeling approach to estimate the dynamic viscosity of CO_2-hydrocarbon mixtures: CO_2 + normal paraffins (from methane up to decane), over wide temperature and pressure ranges (those typically encountered in oil reservoirs: from 310 to 403 K, and from 7 up to 30 Mpa).

(278g) A Modified Capillary Viscometer for Dynamic Viscosity Measurements: Preliminary Results for Octane and Nonane Up to 30 Mpa and 353 K

Galicia-Luna, L. A., INSTITUTO POLITECNICO NACIONAL-ESIQIE
Pimentel-Rodas, A., INSTITUTO POLITECNICO NACIONAL-ESIQIE
Castro-Arellano, J. J., INSTITUTO POLITECNICO NACIONAL-ESIQIE
Originally presented on: 11/5/2013 10:36:00 - 10:57:00

In this work, results of dynamic viscosities measurements for Octane and Nonane are presented. The viscosity have been measured with a modified capillary viscometer designed and built to operate at pressures up to 40 MPa and temperatures up to 473.15 K. The dynamic viscosity was measured at pressures up to 30 MPa and temperatures up to 353 K. The total uncertainties of the dynamic viscosity measurements were estimated to be less than 1%. The measured dynamic viscosities were compared with data, predictions and correlations previously reported in the literature, where the differences between the literature and the measurements reported here are less than 0.5%.

* To whom correspondence should be addressed

E-mail: lgalicial@ipn.mx

Tel: (52) 55 5729-6000, Ext. 55133 Fax: (52) 55 5586-2728

(279a) Experimental Investigation to Characterize the Influence of Single Components On a Complex Multicomponent Liquid-Liquid-Equilibrium

Lorenz, H. M., TU Bergakademie Freiberg
Staak, D., Technical University of Berlin
Rijksen, C., Lonza AG
Repke, J. U., TU Bergakademie Freiberg
Originally presented on: 11/5/2013 8:30:00 - 8:45:00

The most commonly used separation process in chemical industry is distillation, followed by liquid-liquid-extraction. In general, the separation process can be modeled using the equilibrium method. While vapor-liquid-equilibria (VLE) are the basis for distillation processes, extractive separations can be described using liquid-liquid-equilibria (LLE). The NRTL-equations are a common model used to account for the real behavior of the respective liquid phases by activity coefficients. Experimentally determined binary interaction parameters are commonly used to describe the behavior of multicomponent systems. Furthermore, predictive methods, like UNIFAC, often estimate the VLE behavior of simple mixtures with a satisfactory accuracy. The complexity increases with two liquid phases in equilibrium (VLLE or LLE). Due to strong interactions in the liquid phases, the thermodynamic modeling of such industrial relevant systems is very complex. The exact description of these systems especially when polar molecules or electrolytes are included is an enormous challenge. A theoretical or predictive behavior of such complex LLE systems often does not reflect the reality since there is a huge number of influencing parameters. In order to reduce the experimental effort, a systematic approach, based on a minimal necessary number of fundamental experiments is needed to describe such complex mixtures.

This systematic approach is required to understand the influence of every component on the thermodynamic equilibrium. The characterization of a multicomponent liquid-liquid-mixture can be realized in four steps that are schematically shown in Figure 1. In the *first step* the components which are forming two liquid phases have to be identified. Their binary or ternary mixing behavior has to be investigated to a sufficient extent. The *second step* includes the investigation of the influence of single additives on the thermodynamic equilibrium. Therefore, the respective components are added separately in different quantities to a binary or ternary model mixture in order to characterize their influence on the equilibrium by use of the partition coefficient. In the *third step* of the routine, the additive interactions are investigated. All relevant components are combined into a multicomponent reference system, with which the combined effect can be compared to the influence of the single components. In the *fourth step,* the key components that show the largest impact on the equilibrium have to be identified evaluating the results of steps one to three. The resulting reduced mixture of key components is less complex but shows equivalent characteristics compared to the real process solution including even unknown components.

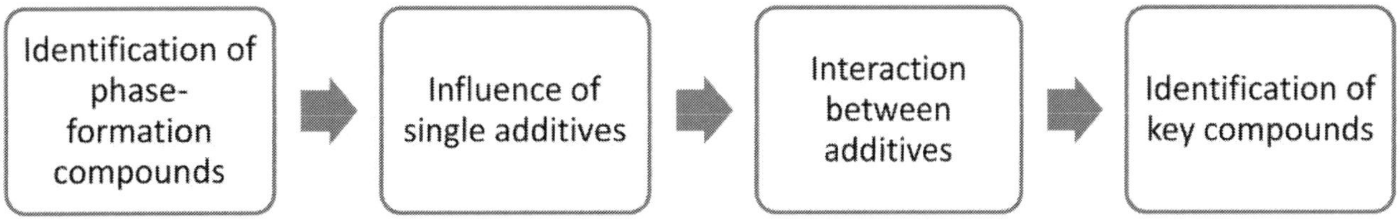

Figure 1:Approach for the identification of key compounds

This practical approach is illustrated in an example where the optimization of a complex industrial process is subsequently accomplished. A liquid-liquid extraction process is the key operation of the entire process. The investigated mixture consists of water, different alkyl-pyridines, an ammonium salt, other organic nitrogen compounds and unknown components, where water has to be separated from the pyridines. In the *first step,* water and two alkyl-pyridines have been identified as main components. The isothermal LLE of the three main components is presented in Figure 2. Pyridine A and water form two liquid phases. The closed miscibility gap has its critical point near the water/pyridine B binary system. In the *second step,* components were added to model solution. The thermodynamic behavior shows a strong dependence on the species of the additives and the added amount. A correlation between the polarity of the additives and the impact on the equilibrium can be seen. Interactions between additives in a multicomponent model mixture and the process solution will be investigated in *step three. Step four* ends with the definition of the resulting key components. The thorough application of the systematic approach will help to optimize the extractive separation unit incorporated in this complex industrial process and to reduce the complexity of the downstream processing.

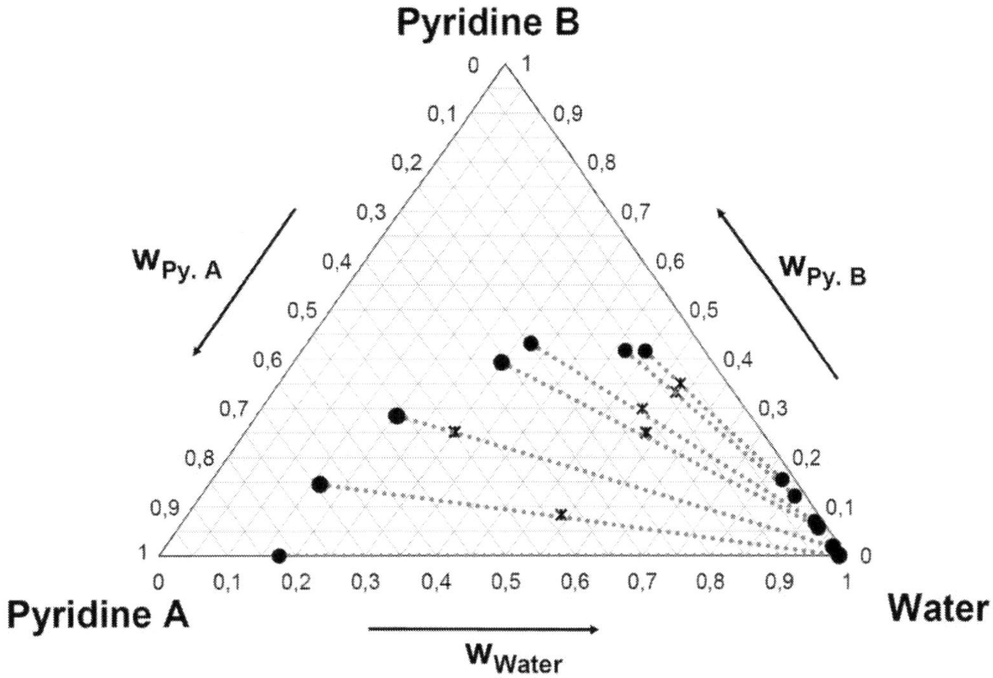

Figure 2: Ternary LLE of Pyridine A, Pyridine B and water, experimental results, T= 60Â°C

(279b) Experimental Measurement and Modeling of Type 3 Ternary System Containing n-Decane + Water + Triacetin

Revellame, E., Mississippi State University
Holmes, W., Mississippi State University
Hernandez, R., University of Louisiana at Lafayette
French, W. T., Mississippi State University
Est vez, L. A., University of Puerto Rico
Originally presented on: 11/5/2013 8:45:00 - 9:00:00

Experiments were conducted to establish the phase diagram for the ternary system containing n-decane + water + triacetin at 23°C and ambient pressure. Results indicated that this system belongs to type 3 Treybal classification with a LLLE region. Modeling indicates that the NRTL equation can satisfactorily predict the ternary behavior of the system and could accurately correlate ternary data.

(279c) Thirty Years With Association Models - What Have We Learnt ?

Kontogeorgis, G., Technical University of Denmark
Originally presented on: 11/5/2013 9:00:00 - 9:15:00

Demanding applications in the oil & gas and chemical industries require advanced models – better than the cubic equations of state. Some of the most successful such models are the so-called association theories, which account explicitly for hydrogen bonding and other complex interactions.

Many such models have been developed over the last 30 years based on chemical, lattice and perturbation theories. Most are expressed in the form of equations of state. They have been used extensively for describing a wide range of complex systems, including associating fluids, polymers, electrolytes and biomolecules. Thus, they have been applied to petroleum, chemical and pharmaceutical industries. Recent books [1-4] have summarized many of these models.

In this presentation, we will present capabilities and limitations of some of these association theories based on our research with the CPA, PC-SAFT and NRHB equations of state. We will present similarities and differences in their performances and emphasize applications in some of the following fields:

- Petroleum industry especially oil – water – gas hydrate inhibitor chemicals
- CO_2 capture and storage
- Chemical industry especially mixtures with organic acids and biofuels
- Electrolytes
- Derivative properties
- Polymers and pharmaceuticals

We will clarify where today association theories perform better than the state of the art models of the previous generation and we will also illustrate what is still challenging almost 30 years after the first systematic applications of association models.

References

[1] G.M. Kontogeorgis, G.K. Folas. Thermodynamic models for industrial applications. From Classical and Advanced mixing rules to association theories, Wiley, 2010.

[2] A.R.H. Goodwin, J. V. Sengers, C.J. Peters Eds, Applied Thermodynamics of Fluids. RSC Publishing, 2010.

[3] J-Ch. de Hemptinne, J-M. Ledanois, P. Mougin, A. Barreau, Select Thermodynamic models for process simulation. A practical guide using a three steps methodology. Editions Technip, 2012.

[4] J.M. Prausnitz, R.N. Lichtenthaler, E.G. de Azevedo. *Molecular Thermodynamics of Fluid-Phase Equilibria* (3rd edition). Prentice Hall Int., 1999.

(279d) A Combined Molecular Simulation and Group Contribution Method to Efficiently Predict the Solubility of Nonelectrolyte Solids

Paluch, A. S., University of Notre Dame
Maginn, E. J., University of Notre Dame
Originally presented on: 11/5/2013 9:15:00 - 9:30:00

Knowledge of the solubility of pharmaceutical solids in a wide range of solvents is essential for all steps of the drug discovery and development process. For drug discovery, one is interested in the design of novel compounds, for which experimental data does not exist, complicating the development process. In the present talk, our efforts to predict the solubility of nonelectrolyte solids in the absence of experimental data will be discussed. The solution phase behavior of the solute is modeled in atomistic detail via molecular simulation. Further, we will show how established solution theories may be used to supplement molecular simulation. The necessary pure solid phase properties of the solute are calculated using group contribution methods. By complementing molecular simulation with solution theories and group contribution methods, solubility predictions may be may in an extremely efficient fashion. In addition, the molecular simulations enable insight into the underlying driving forces behind the solubility process.

Results will be presented for acetaminophen in a range of pure and mixed solvents. Comparison will be made to conventional MOSCED and UNIFAC based methods. Overall, the molecular simulation based predictions are in good quantitative agreement with experimental data, and are comparable to conventional methods.

(279e) Predicting the Extraction Behavior of Pharmaceuticals

Laube, F., Technische Universit t Dortmund
Sadowski, G., Technische Universit t Dortmund
Originally presented on: 11/5/2013 9:30:00 - 9:45:00

Liquid-liquid extraction is a potential separation method for the purification and isolation of pharmaceuticals. Driving force of the liquid-liquid extraction is the different solubility of a pharmaceutical in two immiscible solvents (extraction system). Choosing the adequate extraction system and optimizing this system however usually requires a high experimental effort.

Therefore, within this work the PC-SAFT Equation of State [1] is used to predict the extraction behavior based on a limited number of solid-solubility data (SLE data) of the pharmaceutical in pure solvents. These data were used to determine the PC-SAFT pure-component parameters of the pharmaceutical as well as binary interaction parameters for the pharmaceutical/solvents systems. Solvent 1/solvent 2 binary parameters were fitted to binary liquid-liquid equilibria (LLE). Since binary LLE data for extraction systems can usually be found in literature, only the binary SLE data need to be measured. However, due to the fact that SLE data is also required for drug licensing and registration, this kind of data is in most cases available for pharmaceuticals.

Within this work the separation of different pharmaceuticals was investigated for varying extraction systems. Additionally the influence of pH and temperature on the extraction behavior was predicted. It was found that the modeling results of the extraction behavior are in good agreement with the experimental data. The sequence of solvent selectivity found from the modeling results was the same as the experimentally-determined one. Moreover the comparison of experimental data and modeling results shows that the influence of pH and temperature can be predicted correctly.

Hence, it can be concluded that the extraction behavior of pharmaceuticals can be successfully predicted by PC-SAFT based on their solubility in pure solvents, only. This drastically reduces the experimental effort for the selection of the solvent best suitable for extraction purposes.

1. Gross, J. and Sadowski, G., **Industrial & Engineering Chemistry Research** 40 (2001) 1244-1260

(279f) Modelling the Thermodynamic Properties of Complex Organic Molecules of Pharmaceutical Interest With the SAFT-γ Mie Group Contribution Approach

Papaioannou, V., Imperial College London
Sadeqzadeh, M., Imperial College London
Adjiman, C. S., Imperial College London, Centre for Process Systems Engineering
Jackson, G., Imperial College London, Centre for Process Systems Engineering
Galindo, A., Imperial College London, Centre for Process Systems Engineering
Originally presented on: 11/5/2013 9:45:00 - 10:00:00

The accurate modelling of complex organic compounds typically present in systems of pharmaceutical interest is of great importance for the efficient design of drug processing and production. The major challenge in this case is the characterisation of these multifunctional compounds polar and apolar chemical moeties for which experimental data are often scarce. A number of correlative approaches have been presented, including work based on the PC-SAFT[1] equation of state (EoS) and the NRTL-SAC method[2]. Such approaches have prove successful in modelling of the solubility of active pharmaceutical ingredients (APIs) in pure and mixed solvents, though their applicability is limited since a significant amount of experimental solubility data is required for the parameterisation of the models. A way to overcome this limitation is the application of predictive thermodynamic methodologies, such as those based on activity coefficient models, e.g., COSMO-SAC[3] and the UNIFAC[4] methods. In this work we present the application of the group-contribution based SAFT-g EoS[5,6] in the study of complex molecules, with a main focus on the use of the extension of the new methodology for Mie potentials[7] to describe APIs. The advantage of using a predictive approach, such as the SAFT-g Mie EoS, is that the required parameters for the characterisation of the systems of interest are fully transferable and can thus be obtained by the study of "simpler" systems, for which experimental data is more readily available. No experimental information for the specific APIs under study are required for the parameterisation of the model. Our SAFT-g Mie GC methodology is applied here for a range of APIs both in the study of the solubility of APIs in organic solvents and solvent mixtures, and the study of their partition coefficients in systems of solvents that form two or more liquid phases.

[1] F. Ruether and G. Sadowski, *J. Pharm. Sci.* 98 (2009), p. 4205

[2] C.-C. Chen and P. A. Crafts, *Ind. Eng. Chem. Res.*, 45 (2006), pp. 4816-4824

[3] C.-C. Shu and S.-T. Lin, *Ind. Eng. Chem. Res.*, 50 (2011), pp. 142-147

[4] A. Diedrichs and J. Gmehling, *Ind. Eng. Chem. Res.*, 50 (2011), pp. 1757-1769

[5] A. Lymperiadis, C. S. Adjiman, A. Galindo and G. Jackson, *J. Chem. Phys.*, 127 (2007), p. 234903

[6] A. Lymperiadis, C. S. Adjiman, A. Galindo and G. Jackson, *Fluid Phase Equilib.*, 274 (2008), pp. 85-104

[7] C. Avendano, T. Lafitte, C. S. Adjiman, A. Galindo, E. A. Muller and G. Jackson, *J. Phys. Chem. B*, 117 (2013), p. 2717

(279g) Assessing the Solubility of Pharmaceuticals in Supercritical Carbon Dioxide By Using Sublimation Enthalpies

Tabernero, A., University of Salamanca
Galan, M. A., University of Salamanca
Mart n del Valle, E. M., University of Salamanca
Originally presented on: 11/5/2013 10:00:00 - 10:15:00

Over the last few years, research has been conducted regarding the use of environmentally friendly technologies for the pharmaceutical industry. In this context, supercritical fluid technologies (mainly with supercritical CO_2 (sc-CO_2)) can take advantage of their special properties for developing alternatives to different conventional processes. As an example, it is well known the importance of the use of supercritical fluids to extract drugs from seeds or plants, or to obtain non-solvent particles with controlled particle size distribution.

The determination of the solubility of the solid in sc-CO_2 must be the first step prior to the use of these processes. In supercritical fluid extraction, this solubility provides information about the limit amount of the compound that can be extracted. On the other hand, the chosen supercritical fluid technique for drug processing (supercritical fluid as a solvent or as an antisolvent) depends strongly on that solubility.

However, the determination of the solubility of the pharmaceuticals in sc-CO_2 is not an easy task. Due to the common polarity difference between the drug and the sc-CO_2, the solubility is usually between 10^{-5}-10^{-8}. That means that a small difference in solid purity, or in the protocol can provide high deviations. Moreover, depending on the methodology, experimental differences around one order of magnitude can be obtained. Due to the previous reasons, discrepancies are often observed between experimental solubility data for the same drug.

The previous facts point out the importance of the development of consistency tests to assess the solubility data of highly polar compounds in sc-CO_2. So far, MÃ©ndez Santiago and Teja test is the most used option, although it needs at least three different studies to distinguish the data accuracy. Other theoretical tools, such as cubic equations of state or semiempirical models need adjustable parameters, and consequently they cannot be used for that aim.

An empirical analysis of the solubility of solids in sc-CO_2 has been performed in this work by using only a group contribution method to estimate the sublimation enthalpy of the solid, and by defining a modified enhancement factor. After studying more than 2300 solubility data of pharmaceuticals in sc-CO_2 at different temperatures, polar solids with similar sublimation enthalpies and at similar experimental conditions showed similar solubility behaviour in terms of the previously proposed modified enhancement factor. Based on these results, several equations to estimate that modified enhancement factor (depending only on the sublimation enthalpies) at different temperatures have been defined, providing with an average absolute relative deviation of less than 5%.

Therefore, by knowing only the molecular structure of the compound (as long as there is a group contribution method to calculate the sublimation enthalpies), those equations can be used as a helpful tool to identify if the experimental solubility data follow the usual solubility tendency, and as a consequence their accuracy.

(279h) Prediction of Drug Solubility in Supercritical Carbon Dioxide Using Equation of State Based On Hole Theory With Molecular Information

Sakabe, J., Tokyo Institute of Technology
Shimoyama, Y., Tokyo Institute of Technology
Uchida, H.,
Originally presented on: 11/5/2013 10:15:00 - 10:30:00

There are great interests in preparation of nanometer size materials because of their unique properties in the filed of pharmaceutical, coating, environmental, chemical processing, electronic, and sensing applications. In the pharmaceutical industry, their particle sizes are important for the bioavailability. The bioavailability can be improved by the reduction of the drug particle size. The development of methods for the preparation of nanoparticles for drug has received a great deal of attention. Various crystallization techniques using supercritical fluids have recently been proposed as preparation method of drug nanoparticles. Rapid expansion from supercritical solution method (RESS) is the most well-known crystallization technique using supercritical fluid. We proposed the strategies for nanoparticle design of drugs using RESS technique in our previous work. The mean sizes of the particles micronized by RESS process can be correlated with the supersaturation defined as the difference between the solubilities in supercritical carbon dioxide and the sublimation pressure on the particle collection part. To develop and design the RESS process, a knowledge of drug solubility in supercritical carbon dioxide is necessary and important. In this work, the solubility of several pharmaceutical compounds in supercritical carbon dioxide were predicted using an equation of state. The coordination numbers around the molecules in mixture were estimated from the molecular shape factor which is given by the molecular volume and surface area calculated an quantum calculation based of conductor-like screening model. The prediction performances were evaluated by the comparison with the experimental data It is found that the prediction model proposed in this work is able to represent the experimental data within the same order of the magnitude of the solubility without data fitting.

(302a) Experimental and Theoretical Investigation of Rejection Coefficients of Nanoparticles in Porous Membranes

Baltus, R. E., Clarkson University
Agasanapura, B. N., Clarkson University
Originally presented on: 11/5/2013 12:30:00 - 12:45:00

Nanoparticles (NPs) are often detected in wastewater effluents due to their extensive usage in the consumer product, pharmaceutical and NPs manufacturing industries. The particles require unique membrane separation systems due to their size, shape and surface charges.

In the present work, a series of experiments were designed to steady the rejection of gold NPs. Spherical and capsule shaped gold-particles were synthesized and functionalized using organic compounds to yield negative charges on the particle surface. These gold-particle dispersions were filtered through polycarbonate track-etch membranes with pore size ranging from 0.015 to 0.8 micron. In addition, a computational model was developed to predict the combined effect of steric, hydrodynamic and electrostatic interactions on the rejection coefficient of NPs in a porous membrane. The non-linear Poisson-Boltzmann equation was used to determine the electrostatic particle-wall interaction energy, and a fluid dynamic model that includes particle shape was developed to determine the hydrodynamic interactions. These were then incorporated into a membrane rejection model.

Rejection coefficients from experiments and theoretical modeling for both spherical and capsule particles with an aspect ratio (length/diameter) = 2 and relative particle diameters (particle diameter/ pore diameter) ranging from 0.1 to 0.8 will be presented. When particle and pore have the same charge, theoretical results predict that the rejection coefficient exceeds that of an uncharged system due to the stronger exclusion resulting from the electrostatic repulsive interactions between the particle and the pore. Model results also revealed that 99.99% particle rejection is predicted for a membrane with pore size 1.66 times larger than the particle diameter.Theoretical predictions and experimental results will be compared in this presentation.

(302b) Morphological Stabilization of Solid Surfaces Induced By the Combined Action of Electric Fields and Thermal Gradients

Dasgupta, D., University of Massachusetts Amherst
Du, L., University of Massachusetts Amherst
Sfyris, G. I., University of Massachusetts Amherst
Maroudas, D., University of Massachusetts Amherst
Originally presented on: 11/5/2013 12:45:00 - 13:00:00

Surface morphological instability of stressed solids underlies various reliability problems in technologically important materials, which have a broad range of applications in microelectronics and nanotechnology. In stressed elastic solids, the competition between elastic strain energy and surface energy can destabilize the surface through the Asaro-Tiller or Grinfeld (ATG) instability. In this presentation, we analyze the surface morphological stability of electrically and thermally conducting crystalline elastic solids in uniaxial tension under the simultaneous action of an electric field and a temperature gradient. We use linear stability analysis of a surface mass transport model that accounts for surface electromigration and thermomigration induced by the applied fields and for surface diffusional anisotropy. We find that a properly oriented applied thermal gradient can reduce the critical electric-field strength requirement for stabilization of the planar surface morphology. We validate the conclusions of the linear stability theory based on self-consistent dynamical simulations of surface morphological evolution according to the fully nonlinear model, coupled with numerical solutions of the elastostatic, electrostatic, and thermal boundary-value problems (BVPs) using a Galerkin boundary-integral method.

We also study the surface morphological stability of a coherently strained thin film grown epitaxially on a substrate and subjected to an external electric field and temperature gradient. Due to its lattice mismatch with the substrate, the film may undergo a Stranski-Krastanow (SK) instability, resulting in formation of islands on its surface. We consider various types of substrates placing emphasis on compliant substrates that partly accommodate elastically the lattice-mismatch strain in the epitaxial film. To examine the morphological stability of the film's planar surface state, we conduct a linear stability analysis based on a three-dimensional model of driven film surface morphological evolution. We derive an evolution equation for the film's height and solve analytically the elastostatic, electrostatic, and thermal BVPs that are coupled with it. We find that the action of properly applied and sufficiently strong external fields is necessary to stabilize the planar film surface morphology; in such cases, surface electromigration and thermomigration can inhibit SK-type instabilities and control the onset of island formation on the film surface. We derive the conditions for synergy and competition of the two external fields for surface stabilization and demonstrate the beneficial effects of the thermal field on reducing the critical electric-field strength required to stabilize the planar film surface morphology.

In both cases, we examine the effects on the surface morphological stability of the temperature dependence of the surface diffusivity and the thermal conductivity of the film and the bulk solid, as well as effects of Joule heating of the electrically conducting materials. Our analysis generates experimentally testable hypotheses and motivates experimental measurements that can be compared directly with the theoretical predictions.

(302c) Theoretical and Experimental Study of Gas Permeability and Klinkenberg Effects

Firouzi, M., Stanford University
Alnoaimi, K.,
Kovscek, A., Stanford University
Wilcox, J., Stanford University
Originally presented on: 11/5/2013 13:00:00 - 13:15:00

To predict accurately the transport properties of coals and shales, it is important to study a non-adsorptive gas such as helium to investigate the slippage phenomenon independently of adsorption effects. Non-equilibrium molecular dynamics (NEMD) simulations are carried out with an external driving force imposed on the system to investigate the transport of helium and predict Klinkenberg parameters. The structure is modeled using a three-dimensional pore network, generated atomistically using the Voronoi tessellation method, to represent the carbon-based porous structure of natural systems. Simulations are performed to determine the effect of the pore structure and pressure gradient on gas permeability. In addition, experimental analysis is conducted using the classical Brace pressure solution to measure helium permeability and Klinkenberg parameters of a shale core plug at similar conditions. Simulation results are compared with experiments to understand both the merit and limitations of the simulation approach.

The results indicate the gas permeability, obtained by the pulse decay experiment, is about a magnitude greater than that calculated by the molecular dynamics simulations. The estimated experimental permeability, however, is extracted from the early-time pressure profile that corresponds to macropores as indicated by the estimated apparent pore diameter and referred to as Darcy permeability. Crushing the intact core and sieving the sample to control the grain size could bridge the gap in scale between two methods in a way that the permeability measurements are comparable to the simulations. More complex molecular models of graphite structures containing fractures, local charge and defect sites within the pores, in addition to the inclusion of chemical functional groups to generate a more realistic model of natural systems of interest such as coal and shales could also bridge the gap.

The results have potentially important implications on gas transport in carbon-based materials and geologic formations and may provide an understanding of the limitations of the use of continuum fluid to model transport properties for confined fluids.

(302d) What Really Happens in Diffusion With Reversible Trapping, and What Are the Implications for Biological Diffusion Models?

Nitsche, J. M., University at Buffalo, The State University at New York

Originally presented on: 11/5/2013 13:15:00 - 13:30:00

A classical question in homogenization/coarse graining is, what are the effective (average) diffusion coefficient D_{eff} and rate coefficient k_{eff} characterizing the macroscopically observable transport and reaction of a solute molecules/particle that diffuse through an array (periodic or random) of perfectly or partially absorbing traps (reactive or sticky grains)? Most analyses to date have addressed the irreversible case where?once trapped?the diffusing solute is effectively destroyed or bound forever (see representative references[1-12]). This case is directly relevant to such technological applications as aerosol deposition. In biological systems, however, reversibility of binding/trapping is the rule rather than the exception. Indeed, any irreversibly bound species would necessarily accumulate without bound, and spell disease or death in the absence of a mechanism for removal/clearance. Interestingly, there is a dearth of homogenization theory for the reversible case.

This talk develops a comprehensive answer to the following question. Suppose a solute diffuses through a medium with two-phase microstructure, so that its free concentration $C(x,y,z,t)$ and bound concentration $B(x,y,z,t)$ are governed by the equations

$$\partial C / \partial t = D_a \nabla^2 C \,?\, [(k_{on})_a C \,?\, (k_{off})_a B]$$

$$\partial B / \partial t = (k_{on})_a C \,?\, (k_{off})_a B \qquad (1a)$$

in phase a, and

$$\partial C / \partial t = D_b \nabla^2 C \,?\, [(k_{on})_b C \,?\, (k_{off})_b B]$$

$$\partial B / \partial t = (k_{on})_b C \,?\, (k_{off})_b B \qquad (1b)$$

in phase b at the microscopic scale. What is the equation governing the macroscopically observable behavior? The answer turns out *not* to be of the same form with homogenized coefficients, namely

$$\partial C / \partial t = D_{\text{eff}} \nabla^2 C \,?\, [(k_{on})_{\text{eff}} C \,?\, (k_{off})_{\text{eff}} B]$$

$$\partial B / \partial t = (k_{on})_{\text{eff}} C \,?\, (k_{off})_{\text{eff}} B, \qquad (2)$$

as might be expected. Rather, the correct macroscopic description generally involves a free concentration $C(x,y,z,t)$ and two effective bound concentrations $B^{[1]}(x,y,z,t)$ and $B^{[2]}(x,y,z,t)$ that evolve collectively according to the more complex system of transport equations

$$\partial C / \partial t = D_{\text{eff}} \nabla^2 C \,?\, [(k_{on}^{[1]})_{\text{eff}} C \,?\, (k_{off}^{[1]})_{\text{eff}} B^{[1]}] \,?\, [(k_{on}^{[2]})_{\text{eff}} C \,?\, (k_{off}^{[2]})_{\text{eff}} B^{[2]}]$$

$$\partial B^{[1]} / \partial t = (k_{on}^{[1]})_{\text{eff}} C \,?\, (k_{off}^{[1]})_{\text{eff}} B^{[1]}$$

$$\partial B^{[2]} / \partial t = (k_{on}^{[2]})_{\text{eff}} C \,?\, (k_{off}^{[2]})_{\text{eff}} B^{[2]}. \qquad (3)$$

A general prescription is derived to calculate the effective coefficients $(k_{on}^{[1]})_{\text{eff}}$, $(k_{off}^{[1]})_{\text{eff}}$, $(k_{on}^{[2]})_{\text{eff}}$ and $(k_{off}^{[2]})_{\text{eff}}$, involving an eigenvalue problem posed in a unit cell of the microstructure. Asymptotic analysis yields an explicit formula that approximates the requisite eigenvalue very accurately, thereby allowing the effective binding rate constants to be determined without solving a diffusion problem. Although the analysis is originally carried out for spatially periodic microstructures, the asymptotic reveals a broader applicability of the results, which are verified vis-à-vis a brute-force numerical calculation. The analysis is generalized to an N-phase microstructure for any $N \geq 2$.

The homogenization paradigm has important implications for diffusion models of biological tissues, for which: (ii) multiphase microstructure invariable produces two or more microscopic sets of on- and off binding rate constants $[(k_{on})_a, (k_{off})_a]$, $[(k_{on})_b, (k_{off})_b]$, ?; and (ii) Eq. (2) is invariably assumed to apply at the macroscopic scale. Practical application of the analysis is demonstrated within the context of drug/chemical diffusion through the surface (barrier) layer of skin, the stratum corneum, in which reversible binding to keratin protein is known to occur.[13-16]

1. Ochoa JA, Stroeve P, Whitaker S. 1986. Diffusion and reaction in cellular media. *Chem Eng Sci* **41**(12):2999-3013.

2. Shapiro M, Brenner H. 1988. Dispersion of a chemically reactive solute in a spatially periodic model of a porous medium. *Chem Eng Sci* **43**(3):551-571.

3. Dungan SR, Shapiro M, Brenner H. 1990. Convective-diffusive-reactive Taylor dispersion processes in particulate multiphase systems. *Proc R Soc London Ser A - Math Phys Eng Sci* **429**(1877):639-671.

4. Torquato S. 1991. Diffusion and reaction among traps: some theoretical and simulation results. *J Stat Phys* **65**(5-6):1173-1206.

5. Torquato S, Avellaneda M. 1991. Diffusion and reaction in heterogeneous media: pore size distribution, relaxation times, and mean survival time. Journal of Chemical Physics **95**(9):6477-6489.

6. Mauri R. 1991. Dispersion, convection and reaction in porous media. *Phys Fluids A ? Fluid Dynam* **3**(5):743-756.

7. Brenner H, Edwards DA. 1993. Macrotransport Processes. Boston: Butterworth-Heinemann.

8. Lu SY. 1998. Diffusion and reaction in regular arrays of spheres. *J Chem Phys* **109**(12):4985-4989.

9. Lu SY, Yen YM. 2002. A first-passage scheme for determination of overall rate constants for non-diffusion-limited suspensions. *J Chem Phys* **116**(7):3128-3133.

10. Kansal AR, Torquato S. 2002. Prediction of trapping rates in mixtures of partially absorbing spheres. *J Chem Phys* **116**(24):10589-10597.

11. Vaughn MW. 2004. Diffusion and trapping in a suspension of spheres with simultaneous reaction in the continuous phase. *J Chem Phys* **120**(19):9351-9358.

12. Vaughn MW. 2005. First-passage approach for permeable traps. *J Chem Phys* **123**(13):134905 (7 pp.).

13. Anissimov YG, Roberts MS. 2009. Diffusion modelling of percutaneous absorption kinetics: 4. Effects of a slow equilibration process within stratum corneum on absorption and desorption kinetics. *J Pharm Sci* **98**(2):772-781.

14. Frasch HF, Barbero AM, Hettick JM, Nitsche JM. 2011. Tissue binding affects the kinetics of theophylline diffusion through the stratum corneum barrier layer of skin. *J Pharm Sci* **100**(7):2989-2995.

15. Hansen S, Selzer D, Schaefer UF, Kasting GB. 2011. An extended database of keratin binding. *J Pharm Sci* **100**(5):1712-1726.

16. Seif S and Hansen S. 2012. Measuring the stratum corneum reservoir: desorption kinetics from keratin. *J Pharm Sci* **101**(10):3718-3728.

(302e) Influence of Immersed Objects On Liquid Pool Boil-Over

Shi, X., Worcester Polytechnic Institute
Rangwala, A. S., Worcester Polytechnic Institute
Originally presented on: 11/5/2013 13:30:00 - 13:45:00

Accidental scenarios involving hydrocarbon pool fires usually involve interaction with objects in the path of the liquid. The interaction depends on object type (steel, concrete), geometry as well as height of the liquid surface. For example an object may be completely immersed or above the liquid surface. In the later case there is also interaction of the flame with the object. In this study, we investigate the influence of nonflammable cylindrical objects in a hydrocarbon (hexane) pool fire. A 10 cm diameter steel pan filled with hexane is used as the base case. As a first part of the investigation, one single cylindrical rod of varying heights with respect to the liquid surface is placed in the pool and the influence on burning rate, flame height and radiative heat flux is explored. Two different materials (glass and aluminum) are chosen because of their low and high thermal diffusivities (5.9×10^{-7} m^2/s and 731 $\times10^{-7}$ m^2/s) but similar densities (2200 kg/m^3 and 2707 kg/m^3). Results show the regression rate is a function of two competing effects. The presence of an obstacle reduces the overall burn area thereby decreasing the mass burning rate of the pool fire. At the same time the flame heats the obstacle and depending on its thermal properties, heat is dissipated into the fuel surface improving the vaporization rate. Occasionally, this causes a recirculating current within the pool eventually resulting in boil-over phenomena.

The second part of study investigates the influence of change in surface area to volume ratio of the obstacle in a pool fire. Several smaller diameter aluminum cylindrical rods are used instead of a single rod (same volume but various total surface area). Cylindrical rods of varying heights with respect to the liquid surface are placed in the pool and the influence on burning rate, flame height and heat fluxes is also explored.

A simple one-dimensional empirical heat transfer model is developed to predict the mass loss rates for various pool sizes, fuel and obstacle sizes. Implications to safety considerations involving cylindrical obstacles in pool fires are explained.

(302f) Computational Fluid Dynamics Modelling of Single and Two-Phase Heat and Mass Transfer in Micro- and Milli-Scale Flows

Kuhn, S., ETH Zurich

Originally presented on: 11/5/2013 13:45:00 - 14:00:00

Computational Fluid Dynamics modelling of single and two-phase heat and mass transfer in micro- and milli-sc

Simon Kuhn

Department of Chemical Engineering, University College London, UK

The investigation and characterization of heat and mass transfer phenomena in single and two-phase flow is im mechanisms influencing chemical reactions in micro- and milli-scale devices, where no (or only little) turbulent interfacial transport processes. This detailed analysis is accomplished by a rational approach, identifying the ph interfacial mass transfer on various length-scales using non-invasive, laser-optical measurement techniques[1-3], results to validate and develop predictive multiphase flow models for computational fluid dynamics (CFD).

For this presentation we will focus on the CFD modelling aspect of the work, where we use the open-source so single and two-phase flow heat and interfacial mass transfer on the micro-scale (i.e. 400 μm microchannels) an geometry for the milli-scale is the Corning Advanced Flow Reactor (AFR) depicted in Fig. 1, which consists of in series.

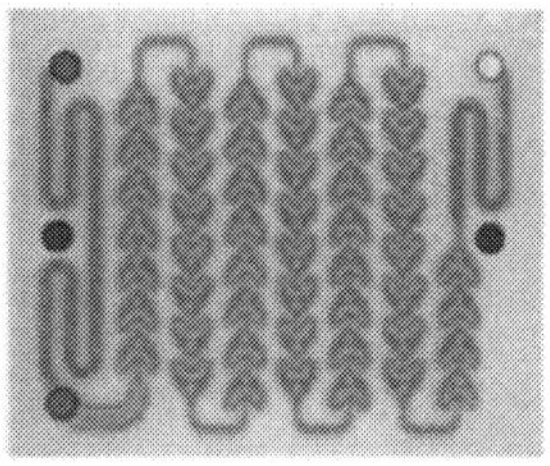

Fig. 1 Representation of the Corning Advanced Flow Reactor geometry. The red dots represent the two inlets for the continuous and dispersed phase, the green dot the outlet. The plate volume is 8.7mL.

The aim of these simulations is to understand interfacial transport processes in more detail, and to use this know milli-scale multiphase flow reactors. The developed models and obtained results will therefore directly impact t and sustainable advanced manufacturing.

The numerical work is conducted using the volume-of-fluid (VOF) approach to simulate two-phase flows, whic describe the volumetric phase fraction in each computational cell. Since the phase interface definition is highly extra terms are needed to account for numerical diffusion, and to provide a sharp interface definition. In the cu OpenFOAM™ solver *interFoam* an interface compression approach is used, and in initial simulations we validat case of bubble generation in a microfluidic T-junction against experimental results. It was found that the standa thus we introduced changes in the interface compression scheme to increase the interface sharpness, which the and proved also successful on the milli-scale (see Fig. 2).

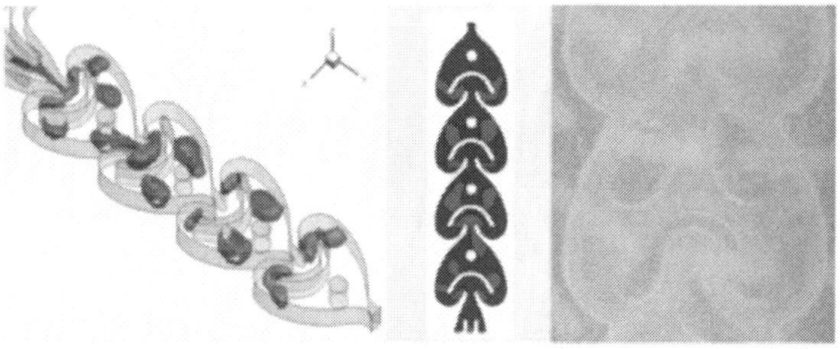

Fig. 2: Comparison of the gas-liquid flow distribution predicted by the modified VOF solver with experimental results from a high-speed camera.

The standard *interFoam* solver does not include scalar transport processes, thus we implemented the energy eq indices denote the gas (G) and liquid (L) phase respectively, and can be transformed to liquid-liquid systems as

$$\frac{\partial(\rho T)}{\partial t}+\frac{\partial(\rho u_j T)}{\partial x_j}=\frac{\partial}{\partial x_j}\left[\frac{\mu}{\mathrm{Pr}}\left(\frac{\partial T}{\partial x_j}\right)\right]$$

where μ denotes the dynamic viscosity of the fluid, and Pr is the Prandtl number defined as $\mathrm{Pr}=\mu c_P/\lambda$, with th heat capacity c_p. This equation can be readily adapted for the VOF approach by calculating the fluid properties fraction α.

The implementation of species concentration in the VOF framework uses the same approach, i.e. we define the coefficient D as

$$c=\alpha c_G+(1-\alpha)c_L; \quad \text{and} \quad D=\alpha D_G+(1-\alpha)D_L\,. \qquad (2)$$

The boundary condition at the interface between the gas and liquid phase is satisfied by Henry's law

$$c_G = Hc_L \qquad (3)$$

with the Henry coefficient H describing the equilibrium solubility of the species in the liquid phase. However, a single species equation this needs to be taken into account as an explicit source term in the scalar transport equ framework

$$\frac{\partial c}{\partial t} + \frac{\partial(u_j c)}{\partial x_j} - \frac{\partial}{\partial x_j}\left[D\left(\frac{\partial c}{\partial x_j}\right)\right] = \frac{\partial}{\partial x_j}\left[D\frac{1}{\alpha H+}\right.$$

. (4)

Figure 3 depicts the predicted temperature field in single and two-phase flow in a microfluidic T-junction. The f temperature and then heat up as they flow downstream in the microchannel. For the single phase case a parabo which is expected for laminar convective heat transfer. The temperature field of the gas-liquid flow case reveal generation process in the entrance region of the microchannel on heat transfer. Initially the gas phase is still in c faster than the liquid phase. After detachment of the gas bubble it is isolated from the wall by the thin liquid filr in the slight temperature difference between the gas and liquid phases further downstream.

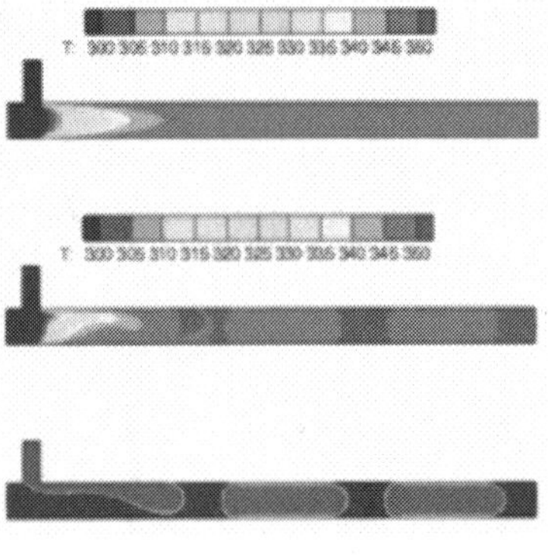

Fig. 3: Comparison of single (top) and two-phase flow (middle) temperature fields in a microchannel. The bottom representation shows the distribution of the gas bubbles.

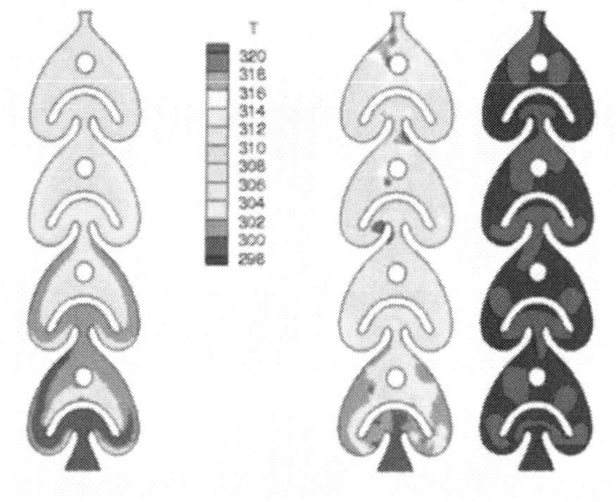

Fig. 4: Comparison of single (left) and two-phase flow (middle) temperature fields in 4 hearts of the Corning AFR. The right representation shows the distribution of the gas bubbles.

Figure 4 depicts the comparison of temperature fields for the Corning AFR. In single phase flow the individual l temperature gradients within a single cell. Elevated temperatures are observed in regions of low fluid velocity, t recirculating fluid motion, which develops between the U-shaped structure and the downstream post. These reg by fluid of lower temperature, thus this particular design leads to limited cross-mixing in single phase flow. The establish in two-phase flow due to the bubble dynamics (constant break-up and merging), and as a consequence much smoother and the heat transfer performance is increased, as seen by the higher average temperature in th phase flow.

We will report in detail on the implemented VOF method, and the obtained heat and mass transfer data are vali results and are further interpreted to characterize the transport processes on the micro- and milli-scale.

References:

[1] Kuhn, S.; Jensen, K.F. (2012) A pH sensitive Laser-Induced Fluorescence technique to monitor mass transf microfluidic devices. Ind. Eng. Chem. Res. 51:8999-9006.

[2] Kuhn, S.; Woitalka, A.; Jensen, K.F. (2012) Mass Transfer in gas-liquid and liquid-liquid multiphase flows. I Conference on Microfluidics.

[3] Nieves-Remacha, M.J.; Kulkarni, A.A.; Jensen, K.F. (2012) Hydrodynamics of Liquid-Liquid Dispersion in Eng. Chem. Res. 51:16251–16262.

(302g) Experimental Studies On the Effect of Surface Roughness On Laminar, Transitional and Turbulent Flow in Microtubes

Chen, Q. L.,
Zhao, C. J., State Key Laboratory of Chemical Engineering, Department of Chemical and Biological Engineering, Zhejiang University
Wu, K. J., State Key Laboratory of Chemical Engineering, Department of Chemical and Biological Engineering, Zhejiang University
He, C. H., State Key Laboratory of Chemical Engineering, Department of Chemical and Biological Engineering, Zhejiang University
Originally presented on: 11/5/2013 14:00:00 - 14:15:00

With the rapid development of microsystem technology, the investigation of liquid flow in microtubes has become one of the most important subjects. Microsystems have been widely used in industry and biological medicine, such as heat exchangers, mixing and separation processes, fuel cells, DNA analysis, artificial blood vessels and so on. To optimize microsystems, a sound understanding of the properties of liquid flow in microtubes is necessary. In the recent 15-20 years many researchers are trying to figure out whether the liquid flow behavior in microscale is the same as classical theory in macroscale. However, different researchers came up with conflicting conclusions. Some researchers found that the friction factor is below the classical theory, some said that friction factor is the same as classical theory, but many recent researchers observed higher friction factor than classical theory. The surface roughness is one of the major parameters leading to higher friction factor, but no one gave a quantitative conclusion about the differences surface roughness made between microscale and macroscale.

In this paper, an experimental study of water flow in microtubes has been performed in 5 stainless steel microtubes whose diameters (D) are 524.7, 353.2, 161.6, 127.5, and 126.3 μm to investigate the effect of surface roughness on characteristics of liquid flow in microscale. The average surface roughnesses (k) of these five microtubes are 2.32, 1.31, 2.80, 2.98, and 4.37 μm, respectively. The average surface roughness was determined by a non-contact 3D surface measurement (Wyko NT9100) using a vertical scanning interferometry mode for a wide variety of topologies, as well as phase-shift interferometry mode for more subtle profile changes.

The experiments were performed in an apparatus including pressure source, test section, pressure transducer, balance, and data acquisition system. The Reynolds number (Re) and friction factor (f) were calculated from experimental data to describe the characteristics of liquid flow.

It was found that the critical Reynolds number (Re_c, the Reynolds number when the friction factor departs from the Poiseuille equation) is a function of the relative roughness (k/D). The Re_c reduces with the increase of the k/D, and the minimum Re_c is 200 when the k/D is 3.46 % in our experiment. The f-Re relationship is the same as macroscale theory in the laminar region. However, the f-Re relationship in the transition region and turbulent region can't be predicted by macroscale theory such as Blasius equation or Colebrook equation any more. A mathematical model was proposed to describe the relationship between Re_c and k/D, and it is also applicable to literature data. A modified Blasius equation was established to calculate f using Re for transition region and a power correlation was established to calculate f using k/D for quadratic resistance region.

The application of the correlations mentioned above is limited to the diameters and surface roughnesses in this work. To obtain more accurate parameters for these models, more experiments need to be carried out in larger and smaller tubes. We considered that Re_c can be one indicator to find the critical diameter (D_c) that distinguish microscale and macroscale (Re_c equals 2300 in macroscale and smaller than 2300 in microscale). More researches will be done to find the D_c.

(302h) Turbulent Dissolution, Drag Reduction and Degradation of Pipe Wall-Injected Polyox

Virk, P. S., Massachusetts Institute of Technology

Originally presented on: 11/5/2013 14:15:00 - 14:30:00

Turbulent dissolution, drag reduction and degradation of wall-injected Polyox P309 solutions was studied by simultaneous axial pressure profile and flow visualization measurements at Reynolds numbers from 15000 to 150000 in a transparent test pipe of ID 16 mm and L/D 430. Friction factors for solvent, deionized water, adhered to the Prandtl-Karman law for fully-developed turbulent flow in smooth pipes. The P309 Polyox, of MW ~ $13x10^6$ with N_{bb} ~ $0.87x10^6$ backbone chain links, was injected as a concentrated solution C_{inj} = ei 500 or 1000 wppm through 6 equi-spaced circumferential ports at friction-normalized injection velocities V_{inj}^+ ~ 1 << pipeline mean velocity V^+ ~ 25. Downstream polymer concentrations C_{line} varied from 0.2 to 20 wppm, exhibiting eventual fractional drag reductions DR_{line} from 0.10 to the asymptotic maximum drag reduction MDR ~ 0.80 possible at the present Re. Polymer dissolution into the flow was assessed from the initial increase in fractional drag reduction towards the eventual downstream DR_{line}, which provided a characteristic development distance $(L/D)_{dr}$ and also from flash photography which provided a characteristic distance $(L/D)_{pv}$ where red-dyed strands of the injected solution disappeared. Finally, at the lowest C_{line} and highest Re, drag reduction was observed to increase, peak, and then decline with increasing downstream distance. Since the peak drag reduction corresponded roughly to the disappearance of the red-dyed strands, that is, to full polymer dissolution, the subsequent downstream decline in drag reduction could be ascribed to polyox degradation and used to estimate its kinetics.

(302i) Heat Transfer Analysis of An Advanced Flow Reactor Using Quantum Dots

Shi, Y., Iowa State University
Kuhn, S., ETH Zurich
Jensen, K. F., Massachusetts Institute of Technology
Originally presented on: 11/5/2013 14:30:00 - 14:45:00

The full-field temperature measurements in a milli-scale CorningÂ®; Advanced FlowTM reactor (AFR) are performed in this work by utilizing a non-intrusive technique that involves quantum dots (QD) as the sensor. Heat removal/addition can be extremely important for chemical reaction systems in terms of selectivity, yield and conversion, in addition to ensuring safety. The detailed temperature information inside the flow field is therefore desirable for reactor design and process optimization. QD's possess various interesting properties that make them a superior choice for such purposes. The photo-luminescence stability of QD's is the key factor to the success of this measurement technique and is addressed by using freshly synthesized QD's. As temperature increases, the intensity of the fluorescence of the QD's decreases. Based on this property, a calibration curve can be constructed. Intensity fields are then recorded at six different flow rates (5, 10, 15, 20, 25 and 30 ml/min) and translated back to the temperature fields for each mixing units in the AFR. From the measured temperature fields, local overall heat transfer coefficients as well as Nusselt numbers are calculated for the mixing unit of interest at each flow rates. From the analysis, it was found the heat transfer performance of the AFR is better than the conventional reactors due to increased ratio of surface area to volume. This unique technique also presents promise for temperature measurements for two-phase or even multiphase flows.

(302j) Diffusion-Driven Colloidal Deposition: From Flat Plate Sensors to Occluded Pipes

Hashmi, S. M., Yale University
Loewenberg, M., Yale University
Firoozabadi, A., Yale University
Originally presented on: 11/5/2013 14:45:00 - 15:00:00

Diffusive driven transport and deposition processes affect real-world applications, from microfluidics based bio-sensors to large scale pipelines in the petroleum industry. Colloidal and molecular deposition can therefore either be a desirable phenomenon to be measured, or a detrimental result to be reversed. In either case, a fundamental understanding of deposition dynamics will both aid applications and inspire experimental design. In this talk, we present an analysis of the deposition of a dispersed-phase fluid onto a flat plate from a suspension in shear flow. According to our model, the profile of the deposit is governed by an inhomogeneous Bergers equation that yields an analytical self-similar solution. We also consider the case of deposition onto a pipe wall under parabolic flow conditions, exploring the parametric dependence of the occluded flow that results. We will discuss applications of this work to industrial settings. In particular, the parametric dependence of the transition between slight and severe clogging is of great interest for flow assurance in the petroleum industry.

(303a) Bubble Point Pressures of Binary Mixtures Wetting Screens Against a Vapor: Implications for Low Gravity Cryogenic Liquid Acquisition

Hartwig, J. W., NASA Glenn Research Center
Mann, J. A., Department of Chemical Engineering
Originally presented on: 11/5/2013 12:30:00 - 12:45:00

Bubble Point Pressures of Binary Mixtures Wetting Screens Against a Vapor: Implications for Low Gravity Cryogenic Liquid Acquisition

Jason Hartwig, Cryogenic Propulsion Engineer, NASA Glenn Research Center, Cleveland, OH 44135

J. Adin Mann Jr., Professor of Chemical Engineering, Case Western Reserve University, Cleveland OH, 44106, USA

Gravity affects many fluidic processes, such as the position of the liquid and vapor phases within a propellant tank. In a standard 1-g environment the density of the individual fluid phases dictates the location of vapor and liquid interface; the heavier fluid (liquid) settles to the bottom and the lighter fluid (vapor) rises to the top. However, in the reduced gravity of a space environment, surface tension becomes the controlling mechanism for this liquid/vapor separation. When transferring cryogenic propellant from a storage tank to a transfer line en route to an engine or depot receiver tank, it is required to transfer vapor free liquid. Multiple liquid acquisition devices (LAD) may be required to ensure the liquid phase sufficiently covers the tank outlet across the widely varying thermal and gravitational environments experienced during a space mission. Of the available choices, the most popular, flexible, and robust type of LAD is the screen channel gallery arm. The channel side that faces the tank wall is covered by a fine mesh screen, which creates a system of micron sized pores that act as a barrier to vapor ingestion. The other channel sides are solid metal. The channels converge and connect at a common location within the tank to position liquid over the outlet. During tank outflow, liquid is wicked across the screen, preventing the pores from drying out when in contact with vapor. Here, the "maximum bubble point pressure" (the pressure that is signaled by bubble formation on the screen) defines the upper limit on the LAD system performance. The total pressure loss across the screen and in the LAD channel must be less than the bubble point pressure to prevent vapor ingestion into the channel. While certain screen channel LAD meshes have proven flight heritage with storable propellants, finer meshes may be required to safely deliver low surface tension cryogenic liquid propellants to an engine.

Experimental results are presented here for the room temperature binary mixture bubble point tests conducted at the Cedar Creek Road Cryogenic Complex, Cell 7 (CCL-7) at the NASA Glenn Research

Center. The purpose of these tests was to investigate the performance of three different stainless steel fine mesh LAD screens in room temperature liquid mixtures which span the intermediate to high surface tension range (25 – 70 mN/m) to provide pretest predictions for cryogenic liquid hydrogen (LH_2) as part of NASA's microgravity LAD technology development program. Bench type tests based on the maximum bubble point method were conducted for a 325x2300, 450x2750, and 510x3600 mesh sample in binary mixtures of methanol and water.

The simplified bubble point model from Hartwig and Mann (2013) for pure liquids is extended to predict bubble point pressures of binary liquid mixtures. A Langmuir adsorption isotherm is used to obtain a curve fit to mixture surface tension data in order to predict the bubble point pressure at any mixture mass fraction. The Gibbs equation using the convention that the excess volume and the excess of the solvent (water) are set to zero in order to satisfy the Gibbs phase rule that specifies the number of independent intensive variables allowed in the representation of the liquid/vapor (L/V) surface tension of the binary solution, . Contact angle between LAD porous screen and binary liquid is measured as a function of methanol mass fraction using the Sessile Drop technique. The solid/vapor and solid/liquid surface tensions are then estimated using the Young-LaPlace equation. From a Zisman plot (work of adhesion vs. liquid/vapor surface tension) of the binary mixture data, one can obtain the so-called "critical" Zisman surface tension for a given liquid/solid pair where contact angle approaches zero. The critical Zisman surface tension defines the point of total wettability at the surface, and for , contact angle is not expected to deviate from , and the surface can be assumed to be completely wetted.

Experimental results are as follows: Contact angle measurements from this work indicate that for the porous LAD screen. Since the surface tension of all cryogenic liquids is less than this critical value, the contact angle can be assumed to be zero for all cryogenic propulsion systems employing stainless steel LAD screens. Bubble point tests conducted in methanol/water mixtures are in agreement with the previous pure fluid tests, in that the middle 450x2750 mesh outperforms both the coarsest 325x2300 and finest 510x3600 meshes. Excellent agreement between experimental data and the model predicted values is obtained for methanol mass fractions greater than 50%; for methanol mass fractions less than 50%, as the mixture moves toward higher concentrations of water, theory and model deviate proportionally. Discrepancies between model and data are due to differences in mixture compositions in the bulk liquid vs. the liquid at the liquid/vapor interface at the LAD screen. As the mixture tends toward higher water concentrations, the low surface tension of methanol and high contact angle of water dominate the breakdown sites within the porous screen, effectively lowering the bubble point from the expectation value. , surface tension data are all consistent with the Langmuir isotherm description of the thermodynamics of adsorption.

The two implications for cryogenic propulsion LAD design from this work are that

1) "Perfect" wettability can be assumed between porous LAD screen and all cryogenic liquids.

2) Predictions for LH_2 bubble point pressures based on experimental data here indicate that bubble point pressures in excess of 750 Pascals may be achievable with the 450x2750 mesh, which represents a significant improvement in performance over the baseline 325x2300 screen.

Results here have implications for all future LH_2 fueled missions, as higher bubble point pressures imply that the LAD can deliver higher flow rates to an engine or receiver fuel depot tank in the micro-gravity conditions of space.

(303b) Dynamic Model of a Water Nanodroplet Under a Transmission Electron Microscope

Leong, F. Y., Institute of High Performance Computing
Mirsaidov, U. M., University of Illinois at Urbana-Champaign
Mahadevan, L., Harvard University
Matsudaira, P., National University of Singapore
Originally presented on: 11/5/2013 12:45:00 - 13:00:00

We report and investigate an experimental phenomenon of torus-shaped water nanodroplets sliding on a hydrophilic substrate in a cyclical stick-slip fashion under a transmission electron microscope. Using continuum long wave theory, we propose a physical model for the stress imposed by electrical charges and explain how the non-uniform distribution of charges on the surface of a sufficiently pinned convex drop could eventually lead to a deformation into a toroidal shaped drop. The depression at the drop center is shown to depend on the balance between the electrostatic stress and the surface tension of the drop. On the other hand, the lateral component of the unbalanced electrostatic stress creates a lateral driving force, which when sufficiently large, could overcome surface heterogeneities in the substrate disjoining pressure, such that the drop slides on the substrate surface through a cyclical stick-slip motion. The present model predicts step-like dynamics in drop displacement and surface area jumps, both qualitatively consistent with experimental observations.

(303c) A Platform for Retaining Native Morphology At Sub-Second Time Scales in Cryo-TEM

Croote, D., Brown University
Godfrin, M., Brown University
Bose, A., University of Rhode Island
Tripathi, A., Brown University
Lee, J., Sungkyunkwan University
Originally presented on: 11/5/2013 13:00:00 - 13:15:00

Introduction

Cryogenic transmission electron microscopy (cryo-TEM) is a widely used tool for morphological analysis in biology, chemistry, and polymer science. Using this technique, samples can be imaged in their native state without fixing or staining, which often cause fundamental conformational changes. Sample preparation for cryo-TEM involves removing the excess liquid deposited on a cryo-TEM grid by blotting with filter paper, which can be seen in Figure 1c-d, and plunging the grid into a cryogen (usually liquid ethane), resulting in vitrification. The blotting step is required to thin the sample, allowing the electron beam to pass through the sample, and to ensure complete sample vitrification. During this process, however, native sample morphology is compromised due to high shear rates inflicted on the sample. Additionally, species may non-specifically adsorb to the filter paper, resulting in a non-native sample concentration when imaged.

Platform Overview

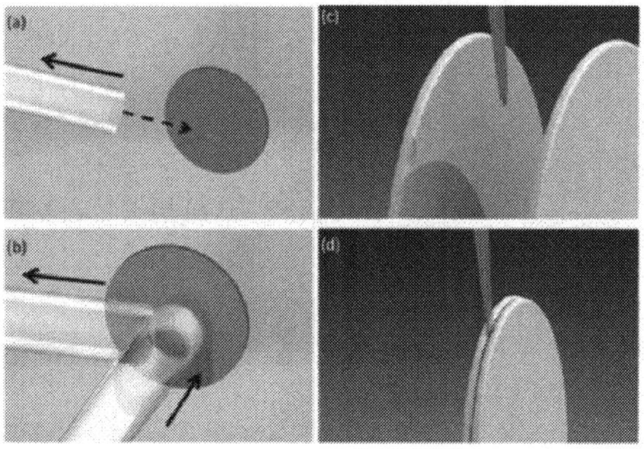

FIG. 1. Capillary positioning for sample deposition and removal for (a) time insensitive study and (b) time sensitive study. Solid arrows portray liquid motion and dashed arrows portray capillary motion. Conventional blotting entails the placement of the sample grid between two pieces of blotting paper (c), which are then pressed against the sample grid for sample thinning (d).

We report the development of a new sample preparation device based on capillary action that overcomes the above limitations. The use of a glass removal capillary aligned parallel to the grid, as shown in Figure 1a-b, results in reduced shear and lower absorption of particulate material from the sample. Furthermore, a deposition capillary aligned perpendicular to the grid coupled with LabView-controlled hardware allows for precise and sub-second resolution for time resolved studies. We demonstrate each of the features of this platform using model samples, and where appropriate, compare our results to those prepared using current vitrification platforms.

Results

The efficacy of our new sample preparation technique is shown through the imaging of CTAB in phenol, Ultra Downy fabric softener, calcium carbonate intermediates, and a carbon black nanoparticle suspension. A mixture of 50 mM CTAB and 50 mM phenol, which forms wormlike micelles in aqueous solution, was utilized in order to compare the morphological differences resulting from choice of sample preparation method. The natural morphology of the wormlike micelles exhibits an entangled structure, which was captured after preparing a sample with our method. A sample prepared using a Vitrobot system with blotting and without relaxation results in wormlike micelles that are oriented parallel to each other, aligned in the direction of flow. Numerical studies were used to quantify the shear causing this alignment; it was found that shear during blotting is one to three orders of magnitude greater than for our method of parallel capillary action. To illustrate the time resolution of the system, the evolution of a mixture of aqueous calcium chloride and sodium carbonate from amorphous calcium carbonate at 700 milliseconds to calcium calcite at 5 was depicted. Another advantage of our system is the capability of imaging higher viscosity fluids. Cryo-TEM images were successfully taken of Downy fabric softener with a viscosity ~100x higher than that of water and a surface tension lower than that of water. Lastly, images of carbon black particle suspension prepared using our method showed species adsorption was reduced by ~20x compared to blotting.

Conclusion

We have developed a novel cryo-TEM sample preparation technique that allows for the vitrification of samples without the shear induced orientation of artifacts that are typical of the conventional cryo-TEM blotting sample preparation. Through the use of a glass removal capillary, we have also effectively eradicated the issue of the altering of sample nature through the adsorption of sample species typical of current preparation methods. Our technique also allows for sub-second time resolution, which is beneficial for the imaging of the intermediates of chemical or biological processes present at sub-second intervals. Our results confirm that this new sample vitrification device opens up previously unattainable regimes for sample preparation and imaging and is a powerful new tool for cryogenic transmission electron microscopy.

(303d) Can Dynamic Contact Angle be Measured Using Molecular Modeling?

Malani, A., Indian Institute of Technology Bombay
Raghavanpillai, A., E.I. du Pont de Nemours,
Wysong, E. B., E.I. du Pont de Nemours,
Rutledge, G. C., Massachusetts Institute of Technology
Originally presented on: 11/5/2013 13:15:00 - 13:30:00

Many natural phenomena, such as, rolling of water droplet on lotus leaf, dryness of bird feather even after immersing in water and skimming of water strider on water surface, have inspired humans to prepare man-made superhydrophobic and self cleaning surfaces. These modified surfaces have potential applications in variety of areas ranging from chemical industry to food appliances. In the design of such surfaces, the key parameter is their wetting behaviour which is characterized by equilibrium and dynamic (advancing and receding) contact angles. Molecular modeling has a potential to provide detail information of contact angle phenomena down to a sub-nanometer length scale resolution, which is essential for success of these materials.

We have proposed a method for determining the dynamic contact angle at the three-phase contact between a solid, a liquid, and a vapor under an applied force, using molecular simulation. The method is demonstrated using a Lennard-Jones (LJ) fluid in contact with a cylindrical shell of the FCC Lennard-Jones solid. We report the advancing and receding contact angles and the contact angle hysteresis for LJ system using this approach. We also evaluate the increase in force required to wet fully an array of solid cylinders (robustness) with decrease in separation distance between cylinders. The dynamic contact angle is characterized by partial slipping of the three phase contact line when a force is applied.

(303e) Molecular Dynamics Simulations of Polymer Electrolyte Membrane Interfaces

Daly, K., Princeton University
Benziger, J., Princeton University
Debenedetti, P. G., Princeton University
Panagiotopoulos, A. Z., Princeton University
Originally presented on: 11/5/2013 13:30:00 - 13:45:00

Polymer electrolyte membranes are an important component of hydrogen fuel cells, and materials for these membranes are selected to maximize proton conductivity. Currently, the material of choice is Nafion, a polyperfluorosulfonic acid. The proton conductivity of Nafion is highly sensitive to the amount of water in the membrane because water molecules aggregate into nano-scale clusters that act as conduits for protons.[1] Consequently, water uptake and retention is a practical concern in fuel cell operation. Recent experiments suggest that the air/membrane interface constitutes a significant transport barrier to water uptake, while the liquid water/membrane interface is a much smaller barrier.[2]The molecular origin of these barriers is not well understood.

Atomistic molecular dynamics simulations of the air/membrane and liquid/membrane interfaces will be presented, revealing local concentrations and configurations of polymers chains that differ substantially from bulk behavior. Mass transfer resistances at these interfaces are directly calculated using a non-equilibrium molecular dynamics technique.

[1] Mauritz, K. A.; Moore, R. B. *Chem. Rev.* **2004**, *104*, 4535–4586.

[2] Majsztrik, P.; Bocarsly, A.; Benziger, J. *J. Phys. Chem. B* **2008**, *112*, 16280–16289.

(303f) A Reactive Molecular Dynamics Research for the Effect of Strain Energy On Hydrolysis Reaction of the Silica-Water Interface

Yeon, J., Pennstate University
van Duin, A. C. T., Pennsylvania State University
Originally presented on: 11/5/2013 13:45:00 - 14:00:00

We performed reactive force field MD simulation to observe the hydrolysis reactions between water molecules and locally strained Si geometry. We established that the ReaxFF potential used in this research accurately describes the water - silica hydrolysis reaction by comparison with DFT data. Number density of silanol-grouups on the surface and properties of the vitreous silica slab obrained from ReaxFF simulation are consistent with experimental results. We demonstrate that the reaction barrier for silanol formation in strained geometry is significantly less than that for non-strained sites. Observation of silanol formation at the high-strain region of a silica nano-rod also supports the concept that the adsorption of water molecule and hydroxyl formation favors the geometry with higher strain energy. Formation of surface hydroxyl in amorphous silica double slit displays a similar tendency: SiOH formation prefers high-strain sites. The most frequent way of surface hydroxyl formation was dissociation of an adsorbed water molecule, creating 2 SiOH on the surface. In addition, we observed proton transfer and proton hopping through the interface, which is also responsible for SiOH formation. A third mechanism observed from our simulation is water or hydronium molecule donating a proton to the surface dangling oxygen, or strained bridge oxygen. This result implies that hydrolysis reaction with water molecules on the silica surface prefers the strained sites of the surface. In addition, we expect that this strain energy - hydrolysis relationship on the surface are relevant to tribology research, as it enable us to predict the attachment points of the lubrication film to the silica surface.

(303g) Reactive Molecular Dynamics Study of Alkylsilane Monolayers On Realistic Amorphous Silica Substrates

Black, J. E., Vanderbilt University
Iacovella, C. R., Vanderbilt University
McCabe, C., Vanderbilt University
Cummings, P. T., Oak Ridge National Laboratory
Originally presented on: 11/5/2013 14:00:00 - 14:15:00

Alkylsilane monolayers are used in many applications, including as lubricants for micro- and nano-electromechanical systems (MEMS and NEMS), where they are employed to reduce stiction and protect surfaces from oxidation and wear. Molecular simulation has begun to play an important role in understanding the molecular-level properties and behavior of these systems and has aided in the understanding and design of monolayer-based nanolubricants (see e.g. references [1,2]). While simulation studies typically use idealized surfaces on which the alkysilane monolayers are attached, previous work has shown that monolayers assembled on crystalline versus amorphous silica exhibit different frictional forces at identical monolayer densities as a result of voids present in the monolayers assembled on amorphous silica [3]. In this work, in an effort to get even closer to the systems studied experimentally, we study monolayers assembled on an amorphous hydroxide surface layer. This closely mimics the post-synthesis processing of amorphous silica with "piranha" solution in order to generate surface hydroxide groups that serve as active sites for the attachment of the alkylsilane chains in the experimental preparation of SAMs.

To examine the impact of silica processing on the properties of monolayers, a synthesis mimetic scheme (SMS) was developed that uses the ReaxFF reactive force field [4]. In this study, amorphous silica substrates were exposed to hydrogen peroxide to create an amorphous hydroxide surface layer to which alkylsilane chains were ultimately bonded. Monolayer properties (including nematic order parameter and tilt angle) were investigated as a function of monolayer density and chain length and compared to the properties of idealized systems (including crystalline substrates and non-processed amorphous substrates). In all cases, the SMS-based systems are considerably less uniform than their idealized counterparts; additional simulations demonstrate that this trend toward disorder results from both a non-regular in-plane arrangement of chains in the monolayer and the processing-induced atomic-scale surface roughness of the silica substrate. This study highlights the importance of considering the synthesis and processing steps associated with the creation of monolayers; including these effects is an important step toward the overall goal of using simulation as a method to screen and predict the next generation of monolayer-based lubricants.

[1] Chandross, M., Grest, G., and Stevens, M. "Friction between Alkylsilane Monolayers: Molecular Simulation of Ordered Monolayers," *Langmuir*, vol. 18, January 2002.

[2] Booth, B., Vilt, S., Lewis, J., Rivera, J., Buehler, E., McCabe, C., Jennings, K. "Tribological Durability of Silane Monolayers on Silicon," *Langmuir*, vol. 27, April 2011.

[3] Lorenz, C., Chandross, M., Grest, G., Stevens, M., and Webb, E. "Tribological Properties of Alkylsilane Self-Assembled Monolayers," *Languir*, vol. 21, September 2005.

[4] van Duin, A., Dasgupta, S., Lorant, F., and Goddard, W. "ReaxFF: A Reactive Force Field for Hydrocarbons," *Journal of Physical Chemistry*, vol. 105. 2001.

(303h) The Influence of Ion Size and Effective Polarizability of the Capacitance of the Electric Double Layer

Minton, G.,
Lue, L., University of Strathclyde
Originally presented on: 11/5/2013 14:15:00 - 14:30:00

Theories for the dynamics of charged colloids in electrolyte solutions under applied electric fields are, for the most part, still based on using the Poisson-Boltzmann theory to describe the electric double layer. This theory is known to neglect several import, including the neglect of ion size, ion polarizability, and charge correlations. These shortcomings can lead to qualitatively incorrect predictions in the electrokinetic motion of charged particles.

In this work, we employ a simple model that includes both ion size and polarizability to examine their effect on the structure and response of the electric double layer. Including either finite ion size or a negative excess ion polarizability leads to a decrease in differential capacitance above a threshold potential. For the case of excluded volume interactions, this is due to the well studied ion crowding effect. In the case of ion polarizability, the decrease can be interpreted in terms of thickening of the double layer due to ion-induced polarizability holes in water. We identify a new length scale which controls the role of excess ion polarizability in the double layer, and show that when this is comparable to the size of the effective Debye screening length, ion polarizability can significantly influence the properties of the double layer [1].

We obtain analytical results for the surface charge and differential capacitance of the double layer as a function of the applied potential on a planar surface. The model is found to be able to quantitatively describe experimental data for the differential capacitance of various systems.

Finally, we discuss how ion correlations can be incorporated into the theory, based on a recently developed variational field theory to account for fluctuations in the electric potential [2].

[1] MM Hatlo, R van Roij, and L Lue, "The electric double layer at high surface potentials: The influence of excess ion polarizability," EPL 97, 28010 (2012).

[2] MM Hatlo and L Lue, "Electrostatic interactions of charged bodies from the weak- to the strong-coupling regime," EPL 89, 25002 (2010).

(303i) In-Situ Observation of Meniscus Tip Behavior During Stripe Pattern Formation in Vertical-Deposition Convective Self-Assembly

Watanabe, S., Kyoto University
Mino, Y., Kyoto University
Miyahara, M., Kyoto University
Originally presented on: 11/5/2013 14:30:00 - 14:45:00

Convective self-assembly is one of the attractive techniques providing one with simple, scalable, and economical routes to the fabrication of patterned structures of particles. In our previous study, we utilized the convective self-assembly technique and successfully fabricated striped colloidal arrays having well-ordered periodicity of the order of micrometers in their line width and spacing. Furthermore, conducting the second stripe formation after the rotation of the substrate with the stripes by 90°, we obtained grid network patterns of colloidal particles. However, the stripe formation mechanism has to be elucidated to further control the formation process. In the present study, we directly observe the stripe formation process in the convective self-assembly of colloidal particles to clarify the formation mechanism. A hydrophilic substrate is vertically immersed in a suspension and kept stationary at a constant temperature. As the solvent evaporates, dispersed particles are carried into the contact line by the convection which compensates for the solvent evaporation from drying region, and deposited on the substrate to form stripe patterned colloidal arrays. We monitor the stripe formation processes by using an optical microscope. The illuminating light is monochromatic with the wavelength of $l = 475 \pm 15$ nm. The light reflected on the meniscus and substrate surfaces interferes with each other, resulting in the interference fringe pattern near the meniscus edge. Calculating the thickness profile of the meniscus from the fringe images, we clarify the stripe formation mechanism as follows. Under a fairly low particle concentration condition, the meniscus is gradually stretched downward to be concave toward a substrate as the solvent evaporation progresses, because the rate of particulate film growth is slower than that of liquid level lowering by the solvent evaporation. The convective flow toward the meniscus tip accelerates the deformation of the meniscus, resulting in the rupture of liquid film at the thinnest point of the meniscus. The rupture propagates in lateral directions, and then the meniscus tip shifts downwards and sticks to a new position which is determined by the equilibrium contact angle.

(303j) On the Propulsion and Saturable Binding Kinetics of a Self-Diffusiophoretic Colloidal Motor

Knapp, E. A., The City College of New York
Maldarelli, C., The City College of New York
Originally presented on: 11/5/2013 14:45:00 - 15:00:00

Colloidal engines are self-propelled particles which find applications as roving sensors and shuttles, as drug delivery methods and in the assembly of nanostructures. We have designed a prototype colloidal engine driven by the bioconjugation of a protein receptor, NeutrAvidin, to its binding ligand, biotin, using diffusiophoresis as a chemo-mechanical transduction mechanism. In this prototype, one micron diameter polystyrene colloids which have been functionalized with Neutravidin on their surfaces are coated on one hemisphere with a gold layer to block the receptors. Immersion of this Janus colloid in a solution of biotin creates a concentration gradient of biotin across the particle. The biotin in solution binds to only one side of the colloid and thus the intermolecular attraction of the biotin with the colloid is unbalanced. This imbalance causes the particle to move over the time scale required for the receptors to saturate. These motors therefore have a clock for self-propulsion.

We report experiments on this colloidal motor which demonstrate that the colloids execute motion along trajectories that are randomized by Brownian rotation but with a velocity obtained by calculating the mean square displacement. The magnitude of the velocity and the duration of the motion are determined by the concentration of the biotin fuel. The larger the fuel concentration, the greater the velocity but the shorter the time scale of the motion. A model is constructed in which the intermolecular interaction of the biotin with the colloid is assumed to be short-ranged, attractive interaction between specifically adsorbed biotin on the colloid surface and the surface NeutrAvidin. The surface adsorbed biotin is obtained by solving the unsteady mass transport equation, and the force exerted on the colloid is obtained by balancing the force of the short range attraction with the hydrodynamic resistance to the colloid motion. The results are compared successfully with the measurements of the diffusiophoretic velocity.

Self propulsion has application in a variety of far reaching technologies. As a result, the ability to manipulate colloidal particles over small length scales is desirable to enhance fields such as material science, drug delivery, and point of care diagnostics. This work expands on earlier studies in self propulsion via the mechanism of self diffusiophoresis in which asymmetry of a particle creates localized concentration gradients of a solute fuel. Here, a biologically friendly binding interaction enables particle propulsion based on the binding of ligand fuel to a target protein on one hemisphere of the particle. Through these studies, particles exhibit directed motion over a finite timescale. Once the receptors on the particles surface saturate, these motor particles will return to the random Brownian rotation and translation as seen by ordinary colloids. Models of this saturable binding mechanism capture the finite time scale for propulsion. As the protein receptor sites saturate, velocity decays and eventually stops. Additionally, the impact of microfluidic confinement is investigated and the resulting drag force felt by propelling particles is assessed.

(309a) Self-Assembly of Pluronic Block Copolymers and Vesicle Formation in a Protic Ionic Liquid

Wagner, N. J., University of Delaware
Li, D., University of Delaware
L pez-Barr n, C. R., University of Delaware
Originally presented on: 11/5/2013 12:30:00 - 12:50:00

Pluronics are triblock copolymers formed by a polyoxypropylene block (PO) and two polyoxyethylene blocks (EO). They self-assemble into micelles when dissolved in protic solvents and these microstructures are excellent candidates for applications such as nano-carriers for drug delivery applications [1] and as templates for nano-structured materials [2-3]. Room temperature ionic liquids (RTIL) are salts with melting point (m.p.) below room temperature. Classified as green solvents due to their negligible vapor pressure, RTILs are good alternatives to volatile organic compounds in synthesis, catalysis, extraction and separation [4-5]. Of interest here is a fundamental, molecular-level understanding of the self-assembly of amphiphilic molecules, such as Pluronic polymers, in a RTIL. This can facilitate the rational design of desired structures and rheological characteristics for specific applications

Previous our studies of Pluronic polymers in a protic ionic liquid, ethyl ammonium nitrate (EAN) have elucidated self-assembled microstructures at intermediate concentrations, including spherical, wormlike micelles [6], gels and liquid crystals [7]. In this study, we systematically investigate the self-assembled structures in a model system, L121/EAN mixture at lower concentrations . These compositions are potential candidates for spontaneous formation of thermodynamically stable vesicles which are useful in drug delivery applications. A broad combination of experimental probes (i.e., ultra-small, small-angle neutron scattering, small-angle x-ray scattering, dynamic light scattering, rheology, conductivity, and fluorescence microscopy) are used to determine the structural, thermodynamic and rheological changes associated to the vesicular structures. The onset unilamellar vesicle formation is found upon increasing the concentration of Pluronic polymer L121 or increasing the temperature, defining a critical vesicle concentration (CVC) or temperature (CVT). Further increases in concentration or temperature leads to the continuous growth of multilamellar vesicles. Such unilamellar-multilamellar vesicle structural transition is shown to be thermoreversible by applying heating-cooling cycles on the sample, and in-situ SANS-temperature study. To date, this is the first experimental report on the spontaneous formation of thermoreversible vesicles in Pluronic-ionic liquid systems.

[1] Batrakova and Kavanov, J. Controlled Release 2008, 130, 98–106

[2] Hess et al., Chem. Mater. 2009, 21, 2125–2129

[3] Sakai and Alexandridis, Nanotechnology 16 (2005) S344–S35

[4] Evans et al., J.Coll. Int. Sci. 1982, 88, 89

[5] Greaves et al., Langmuir 2007, 23, 402

[6] Lopez-Barron, C.R., Li, D., DeRita L., Wagner, N., Macromolecules, in preparation

[7] Lopez-Barron, C.R., Wagner, N., et al., Phys. Rev. Lett. 2012, 108, 258301

(309b) Interaction of Ionic Liquids With Lipid Biomembrane: Implication From Supramolecular Assembly to Cytotoxicity

Jing, B., University of Notre Dame
Lan, N., University of Notre Dame
Zhu, Y. E., University of Notre Dame
Originally presented on: 11/5/2013 12:50:00 - 13:10:00

An explosion in the research activities using ionic liquids (ILs) as new “green” chemicals in several chemical and biomedical processes has resulted in the urgent need to understand their impact in term of their transport and toxicity towards aquatic organisms. Though a few experimental toxicology studies have reported that some ionic liquids are toxic with increased hydrophobicity of ILs while others are not, our understanding of the molecular level mechanism of IL toxicity remains poorly understood. In this talk, we will discuss our recent biophysical study of the interaction of ionic liquids with model cell membranes. We have found that the ILs could induce morphological change of lipid bilayers when a critical concentration is exceeded, leading to the swelling and tube-like formation of lipid bilayers. The critical concentration shows a strong dependence on the length of hydrocarbon tails and hydrophobic counterions. By SAXS and fluorescence microscopic measurement combined with MD computer simulation, we have confirmed that tube-like lipid complexes result from the insertion of ILs with long hydrocarbon chains to minimize the hydrophobic interaction with aqueous media. This finding could give insight to the modification and adoption of ILs for the engineering of micro-organisms.

(309c) Ionic Liquid-Based Supercapacitors: Insight From Molecular Simulations

Bedrov, D., University of Utah
Vatamanu, J.,
Originally presented on: 11/5/2013 13:10:00 - 13:30:00

Electrical double layer capacitors (EDLCs), or supercapacitors, continue to gain an increasing attention as promising electrical energy storage devices that can be complementary to batteries, particularly when high power density and fast power delivery/uptake and long cycling life are required. The EDLCs store energy via the accumulation of ions in a nanometer-thick layer at electrochemically-stable high specific surface area active electrodes. A notable improvement in EDLC performance has been achieved due to recent advances in understanding charge storage mechanisms, development of advanced nanostructured electrodes and electrochemically stable electrolytes such as room temperature ionic liquids. Despite the extensive experimental work on these systems, the specifics of what happens inside and on the surfaces of most promising nanostructured electrodes are still not well understood. In this presentation, we will show how molecular dynamics simulations using state-of-the-art methodologies offer crucial insight into understanding the molecular level phenomena that influence supercapacitor performance and provide guidance for the design of novel materials. Specifically, we systematically investigate performance of various ionic liquids in nanoconfined geometries and on various nanoparticle surfaces. Our simulations show that depending on complex interplay of anion/cation asymmetry in shape, size, details of charge distribution, and specifics of ion-surface interactions, the capacitance on positive and negative electrodes can be very asymmetric. We discuss which structural characteristics of electrode (nanopore size, surface roughness and curvature) are beneficial for the capacitance enhancement as well as examine mechanisms of ion mobility in the nanoconfined geometries during charging/discharging processes. Using the obtained molecular level insight we discuss how to optimize the design and selection of electrolyte/electrode combinations to achieve increased capacitance in ELDC.

(309d) Thickness-Dependent Molecular Arrangement and Topography of Ultrathin Ionic Liquid Films On a Silica Substrate

Li, L.,
Gong, X.,
Frankert, S.,
Wang, Y.,
Originally presented on: 11/5/2013 13:30:00 - 13:50:00

Thickness-dependent molecular arrangement and topography of ultrathin ionic liquid films on a silica substrate

Xiao Gong, Scott Frankert, Yongjin Wang and Lei Li

Department of Chemical and Petroleum Engineering, University of Pittsburgh

Pittsburgh, PA 15261

Room temperature ionic liquids (RTILs), due to their extremely low volatility, non-flammability and other remarkable physicochemical properties, have become a very promising candidate as next-generation green chemicals. Understanding the molecular arrangement and topography of RTILs at RTIL-solid interfaces is critical to many important applications, including catalysis, lubrication, electrochemistry and photovoltaic power generation. To gain this understanding, an effective approach is to investigate the solid-supported ultrathin RTIL films with the thickness around one monolayer or below, where the otherwise buried interface can be more easily characterized. In the current presentation, we report our experimental study of the molecular arrangement and topography on a silica-supported ultrathin film of ionic liquid, 1,2-dimethyl-3-propylimidazolium bis (trifluoromethylsulfonyl) imide (DMPIIm). Angle-resolved X-ray photoelectron spectroscopy (ARXPS) results indicated a transition from an "anion/cation" layered structure to a "drop-on-layer" structure when the DMPIIm film gets thicker. AFM results agreed to the ARXPS results and showed a transition from a smooth to a "rough" surface with the increase of DMPIIm thickness. The observed phenomena have been attributed to the competition between solid/IL attraction and anion-cation cohesion.

(309e) Ionic Liquids Behave As Dilute Electrolyte Solutions

Gebbie, M. A., University of California, Santa Barbara
Valtiner, M., Max-Planck-Institut fÃ¼r Eisenforschung GmbH
Banquy, X., University of California, Santa Barbara
Fox, E. T., North Carolina State University
Henderson, W. A., North Carolina State University
Israelachvili, J. N., University of California, Santa Barbara
Originally presented on: 11/5/2013 13:50:00 - 14:10:00

Ionic liquids are technologically-important materials that have been demonstrated for numerous applications, including safe, high-efficiency electrochemical storage devices, self-assembly media, and lubrication. Since these liquids are composed solely of electrically-charged ions, ionic liquids are typically expected to exhibit electrostatic screening properties that should be comparable to highly-concentrated electrolyte solutions (1, 2). Here, we combine direct surface force measurements, performed across the common ionic liquid [C_4mim][NTf_2], with simple thermodynamic arguments to demonstrate that ionic liquids instead behave as dilute weak electrolyte solutions, with typical effective dissociated ion concentrations of less than 0.1% of the maximum ion density at room temperature. Our results clear up several scientific controversies and paradoxes. We also provide a general, molecular-scale framework to guide the design of novel high free ion density ionic liquids.

1. Weingärtner Hermann (2008) Understanding ionic liquids at the molecular level: facts, problems, and controversies. *Angew. Chem. Int. Ed.* 47:654-670.

2. Kornyshev Alexei A (2007) Double-layer in ionic liquids: paradigm change? *J. Phys. Chem. B* 111:5545-5557.

(309f) Ionic Liquid Electrolytes Confined Inside Model Nanoporous Carbons: A Molecular Simulation Study

Rajput, N. N., Louisiana State University
Monk, J. D., Louisiana State University
Hung, F. R., Louisiana State University
Originally presented on: 11/5/2013 14:10:00 - 14:30:00

A fundamental understanding of the properties of ionic liquids (ILs) inside nanopores is essential to enhance the performance of energy storage devices such as electrochemical double layer capacitors (EDLCs). Molecular dynamics simulations were performed to study the properties of the IL [emim+][NTf2-], as well as its mixtures with an organic solvent, acetonitrile, when confined inside two model materials, an ideal slit graphitic nanopore and a model nanoporous carbon with roughly rectangular pores that are interconnected. The structure and dynamics of the IL electrolyte vary strongly with pore size, amount of electrolyte inside the nanopores, density of electrical charges in the pore walls, and concentration of IL/acetonitrile. The dynamics of the ions inside the pore are highly heterogeneous, and depend strongly on their position with respect to the pore walls, and on variables such as surface charge density and concentration of IL/acetonitrile. The structural properties of ILs inside nanoporous electrodes affect the macroscopic capacitance in EDLCs, and the dynamical properties of the confined IL is one of the factors that determines the macroscopic electrical resistance in electrochemical devices.

(309g) Dynamic and Structural Properties of Room Temperature Ionic Liquids Near Silica and Carbon Surfaces

Li, S., Vanderbilt University
Feng, G., Vanderbilt University
Han, K. S., Oak Ridge National Lab
Hagaman, E., Oak Ridge National Lab
Vlcek, L., Oak Ridge National Laboratory
Cummings, P. T., Oak Ridge National Laboratory
Originally presented on: 11/5/2013 14:30:00 - 14:45:00

The dynamics and structural properties of a room temperature ionic liquid (RTIL) 1-Butyl-3-Methyl-Imidazolium(trifluoromethanesulfonimide) ([C_4mim][Tf_2N]) confined in silica and carbon mesopores were investigated by molecular dynamics (MD) simulations and nuclear magnetic resonance (NMR) experiments. Different interfacial microstructures of confined [C_4mim][Tf_2N] were observed and the temperature-dependent diffusion coefficients of [C_4mim][Tf_2N] confined in silica or carbon mesopore exhibit divergent behavior. The effects of loading fraction on the magnitude of the diffusion coefficient in the silica pore are investigated and is found to display weaker temperature dependence as loading fraction decreases. The diffusion coefficients of mesoporous carbon-confined [C_4mim][Tf_2N] are relatively insensitive to the loading faction and exhibit temperature dependence that is similar to the bulk temperature dependence at all loading levels. The unique surface heterogeneity, dissimilar interfacial microstructures, and interaction potential profile of RTILs near silica and carbon walls are shown to govern the differences exhibited in RTILs confined in mesoporous silica and carbon.

(309h) Electrical Double Layers in Double-Cationic Ionic Liquids

Li, S., Vanderbilt University
Feng, G., Vanderbilt University
Cummings, P. T., Oak Ridge National Laboratory
Originally presented on: 11/5/2013 14:45:00 - 15:00:00

As a novel class of energy storage devices, electrical double layer capacitors (EDLCs), also called supercapacitors, have attracted tremendous attention in recent years as a promising type of electric energy storage (EES) device, due to their advantageous properties, such as high power density, high capacitance, and excellent durability. In view of their exceptionally wide electrochemical stability window, excellent thermal stability, non-volatility and relatively inert nature, room-temperature ionic liquids (RTILs) have emerged as a very advanced class of electrolytes for supercapacitors. In contrast to the tremendous amount of work on single-cationic ionic liquids, few efforts have been devoted to double-cationic ionic liquids used as electrolytes in supercapacitors. We performed molecular dynamics (MD) simulations to investigate the structure and capacitive behaviour of the electrical double layers (EDLs) in double-cationic ionic liquids with varying alkyl linkage chain lengths near planar carbon surfaces, for comparison with that for single-cationic ionic liquids. Different capacitance-potential relations are observed for single- and double-cationic RTILs in supercapacitor applications.

(311b) Systematic Molecular Simulation Investigation of the Influence of the Chemical Structure of Anion and Cation to the Properties of Imidazolium-Based Ionic Liquids and to the Thermodynamic Properties of Light Gases Dissolved Therein

Economou, I. G., National Center for Scientific Research Demokritos
Androulaki, E., National Center for Scientific Research Demokritos
Vergadou, N.,
Originally presented on: 11/5/2013 12:50:00 - 13:10:00

This work focuses on the computational study of the physical properties of pure imidazolium-based tricyanomethanide Ionic Liquids (ILs), [$C_n mim^+$][TCM^-], and bis-trifluoromethylosulphonyl ILs, [$C_n mim^+$][Tf_2N^-], and on the prediction of solubility and diffusivity of light gases (e.g. Ar, CO_2, N_2) therein over a wide temperature range at atmospheric pressure. Molecular simulation plays a vital role in designing ILs with controlled permeability properties for use in a number of industrial processes, such as CO_2 capture, by revealing a wealth of microscopic information for the underlying mechanisms that are responsible for materials' macroscopic properties.

Here, long molecular dynamics (MD) simulations were performed in a wide temperature range and at atmospheric pressure in order to predict the thermodynamic, structural and dynamical properties of the pure ILs. A non-polarizable atomistic force-field was optimized in order to accurately predict the density and the self-diffusion coefficient. The dynamical heterogeneity exhibited by the ILs was investigated through the calculation of the non-Gaussian parameter and the deviation of the self-part of the van-Hove distribution function from the expected normal distribution. Subsets of "fast" and "slow" ions form individual clusters that consist of either mobile or immobile ions in the bulk. The structural relaxation process was analyzed by the self-part of the intermediate scattering function which exhibits strong wave number and temperature dependence. The distinct behavior of anions and cations at various length scales was thoroughly investigated as a function of the alkyl tail length. Detailed analysis of the ions' diffusion reveals preferential motion along the direction of the alkyl tail for the cation and the S-S vector for the [Tf_2N^-] anion. As the alkyl tail length increases, the heterogeneity in the dynamics becomes more pronounced and is preserved for several nanoseconds at low temperatures.

Simulation of light gases (e.g. Ar, CO_2, N_2) in the ILs at infinite dilution was performed in order to calculate the solubility and diffusivity of the gas molecules in the ILs. The solubility was estimated based on the Widom test particle insertion technique. In all cases, the agreement between experimental data and molecular simulation is excellent. The temperature effect was thoroughly studied for all the properties of the pure ILs and the dynamic properties of light gases dissolved in the above mentioned systems. Moreover, the influence of the cation's alkyl chain length and the choice of the anion were investigated in order to study the underlying molecular mechanism that governs the macroscopic behavior of these systems. Diffusivity was estimated through the mean square displacement of the ions.

In conclusion, strong spatial organization with an indication of alkyl tail aggregration phenomena was determined. This organization leads to an inhomogeneous bulk structure and is connected to heterogeneity in the dynamics of these systems which has been also reported by previous studies of similar systems. This inhomogeneity affects the macroscopic physical properties of the system.

(311c) Distinctive Nanoscale Organization of Double- Cationic Versus Single-Cationic Ionic Liquids

Li, S., Vanderbilt University
Feng, G., Vanderbilt University
Banuelos, J. L., Oak Ridge National Lab
Rother, G., Oak Ridge National Lab
Fulvio, P. F., Oak Ridge National Lab
Dai, S., Oak Ridge National Laboratory
Cummings, P. T., Oak Ridge National Laboratory
Originally presented on: 11/5/2013 13:10:00 - 13:30:00

Double-cationic ionic liquids are attracting the increasing research interests recently. In this work, the distinctive structural organization of double-cationic ionic liquids (DILs) with varying alkyl linkage chain lengths is systematically investigated using classical molecular dynamics (MD) simulations. In comparison with their single-cationic ionic liquids (SILs) counterparts, DILs with varying length alkyl chains exhibit structural features that are different from SILs regardless of anion types. SILs are found to exhibit various structural heterogeneities that are more pronounced than DILs. Moreover, the predominant role of anion type in the structure of DILs was verified be analogous to that in MILs. In the end, the different nanoscale organizations of DILs and MILs are rationalized by straight- and folded-chain models proposed for the nanoaggregates in DILs and those in MILs. This study provides molecular insights into the nature of structural heterogeneity in DILs as well as a theoretical model for the interpretation of experimental observations.

(311d) Aggregation in Ionic Liquids: Molecular Dynamics Simulation

Liu, X., Institute of Process Engineering, Chinese Academy of Sciences
Zhang, X., Institute of Process Engineering, Chinese Academy of Sciences
Zhang, S., Institute of Process Engineering, Chinese Academy of Sciences

Originally presented on: 11/5/2013 13:30:00 - 13:50:00

The spatial microheterogeneity in ionic liquids (ILs) is one of the the key features and has been proved useful in understanding the special behavior and interpreting many Physical phenomena of ILs, including the heterogeneous self diffusion, surface layering, as well as their surfactant-like micelles formed in IL/water mixtures.[1,2] As a result, much attention has been paid to the related researches, including both experiment and simulation. [3,4]

The self-aggregation or segregation exists both in IL solution and neat IL composed entirely of ions. In this work, the self-aggregation in ILs composed by $[C_n mim]^+$ (n=2, 4, 6, 8 , and 10) and Br^- were investigated by molecular dynamics simulation at the molecular level based on the all-atom force field. Both the neat ILs and the mixture of $[C_{12}mim]Br$ with water were studied. In order to reduce the computational time, united atom force field was utilized for the mixture of $[C_{12}mim]Br$ with water. Center of mass and site-to-site radial distribution functions (RDFs) were employed to study the ionic pair interactions. To depicture more visual structures of these ILs, the spatial distribution functions (SDFs) that stand for the three-dimensional probability distributions of ions and molecule were studied. A way based on the atomicity density was developed for estimating and understanding the inhomogeneous structure at nanometer scale. Dynamics behaviour and diffusion of ions, molecules and clusters were characterize through the mean-square displacement and self diffusion fuctions.

References:

(1) Wang, Y.; Jiang, W.; Yan, T.; Voth, G. A. *Acc. Chem. Res.* **2007**, *40*, 1193-1199.

(2) Dupont, J. *Acc. Chem. Res.* **2011**, *44*, 1223-1231.

(3) Chen, S. M.; Kobayashi, K.; Kitaura, R.; Miyata, Y.; Shinohara., H. *ACS Nano* **2011**, *5*, 4902-4908.

(4) Liu, X.; Zhou, G.; Zhang, S.; Wu, G. *Mol. Sim.* **2010**, *36*, 1123-1130.

(311e) Modeling Chemical and Physical Interactions in CO2/Ionic Liquids With the Reaxff Force Field

Zhang, B., US DOE-National Energy Technology Laboratory
van Duin, A. C. T., Pennsylvania State University
Johnson, J. K., University of Pittsburgh
Originally presented on: 11/5/2013 13:50:00 - 14:10:00

Ionic liquids (ILs) are promising sorbents for CO_2 due to their essentially zero vapor pressure and the potential to tune the CO_2-IL interaction. Both experimental and computational studies of CO_2/IL systems have been carried out. The ILs can interact with CO_2 through one or more of the following: physical absorption, complexation, chemical reaction. It is a challenge for a computational model to be able to accurately account for all three of these types of interactions. Physical interactions can be described with classical potentials very well, but cannot account for the bond formation and breaking events associated with chemical interactions between CO_2 and ILs. Treatment of complexation and chemical reactions usually requires high-level quantum chemical methods that are computationally expensive. An alternative approach is to use a reactive force field that can capture both physical interactions and chemical reactions. We have developed a new parameterization of the ReaxFF reactive force field for tetrabutylphosphonium glycinate ([$P(C_4)_4$][Gly]) and CO_2. We developed an extensive training set for both the pure [$P(C_4)_4$][Gly] system and the [$P(C_4)_4$][Gly]/CO_2 mixture from periodic and gas phase density functional theory (DFT) calculations. We have validated the optimized ReaxFF force field by comparing the energies from DFT and ReaxFF calculations for the structures in the training sets. We also tested the predictability of the optimized force field by comparing the same configurations of ILs with CO_2 not in the training sets. The trends of relative stability of different configurations based on energies from DFT and ReaxFF methods are in good agreement. The optimized ReaxFF force field has been used in large-scale molecular dynamics (MD) simulation. The calculated density is in very good agreement with experiment measured values at 300 K at ambient pressure. The radial distribution function and transport properties (MSD, diffusivity) of pure IL and IL with CO_2 mixture have been calculated and compared with those from classical potential. The reaction between CO_2 and IL has also been monitored and compared with first principles MD results.

(311f) Prediction of Thermodynamic Properties of Ionic Liquids: Improved Performance From Ab Initio Molecular Dynamics

Zhang, Y., University of Notre Dame
Maginn, E. J., University of Notre Dame
Originally presented on: 11/5/2013 14:10:00 - 14:30:00

Classical molecular dynamics and Monte Carlo simulations have been used extensively over the past ten years to predict the thermodynamic and transport properties of ionic liquids. As more experimental data for these systems becomes available, it is becoming clear that the quality of the force field is essential for making accurate property predictions. In this talk, we show how relatively short ab initio molecular dynamics (AIMD) simulations can be used to improve the accuracies in the prediction of the melting points and electrochemical windows of ionic liquids. On one hand, AIMD was used to better parameterize classical force fields, which can then be used to conduct traditional classical simulations on large systems for long time scales. We show that properties such as melting point can vary tremendously depending on both the magnitude and distribution of partial charges. On the other hand, AIMD was directly combined with classical MD simulations to provide better description of the potential energy surface. It was found that a quick relaxation of snapshots from classical MD simulation using AIMD can improve the calculated electrochemical window significantly.

(311g) Molecular Thermodynamic Characterization of Ionic Liquids for Separation Processes

Vega, L. F., MATGAS Research Center and Carburos Met licos, Air Products Group
Llovell, F., MATGAS Research Center
Oliveira, M. B.,
Vilaseca, O., MATGAS Research Center and ICMAB-CSIC
Marcos, R. M., Universitat Rovira i Virgili (URV)
Coutinho, J. A. P., Universidade de Aveiro
Originally presented on: 11/5/2013 14:30:00 - 14:50:00

Ionic Liquids (ILs) have been widely studied in the last ten years because of their potential applications from a technological point of view as well as their highly non-ideal properties from a fundamental perspective. The fact that they are liquid and have nearly zero vapor pressure values at room temperature, as well as their capacity to modify their adsorption capabilities by modifying the anion-cation pair, convert them in good candidates for gas absorption separation and several other industrial processes. However, the process of thermodynamic characterization, needed before put into a final application, is very expensive and time-consuming in terms of experimental work. Hence, different equations of state have been applied to the estimation of the thermodynamic properties of these ionic liquids, with different degrees of success. The complexity of these molecules has reduced the ability to reproduce some thermophysical properties in good agreement with the measured data, and most of the work done up to now with success has focused only in the description of the temperature-density diagram and the solubility of a particular gas on the ionic liquid.

One of the challenges in these theoretical approaches is to make them as general as possible, reproducing several features of ionic liquids and, in particular, their interfacial and derivative properties, as well as mixture data, with the same degree of accuracy as temperature-density data, instead of focusing on a particular property.

We will present and discuss the work our team has been doing in this field in the last years. This work is intended to fill this gap by proposing and validating a general and simple molecular model with the soft-SAFT equation of state [1] giving physical insight into the development of the model and the accuracy of the parameters for predictive purposes. This is accomplished by including different calculations of thermophysical properties of ionic liquid families with two different cations (imidazolium and pyridinium) and several anions (BF_4, PF_6, Tf_2N, $MeSO_4$, etc.). In all cases, simple but reliable molecular models are presented based on structural information and, also, in results obtained from molecular simulations and ab-initio calculations. As a first step the molecular parameters are fitted to single-phase equilibrium data and used in a transferable manner to predict other properties, such as high pressure density data, enthalpy of vaporization, critical points, interfacial tensions [2] and second order thermodynamic derivative properties. Correlations with the molecular weight are identified. Some derivative properties (heat capacity, isothermal compressibility, etc.) are used to discriminate among sets of parameters. The second step involves the assessment of the predictive ability to reproduce the solubility of several pollutant gases (CO_2, SO_2, NH_3, H_2S) in those ILs, as compared to experimental data [3-5]. In most of the cases, excellent agreement is found using a single temperature-independent binary parameter. The transferability of the binary parameters is also assessed. An estimation of the energetic properties is given in order to determine the effect of enthalpic and entropic effects on the solution. Finally, aqueous mixtures of ionic liquids and their mutual solubilities (liquid-liquid equilibrium) are reproduced with this approach [5-6].

Acknowledgements

F. Llovell acknowledges a TALENT fellowship from the Generalitat de Catalunya. M.B. Oliveira acknowledges her Post-Doctoral grant (SFRH/BPD/71200/2010). This work has been partially financed by the Spanish government (projects CTQ2008-05370/PPQ and CENIT SOST-CO2, CEN2008-01027), the Catalan government (project 2009SGR-666), the Portuguese Government (project PTDC/EQU-FTT/102166/2008) and from Carburos Metálicos, Air Products Group. CICECO is being founded by FCT through project Pest-C/CTM/LA0011/2011.

References

[1] F.J. Blas, L.F Vega, Mol. Phys. 92 (1997) 135.

[2] J.D. van der Waals, Z. Phys. Chem. 13 (1894) 657.

[3] J.S. Andreu, L.F. Vega, J. Phys. Chem. B 112 (2008) 15398.

[4] F. Llovell, R. M. Marcos, N. Mac Dowell, L. F. Vega, J. Phys. Chem. B. 116 (2012), 7709.

[5] M. B. Oliveira, F. Llovell, J.A.P. Coutinho, L. F. Vega, J. Phys. Chem. B. 116 (2012) 9089.

[6] F. Llovell, E. Valente, O. Vilaseca, L.F. Vega, J. Phys. Chem. B 115 (2011) 4387.

(315a) Gas Dissolution in Microscale Segmented Gas-Liquid Flow

Abolhasani, M., University of Toronto
Kumacheva, E., FlowJEM Inc.
Guenther, A., University of Toronto
Originally presented on: 11/5/2013 12:30:00 - 12:45:00

We investigate the scalar transport of gas molecules from uniformly and equally spaced bubbles across the gas-liquid interface and the liquid plugs that separate them (segmented gas-liquid flow). The saturation time, i.e., the time that is required for a gas molecule to achieve a uniform concentration of the gas in the liquid is measured experimentally and validated by a two-dimensional numerical model.

We investigated the dependence of the saturation time while varying the Peclet number, *Pe*, between 2,500 and 30,000, utilizing an image-based control strategy in combination with an automated microfluidic platform. The experimental results confirmed the analytical prediction of the saturation time that scales with a quarter of the microchannel width as the characteristic length scale for the diffusive transport during gas dissolution. We found that at low capillary numbers ($Ca < 0.01$), the saturation time varies with *Pe* to the power -0.67. In a serpentine microchannel we assessed the effect of chaotic advection in reducing the saturation time and length, and in increasing the dissolution rate. We found that 5 microchannel turns increased the dissolution rate by 30% and decreased the saturation length by 17%. No additional enhancement was observed in configurations with a larger number of turns.

(315b) Investigations Of The Coalescence Phenomena In Liquid-Liquid Taylor-FLOW

Praefke, H., RWTH Aachen University
Liauw, M. A., RWTH Aachen University
Originally presented on: 11/5/2013 12:45:00 - 13:00:00

INVESTIGATIONS OF THE COALESCENCE PHENOMENA IN LIQUID-LIQUID TAYLOR-FLOW

The general advantages of micro reaction engineering like a high surface to volume ratio and a resulting efficient high heat- and mass transport can also be observed by using Taylor-Flow a certain flow regime which occurs by combining two immiscible fluids in a capillary. Taylor-Flow can be used for chemical operations like (reactive) extraction. The type of Taylor-Flow differs with the used fluids; there is gas-liquid, liquid-liquid or even liquid-liquid-solid Taylor-Flow. The adjustment of different slug lengths is well investigated in gas-liquid Taylor-Flow both as numerical simulation [1] or even experimentally concerning capillary diameter [2], mixer geometry [3], liquid viscosity [4] and many other parameters. Numerical work has already been done in liquid-liquid Taylor-Flow but capillary lengths in these investigations were always shorter than a few meters. The undesired phenomenon of coalescence which describes an elongation of slugs along the capillary normally occurs at higher capillary lengths. To keep the extraction performance on a high level this coalescence must be avoided.

The aim of the ongoing research is to find parameters which are related to the formation of slugs in Taylor-Flow and their coalescence. This information could be very useful in following extraction experiments.

The used model system of water and 1-octanol is a common system in chemistry to determine the lipophilic and hydrophilic properties of a substance. For the investigation on the stability in liquid-liquid Taylor-Flow in consideration of different parameters a modularly set-up was built. This set-up made changes of the mixer and the capillary easy. To determine the resulting slug length distribution several sensors were tested. One was produced by ultrasonic hot embossing at KEμ RWTH Aachen University [5] the others were commercially available. The commercial sensors work with detection of light transmission and the newly produced sensors measures the electric conductivity directly inside the capillary.

From the commercially available sensors a laser light beam sensor was found to be the most suitable. It detects slugs with a length of at least 0.5 mm. For the detection with the laser light beam it was necessary to dye the aqueous phase with methylene blue which causes no significant change in surface tension or density of the phase. This was not necessary for the sensor from KEμ RWTH due to the measurements via electric conductivity.

To get a deeper understanding of the slug generation in Taylor-Flow many test parameters were defined. The variation of the total volume flow, the volume flow ratio, the material, diameter and length of the capillary and the resulting pressure drop were analyzed first in the research. Following the expectations, the capillary diameter and the volume flow parameters had the strongest impact on the resulting flow pattern.

In flows with a wide slug length distribution coalescence could be observed which could be caused by the different velocities of the slugs with differing lengths. To support this hypothesis a new set-up which can produce a heterogeneous slug length distribution was built. The research is still ongoing.

As already seen in gas-liquid Taylor Flow [2] the resulting Taylor-Flow is much more stable in smaller capillary diameters. This leads to higher adjustable volume flows until the flow regime changes from Taylor-Flow to a non-consistent flow. The influence of the capillary material was negligible which could be explained by the quite even properties of the both materials, fluorinated ethylene propylene (FEP) and perfluoroalkoxy (PFA). The strongest impact on stability of Taylor-Flow was observed by varying the volume flow and the water/1-octanol ratio. A decrease in stability could be observed at high octanol to water values wherein the increase of the water flow had no strong influence beyond a critical value.

For further research the focus will be on the mixer, especially its geometry and the orientation of the outlet. Also the geometry of the capillary will be further investigated.

[1] R. Gupta et al., *J. Comput. Multiphase Flows* **2010**, *2*, 1.

[2] J.W. Coleman, S. Garimella, *Int. J. Heat Mass Transf.* **1999**, *42*, 2869.

[3] A. Leclerc et al., *Chem. Eng. J.* **2010**, *165*, 290.

[4] S. Waelchli, P. Rudolf von Rohr, *Int. J. Multiphase Flow* **2006**, *32*, 791.

[5] W.K. Schomburg et al., *Micromachines* **2011**, *2*, 157.

(315c) Confined Droplets At T-Junctions

Hoang, D. A., Delft University of Technology
van Steijn, V., Delft University of Technology
Portela, L. M., Delft University of Technology
Kleijn, C. R., Delft University of Technology
Kreutzer, M., Delft University of Technology
Originally presented on: 11/5/2013 13:00:00 - 13:15:00

We analyze droplet breakup in confined T-junction geometries. The aims of this talk are to elucidate the break-up mechanism, to use this understanding to predict whether a droplet will break up or drift away to one of the branches, and to describe behavior close to the critical transition from breaking to drifting.

Earlier work (Link et al., 2004) had pointed to a breakup mechanism akin to Rayleigh-Plateau instability, whereas Leshansky and coworkers found (2009, 2012) that experimental data could equally well be explained without any capillary instability using a two-dimensional steady-state analysis. We show that a capillary instability is crucial in the description of the final stages of collapse. This instability is compared to end-pinching and to Rayleigh-Plateau instability found in unconfined droplets. Only for the early stages of breakup, three-dimensional effects are not important. Predicting when the droplet breaks needs the full 3D analysis.

Whether a droplet breaks is an unresolved fundamental problem. All experimental data to date indicate that there is a certain droplet length, beyond which all droplets will always break, whatever the flow speed. We present careful experiments that show that this is not the case, with long droplets that do not break at sufficiently small flow speeds. It may take a long time to drift away to one side (> 30 minutes in some experiments), but drift away they do. This experimental phase diagram that spans many decades of the capillary number allows the further development of theory.

Many practical applications will want to operate with rather short droplets at rather slow speeds, i.e. close the the critical line that marks the transition from drifting to breaking. We analyze how close to this transition, the timescale for drifting and the timescale for breakup become comparable, leading to a good description of the asymmetric breakup that occurs under these conditions.

This work provides insight into the behavior of confined droplets and predicts traffic of droplets in microfluidic networks for lab-on-a-chip applications.

D.R. Link et al, PRL 2004

Leshansky and Pismen, Phys. Fluids, 2009

Leshansky and Pismen, PRL, 2012

Hoang et al, Journal of Fluid Mechanics, 2013

(315d) Microfluidic Generation of Pickering Drops With Independently Controlled Drop Volume and Particle Surface Concentration

Shi, X., Texas Tech University
Christopher, G. F., Texas Tech University
Originally presented on: 11/5/2013 13:15:00 - 13:30:00

Particle stabilized emulsions, known as Pickering Emulsions, are found in food, energy applications, cosmetics, and pharmaceuticals. Pickering emulsions' stability occurs due to mechanical protection provided by and the enhanced interfacial rheology of individual drop's particle laden interface. Interfacial rheology is a strong function of particle surface concentration; it also affects drop deformation, which impacts bulk emulsion rheology. However, due to the difficulty in creating Pickering drops with controlled surface concentration, the relationship between surface concentration, interfacial rheology and droplet deformation has not been explicitly characterized.

We present a new microfluidic method to create Pickering drops with controlled particle surface concentration, which are then used to examine the role of surface concentration on deformation. A microfluidic co-flow geometry is used to generate drops; the design is modified to introduce an inner dispersed phase flow composed of a particle dispersion sheathing flow which surrounds an inner aqueous flow. Due to the narrow width of the sheathing flow, particles are trapped in a region close to the interface during breakup, which ensures fast diffusion of all particles to the interface. The Pickering drop surface concentration is a function of droplet volume and particle number within a drop. Droplet volume is controlled by the Capillary number defined by an outer oil phase flow rate, and a Weber number defined by the total dispersed phase flow rate. Particle number is controlled by bulk particle concentration in the sheathing flow and the flow rate ratio between the sheathing flow and the inner aqueous phase. By adjusting Capillary number, Weber number, and inner phase flow rate ratio, we can independently control both particle surface concentration and drop volume in real-time. By using mass conservation, particle surface concentration can be predicted; it is measured experimentally by polymerizing Pickering drops and imaging them using confocal microscopy. Experimental results are compared to predicted surface concentration. The role of particle surface concentration on droplet breakup regimes is also examined. Using these drops, we present initial results on the effect of particle concentration on droplet deformation in uniaxial extension; microfluidics are particularly advantageous to this work due to their ability to allow tracking of multiple length and time scales associated with the particles, their organization, surface flow, and drop deformation.

We present a systematic study the generation of Pickering drops of controlled surface concentration and examine the effect of particle laden interfaces on generation and the deformation of these drops.

(315e) Generating Single Cell Samples As Droplets Using a DVD Pickup Based Optical-Tweezer in a Microfluidic Platform

Kasukurti, A., Colorado School of Mines
Marr, D. W. M., Colorado School Of Mines
Desai, S. A., National Institute of Allergy and Infectious Disease
Eggleton, C., University of Maryland Baltimore County
Originally presented on: 11/5/2013 13:30:00 - 13:45:00

Single cell detection, handling and isolation are vital for the generation of samples for single cell-based diagnostics and purifying rare cells such as stem cells, circulating tumor cells, and cells for genetic testing. Commercial instruments available for single-cell deposition into 96 well plates are often bulky and expensive. Micro-fluidic platforms have addressed these issues but have often had the limitation in sample delivery as the generated samples are often inaccessible or become contaminated as you try to extract them. There is a need for a microfluidic chip that can deliver the generated samples off the chip in the form of droplets that can be collected and used readily. We have developed a compact (scale of a handheld device) and inexpensive platform to generate single cell samples as individual droplets. Our device is powered by a gravity driven microfluidic chip coupled with an optical trapping DVD optical pickup. Our device exploits the laminar nature of flow in microfluidics to deterministically guide cells to separate exits with an optical trap. Directional vaults in the microfluidic channel roof move sorted cells away from slow streamlines close to the wall into faster streamlines at the channel center. This reduces cell residence time in the device and improves the throughput of the device manifold. Holes were drilled in the glass substrate for the fluid to leave the device. Selective hydrophobic treatment of the underside of glass slides causes samples to form droplets. Consistency of droplet generation is crucial for the performance of the continuous flow chip that delivers single-cell samples as droplets. We have designed a device where droplets are generated at regular intervals and have a tight distribution around 53.72 μl ± 0.07μl. This small deviation gives us the confidence that a particle sorted at a regular interval upstream will end up in a new droplet every time. However, falling droplets cause significant pressure fluctuations upstream. Compliance reservoirs were added upstream of the droplet outlet to absorb/dampen pressure waves from falling droplets and shield the sorting section. The compliance reservoirs dampen the large fluctuations effectively. We believe we have a compact sorting and delivery module that can be added downstream of many common detection mechanisms based on viscoelasticity, electrical impedance, size or florescence.

(315f) Bubble-Propulsion of Colloidal Dimers in Hydrogen Peroxide

Wang, S., Colorado School of Mines
Wu, N., Colorado School of Mines
Originally presented on: 11/5/2013 13:45:00 - 14:00:00

Platinum-coated Janus spheres have shown self-propelling behavior in hydrogen peroxide. However, they were typically produced by template-assisted method with simple symmetry. By utilizing the surface anisotropy of colloidal dimers, we have successfully coated one lobe with platinum nanoparticles via a scalable synthetic route. More interestingly, this type of dimers have shown much stronger propulsion than Janus spheres in hydrogen peroxide solutions due to bublle propulsion. The speed of dimers can be as high as ~53 μm/s (~13 body length/s). Both linear and circular movement were observed, depending on the position where bubbles are generated. For example, when the center of the bubble is along with the long axis of the dimer, it moves in a linear trajectory. Otherwise, circular movement is observed. The rotational direction (either clockwise or counter-clockwise) can also be predicted, based on the torque that a dimer experiences. Analyses from high speed camera images also reveal intriguing modes of bubble propulsion. Both pull-back and push-out have been observed. Our study is the first report for the bulk-synthesis of catalytic micro-motors. The rich propelling behavior reported here could help gain deeper understanding on the propulsion mechanisms of microparticles in solution, which are important for applications in mutli-tasking drug delivery agents and sensors.

(315g) Self-Organizing Microfluidic Crystals

Uspal, W. E., Massachusetts Institute of Technology
Eral, H. B., MIT
Doyle, P. S., Massachusetts Institute of Technology
Originally presented on: 11/5/2013 14:00:00 - 14:15:00

We consider how to design a microfluidic system in which suspended particles spontaneously order into flowing crystals when driven by external pressure. Via theory, numerics, and experiments, we find that particle-particle hydrodynamic interactions drive self-organization under suitable conditions of particle morphology and geometric confinement. Small clusters of asymmetric "tadpole" particles, strongly confined in one direction and weakly confined in another, spontaneously order in a direction perpendicular to the external flow, forming one dimensional lattices. Large suspensions of tadpoles exhibit strong density heterogeneities and form jammed aggregates. By rationally tailoring particle shape, we tame this aggregation and achieve formation of large two-dimensional crystals. Key aspects of the theoretical/numerical results are confirmed by Stop Flow Lithography experiments.

(315h) Coalescing Drops in Microfluidic Parking Networks: A Multifunctional Platform for Drop-Based Microfluidics

Bithi, S. S., Texas Tech University
Wang, W. S., Texas Tech University
Sun, M., Texas Tech University
Blawzdziewicz, J., Texas Tech University
Vanapalli, S. A., Texas Tech University
Originally presented on: 11/5/2013 14:15:00 - 14:30:00

Automated multiwell plates and pipetting systems have revolutionized modern biological analysis. Multiwell storage allows easy barcoding for identification and retrieval of samples, and pipettes can dispense and remove known volumes of fluids into wells, allowing operations such as dilutions and washing. However, it is difficult to downscale wellplates and pipettes to manipulate fluids at the sub-microliter scale.

Droplet-based microfluidics has the potential to bring a paradigm shift in liquid handling at the sub-microliter-scale because of the ability to compartmentalize reactions in oil-isolated drops of volumes as small as picoliters. In this study, we report that a simple switch from surfactant-covered to surfactant-free drops and exploiting the coalescence between moving drop(s) and stationary drop(s) in microfluidic parking network allows (i) robust storage of arrays of monodisperse drops at prescribed coordinates, (ii) arraying of drops with gradually varying volumes in a single step, (iii) dilutions producing mondisperse droplet arrays with fine gradation in reagent concentration from drop to drop, (iv) multiplexed storage of drops of different compositions, and (v) fluid removal from static drops while retaining particles and cells. These functionalities can be achieved over a wide system parameter space, lending flexibility to our approach. We will discuss the hydrodynamic mechanisms underlying these capabilities and our method's insensitivity to system parameters.

In summary, the functionalities delivered by our platform are akin to those provided by wellplates and pipettes, and they can be achieved using one or two pressure sources, simplifying significantly the world-to-chip interface. Thus, our simple platform is a promising inexpensive candidate for performing biological and biomedical assays with low sample/reagent volumes and parallelization capabilities.

(315i) Synthesis of Magnetic Glass Microcomponents Using the Stop-Flow Lithography and Sol-Gel Process

Hsieh, I. M., National Taiwan University
Hsieh, C. C., National Taiwan University
Originally presented on: 11/5/2013 14:30:00 - 14:45:00

We have synthesized magnetic glass microcomponents by using the stop-flow lithography and sol-gel process. Polymeric microparticles with complex shape were first photopolymerized in a PDMS microchannel by photolithography. The prepolymer solutions consisted of PEGDA as the monomer and the crosslinker, the photoinitiator, the precursor of silicon-dioxide (TEOS) and the bridging agents that maintain the affinity between PEGDA and TEOS molecules. The polymeric particles were then heated to generate glass microparticles. The heating step does not only condense TEOS to SiO_2 but also burn out the polymer and leave a whole glass structure. The resulting glass microparticles were found shrinking significantly from their original size but having much greater mechanical strength. By adding magnetic particles into the prepolymer solutions, we were also able to manufacture magnetic glass particles that can be controlled remotely by magnetic field. We have made magnetic glass microgears and explored the possibility of using them to serve as the power source of micromechanical machines.

(315j) Long-Range Interparticle Forces in Complex Particle Arrays Assembled By Inertial Microfluidic Flows

Oakey, J., University of Wyoming
Reece, A., University of Wyoming
McConnell, J. T., University of Wyoming
Originally presented on: 11/5/2013 14:45:00 - 15:00:00

Microfluidic inertial focusing provides the passive alignment of small particles without the need for external actuation or sheath fluid by predictably positioning particles to well-defined lateral and longitudinal locations. The benefits of inertial focusing have quickly enabled the development of miniaturized flow cytometers, size-selective sorting devices and other high-throughput particle screening tools. Rudimentary design of inertial focusing devices requires only knowledge of the fluid properties and particle to channel size ratio. Equilibrium behavior of inertially focused particles has been extensively characterized and the constitutive phenomena described by scaling relationships for straight channels of square and rectangular cross-section. At the limits of inertial focusing, however, interparticle repulsions give rise to complex particle ordering that, while interesting and potentially useful, dramatically diminishes the technique's effectiveness for high-throughput particle handling applications. We have identified three scaling regimes with increasing degrees of geometrical ordering between focused particles. To explore the limits of inertial focusing and identify the origins of these long range interparticle forces, we have explored both equilibrium focusing behavior and particle dynamics during focusing. Experimental results for highly concentrated particle solutions identify equilibrium thresholds to focusing that scale weakly with concentration and strongly with channel geometry. These results have in turn motivated experiments to quantify geometry-dependent inertial forces that drive lateral particle migration at the dilute particle limit. Balancing inertial forces with interparticle repulsive forces now provides a complete picture of pattern formation within inertially focused particles and enhances our understanding of the fundamental limits of inertial focusing for technological applications.

(324a) Invited Talk: Molecular Simulation of the Dynamics in Thin Polymer Films

Riggleman, R. A., University of Pennsylvania

Originally presented on: 11/5/2013 12:30:00 - 13:00:00

After more than 15 years of study since the original article by Keddie et al. demonstrating the effects of confinement on the glass transition temperature (Tg) in nanoscopic polymer films supported on a silicon substrate, there is not yet a consensus on the origins of the Tg shift. Understanding and controlling the effects of confinement on glass-forming polymers is essential to further development of photolithography and semiconductor manufacturing, as well as several emerging technologies that will depend on the properties of confined glasses, such as stable glasses, flexible displays, and responsive materials. A growing body of experimental literature exists that suggests that the dynamics near a free surface are not only enhanced but are fundamentally different in their nature compared to a bulk glass-forming material. However, despite the fact that the experimentally-relevant length scales can be easily captured by molecular simulation, here are few simulation studies examining the dynamics of glass-forming polymers in confined geometries compared to the extensive experimental work. In this talk, I will describe some of our recent efforts to understand how the dynamics of glass-forming polymers change near free surfaces. We have identified several cases where the magnitude of the confinement effect can depend strongly on the property under investigation, which can be due to changes in the molecular packing near an interface. In addition, we have shown how physical aging can vary depending on the property measured. Finally, I will describe some of our work trying to understand how the view of confinement effects found in model systems studied with molecular dynamics compares with recent experiments.

(324b) Fragility Plays a Key Role in Determining the Magnitude of the Tg-Confinement Effect in Polymer Films

Torkelson, J. M., Northwestern University
Evans, C. M., Northwestern University
Deng, H., Northwestern University
Originally presented on: 11/5/2013 13:00:00 - 13:20:00

Polymer films with one or more free surfaces and no substantial attractive substrate interaction report a reduced glass transition temperature, Tg, upon confinement by various thermodynamics or pseudo-thermodynamic experimental methods including the following: ellipsometry, dielectric relaxation spectroscopy, differential scanning calorimetry, fluorescence spectroscopy, Brillouin light scattering, x-ray reflectivity, and neutron reflectivity. Despite the many theories and models which seek to explain why Tg decreases upon confinement in supported films with no substantial substrate interaction, rarely has the question been considered regarding why the Tg-confinement effect is so dependent on polymer species.

To understand what aspect of polymer structure or dynamics contribute to the strength of Tg-confinement effects, we consider the well-documented reduced surface Tg. If the surface is a region where the requirements for cooperative segmental relaxation are reduced, and thus Tg is reduced, polymers which relax in more cooperatively should experience a greater free-surface effect. The ability of a polymer species to undergo cooperative relaxation is connected to the fragility (m), which is a parameter that reflects the structure and intermediate range order of a glass former. According to Angell (J. Non-Cryst. Solids, 1985, 73, 1-17) who first defined fragility and classified glasses as strong or fragile, "The temperature dependence of the average relaxation time as well as the detailed relaxation function seem to be closely connected with the nature of the intermediate range order." Materials with higher fragilities possess less intermediate range order while stronger glass formers are typically characterized by an open network structure. Strong glasses possess a nearly Arrhenius temperature dependence of alpha-relaxation times when heated above Tg while fragile glass formers exhibit a sharp, non-Arrhenius decrease in structural relaxation times. This non-Arrhenius behavior reflects the presence of cooperative behavior as fragile glasses require a larger number of neighboring units to participate in structural relaxation. Thus, fragile glass formers may experience a greater modification to Tg by the presence of a free surface in a film. We hypothesize that more fragile glass formers will experience larger Tg-confinement effect in the absence of substantial polymer-substrate interactions.

We combine some of our group's past, published research data as well as new experimental results of Tg-confinement effects in seven polymer film systems: polystyrene (PS), PS with 2 wt% dioctyl phthalate (DOP), PS with 4 wt% DOP, poly(4-methyl styrene) (P4MS), polycarbonate (PC), polysulfone (PSF), and poly(vinyl chloride) (PVC). All Tg measurements were performed by fluorescence so that issues associated with comparing results from different techniques are circumvented. We also present measurements of fragility using a single technique, a differential scanning calorimetry based approach, first outlined by Angell, and investigate the connection between fragility and confinement effects. We restrict discussion to single-layer films with no substantial polymer-substrate interaction. This geometry represents the simplest case of a confined polymer glass former and is an appropriate starting point for understanding the role of polymer structure and properties on the strength of confinement effects. For confined films of linear chains with a free surface and no substantial interactions, we observe a one-to-one correlation between higher fragility and larger Tg reductions upon confinement. In particular, the magnitude of the Tg-confinement effect increases in the following order: PS + 4 wt% DOP < PS + 2 wt% DOP < PS < P4MS ~ PC < PSF < PVC. Fragility values increase in the following order: PS + 4 wt% DOP < PS + 2 wt% DOP < PS < P4MS ~ PC < PSF < PVC.

The fact that the steepest reduction in Tg with confinement occurred with PVC, which is the most fragile known polymeric glass former, makes evident that the magnitude of the Tg-confinement is not correlated with increasing Tg, because PVC has the lowest Tg of any of the polymer systems tested. As an example of the extreme strength of its Tg-confinement effect, Tg is reduced from its bulk value by more than 40 K in a 47-nm-thick film. Polysulfone also exhibit very strong confinement effects, albeit weaker than PVC, with a 24-nm-thick films reporting a Tg reduced from bulk response by 53 K. At the opposite extreme, PS + 4 wt% DOP exhibits not Tg-confinement effect in a film of 13 to 14 nm thickness.

Our study demonstrates that fragility is a key variable which is a reflection of the local medium range order in a polymer glass which determines the susceptibility of a glass former to perturbations (in this case induced by confinement). Thus, a picture for Tg-confinement effects emerges in terms of the molecular level packing of the polymer chains which are in turn related to the fundamental vibration time scale of dynamics. Fragility can explain the magnitude of Tg-confinement effects in neat polymer as well as plasticized, antiplasticized, or solvent sorbed films provided there are no attractive interactions with the supporting substrate and the polymers are linear chains. Routes for further testing of the fragility hypothesis will be suggested and caution will be issued for cases where additional factors may influence the Tg-confinement effect.

(324c) Image Formation in Thin Films of Glassy Photoresists

Patil, A. A., University of Kentucky
Doxastakis, M., University of Houston
Stein, G. E., University of Houston
Originally presented on: 11/5/2013 13:20:00 - 13:40:00

Image formation in modern lithographic processes is based on the acid-catalyzed deprotection of glassy polymer films. It is well establish that slow acid diffusion controls the reaction kinetics, but models based on Fickian transport cannot describe experimental data. We studied the acid catalyzed deprotection of glassy poly(hydroxystyrene-co-tertbutyl acrylate) films using infrared absorbance spectroscopy, scanning electron microscopy, and stochastic simulations. Experimental data were interpreted with a model that explicitly accounts for acid transport, where heterogeneities at local length scales are introduced through a non-exponential distribution of waiting times between successive hopping events. Subdiffusive behavior predicts key attributes of the observed deprotection rates, such as fast reaction at short times, slow reaction at long times, and a non-linear dependence on acid loading. This transport model is remarkably consistent with other literature studies of probe diffusion in inert glasses, which suggests that the same underlying physics (i.e., dynamic heterogeneities) are responsible for the observed anomalous behavior in our reacting system. In our ongoing work, we highlight the complex behavior in photoresists by changing the size of acid-counterion pairs, incorporating plasticizers, and tuning the film thickness. These data can facilitate the development of predictive lithography models that reflect the properties of confined polymer glasses.

(324d) The Calorimetric Glass Transition and Structural Relaxation of Single Polystyrene Films

Koh, Y. P., Texas Tech
Gao, S., Texas Tech Univeristy
Simon, S. L., Texas Tech University
Originally presented on: 11/5/2013 13:40:00 - 14:00:00

The behavior of materials confined at the nanoscale has been of considerable interest over the past two decades and considerable controversy has arisen. In particular, calorimetric measurements on utlrathin polymeric films using fast scanning or nanocalorimetry have been interpretted as showing no depression in the glass transition temperature (T_g), whereas mulitple other techniques have generally shown decreases in T_g. We attempt to resolve this discrepancy by using a commercial rapid-scanning chip calorimeter to measure T_g for single polystyrene thin films as a function of cooling rate and film thickness. Films were prepared in two ways: spincast films placed on a layer of inert oil or grease and films directly spincast on the back of the calorimetric chip. Results for thick films on oil or grease and directly spicast show results consistent with those of a bulk sample measured by conventional DSC. On the other hand, the 47 nm thick film on oil and 71 nm thick films both on oil and on grease show a T_g depression which decreases with increasing cooling rate. For films directly spincast onto the sensor, the magnitude of the T_g depression is smaller and is only observed for our thinnest film, 16 nm thick. All of our data are within the range of the results reported in the literature on supported films and are consistent with prior findings that T_g decreases with decreasing film thickness and substrate surface energy. The differences in our data and those of prior researchers using nanocalorimetry are attributed to rate effects and differences in data analysis. Our results with respect to high temperature annealing also conclusively demonstrate that annealing results in film dewetting and thickening and is accompanied by a reversion of T_g to the bulk value. Thus, the T_g depression observed is neither due to degradation nor to plasticization effects. In addition to measurements of T_g as a function of cooling rate, the structural recovery of single thin films will be discussed.

(324e) Confinement Effects On the Glass Formation Behavior of Nanolayered Polymers

Lang, R., Ohio University
Mackura, M.,
Simmons, D. S., University of Akron
Originally presented on: 11/5/2013 14:00:00 - 14:20:00

Confinement to nanoscale dimensions produces large changes in the engineering properties of glass-forming materials. Until recently, the majority of studies on nanoconfinement effects on the glass transition have focused on freestanding films and films supported by rigid substrates. In contrast, many real-world materials exhibit internal nanoconfinement: they possess distinct polymeric domains of nanoscale dimension confined between mutual interfaces. Examples include block copolymers, multi-nanolayered polymers, polymer nanoblends, and bulk heterojunction organic photovoltaics. In this presentation, I will describe recent and ongoing work in our group employing computer simulations to characterize and understand nanoscale confinement effects in such a system – a nanolayered polymer.

Our results show that the glass formation properties of nanostructured polymers depend strongly on both the relative T_g of the nanoscale polymer domains and the degree of immisibility of the polymers. We also observe confinement-induced changes in the 'fragility' of glass formation, which quantifies the 'abruptness' with which the glass forms on cooling. Results indicate that these changes are neither a direct result of interfacial mixing nor of a change in free volume at the interface, but rather emerge from changes in interfacial mobility parallel and normal to the interface. These changes are hypothesized to be controlled by dynamic coupling across the interface, resistance of the interface to penetration by chains, and the inherent susceptibility of each polymer to dynamical perturbations as quantified by their fragilities of glass formation. These studies suggest a new basis for understanding and controlling glass formation behavior in nanostructured polymers via rational selection of polymer bulk and mixing properties.

(324f) Temperature-Step Dewetting of Ultrathin Polymer Films: A Rapid Means to Tg Determination

Wang, J., Texas Tech University
Torres Arellano, A.,
McKenna, G. B., Texas Tech University
Originally presented on: 11/5/2013 14:20:00 - 14:40:00

The determination of the glass transition temperature of polymer films as a function of film thickness can be tedious and time consuming because the general methods of measurement of the properties of ultrathin polymer films have required a new sample for every determination of Tg as a function of thickness h. Here we show a novel temperature-step rheological method that overcomes this difficulty. We build upon the methods of Bodiguel and Fretigny [1] for determination of the viscoelastic response of ultrathin polymer films to develop a temperature-step dewetting procedure that gives the Tg-h relationship for multiple values of h from a single measurement. In principle, one can make a single measurement that provides the same information as do 10 to 15 individual measurements with current methods. We will demonstrate the method and provide new results for polystyrene and polycarbonate dewetting from glycerol and from an ionic liquid. The results show that the Tg reduction of films floating on glycerol is greater than that for films floating on the [BMIM][OTf] ionic liquid, but that in both instances the reductions are less than observed for the same materials supported on SiO_2 rigid substrates. The results suggest that the T_g - thickness effects observed in ultrathin films are very subtley dependent on surface energy but not on the surface mobility or rigidity.

[1] H. Bodiguel and C. Fretigny, Phys. Rev. Lett., 97, 266105 (2006).

This work is supported by the J.R. Bradford Endowment at TTU and by the National Science Foundation under grants DMR-0804438 and DMR-1207070.

(324g) Effects of Nanoscale Confinement On the Physical Aging of Glassy Polymers

Shavit, A., University of Pennsylvania
Riggleman, R., University of Pennsylvania
Originally presented on: 11/5/2013 14:40:00 - 15:00:00

Despite nearly twenty years of active research, the effects of nanoscale confinement on the properties of glass forming polymers remain poorly understood. Furthermore, molecular simulations have so far only played a limited role in our understanding of these confinement effects, which are important for applications in membrane separation, semiconductor manufacturing, and fabrication of novel materials (i.e., stable glasses). We have used molecular dynamics simulations to investigate the effects of aging on bulk and free-standing thin-film glass-forming polymers. We demonstrate that in the vicinity of the bulk glass transition temperature, there are regimes where the free surface is in equilibrium while the center of the film exhibits bulk-like aging. We also show that two measures of dynamics can exhibit significantly different local aging. We compare our results with those published from recent experiments, and we provide a microscopic picture on the differences in physical aging in bulk and free-standing polymer films.

(334a) Evaluation of Thermodynamic Properties of Gas Mixtures Via the Virial Equation of State With Accurate Molecular Models

Yang, S., University at Buffalo, The State University of New York
Schultz, A. J., University at Buffalo, The State University of New York
Kofke, D. A., University at Buffalo, The State University of New York
Originally presented on: 11/5/2013 12:30:00 - 12:50:00

Experimental measurements of vapor pressures of heavy organic molecules can be performed via a gas saturation technique in a light carrier gas. The apparent vapor pressure based on an ideal-gas treatment of vapor mixture is found to depend on the carrier gas, indicating that treatment of non-ideality is needed to properly interpret the experimental data. Semiempirical methods can be applied, but a successful outcome is more likely if based on treatment that considers the specific molecular features of the component molecules. The virial equation of state (VEOS) for mixtures presents a convenient approach to develop such an approach.

The VEOS for mixtures is given as a power series in density and mole fraction of each component, with coefficients relating to the interactions of corresponding number of molecules of each species. The pure virial coefficient of each component can be calculated via evaluation of cluster integral(s) dependent on the intermolecular interactions among molecules of the same species, whereas the cross virial coefficients are given as cluster integral(s) involving the interactions among unlike molecules species.

In previous work [1] we explored the Lennard-Jones binary mixtures using Mayer sampling Monte Carlo method [2], computing virial coefficients and examining vapor-liquid critical behavior. In the present work, we focus on more realistic molecular models, which requires us to re-develop the virial series for mixtures of non-rigid molecules. For such systems, it is necessary to consider cluster integrals that are not present in conventional formulations of the virial series (applicable to rigid molecules), because molecular flexibility prevents cancellation of cluster integrals that is usually assumed in the conventional development [3,4].

We examine coefficients for nitrogen(N_2) and carbon dioxide(CO_2), which are sometimes used as carrier gases in experimental measurements. For the heavy organic, we examine n-alkanes as prototype molecules. Both the pure virial coefficients of every species and the cross virial coefficients of N_2/n-alkanes and CO_2/n-alkanes binaries are calculated up to the 5^{th}order. We examine the convergence behavior of the VEOS for these binaries, and compare the thermodynamic properties of with available simulation and experimental data.

[1] A. J. Schultz and D. A. Kofke, J. Chem. Phys. 130, 224104 (2009).

[2] J. K. Singh and D. A. Kofke, Phys. Rev. Lett. 92, 220601 (2004).

[3] S. Caracciolo, B. M. Mognetti, and A. Pelissetto, J. Chem. Phys. 125, 094903 (2006).

[4] K. R. S. Shaul, A. J. Schultz, and D. A. Kofke, J. Chem. Phys. 135, 124101 (2011)

(334b) Modeling of CO2 Scrubbing With Hyperbranched Polymers Solutions Using Lattice Cluster Theory

Walowski, C., TU Berlin
Langenbach, K., TU Berlin
Enders, S., TU Berlin
Originally presented on: 11/5/2013 12:50:00 - 13:10:00

Concerning the global warming effect which is caused by the increase of the environmental CO_2 concentration, large efforts have been made to optimize the CO_2 sequestration. Until now, CO_2 scrubbing using monoethanolamin solutions is the state of the art chemical adsorption process. However, this process has several disadvantages1, like high energy costs, solvent loss and salt formation. Solving the drawbacks of the classical amine systems, Rolker at al.[1] suggested the application of hyperbranched polymers to achieve a physical absorption with a high selectivity and capacity for CO_2. The polyamines offer very high CO_2 loading because of the presence of different kinds of amino groups per molecules[1]. CO_2 capture potential of hyperbranched polymers, like polyglycerol or polyethylene imine, were studied in the literature2[,3,4]. The hyperbranched polymers can by tailored by different functional terminal groups and the architecture of the polymer in terms of degree of branching or molecular weight. Until now, the modeling of the gas solubility was performed using the classical approaches, like UNIFAC-FV[1,4] or PC-SAFT[2,3]. Both methods are not able to take the architecture of the hyperbranched polymer into account. The purpose of this contribution is the investigation of the impact of the polymer architecture on the gas solubility in hyperbranched polymer solutions.

Recently, phase equilibria of mixtures containing hyperbranched polymers could be modeled using Lattice Cluster Theory (LCT), developed by Freed and coworkers5[,6], in a compressible (LCT-EOS)7[,8,9] or in a non-compressible1[0] fashion. In order to model the gas solubility the LCT-EOS is applied. This approach allows the systematic investigation of impact of the polymer architecture on the gas solubility. This is possible, using a cluster expansion of the partition function that incorporates a molecule's structure through a set of combinatorial numbers describing its united-atom architecture. For a given architecture of a hyperbranched polymer the combinatorial numbers involved in the LCT-EOS can be obtained by a graph theoretical approach[7]. The theoretical results were compared with experimental data taken from the literature[1,2].

Literature

[1] J. Rolker, M. Seiler, L. Mokrushina, W. Arlt, Potential of Branched Polymers in the Filed of Gas

Absorption: Experimental Gas Solubilities and Modeling. Ind. Eng. Chem. Res. 46 (2007) 6572-6583.

2 M.K. Kozlowska, B.F. Jürgens, C.S. Schacht, J. Gross, T.W. de Loos, Phase Behavior of Hyperbranched Polymer Systems: Experiments and Application of the Perturbed-Chain Polar SAFT Equation of State. J. Phys. Chem. B 113 (2009) 1022-1029.

3 C.S. Schacht, S. Bahramali, D. Wilms, H. Frey, J. Gross, T.W. de Loos, Phase Behavior of the system hyperbranched polyglycerol + methanol + carbon dioxide. Fluid Phase Equilibria 299 (2010) 252-8.

4 S. Stünkel, W. Martini, H. Arellano-Garcia, G. Wozny, Entwicklung eines optimalen CO_2-Abtrennungsprozesses. Chemie Ingenieur Technik *83 (*2011) 448 – 495.

5 J. Dudowicz, K.F. Freed, Effect of Monomer Structure and Compressibility on the Properties of Multicomponent Polymer Blends and Solutions: 1. Lattice Cluster Theory of Compressible Systems. Macromolecules 24 (1991) 5076-5095.

6 J. Dudowicz, M.S. Freed, K.F. Freed, Effect of Monomer Structure and Compressibility on the Properties of Multicomponent Polymer Blends and Solutions: 2. Application to Binary Blends. Macromolecules 24 (1991) 5096-5111.

7 K.O. Langenbach, S. Enders, Development of an Equation of State Based on Lattice Cluster Theory for Pure Components. Fluid Phase Equilibria 331 (2012) 58–79.

8 K. Langenbach, S. Enders, C. Browarzik, D. Browarzik, Calculation of the high pressure phase equilibrium in hyperbranched polymer systems with the Lattice-Cluster Theory. J. Chem. Thermodynamics 59 (2013) 107–113.

9 K. Langenbach, D. Browarzik, J. Sailer, S. Enders, New Formulation of the Lattice Cluster Theory Equation of State for Multi-Component Systems. in preparation.

10 T. Zeiner, D. Browarzik, S. Enders, Calculation of the liquid-liquid equilibrium of aqueous solutions of hyperbranched polymers. Fluid Phase Equilibria 286 (2009) 127-133.

(334c) A Predictive Modeling Framework for Structure I Gas Hydrates

Lucia, A., University of Rhode Island
Henley, H., University of RHode Island
Originally presented on: 11/5/2013 13:10:00 - 13:30:00

Gas clathrate hydrates are non-stoichiometric, crystalline compounds that form under ambient temperatures and high pressures. Modeling their physical and thermodynamic properties is of great interest to the scientific community, yet current methods such as van der Waals-Platteeuuw (vdW-P) theory rely heavily on empirical correlation. In addition, there is some controversy in the open literature regarding the vdWP approach; some studies (Chialvo et al., 2002) argue that it is flawed because it does not appropriately account for guest-guest interactions while other studies (Wierzchowski and Monson, 2007) suggest it is only long range forces that are important for accurate modeling.

In this presentation, a new modeling framework for gas hydrate systems is described. Our approach is based on the use of the recently developed multi-scale predictive Gibbs-Helmholtz Constrained (GHC) equation of state (Lucia et al., 2012). The key to the GHC EOS is the novel up-scaling equation that communicates information from the molecular to the bulk phase length scale, and, in the case of hydrates, allows the inclusion of all necessary guest-guest, guest-host, or host-host interactions and structural contributions. Results for GHC predicted gas hydrate densities and phase equilibria are compared to vdW-P theory and experimental data.

(334d) A Practical Molecular Thermodynamic Model for Asphaltene Aggregation

Chen, C. C., Aspen Technology, Inc.

Originally presented on: 11/5/2013 13:30:00 - 13:50:00

One of the most challenging properties of petroleum fluids is crude oil compatibility. Asphaltene molecules precipitate or aggregate when one blends "incompatible" crude oils (Shen, 2002). Such crude oil "incompatibility" often results in significant negative economic impact. Therefore it is highly desirable to advance theoretical understanding of crude oil compatibility and develop robust thermodynamic models to describe and predict crude oil compatibility.

Recent advances in molecular characterization of petroleum fluids make it possible to derive optimal chemical compositions of representative hydrocarbon constituent molecules for crude oils and petroleum fractions from correlating readily available petroleum assay data. From such chemical compositions one could then estimate and predict wide ranges of physical and chemical properties of petroleum fluids rigorously and consistently.

In this work we present a novel and practical molecular thermodynamic framework for modeling crude oil compatibility. Chemical compositions of representative hydrocarbon constituent molecules, including asphaltene molecules and their compositions, are first identified from readily available assay data for crude oils of interest. Aggregation of asphaltene molecules is described with the Yen-Mullins molecular model (Mullins, 2010). Liquid phase nonideality of molecular species in petroleum fluids is modeled with the NRTL-SAC conceptual segment activity coefficient model of Chen and Song (2004). Model parameters including asphaltene aggregation constant and conceptual segment numbers for asphaltene molecules are then derived from appropriate literature data.

Shen, E.Y., Petroleum Asphaltene-Properties, Characterization, and Issues, *Energy & Fuels*, **2002**, *16*, 74-82

Mullins, O.C., The Modified Yen Model, *Energy & Fuels*, **2010**, *24*, 2179-2207

Chen, C.-C. and Song, Y., Solubility Modeling with a Nonrandom Two-Liquid Segment Activity Coefficient Model, *Ind. Eng. Chem. Res.* **2004**, *43*, 8354-8362

(334e) Modeling Asphaltene Aggregation With a Single Compound

Bruere, B., The Petroleum Institute, Abu Dhabi
Subramanian, D., University of Maryland
Leys, J., KU Leuven
Peters, C. J., Petroleum Institute
Anisimov, M. A., University of Maryland
Originally presented on: 11/5/2013 13:50:00 - 14:10:00

Modeling Asphaltene Aggregation with a Single Compound

Bianca Breure[1], Deepa Subramanian[1,2], Jan Leys[2,3], Cor J. Peters[1], and Mikhail A. Anisimov[2,3]

[1]Chemical Engineering Department, The Petroleum Institute, Abu Dhabi, United Arab Emirates

[2]Institute for Physical Science and Technology, and Department of Chemical and Biomolecular Engineering, University of Maryland, College Park, Maryland 20742, United States

[3]Laboratorium voor Akoestiek en Thermische Fysica, Departement Natuurkunde en Sterrenkunde, KU Leuven, B-3001 Leuven, Belgium

ABSTRACT

Aggregation of a model asphaltene compound hexa-tert-butylhexa-perihexabenzocoronene (HTBHBC) in toluene/heptane mixtures was monitored in real time using dynamic light scattering. The observed aggregation behavior was compared to that of real asphaltenes. It was found that heptane induces aggregation of HTBHBC and that the growth rate of the aggregates depends upon the toluene/heptane ratio, closely resembling the aggregation kinetics of real asphaltenes. For the conditions studied in this work, only diffusion-limited aggregation was observed. We thus confirm that a single well-defined compound can be successfully used for modeling the process of asphaltene aggregation.

(334f) Experimental and Modeling Study of the Phase Behavior of (Synthetic Crude Oil + CO2)

Al Ghafri, S., Imperial College London
Maitland, G., Imperial College London
Trusler, J. P. M., Imperial College London
Originally presented on: 11/5/2013 14:10:00 - 14:30:00

Experimental and Modeling Study of the Phase Behavior of (Synthetic Crude Oil + CO_2)

*Saif Zahir Al Ghafri, Geoffrey C. Maitland, and J. P. Martin Trusler**

Qatar Carbonates and Carbon Storage Research Centre, Department of Chemical Engineering, Imperial College London, South Kensington Campus, London SW7 2AZ

* Corresponding author e-mail: m.trusler@imperial.ac.uk.

<>1. Introduction

A quantitative understanding of the phase behavior of CO_2-hydrocarbon mixtures at reservoir conditions is essential for the proper design, construction and operation of carbon capture and storage (CCS) and enhanced oil recovery (EOR) processes. In order to model these processes quantitatively, it is necessary to know the phase behavior and physical properties of the mixtures formed between CO_2 and reservoir fluids under the conditions prevailing in the reservoir and, for this purpose, compositional thermodynamic models are required. Unfortunately, the phase behavior and other thermophysical properties of petroleum fluids are very difficult to predict *a priori*. Accordingly, there is a great deal of interest in developing improved models and computational packages to predict the phase behavior and thermodynamic properties of such mixtures with the least inputs of experimental data.Â In pursuit of this objective, a wide variety of approaches have been developed, ranging from empirical and semi-empirical equation of state (EoS), such as cubic equations, to molecular-based models such as the Statistical Associating Fluid Theory (SAFT). Nevertheless, experimental data are still required (and are in fact vital) to tune the interaction parameters of such models, to help in developing predictive approaches, and to assess the predictive capabilities of all models. While equilibrium data for binary CO_2-hydrocarbon mixtures are plentiful, equilibrium data for ternary and multi-component CO_2-hydrocarbon, CO_2-brines and CO_2-hydrocarbon-brines mixtures are limited, especially for systems containing heavy hydrocarbons and/or hydrocarbons other than alkanes. Therefore providing new experimental data for multi-component mixtures at reservoir conditions is of great value.

<>2. Experimental Design and Measurements

In this work, a new experimental apparatus was designed and constructed to measure the phase equilibria of systems containing CO_2 and hydrocarbons at reservoir temperatures and pressures. The apparatus involved a high-pressure high-temperature variable-volume view cell, with wetted parts fabricated from Hastelloy C-276, driven by a computer-controlled servo motor system, and equipped with a sapphire window for visual observation. Two calibrated syringe pumps (Quizix, model Q5210) with a maximum service pressure of 70 MPa were used for quantitative fluid injection. A low-dead-volume pressure transducer (Model DF2, DJ Instruments Ltd) was used for the pressure measurements while the cell temperature was controlled by a heating jacket, and the temperature was measured by means of a Pt100 sensor. The phase behavior of the mixture in the cell was observed with the aid of a CCD camera operating with both front and rear illumination. The maximum operating pressure and temperature of the entire setup were 40 MPa and 473.15 K, respectively. The apparatus was validated by comparison with published isothermal vapor-liquid equilibrium data for the binary system (CO_2 + heptane).

In this work, we report experimental measurements of the phase behavior of mixtures of CO_2 with a synthetic crude oil. The composition of the synthetic oil was chosen to match the physical and chemical properties of a bottom-hole crude oil sample from a Qatari field. The ?dead' oil contained a total of 17 components, each of between 5 and 30 carbon atoms, including alkanes, branched-alkanes, cyclo-alkanes, and aromatics. Solution gas (0.81 methane + 0.13 ethane + 0.06 propane) was added to obtain live synthetic crudes with various gas-oil ratios (GORs).Â Phase equilibrium measurements are reported for the ?dead' oil and also for ?live' oils, with GORs of 58 and 200 [1], under the addition of CO_2. The measurements were carried out at temperatures of (298.15 K, 323.15 K, 373.15 K and 423.15 K) and at pressures up to 36 MPa and include vapor-liquid, liquid-liquid and vapor-liquid-liquid equilibrium conditions.

<>3. Modeling

Traditional EoS models contain large numbers of adjustable binary parameters which require for their determination not only a compositional analysis but also experimental phase equilibrium data on each binary sub-system. In this work, we tested the capabilities of a recently-developed predictive model [2] based on the Peng-Robinson 78 EoS [3]. In this approach, the binary parameters are obtained from a group contribution scheme. Careful attention was paid the critical constants and acentric factor of high molar-mass components for which these properties are subject to significant uncertainty. Since the mixture also involved several light substances with critical temperatures lower than the some or all experimental temperatures, we also investigated the use of different alpha functions in the PPR78 EoS that have been developed specifically for super-critical components. The results showed that the model can predict the vapor-liquid equilibria of CO_2-hydrocarbon mixtures with reasonable accuracy without the need to regress binary parameters against experimental data.

References and notes

1. The GOR is the ratio of the volumes of gas and liquid when the mixtures is flashed at standard conditions of T = 288.15 K and p =101.3 kPa.

2. Vitu, S.; Privat, R.; Jaubert, J.-N.; Mutelet, J. Supercrit. Fluids 45 (2008) 1-26

3. D.Y. Peng, D.B Robinson, Ind. Eng. Chem. Fundam. 15 (1976) 59-64.

Acknowledgment

We gratefully acknowledge the funding of QCCSRC provided jointly by Qatar Petroleum, Shell, and the

Qatar Science and Technology Park, and their permission to publish this research.

Keywords: Phase behavior; modeling; CO_2; hydrocarbon; PPR78; variable volume cell, SAFT, heptane, toluene.

(334g) Phase Behavior of PS/Tetramethyl Orthosilicate/CO2 Ternary System

Matsukawa, H., Tokyo University of Science
Hamada, T., Tokyo University of Science
Otake, K., Tokyo University of Science
Shono, A., Tokyo University of Science
Shimada, Y., Tokyo University of Science
Kobayashi, D., Tokyo University of Science
Yoda, S., National Institute of Advanced Industrial Science and Technology (AIST)
Furuya, T., National Institute of Advanced Industrial Science and Technology (AIST)
Takebayashi, Y., National Institute of Advanced Industrial Science and Technology (AIST)
Okawa, Y.,
Originally presented on: 11/5/2013 14:30:00 - 14:50:00

Phase Behavior of PS/Tetramethyl orthosilicate/CO_2 Ternary System

In recent years, for the effective energy usage, development of high performance heat insulator is the subject of research for engineers. Polymer foam that inside of the foam filled with silica aerogel is regarded as one of the promising candidate for high performance heat insulator. This composite material will be commercially produced with a supercritical carbon dioxide (CO_2) assisted extruding of the mixture of the polymer and tetraalkoxysilane. As the foaming pressure depend on the phase separation, phase behavior of the systems is very important. In the previous work, we examined Tetramethyl orthosilicate (TMOS)/CO_2 binary systems and poly(methyl methacrylate) (PMMA)/TMOS/CO_2 ternary systems.

In this work, phase diagrams of poly(styrene) (PS)/ TMOS/CO_2 ternary systems were measured. PS/TMOS system is mutual solubility system at atmospheric pressure. Phase equilibrium measurements were performed by visual observation of the Vapor-Liquid (VL) separation, Liquid-Liquid (LL) separation and Vapor-Liquid-Liquid (VLL) separation with a synthetic method at wide range of temperatures and Pressures. Experimental conditions for the ternary systems were at PS weight fraction from 10 wt % to 40 wt % and PS molecular weight from 35,000 to 250,000. The results of the measurements were compared with the results of the TMOS/CO_2 binary systems and PMMA/TMOS/CO_2 ternary systems.

The PS/TMOS/CO_2 ternary system showed two types of different phase behavior with the amount of the CO_2 mass fraction. When CO_2 mass fraction was small, it showed the behavior of the Vapor-Liquid (VL) separation, and the VL equilibrium line appeared. On the other hand, when CO_2 mass fraction was large, as CO_2 acts as a poor solvent for PS/TMOS, Liquid-Liquid (LL) phase separation occurred.

Consequently the VLL and LL equilibrium line appeared. As PS molecular weight was increased, the LL equilibrium line was shifted to the lower CO_2 mass fraction. It seems that because the affinity of the PS and CO_2 is reduced due to the increase in molecular weight, it became easier to LL phase separation. Effect of molecular weight was not observed in the case of the VL and VLL equilibrium line.

Further, with the increase in temperature, the VL equilibrium line shifted to the high pressure. In contrast, the LL lines of two temperatures (40 °C and 80 °C) crossed each other. It seems that this phenomenon is the same effect as the retrograde crystallization. The VL and VLL equilibrium lines of the ternary systems and the VL equilibrium lines of the TMOS/CO_2 binary systems were almost identical. Thus, the polymer is not involved in to the separation of carbon dioxide from the mixture.

In addition, this work was compared with PMMA/TMOS/CO_2 ternary system. The impact of differences in the polymer was not found in the VL and VLL equilibrium line, but in the LL equilibrium line, the equilibrium line with PS showed a rise in the region of lower CO_2 mass fraction than that with PMMA. This reason is considered that the affinity with TMOS is good towards PS than PMMA.

(334h) Experimental Measurementsof Bubble Point (V-L) and Three-Phase Hydrate Equilibrium Condition (H-Lw-V) for CO2 Rich Gas Mixtures

Nasir, Q., University Tecnology Petronas
Sabil, K. M., Universiti Teknologi Petronas
Partoon, B., Universiti Teknologi Petronas
Riduan, M., Universiti Teknologi Petronas
B Abu Seman, A.,
Originally presented on: 11/5/2013 14:50:00 - 15:10:00

Bubble point estimation in petroleum industry is the most crucial pressure-volume-temperature (PVT) property especially for strategies concerning the oilfield development such as Enhanced oil Recovery (EOR).Gas hydrate on the other hand plays a significant role in the natural gas production and transmission lines. Its formation in pipelines could prompt to safety hazards to both production and flow assurance problem which leads to economic intimidation. Green house gas like CO_2 is an important constituent in natural gas and its significance in hydrate formation is inevitable. The purpose of this work is to perform systematic measurements for bubble point (V-L) and hydrate-liquid-vapor (H-Lw-V) equilibrium for CO_2-riched synthetic natural gas (SNG) mixtures. Four different SNG mixture were used and purchase from gas walkers SDN BHD Malaysia, with a cylinder capacity of 10L having fill pressure of 45 bar and $0.6/m^3$ gas volume. The four SNG used in this study contains components ranges from minimum two to maximum seven, with CO_2 content varies from 69.14% to 72.5%. Data were generated by using PVT equipment for bubble point measurement with a working temperature ranges from -15°C to °C and pressure from 0 MPa to 10 Mpa , The equipment can be operated with a pressure ranges from 68 Mpa to 103 Mpa and temperature from -20°C to 175 °C, Equilibrium cell enclosed in the equipment has a volume of 200 cc with a visual volume of 100 cc. Temperature , pressure and volume is recorded with accuracy of 0.5°C , 0.01% full scale and 0.01 ml. Fluid Eval PVT equipment offers full sample visibility from front and back windows and attached camera device is used for recording sample picture in regular intervals. Similarly for hydrate equilibrium measurements equilibrium cell provided in Hydreval was used to investiagte the thermodynamic equilibrium conditions of hydrate. The measurement is done using step heating tecnique of Tohidi et al with working temperature ranges from 4°C to 16°C and pressure 2.55 MPa to 14.61 Mpa. The hydrate cell has a maximum capacity of $80cm^3$ with maximum operating pressure is 20MPa with temperature ranges from -20°C to 150°C. Also magnetic driven stirrer is provided for homogenous agitation of the fluid sample. Temperature, pressure and volume conditions are measured and recorded in every 2 sec with accuracy of ±0.1K, ±0.01 MPa and ±0.001 cm^3 respectively. Experiment results obtained in this study are then compared with those obtained from PVT modeling softwares. The results attained are in good agreement with commerical PVT software with minimum deviation of -0.22 with bubble point prediction and 0.98 for hydrate equilibrium prediction. The hydrate equilibrium data generated for SNG mixture are also compared with pure methane and CO_2 and results obtained with SNG mixture having rich CO_2 content fall in between the pure component vapor-liquid water-hydrate ($V\text{-}L_w\text{-}H$) equilibrium data attained in this work.

(336a) John Prausnitz, Humanist and Epistolarian

Debenedetti, P. G., Princeton University

Originally presented on: 11/5/2013 12:35:00 - 12:55:00

John Prausnitz's towering influence on our profession has helped shape modern Chemical Engineering. Undoubtedly this influence can be traced in large part to his creation of the discipline known as molecular thermodynamics, which has enabled engineers to make accurate predictions of important physical properties through a judicious blend of intuition and rigor. Remarkably, John has also shaped the careers of many practitioners of Chemical Engineering by reminding us of the profoundly human dimensions of our craft and discipline. He has done this through his scholarly writings, his formal speeches, and, most notably, his personal correspondence with countless colleagues, disciples and former students. Drawing on examples from his papers and letters, I will pay tribute to John Prausnitz the eloquent humanist and virtuoso epistolarian, whose insatiable intellectual curiosity has enabled him, in the course of a remarkable career spanning more than half a century, to enlighten, inspire, and educate an entire profession.

(336b) Berkeley in the Early 60's

Eckert, C. A., Georgia Institute of Technology

Originally presented on: 11/5/2013 12:55:00 - 13:15:00

First of all I was a maverick; I had done two degrees at MIT, and everyone who was able to do so remained there for their PhD. My advisor and department head, Ed Gilliland told me I was crazy to leave MIT and that Berkeley was "no good." But Tom Sherwood told me about a hot young faculty member there -- John Prausnitz. I had no intention of doing thermodynamics, but his energy, enthusiasm, and creativity seduced me. Thermodynamics was dull rote stuff that would get properties from other properties. But Molecular Thermodynamics could get something from nothing, using "enlightened empiricism."

Really what appealed to me most is that it seemed that every day John came into the lab with five new ideas. He was so prolific that we, his students, formulated a defense, and that was to have more new ideas. I don't believe that at the time we realized that that was his plan all along, but it worked and we opened lots of new doors. Another technique he used was to put his students into competition with each other. He would tell me that another student was really doing better than I was, so I had better work harder. Eventually I compared notes with the other student, who said he was being told that I was surpassing him. So we knew what was going on, but it it elevated us both.

John's great skill was motivating students and colleagues to achieve important results. For example, he was the antithesis of an experimentalist, but he knew what experiments could do, and he knew to whom to send us to work out the details. For example, we were using IR at the time, ages before the advent of Fourier transform. There was no quantitative chromatograph commercially available, so we built one and used it successfully.

Another case was computers. Fortran had just come out and it seemed that a whole class of new problems had become tractable. But John would not even look at computer output – we had to make graphs. The department head, Don Hanson, came up with a plot to get John computer-active, and outlined the possibility of setting up computer programs for calculating separations. Several of us worked on it, and the upshot was that we came up with what was really the first monograph on applications to separations, full of computer programs. John coauthored the book and made very substantial contributions to the text, but still never looked at the output.

One of the outstanding applications of computers was the opportunity to use novel excess Gibbs energy expressions based on entropy more than enthalpy. These really quite inaccessible with slide rules, but facile with Fortran. At one point George Scatchard from MIT came to Berkeley and gave a talk of such an equation, developed with his student Wilson. It had not yet been published (and would not be published until some years later), but John's students copied the form and started using it. Soon they had demonstrated clearly its superiority to the enthalpic VLE models such as Margules and van Laar, and they published the results (with attribution) before Wilson actually published what we now know as Wilson's Equation. This led to improved entropic equations still used, such as NRTL and UNIQUAC, but still without John becoming computer-active.

John did always want to learn. I remember when he decided to learn statistical mechanics. He assigned three of his students to go learn about it and teach him. One took the course in Chemistry, one in Physics, and I took the Mechanical Engineering course from the youngest (and youngest-looking) professor at Berkeley, Prof Tien. He was a wonderful teacher, and later became the first nonwhite Chancellor of Berkeley. But then I digress; the three of us started a class for John and the rest of the group, and he went on to use the results most powerfully.

One of the most striking things I learned about John early on was his ability to communicate to a group. Other seminar speakers would inundate the audience in equations attempting to be paragons of elegance. John was different – I always emerged from his presentations saying "My goodness that was easy. I wish I had thought of that." He did not try to appear elegant; rather he solved important problems simply.

I went to Berkeley thinking I might want to be a university professor, and by his example and guidance, John intensified that attitude. He has stayed in touch with me, and with his other academic progeny, for now nearly half a century, always challenging and elevating us.

(336c) Fluctuation Solution Theory and Messages Heard From A Master: Reprise

O'Connell, J. P., University of Virginia

Originally presented on: 11/5/2013 13:15:00 - 13:35:00

The objective of physical properties modeling is to judiciously use fundamentals, data, approximation, computation, and generalization to optimize descriptions of chemical systems, especially for process design. John Prausnitz masterfully devised, demonstrated, applied, and taught - to generations of students and professionals - some extraordinarily successful strategies for comprehending and treating an incredible richness of important systems, substances, and conditions. Further, he has shown their value in many spheres of intellectual work and human endeavor.

Some time ago, I presented a talk that shared my appreciation of John's influence on my 3 decades of research on extending and applying fluctuation solution theory, the molecular theory of solutions due to Kirkwood and Buff. I phrased a set of "messages" that I could imagine hearing from him about how to proceed at moments of decision along the way. In the last 15 years, our work has made further progress that was aided by more messages I heard from within me because of John's impact.

The talk will reprise the original theme and format, describe what we have achieved in recent years, and articulate additional messages.

(336d) Prausnitizian Phase Equilibria in Soft-Contact-Lens Copolymer Hydrogels

Radke, C. J., University of California Berkeley
Prausnitz, J., University of California at Berkeley
Originally presented on: 11/5/2013 13:35:00 - 13:55:00

Prausnitizian Phase Equilibria in Soft-Contact-Lens Copolymer Hydrogels

C.J. Radke[1,2] and J.M. Prausnitz[1]

[1] Chemical and Biomolecular Engineering Department, University of California, Berkeley, CA, 94720-1462

[2] Vision Science Group, University of California, Berkeley, CA, 94720-2002

85th Birthday Celebration of J. M. Prausnitz

Thermodynamics and Transport Properties

AIChE San Francisco, 2013

Nearly 50 % of first-time SCL wearers drop out because of discomfort. Wear comfort increases with increasing lens water content making water saturation a critical variable. John pioneered the language to tackle such problems and continues to apply them toward solution of practical problems. We report studies with undergraduate students on the swelling of 2-hydroxyethyl methacrylate (HEMA)/methacrylic acid (MAA) hydrogels over the complete range of MAA content. Gel water content strongly increases near the pK of MAA. At low pH, water content is insensitive to the MAA/HEMA ratio, whereas at high pH, small amounts of MAA increase water saturation dramatically, but less so as MAA copolymer fraction increases. Application of multipolymer Flory-Huggins-Rehner theory proves inadequate as do Donnan electrostatics. We propose specific complexation of water with charged acid groups of MAA along with charge-dependent ion-binding of salt counterions. John's appetitive for mentoring students through insightful research remains alive and well at Berkeley.

(336e) The Chemical Engineers' Role in Energy and Economic Security
Trainham, J.,
Originally presented on: 11/5/2013 13:55:00 - 14:15:00

The Chemical Engineers' Role in Energy and Economic Security

James A. Trainham

RTI International and NC State University

Transportation—not power generation—is the real energy challenge.

Energy Independence for the U.S. is really North American Energy Independence because Canada, Mexico and the U.S. are interconnected, linked by geography, geology, infrastructure and economics. Together we can detox North America from OPEC's monopoly.

Until now no substitutes for oil existed. Unconventional natural gas (UG) reserves are plentiful and increasing. The opportunity is for UG to diversify the fuel choices. UG will change the geopolitics of the world as UG recovery and use technology, applied globally, substantially reduce oil dependencies.

Chemical engineers must lead this migration to UG, reducing the investment intensity of UG based fuels in the short-term and in the longer-term reduce the carbon footprint of UG usage..

These circumstances and the leadership legacy of Professor Prausnitz demand that we chemical engineers develop the fundamental understanding to make the new technologies economic reality and eliminate transition barriers.

(336f) How to Transform a Field of Study and Define a Discipline

de Pablo, J. J., University of Wisconsin-Madison

Originally presented on: 11/5/2013 14:30:00 - 14:50:00

In a career spanning over five decades John Prausnitz transformed chemical engineering. Through his pioneering research efforts and visionary studies, the field of molecular thermodynamics gradually evolved into a discipline where some of the most fundamental experimental observations and theoretical constructs found an outlet in engineering practice. Indeed, one of John's lasting imprints on chemical engineering and on his students' psyche has been that, to the extent possible, research should lead to solution of important problems. That evolution continues today, and builds on John Prausnitz's text books and research monographs, his nearly 700 papers, many of them seminal, and those of hundreds of students and collaborators around the world that have been influenced by his ideas. In this presentation I will describe several recent, exciting developments in chemical engineering that can be traced back to John Prausnitz's original contributions and those of his co-workers, and I will attempt to anticipate how those developments might shape engineering for years to come.

(336g) Nature-Inspired Chemical Engineering - Pathways to Innovation and Sustainability
Coppens, M. O., Rensselaer Polytechnic Institute
Originally presented on: 11/5/2013 14:50:00 - 15:10:00

Some of the most urgent challenges facing us today are related to energy, lack of fresh water, and how to translate discoveries at the nanoscale to macroscopically relevant scales for practical applications, as in scalable manufacturing.

My presentation will give a brief overview of our nature-inspired chemical engineering (NICE) approach, which addresses these challenges by a rigorous approach that takes guidance from Nature. In particular, we seek to discover fundamental mechanisms (rather than imitation out of context) behind desirable traits in natural systems – like scalability, robustness and efficiency – and use this insight to guide the design and synthesis of engineered systems that have the same properties.

Particularly promising for innovation are three fundamental and ubiquitous natural mechanisms, which we see as most timely for exploration: hierarchical transport networks (such as in lungs and trees), force balancing (e.g., steric vs. electrostatic effects), and dynamic self-organisation (as in regular pattern formation induced by perturbations, and emergence of robust properties as a result of collective phenomena behind much simpler entities).

The NICE methodology allows us to discover innovative solutions to technological problems. In my presentation, I will illustrate this thematic approach via a few examples, e.g., how the structure of trees and leaves guides the design of fractal fluid distributors and hierarchically structured catalysts; how chaperones present us with ideas on how to stabilize enzymes for therapeutics; and how bacterial communities teach us lessons on how to design adaptive, robust materials.

Taking ideas from biology to architecture, and cognizant of the context in product and process design as a boundary condition to engineering practical solutions, nature-inspired chemical engineering brings the humanities back to engineering.

(336h) A Variational Field Theory for Ion/Charge Correlations in Electrolyte Systems

Lue, L., University of Strathclyde

Originally presented on: 11/5/2013 15:10:00 - 15:30:00

Many biological systems and industrially important formulations consist of solutions containing highly charged colloidal particles. For these solutions, electrostatic and dispersion forces typically play a major role in determining their structure and thermodynamic properties. Much of our understanding of the role of electrostatic interactions in these systems is based on Poisson-Boltzmann theory which, although quite successful, has many well documented limitations, such as its failure for systems with multivalent ions where ion-ion correlations are significant. In this work, a variational field theory approach is developed, which removes many of
these limitations.

The key physical motivation behind the theory is to treat the short and long wavelength fluctuations in the system within different approximation schemes. At short range, the systems are strongly coupled, and at long range the systems are weakly coupled. To describe both these regimes within a single theory, we split the interaction into a short and long range contribution. The long range
behavior is often well approximated by a mean field theory, or including first order fluctuation corrections, while the short range behavior can be captured by a virial expansion, or other liquid state methods suitable to describe particles with short range pair interactions.

This approach is able to properly account for the many-body effects of
the electrostatic interactions, even for strongly charged particles, and it also naturally handles the coupling between electrostatic, induction, and dispersion interactions. For weakly coupled systems,
this theory approaches the Poisson-Boltzmann theory, while for strongly coupled systems, the theory resembles the strong coupling

expansion. The theory also performs well for intermediate couplings.
In addition, the accuracy of the theory can, in principle, be systematically improved.

The theory can be applied to a wide variety of problems, such as systems with different geometries and conditions (e.g., dielectric interfaces) and particles with different shapes and charge distributions. We examine its application to determine the effective
interaction between charged, low dielectric particles in electrolyte solutions. This fairly simple system exhibits a range of interesting phenomena, such as like-charge attraction and electrostatic depletion,
that can not be captured with mean field approximations. The resulting theory is shown to gives accurate predictions in comparison
to Monte Carlo simulation results.

(336i) Becoming a Prausnitzian

Wu, J., Delft University of Technology
Originally presented on: 11/5/2013 15:30:00 - 15:50:00

Becoming a Prausnitzian

Jianzhong Wu

Department of Chemical and Environmental Engineering and Department of Mathematics

University of California, Riverside, California 92521

On several occasions after a technical presentation someone would come to tell me in private that my talk sounds very Prausnitzian. I take it as a great compliment and enjoys a sense of pride for being affiliated with the big name in chemical engineering. I once shared this experience with the master. While he was apparently pleased with the comment, he told me that he was not quite sure whether it means anything because a Berkeley professor always has a strong influence on his/her graduate students. I worked with John Prausnitz first as a visiting fellow then as a graduate student and also as a postdoctoral researcher. His influence on my personal and professional developments has been enormous. I will share some of that experience in this symposium to celebrate his Life Time Accomplishment Award from the American Society of Engineering Education (ASEE).

(336j) Prausnitz Talks About Prausnitz

Prausnitz, M. R., Georgia Institute of Technology

Originally presented on: 11/5/2013 15:50:00 - 16:00:00

Prausnitz Talks About Prausnitz

Mark R. Prausnitz, Chemical and Biomolecular Engineering, Biomedical Engineering, Georgia Institute of Technology, Atlanta, GA

This talk will highlight a series of professional accomplishments in the career of John Prausnitz and describe their impact and their context from the perspective of a family member using anecdotes and photos.

(352a) Estimation of the First Normal Stress Difference From Shear Viscosity Data for Polymer Melts and Solutions

Sharma, V., Massachusetts Institute of Technology
Originally presented on: 11/5/2013 15:15:00 - 15:30:00

The relevant stress distribution for Non-Newtonian fluids in response to a steady simple shear flow is expressed using three viscometric measures; the shear stress and two normal stress differences. The normal stress differences are associated with nonlinear viscoelastic effects, and become vanishingly small in linear viscoelastic measurements. Normal stress measurements are relatively difficult to perform and require specialized equipment; by contrast, measurements of the complex viscosity/modulus (in oscillatory shear) or steady flow curves are relatively straightforward and far more accurate. The Cox-Merz rule and Laun's rule are two empirical relations that somewhat intriguingly allow the estimation of steady shear viscosity and first normal stress difference (nonlinear response), respectively, using small amplitude oscillatory shear measurements (of the linear viscoelastic response). Recently, by using a lesser known relationship also proposed by Cox and Merz, in conjunction with Laun's rule, Sharma and McKinley deduced an empirical relationship between the rate-dependent steady shear viscosity and the first normal stress difference. The new rule enables *a priori* estimation of the first normal stress difference using only the steady shear viscosity vs shear rate data of entangled polymeric solutions and melts. In this talk, we rigorously examine the dissipative and recoverable contributions to the measured nonlinear rate-dependent shear viscosity response, in pursuit of a self-consistent physical description of the correlation between the three viscometric functions.

(352b) Crooks Fluctuation Theorem for Flowing Complex Fluids

Latinwo, F., University of Illinois at Urbana-Champaign
Schroeder, C. M., University of Illinois at Urbana-Champaign
Originally presented on: 11/5/2013 15:30:00 - 15:45:00

In this work, we apply the Crooks fluctuation theorem to determine fundamental nonequilibrium steady state properties of polymer molecules driven by fluid flow. In general, nonequilibrium work relations provide a new framework to study the dynamics of complex fluids. Recently, by using the Jarzynski equality, we demonstrated the direct determination of equilibrium properties such as polymer chain elasticity from transient stretching trajectories of polymer molecules in flow. Here, we report the determination of steady state energies of polymers in flow using Brownian dynamics simulations of dsDNA molecules transitioned between strain rates in forward and reverse directions. Interestingly, the steady state energy allows for the determination of rheological quantities such as polymer relaxation time and single chain elasticity. In addition to free energies, we observe a polymer stretch hysteresis in flow that exists in the absence of intra-chain hydrodynamic interactions. We find that the ensemble-average stretching trajectories of polymer molecules strongly depend on the steady state history. Overall, analyzing complex fluids in the context of Crooks fluctuation theorem allows for the determination of fundamental flowing materials properties and the observation of history-dependent phenomena.

(352c) "Not So" Simple Shear Flow of Polymer Solutions (Shear Banding)

Leal, L. G., University of California, Santa Barbara
Cromer, M.,
Villet, M. C., University of California, Santa Barbara
Fredrickson, G. H., University of California, Santa Barbara
Originally presented on: 11/5/2013 15:45:00 - 16:00:00

We consider the problem of shear banding. Recent experimental work by Wang and coworkers suggests that the phenomenon of shear banding can occur in polymer solutions. These observations are, however, at odds with the usual expectation that shear banding is due to a constitutive instability associated with a range of shear rates where the shear stress decreases with increasing shear rate. This behavior is believed to occur in solutions of wormlike micelles, and these fluids do exhibit shear banding. However, polymer solutions exhibit a monotonic increase of shear stress with shear rate, and thus (according to common wisdom) should not exhibit shear banding. Indeed, other investigators have had difficulty reproducing Wang's experimental results. In this talk, we show that a commonly accepted, monotonic constitutive model for polymers (the "Rolie-Poly" model) exhibits a linear instability that leads to a steady shear banding solution, provided that we treat the polymer concentration profile as an unknown, in addition to the velocity profile. We further show that many of the observations of Wang and coworkers are at least qualitatively reproduced, such as multiple steady states at certain shear rates, and elastic recoil during the startup flow that leads to the steady shear banded solutions.

(352d) Hi-Fidelity Dissipative Particle Dynamics (DPD) Simulation of Shear Banding in Entangled Polymer Melts

Mohagheghi, M., University Of Tennessee
Khomami, B., University of Tennessee
Originally presented on: 11/5/2013 16:00:00 - 16:15:00

In recent years, the existence of spatially inhomogeneous shear flow in entangled polymeric solutions and melts has been demonstrated. Specifically, in steady or transient shear flows, this class of fluids could exhibit shear banding, i.e., the flow separates into fast and slow moving regimes. In this presentation, the result of the first hi-fidelity and coarse grained DPD simulations of shear flow of entangled polymeric melts will be presented with the aim of paving the way for a mechanistic understanding of shear banding in entangled polymers. Specifically, large-scale DPD simulations of entangled polymers over a broad range of deformation rates and entanglement densities have been performed. Overall, our results clearly demonstrate the existence of shear banding and the associated macromolecular dynamics. In this presentation, these results will be discussed in the context of recent coarse-grained molecular dynamics of moderately entangled polymers as well as continuum level models devoted to phenomenology of shear banding.

(352e) Dynamics of Polymer Adsorption From Dilute Solution in Shear Flow Near a Planar Wall

Dutta, S., University of MInnesota
Dorfman, K. D., University of Minnesota
Kumar, S., University of Minnesota
Originally presented on: 11/5/2013 16:15:00 - 16:30:00

Adsorption of polymers from dilute solution subject to shear flow near a planar wall is studied using kinetic theory. A dumbbell model consisting of two beads connected by a nonlinear spring is used to describe the polymer molecules, and the beads interact with the wall via a short-range exponential potential. Bead-bead and bead-wall hydrodynamic interactions are also included in the theory. The kinetic theory yields an advection-diffusion equation which describes the probability density of the dumbbell center-of-mass as a function of the distance normal to the adsorbing surface and time. This equation is solved numerically via a spectral method for cases where polymer is initially present or absent on the adsorbing surface. For an initially bare surface, it is found that the quantity of polymer adsorbed decreases with an increase in polymer molecular weight at a given shear rate and point in time. In addition, for a given molecular weight and point in time, the quantity adsorbed decreases with an increase in shear rate. When adsorbed polymer is initially present, similar trends are observed. Furthermore, complete desorption can be achieved at a sufficiently high shear rate. In all cases, the time required to approach a steady value of the adsorbed amount is many orders of magnitude larger than the dumbbell relaxation time. The above findings are in qualitative agreement with experimental measurements reported nearly three decades ago by Lee and Fuller [J. Colloid Interface Sci. 103 (1985) 569]. Our findings also suggest that the physical mechanism underlying the long-standing observation that shear flow inhibits polymer adsorption and assists polymer desorption is hydrodynamic interaction between stretched polymer molecules and the adsorbing surface.

(352f) On the Reversibility and Chaos in Microscopic Fluid Systems

Rosenfeld, L., Stanford University
Fan, L., Stanford University
Tang, S. K. Y., Stanford University
Originally presented on: 11/5/2013 16:30:00 - 16:45:00

Soft amorphous materials such as foams, emulsions, granular systems, or colloidal suspensions often display complex flow properties. These complex materials have many applications in chemical, biological and industrial processes. The central quantity of interest for determining complex fluids rheology is the spatial distribution of the particles. Under imposed flow, the ability of the microstructure to rearrange to accommodate flow and interparticle forces determines its macroscopic rheological response. The many-body, shear-history-dependent nature of the microstructure renders the prediction of the complex material's dynamics and rheology highly nontrivial. Hence a fundamental understanding of the microscopic rearrangements and specifically the transition from hydrodynamic reversible to irreversible behavior is of central importance.

In this study we explore the transition from reversible to chaotic behavior in the oscillatory shear flow of water-in-oil emulsions. The emulsion is flown through a microchannel and is forced to rearrange due to a central constriction in the channel. We study the motion of the individual droplets and their neighbors in order to determine their ability to retain their original position after several cycles of oscillations.

We have found that while at the Stokes flow limit, the emulsion exhibit behaviors that vary from complete reversible to complete irreversible depending on the volume fraction, velocity and strain amplitude. We provide the first direct visualization of this phenomenon. This work is an important step in understanding the microscopic rearrangements of droplets and particles near jamming.

(352g) Yielding and Microstructure in a 2D Jammed Material Under Shear Deformation

Arratia, P. E., University of Pennsylvania
Keim, N. C.,
Originally presented on: 11/5/2013 16:45:00 - 17:00:00

In this talk, we present an experimental investigation on the rheology and microstructure of two-dimensional colloidal suspensions undergoing oscillatory shear. In the experiments, colloidal particles are adsorbed at an oil-water interface and form a dense disordered monolayer. Rheological measurements are performed in a custom-built interfacial stress rheometer, using a magnetic needle within the material. This configuration allows for the simultaneous characterization of both fluid microstructure (tracking ~10^5 particles) and bulk rheology. Results show that below a certain strain amplitude, the microstructure becomes reversible after a transient. Above this strain amplitude, the microstructure continues to evolve through many irreversible events. We argue that this boundary between a reversible and irreversible steady state is a yielding transition, and that our experiments measure a meaningful yield stress.

(352h) Structure and Rheology of Micelle-Nanoparticle and Polymer-Nanoparticle Networks From Molecular Dynamics Simulations

Dhakal, S., Syracuse University
Sambasivam, A., Syracuse University
Sureshkumar, R., Syracuse University
Originally presented on: 11/5/2013 17:00:00 - 17:15:00

Self-assembly of nanoparticles (NPs) with cylindrical micelles or polymers in an aqueous solution has generated much interest recently [1-3]. Such "nanofluids" can be tailored to respond to specific optical, magnetic, thermal or electrical stimulus depending on the NP characteristics. For instance, we have shown previously that by utilizing noble metal NPs, plasmonic fluids with tunable optical properties could be synthesized by controlling the particle shape, size, and surface functionalities [4]. Although these experiments have shed light into the plausible mechanisms of self-assembly, a quantitative description of the role of NP shape and size, ratio of the surfactant to the salt concentrations, and surfaces charge on the structure, stability and rheology of such fluids is still lacking. Extending our previous coarse grained molecular dynamics (MD) simulations for surfactant self-assemblies [5, 6] and their interactions with a single NP [7-8], we investigate the structure and rheological properties of particulate networks consisting of multiple (O(10)) NPs. In these simulations, explicit solvent/salt mediated interactions among surfactants are considered while the NP-surfactant interactions are represented through a mean field potential inspired by the predictions of fully explicit MD simulations of NP-micelle interactions. We find that the stability and structure of the network are highly dependent on the ratio of surfactant/salt concentration. Effects of NP size, shape, volume fraction and surface charge density on the structure and rheological properties of the network will be discussed. Extension of the simulation framework to study polymer-colloidal particle networks as a mechanical model for bacterial biofilms will also be presented.

(1) F. Nettesheim, M.W. Liberatore, T.K. Hodgdon, N.J. Wagner, E.W. Kaler & M. Vethamuthu, Langmuir, 24, 7718 (2008).

(2) M.E. Helgeson, T.K. Hodgdon, E.W. Kaler & N.J. Wagner, Langmuir, 26, 8049 (2010).

(3) A. C. Balazs, T. Emrick & T. P. Russell, Science, 314 (5802) 1107– 1110 (2006).

(4) T. Cong, S.N. Wani, P.A. Paynter & R. Sureshkumar, Applied Physics Letters, 99, 043112 (2011).

(5) A.V. Sangwai & R. Sureshkumar, Langmuir, 27, 6628 (2011).

(6) A.V. Sangwai & R. Sureshkumar, Langmuir, 28, 1127 (2012).

(7) S. Nangia & R. Sureshkumar, Langmuir, 28, 1766 (2012).

(8) T. Cong, G. Zhou, A. Sambasivam & R. Sureshkumar, Structure, Rheology and Optical Properties of Self-Assembled Micelle-Nanoparticle Plasmonic Nanogels: Experiments and Molecular Dynamics Simulations, Paper 413c, AIChE Annual Meeting, Pittsburg, PA, October 28-November 2 (2012).

Acknowledgements: We acknowledge National Science Foundation grants 1049489 and 1049454 for the support of this research.

(352i) Time-Strain Superposition in Soft Glasses

Srivastava, S., Cornell University
Archer, L. A., Cornell University
Originally presented on: 11/5/2013 17:15:00 - 17:30:00

Soft glassy materials are typified by large relaxation times and thus it is imperative to probe their long time responses. Most of these systems are athermal, and thus the conventional time-temperature superposition (TTS) approach has not proven to be suitable for accessing long time behaviors. We report a simple time-strain superposition (TSS) procedure that can be used instead to create rheological maps of soft glassy dynamics spanning over unprecedented time-scales. Experimental results from model soft nanoparticle suspensions as well as self-suspended nanoparticle fluids are presented. The roles of temperature and strain in assisting the relaxation in materials that exhibit TTS and TSS, respectively, are found to be analogous; the shift factors are shown to exhibit a WLF like dependence on strain in TSS. The generality of the approach is confirmed by theoretical calculations using the soft glassy rheology (SGR) model that also predict similar behavior, thus suggesting that time-strain superposition is a generic feature of all soft glasses.

(352j) Shear Thickening and Defect Formation of Fumed Silica CMP Slurries Using Rheo-Polishing

Liberatore, M. W., Colorado School of Mines
Crawford, N., Colorado School of Mines
Williams, K., Colorado School of Mines
Boldridge, D.,
Originally presented on: 11/5/2013 17:30:00 - 17:45:00

During the chemical mechanical polishing (CMP) process, it is believed that shear thickening of the slurry, caused by particle agglomeration, results in a significant increase in surface defects (i.e, scratches, gouges, pits, etc.). These CMP-induced defects then cost the semiconductor industry billions of dollars annually in lost production. We have developed a methodology for the synchronized measurement of rheological behavior while polishing a semiconductor wafer, the first of its kind (a technique termed rheo-polishing). We investigated the shear thickening of a 25 wt% fumed silica slurry with 0.15 M added KCl (100-200 nm particle diameter). The thickened slurry displayed a ~5-fold increase in viscosity with increasing shear rate. As the shear rate was reduced to zero, the slurry continued to thicken showing a final viscosity that was ~100x greater than the initial viscosity. Optical microscopy and non-contact profilometry were then utilized to directly link slurry thickening behavior to more severe surface scratching. The thickened slurry generated up to 7x more surface scratches than a non-thickened slurry. Both slurry thickening and surface scratching were associated with a dramatic increase in the population of "large" particles (>300 nm) which were undetectable in the non-thickened slurry. These "large" and potentially scratch-generating particles are believed to instigate measurable surface damage.

(363a) A New Method for Determining the Hamaker Constant of a Solid With Atomic Force Microscopy

Dong, J., Purdue University
Zhao, Y., Hewlett-Packard Laboratories, Hewlett-Packard Co.
Ng, H. T., Hewlett-Packard Laboratories, Hewlett-Packard Co.
Franses, E. I., Purdue University
Corti, D. S., Purdue University
Originally presented on: 11/5/2013 15:15:00 - 15:30:00

The determination of the Hamaker constant, A, of a solid material with a flat surface using atomic force microscopy (AFM) is often done based on the following steps. First, data of the dynamic AFM tip deflection under the influence of the attractive force between the tip and the solid and for a given cantilever approach speed are obtained. Second, a quasi-static analysis is used to fit the dynamic data, which enables one to infer the jump-into-contact distance Δd and hence the value of A. Here, we evaluate the validity of this quasi-static method and develop an improved procedure for determining A which acknowledges explicitly that the tip motion is dynamic, and not quasi-static.

We begin by correcting an earlier quasi-static method based on the "jump-into-contact" distance for the simplest sphere-plane model. This analysis is further extended to three additional tip-surface models. The sensitivity of these models in the quasi-static analysis is investigated. Moreover, a mechanically equivalent energy-based analysis is reported. The energy approach offers additional insights on the mechanical stability of the tip as it approaches the solid surface quasi-statically. Next, a rigorous dynamic analysis of the tip motion is developed to determine the time-dependent tip deflection. For close contact distances with the surface, the tip motion is found to deviate from the predictions of the quasi-static analysis except at the limit of very low or zero cantilever speeds. This dynamic analysis also indicates that a "jump-into-contact" distance, which is defined in the quasi-static analysis, cannot be uniquely determined from the dynamic data. Hence, a new method is needed if the quasi-static model is still to be used when interpreting the dynamic data.

We introduce a novel approach whereby an energy analysis is used to approximate the dynamic data with a quasi-static model. An iterative scheme is introduced, in which an apparent Hamaker constant, A_{app}, is determined from the dynamic data and the quasi-static model. This A_{app} depends on the cantilever speed and the spatial resolution δ of the data points. In the limit of zero cantilever speed, A_{app} should converge to the value of A obtained from the quasi-static model. For proper use of the new method, one therefore needs to obtain dynamic data at resolutions of $\delta < \Delta d$ and for several decreasing cantilever speeds. We test the method using simulated data, and investigate the effects of the number of data points sampled from the dynamic curve of the tip motion. High sampling rates, small values of δ and low cantilever speeds are shown to yield the most accurate estimates of A. Finally, the new method is applied to dynamic AFM data for a model system of glass (amorphous SiO_2).

(363b) Large-Scale Molecular Dynamics Study of Dewetting of Thin Liquid Films On Solid Substrates

Nguyen, T. D., Oak Ridge National Laboratory
Fuentes-Cabrera, M., Joint Institute for Computational Science, University of Tennessee and Oak Ridge National Laboratory
Fowlkes, J. D., Oak Ridge National Laboratory
Diez, J. A., Universidad Nacional del Centro de la Provincia de Buenos Aires
GonzÃ¡lez, A. G., Universidad Nacional del Centro de la Provincia de Buenos Aires
Kondic, L.,
Rack, P. D., The University of Tennessee
Originally presented on: 11/5/2013 15:30:00 - 15:45:00

The initial rupture of thin liquid films on a solid substrate has been extensively investigated in experiment and continuum theory regarding the interplay between two spontaneous mechanisms, i.e. spinodal instability and nucleation, at the early stage of dewetting. Molecular simulation becomes a powerful means for investigating the dewetting dynamics and structural changes at such a short time scale while ruling out heterogeneous nucleation, which would be almost inevitable in experiment. The enormous challenges, however, stem from the system size that should be sufficiently large to accommodate the tempo-spatial growth of unstable undulate modes in the film. Using LAMMPS GPU, a massively parallel Molecular Dynamics software package with graphics processing units acceleration, on Titan, we are now able to study metallic thin films at close-to-experiment length scales (i.e. hundreds of nanometers and micrometers), consisting of up to tens of millions of atoms. We found distinctive signatures of spinodal instability and thermal nucleation in these systems in remarkable agreement with previous experiments and theoretical predictions. Importantly, our results reveal insights into the early stage of dewetting at the atomistic level, which are inaccessible with continuum models and experiment.

(363c) Water Structure and Dynamics in Thin Films On Mineral Surfaces

Sarupria, S., Princeton University

Originally presented on: 11/5/2013 15:45:00 - 16:00:00

Understanding water behavior near surfaces is of primary importance to several processes, including heterogeneous ice nucleation and water-mediated self-assembly processes. The structure and dynamics of water molecules in films on solid surfaces are strongly influenced by several factors including the structure of the surface, surface-water interaction energy, the thickness of the water film and temperature. This directly affects the phase transitions that can occur within the film. In our work we use molecular dynamics simulations to characterize the structure and dynamics properties of water molecules in thin films of water near kaolinite surfaces. We further investigate the effect of lattice spacing on water behavior by modifying the bond distances of the kaolinite surface. We find that the lattice spacing influences the overall arrangement of water molecules especially in single monolayer thick water films. Our work provides insights to understand the ice nucleating efficiency of mineral surfaces.

(363d) Effect of External Vibrations On Non-Isothermal Ice-Nucleation Rates

Dursch, T. J. Jr., University of California, Berkeley

Trigub, G.,

Liu, J.,

Radke, C. J., University of California Berkeley

Weber, A., Lawrence Berkeley National Laboratory

Originally presented on: 11/5/2013 16:00:00 - 16:15:00

Effect of External Vibrations on Non-isothermal Ice-Nucleation Rates

T.J. Dursch,[1,2] G.J. Trigub,[1] J.F Liu,[1] C.J. Radke,[1,3] A.Z. Weber[2]

[1] Chemical and Biomolecular Engineering Department, University of California, Berkeley, CA

[2] Environmental Energy Technology Division, Lawrence Berkeley National Laboratory, Berkeley, CA

[3] Earth Sciences Division, Lawrence Berkeley National Laboratory, Berkeley, CA

AICHE 2013: Session 01C02 Fundamentals of Interfacial Phenomena I

Nucleation stochastically generates aggregates of a new and more stable phase within a metastable mother phase, capable of growing rapidly into macroscopic crystallites. Controlled nucleation is critical to the success of numerous applications, including: manufacturing of bioactive drugs, thermal energy storage in phase-change materials, zeolite synthesis, and freeze drying. Our particular focus lies in ice crystallization during cold-start of a proton-exchange-membrane fuel cell[1]. Commonly, nucleation rates are obtained from induction-time measurements[1]. In many cases, especially in ice crystallization, large driving forces (e.g., subcoolings of 10-30 °C) are required for sufficiently large nucleation rates and short induction times. In this work, we examine the effect of an external driving force, specifically, applied vibrations, on ice-nucleation rates and induction times.

Distilled/deionized (DDI) water, placed in a plexiglass (PMMA) cylinder, is cooled from beneath by a copper heat exchanger positioned on a vibration generator. Induction times are obtained using both infrared thermography and thermocouples as a function of vibration frequency (0 to 50 Hz) and amplitude (1 to 10 mm). Figure 2 plots 10 repeated induction-time measurements for 30-μL of DDI water at frequencies of 0 (blue), 10 (green), 20 (purple), and 50- Hz (light blue). Induction times decrease with increasing vibration frequency (i.e., a factor of 3 from 0 to 50 Hz). Additionally, induction-time Poisson distributions[1] narrow with increasing vibration frequency, yielding well-controlled nucleation rates. No theory is currently available to predict the role of vibrations in ice nucleation. We find that accounting for work input by mechanical vibrations in classical nucleation theory provides qualitative agreement with the new results.

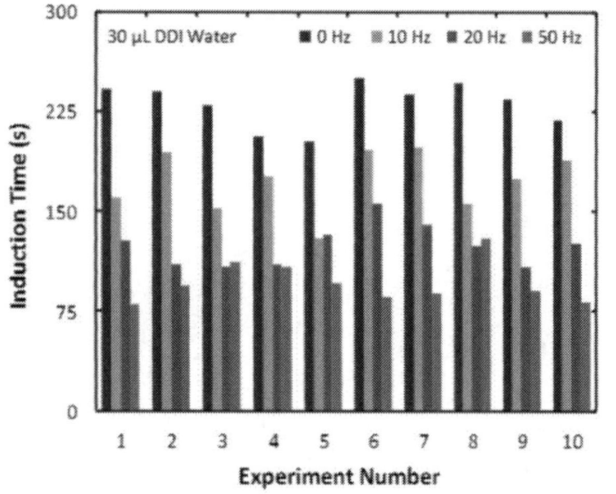

Figure 1. 10 repeated induction-time measurements for 30-μL of DDI water at frequencies of 0 (blue), 10 (green), 20 (purple), and 50 Hz (light blue).

References

[1] T.J. Dursch, M.A. Ciontea, C.J. Radke, A.Z. Weber, Isothermal ice-crystallization kinetics in the gas-diffusion layer of a proton-exchange-membrane fuel cell, Langmuir 28 (2012) 1222-1234.

(363e) Surface Corrugation Effects On the Behavior of Water Under Extreme Graphene Confinement

Chialvo, A. A., Oak Ridge National Laboratory
Vlcek, L., Oak Ridge National Laboratory
Cummings, P. T., Oak Ridge National Laboratory
Originally presented on: 11/5/2013 16:15:00 - 16:30:00

Significant theoretical, experimental and simulation efforts have been aimed at understanding the behavior of water in contact with solid surfaces. The emerging picture from these studies confirms the expectation that the structural and dynamical properties of water are significantly perturbed at the fluid-solid interfaces. The structural perturbation is characterized by an inhomogeneous local density distribution whose dynamical counterpart translates into anisotropic behavior of the resulting translational and rotational molecular motions.

An interfacial phenomenon of great relevance in biochemical and geochemical processes as well as energy-related industrial applications, is solid surface wettability. In fact, current evidence points to the strong effect of surface topography (i.e., curvature and corrugation) on the surface wettability, whose most obvious macroscopic evidence is the modification of the fluid's contact angle, a phenomenon frequently described in terms of the meso-scale models of Wenzel [1] and Cassie [2].

Molecular-based simulation of solid-water interfacial phenomena at nanoscopically-flat surfaces show that the interfacial regions extend only a few molecular diameters, therefore, any surface corrugation-driven interfacial phenomenon might be significantly enhanced as the surface corrugation becomes commensurate to the solvent molecular characteristic length-scale, i.e., a few angstroms for water [3].

Our current work has been motivated by the experimental evidence of the strong influence of surface roughness on the wettability of carbon surfaces in contact with aqueous environments, as well as the macroscopic and controversial Wenzel's prediction of an enhancement of the solid-fluid hydro-philicity (phobicity) with increasing roughness (corrugation) over that of the corresponding flat surface.

Therefore, the aim of this work is to provide essential molecular-based insights into the hydro-phobic/philic nature of the water-(corrugated graphene) surface interactions, by addressing the following central issues: (a) the link between the corrugation-driven hydration free energy changes in the association process involving graphene plates and the resulting water-graphene interfacial tension, (b) the simultaneous effects of surface corrugation plus confinement on the water's thermodynamic response functions and the resulting macroscopic modeling implications, (c) the ill-defined nature of Wenzel formula to describe the water-corrugated graphene interfacial behavior at the nanoscale.

For that purpose, we performed isobaric-isothermal (locally grand canonical) molecular dynamics simulations to study the behavior of water under extreme confinement between uncharged finite-sized graphene plates, to simultaneously characterize the behavioral differences between water at interfacial and under confinement, while in equilibrium with its own bulk [4,5]. Using the slit-pore (graphene flat-plates) configuration as a reference system, we introduce precisely defined corrugation patterns that preserve the original pore volume, a feature that facilitates the interpretation of simulation results. The characterization involves the axial profiles of the potential of mean force (PMF), local isothermal compressibility, diffusivity, and thermal expansivity. Moreover, the profiles of the solvent contribution to the PMF in conjunction with that of the local isothermal compressibility are then used to assess the surface-corrugation effects on the hydro-philic/phobic nature of the water-graphene interactions and on the (oscillatory) onset of wet-dry transition. We also make contact with quasielastic neutron scattering (QENS) experiments by calculating the simulated scattering spectra and interpreting them in terms of microscopic information from the corresponding molecular dynamics simulation [6].

[1] Wenzel, R. N. Industrial & Engineering Chemistry 1936, 28, 988-994.

[2] Cassie, A. B. D.; Baxter, S. Transactions of the Faraday Society 1944, 40, 546-551.

[3] Striolo, A. Adsorption Science & Technology 2011, 29, 211-258.

[4] Chialvo, A. A.; Cummings, P. T. The Journal of Physical Chemistry A 2011, 115, 5918-5927.

[5] Chialvo, A. A.; Cummings, P. T. The Journal of Physical Chemistry C 2013 Submitted for publication.

[6] Mamontov, E.; Wesolowski, D. J.; Vlcek, L.; Cummings, P. T.; Rosenqvist, J.; Wang, W.; Cole, D. R. Journal of Physical Chemistry C 2008, 112, 12334-12341.

(363f) Shape and Roll-Off Angle of a Pinned Droplet On a Progressively Inclined Plate

Musterd, M., Delft University of Technology
van Steijn, V., Delft University of Technology
Kreutzer, M. T., Delft University of Technology
Kleijn, C. R., Delft University of Technology
Originally presented on: 11/5/2013 16:30:00 - 16:45:00

Droplets hanging on a window are a common sight. A long standing problem is to predict the maximum size of a droplet that can hang on a window without rolling down. In this work we contribute to the solution of this problem for the model system of a droplet on an incline. We construct a model that allows us to a priori predict how a droplet deforms with increasing inclination angle and for what angle it rolls o .

The general approach to model this roll-o angle is to resolve the force balance between gravitational and pinning forces: $Bo \sin \alpha = kr(\cos \theta_r - \cos \theta_a)$. In this description, k is currently predicted to be a constant in contrast to recent experimental and numerical findings [1, 2]. In our model we take a more fundamental approach to determine the pinning force by predicting the shape of the droplet base and the contact angle distribution around the base. From this approach we predict the variable k. We find that it depends on the initial contact angle of the droplet and, surprisingly, also on the value of the Bond number and contact angle hysteresis ($\cos \theta_r - \cos \theta_a$) itself. Additionally, the model allows us to understand more generally how a pinned droplet will behave under given forcing.

[1] V. Berejnov and R. E. Thorne. E ect of transient pinning on stability of drops sitting on an inclined plane. Phys. Rev. E, 75:066308, 2007.
[2] M. J. Santos, S. Velasco, and J. A. White. Simulation analysis of contact angles and retention forces of liquid drops on inclined surfaces. Langmuir, 28(32):11819–11826, 2012.

(363g) Electrostatic Adsorption of Polyelectrolyte Brush-Grafted Nanoparticles At the Solid/Liquid Interface

Riley, J. K.,
Tilton, R. D., Carnegie Mellon University
Originally presented on: 11/5/2013 16:45:00 - 17:00:00

Nanoparticulate polymer brushes consist of densely packed, end-attached polymer chains extending from a nanoscale core. Examples include multi-arm star polymers with cross-linked cores or inorganic nanoparticles with chains grafted at high density from their surfaces. Similar to block copolymer micelles, they have a core/corona structure, but unlike micelles they are permanent, non-dissociating objects. When block copolymer micelles adsorb, they can disassemble so that all chains found at the surface are in direct contact with it. Adsorbed nanoparticulate brushes cannot do this, and a large fraction of their chains may have no direct contact with the surface. Adsorption of nanoparticulate brushes may therefore be advantageous for surface conditioning technologies that benefit from dense layers of highly extended polymer chains, such as steric stabilization or boundary lubrication.

Adsorption thermodynamics differ in significant ways for nanoparticulate brushes relative to linear polymer chains, particularly with regard to the strong inter-segment repulsions that exist within the brush and how they influence the configurational part of the adsorption entropy. Here we experimentally compare adsorption of nanoparticulate brushes and chemically similar linear polymer chains to the solid/liquid interface. This presentation focuses mainly on an annealed nanoparticulate polyelectrolyte brush, consisting of the weak polybase poly(2-(dimethylamino)ethyl methacrylate) grafted at high density from silica nanoparticles by surface-initiated atom transfer radical polymerization (PDMAEMA-Si). These are adsorbed from aqueous suspensions to negatively charged silica surfaces and compared to linear PDMAEMA adsorption. For both systems, adsorption balances electrostatic attraction to the surface with lateral electrostatic repulsions among neighboring species or intramolecular/intra-brush electrostatic repulsions. These are readily tuned by pH and ionic strength changes. The main points of discussion will concern the effects of altered electrostatic conditions on the extent of adsorption. In general, the extent of adsorption is significantly larger at any pH for the PDMAEMA-Si nanoparticulate brush than for linear PDMAEMA, due to the high density packing of polymer segments in the former. Consistent with a prior investigation of linear PDMAEMA adsorption to silica, a critical pH threshold is observed, above which little or no adsorption occurs. This happens because the weakly charged chains have insufficient attraction to the surface at elevated pH to offset the unfavorable adsorption entropy. In contrast, no such pH threshold is observed for PDMAEMA-Si nanoparticulate brushes, since the adsorption entropy is less unfavorable for this system, and even the weak residual attraction to the surface is sufficient to allow adsorption. This is consistent with prior models of star polymer adsorption. The overall structure of the adsorbed layers is interpreted in terms of pH effects on brush charge and swelling, aided by electrophoretic mobility and dynamic light scattering measurements on bulk suspensions, as well as streaming potential measurements and atomic force microscopy images of PDMAEMA-Si covered surfaces. When possible, this electrostatic-dominated system will be compared and contrasted with behaviors displayed by nonionic poly(ethylene oxide) star polymers and linear polymers.

(363h) Interfacial Activity of Polymer Grafted Nanoparticles

Kim, D., University of Houston
Krishnamoorti, R., University of Houston
Originally presented on: 11/5/2013 17:00:00 - 17:15:00

The oil-water interfacial tension is reduced when nanoparticles segregate to the oil-water interface and this segregation is governed by the interparticle interactions of the nanoparticles. In this study, hydrophilic polymers or hydrophilic-oleophilic copolymers were grafted from silica nanoparticles using a living radical atom transfer polymerization in order to tune the interparticles interaction. We have studied the nature of interparticle interactions of such nanoparticles using interfacial tension measurements, and their structure using small angle X-ray scattering. Amphiphilic polymers grafted from nanoparticles promote their segregation to the interface and make it possible to reduce the hexane-water interfacial tension significantly. Moreover, the hybrid particle size was the dominant variable in determining the critical micelle concentration and could be manipulated by changing the molecular weight and grafting density of polymer brushes. The results of this study can be applied to the development of particle-based oil dispersants or enhanced oil recovery agents to improve their efficiency and performance.

(363i) Study of the Formation and Morphology of Biodegradable Polymeric Janus Particles Via Atomistic Simulation of Interfacial Interactions

Tomassone, M. S., Rutgers University
Zhu, W., Rutgers University
Winkler, J., Rutgers University
Originally presented on: 11/5/2013 17:15:00 - 17:30:00

Janus particles are defined as isotropic amphiphilic particles having different chemical or physical properties in two surface regions. Biocompatible or biodegradable polymeric Janus particles have many potential applications in drug delivery, such as acting as a stabilizer for nanocrystal drug suspensions or site-specific drug delivery vehicles. The so-called internal phase separation method is widely used to produce polymeric Janus particles because it is applicable to most polymers and amenable to scale up. In this method, when the volatile organic solvent is completely evaporated, Janus particles are formed in the aqueous system. The morphology and stability of such particles depend on various interfacial tensions. It is inconvenient and costly to screen good combinations of biocompatible/biodegradable polymers capable of producing Janus particles by measuring surface/interfacial tensions in polymeric Janus particle aqueous system. Little atomistic simulation work has been done to explore the interfacial interactions governing Janus particle formation. Interfacial tensions and binding energies of various systems were approximated using atomistic simulations to determine whether Janus particle formation is favorable under the given conditions. The simulated systems consisted of poly(lactic-co-glycolic acid) (PLGA 50:50) and poly-e-caprolactone (PCL) as the matrix polymers and SDS as surfactant. Wang et al. developed a geometry map for predicting the final morphology of polystyrene (PS)/poly(methylmetacrylate) (PMMA) composite particles prepared via internal phase separation in which the particle morphology is dependent upon the interfacial tensions of the system. Specifically, the ratio of the interfacial tension between the aqueous phase and polymer phase to the interfacial tension between the oil phase and aqueous phase normalized by the interfacial tension between the oil and polymer phases corresponds to a region on the geometry map. Inside the region for Janus particle formation, there is a range of possible architectures ranging from acorn-like to crescent moon. When applied to our PLGA/PCL system, this map accurately predicted the resulting morphology of PLGA/PCL Janus particles formed by internal phase separation in an oil-in-water emulsion template based on interfacial tensions calculated for PLGA-PCL, PLGA-water, and PCL-water with and without surfactant in the aqueous phase. Additionally, the presence of drug was explored via simulations. In some cases, the inclusion of drug affects the thermodynamics of Janus particle formation. For example, the introduction of 5% w/w griseofulvin was found to significantly alter the interfacial tensions of the system, whereas the same mass fraction of curcumin does not. Simulation results were experimentally validated. The multitude of surfactant choices and drug compounds currently available makes it impossible to experimentally test each one to find the optimal formulation. Determining the effects of different drugs and surfactants on interfacial tensions and corresponding particle morphology via atomistic modeling prior to experimentation enables rational design of Janus particles.

(363j) Perspective of Spectroscopy in Structure, Function, and Formation Mechanism of Polymer Nanocomposites

Wei, S., Lamar University
Zhu, J., Lamar University
He, Q., Lamar University
Gu, H., Lamar University
Guo, Z., Lamar University
Originally presented on: 11/5/2013 17:30:00 - 17:45:00

Polymer nanocomposites have proven enormous contributions to many noteworthy technological fields. The rational design of such materials toward desired structure and properties requires powerful tools that can help elucidate the correlation between starting chemical components and the final products. Nevertheless, it is challenging especially for multifunctional polymer nanocomposites that integrate both functionalities and properties into one entity. Here we discuss the application of two particular spectroscopic techniques which have helped elucidate the formation mechanism of the PNCs, and meanwhile infer the function mechanism such as in the heavy metal remediation process. One is the surface molecular spectroscopy XPS that has been used to identify elemental composition and chemical speciation in the nanoscale region of the PANI-based nanocomposites before and after heavy metal remediation, which provided very insightful mechanism for the redox process. Another is the 57Fe Mössbauer spectroscopy that has been applied to probe the varied iron species in the PNCs, which indicated the interfacial interactions between the macromolecule chain and the iron nanoparticle surfaces and further attributed to the unique nanofiller structure formation of the polyolefin nanocomposites.

(367a) Phase Behavior of Gases in Ionic Liquids

Shiflett, M. B., DuPont Company
Yokozeki, A., DuPont
Originally presented on: 11/5/2013 15:15:00 - 15:35:00

This presentation will emphasize the importance of characterizing the global phase behavior of gases in ionic liquids and how this can provide insight into new applications. Solubility measurements of several gases in ionic liquids are discussed and important experimental details are highlighted when measuring the vapor-liquid equilibria using a gravimetric microbalance and vapor-liquid-liquid equilibria using a mass-volume technique. The global phase behavior has been well correlated with a modified cubic equation and many of the systems described exhibit type-III and type-V phase behavior according to the van Konynenburg-Scott classification. We have found that gases can exhibit different solubility behaviors in ionic liquids (*i.e.* physical and chemical absorption) and that these behaviors can be analyzed with the present equation of state using a simple association model and excess thermodynamic functions. Gas separation using ionic liquids is discussed which includes examples of ternary phase behavior (experiments and model calculations) and what will be required to develop future commercial processes.

(367b) The Viscosity of Ether-Functionalized Ionic Liquids

Chen, Z., Nanyang Technological University
Xue, T., Nanyang Technological University
Lee, J. M., Nanyang Technological University
Originally presented on: 11/5/2013 15:35:00 - 15:55:00

The development of low viscous and highly conductive ionic liquids is always being driven by the ambition to improve their performance in related fields, like heterogeneous catalysis and supporting electrolyte. Generally, a low-viscosity ionic liquid is characteristic of well-delocalized positive and negative charges, low-symmetry ions and highly flexible side chains, for example ether groups. Recently, hole theory indicated that: 1) the ions in ionic liquids cannot pack closely and lots of holes may come into existence; 2) the bulk volume consists of the inherent volume and the total volume of holes; and 3) a large amount of total hole volume is of advantage to a high rate of mass transfer and low viscosity.[1,2]

In this study, 12 ether-functionalized ionic liquids and their alkyl counterparts short of an alkoxy O atom were prepared and characterized to determine the reason for the alkoxy chain effect on viscosity, as shown in Figure 1, based on the ether chains' contributions to viscosity (η_O/η), the activation energy of viscosity (ΔE_η) and the total free volume (ΔFV_O), which was calculated from density and surface tension [$FV = 2N \times 0.6791(kT/\gamma)^{3/2}$].[2] For example at 25 °C, when the C_3H_7- group in [C_3H_7-MPyrr]NTf_2 was replaced by $C_2H_5OCH_2$-, the viscosity and its activation energy would be reduced from 62.4 cP and 28.83 kJ mol^{-1} to 38.7 cP and 25.29 kJ mol^{-1} respectively, yielding $\eta_O/\eta = 0.620$ (38.7/62.4) and $\Delta E_\eta = -3.54$ kJ mol^{-1} (25.29 - 28.83) respectively, accompanied by $\Delta FV_O = 4.76$ mL mol^{-1}, while the total free volume in [$C_2H_5OCH_2$-MPyrr]NTf_2 is 32.25 ml mol^{-1}. Although the obtained η_O/η, ΔE_η and ΔFV_O for the 12 pairs of ionic liquids are highly variable, both η_O/η and ΔE_η show a decreasing dependence on ΔFV_O. As a result, the alkoxy chain effect on reducing viscosity was ascribed to the ability of the highly flexible alkoxy chains to increase the total free volume for the convenience of mass transfer. In order to maximize the effect on decreasing viscosity, the alkoxy group was supposed to contain a rod-like alkyl tail and a spacer length as short as only one methylene between O atom and the cationic N atom, and be in a non-hydrogen-bonding environment.

[1] A. P. Abbott, *ChemPhysChem*, **2004**, *5*, 1242-1246.

[2] J. Z. Yang, X. M. Lu, J. S. Gui and W. G. Xu, *Green Chem.*, **2004**, *6*, 541-543

[3] Z. J. Chen, T. Xue, J. M. Lee, *Rsc Adv.*, **2012**, *2*, 10564-10574.

(367d) Time-Dependent Diffusion and Chemical Exchange in Mixtures of An Amine-Functionalized Ionic Liquid and Carbon Dioxide By NMR

Hazelbaker, E., University of Florida
Wang, H., University of Florida
Budhathoki, S., University of Notre Dame
Shah, J. K., University of Notre Dame
Maginn, E. J., University of Notre Dame
Vasenkov, S., University of Florida
Originally presented on: 11/5/2013 16:15:00 - 16:35:00

Task-specific ionic liquids (TSILs) containing amine functional groups are promising materials for carbon dioxide capture. In this work a novel diffusion NMR technique, which combines advantages of high magnetic field and high magnetic field gradients, was applied to investigate self-diffusion in mixtures of an amine-functionalized TSIL and carbon dioxide. The measured diffusion data reveal time-dependent diffusivities of carbon dioxide in the mixtures on the millisecond time scale. This dependence was attributed to the existence of chemical exchange of carbon dioxide, i.e. exchange of carbon dioxide between the chemisorbed and physisorbed states in the mixtures under conditions of diffusion studies. At the same time, ion diffusivities were found to be diffusion time independent. High field exchange NMR was used to confirm the existence of the chemical exchange process and to obtain the mean residence times of carbon dioxide in the chemisorbed and physisorbed states. The diffusion and exchange NMR studies were performed for different molar ratios of carbon dioxide and at different temperatures. The experimental results will be compared with the predictions of computer simulations performed for the same or similar mixtures of TSIL and carbon dioxide.

(367e) Investigation of the Feasibility of Flue Gas Separations Using An Ionic Liquid From a Thermodynamic Point of View

Raeissi, S., Shiraz University
Florusse, L. J., Delft University of Technology, Faculty of Applied Sciences, DelftChemTech (PCMT)
Peters, C. J., Petroleum Institute
Originally presented on: 11/5/2013 16:35:00 - 16:55:00

Just one decade ago, the term "ionic liquid" was unknown to the majority of chemical engineers. Today, this unique family of substances is being investigated so intensely that over four-thousand journal articles are being published on ionic liquids in a period of just one year. One of the areas that have attracted great research interest is the system of CO_2 and an ionic liquid (IL). A Scopus search on the combination of words "ionic liquid" and "carbon dioxide" leads to 250 results for the year 2012. Potential applications involving such systems include CO_2 absorptions and separations by ionic liquids. Due to their importance in the design of such processes, numerous publications have appeared, and are continuing to do so, on the phase behavior of CO_2 + ionic liquid systems. However, the number of articles that present the solubility of carbon monoxide (CO) in ILs is an order of magnitude less than the corresponding papers on CO_2. While the solubility of CO in ILs is not as high as that of CO_2, it is still significant enough to deserve further investigation. Potential applications include, for example, IL-enhanced reactions involving CO, or gas purifications with ILs. Flue gas, in particular, consists mainly of N_2, CO, H_2, CO_2, SO_2, and N_2O. With environmental laws becoming increasingly strict, the purification of gas effluents is gaining even more urgency. In this respect, various routes have been investigated. One potential path that can be investigated regarding its feasibility, is the separation of flue gas pollutants such as CO, CO_2, SO_2, N_2O, and CH_4 by ionic liquids, for example, in the form of supported ionic liquid membranes.

In this respect, this study investigates the high-pressure solubility of CO in 1-butyl-3-methylimidazolium bis(trifuoromethylsulfonyl)amide ([bmim][Tf_2N]) up to temperatures and pressures of 460 K and 10 MPa, respectively, using a synthetic method. The results obtained show that the extent of solubility of CO in this ionic liquid is not great. For example, the single gas solubility of CO at 323 K and 2 MPa is about 0.02 in mole fraction, while the corresponding values for CO_2, N_2O, and CH_4 at the same temperature and pressure are 0.3, 0.3, and 0.04, respectively. A very rough estimate indicates that at 323 K and 2 MPa, the ratios of single-gas equilibrium molar concentrations in [bmim][Tf_2N] of carbon dioxide to nitrogen is roughly $CO_2/N_2 \approx 17$, and the corresponding values for other gases are: $SO_2/N_2 >> 17$, $N_2O/N_2 \approx 17$, $CO/N_2 \approx 1$, $CH_4/N_2 \approx 2$. The solubilities of the different gases are higher at higher pressures, much more so for the highly soluble gases such as SO_2, N_2O, and CO_2 and much less for the low-soluble gases such as N_2 and O_2.

Therefore, flue gas purifications can be more efficient at higher pressures. Since the above estimates indicate single-gas solubility differences of up to one order of magnitude with one another, it should be possible, at least from a thermodynamic point of view, to absorb some of the impurities of flue gases in this ionic liquid in order to separate them from N_2 and O_2, before flue gas is vented into the atmosphere.

However, if one focuses on CO capture alone, the results of this study may seem discouraging at a first glance. Although CO was shown to dissolve better in [bmim][Tf_2N] than in most conventional organic solvents, it still does not have the high solubility that is desired in order to achieve easy and cheap separations. However, if one focuses on the common liquid absorbents that have been investigated for CO separations, it is seen that they are not capable to simultaneously remove other impurities from flue gas. On the other hand, ionic liquids are very good absorbents for CO_2, SO_2, and NO_x. Organic solvents are simply not good solvents for volatile gases such as CO, therefore to increase the solubility in a liquid, researchers have mainly focused on absorbents which form complexes with CO. Examples include COSORB, ammoniacal copper liquor, Fe(TIM), palladium complex, and selenium and secondary amine. However, such complex-forming absorbents do not make complexes with other flue gas pollutants. For example, in the COSORB process CO_2 and CH_4 are chemically inert to the solvent, CO_2 does not react with the Fe(TIM) solvent, the palladium complex is good for separating CO_2 and CH_4 from CO, and the selenium and secondary amine system can be used to separate CH_4 from CO. Therefore, using such absorbents for CO removal would require additional separation units in order to remove the other impurities of flue gases. But ILs can simultaneously dissolve many gases, and additionally, since gas solubility in an IL is a purely physical process with no chemical interactions involved, the ILs can be cheaply and very easily regenerated by pressure drop or temperature increase. One further advantage of ILs, compared to other liquid absorbents, is that upon regeneration, the separated gases are not polluted with the solvent, because an ionic liquid has insignificant vapour pressure. Because of all these considerations, flue gas purifications with ionic liquids deserve further investigations. However, it is suggested that future investigations should consider mixed gases and real flue gases instead, since the presence of other solutes in an IL have been shown to increase/decrease the extent of solubilities compared to their corresponding binary-pair values, i.e., to show cosolvency or antisolvency effects.

(367f) Free Volume Basis of Gas Solubility and Selectivity in Functionalized Imidazolium-Based Ionic Liquids

Shannon, M. S., University of Alabama
Bara, J. E., University of Alabama
Horne, W. J., University of Alabama
Originally presented on: 11/5/2013 16:55:00 - 17:15:00

Due to the tunable structures and functionalities of ionic liquids (ILs), current research efforts have introduced several unique materials for industrial applications, including solvents and media for CO_2 capture. Fundamental conclusions and aspects of these IL materials and their physical properties are imperative to understanding governing driving-forces for gas separations and underlying molecular structures. Recently, we investigated fractional free volume (FFV)-dependence of gas solubility in 165 existing and theoretical 1-*n*-alkyl-3-methylimidazolium-based ionic liquids (ILs) via COSMOtherm, a very powerful and rapid software utilized commonly now for predicting IL thermophyscial properties.

Our present study investigates the effects on gas solubility of novel functional groups by replacing *n*-alkyl chains with nitriles ($C_nH_{2n}C{\equiv}N$) and ethers or oligo(ethylene glycols) ($(-C_2H_4O-)_n$). Initial computational results with these new functional groups via COSMOtherm reflect deviations in FFV from previously studied [$C_n$mim] ILs. Polar substituents provide a higher CO_2 binding affinity and repulsion of non-polar gases (i.e. CH_4, N_2, etc.) leading toward improved CO_2 separation performance. Initial results suggest decreased FFV values for appended nitrile and ether groups when compared to previous [$C_n$mim] IL data, which would also imply decreased gas solubility. However, with this observed decrease in FFV due to chain aggregation of polar entities, ideal gas selectivity is shown to increase with these materials being more size selective within these microdomain/voids and unable to accommodate the larger CH_4 and N_2 molecules.

Previous models presented for gas solubility in ILs consist of a molar volume-basis for families of 1-*n*-alkyl-3-methylimidazolium ILs. Free volume within these ILs as well as the currently studied nitrile- and ether-based ILs was shown to be the underlying property driving gas solubility and selectivity. For light gases (i.e. CH_4 and N_2) considered in CO_2 separations, solubility increases with increasing FFV implying direct proportionality with free volume. However, solubility of CO_2 was observed to decrease with increasing FFV, with CO_2 selectivity being a function of free volume to the (-3/2) power. Controlling IL FFV via cation/anion combinations, as well as, cation functional groups can lead to improvements in CO_2 solubility and selectivity and permeability in IL-based membranes.

(367g) The Effects of Water Concentration On the Properties of Imidazolium-Based Ionic Liquids

Niazi, A. A., RWTH Aachen University
Crothers, A., RWTH Aachen University
Rabideau, B. D., RWTH Aachen University
Ismail, A. E., RWTH Aachen University
Originally presented on: 11/5/2013 17:15:00 - 17:35:00

In the conversion process of production of tailor-made biofuels from waste biomass, cellulose is solvated and broken down into glucose units using solvents such as ionic liquids (ILs). Because this process releases trapped water molecules from the cellulose matrix, and because water is known to inhibit the dissolution of cellulose in IL's, study of the effect of water content on local interactions between individual components of the system has an important role in understanding the details of the mechanism of cellulose dissolution using ILs. In this work, we have analyzed the results of series of molecular dynamics (MD) simulations of twelve different commercially available imidazolium-based ILs: a combinatorial set of four cations: [dmim], [emim], [pmim], and [bmim] and three anions: [DMP], Cl, and [Ac]. For each IL, we have tested the effect of presence of water molecules on structural properties and dynamics of systems with water mole fractions ranging from 0.0% to 99.9%. Particularly, we have examined bulk properties (density), structural properties (radial distribution functions, water clustering and hydrogen bonding) and dynamical properties (shear viscosity and diffusion coefficients) of those mixtures and have made direct comparisons with experimental works whenever possible. We have observed for each IL a critical concentration, usually at water mole fractions between 70% and 80%, at which the mixture undergoes a transition from water dispersed in IL to an aqueous solution.

(368a) Asymptotic Theory and Numerical Analysis for Unraveling the Stokes Flow Singularity At the Junction Between Solid and Porous Walls With Arbitrary Wedge Angle

Nitsche, L. C., University of Illinois at Chicago
Bernal, B.,
Originally presented on: 11/5/2013 15:15:00 - 15:30:00

We consider the two-dimensional Stokes flow singularity that arises at the vertex of an angular wedge with one solid wall and one porous wall. In the practical context of spiral-wound membrane modules, such discontinuous transitions in wall permeability occur between the membrane and either (i) the initial and final impermeable sections of the channel[1,2] (wedge angle 180 degrees), or (ii) transverse filaments of a ladder-type spacer[3,4] (wedge angle 90 degrees). On a macroscopic, outer length scale the well-known similarity solution[5-7] appears like a jump discontinuity in the normal velocity, characterized by a non-integrable $1/r$ divergence of the pressure. Our goal is to resolve the fine structure, particularly because reverse osmosis (e.g., desalination of seawater) is sensitive the local transmembrane pressure drop. Previous results for a 180 degree wedge[8] are here extended to arbitrary wedge angle Q.

The porous wall is characterized by a Darcy permeability for the seepage velocity and a Beavers-Joseph condition[9] for the slip velocity. Preliminary numerical solutions with a least-squares variant of the method of fundamental solutions indicate a continuous velocity coupled to a weaker, integrable singularity in the pressure field: $p \sim r^{a}$ with $a > -1$. However, inconsistencies in the numerically imposed outer boundary condition indicate a very slow radial decay, which becomes worse with decreasing angle Q. Thus, we undertake asymptotic analysis to: (i) understand the radial decay behavior; and (ii) find a more accurate far-field solution to impose as the outer boundary condition. Similarity solutions (involving a stream function that varies like some power of r) must be augmented with logarithmic terms to facilitate iterative satisfaction of the boundary conditions along the walls. We then use a hybrid computational scheme in which the numerics are iteratively patched to the outer asymptotics, thereby determining free coefficients in the latter. We also derive an inner asymptotic series and fit its free coefficient to the numerics at small r.

The flat wall (Q = 180 degrees) emerges as a very special, degenerate case — in which the slip condition plays no role. For any smaller wedge angle, the slip length appears as an important, intermediate length scale in addition to the inner Darcy length scale. Based upon a determinant criterion for the leading inner similarity solution, we devise a way to determine the power-law exponent $a(Q)$ independent of the numerics. An unexpectedly rich mathematical structure emerges, with multiple solution branches arising at both small and large Q.

Our results highlight the fruitful interplay between asymptotic theory and numerical analysis, both of which were needed to make progress on this problem.

REFERENCES

[1] Miranda, J. M. & Campos, J. B. L. M. 2001 An improved numerical scheme to study mass transfer over a separation membrane. *J. Membr. Sci.* **188**, 49–59.

[2] Geraldes, V., Semiao, V. & De Pinho, M. N. 2002 The effect on mass transfer of momentum and concentration boundary layers at the entrance region of a slit with a nanofiltration membrane wall. *Chem. Engng Sci.* **57**, 735–748.

[3] Geraldes, V., Semiao, V. & De Pinho, M. N. 2002 The effect of the ladder-type spacers configuration in NF spiral-wound modules on the concentration boundary layers disruption. *Desalination* **146**, 187–194.

[4] Geraldes, V., Semiao, V. & De Pinho, M. N. 2002 Flow management in nanofiltration spiral wound modules with ladder-type spacers. *J. Membr. Sci.* **203**, 87–102.

[5] Taylor, G. I. 1960 Similarity solutions of hydrodynamic problems. In *Aeronaut. Astronaut.*(Durand Anniv. Vol.), pp. 21–28. Pergmon.

[6] Taylor, G. I. 1962 On scraping viscous fluid from a plane surface. Reprinted in *The Scientific Papers of Sir Geoffrey Ingram Taylor*, vol. IV (ed. G. K. Batchelor), pp. 410–413. Cambridge University Press.

[7] Moffatt, H. K. 1964 Viscous and resistive eddies near a sharp corner. *J. Fluid Mech.* **18**, 1–18.

[8] Nitsche L. C. & Parthasarathi P. 2012 Stokes Flow singularity at the junction between impermeable and porous walls. *J. Fluid Mech.* **713**, 183-215.

[9] Beavers, G. S. & Joseph, D. D. 1967 Boundary conditions at a naturally permeable wall. *J. Fluid Mech.* **30**, 197–207.

(368b) Onset of Buoyancy-Driven Convection in Cartesian and Cylindrical Geometries

Myint, P. C., Lawrence Livermore National Laboratory
Firoozabadi, A., Yale University
Originally presented on: 11/5/2013 15:30:00 - 15:45:00

We perform a linear stability analysis to examine the onset of buoyancy-driven convection relevant to subsurface carbon dioxide sequestration in confined, porous Cartesian and cylindrical domains. We consider Cartesian geometries where the aspect ratio between the two horizontal dimensions is not necessarily equal to one. Two key elements of the stability analysis are the critical time and the critical wavenumber. Lateral boundaries have a much greater effect on the critical wavenumber than on the critical time. The confinement due to these boundaries impedes the onset of convection to the extent that convection cannot even occur in domains that are smaller than a certain size. Large aspect ratios can significantly reduce boundary effects. Patterns of the earliest-growing perturbation mode in the horizontal plane reveal many interesting features. We illustrate several differences between patterns in Cartesian geometries and patterns in cylindrical geometries. Our results may help with understanding the onset of buoyancy-driven convection in real systems where lateral boundary effects are significant.

(368c) Modeling of Particle-Particle and Particle-Wall Contact Points in the 3D CFD Simulation of Transport in Fixed Beds of Spheres

Dixon, A. G., Worcester Polytechnic Institute
Nijemeisland, M., Johnson Matthey Catalysts
Stitt, E. H., Johnson Matthey Catalysts
Originally presented on: 11/5/2013 15:45:00 - 16:00:00

The use of computational fluid dynamics (CFD) as a tool for obtaining detailed flow and heat and mass transport information in packed beds is growing, despite the apparent intractability of the geometry of the packing. In particular, the simulation by CFD of fixed beds at higher particle Reynolds number (Re > 1000) and low tube-to-particle diameter ratio (2 < N < 8) is of interest as highly-exothermic partial oxidation reactors and highly-endothermic steam reforming reactors operate under these conditions. Recently, studies that presented simulations of beds of a large number of particles (usually between 100 – 1000 spheres) have appeared in the literature, for flow and heat transfer simulations.

In the development of meshes for CFD simulations of transport in fixed beds of spheres, particle-particle and wall-particle contact points often present difficulties. This paper presents a comparison between four methods of dealing with contact points in CFD simulations of both fluid flow and heat transfer in fixed beds of spheres, at commercially-relevant flow rates. These methods represent unselective (global) and selective (local) geometric manipulation of the spherical particles through either reducing or enlarging them. We give results for drag coefficient (C_D) and heat flow (Q) for flow past sphere-sphere and wall-sphere contact points, focusing on higher flow rates typical of industrial steam reformers (500 < Re < 10,000). Global methods, in which all particles in a bed are either shrunk or enlarged uniformly, change bed voidage giving erroneous results for C_D. Local methods, in which bridges are inserted or spherical caps are removed only at the points of contact, give much better results for C_D. The bridges approach is preferable for heat transfer, as fluid gaps reduce heat transfer too much, and particle overlaps increase it. A set of graphs is presented to allow estimation of the error introduced by the various methods of dealing with contact points. Examples are presented of the application of these methods to fixed bed transport simulations.

(368d) Multicomponent Fuel Spray Evaporation in High Pressure Hot Gas Flows

Samimi Abianeh, O.,
Chen, C. P., University of Alabama in Huntsville
Mahalingam, S.,
Originally presented on: 11/5/2013 16:00:00 - 16:15:00

In this study, we investigate multicomponent fuel droplet dispersion and evaporation in a hot gas flow with elevated pressure environment. A new modeling methodology to simultaneously account for finite thermal conductivity, finite mass diffusivity and turbulence effects within atomizing multicomponent liquid fuel sprays has been developed. This new model incorporated the liquid turbulence effect in modeling the boundary layer heat and mass resistance during multi-component droplet evaporation. The atomization/spray model used is the T-blob/T-TAB model developed by the authors. This model supplies relevant turbulence characteristics within the atomizing droplets from the injector. The finite conductivity and mass diffusivity models are based on two-layer film theory, where the turbulence characteristics of the droplet are used to estimate the effective thermal conductivity and mass transfer diffusivities. Real fuels such as gasoline, diesel and bio-derived fuels are replaced by surrogate fuels based on a newly developed methodology using the inversed batch distillation concept. In the droplet model four regions are considered: the interior region of the droplet, droplet-side interface, gas-side interface, and the surrounding gas phase. Approximate solutions to the quasi-steady energy and mass transfer equation were used to derive an explicit expression for the heat and mass flux from the surrounding gas to the droplet–gas interface, and within the multi-component droplet. At the droplet/gas interface, extension of the phase equilibrium model to high pressures using the Peng-Robinson equation of state and activity coefficient is also considered. So far, the model has been validated with one-way coupled numerical calculations. The purpose of this study is to incorporate the KIVA-3V computational fluid dynamics (CFD) code utilizing hybrid Large Eddy Simulation for two-way coupled droplet-fluids interactions. Validations against available experimental data will be performed and discussed.

(368e) Multicomponent Mass Transfer in the Compressible Flow of Semi-Continuous Mixtures Using Adaptive Characterization Method

Jatob , L., Universidade Federal do Rio de Janeiro
Lage, P., Universidade Federal do Rio de Janeiro
Silva, L., Universidade Federal do Rio de Janeiro
Originally presented on: 11/5/2013 16:15:00 - 16:30:00

Continuous mixtures are characterized by a large amount of components with similar properties which makes the determination of its exact composition unfeasible. These mixtures can be represented by a continuous component using a distribution function to characterize its composition. The distribution variable can be the molar mass or any other convenient property. Examples of continuous mixtures are oil fractions and polymer solutions. If the mixture has some known components, it is treated as a semi-continuous mixture. In order to solve the mass transport equation of a continuous component using the conventional discrete component approach, the distribution function has to be described by pseudo-components which are usually determined by quadrature based methods. However, the elevated number of pseudo-components required to capture the composition changes during mass transfer process still represents a drawback for CFD simulations. The present work developed the mathematical model and numerical solution of the multicomponent mass transport equations for the flow of semi-continuous mixtures. The model is an extension of the adaptive Quadrature Method of Moments (QMoM) for continuous thermodynamics to field problems. The method is called Direct QMoM for continuous thermodynamics. It was implemented in a CFD open-source package (OpenFOAM) as a compressible ideal gas flow solver. The DQMoM for continuous thermodynamics was compared with the conventional discrete component model (DCM) in a mixing flow of two mixtures consisting of 57 hydrocarbons and nitrogen. It was shown that 4 adaptive DQMoM pseudo-components were enough to reproduce the properties of the DCM mixture with a 2% accuracy. Furthermore, the DQMoM CFD solution was approximately two times faster than the DCM solution for the mixture flow.

(368f) A Modified Dynamic Method for Measuring kLa

Damiani, A., Auburn University
Kim, M. H., Auburn University
Liang, M., Auburn University
Wang, J., Auburn University
Originally presented on: 11/5/2013 16:30:00 - 16:45:00

The oxygen transfer rate (OTR) is a crucial parameter in aerobic and microaerobic processes, since it is often the rate-limiting step due to the low solubility in the broth. Therefore, it is essential parameter to accurately measure the mass transfer coefficient for process control, design, and scale-up of bioreactors. Both physical and chemical methods have been proposed for the measurement of the mass transfer coefficient,. Among them, a physical method named dynamic method[1]has been widely used to measure the with the presence of cells.

In general, the dynamic method consists of two steps: consumption and absorption. During consumption, aeration is stopped and dissolved oxygen (DO) decreases due to cell respiration, therefore the graphical slope of DO can be used to estimate the oxygen uptake rate; during absorption, aeration is resumed, and the DO increases until a steady-state is reached. With determination of the oxygen uptake rate from the consumption process, the can now be calculated from the absorption process.

Despite the many successful applications of the dynamic method, it has difficulties to obtain accurate measurement in at least two common cases. First, when cell density is low while agitation is high, a linear decreases of DO during consumption process cannot be obtained, which makes the accurate estimate of oxygen pick up rates by the cells impossible. The departure from the linearly decreased DO is caused by the mass transfer between head-space and liquid phase, which is not considered in the dynamic methods. Second, when cell density is high while agitation is low, DO barely increases during the absorption process, which makes the estimate of impossible. This is caused by the limited mass transfer between the bubbles and the liquid phase which is lower than the oxygen pick up rate.

In this work, a modified dynamic method is developed and tested to circumvent these difficulties. In the modified dynamic method, both mass transfer between the headspace and liquid phase, and mass transfer between bubbles and liquid phase are considered; in addition, during the consumption process, nitrogen gas is bubbled through the broth which not only expedites the desorption process, but also enables the measurement this step alone.

We used *Scheffersomyces stipitis* as the model system to test our method. Since *S. stipitis* is highly sensitive to oxygen supply, the determination of an optimal OTR is very important for improving product yield[2]. We have conducted a series of experiments with different cell density and agitation speeds to validate the modified dynamic method. Our results show that the parameters estimated from the consumption and absorption processes agree very well with each other; in addition, parameters estimated from liquid phase and headspace measurements also agree very well with each other.

Our experimental results show that modified dynamic method allows researchers in the biotechnology field to measure more easily and accurately. In addition, we found that S. stipites achieved a maximum value at OD_{600}= 3.0 and then proceeded to steadily decay with cell density increases, which might be caused by the changing interfacial properties of the fermentation broth. The headspace mass transfer coefficient increased with agitation speed, while oxygen uptake rate increased with cell density and decreased with agitation speed.

References

1. Biotechnology Advances (2009) 27: 153–176
2. Biotechnology and Bioengineering (1989) 34:398-402

(368g) Asymptotic Analysis of Double-Carrier, Space-Charge-Limited Transport in Organic Light-Emitting Diodes

Feicht, S.,
Schnitzer, O., Technion
Khair, A. S., Carnegie Mellon University
Originally presented on: 11/5/2013 16:45:00 - 17:00:00

In organic light-emitting diodes (OLEDs), an applied voltage drives the injection and transport of charge carriers, electrons and holes, across an organic semiconducting thin film; the carriers then recombine to generate light emission. We analyze electron and hole transport in OLEDs via the drift-diffusion equations. We focus on space-charge-limited transport, in which rapid variations in charge carrier density occur near the injecting electrodes, and in which the electric field is highly nonuniform. This motivates our application of singular asymptotic analysis to the drift-diffusion equations. In the absence of electron-hole recombination, our analysis reveals three regions within the OLED: (i) "space-charge layers" near each electrode whose width λ are much smaller than the device width L, wherein carrier densities decay rapidly and the electric field is intense; (ii) a "bulk" region whose width is on the scale of L, where carrier densities are small; and (iii) intermediate regions bridging (i) and (ii). Our analysis shows that the current J scales as $J \sim \varepsilon\mu V^2/L^2\lambda$, where V is the applied voltage, ε is the permittivity, and μ is the electric mobility, in contrast to the familiar diffusion-free scaling $J \sim \varepsilon\mu V^2/L^3$. Thus, diffusion is seen to lead to a large $O(L/\lambda)$ increase in current. Finally, we derive an asymptotic recombination-voltage relation for a kinetically-limited OLED, in which charge recombination occurs on a much longer timescale than diffusion and drift.

(368h) Modeling Reactive Heat and Mass Transfer in Biogas Production Unit

Horgue, P., Universit de Toulouse ; INPT, UPS ; IMFT (Institut de M canique des Fluides de Toulouse)
Debenest, G., Universit de Toulouse ; INPT, UPS ; IMFT (Institut de M canique des Fluides de Toulouse)
Originally presented on: 11/5/2013 17:00:00 - 17:15:00

Biogas production is one of the existing ways to produce renewable energy from wastes. This is mainly governed by multiphase transport processes and biological reactions without oxygen. Consequently, a fine description of those phenomena at the production unit scale is required in order to characterize the biogas production. It is necessary to take into account the various heterogeneities (phase distribution, gas composition, temperature) inside the production unit to accurately predict the installation efficiency.

The goal of this work is the development and the validation of a CFD software solving the complete set of equations constituting the model. The code is developed using the OpenFOAMÃ‚® platform, an open-source CFD toolbox highly programmable, which enables parallel computation and disposes powerful pre- and post-processing tools.

The set of equations is solved sequentially, decoupling the hydrodynamics part from component transport and heat transfer. First, gas and liquid flows in the porous medium are solved using the IMPES method (Implicit Pressure and Explicit Saturation) with, for each equation, an adaptive independent time-step. Then, the reactive mass transfer is solved decoupling the purely reactive part from the transport part. Finally, heat transfer is computed using the fields previously updated.

Results on the hydrodynamic part are first validated on the Buckley-Leverett case by comparison with the analytical results. Then, results of the complete model are compared with experimental data obtained on a pilot unit.

(368i) Modeling of Hydrodynamic Interactions of Large Groups of Swimming Microorganisms in Viscoelastic Fluids

Bozorgi, Y., Rensselaer Polytechnic Institute
Underhill, P. T., Rensselaer Polytechnic Institute
Originally presented on: 11/5/2013 17:15:00 - 17:30:00

Hydrodynamic interactions of swimming microorganisms can lead to coordinated behaviors of large groups. The impact of viscoelasticity on the collective behavior of active particles driven by hydrodynamic interactions has been quantified with the inclusion of rotational diffusion. Oldroyd-B, Maxwell, and generalized linear viscoelastic modeled are considered as the constitutive equation of the suspending fluid, inspired by some biological fluids. A mean field assumption is used to model the suspension dynamics near an isotropic state. The onset of instability has been quantified by a linear stability analysis in terms of wavenumber, diffusivities, and constitutive equation parameters. Some key results are in contrast to suspensions in Newtonian fluids. The maximal growth rate can occur at a particular wavelength, and diffusion can act to make the system more unstable. Viscoelasticity can also affect the long time dynamics of the continuum equations.

(368j) Equivalents - A New Concept for the Prediction of Transport

Churchill, S. W., The University of Pennsylvania

Originally presented on: 11/5/2013 17:30:00 - 17:45:00

I recently received a written communication that asked why the prediction by an expression of Churchill and Chu in 1975 of the Nusselt number for free convection from a cylinder in the limit of $Ra \rightarrow 0$, namely 0.60, was twice that by an expression of Churchill and Bernstein in 1977 for forced convection from a cylinder in the limit of $Re \rightarrow 0$, namely 0.30. That discrepancy was exaggerated by a misinterpretation ? the limiting value in the correlating equation for free convection was actually $(0.60)^2 = 0.36$, but there should not have been any difference. Both 0.30 and 0.36 were artifacts of independent ill-conceived attempts to extend by means of an additive constant the applicability of the predictive expressions to values of *Gr* and *Re* below the limits of validity of thin laminar boundary layer theory. A better means of extension is presented herein, but attention is first turned to a matter of broader scope that was prompted by the inquiry, namely the possible implications and applications of a quantitative equivalence between free and forced convection.

The dimensionless grouping which determine the relative rates of combined free and forced convection in the thin laminar boundary layer regime in the asymptotic limits of $Pr \rightarrow 0$ and ∞ were noted by Acrivos in 1966, by Morgan in 1975, and by myself in 1983, but practical applications were not identified by any of us. The current investigation has revealed that the equivalence is functionally independent of but numerically dependent on the Prandtl number, that it may be extended for other thermal boundary conditions and for other geometries, that it may contribute to understanding in the classroom, and that it has the potential for prediction in a practical sense.

Analogies have long had a prominent role in chemical engineering design and analysis despite their flaws in both a fundamental and a practical sense. For example, that between momentum and energy transfer allows the heat transfer coefficient to be predicted from a correlating equation for the friction factor. Unfortunately, the predictions of the most widely used analogy are in error by as much as 40% for the very conditions for which it was devised. The newly identified equivalences are found to have a sounder rationale than the analogies, and their predictions to be more accurate.

The Grashof number that produces the same value of the local Nusselt number as the Reynolds number for the same geometrical configuration and the same Prandtl number was chosen as a marker of the equivalence rather than the Rayleigh number because that choice simplifies the ultimate expressions.

The first equivalence to be examined was that for an isothermal plate in the thin laminar boundary layer regime as represented by expressions in the form of the power-mean of asymptotes for limiting values of the Prandtl number. Fortuitously, the first description of that methodology for the construction of correlative/predictive expressions by Churchill and Usagi in 1972 included the following expression for the local Nusselt number for forced convection from an isothermal plate:

(1)

The coefficients 0.5027 and 0.4914 have a theoretical basis. The value of -9/4 for the combining exponent was determined empirically but its ubiquity in such expressions implies the existence of an as yet unidentified theoretical rationale.

Although this is not the place for a detailed discussion of the idealizations of thin laminar boundary layer theory, it should at least be mentioned that some limited range of flow is presumed. In the case of forced convection that implies a particular range of the Reynolds number and in the case of free convection a particular range of the Grashof number.

Churchill and Ozoe in 1973 devised a correlative/predictive expression for the local Nusselt number for forced convection from an isothermally plate in the thin laminar boundary layer regime. They derived the asymptotes and used their own numerical solutions for several finite values of *Pr* to evaluate a combining exponent of -1. The result is

(2)

Just as with Eq. 1, the coefficients have a theoretical basis and, again by virtue of its ubiquity,

the combining exponent probably has an as yet unidentified rationale.

Equating the right-hand sides of Eqs. 1 and 2 results in

= (3)

which can be re-arranged and simplified as

$$Gr_x = A\{Pr\}Re \quad (4)$$

where

$A\{Pr\}$ (5)

According to Eq. 4, Gr_x is equivalent to *Re*times a function of *Pr*. This result suggests that the concept of equivalence has potential for prediction and for understanding. As an example of the latter, it may be inferred that free convection generates an equivalent velocity equal to $(xg\beta\Delta T/A\{Pr\})^{1/2}$.

The combination of Eqs. 4 and 5 is apparently the first expression to be identified for an equivalence between free and forced convection for all values of *Pr,* and its generality exceeds expectations in several respects. First, for any finite value of *Pr*, the function $A\{Pr\}$becomes simply a numerical value. Second, an equivalence for the integrated-mean Nusselt number, , can be derived simply by multiplying the leading coefficient for free convection by 4/3 and that for forced convection by 2. The net result is that the leading coefficient of Eq. 5 is replaced by $0.2061(3/2)^4 = 1.043$. Third, insofar as the Grashof number is defined as it is for a uniform wall temperature, that is in terms of the difference between the local wall temperature (now a dependent variable) and the ambient temperature, the expressions for uniform heating are identical to those for a uniform wall temperature except for the numerical values of the three coefficients. Fourth, the corresponding expressions for a horizontal cylinder and for a sphere are also identical to those for a flat plate except for the numerical values of the three coefficients.

If Eq. 4 is to be predictive in a quantitative sense for other geometries and thermal boundary conditions, the function $A\{Pr\}$ must be generalized and/or predictable. Considerable progress has been made in those respects and is to be reported in the CDROM version and the presentation.

The complete differential models for steady flow over a long horizontal cylinder and for heat transfer by

***** Formatting Issue - Best Available Paper/Graphic *****

thermal conduction from a long horizontal cylinder are ill-posed in a strict mathematical sense but the postulate of a thin laminar boundary layer and the neglect of the effect of the wake and plume on heat transfer have permitted the derivation of the afore-mentioned approximate solutions.

The various attempts to encompass the regime of a thickening boundary layer by the addition of a limiting value to the solutions for a flat plate and a cylinder failed because the experimental values of the Nusselt number appear to approach zero as does *Gr* for free convection and *Re* for forced convection. Langmuir in 1912, in the process of modeling heat transfer within a partially evacuated light-bulb, devised an approximate solution for the effect of the cylindrical curvature of the filament that can be adapted for the regime of thickening. He began by postulating that the heat loss from the filament by free convection could be represented by thermal conduction across a hypothetical "sheath" (annular ring) of stagnant gas of thickness $d = (D_o - D_i.)/2$. He correctly expressed the ensuing heat flux density at the surface of wire in terms of the logarithmic-mean area and thereby obtained

(6)

Equation 6 can be expressed more simply and generically as

(7)

Here represents the heat transfer by conduction through a flat "sheath" of gas and

through a cylindrically curved one. Equation 7 is promising in that the hypothetical thickness of the sheath does not appear explicitly, but the prediction of the Nusselt number for convection from a cylinder from that for a flat plate is highly inaccurate. If, however, Eq. 7 is re-expressed as

(8)

it predicts values in good agreement with experimental data for the entire laminar boundary Nu_l, from those for thin laminar boundary layer theory as symbolized Nu_{tlbl} . Equation 8 proves to be applicable for both flat plates and cylinders and for both free and forced convection. The latter duality resolves the discrepancy that prompted this investigation.

Theoretically based correlating equations for the integrated-mean Nusselt number for an isolated sphere in laminar flow generally consist of the sum of a closed-form solution for convection in the thin laminar boundary layer regime and the exact solution for pure conduction to unbounded surroundings. That is

$+ 2$
(9)

Curiously. the application of Langmuir's concept of approximating convection by conduction across a stagnant film in a spherical annulus, abetted by the geometric-mean area, results in

$+ 2$ (10) The forgoing is only a sample of the potential applicability of the concept of equivalence

between free and forced convection, which is also found to be applicable to turbulent and confined flows.

AUTHOR INDEX

AUTHOR INDEX

AUTHOR INDEX

AUTHOR INDEX

AUTHOR INDEX

AUTHOR INDEX

AUTHOR INDEX

AUTHOR INDEX

AUTHOR INDEX

AUTHOR INDEX

AUTHOR INDEX

AUTHOR INDEX

AUTHOR INDEX

AUTHOR INDEX

AUTHOR INDEX

AUTHOR INDEX

AUTHOR INDEX

AUTHOR INDEX

AUTHOR INDEX

AUTHOR INDEX

AUTHOR INDEX

AUTHOR INDEX

AUTHOR INDEX

AUTHOR INDEX

Engineering Sciences and Fundamentals 2013

Core Programming Area at the 2013 AIChE Annual Meeting: Global Challenges for Engineering a Sustainable Future

San Francisco, California, USA
3 – 8 November 2013

Volume 2 of 2

ISBN: 978-1-63439-039-2

Printed from e-media with permission by:

Curran Associates, Inc.
57 Morehouse Lane
Red Hook, NY 12571

Some format issues inherent in the e-media version may also appear in this print version.

Printed by Curran Associates, Inc. (2014)

For permission requests, please contact AIChE
at the address below.

AIChE
3 Park Avenue
New York, NY 10016-5991

Phone: (203) 702-7660
Fax: (203) 775-5177

www.aiche.org

Additional copies of this publication are available from:

Curran Associates, Inc.
57 Morehouse Lane
Red Hook, NY 12571 USA
Phone: 845-758-0400
Fax: 845-758-2634
Email: curran@proceedings.com
Web: www.proceedings.com

TABLE OF CONTENTS

VOLUME 1

VOLUME 2

(379a) Structure Formation During Vibration of Suspensions of Hard Spheres

Staton, J.,
Dungan, S. R., University of California, Davis
Phillips, R. J., University of California, Davis
Originally presented on: 11/5/2013 15:15:00 - 15:30:00

Complex fluids, when subjected to vertical vibration, have been shown to exhibit a variety of interfacial morphologies considerably different from those observed in Newtonian fluids. Currently there is no accepted model or theory to explain many of the observed phenomena. Here we investigate the surface instabilities in thin layers of vertically vibrated, highly-filled dispersions of glass microspheres in aqueous solutions of varying density. In contrast to what has been reported previously, we find that the behavior is similar to that of a Newtonian fluid when the fluid and particle densities are approximately equal. Large fluid-particle density differences and low particle volume fractions reveal standing waves around the outer edge of the dispersion. At high particle volume fraction, spontaneous craters form and grow into protrusions that quickly collapse. Under some conditions, the suspension gathers itself into a single, spherical ball floating on a thin fluid layer. Fluid density, volume fraction and particle size impact the presence and topology of the observed phenomena.

(379b) Particle Sedimentation in Yield Stress Fluids With Different Microstructures

Emady, H. N., Purdue University
Caggioni, M., Procter and Gamble Co.
Spicer, P. T., Procter and Gamble Co.
Originally presented on: 11/5/2013 15:30:00 - 15:45:00

Yield stress fluids are widely used in practical applications and deeply studied by the soft matter community. Conceptually, a yield stress fluid is easy to describe: A solid that can be transformed into a liquid by applying a minimum stress on it and will re-solidify once the applied stress is removed. Yield stress is useful for stably suspending particles against sedimentation if the fluid yield stress is greater than the particles' stress. Current methods to predict particle sedimentation are based on the calculation of a single dimensionless yield stress parameter that compares this stress ratio. However, wide variations have been observed in the critical yield stress parameter. We begin to explain these variations by examining the fluid microstructure. Based on the microstructure at the origin of the yield stress, we propose a distinction between yield stress provided through colloidal glasses and yield stress provided through colloidal gels. We compare the sedimentation behavior of a single sphere in commonly-studied Carbopol, a colloidal glass, and in microfibrous cellulose, a colloidal gel, and show that fluids with the same yield stress can differ widely in their ability to suspend a particle. Therefore, in addition to yield stress, fluid microstructure must be characterized in order to better predict sedimentation in these fluids.

(379c) Endothelial Cells Migrate Upstream and Align Against the Shear Stress Field Created By Impinging Flow

Ostrowski, M. A., Stanford University

Huang, N. F., Stanford University

Walker, T. W., South Dakota School of Mines and Technology

VerWijlen, T.,

Poplawski, C.,

Khoo, A. S., Stanford University

Cooke, J. P., Stanford University

Dunn, A. R., Stanford University

Fuller, G. G., Stanford University

Originally presented on: 11/5/2013 15:45:00 - 16:00:00

Endothelial Cells Migrate Upstream and Align Against the Shear Stress Field Created by Impinging Flow

Spatial gradients in wall shear stress have emerged as an important factor for cardiovascular development and disease. High shear and vortical flow were required for normal heart development in zebrafish, where occluded flow results in an abnormal third chamber.[1] Large shear stress gradients likewise exist in the heart tube of the developing quail embryo, where high localized shear stresses in the outflow tract coincide with the future locations of aortic and pulmonary valve formation.[2,3] ECs lining the aortic valve leaflets, where high shear stress gradients occur, align perpendicular to flow and may reflect a response to the high shear stress present in the left ventricle.[4] Importantly, the mechanisms underlying ECs response to spatial gradients in wall shear stress remain essentially unknown. In order to study the influence of shear stress gradients on endothelial cell response, we built an impinging flow cell that exposes endothelial cells to a gradient of shear stress whose slope can be tuned by changing the flow rate. We investigated the response of endothelial cells to shear stress gradients that ranged from 0 to a peak shear stress of between 9-220 dynes/cm^2.

Materials and Methods

We built an impinging flow cell that exposes endothelial cells to a gradient of shear stress. The impinging flow cell device is used to produce gradients in shear stress while imaging the cellular migration and reorientation over a 21-hour period. A photo of the device is shown in Fig. 1a. The device was fabricated out of acrylic in order to enable live-cell imaging by allowing light to pass through the transparent device and illuminate the cells while tracking their response to impinging flow. The device is mounted on an x-y-z translation stage suspended at a fixed height of 1 mm over a cell culture dish seeded with ECs. The impinging flow exit orifice is oriented normal to the EC surface and dispenses fluid (cell culture media) onto the EC monolayer seeded on a cell culture dish The orifice radius is 0.35 mm. A pump is used to recirculate cell culture media onto the cell surface.

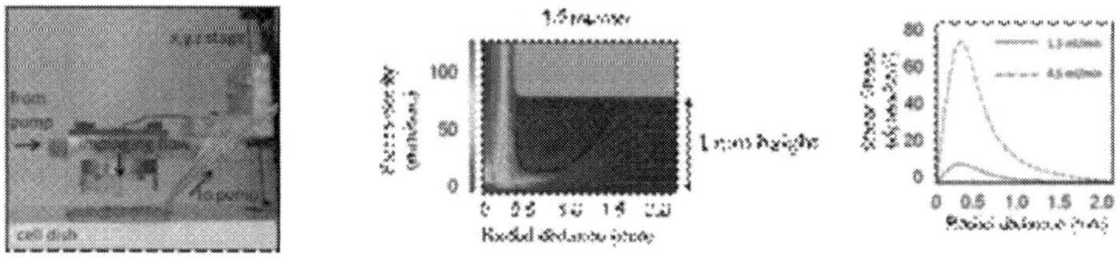

Figure 1. Device design and modeling of impinging flow cell. (a) Impinging flow cell device is used to produce gradients in shear stress while imaging the cellular migration and reorientation. The fluid flow enters the impinging flow chamber shown at the left and impinges onto the cell culture dish normal to the cell surface. Fluid leaves the dish at the same rate through a tube located at the edge of the dish via recirculation by a peristaltic pump (b) Velocity flow field from finite element analysis performed in Comsol shown in vertical cross section for flow through an impinging flow cell. Flow rate of 1.5 mL/min for an orifice with a 0.35 mm radius located 1 mm above the cell surface is modeled. Fluid velocity magnitude is represented with color, fluid direction with arrows and flow streamlines with black lines. The conditions show a low flow region (depicted in blue) at $r = 0$ and far from the orifice. The inlet flow rate corresponds to jet Reynolds numbers of 67 and demonstrates that the flow is in the laminar regime. (c) Wall shear stress on the cell culture dish is plotted as a function of radius from the impinging flow. Shear stress ranges from zero (at $r = 0$) directly under the exit of the impinging flow cell orifice (the stagnation point) and increasing symmetrically and radially outward to a maximum at a distance of 0.350 mm. The region of maximum in shear stress is radially symmetric and corresponds to the radius of the impinging exit orifice. Shear stress values from the analysis are plotted for flow rates of 1.5 mL/min and 4.5 mL/min.

Results

At high confluency, endothelial cells migrate *against* the direction of fluid flow for all flow rates tested and concentrate in the region of maximum wall shear stress, while sub-confluent endothelial cells migrate *with* the flow direction. In addition, endothelial cells align parallel to the flow at low wall shear stresses but orient perpendicular to the flow direction and in a densely packed arrangement above a critical threshold in local wall shear stress as shown in Fig. 2b, within region 2.

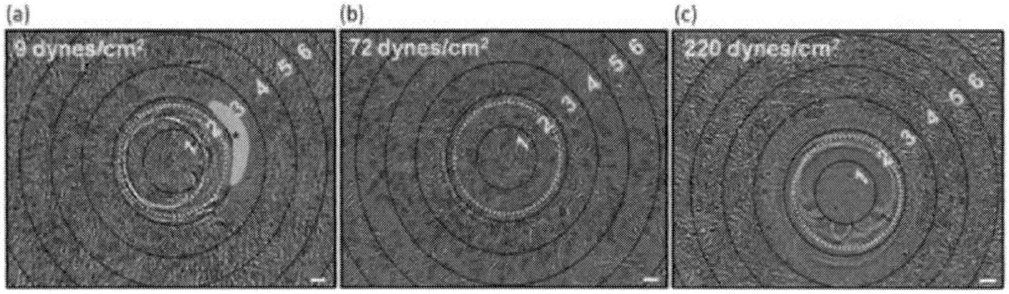

Figure 2. Orientation analysis of HMVEC's exposed to flow in the impinging flow cell for 21 hours. The peak location in shear stress is shown as a yellow dashed ring. Cells have been colored to clearly observe their long axis with respect to the fluid flow direction. *Blue* indicates EC alignment parallel with flow, *red* is perpendicular to flow, and *green* is intermediate orientation. Blank lines indicate concentric rings with a spacing of 0.185 mm to show axisymmetric cell response. (a) Image of HMVECs exposed to 1.5mL/min flow producing peak WSS of 9 dynes/cm^2. Cells beyond the shear stress maximum are primarily oriented with flow (regions 3, 4 and 5), while cells in the stagnation region (region 1) are predominantly unaligned. Gray region with '*' represents an area where valid EC images were not obtained because of poor contrast. (b) HMVECs exposed to impinging flow rate of 4.5 mL/min with maximum shear stress of 72 dynes/cm^2.

Cells in the vicinity of the wall shear stress maximum (region 2) are predominantly oriented perpendicular to flow (red), while cells beyond this region are oriented parallel (blue) to the flow. (c) HMVECs exposed to 9 mL/min flow with peak WSS of 220 dynes/cm^2. Cells in regions 1, 2, and much of 3 were ablated due to high shear stress. Cells in regions 3-5 oriented perpendicular to flow. Scale bars equal 100 μm.

Our observations suggest that endothelial cells are exquisitely sensitive to both the magnitude and spatial gradients in wall shear stress. The perpendicular reorientation of ECs at high shear stress values agrees with *in vivo* data where ECs are observed to orient perpendicular to flow at the valve leaflet where shear stress is high. Their resulting migration and polarization may play presently unrecognized roles both during cardiovascular development and disease, particularly in regions of complex flow. The mechanisms underlying these responses may additionally play important roles during development, where high shear and large gradients exist transiently during heart development.

1. Hove, J. R. *et al.* Intracardiac fluid forces are an essential epigenetic factor for embryonic cardiogenesis. *Nature* (2003).

2. Peterson, L. M. *et al.* 4D shear stress maps of the developing heart using Doppler optical coherence tomography. *Biomed Opt Express* **3**, 3022-3032 (2012).

3. Garita, B. *et al.* Blood flow dynamics of one cardiac cycle and relationship to mechanotransduction and trabeculation during heart looping. *Am J Physiol Heart Circ Physiol* **300** (2011).

4. Deck, J. D. Endothelial cell orientation on aortic valve leaflets. *Cardiovasc Res* **20**, 760-767 (1986).

(379d) Dynamics of Repulsive and Attractive Colloidal Glasses Under Constant-Stress Deformation

Chan, H. K., University of California, Irvine
Mohraz, A., University of California, Irvine
Originally presented on: 11/5/2013 16:00:00 - 16:15:00

External stresses can cause dense colloidal suspensions to exhibit shear melting, delayed yielding, creep, and other complex rheological behavior. Such phenomena are not only of utter practical and technological importance for the use and processing of colloid-based materials, but also play a role in the physics of dynamical arrest and jamming transitions in colloidal glasses and gels. However, the underlying origins of these phenomena remain poorly understood without the ability to characterize in detail how the microstructure of these materials evolves during deformation under externally applied stresses. In this study, we use a custom-built shear cell coupled to a fast-scanning confocal microscope to directly visualize the creep deformation and yielding behavior of a dense colloidal suspension in which the interparticle interactions can be varied systematically from nearly hard-sphere to attractive. Investigation of particle rearrangements in response to different applied stresses allows us to capture and contrast the dynamic events that trigger the transition from slow, prolonged creep to liquid-like flow in both repulsive and attractive glasses. We will discuss our findings in light of the widely accepted bond- and cage-breaking yielding mechanisms in these systems.

(379e) Effect of Particle Stiffness On Glassy Dynamics of Dense Colloidal Liquids in the Bulk and Confined Regimes

Seekell, R. III, Villanova Univesity
Sarangapani, P., University of Notre Dame
Zhu, Y. E., University of Notre Dame
Originally presented on: 11/5/2013 16:15:00 - 16:30:00

"Fragile" glassy materials show a non-Arrhenius dependence of relaxation time with temperature close to the glass transition and have been extensively studied for molecular glass formers as model "hard-sphere" colloidal suspensions, but we lack a complete understanding of "strong" glass formers which show an Arrhenius dependence on temperature approaching the glass transition. In this work, we investigate the glassy dynamics of microgels of varied particle stiffness in dense aqueous suspensions using confocal microscopy. Poly(N-isopropylacrylamide) (PNIPAM) microgel particles of variable stiffness in aqueous media are synthesized by free radical polymerization varying the cross-linking density. We investigate five separate crosslinking densities to fully encompass the "soft" to "stiff" transition. Our results show that "soft" particles and "stiff" particles behave as ideal strong glass formers and fragile glass formers, respectively. By exploring the intermediate crosslinking densities, we offer insight into the transition from a strong glass former to a fragile glass former, specifically that dynamic heterogeneity, specifically string-like motion, is more pronounced as stiffness increases. Furthermore, when these glasses are confined to a thickness of several particle diameters, these suspensions show arrested particle motion at smaller volume fractions while the dynamic heterogeneities show notable increase.

(379f) From Transition to Fully Developed Particulate Turbulent Flows: the Effect of Finite Size Particles

Climent, E., IMFT
Loisel, V., Laboratoire Genie Chimique
Abbas, M., Laboratoire de Genie Chimique
Masbernat, O., Laboratoire de Genie Chimique
Originally presented on: 11/5/2013 16:30:00 - 16:45:00

The effect of particles on turbulence is a key phenomenon in many practical industrial flows (slurries, fluidized beds, ...). Depending on the size ratio between the particles and the smallest scale of turbulence, the particle feedback on the flow can either enhance or suppress turbulence. In this communication, we will focus on two generic flow configurations: homogeneous isotropic turbulence seeded with particles and a transitional plane channel flow.

Fully-coupled numerical simulations are used to investigate the interactions between neutrally-buoyant finite-size particles and the fluid flow. The numerical method chosen for this work is the Force-Coupling Method (FCM) (Climent and Maxey, 2009). It is fully-resolved in the sense that the fluid equations are solved at a length-scale smaller than the particle radius. The FCM is based on a low-order, finite multipole expansion of the velocity disturbance induced by the presence of the particles. The presence of the dispersed phase in the fluid is then represented by a body force distribution written as a multipole expansion where the first term is the monopole representing the force that the particle applies on the fluid (due to an external forcing or particle-to-particle contact forces). The second term is the dipole tensor. Its anti-symmetric part is related to external torques applied on the particle. The symmetric part is set through an iterative procedure to ensure that the strain-rate within the fluid volume occupied by the dispersed phase is zero (enforcing solid body response).

Particles in turbulence are experiencing shear flows at different scales. The method has been validated on low Reynolds number shear flows (Abbas et al., 2006) and extension to finite Reynolds will be commented in this communication. Second, the numerical method has been validated in the case of inertia-induced particle migration in a plane Poiseuille flow. In laminar channel flows, neutrally buoyant particles are submitted to a lift force induced by the interaction between the finite-size particle and the parabolic velocity profile when the particle Reynolds number is finite.

Homogeneous isotropic turbulence seeded with finite size neutrally buoyant particles (typically few Kolmogorov length scales) is investigated (Yeo et al., 2010). Results are given on the modulation of the turbulence at volumetric concentrations of 6%. We analyzed the Lagrangian statistics for the velocity and acceleration of the dispersed phase. The turbulent fluctuations are damped at mid-range wavenumbers by the particles while the small scale kinetic energy is significantly enhanced. The pivoting wavenumber characterizing the transition from damped to enhanced energy content is shown to vary with the size of particles.

For neutrally buoyant particles seeded in transitional pipe flows, Matas *et al.*(2003) observed changes in the values of the critical Reynolds numbers depending on both the solid volume fraction and the particle-to-pipe size-ratio. Typically, the transition occurs at lower Reynolds numbers when the flow carries macro-sized particles at dilute to moderate concentrations (up to 25%). On the contrary, the critical Reynolds numbers of the onset of transition is shifted towards greater values when particles are micro-sized and their concentration is higher. In this communication, we aim at understanding the mechanisms lying behind the shift of the laminar-turbulent transition regime down to lower critical Reynolds numbers in suspension flows of macro-sized particles. Particles are randomly seeded into a fluctuating channel flow at a solid volume fraction of 5%, the size ratio of particle diameter to channel height being 1/16. After a transient regime, particles are homogeneously distributed in most of the channel flow. A larger concentration occurs in the near-wall region mainly due to inertial particle transversal migration (SÃ©grÃ©-Silberberg effect). The wall-normal and spanwise flow velocity fluctuations are significantly changed compared to the single-phase flow case. They are larger in the wall vicinity due to larger particle concentration in this region, and they are slightly larger in the channel core due to energy redistribution from the streamwise towards the transverse directions. Hence in the suspension flow, the particles are responsible for the significant agitation growth (in transverse directions) compared to the single-phase case. In our simulations of suspension flows, values of the wall friction coefficient are larger than in the single-phase case. The fluctuating flow exhibits a nearly constant friction coefficient in the transition regime whereas the friction coefficient increases when the Reynolds number is decreased below 1200. Substituting the Reynolds number by a mixture Reynolds number based on the effective suspension viscosity, the evolution of the friction coefficient almost collapses with the laminar law.

These results tend to show that compared to single-phase flows, even at low concentration, finite Reynolds particulate flows relaminarize at lower Reynolds number, in accordance with previous experimental observations of Matas et al. (2003).

References

Matas, J.-P., Morris, J. F., Guazzelli, E., 2003. Transition to turbulence in particulate pipe flow. Phys. Rev. Lett. 90, 014501.

Abbas, M., Climent, E., Simonin, O., Maxey, M. R., 2006. Dynamics of bidisperse suspensions under Stokes flows: Linear shear flow and sedimentation. Phys. Fluids 18.

The Force Coupling Method: A flexible approach for the simulation of particulate flows, E. Climent & M.R. Maxey, (2009) inserted in "Theoretical Methods for Micro Scale Viscous Flows", Ressign Press, Eds F. Feuillebois and A. Sellier (ISBN: 978-81-7895-400-4).

Modulation of homogeneous turbulence seeded with finite size bubbles or particles (2010) K. Yeo, S. Dong, E. Climent, M.R. Maxey, Int. J. of Multiphase flows, 36, 221–233.

(379g) Adapting Data Processing to Compare Model and Experiment Accurately: A Discrete Element Model and Magnetic Resonance Measurements of a 3D Cylindrical Fluidized Bed

Boyce, C. M., University of Cambridge
Holland, D. J., University of Cambridge
Scott, S. A., University of Cambridge
Dennis, J. S., University of Cambridge
Originally presented on: 11/5/2013 16:45:00 - 17:00:00

Abstract:

Despite the widespread practical importance of fluidized beds, the underlying physics of these systems is still not well understood [1]. This lack of understanding is due in part to the fact that fluidized beds are difficult to measure because they are opaque, and non-intrusive measurement techniques such as magnetic resonance (MR) must be employed. Additionally, granular matter in fluidized beds can possess gas-, liquid-, and solid-like properties [2], making it very difficult to model accurately. In order to predict the behaviour of fluidized beds, two characteristics must be modelled properly: (1) the contact forces between the particles and (2) the interaction forces between the particles and the fluid. The difficulties in measuring and modelling granular systems potentially lead to experimental results and simulations with a combination of known and unknown inaccuracies. Thus, in order to cross-validate these models and measurements, it is important to compare their results in as direct a manner as possible. This paper describes the cross-validation of a computational model and magnetic resonance measurements of a 3D cylindrical bubbling fluidized bed; the model was processed in a manner that simulated the acquisition process of the experiment to provide a direct comparison between the two.

A discrete element model (DEM) to simulate individual particles combined with computational fluid dynamics (CFD) to model fluid flow was used here to simulate a bubbling fluidized bed. The authors developed a 3D cylindrical DEM-CFD model by integrating a discrete element model in which particles are tracked in rectangular coordinates and contained by tubular walls with a CFD grid modelled in cylindrical coordinates. A novel structured CFD grid, shown in Figure 1, was setup such that the volumes of the fluid cells were kept constant and of similar shape, to ensure the volume-averaged fluid equations for modelling unresolved fluid flow around particles [3], were satisfied. The accuracy of a DEM-CFD model is unclear because of inherent assumptions and inaccuracies in parameters describing the contact mechanics between particles as well as a drag law describing the force of interaction between fluid and particles.

Thus, the DEM-CFD model required validation by experiment. Magnetic resonance (MR) was chosen for experimental comparison because it has proven effective at measuring a variety of physical aspects of fluidized beds [4,5,6], it does not require tracer particles, and it can make measurements on 3D cylindrical fluidized beds. An inherent problem of MR is the complex and indirect tomographic measurement techniques it requires. It is unclear how the mathematical sampling and post-processing methods used to convert the raw MR measurements into images, and derived quantities such as particle velocity, affect the results of measurements on dynamic systems such as fluidized beds. Thus, it is desirable to cross-validate MR measurements by comparison with computational models.

In this study, the DEM-CFD model developed was used to simulate an MR experiment measuring time-averaged particle velocity in a bubbling fluidized bed, conducted by Holland *et al.* [7]. The time averaged velocity was measured in the x-z plane with a 5 mm slice in the y-direction at the centre of a 44 mm diameter bed, as shown in Figure 2. The bed contained 1.2 mm diameter particles with a settled height of 30 mm and was fluidized at twice the minimum fluidization velocity, 2 U/U_{mf}. In the experiment, a pulse sequence of magnetic field gradients, shown in Figure 3, was used to manipulate MR signal from oil contained in poppy seeds and ultimately measure the time-averaged velocity of the particles over an acquisition time of ~30 minutes. This same pulse sequence was simulated on the particle positions output from the DEM-CFD model over a timespan compressed to 20 seconds, in order to directly compare the results measured by model and experiment.

Additionally, other, simpler averaging techniques were applied to the particle positions and velocities output from the DEM-CFD simulation to provide a comparison with the MR-type acquisition and averaging procedure. A frame-based technique averaged together the average particle velocity in individual pixels at ?frames? in time, separated by 10 ms intervals, over the total simulation time. In contrast, particle-based averaging techniques gave equal weighting to the velocity of each particle which passed through a pixel. The MR-type average was expected to be closer to a particle-based average, since MR signal is proportional to the volume of the particle from which it is produced. However, it was unclear how far other complications in the MR acquisition process would cause the MR-type average to deviate from a particle-based average in a dynamic fluidized bed system and how far this average would deviate from a frame-based average. Additionally, since technical complications in MR required signal from particles to be weighted by a 5 mm Gaussian slice, shown in Figure 3, as compared to being bounded by a simple 5 mm ?top-hat? slice, a particle-based average was conducted with both of these slices to see how the different slices affected the final average.

Figure 4 shows a comparison between the experimental MR results for time averaged particle velocity and those from the DEM-CFD simulation with the simulated MR pulse sequence. The model and measurements are cross-validated by the similarity in velocity profiles. Notably, the experimental profile extends further upward in the bed and has a higher maximum particle speed than the DEM-CFD velocity profile. These differences can be attributed to inaccuracies in modelling the fluid-particle interaction force: spherical particles were used in the DEM-CFD simulation, giving them a lower surface-area-to-volume ratio than the non-spherical poppy seeds used experimentally and thus a lower drag force. Additionally, the drag correlation of Beetstra *et al.* [8] used here has been demonstrated to significantly underestimate the average drag force on particles in a fluidized bed [9].

Figure 5 provides a comparison between the different averaging techniques used to time-average particle velocity from the DEM-CFD data. The comparison shows very little difference between MR-type averaging, particle-based averaging with a 5 mm top-hat slice, and particle-based averaging with a Gaussian-weighted 5 mm slice. Thus, this comparison shows that, despite the complex acquisition and processing procedure, MR measurements do provide a particle-based time average in dynamic systems and a 5 mm Gaussian slice yields equivalent results to a 5 mm top-hat slice in these systems. Additionally, the frame-based average provides a similar velocity profile, yet much greater upwards and downwards particle speeds than the other averages. This demonstrates that different averaging techniques on the exact same data can yield vastly different results. Thus, great care should be taken when comparing results from different modelling and measurement techniques for cross-validation, as it may be the averaging technique, as opposed to validity of the techniques, which causes discrepancies.

References

[1] P. G. de Gennes, Rev. Mod. Phys. **71**, S374 (1999).

[2] H. M. Jaeger, S. R. Nagel, and R. P. Behringer, Rev. Mod. Phys. **68**, 1259 (1996).

[3] T. B. Anderson, and R. Jackson, Industrial & Engineering Chemistry Fundamentals **6**, 527 (1967).

[4] C. R. MÃ¼ller, J. F. Davidson, J. S. Dennis, P. S. Fennell, L. F. Gadden, A. N. Hayhurst, M. D. Mantle, A. C. Rees, and A. J. Sederman, Chemical Engineering Science **62**, 82 (2007).

[5] C. R. MÃ¼ller, J. F. Davidson, J. S. Dennis, P. S. Fennell, L. F. Gadden, A. N. Hayhurst, M. D. Mantle, A. C. Rees, and A. J. Sederman, Phys. Rev. Lett. **96**, 154504 (2006).

[6] D. J. Holland, C. R. MÃ¼ller, J. F. Davidson, J. S. Dennis, L. F. Gladden, A. N, Hayhurst, M. D. Mantle, and A. J. Sederman, Journal of Magnetic Resonance **187**, 199 (2007).

[7] D. J. Holland, C. R. MÃ¼ller, J. S. Dennis, L. F. Gladden, and A. J. Sederman, Powder Technology **182**, 171 (2008).

[8] R, Beetstra, M. A. van der Hoef, and J. A. M. Kuipers, Chemical Engineering Science **62**, 246 (2007).

[9] S. H. L. Kriebitzsch, M. A. van der Hoef, J. A. M. Kuipers, Chemical Engineering Science **91**, 1 (2013).

Figures:

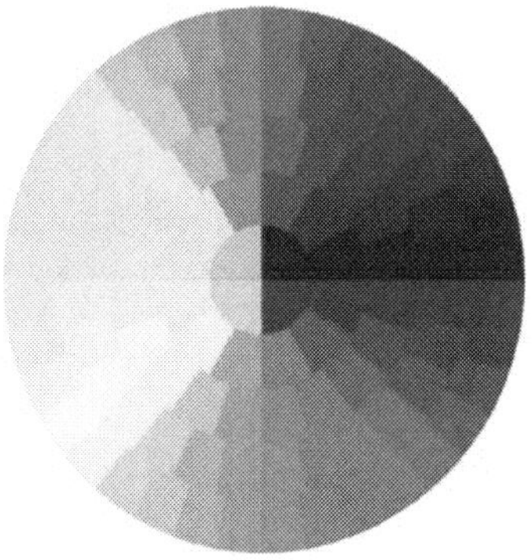

Figure 1 (Colour online) Horizontal cross section of novel CFD grid. Different colours denote different CFD cells. More CFD cells are used in the annuli further from the centre, such that the fluid cells have a constant volume to ensure the same accuracy in the volume-averaged fluid equations.

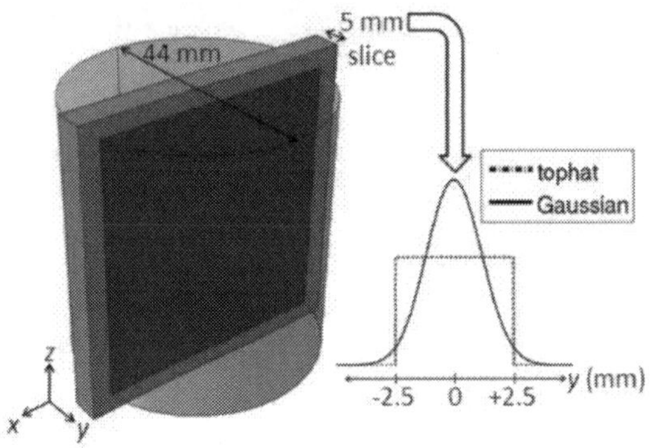

Figure 2 (Colour online) Plot of 5 mm y-slice in fluidized bed (left). Plot of Gaussian and top-hat signal weighting functions versus y-position in the fluidized bed (right).

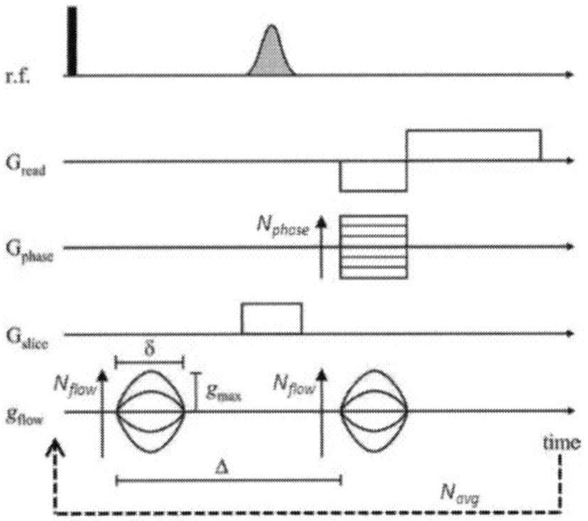

Figure 3 Pulse sequence modelled for MR-type post processing of time averaged axial velocity measurements (Reproduced from Holland *et al.* [7]).

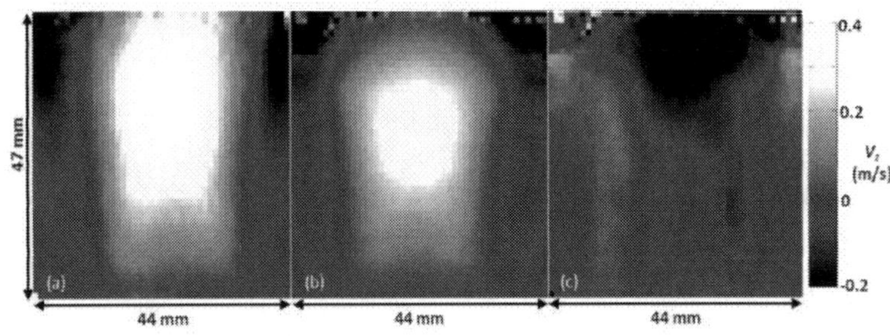

Figure 4 (Colour online) Comparison of average axial velocity images from: (a) experimental MR imaging [5] and (b) MR-type post-processing of the DEM simulation. Image (c) gives the difference map between (b) and (a). The resolution of each image is 1.04 mm (z) by 0.94 mm (x).

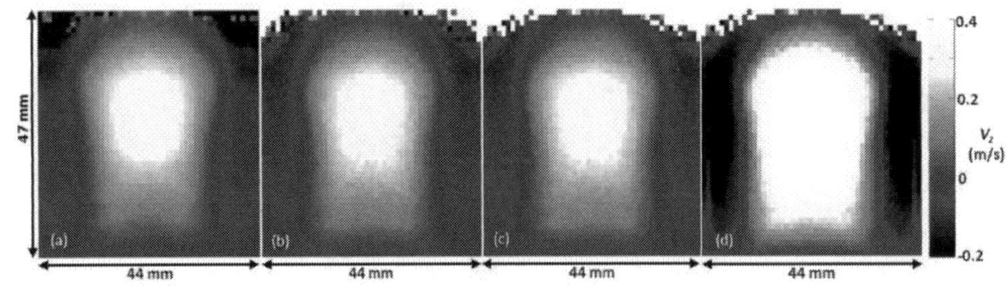

Figure 5 (Colour online) Comparison of average axial velocity images of DEM simulation with different averaging methods: (a) MR-type post-processing, (b) Gaussian-slice particle-based average, (c) ?top-hat? slice, particle-based average, and (d) frame-based average. The resolution of each image is 1.04 mm (z) by 0.94 mm (x).

(379h) Collision Dynamics and Impaction of Inelastic Nano Particles

Ahuja, S., Xerox Corporation

Originally presented on: 11/5/2013 17:00:00 - 17:15:00

Analysis of flow of granular material is useful in designing storage of food grains, ores, and coal to pharmaceutical powders and to digital printing. High definition digital imaging is enabled in printers by using small micron and nano toner particles. Use of small particles in a container gives rise to complex particulate flow as small particles tend to aggregate and large particles break on normal and oblique collisions with each other or with walls of the container. Particulate solids undergo brittle- ductile transition which depends on the size of a particle and its shape, its inelastic behavior and strain rates under which it is deformed.

Particulate flow can result in jamming which has been analyzed using Lattice- Boltzmann (LB) or DEM (Discrete Element Methods) on deformation from collisions during hydrodynamic forces in fluid-particle flows, An elastic-plastic model is developed for particle collisions during a particulate flow that depends on particle radius, incident particle speed and yield stress of the material. The model predictions are compared with experimental data on particles prepared by reactive extrusion and subjected to time dependent deformation in a contained cavity.

(379i) Water Slug Formation and Motion in Gas Flow Channels

Benziger, J., Princeton University
Cheah, M. J.,
Kevrekidis, I. G., Princeton University
Originally presented on: 11/5/2013 17:15:00 - 17:30:00

Water emerging from 100 micron pores into mm size gas flow channels forms drops that grow and become slugs which span the flow channel. Flowing gas causes the slugs to detach and move down the channel. The effect of channel geometry, surface wettability and gravity on the formation and motion of water slugs has been analyzed using high-speed video images of the drops and differential pressure-time traces. Drops grow assuming a sequence of shapes that minimize the total interfacial energy of the gas/liquid and liquid/solid interfaces. The drops are initially spherical caps centered on the pore (liquid contacts one wall). Above a certain size the drops move to the corner, forming "corner drops" (liquid contact two walls). Corner drops grow across the channel, evolving into partial liquid bridges (drops confined by three walls) and finally the drops span the channel cross-section forming slugs (contacting all four walls). Smaller slugs are formed in channels with hydrophobic walls than in channels with hydrophilic walls. Smaller slugs are formed in channels with curved walls than in square or rectangular channels. Slugs move when the differential gas pressure overcomes the force to move the advancing and receding gas/liquid/solid contact lines of the slugs. Residual water left behind in corners by moving slugs reduces the barriers for drops to form slugs, causing the steady state slug volumes to be smaller than those seen at start-up in dry channels.

(379j) A Generalized Oldroyd's Model for Dilute Emulsions of Deformable Drops

Martin, R.,
Zinchenko, A., University of Colorado
Davis, R. H., University of Colorado
Originally presented on: 11/5/2013 17:30:00 - 17:45:00

Formulating a constitutive equation for a non-Newtonian fluid valid in a broad class of kinematic conditions is an outstanding problem of great practical relevance. The focus in theoretical rheology so far has been on viscometric (shear) flows, and, to a lesser extent, on extensional flows (typically, uniaxial extension/compression). However, the information about the stress tensor in these types of flow alone cannot predict the material behavior for an arbitrary history of deformation. We propose a novel theoretical approach to constitutive modeling for non-Newtonian fluids based on a generalized Oldroyd's equation. In its traditional form [1], this constitutive equation of differential type relates the instantaneous stress and its corotational time derivative to the rate-of–strain tensor and its derivate; all seven material parameters in this equation are assumed constant. We work instead with a generalized model, in which these parameters are allowed to be functions of the instantaneous second invariant of the rate-of-strain tensor; such a model still obeys the objectivity principle. For a deviatoric stress, the number of material parameters reduces to five. The basic idea of the present approach is to consider two types of flow: (1) steady shear and (2) planar extension (PE) or hyperbolic flow, and then match all five theoretically-obtained rheological functions (three for shear flow , and two for PE) to those predicted by Oldroyd's equation. This match gives five non-linear equations for five material parameters solved by Newton iterations at every value of the second invariant. The choice of shear and PE is crucial, since these are the only two types of flow with a reproducible lattice [2] that allows for long-time simulations of concentrated systems (e.g., emulsions) with periodic boundaries.

A dilute emulsion of isolated deformable drops gives a rare opportunity to verify this approach to constitutive modeling, since the single-drop contribution to the average stress can be found precisely by boundary-integral (BI) solutions for arbitrary ambient-flow kinematics. The generalized Oldroyd's parameters were found by this approach for drop-to-medium viscosity ratios of 1 and 5; the solution was found to be always unique, regardless of an initial guess in the matching algorithm. The second invariant enters the equations through the effective capillary number Ca; the range of Ca is limited by breakup conditions for PE ($Ca<Ca\_cr$). Unlike usual shear-thinning, the dilute emulsion is shown to be tension-thickening in PE.

Predictions using the generalized Oldroyd model were compared with exact BI results for several types of flow kinematics. For a mixed linear flow between simple shear and PE and equal viscosities, the model predictions for all intrinsic stress components are practically indistinguishable from exact BI results for all $Ca<Ca\_cr$. For viscosity ratio of 5, larger deviations are observed near Ca_cr, but our model is still much more accurate than small-deformation theories. The second example is the rheological response of a small

drop submerged in a Stokes flow past a macroscopic sphere. Here, a fluid element trajectory resembles uniaxial compression near the stagnation point and extension in the wake, and is akin to unsteady shear flow at intermediate times. Again, the model has excellent accuracy, far exceeding the accuracy of small deformation theories. Another type of kinematics considered is finite-Reynolds-number flow in a rectangular cavity with a moving wall. Here, the history of deformation along a fluid element trajectory is time-periodic, different from simple shear or extension. Most importantly, our approach to constitutive modeling can be extended to large-strain flows of concentrated emulsions, unlike small deformation theories limited to single drops.

Acknowledgment: This work was supported by NSF Grant No. CBET 1064132

1. Oldroyd J.B. 1958 Non-Newtonian effects in steady motion of some idealized elastico-viscous liquids. Proc. Roy. Soc. Lond. A**245** , 278-297.

2. A.M. Kraynik and D.A. Reinelt 1992 Extensional motions of spatially periodic lattices. Int. J. Multiphase Flow 18, 1045-1059.

(382a) Comb Polymers: Film Surface Fluctuation Dynamics and Surface Segregation in the Limit of Dense Branching

Liu, B., The University of Akron
Narayanan, S., Argonne National Laboratory
Hu, R., Colorado School Of Mines
Wu, D. T., Colorado School Of Mines
Foster, M. D., The University of Akron
Originally presented on: 11/5/2013 15:15:00 - 15:35:00

The surface height fluctuations of melt films of densely branched comb polystyrenes of thicknesses greater than 55nm and at temperatures more than 23 °C above the $T_{g,bulk}$ can be rationalized using the hydrodynamic continuum theory (HCT) known to describe melts of linear and unentangled cyclic chains. Film viscosities (η_{XPCS}) for three combs inferred from fits of the HCT to X-ray Photon Correlation Spectroscopy (XPCS) data are the same as bulk viscosities (η_{bulk}) measured with rheometry. For the comb most like a star polymer and the comb closest to showing bulk entanglement behavior, $\eta_{XPCS} > \eta_{bulk}$. However, the values of η_{XPCS} - η_{bulk} are much smaller than those seen for less densely branched polystyrenes. We conjecture that the smaller magnitude of η_{XPCS} - η_{bulk} for the densely grafted combs is due to a lack of interpenetration of the side chains when branching is very dense. While data of relaxation time versus T for cyclic chains virtually collapse to a single curve when $T_{g,bulk}$ is accounted for, that is not the case for combs. $T_{g,bulk}$ and specific chain architecture both play important roles in determining the surface fluctuations.

We have also considered the effect of the dense branching on surface segregation in blends of linear and comb-branched polystyrenes. The comb-branched chains are strongly preferred at the surface and the experimentally determined surface segregation can be compared with self-consistent field theory calculations.

Acknowledgements: NSF CBET 0730692 & CBET-0731319 and DURIP W911NF-09-1-0122.

(382b) Examining Hydration and Mineralization in Polyelectrolyte Brushes

Kilbey, S. M., University of Tennessee
Deodhar, C., University of Tennessee - Knoxville
Ankner, J. F., Oak Ridge National Laboratory

Originally presented on: 11/5/2013 15:35:00 - 15:55:00

Layers of end-grafted polymer chains immersed in good solvents – often referred to as "polymer brushes" – are useful for tailoring interfacial properties and are often viewed as archetypical systems for teasing apart the inherent links between organization, structure, and properties of confined, soft materials. Polyelectrolyte brushes are of particular interest due to their relevance to biology and the possibilities they offer as constructs for lubricious biomaterial coatings or scaffolds, electro-actuators (synthetic muscles), biosensors, or high affinity separation agents. Despite this potential, polyelectrolytes remain perhaps the least understood form of soft matter due to their complex connections between charge and conformation. In this presentation, we describe recent efforts aimed at understanding how responsive polyelectrolyte brushes regulate their nanoscale structure as a function of their design and environmental conditions. We have synthesized homopolymer and copolymer brushes based on methacrylic acid using controlled (free) radical polymerization methods, and characterized their surface organization and structure using in situ ellipsometry and neutron reflectometry. In addition to useful insights into how polymer design impacts the swelling behavior of the brush, we find impactful consequences in terms of how we think about these systems: In short, our studies reveal a general misunderstanding in terms of the thickness and areal density of water-soluble brush systems, and we find that structural transitions are shifted or stretched compared to behaviors expected for unconfined systems because of how the chains regulate their structure. In addition, we describe the use of neutron reflectivity measurements to understand the in situ growth of calcium carbonate mineral phases in these ionotropic matrices. These findings underscore not only the complexity of these materials, but also how sophisticated characterization methods such as neutron reflectometry facilitate decoding synthesis-structure-property relationships.

(382c) Patterning Nano-Square Arrays Using Shear-Aligned Block Copolymer Thin Films

Kim, S. Y., Princeton University
Davis, R. L., Princeton University
Michal, B. T., Princeton University
Gwyther, J.,
Manners, I.,
Chaikin, P. M., New York University
Register, R. A., Princeton University
Originally presented on: 11/5/2013 15:55:00 - 16:15:00

Block copolymers can microphase-separate, at sufficiently high values of the interaction parameter and degree of polymerization (N), and show different morphologies depending on the volume fraction of each block. Dimensions are easily controllable over the range of 10-100 nm by varying N. In thin films, due to their ability to create dense periodic patterns, block copolymers have been highlighted in nanolithographic applications expected to provide fine resolution and scalability [1, 2]. While many techniques have been introduced to impart long-range order to arrays of block copolymer microdomains, such as thermal annealing, solvent annealing, graphoepitaxy, surface chemical patterning, and shear stress [3], only a few symmetries have been demonstrated in such arrays [3, 4]: hexagonally packed arrays of dots can be formed from spherical or cylindrical phase block copolymers, while periodic stripes can be created from cylindrical or lamellar phase block copolymers. There is increasing interest in creating square or rectangular patterns since these patterns are compatible with semiconductor integrated circuit design standards [1]. Nevertheless, creating square or rectangular arrays-which do not naturally form by spontaneous self-assembly of simple diblock copolymers- is more challenging [5]. A few attempts have been made to create rectangular patterns via directed assembly of block copolymers via electron-beam lithography or by fabricating arrays of posts on the substrate that guide the block copolymer microdomains into a square array; however, these methods are often time consuming and expensive [6, 7].

In this study, we demonstrate the generation of nanoscale rectangular patterns over large areas, using cylinder-forming block copolymers aligned in a cross-pattern. The cross-pattern is created from two monolayers of cylinder-forming block copolymers, where each layer is spin-coated and shear-aligned sequentially; shear stress can rapidly produce alignment over centimeter-scale areas [8]. After the first block copolymer layer is shear-aligned, the film structure is fixed with UV irradiation; the second block copolymer layer is then spin-coated and shear-aligned in an orthogonal direction. Reactive ion etching (RIE) is used to selectively remove the matrix polymer surrounding the cylinders, yielding the double-layer cross-pattern which serves as a template for square arrays.

Diblocks of polystyrene-b-poly(ferrocenylisopropylmethylsilane) (PS-PFS), forming iron-containing PFS cylinders, and polystyrene-b-polyhexylmethacrylate (PS-PHMA), forming PS cylinders, were employed to create patterns in double-layer thin films. A spin-coated monolayer film was first aligned by shearing a 1×1 cm^2area with a cured polydimethylsiloxane pad at 150 °C, well above the glass transition temperatures of all the blocks. In the PS-PFS case, after shear alignment, the majority PS block was cross-linked using UV irradiation. In the PS-PHMA case, UV irradiation both cross-linked the minority PS block and partially degraded the PHMA matrix; degraded polymer residues were removed by cleaning with toluene briefly, reducing the total film thickness by 2-5 nm. Thus, in both cases, UV irradiation fixed the shear-aligned cylinder structure of the monolayer film. Then, a second uniform monolayer of block copolymer was spin-coated on top of the first layer. A second shear stress was then applied to the film perpendicular to the first shear direction. While the second shear produced alignment of the top layer perpendicular to the first alignment direction, the cylinders in the first layer retained their orientation due to the cross-linked PS blocks.

After the second alignment, RIE was employed to selectively remove the PS or PHMA blocks in PS-PFS and PS-PHMA, respectively, so that only the cylinder-forming minor blocks remained. For PS-PFS, oxygen was used to remove the PS block; PFS is strongly resistant to oxygen RIE due to the iron and silicon it contains, while the all-organic PS blocks are completely removed by this process [9]. After RIE, the stacked PFS cylinders yielded square patterns; the pitch of the nano-square grid was about 35 nm, set by the intercylinder spacing of the PS-PFS (50k-16.5k block molecular weights). The lines constituting the grid each had a width of approximately12 nm. The strong etch resistance of PFS allows the pattern to be transferred into other substrates, creating nano-square wells. In one example, after oxygen RIE of the double-layer film, additional Bosch etching [10] was used to create a regular array of nano-square wells in the silicon wafer substrate. The depth of the wells could be increased up to 60 nm, or an aspect ratio of approximately 2, depending on the etch time. For pattern transfer from PS-PHMA diblocks, CF_4 was used to preferentially remove PHMA blocks, as CF_4RIE etches PHMA twice as fast as PS. When the etch time was carefully controlled, most of the PHMA matrix could be removed while most of the PS in both layers remained, yielding square grids from the stacked PS cylinders. The pitch of the array was about 40 nm (from PS-PHMA 33k-126k), while the width of each line was again about 12 nm. We synthesized PS-PHMA diblocks with various molecular weights so that the pitch of the grid could be varied from 25 nm to 40 nm. If block copolymers of different molecular weights are used for the two layers, rectangular-shaped (vs. square) patterns could be generated. Also, since the second alignment direction can be arbitrarily selected, a parallelogram-shaped pattern with an arbitrary angle can also be created.

To conclude, we present a simple way to create nano-square arrays by building up a double-layer film of a cylinder-forming diblock, where each layer is sequentially deposited, shear-aligned, and cross-linked. The pitch of the nano-square arrays is tunable by varying polymer molecular weight, and this method capitalizes on the fact that shear alignment is a simple and cost-effective approach to generate ordered arrays over macroscopic areas.

This work was supported by the National Science Foundation (MRSEC Program) through the Princeton Center for Complex Materials (DMR-0819860).

[1] H.C. Kim, S.M. Park, W.D. Hinsberg, Chem. Rev., 110 (2010) 146-177.

[2] C.J. Hawker, T.P. Russell, MRS Bull., 30 (2005) 952-966.

[3] A.P. Marencic, R.A. Register, Ann. Rev. Chem. Biomol. Eng., 1 (2010) 277-297.

[4] R.A. Segalman, Mater. Sci. Eng. R., 48 (2005) 191-226.

[5] C.G. Hardy, C.B. Tang, J. Polym. Sci. Pt. B-Polym. Phys., 51 (2013) 2-15.

[6] R. Ruiz, E. Dobisz, T.R. Albrecht, ACS Nano, 5 (2011) 79-84.

[7] A. Tavakkoli K G, K.W. Gotrik, A.F. Hannon, A. Alexander-Katz, C.A. Ross, K.K. Berggren, Science, 336 (2012) 1294-1298.

[8] D.E. Angelescu, J.H. Waller, D.H. Adamson, P. Deshpande, S.Y. Chou, R.A. Register, P.M. Chaikin, Adv. Mater., 16 (2004) 1736-1740.

[9] A. Nunns, J. Gwyther, I. Manners, Polymer, 54 (2013) 1269-1284.

[10] C.L. Chang, Y.F. Wang, Y. Kanamori, J.J. Shih, Y. Kawai, C.K. Lee, K.C. Wu, M. Esashi, J. Micromech. Microeng., 15 (2005) 580-585.

(382d) Silicon-Containing Block Copolymers for Sub-20 Nm Pitch Patterning

Cushen, J. D., University of Texas at Austin
Wan, L.,
Ruiz, R.,
Willson, C. G., The University of Texas at Austin
Ellison, C. J., The University of Texas at Austin
Originally presented on: 11/5/2013 16:15:00 - 16:35:00

Block copolymers demonstrate potential in next-generation lithography as a solution for overcoming the limitations of conventional lithographic techniques. Ideal block copolymer materials for this application can be synthesized on a commercial scale, have high χ-parameters, promoting self-assembly into sub-20 nm pitch features, and have high etch contrast between the domains for facilitating pattern transfer. Block copolymers that contain silicon in one domain are attractive materials for nanopatterning since they often fulfill these requirements. Here, we describe a new silicon-containing block copolymer system, poly(trimethylsilyl styrene-*block*-D,L lactide) (PTMSS-*b*-PLA), synthesized by a combination of anionic and ring-opening polymerization techniques. Due to the large difference in surface energies between the blocks in silicon-containing materials, a neutral surface treatment and solvent annealing or top coat materials are generally required to achieve perpendicular orientation. The processing conditions required to promote perpendicular orientation of sub-20 nm pitch PTMSS-*b*-PLA, direct the self-assembly, and transfer the block copolymer pattern into the underlying substrate will be presented.

(382e) High χ Block Copolymers Based On Hydrogen Bonding Polymers

Lawson, R. A., Georgia Institute of Technology
Cheng, J., Georgia Institute of Technology
Jarnagin, N., Georgia Institute of Technology
Tolbert, L. M., Georgia Institute of Technology
Henderson, C. L., Georgia Institute of Technology
Originally presented on: 11/5/2013 16:35:00 - 16:55:00

Directed self-assembly (DSA) of block copolymers (BCPs) is a promising technique for producing sub-30 nm pitch patterning by allowing for pitch multiplication of lithographically derived patterns. However, to achieve pattern pitches below 20 nm, materials with high χ (i.e. the Flory-Huggins interaction parameter) are required. This is because materials with lower χ values, such as polystyrene-block-polymethylmethacrylate (PS-b-PMMA), phase mix at the low degrees of polymerization that would be required to obtain such a small dimension. Our approach to designing high χ block copolymers is based on polystyrene blocks attached to blocks which contain strong hydrogen bonding moeties such as alcohols, phenols, and carboxylic acids. This produces BCPs where each block has relatively strong cohesive interactions with itself, but which are orthogonal to the interactions the other block has with itself. For example, in polystyrene-block-poly(acrylic acid) (PS-b-PAA), the PS block has strong aromatic and pi-pi interactions with itself, while the PAA block has strong hydrogen bonding, but little-to-none aromatic interactions. Within this framework, three different polymers have been synthesized using either atom transfer radical polymerization (ATRP) or nitroxide-mediated radical polymerization (NMP): polystyrene-block-poly(acrylic acid) (PS-b-PAA), polystyrene-block- poly(2-hydroxyethyl-methacrylate) (PS-b-PHEMA), and polystyrene-block-poly(4-hydroxystyrene) (PS-b-PHOST). PS-b-PHEMA and PS-b-PHOST have both shown sub-20 nm pitches with the possibility for sub-10 nm pitches. The challenges of directed self-assembly of each system and their unique properties will be discussed.

(382f) Controlled Long Range Orientation of Block Copolymer Nanostructures Through Soft Confinement Solvent Vapor Annealing

Vogt, B. D., University of Akron
Qiang, Z.,
Cavicchi, K. A.,
Originally presented on: 11/5/2013 16:55:00 - 17:15:00

Long range alignment of block copolymer domains in thin films is generally obtained through relatively complex processing involving pre-patterned substrates,[1] electric fields,[2] crystal faults,[3] or slow thermal zone annealing.[4] In this work, we describe a simple method that is similar to typical solvent vapor annealing (SVA) strategies except that the surface of the block copolymer film is capped with a swellable elastomer. By controlling the shape and size of the elastomer on the block copolymer film during SVA, the orientation of perpendicular lamellae can be controlled over 180 degrees. We will discuss how the thickness of the film, block copolymer composition, and solvent using in SVA impact the alignment and orientation of block copolymer domains. From GISAXS, the Herman's orientation parameter of the films can exceed 0.95 (second order parameter) under typical SVA conditions for less than 2 h.

1. Segalman, R. A.; Yokoyama, H.; Kramer, E. J., Graphoepitaxy of spherical domain block copolymer films. *Advanced Materials* **2001,** 13, (15), 1152-1155.

2. Morkved, T. L.; Lu, M.; Urbas, A. M.; Ehrichs, E. E.; Jaeger, H. M.; Mansky, P.; Russell, T. P., Local control of microdomain orientation in diblock copolymer thin films with electric fields. *Science* **1996,** 273, (5277), 931-933.

3. Hong, S. W.; Voronov, D. L.; Lee, D. H.; Hexemer, A.; Padmore, H. A.; Xu, T.; Russell, T. P., Controlled Orientation of Block Copolymers on Defect-Free Faceted Surfaces. *Advanced Materials* **2012,** 24, (31), 4278-+.

4. Berry, B. C.; Bosse, A. W.; Douglas, J. F.; Jones, R. L.; Karim, A., Orientational order in block copolymer films zone annealed below the order-disorder transition temperature. *Nano Letters* **2007,** 7, (9), 2789-2794.

(382g) Phase Evolution of PS/PMMA Blend Films Between Two Parallel Rigid Plates With Nano-Scale Gap

Wang, Z.,
Zhang, Z., University of Colorado at Boulder
Wang, L., University of Colorado at Boulder
Ding, Y., University of Colorado at Boulder
Originally presented on: 11/5/2013 17:15:00 - 17:35:00

We present the phase evolution of PS/PMMA blend film strongly confined between two parallel rigid plates with identical surface energies and a gap distance ranging from 50 nm ~100 nm. This symmetrical configuration is achieved by pressurizing a spun-cast PS/PMMA film on a silicon wafer against another silicon wafer. In the absence of pressurization, phase inversion process occurred on both substrates, with PMMA wetting both substrates and PS breaking into correlated droplets. The phase inversion was completely suppressed by pressurizing the plates, as the pressure overcame the capillary pressure of confined PS domains.

(388a) Second Osmotic Virial Coefficients for "Patchy" Proteins

Roberts, C. J., University of Delaware
Blanco, M. A., University of Delaware
Originally presented on: 11/5/2013 15:15:00 - 15:33:00

Proteins and anisotropic nanoparticles often have surfaces composed of neutral hydrophilic, hydrophobic, and charged "patches". In some cases it has been suggested that complementary patch-patch interactions dominate the thermodynamics of protein and nano-particle solutions if the attractive energies are sufficiently large and short-ranged, although criteria for achieving this scenario have not been quantified systematically. A series of models are used here to evaluate the balance of the favorable energy gain and the entropic penalty for forming such patch-patch contacts, and tested for conditions where experimentally realistic, quantitative values of the second osmotic virial coefficient (B22) are recovered. Using estimates of the strength of interactions between hydrophobic (HP) patches on proteins and how they scale with patch size, along with a patch-patch model akin to that of Kern and Frenkel (J. Phys. Chem. 118, 9882 (2003)), the results indicate that one of three scenarios is realistic for highly "patchy" proteins or similarly sized nanoparticles: (i) patches are too small, and attractions are not sufficiently strong to overcome the entropic penalty of forming patch-patch pairs; (ii) patches fall within a narrow range of sizes, and contribute significantly to B22; (iii) patches are so large that solutions are effectively unstable (i.e., unphysically large B22 values). In addition, the models illustrate that repulsive interactions contribute significantly to B22 for only those conditions where the repulsions are long-ranged. Finally, the importance of the balance between energetic and entropic contributions is also highlighted by reformulating B22 in terms of the orientation-space density of states for a given energy and protein-protein distance. Doing so provides insight into why B22 is indicative of protein phase separation and aggregation propensity for some proteins / solvent conditions but not others. The results also suggest experimental signatures that can be used to identify whether B22 should be expected to be a useful predictor of aggregation or phase separation for a given protein/solvent system, with examples such as eye lens crystallin(s), antibodies, and model proteins.

(388b) Effects of Macromolecular Crowding On Amyloid Beta (16-22) Aggregation Using Coarse-Grained Simulations

Latshaw, D. C. II, North Carolina State University
Hall, C. K., North Carolina State University
Originally presented on: 11/5/2013 15:33:00 - 15:51:00

In Alzheimer's Disease (AD), the amyloid β (Aβ) peptide aggregates in the human brain and forms oligomer and fibril structures, causing a decline in thought processes and memory, ultimately resulting in dementia and death. The intracellular and extracellular environments of the brain contain a variety of macromolecules and structures that can occupy anywhere from 7% to in excess of 40% of the total volume of their respective mediums with cellular cytoplasm typically falling within the 20% to 30% range. Typically studies of protein aggregation do not account for the fact that both intracellular and extracellular regions of the body are crowded and while the observations made in these experiments lead to a better understanding of protein aggregation, they do not accurately depict how the process occurs *in vivo*. Since the rate and extent of aggregation in a crowded environment such as the human brain could differ by orders of magnitude from that in dilute solution, an investigation of aggregation rates and mechanisms in the presence of crowding spheres using computer simulations could provide unique insight into what causes differences between *in vitro* and *in vivo* protein aggregation. To examine the effect of crowding on protein aggregation, discontinuous molecular dynamics (DMD) simulations combined with an intermediate resolution protein model developed in the Hall group, PRIME20, were applied to a peptide/crowder system. The systems contained 192 Aβ(16-22) peptides and crowders of diameters 5Å, 20Å, and 40Å, represented here by simple hard spheres, at volume fractions of φ_c=0.00, 0.05, 0.10, 0.15, and 0.20. The effects of varying the number and size of the spheres were examined. Preliminary results indicate that aggregation is promoted when the crowder volume fraction is increased, leading to an increase in effective concentration. An additional effect of increased crowder volume fraction is more rapid oligomer formation and fibril seeding, ultimately decreasing or eliminating the lag phase of aggregation and altering the mechanism of aggregation. The aggregation mechanism appears to shift from the reaction limited dock-lock mechanism in dilute solution to a mechanism more characteristic of hydrophobic collapse followed by subsequent rearrangement into fibrils in a crowded environment. When the size of the crowding spheres at fixed crowding volume fraction is increased, aggregation is decreased because the increased mass decreases peptide diffusion.

(388c) Molecular Dynamics Simulations of Self-Assembled DNA Nanostructures

Markegard, C. B., University of California, Irvine
Fu, I. W., University of California, Irvine
Nguyen, H. D., University of California, Irvine
Originally presented on: 11/5/2013 15:51:00 - 16:09:00

Elucidation of the molecular-level mechanisms by which DNA single strands self-assemble into nanostructures might provide new informative guidelines in the engineering of novel biomaterials. Our studies examine the role of DNA sequences and the environmental conditions on the morphology and mechanical properties of DNA assemblies by performing molecular dynamics simulations. In this work, a newly developed coarse-grained model that bridges the gaps between atomistic models and extremely simplified models by capturing geometric and energetic details yet it is sufficiently simple to allow simulations of the spontaneous self-assembly of many DNA strands simultaneously. First, the dependence of the persistence length as a function of the ionic strength is examined. In addition, the melting temperature of double-stranded DNA is examined as a function of sequence, ionic strength, and chain length. Our results from single-molecule simulations agree qualitatively with experimental data. To examine the kinetic mechanisms involved in DNA self-assembly, we perform constant-temperature simulations to observe the whole process starting from random configurations of many single-strands of DNA. In particular, we investigate the self-assembly of homogenous and random sequences to determine the kinetic mechanisms of nanostructure formation. The findings of this research will guide experimentalists to identify systems of novel biomaterials with advantageous morphological properties.

(388d) Strategies to Stabilize Proteins On Surfaces From Improved Modeling of Protein-Surface Interactions

Loong, B., Brigham Young University
Schinn, S. M., Brigham Young University
Bush, D. B., Brigham Young University
Knotts, T. A. IV, Brigham Young University
Originally presented on: 11/5/2013 16:09:00 - 16:27:00

Protein arrays are a potentially disruptive technology—with applications in a variety of fields including medicine, defense, and biology—that have yet to find mainstream use due to unreliability. Protein arrays function on the principle of molecular recognition between the protein on the surface and a ligand molecule, so the technology can only be as reliable as the recognition is robust. But tethering proteins to surfaces during fabrication can affect the stability of proteins in ways that compromise the recognition, and a complete understanding of the relevant phenomena remains elusive. This lack of knowledge not only limits the performance of current microarrays, but also prevents rational design of next-generation platforms.

Studying the stability of proteins on surfaces is difficult because typical experimental techniques to ascertain protein structure (such as NMR and X-ray crystallography) are not transferrable to inhomogeneous environments. As such, molecular simulation has emerged as the primary method to investigate protein-surface interactions. Simulation has provided many insights into these processes, but the vast majority of the studies have been done with relatively idealized systems. The idealities include studying the stability of only one protein on the surface at a time rather than many proteins as would be found in experiment, binding the protein to the surface using only a single tethering molecule attached to a single residue in the protein, and modeling the surface in a rudimentary way such that it interacts with each protein residue with the same energy regardless of the specific chemistries involved.

This presentation will summarize several efforts to move away from the "ideal" case and model the system in a manner that more closely approximates the real system. Results will be presented that demonstrate: 1) how multiple molecules tethered to the surface interact with each other to change the degree of folding found on the surface at a given temperature, 2) how binding the protein to the surface using multiple tethers on different residues can drastically alter the folding mechanism and lead to either stabilization or destabilization of protein structure, and 3) how modeling the protein-surface interactions with appropriate chemical detail changes many of the prior results seen when using more rudimentary surface models. This last topic is made possible by using an advanced coarse-grain model that captures chemically-specific, protein-surface interactions in a manner that produces remarkable agreement with experiment. As a whole, the results offer hope that rational design of future protein arrays is possible.

(388e) Novel Biclustering Framework for Tertiary Contact Prediction Using Low-Homology Protein Templates

Smadbeck, J., Princeton University
Kieslich, C. A., Princeton University
Khoury, G. A., Pennsylvania State University-University Park
Floudas, C. A., Princeton University
Originally presented on: 11/5/2013 16:27:00 - 16:45:00

The prediction of the three-dimensional structure of a protein from its amino acid sequence remains an open question in molecular biology, with important implications in biological engineering. For sequences with low-homology template structures, the ranking of template structures and generation of predicted structures can be particularly difficult. The development of accurate prediction of long-range amino-acid contacts can be important in ranking these low-homology templates for modeling. Additionally, these contacts can be useful in constraining dynamic simulations of protein structure of low-homology targets. To this end we have developed a novel tertiary contact prediction method based on biclustering analysis for extracting C_α distance constraints from low homology templates.

The initial step of our procedure is the identification of template structures for the target sequence using a modified threading algorithm of SPARKS-X [1]. Preliminary models, based on the top template structures, are generated using CYANA [2] in order to remove gaps in the structures resulting from unmapped regions of the alignments. To identify persistent structures and topologies within the templates, hierarchical clustering based on pair-wise GDT (a structure similarity measure) is performed using the initial CYANA models. The template structures belonging to the largest three clusters of the GDT-based dendrogram tree are selected for C_α-C_α distance calculation. This matrix of C_α-C_α distances then serves as input to OREO[3-5], which is an iterative framework for biclustering dense and sparse data matrices via optimal re-ordering of rows and columns.

The final step is to filter the clustered distances to exclude distances of low confidence. We apply three filters: (i) a variance filter, based on the mean/standard deviation of each distance; (ii) a sequence mapping filter, in which contacts involving poorly mapped positions are excluded, according to an accumulation of the position-specific scoring matrix (PSSM) produced by SPARKS-X; (iii) a structure-based filter, which removes distance constraints that deviate significantly from the predicted values during CYANA structure generation. The structure-based filter includes additional constraints derived from CONCORD[6] predicted secondary structure and predicted beta-sheet topology[7], which is essential for identifying conflicting constraints. The remaining clustered distances are considered strong candidates for conserved contacts and are used to generate C_α distance constraints for a final structure generation.

We present results on a series of free-modeling targets from the Critical Assessment of techniques for protein Structure Prediction 10 (CASP10) competition. This method demonstrates a superior performance over other low-homology template-based contact prediction methods in prediction short, medium, and long-range contacts for difficult targets. The contacts are used in the generation of structures through a constrained molecular dynamics (MD) run to demonstrate how such contacts are important for the accurate structural fold determination.

[1] Yang Y, Faraggi E, Zhao H, Zhou Y. Improving protein fold recognition and template-based modeling by employing probabilistic-based matching between predicted one-dimensional structural properties of the query and corresponding native properties of templates. *Bioinformatics* 27:2076-82 (2011)

[2] López-MÃ©ndez B, Güntert P. Automated protein structure determination from NMR spectra. *J. Am. Chem. Soc.* 128:13112–13122 (2006)

[3] DiMaggio PA, McAllister SR, Floudas CA, Feng XJ, Rabinowitz JD, Rabitz HA. Biclustering via optimal re-ordering of data matrices in systems biology: rigorous methods and comparative studies. *BMC Bioinformatics*, 9:458 (2008)

[4] McAllister SR, DiMaggio PA, Floudas CA. Mathematical modeling and efficient optimization methods for the distance-dependent rearrangement clustering problem. *J. Global Optim.* 45:111-129 (2009)

[5] DiMaggio PA, McAllister SR, Floudas CA, Feng XJ, Li G, Rabinowitz JD, Rabitz HA. Enhancing molecular discovery using descriptor-free rearrangement clustering techniques for sparse data sets. *AIChE J.* 56(2):405-418 (2010)

[6] Wei Y, Thompson J, Floudas CA. CONCORD: a consensus method for protein secondary structure prediction via mixed integer linear optimization *Proc. R. Soc. A* 468(2139):831-85 (2012)

[7] Subramani A, Floudas CA. β-sheet Topology Prediction with High Precision and Recall for β and Mixed α/β Proteins. *PLoS ONE* 7(3):e32461 (2012)

(388f) Tunable Nucleation Time of Functional Sphingomyelinase-Lipid Complex

Lin, C. Y.,
Chao, L., National Taiwan University
Originally presented on: 11/5/2013 16:45:00 - 17:03:00

Cell membranes are composed of various types of lipids and proteins. Phase separation, or called lipid membrane heterogeneity, exists in this multi-component system. The phase dynamics has been suggested to influence many cellular processes. We report that a type of peripheral membrane protein, sphingomyelinase (SMase), can form a functional complex with its lipid substrate and product on the lipid membrane, and has a tunable formation time based on classical nucleation theory. The enzyme's hydrolysis rate is found to significantly increase after the complex is formed. We found that the nucleation rate of the complex can be tuned or predicted by the supersaturation of the enzyme, the lipid substrate and the lipid product. The complex nucleation is found to be influenced by the component concentrations in the fluid phase, where the complex nucleates, suggesting that the property in the interested phase instead of the apparent property in the entire system is important in the phase segregated lipid membrane. More generally, our finding suggests that nucleation could serve as a time lag control mechanism in this enzymatic system, while many of the time lags in biology are controlled by diffusion time or reaction kinetic delay. Ways to reduce nucleation energy barrier could be used to shorten the aggregation time lag and vice versa.

(388g) Interaction of Room Temperature Ionic Liquids With Lipid Membrane: A Molecular Simulation Study

Shah, J. K., University of Notre Dame
Maginn, E. J., University of Notre Dame
Originally presented on: 11/5/2013 17:03:00 - 17:21:00

Ionic liquids exhibit extremely low vapor pressure suggesting that they are attractive alternatives to volatile organic compounds. However, elimination of air emissions does not necessarily imply that ionic liquids do not have environmental footprints and they do not pose health concerns. In fact, ionic liquids have been shown to be toxic to varying extents to different organism. However, a little progress has been achieved towards the molecular level understanding of the origins of ionic liquid toxicity.

Driven by the hypothesis that toxic effects of ionic liquids arise as a result of their interactions with the cell membranes, molecular dynamics simulations of ionic liquids in contact with a model lipid bilayer are carried out. Three ionic liquids with varying hydrophobicity are examined: 1-ethyl-3-methylimidazolium [emim] chloride, [emim] ethylsulfate and [emim] bis(trifluoromethanesulfonyl)imide. In this talk, the effect of these ionic liquids as a function of concentrations on the structure and dynamics of the lipid bilayer represented by 1-pamiltoyl-2-oleoyl-sn-glycero-3-phosphotidylcholine (POPC) will presented. Based on the simulation results, a mechanism of ionic liquid toxicity is proposed.

(388h) A Comparative Study of Gromos- and Charmm-Based Force Fields for the Simulation of Non-Hydroxy Ceramide Bilayers

Iacovella, C. R., Vanderbilt University
Guo, S., Vanderbilt University
Moore, T. C.,
McCabe, C., Vanderbilt University
Originally presented on: 11/5/2013 17:21:00 - 17:39:00

The ability of the skin to regulate bodily functions and act as an effective barrier to chemical penetrants is controlled by the thin, outermost layer, known as the stratum corneum. The stratum corneum is predominantly composed of ceramides, cholesterol, and free fatty acids, however, the molecular level organization of the species and their role in maintaining the barrier function of the skin is not well understood [1]. While molecular simulation has been used extensively to study phospholipid bilayers and has provided a wealth of structural information, in comparison far less is known about the atomic level structure and interactions of bilayers involving ceramides. Prior simulation studies of non-hydroxy sphingosine ceramides (CER-NS) have demonstrated important differences with other lipids, e.g., sphingomyelin bilayers, including reduced ordering [2] and reduced residence time [3] of water molecules at the interface, due to the absence of a large polar head group. However, most work to date has examined pre-assembled ceramide bilayers under a very limited timescale (~20 ns simulation time), making it possible that these studies are adversely influenced by the initial, assumed structure. Additionally, only one ceramide (CER-NS) of the twelve distinct ceramides known in the stratum corneum [1] has thus far been considered in the literature. Furthermore, while common force fields (e.g., CHARMM and GROMOS) have been tuned and validated for phospholipid bilayers, it is not yet known how these perform when applied to ceramide-based systems.

Here, the structural properties of CER-NS bilayers are examined using models derived from both the CHARMM and GROMOS force fields. Parameters from CHARMM36 [4] are used with missing parameters and headgroup charges determined from density functional theory calculations; GROMOS-based simulations are performed using the model adapted by Berger [5] that has been employed in prior ceramide studies [6]. Here, it is observed through examination of long simulation runs (>100 ns) that both force fields provide similar results at physiological conditions (T=305K), in good agreement with experimental observations. However, examination of the thermophysical behavior demonstrates that the CHARMM-based force field is better able to reproduce the experimentally determined gel-to-liquid bilayer transition temperature, differing by only ~10K, whereas GROMOS overpredicts the transition temperature by > 50K. The CHARMM-based force field is also able to resolve a secondary transition ~345K, in close agreement with experiment, where evidence of such a transition is absent in the GROMOS-based force field. This suggests that the GROMOS-based force field over-estimates the stability of the gel phase for the ceramide bilayer, which may alter results for simulations of mixtures more representative of the stratum corneum, where structural heterogeneity may be present. Finally, using the validated CHARMM-based force field, simulations of CER NS are compared to simulations of the non-hydroxy phytospingosine ceramide (CER NP) to begin to examine the role of ceramide structure; distinct structural differences are observed with regards to the prefered orientations of the C-O bonds and the structural ordering of nitrogen atoms in the lipid headgroups of the two difference species.

[1] Wertz, P. W., Lipids and barrier function of the skin. Acta Derm.-Venereol. 2000, 7-11.

[2] Pandit, S. A.; Scott, H. L., Molecular-dynamics simulation of a ceramide bilayer. J. Chem. Phys. 2006, 124 (1), 014708.

[3] Imai, Y.; Liu, X. L.; Yamagishi, J.; Mori, K.; Neya, S.; Hoshino, T., Computational analysis of water residence on ceramide and sphingomyelin bilayer membranes. Journal of Molecular Graphics & Modelling 2010, 29 (3), 461-469.

[4] Klauda, J. B.; Venable, R. M.; Freites, J. A.; O'Connor, J. W.; Tobias, D. J.; Mondragon-Ramirez, C.; Vorobyov, I.; MacKerell, A. D.; Pastor, R. W., Update of the CHARMM All-Atom Additive Force Field for Lipids: Validation on Six Lipid Types. The Journal of Physical Chemistry B 2010, 114 (23), 7830-7843.

[5] Oliver Berger, O. E., Fritz Jahnig, Molecular Dynamics Simulations of a Fluid Bilayer of Dipalmitoylphosphatidylcholine at Full Hydration, Constant Pressure, and Constant Temperature. Biophys. J. 1997, 72, 2002-2013.

[6] Notman, R.; den Otter, W. K.; Noro, M. G.; Briels, W. J.; Anwar, J., The permeability enhancing mechanism of DMSO in ceramide Bilayers simulated by molecular dynamics. Biophysical Journal 2007, 93 (6), 2056-2068.

(388i) Identifying Ligand Binding Sites and Poses Using Hamiltonian Replica Exchange Molecular Dynamics: Model Systems and a Validation Study Using the Astex Diverse Set

Wang, K., University of Virginia
Yang, Y.,
Chodera, J. D., University of California, Berkeley
Shirts, M. R., University of Virginia
Originally presented on: 11/5/2013 17:39:00 - 17:57:00

Fast and accurate identification of protein-ligand binding locations is crucial in drug discovery. Though in theory, MD simulations of such systems will converge on the true bound structure if run long enough, such simulations can get stuck in local minima which leads to insufficient sampling. We report a methodology to accelerate and improve sampling between minima by modifying the Hamiltonian replica exchange molecular dynamics, including using multiple fully coupled and completely uncoupled states as well as Monte Carlo simulation techniques, and using GPU-accelerated code from the OpenMM toolkit.

The methodology was tested on the T4 lysozyme model system, a simple test system with an enclosed, hydrophobic pocket, and validated using 85-system Astex Diverse Set, a well-curated data set with druggable proteins and drug-like ligands, which has been widely used in industrial and academic docking studies. For a 10-ns simulation at each intermediate state for each system, in most cases taking only 1-2 days for each system, all possible binding locations and poses are identified and ranked by statistically analyzing and clustering an ensemble of configurations. Multistate Bennett acceptance ratio (MBAR) is used to calculate the binding free energies of all locations.

In the model system, the true binding site is robustly identified as the top site for three binders, but not a non-binder, demonstrating the power of our methodology to not only identify binding sites, but also differentiate binders from non-binders. Preliminary results for the validation study show that the true binding site is identified as the top three possible sites by our study for about 70% of the cases, with a 38% chance of being the top size. This systematic methodology has potential to substantially help early-stage drug discovery researchers to accurately identify binding sites and poses. The fact that multiple competing binding sites with comparable binding free energies are identified for many cases may provide insights for experimentalists designing drugs for novel targets.

(414a) Photocatalytic Conversion of Carbon Dioxide With Water On Ti-Based MOFs: An Experimental Investigation

Ngo, H. T., The University of South Florida
Ma, S., The University of South Florida
Joseph, B., Clean Energy Research Center, University of South Florida
Gao, W. Y.,
Bhethanabotla, V. R., University of South Florida
Originally presented on: 11/6/2013 8:30:00 - 8:50:00

Since Fujishima and Honda first demonstrated photocatalytic water splitting in 1972[1], numerous investigators have extended the idea into many other semiconductors and photoreactions. Utilizing solar energy to facilitate photocatalytic conversion of carbon dioxide (CO_2) with water (H_2O) to fuels and chemicals such as CO, CH_4, CH_3OH holds great potential as a sustainable alternative energy source and as a promising method to reduce pollution and global warming.

In spite of the large number of studies and materials investigated for photocatalytic CO_2 conversion, efficiencies achieved to date have been very small and much work remains to be done to overcome the limitations. Metal-organic frame works (MOFs) are perhaps uniquely suited to address CO_2 capture and photocatalytic conversion because of their exceptional porosities, tunable pore sizes and chemical structure. The principle of reducing CO_2 in a MOF using UV radiation has recently been demonstrated by doping a photocatalyst moiety into a MOF[2]. Researchers at USF have been developing Ti-based semi-conductive MOFs that are photo-active in the visible light range.

In this study, we evaluated the photocatalytic activity of a Ti-based metal-organic framework in a gas-phase CO_2 reduction reaction using a robust photoreactor system. The reactor system has a small volume (<15ml) with the ability to control temperature, pressure and reaction type (batch or flow reactor). Photoreaction efficiency, product conversion and selectivity of MOF photocatalysts are analyzed in comparison to conventional TiO_2 (P-25). Reactor conditions, including temperature and pressure, were optimized to improve reaction kinetics and the conversion of CO_2 and water to the desired products.

References:

[1] Fujishima, A., & Honda, K. (1972). Electrochemical Photolysis of Water at a Semiconductor Electrode. [10.1038/238037a0]. *Nature, 238*(5358), 37-38.
[2] Wang, C., Xie, Z., deKrafft, K. E., & Lin, W. (2011). Doping Metal–Organic Frameworks for Water Oxidation, Carbon Dioxide Reduction, and Organic Photocatalysis. *Journal of the American Chemical Society, 133*(34), 13445-13454. doi: 10.1021/ja203564w

(414b) Electrocatalytic and Materials Properties of Iron-Doped Barium Zirconates for Oxygen Membrane and CO2 Reduction Applications

Zhang, H., Texas A&M University
Wilhite, B. A., Texas A&M University
Originally presented on: 11/6/2013 8:50:00 - 9:10:00

Recently, interest in solid-oxide electrolysis cells (SOECs) with platinum or nickel cermet electrodes capable of reducing combustion effluents to synthesis gas has driven the search for mixed ionic-electronic conductors with suitable catalytic activity for reduction of carbonaceous species. There remains a need for discovering chemically stable electrode materials combining electronic and oxygen-ionic conductivities with suitable catalytic activity for oxidation or reduction. Our research group has focused upon identifying ambivalent transition-metal dopants capable of imparting both electrochemical and catalytic properties to perovskite electrolyte materials. In this study, the authors present a combined electrochemical and catalytic investigation of iron-doped barium zirconium perovskite in the context of the electrochemical reduction of combustion gas (CO_2, H_2O) to synthesis gas (CO, H_2).

Powders of $BaZr_{0.90}Fe_{0.10}O_{3-\delta}$ (BZF10) were synthesized via solid-state reaction and characterized by synchrotron X-ray diffraction (XRD) and transmission electron microscopy (TEM) coupled with energy dispersive X-ray spectroscopy (EDS). The powders were pressed and sintered into 13mm dense button cells, and a porous Pt electrode and a dense Pt blocking electrode were attached to either face of the pellet. The resulting system was used to conduct electrical conductivity relaxation (ECR), electrochemical impedance spectroscopy (EIS), and reaction analysis over a wide range of dry O_2, CO_2, H_2O, and humidified CO, H_2 partial pressures at temperatures of 600 – 800°C. Analysis of equilibrium electrical conductivities (σ_∞) vs. temperature and partial pressures provides critical insight into the underlying electrochemical mechanisms for surface reaction. Non-linear regression of conductivity vs. time data allows deduction of surface electrochemical reaction rates and solid-state diffusivities of charge carriers, and probing of surface oxygen-acceptor/donor kinetics. This information will be used to gain further insight into the underlying mechanisms for oxygen acceptor activity in the presence of oxygen as well as the surface reaction mechanisms for CO_2 reduction to CO and O_2. BZF10 was verified as an electrochemical stable, mixed ionic-electronic conducting ceramic suitable for SOEC application. The σ_∞ vs. O_2 partial pressure indicates a high oxygen-ion acceptor activity for $Ba_{0.9}Fe_{0.1}ZrO_{3+\delta}$. Equivalent experiments conducted over a range of CO_2 partial pressures (pCO_2) indicate a net decrease in σ_∞ with respect to pCO_2, coinciding with up to a 16% conversion of CO_2 to CO and O_2 at steady-state. Reduction of the Fe-doped perovskite suggests the formation of a CO_3^{-2} intermediate during the overall conversion of CO_2 to O_2 and probing of CO_2 reduction activity indicates significant catalytic activity, which is attributed to the oxygen acceptor-donor activity of the Fe-doped perovskite. Experimental results indicate that the material behaves as a mixed oxygen ionic-electronic conductor, with significant catalytic activity for CO_2 reduction.

(414c) Design of Autonomous Multifunctional GaN:ZnO Based Photocatalysts for Artificial Photosynthesis

Matsubu, J. C., University of California Riverside
Christopher, P., University of California, Riverside
Originally presented on: 11/6/2013 9:10:00 - 9:30:00

The global reliance on fossil fuels as a primary energy source is unsustainable due to massive CO_2 emissions. An extremely appealing solution is to develop Artificial Photosynthesis (AP) technologies that utilize solar energy to drive the conversion of water and negatively valued CO_2 into positively valued chemicals and fuels. Previous research in this field has focused on photocatalytic AP reactions at near ambient conditions, high temperature dissociation of H_2O and CO_2 through metal-oxide looping, or photoelectrochemical approaches. These approaches are limited by low rates, high temperatures, and/or complicated and expensive process design.

We will discuss an approach that exploits full solar spectrum conversion to split water and reduce CO_2 at moderate temperatures on autonomous multifunctional materials. The hypothesis driving this work is that a single material can be designed that utilizes high-energy photons to drive the endothermic water splitting reaction and low energy photons to heat the catalyst and drive the exothermic CO_2 reduction process. Multifunctional catalysts are designed by combining autonomous GaN and ZnO based water-splitting photocatalysts with co-catalysts for thermocatalytic CO_2 reduction.

We have explored the effect of co-catalyst metal type and oxidation state, CO_2:H_2 ratio, reaction temperature, and light intensity on thermocatalytic CO_2 reduction. We tested various GaN:ZnO supported methanation co-catalysts (Ni, Ru, Rh and their oxide counterparts) for their ability to drive CO_2 reduction and suppress oxidation back reactions. Results have shown a difference in catalytic activity and selectivity for CO_2 reduction based on co-catalyst composition, oxidation state, and CO_2:H_2 feed ratio. In addition, an increase in activity of CO_2 reduction was observed in the presence of light at elevated temperatures. These results will be discussed in the context of designing an overall H_2O and CO_2 conversion process based on autonomous, multifunctional materials.

(414d) Tandem Reaction for CO2 Hydrogenation to Methanol At Low Temperature

Chen, Y., University of Michigan
Thompson, L. T., University of Michigan
Originally presented on: 11/6/2013 9:30:00 - 9:50:00

Carbon dioxide has attracted much attention as a promising abundant and sustainable feedstock for the synthesis of fuels and commodity chemicals. The hydrogenation of CO_2 has the potential to become a carbon-neutral process, if hydrogen is obtained from renewable energy resources (e.g. H_2O). Methanol is an interesting first product, because it can be utilized as a fuel or precursor for other chemicals. State-of-the-art catalysts for CO_2 hydrogenation usually requires high operating temperatures (> 200 °C), limiting the single-pass yield to methanol, which is thermodynamically favored at low temperatures [1].

An alternative route to produce methanol from CO_2 via tandem reactions has recently been reported, with the promise of improving the methanol yield at milder temperatures (100-150 °C) [2,3]. Methanol is synthesized via three sub-steps in tandem: i) Hydrogenation of CO_2 to formic acid; ii) esterification of formic acid to alkyl formate; iii) hydrogenolysis of alkyl formate to methanol. (Reaction scheme is shown in Figure 1.) An all-homogeneous cascade catalyst, consisting of viable catalyst for each single step, has recently been developed and demonstrated for this tandem, one-pot reaction [2]. However, the overall turnover number (TON) appeared to be limited by hydrogenolysis catalyst, due to its incompatability with step ii catalyst. This tandem reaction was also achievable using a single heterogenous catalyst, with hydrogenolysis being the rate-determining step [3]. Therefore, identifying more viable catalyst for the final step is crucial to enhance the overall TON of this triple-tandem reaction.

This talk will demonstrate ethyl formate hydrogenolysis over a series of copper-based catalysts in a slurry-phase batch reactor. Several process parameters, including H_2 partial pressure and catalyst pretreatment protocol, were studied to identify favorable conditions for methanol formation. Figure 2 shows the reactant consumption and products formation profiles over a barium promoted-copper chromite catalyst. A total yield of 45.3% methanol was achieved at an ethyl formate conversion of 87.4% (8 hr, 135 °C, 30 bar H_2). We also evaluated this catalyst for CO_2 hydrogenation under similar conditions; with just the Cu-based heterogeneous catalyst, an overall yield of 35.8% was achieved at a conversion of 41.4% (20 hr, 135 °C, 10 bar CO_2, 30 bar H_2). The rates are likely to be improved when the appropriate homogenous catalysts are combined with this heterogneneous catalyst.

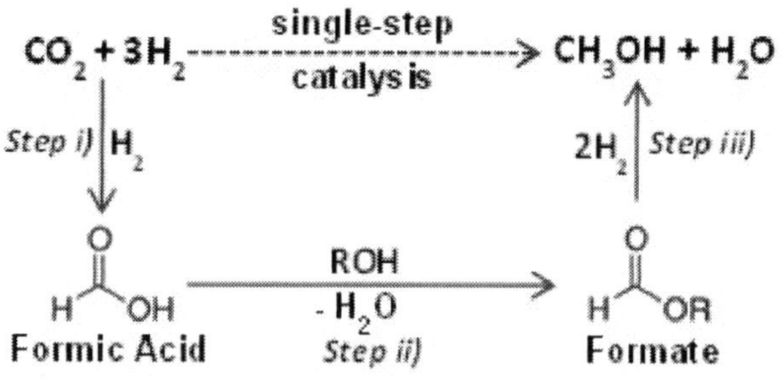

Figure 1: Tandem reaction scheme for CO_2 hydrogenation to methanol

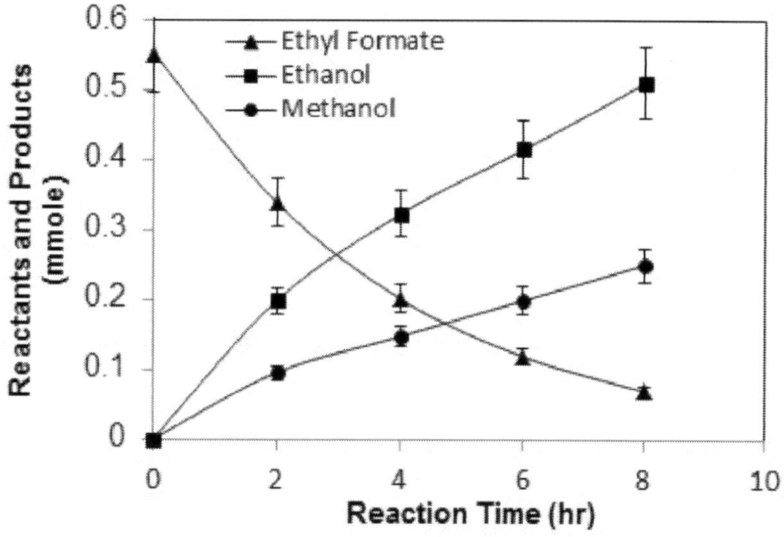

Figure 2: Reactant consumption and products formation profiles of ethyl formate hydrogenolysis over barium promoted-copper chromite. 135 °C, 30 bar, 200 rpm, 200 mg catalyst, 37.5 ml p-Dioxane, 0.55 mmole formate, 4.66 mmole H_2 (based on solubility in p-Dioxane).

References

1. S. Natesakhawat, J. Lekse, J. Baltrus, P. Ohodnicki, Jr. B. Howard, X. Deng, C. Matranga, *ACS. Catal. 2* **(2012)** 1667.

2. C. Huff, M. Sanford, *J. Am. Chem. Soc. 133* **(2011)** 18122.

3. L. Fan, Y. Sakaiya, K. Fujimoto, *Appl. Catal. A: Gen. 180* **(1999)** L11.

(414e) Effects of Surface Modifications On the Electrochemical Reduction of CO2 On Silver-Based Catalysts

Hatsukade, T., Stanford University
Kuhl, K. P., Stanford University
Cave, E. R., Stanford University
Abram, D. N., Stanford University
Jaramillo, T. F., Stanford University
Originally presented on: 11/6/2013 9:50:00 - 10:10:00

Research on the electrochemical reduction of CO_2 has focused on gathering a fundamental understanding of the catalytic process through looking at a range of transition metals. Carbon monoxide is believed to be a key intermediate and the CO binding energy (E_B[CO]) of the transition metal catalyst determines its activity for carbon dioxide reduction reaction (CO_2RR). We investigated CO_2 electroreduction on silver-catalyzed systems for two main reasons: (1) silver has high activity for CO_2RR to CO, and (2) it has a late onset for hydrogen evolution, an unwanted byproduct of electrochemical reduction in an aqueous environment. Silver is known to have a weak, sub-optimal E_B[CO]; by exploring ways to increase the E_B[CO] of silver, a higher activity of the CO_2RR may be attained, confirming the significance of the CO intermediate species.

Initial investigations using a polycrystalline silver surface were performed in a custom electrochemical cell, which offers high sensitivity for minor products of CO_2RR. As expected from literature, CO and H_2 were observed as major products, along with formate as a minor product. In addition to these products, methane, methanol, and ethanol were also detected as minor products at high overpotentials, illustrating the ability of silver to produce hydrocarbon/oxygenate products even with its weak E_B[CO]. The silver surface was then nanostructured through an electrochemical oxidation/reduction process with the aim of strengthening the E_B[CO]. This enhanced the production of CO_2RR products compared to the unmodified surface. The enhanced activity is attributed to the possible shift in the E_B[CO] of the surface as well as the increase in the surface area.

(414f) Design, Characterization and Testing Of Catalysts For Electrochemical Reduction Of CO2 To CO

Jhong, M., University of Illinois at Urbana-Champaign
Kenis, P. J. A., University of Illinois at Urbana-Champaign
Originally presented on: 11/6/2013 10:10:00 - 10:30:00

During the past few decades, the increasing level of CO_2 in the atmosphere has led to a number of undesired climate effects, including global warming, rising sea levels, and more unpredictable weather patterns. The amounts of CO_2 being produced are so large that multiple approaches need to be implemented simultaneously, including switching to renewable energy sources, increasing the energy efficiency of buildings, switching from coal to the less carbon-producing natural gas, natural gas, and underground carbon sequestration to curb the increase in atmospheric CO_2 levels.[1] An additional approach that can be employed to overcome this daunting challenge is the electrochemical reduction of CO_2 into useful chemicals such as formic acid, carbon monoxide, hydrocarbons, or alcohols.[2-4] This process can be driven, for example, by the vast amount of excess renewable power that frequently is available from intermittent sources such as solar and wind. Furthermore, by utilizing CO_2 as the starting material for chemical production, this process reduces our dependency on fossil fuels.

Despite the potential of electrochemical reduction of CO_2, current performance levels are insufficient for commercialization.[5] The catalysts, electrode structures, as well as operation conditions need to be improved such that sufficient selectivity for the desired product (Faradaic efficiency >95%), sufficient energetic efficiency (>60%), and sufficient conversion (current density >250 mA/cm^2) can be achieved.[6-8]

This paper will report some of our recent efforts to understand and improve catalysts, electrode structure,[9] and operation conditions.[10] For example, we have developed organometallic catalysts that exhibit identical performance to Ag nanoparticles, but at a significantly lower loading, as well as multiwall carbon nanotube-supported catalysts and metal-free catalysts that exhibit current densities larger than 180 mA/cm^2, thus vastly exceeding state-of-the-art performance of Ag nanoparticles. The results of these efforts start to approach the aforementioned specified performance metrics needed for an economically viable process.

[1] S. Pacala, R. Socolow, *Science* **2004**, *305*, 968-972.
[2] D. T. Whipple, E. C. Finke, P. J. A. Kenis, *Electrochem. Solid-State Lett.* **2010**, *13*, B109-B111.
[3] Y. Hori, in *Handbook of Fuel Cells, Vol. 2*, John Wiley & Sons, Ltd, **2010**, pp. 720-733.
[4] H.-R. M. Jhong, S. Ma, P. J. A. Kenis, *Curr. Opin. Chem. Eng.* **2013**, *accepted.*
[5] E. J. Dufek, T. E. Lister, M. E. McIlwain, *Electrochem. Solid-State Lett.* **2012**, *15*, B48-B50.
[6] C. E. Tornow, M. R. Thorson, S. Ma, A. A. Gewirth, P. J. A. Kenis, *J. Am. Chem. Soc.* **2012**, *134*, 19520-19523.
[7] A. Salehi-Khojin, H.-R. M. Jhong, B. A. Rosen, W. Zhu, S. Ma, P. J. A. Kenis, R. I. Masel, *J. Phys. Chem. C* **2012**, *117*, 1627-1632.
[8] B. A. Rosen, A. Salehi-Khojin, M. R. Thorson, W. Zhu, D. T. Whipple, P. J. A. Kenis, R. I. Masel, *Science* **2011**, *334*, 643-644.
[9] H.-R. M. Jhong, F. R. Brushett, P. J. A. Kenis, *Advanced Energy Materials* **2013**, *accepted.*
[10] M. R. Thorson, K. I. Siil, P. J. A. Kenis, *J. Electrochem. Soc.* **2013**, *160*, F69-F74.

(414g) Trends in the Catalytic Reduction of CO2 By Hydrogen Over Supported Monometallic and Bimetallic Catalysts

Porosoff, M. D., University of Delaware
Chen, J. G., Columbia University
Originally presented on: 11/6/2013 10:30:00 - 10:50:00

Rising atmospheric concentration of CO_2 is forecasted to have potentially disastrous effects on the global climate due to its role in global warming and ocean acidification. A catalytic process that utilizes CO_2 as a feedstock to make carbon monoxide, methanol, and methane, is potentially more desirable than sequestration. CO is the most desired product because it can be integrated into down-stream processes to produce synthetic fuels via Fischer-Tropsch.

The reduction of CO_2 by hydrogen has been conducted on supported catalysts in batch and flow reactors. Catalysts synthesized on a reducible support (CeO_2) showed higher activity than on an irreducible support (γ-Al_2O_3). The active metal also played an important role in controlling the selective reduction of CO_2 to CO instead of CH_4. Extended X-ray absorption fine structure (EXAFS) and transmission electron microscopy (TEM) confirmed the formation of uniform, bimetallic particles. Among the monometallic and bimetallic catalysts evaluated in a batch reactor, PdNi/CeO_2 was found to be the most active bimetallic catalyst, but formed the greatest amount of CH_4. PtCo/CeO_2 showed the highest selectivity to CO with very low selectivity to CH_4. The selectivity of each catalyst correlated with the d-band center value of each bimetallic surface. Among the catalysts investigated, bimetallics with values of d-band center farther from the Fermi level produced more CO and less CH_4 than catalysts with values closer to the Fermi level.

Batch reactor experiments were verified in a flow reactor under corresponding conditions. Results in the flow reactor were consistent with the batch reactor, with Pt–Co showing the highest selectivity to CO and Pd–Ni the lowest. Establishing a selectivity trend for CO_2 activation provides a facile means to choose effective catalysts from the large database of d-band center values for a variety of monometallic and bimetallic catalysts.

(429a) A Theoretical-Experimental Analysis of the Precipitation of Acetaminophen and Tretinoin By Using Supercritical Enhanced Dispersion of Supercritical Fluids (SEDS)

Tabernero, A., University of Salamanca
Galan, M. A., University of Salamanca
Martin del Valle, E. M., University of Salamanca
Originally presented on: 11/6/2013 8:30:00 - 8:55:00

Over the last decades, supercritical fluids (SCFs) technologies have been investigated as environmentally friendly techniques for pharmaceutical particle engineering. In this context, SCFs can be mainly used as solvents or antisolvents with the aim of micronizing drugs.

Solution enhanced dispersion by supercritical fluids (SEDS) is a SCF antisolvent technique in which the solution is atomized in a supercritical atmosphere (in this case supercritical CO_2). If the solid is poorly soluble in the SCF, and at the same time the SCF is highly soluble in the solvent, a great degree of supersaturation is achieved and an antisolvent effect is produced. Therefore, the solid precipitates in a reduced size without non-solvent traces. The main innovation of the SEDS process in relation with other SCFs technologies is the nozzle design. In this case, a coaxial nozzle with a mixing length is used in order to provide a previous mixture of the SCF and the solution, enhancing the mass transfer.

Thermodynamics, hydrodynamic, mass transfer and crystallization kinetics are therefore involved in that technique and they should be studied to get a complete understanding of the precipitation process.

A theoretical-experimental study of the precipitation of acetaminophen and tretinoin with SEDS has been performed in this work. Phase equilibria of the systems, atomization regimes and mass transfer coefficients of liquid and vapor phases have been separately studied.

According to the experimental results, the smallest particles without aggregates for both systems have been obtained at experimental conditions just above the mixture critical point of the antisolvent-solvent system with a high antisolvent-solvent mass flow rate ratio. On the other hand, according to the theoretical results, a high degree of supersaturation and a high mass transfer coefficient of the liquid phase are produced at these conditions. That means a great antisolvent effect because of the low liquid side mass transfer resistance and a small particle size because of the high supersaturation. Moreover, results indicate that a small initial droplet size provides a higher antisolvent effect.

(429b) Synthesis of Biodegradable Polymers Using Supercritical Carbon Dioxide As Reaction Medium

Beuermann, S., Clausthal University of Technology

Schmidt, C.,

Originally presented on: 11/6/2013 8:55:00 - 9:20:00

Biodegradable polymers are of high interest for applications in the medical sector due to their decomposition into safe compounds under physiological conditions.[1-3] Applications in the biomedical sector require medical grade polymers. Therefore, polymerization processes that avoid the use of toxic substances are mandatory. The use of supercritical carbon dioxide ($scCO_2$) as reaction medium is particulary interesting. Firstly, CO_2 is non-toxic and gaseous at ambient conditions. Consequently, it may easily be removed from the product without chromatographic techniques and/or any energy intensive drying processes. Secondly, the physico-chemical properties may be tuned in a wide range by changes in temperature and pressure. Of particular interest is the density and the ability to dissolve other chemical compounds, which is directly related to the phase behavior of the systems. Another aspect is the processing of the material following the synthesis. Various processes are already established for the generation of particulate material.[4] Recently, it was also shown that the type of crystalline phases may be influenced.[5] In addition to the choice of reaction medium the selection of catalyst is crucial.

The focus of this contribution is on the ring-opening polymerization of diglycolide. It will be shown that $scCO_2$ has various advantages for the synthesis of polyglycolide (PGA). Besides using metal-organic compounds as catalyst enzymatic polymerizations were carried out. The polymerizations was performed in an optical high pressure cell, allowing for reaction pressures up to 1500 bar. Since the decomposition rate of the polymer strongly depends upon its molecular weight, molecular weight control during polymerization is of high importance. Thus, the reaction parameters such as temperature, pressure, monomer concentration, and reaction time were varied to obtain polymer material with well-defined molecular weights. In addition, the degradation rate of the polymer is strongly affected by the chemical structure of the polymer.[3] Therefore additional monomers such as ε-caprolactone and ω-pentadecalactone were employed. Due to the low ring strain in ω-pentadecalatone enzymatic polymerizations of this monomer are particularly attractive. The enzymes *Candida antarctica* and *Burkholderia cepacia* were used.

[1] K. A. Athanasiou, G. G. Niederauer, *Biomaterials* **1996,** *17*, 93.

[2] W. D. Hovis, R. W. Bucholz, *Foot Ankle Int.* **1997**, *18*, 128.

[3] D. Hofmann, M. Entrialgo-Castaño, K. Kratz, A. Lendlein, *Adv. Mater.* **2009**, 21, 3237.

[4] S.-D. Yeo, E. Kiran, *J. Supercrit. Fluids* **2005**, *34*, 287.

[5] D. Bolten, M. Türk, *J. Supercrit. Fluids* **2012**, *66*, 32.

[4] T. Meyer, M. F. Kemmere, ?Supercritical Carbon Dioxide- in Polymer Reaction Engineering?, **2005**, Wiley-VCH, Weinheim.

(429c) Foaming Behavior of Polymers With Using Mixtures of CO2 and N2 As Foaming Agent

Kato, T., Tokyo University of Science
Otake, K., Tokyo University of Science
Shono, A., Tokyo University of Science
Kobayashi, D., Tokyo University of Science
Shimada, Y., Tokyo University of Science
Originally presented on: 11/6/2013 9:20:00 - 9:45:00

Polymer foams are well known to have high shock absorbing ability, thermal insulating ability, good handling ability and many other good characteristics. Due to their properties, polymer foams are used as products for daily use like sponge, housing materials, car bumper and so on.

Recently, due to the interest in environmental issues, non-CFCs such as carbon dioxide (CO_2) or nitrogen (N_2) are used for blowing agent for the production of the polymeric foams. It is well known that foams with CO_2 have large but non-uniform cells. On the other hand, foams with N_2have a lot of small and uniform cells.

In this work, to control the final foaming structure, use of the mixtures of CO_2 and N_2 as blowing agent was examined through the foam morphology and visualization of the initial stage of foaming. Poly(methyl mathacrylate) (PMMA), Polystyrene (PS), CO_2 and N_2 were used for the foaming experiments. Molar fraction of CO_2/N_2 was varied in the range 0-100 mol% to observe the foam morphology of PMMA/CO_2/N_2 and PS/CO_2/N_2systems. The densities of the foamed polymer samples were determined by the Archimedean principle using a picnometer. For microstructural analysis, the foamed samples were observed with scanning electron microscope.

Initial stage of foaming was observed with batch foaming visualization system. In this study, foaming behavior, or the leading time of foaming after pressure release and cell diameters growth rate as well as number density increase of cells, was examined with PMMA/CO_2/N_2systems.

In the foam morphology, we found the difference in density, cell number density, and cell size, following the change in CO_2/N_2 molar fractions. The lowest density (0.047 g/cm^3) was obtained with PMMA/CO_2/N_2 system at the CO_2/N_2 ratio of 40 mol%. It is 4 to 8 times lower than PMMA/CO_2 or PMMA/N_2 binary systems. Also PS/CO_2/N_2 system had the lowest density (0.038 g/cm^3) at the CO_2/N_2 ratio of 83 mol%, which is different from PMMA/CO_2/N_2system. Polymer foams with the lowest density had larger cells than those obtained with the single gas foaming.

The initial stage of foaming with mixtures of CO_2 and N_2 systems exhibited somewhat different behavior compared with the single gas foaming systems. At certain molar ratio of CO_2 and N_2, growth rate of the initially formed bubble changed drastically. This behavior might be caused by the plasticization effect from increasing the CO_2 molar fraction. Decrease in the interface tension of polymer/gas with the increase in the CO_2 molar fraction will encourage the cell diameters to grow. However, the plasticization also decreases viscosity, which will cause unification and collapse of cells in the high CO_2 molar fraction range. The best CO_2/N_2molar fraction for low density of foam might be determined by the balance of the cell diameter growth with unification and collapse of cells.

It could be concluded that use of the mixtures of CO_2 and N_2 as blowing agents is useful to control the foaming structure.

(429d) Formation of "Vespula-Like" Structure By the Foaming of Poly(methyl methacrylate)/ Carbon Dioxide / Tetramethoxysilane Ternary System

Yamaji, Y., Tokyo University of Science
Otake, K., Tokyo University of Science
Shono, A., Tokyo University of Science
Kobayashi, D., Tokyo University of Science
Shimada, Y., Tokyo University of Science
Yoda, S., National Institute of Advanced Industrial Science and Technology (AIST)
Furuya, T., National Institute of Advanced Industrial Science and Technology (AIST)
Takebayashi, Y., National Institute of Advanced Industrial Science and Technology (AIST)
Originally presented on: 11/6/2013 9:45:00 - 10:10:00

In recent years, for the effective energy usage, development of high performance heat insulator has been the subject of research for engineers. Polymeric foams have been used as thermal insulator due to their inherent advantages, such as low density, high mechanical strength, and low thermal conductivity. The use of supercritical carbon dioxide ($scCO_2$) as an alternate foaming agent has been gathering a great attention. With $scCO_2$ as a foaming agent, foams that have high cell density and mechanical strength could be obtained. However, $scCO_2$ also has the weak point that its adiabaticity is comparatively lower than that of other foaming agents. To overcome this, addition of third component was examined. Foam that inside of the cells filled with silica aerogel, or the "vespula-like" structure, is regarded as one of the promising candidate for high performance heat insulator.

In this study, general conditions for the forming of the "vespula-like" structure was investigated. Tetramethoxysilane (TMOS) and Poly(methyl methacrylate) (PMMA) were chosen as a model system because PMMA/TMOS/CO_2 system is known to form the "vespula-like" structure. PMMA/silica aerogel composite with "vespula-like" structure was prepared under various condition and the morphology of those was observed with scanning electron microscope (SEM).

The PMMA/silica aerogel composite foams were prepared by batch foaming process followed by gelation process. In the batch foaming process, PMMA samples and TMOS were placed in high pressure vessel. CO_2 was fed to the pressure vessel at constant temperature and pressure. After saturating PMMA with CO_2 and TMOS, the pressure was released at constant decompression rate. The experimental parameters were saturation temperatures, pressures, and decompression rates. After the foaming process, in order to fill cells with silica aerogel, gelation was conducted under saturated conditions of water or aqueous 3 wt% NH_3 solution. In the cells, TMOS that was separated from PMMA by releasing pressure was hydrolyzed and polymerized by the vapor to obtain the solid silicone earogel.

It was found that the number of cells filled with silica aerogel was increased by foaming at higher temperatures. TMOS likely have a relatively superior solubility at high temperature. Therefore, there may be more TMOS in PMMA at high temperature. At the same temperature, the number of cells filled with silica aerogel was increased by the increase in the saturation pressure followed by faster depressurization rate. It is known that high pressure saturation and fast depressurization rate lead to smaller average cell size. Due to the fact that the cells prepared at these conditions had small volume, it may be easily filled with silica aerogel.

To obtain the foaming structures that most of the cells are filled with silica aerogel, controlling of the foaming conditions, PMMA/silica aerogel compositions are important. Further experiments should be conducted for the full understandings of the foaming structure of this system.

(429e) Determination of Supersaturation Fields By CFD in Supercritical Co-Crystallization Process

Erriguible, A., I2M - TREFLE - IPB - University of Bordeaux
Revelli, A. L.,
Laugier, S.,
Subra-Paternault, P.,
Originally presented on: 11/6/2013 10:10:00 - 10:35:00

In the pharmaceutical sector, a new strategy at nanoscale for modulating the physical and chemical properties of drugs at macroscale is the formation of cocrystals, i.e. multicomponent assemblies of drug (API) and coformer held together by noncovalent interactions. Meanwhile, the need for improved manufacturing technologies is getting more and more crucial, notably because of the emergence of the generics that are less expensive than the original drugs and of the growing social concerns for environment protection and global change that motivate the emergence of alternative products and manufacturing routes. Crystallization assisted by CO_2, and especially the so-called SAS (Supercritical Antisolvent) process has emerged years ago as a promising alternative route for a better control of the particle size. Experimentally, the task is fastidious, and consequently promotes the development of simulation to better identify important parameters and to go off the case-by-case approach. A complete description of the Supercritical Antisolvent Process requires the modeling of several complex physical phenomena such as heat and mass transfer, hydrodynamic, phase equilibria of ternary or quaternary, and nucleation /growth of the species to be precipitated. Our team is currently developing a CFD code that takes in consideration all phenomena cited above, in case of the spray version of the antisolvent process, i.e. when the liquor solution is injected into a vessel pressurized with CO_2 at given temperature, pressure and flow rate. Simulations are performed in 3D configuration with the home-made Computational Fluid Dynamic code “Thetis”, by solving the classical Navier-Stokes model coupled with the Fick’s law to take into account the diffusion of the species. Let us mention however that due to the importance of the micromixing, turbulence was considered using a Large Eddy simulation model. By implementing further the equilibrium data for the solute to recrystallize, the spatial and time-dependent supersaturation field is analysed. As a first step, we propose simulations for the systems naproxen/acetone and nicotinamide/acetone in classical SAS conditions, i.e. when the liquor solution (solute+solvent) is injected in the pressurized CO_2 reactor. Next, the supersaturation fields obtained in the previous configurations are compared with the ones in the case of SAS co-crystallization at the same pressure/temperature conditions, i.e. when the API (naproxen), the co-former (nicotinamide) and the solvent (acetone) are injected simultaneously in the pressurized reactor. The analysis of supersaturation cartographies of the species to crystallize show that the supersaturation levels are particularly modified in the case of co-crystallization.

(429f) A Novel Method of Making Stamps in Soft Lithography Via Compressed Carbon Dioxide

Cang, Y., East China University of Science and Technology
Zhang, R., East China University of Science and Technology
Liu, L., East China University of Science and Technology
Zhang, J., East China University of Science and Technology
Yang, T., East China University of Science and Technology
Guo, X., East China University of Science and Technology
Originally presented on: 11/6/2013 10:35:00 - 11:00:00

A novel method of making stamps in soft lithography via compressed carbon dioxide

Yu Cang, Rui Zhang*, Lixiao Liu, Jianchao Zhang, Tao Yang, Xuhong Guo*

State Key Laboratory of Chemical Engineering, East China University of Science and Technology, 130 Meilong Road, Shanghai 200237, China

**Corresponding authors. Tel.: +86-21-64253491; E-mail: r.zhang@ecust.edu.cn*

guoxuhong@ecust.edu.cn

Abstract: Soft lithography (SL) is one newly developed and high precise microfabrication technology which can fabricate various microstructures of materials costless with the help of medium pattern-transfer element or elastomeric stamp. Polydimethylsiloxane(PDMS) owns excellent viscoelasticity that become widely used in making different-microstructure stamps. PDMS stamps are produced by casting and curing PDMS precursors on the Si master by photolithography or swelled by some organic solvent traditionally. However, these preparation methods are expensive and may distort microstructures which limited their applications. We propose a new method of making stamp that imprint microstructure on PDMS with compressed carbon dioxide due to the good solubility of PDMS in compressed carbon dioxide. The patterns are sandwiched between two PDMS moldings under certain pressure and temperature. We use gauze, silica gel particle and nanometer-size chip as patterns. In the whole process, the reaction pressure is up to 5MPa and temperature is 318.15K, which can soft PDMS well. However, after the reaction the quench temperature should be decreased obviously under constant pressure to prevent the collapse of stamp. The patterns of different dimensions from micro-meters to nano-meters can be imprinted perfectly after about 4 hours. The simply and precise way of making stamp can expand the use of SL in chemistry, biology, physics, semiconductor fields and so on.

Keywords: soft lithography, polydimethylsiloxane, carbon dioxide

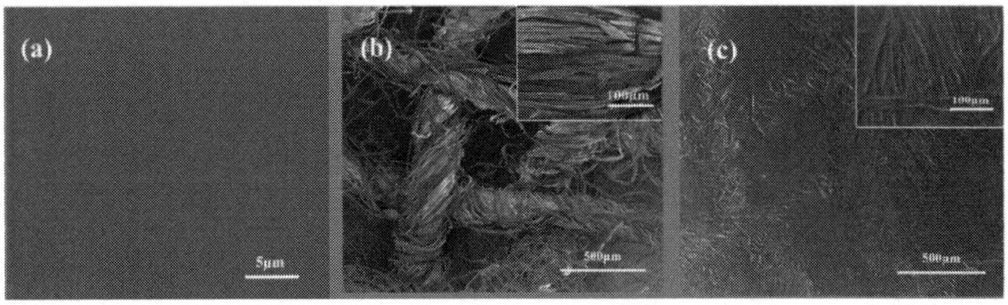

Figure. (a) SEM image of PDMS before imprinted (b) SEM image of gauze pattern (c) SEM image of PDMS after imprinted gauze pattern

(439a) Metal Nanoparticle Morphogenesis and Stabilization Mediated By Amphiphilic Polymers

Sakai, T.,
Gkikas, E.,
Iatrou, H.,
Alexandridis, P., University at Buffalo - The State University of New York (SUNY)
Originally presented on: 11/6/2013 8:35:00 - 8:53:00

Metal nanoparticles exhibit novel optical and catalytic properties, and attract considerable interest in a range of applications, e.g., photonics, diagnostics. The morphology (size and shape) of the nanoparticles and their surface/colloidal properties are very important in the various applications. A methodology for the synthesis in aqueous media of metal nanoparticles with controlled size and shape and exceptional colloidal stability is reviewed. This methodology is based on designer polymers that can exhibit multiple functions on the basis of the polymer intramolecular and supramolecular organization. In addition to being water based, this methodology requires no external energy input and employs commercially available polymers, e.g., poly(ethylene oxide) containing Pluronics or Poloxamers, resulting in low cost and potential environmental benefits. Recent results are presented on the use for gold nanoparticle synthesis and stabilization of PEO-based polymers with different end-functionality, and of poly(L-proline).

Alexandridis, P.; Sakai, T. Preparation of Metallic Nanoparticles. *United States Patent 7,718,094*; publication date: May 10, 2010.

Alexandridis, P. Gold Nanoparticle Synthesis, Morphology Control, and Stabilization Facilitated by Functional Polymers. *Chem. Eng. Technol.* **2011**, *34* (1), 15-28. DOI: 10.1002/ceat.201000335

Sakai, T.; Horiuchi, Y.; Alexandridis, P.; Okada, T.; Mishima, S. Block Copolymer-mediated Synthesis of Gold Nanoparticles in Aqueous Solutions: Segment Effect on Gold Ion Reduction, Stabilization and Particle Morphology. *J. Colloid Interface Sci.* **2013**, *394*, 124-131. DOI: 10.1016/j.jcis.2012.12.003

Gkikas, M.; Timonen, J.; Ruokolainen, J.; Alexandridis, P.; Iatrou, H. Facile Aqueous Synthesis and Stabilization of Nearly Monodispersed Gold Nanospheres by Poly(L-proline). *J. Polym. Sci. Part A: Polymer Chemistry* **2013**, *51* (6), 1448-1456. DOI: 10.1002/pola.26517

(439b) Flow Synthetic Process of Silica@Au Core-Shell Nanoparticles By Using Microreactor

Watanabe, S., Kyoto University
Hiratsuka, T., Kyoto University
Asahi, Y., Kyoto University
Mae, K., Kyoto University
Miyahara, M., Kyoto University
Originally presented on: 11/6/2013 8:53:00 - 9:11:00

SiO_2@Au core-shell type particles are quite attractive because of their unique optical properties and potential biomedical applications including imaging, sensing, photothermal therapy, and gene-silencing technology. A widely-used preparation technique of Au nanoshells is the seed-mediated growth, which is composed of three steps: surface modification of core silica particles, Au seeds decoration of the modified silica surface, and the shell growth by reduction of Au ions on the Au seeds. Although this technique allows good control in the shell thickness, it takes a long time on the order of days to complete the preparation process. In the present study, we develop a flow synthetic process by applying a microreactor separately to the Au seeds decoration and the shell growth processes. In the decoration process, we demonstrate that the microreactor enables one-step fabrication of gold-decorated silica particles by directly reducing Au ions with a strong reducing agent in the presence of silica particles, which can remarkably shorten the preparation period. The microreactor flow synthesis is also demonstrated to be effective for the shell growth process when a mild reducing agent is used to suppress the nucleation in the solution phase instead of the Au seeds surface. We will also discuss the stability of Au nanoshells depending on the type of the reducing agent.

(439c) Sub 50 Nm Au Nanoclusters With Absorbance Up to 1200 Nm Made By Favoring Kinetically Versus Equilibrium Favored Morphologies

Borwankar, A. U., University of Texas at Austin
Willsey, B. W., University of Texas at Austin
Twu, A., University of Texas at Austin
Stover, R.,
Hung, J., University of Texas at Austin
Truskett, T. M., The University of Texas at Austin
Johnston, K. P., University of Texas at Austin
Originally presented on: 11/6/2013 9:11:00 - 9:29:00

Plasmonic nanoparticles with high NIR absorbance are of great interest in fields including biomedical optical imaging, microelectronics, plasmonic sensors and catalysis. Herein we synthesize 35 to 50 nm Au nanoclusters coated with carboxymethyldextran (CMD) with high NIR absorbance at wavelengths ranging from 700 nm to 1200 nm. Nanoclusters, composed of Au primary particles with diameters of ~ 10 nm, were assembled by finely tuning the rate of reduction of Au with pH in the presence of a polymeric stabilizer. To limit extemely fast nucleation and growth, we utilize a low Au precursor concentration together with a mild reducing agent. Strong effects of pH on NIR absorbance are observed and can be understood in terms of growth processes (kinetic vs. thermodynamic) and the resulting particle morphologies. As pH is decreased, the growth becomes more kinetically controlled leading to "finger-like" protrusions that increse the NIR extinction markedly. At higher pH with slower growth, a more thermodynamically controlled parthway is dominant with much smaller protrusions and thus much weaker NIR extinction. In the case of two iterations of growth, the additional protrusions and asymmetry from the sperhical boundary condition lead to even greater NIR absorbace from multi-polar interactions. The particles also have a high negative zeta potential, providing good colloidal stability.

(439d) Templated Synthesis of Multifunctional Plasmonic Nanoparticles

Linn, N., University of Florida
Jiang, P., University of Florida
Originally presented on: 11/6/2013 9:29:00 - 9:47:00

Hollow nanoparticles play a critical role in microencapsulation – a process that has been widely exploited for controlled drug delivery. Here we report the fabrication of hollow flask-shaped plasmonic nanoparticles by a colloidal templating approach. The resulting nanoparticles exhibit tunable plasmonic properties which can be easily tuned by changing the templating particle size and the shell thickness. The experimental optical measurements match with the theoretical simulations using finite-difference time-domain (FDTD) model. We observe strong in-situ optical adsorption and heating by the templated plasmonic nanoparticles. The interior of the hollow nanoparticles can be filled up with chemicals and the flask opening can then be sealed by a thin layer of heat-sensitive polymer. We have demonstrated that the encapsulated chemicals (e.g., a laser dye) can be delivered in a controlled manner when a near-IR laser beam is illuminated on these composite nanoparticles. Additionally, the plasmonic nanoparticles exhibit strong surface-enhanced Raman scattering (SERS), promising for in-situ biosensing.

(439e) Stable Polymer Grafted Iron Oxide Nanoclusters in High Salinity Brine for Transport and Imaging in Subsurface Reservoirs

Xue, Z.,
Bagaria, H. G., Rice University
Neilson, B., The University of Texas
Worthen, A., The University of Texas
Yoon, K., The University of Texas
Nayak, S.,
Cheng, V., The University of Texas
Lee, J., The University of Texas
Bielawski, C. W., The University of Texas
Johnston, K., University of Texas Austin
Originally presented on: 11/6/2013 9:47:00 - 10:05:00

Magnetic nanoparticles with high mobility in subsurface reservoirs at high salinities and temperatures are expected to have a major impact on enhanced oil recovery, carbon dioxide sequestration, and electromagnetic imaging. Herein we present a rare example of elelctrosteric stabilization of iron oxide (IO) nanoparticles (NPs) grafted with poly(2-acrylamido-2-methylpropanesulfonate-co-acrylic acid) (poly(AMPS-co-AA)) that not only display colloidal stability in standard American Petroleum Institute (API) brine (8% NaCl + 2% CaCl2 by weight) at 90 °C for 1 month but also resist undesirable adsorption on silica surfaces (0.4% monolayer NPs). In this copolymer coating, the AMPS groups interact weakly with Ca2+, so that the poly(AMPS-co-AA) chains grafted onto IO NPs surface remain sufficiently well solvated and negatively charged to provide electrosteric stabilization. The AA groups, in addition, enable covalent grafting of the poly(AMPS-co-AA) chains to amine-functionalized IO NPs via formation of multiple amide bonds and prevent polymer desorption even after a 40000-fold dilution. The remarkable stability and transport properties of poly(AMPS-co-AA) grafted IO NPs is attributed to the high stability of poly(AMPS-co-AA) chains as revealed by dynamic light scattering and electrokinetic measurements. The low equilibrium silica adsorption capacity of poly(AMPS-co-AA) grafted IO NPs correlates well with the low retention in 1D column flow experiments. The aforementioned methodology may be readily adapted to stabilize a variety of other functional inorganic and organic NPs at high salinities and temperatures.

(439f) Effects of Water Droplet Size On Polypyrrole Particle Synthesis Using Water in Oil Emulsion System

Shono, A., Tokyo University of Science
Indo, K., Tokyo University of Science
Nakada, H., Tokyo University of Science
Kobayashi, D., Tokyo University of Science
Shimada, Y., Tokyo University of Science
Takahashi, T., Tokyo University of Science
Otake, K., Tokyo University of Science
Originally presented on: 11/6/2013 10:05:00 - 10:23:00

Conducting polymers such as polyacetylene, polyaniline, and polythiophene have been investigated widely for use in various applications. Polypyrrole is one of the most studied conducting polymers because of its environmental stability and relative ease of synthesis. However, the poor processability of PPy is often a problem. Particularly for this reason, the synthesis of a fine-particle conducting polymer by chemical oxidative polymerization is a valuable synthetic goal for applying the conductive nano filler. Normally, PPy particles are obtained by chemical polymerization in various organic solvents and aqueous media, into which an oxidative agent such as ferric chloride, ferric perchlorate, potassium persulfate, or ammonium persulfate is added. However, this process yields not only spherical particles, but also particles that are hexagonal, cubic, wire-like, and ribbon-like. As the rate of reaction in chemical oxidative polymerization for the synthesis of conducting polymers is very rapid, it is important to prepare spherical particles of the conducting polymer to maintain the reaction temperature at approximately 278 K. In this study, control of the reaction rate was investigated, and a new method for the synthesis of polypyrrole particles using water in oil emulsion is proposed. In this system, the diffusion of the monomer into a water droplet as the dispersed phase is the rate-determining step, and controls the reaction rate at room temperature. Monodispersed spherical polypyrrole particles were synthesized at room temperature using this method. The effects of organic solvent as the continuous phase on the synthesized polypyrrole particle size were investigated. The diameter of the polypyrrole particle decreased with increasing viscosity of the organic solvent, and an empirical formula connecting the diffusion coefficient and average particle size is proposed. In addition, the effects of the volume fraction of the dispersed phase on the water droplet size of emulsion and synthesized polypyrrole particle size were investigated. The water droplet size increased with increasing volume fraction of the dispersed phase. On the other hand, the tendency of the particle size differs from water droplet size. In the condition that the volume fraction of dispersed phase was smaller than 10%, the water droplet size of emulsion was around 10 nm, and it is considered reversed micelle. Therefore, the polypyrrole particle formation mechanism changes with water droplet size. From these results, we reveal that monodispersed spherical polypyrrole particles were synthesized at room temperature by chemical oxidative polymerization using W/O emulsion system. And, the polypyrrole particle formation mechanism was proposed.

(439h) Kinetic Model for Stable Amorphous Calcium Carbonate

Harris, D. D., BYU
Lignell, D. O., Brigham Young University
Originally presented on: 11/6/2013 10:41:00 - 10:59:00

Amorphous calcium carbonate (ACC) is important to many industrial processes as well as biological environments. Calcium carbonate has recently attracted a lot of attention due to its potential as a carbon sink for sequestering carbon dioxide. ACC has been found to play a role in the formation of biominerals in aquatic organisms due to their ability to manipulate ACC. ACC is also a common precursor to anhydrous crystalline calcium carbonates in highly supersaturated solutions. The precipitation kinetics of calcium carbonate and consequent transformation mechanism into solid phases is not fully understood.
Intermolecular interactions and steric effects exhibited by aqueous systems make the application of nucleation theory challenging. In this work, classical and novel nucleation mechanisms are applied to an experiment conducted by Gebauer et al. (D. Gebauer, A. Volkel and H. Colfen. Stable prenucleation calcium carbonate clusters. Science, 322:1819-1822, 2008) for comparison. Four solid-phase kinetic models are coupled to the equilibrium chemistry model to simulate results which are compared with data from Gebauer et al. Each of the models have a tuning parameter, optimized at a pH of 10.0. The solid-phases kinetic models consist of two classical models and two meso-crystallization models which operate within the framework of direct quadrature method of moments (DQMOM). The four models each capture various aspects of the data, and operate on the assumption that the calcium carbonate complex from legacy equilibrium expressions, CaCO3 , is representative of stable prenucleation clusters.

(440a) Rising Motion of a Swarm of Drops in a Linearly Stratified Fluid

Bayareh, M.,
Dabiri, S., Notre Dame University
Ardekani, A. M., University of Notre Dame
Originally presented on: 11/6/2013 8:30:00 - 8:45:00

The motion of drops and bubbles in stratified fluids has several natural and industrial applications such as bubbles rising across pycnoclines in marine and aquatic environments, oil spills, ocean sequestration of CO_2, and bubble mixers used for lake/reservoir destratification and aeration. Even though the motion of bubbles and drops in a homogenous fluid has been extensively studies, their motion in a stratified fluid is poorly understood. Here, the three-dimensional motion of a swarm of buoyant drops rising in a linearly stratified fluid is studied using a finite volume/front tracking method. Numerical simulations are performed to investigate the effects of stratification on the cluster formation and the mean rise velocity of the swarm of drops. The pair correlation function is calculated to characterize the microstructure of the dispersed phase. The simulations show stronger tendency to form clusters as stratification increases. The mean rise velocity of the swarm of drops is smaller compared to the one in a homogenous fluid.

(440b) Simulations of Solid-Liquid Scalar Transfer in Turbulent Flow

Derksen, J., University of Alberta

Originally presented on: 11/6/2013 8:45:00 - 9:00:00

We numerically study scalar transfer from a single solid sphere into a surrounding fluid phase that undergoes turbulent flow. The conditions are such that the diameter of the sphere is an order of magnitude larger than the Kolmogorov length scale so that the sphere experiences an inhomogeneous and strongly time varying flow.

In the simulations, the turbulence as well as the sphere are fully resolved. This is achieved by means of the lattice-Boltzmann method. Its uniform, cubic lattice has a spacing close to the Kolmogorov length scale. The no-slip condition on the surface of the translating and rotating sphere is imposed by an immersed boundary method. Turbulence in the fully periodic, three-dimensional domain has been created through linear forcing.

In addition to the flow dynamics, we solve a convection-diffusion equation for the passive scalar c with c=1 at the solid surface and initially c=0 in the rest of the domain. Since we deal with liquid systems, the Schmidt number is high and we anticipate issues with the spatial resolution of the scalar concentration field, at least close to the sphere. To mitigate these we solve the scalar transport equation on two connected grids: an inner spherical grid that moves with the sphere, and an outer cubic grid that is the same as the grid on which the flow is solved (see Figure 1). For solving the scalar transport equation a finite volume method with a TVD scheme has been used.

This dual-grid approach has first been benchmarked for the case of a sphere suspended in simple shear. Reference data for this case are available due to Yang et al (2011). This benchmark provides us with notions as to the spatial resolutions required as a function of Schmidt and Peclet numbers.

First results with the sphere immersed in turbulent flow (illustrated in Figure 2) show the erratic structure of the mass transfer process close to the sphere with strong inhomogeneities due to the turbulent flow. Further analysis of the results gives credit to mass transfer models based on surface renewal with time scales of the order of 50 Kolmogorov times.

Chao Yang, Jingsheng Zhang, Donald L. Koch, Xiaolong Yin. Mass/heat transfer from a neutrally buoyant sphere in simple shear flow at finite Reynolds and Peclet numbers. AIChE J. 57, p1419, 2011.

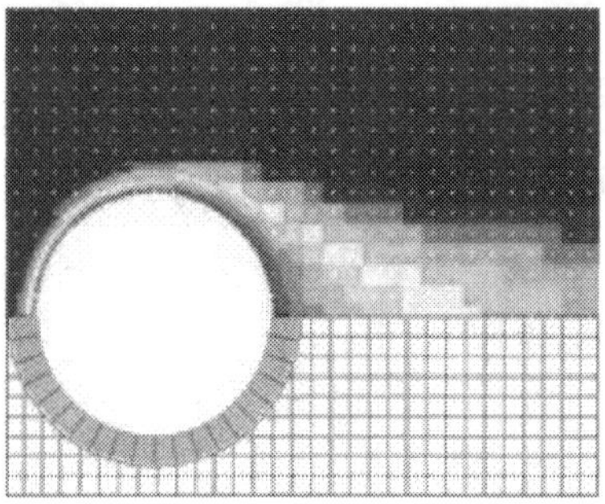

Figure 1

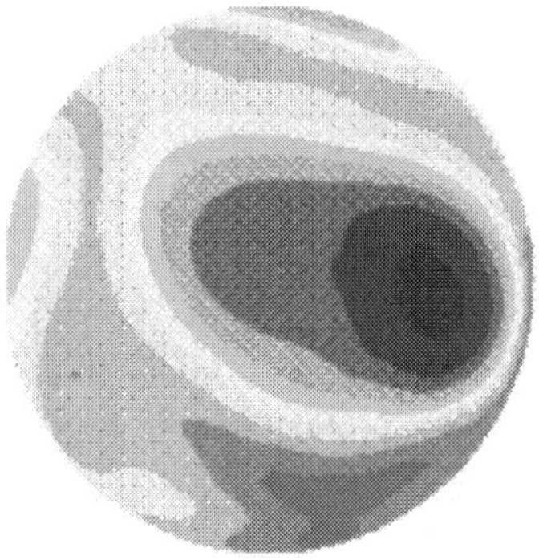

Figure 2

(440c) Pressure Drop Reduction of Stabilized Water-in-Oil Emulsions Using Organoclays

Hussein, I., King Fahd University of Petroleum & Minerals
Al-Yaari, M.,
Al-Sarkhi, A.,
Originally presented on: 11/6/2013 9:00:00 - 9:15:00

In this study, the influence of nano-additives on the pressure drop of surfactant-stabilized water-in-oil (W/O) emulsions was measured. Organoclays as were tested as pressure loss reducing agents for stable W/O emulsions with 0.7 and 0.7 water volume fraction. Pressure drop measurements were conducted in 2.54-cm and 1.27-cm horizontal pipes. The results showed a significant reduction in the emulsion viscosity with the addition of nano-additives and this effect was enhanced as the nano-additive concentration increased. In addition, for the case of concentrated W/O emulsions, the addition of organoclays resulted in 25 % reduction in the emulsion pressure drop in both test sections. Also, for stable W/O emulsion with only 0.3 water fraction no pressure drop reduction was observed in the laminar region. However, a pressure drop reduction was detected in the turbulent region and such effect was pronounced at high Reynolds numbers and organoclay concentration. The observed results were explained in terms of emulsion dispersed phase droplet size. In the laminar regime, the friction factor for stable W/O emulsions with 0.3 volume fraction of the dispersed phase was in a good agreement with single phase predictions. However, in the turbulent regime the friction factor for the multiphase system was below the predictions of single phase flow. Organoclays proved to have a good potential as drag reducing agents by producing smaller and stable emulsion droplets hence suppressing the Reynolds' stresses.

(440d) Growth of Multiparticle Clusters in Sedimenting Suspensions

Zinchenko, A., University of Colorado
Davis, R. H., University of Colorado
Originally presented on: 11/6/2013 9:15:00 - 9:30:00

Sedimentation of multiparticle clusters (aggregates) in a quiescent fluid is relevant to differential phase separation, and to natural processes in marine systems. All computer simulations so far assumed that a rigid multiparticle aggregate is formed by a version of the Monte Carlo (MC) method (typically, in the limits of diffusion-limited cluster aggregation (DLCA) or reaction-limited cluster aggregation (RLCA)), when already existing clusters are moved randomly to meet and form larger clusters (with some sticking probability). The library of clusters built in this manner is then used to simulate aggregate sedimentation by Stokesian Dynamics, Lattice-Boltzmann or multipole methods. However, these purely geometrical MC methods for cluster formation are entirely *ad hoc* and they do not correspond to the physical situation in sedimenting suspensions, at least for non-Brownian particles; there is obviously a disconnection between these methods and subsequent sedimentation simulations.

In contrast, the present work offers, for the first time, rigorous and highly-accurate, first-principle simulations of multiparticle aggregation as it naturally occurs in dilute sedimenting suspensions. The primary particles are monodisperse spheres; Brownian and inertial effects are neglected. For the cluster growth to start, some rigid vertically-aligned doublets are assumed to be already present (their concentration is much smaller than that for primary particles and does not affect the results). A doublet, in its downward motion, captures a new primary particle to form a rigid triplet. In general, once we have a rigid aggregate of N particles, a new primary particle is randomly generated far upstream from the cluster with uniform probability density (a suitable assumption for locally-homogeneous suspensions, prior to aggregation), and the particle-aggregate relative trajectory is computed. If no capture with the aggregate occurs, a new random initial particle position is tested, etc., until a colliding trajectory is found. The new rigid-body aggregate is then allowed to reach a steady sedimentation regime (which, for $N>3$, is not a simple downward translation, but a spiral motion about some axis of steady rotation) before a new primary particle is generated upstream, etc. Particle-aggregate collision requires singular van der Waals attraction (to overcome lubrication) which we assume to be present, but short-range, not appreciable until the particle-cluster surface clearance is much smaller than the particle radius. Technically, the algorithm consists of several parts:

1. A semi-analytical solution to determine the regime of steady sedimentation for a random isolated cluster. Tensor representations for the aggregate translational and rotational velocities via the total applied force reduce the Stokes flow problem to only three necessary solutions at the initial cluster position, resulting in very efficient time integration. The steady-state rotational velocity (about the vertical axis) is found to be an eigenvector of the appropriate coupling tensor but, in general, there are up to three eigenvalues; the only stable solution is selected by time integration. The axis of steady rotation for the cluster is found to be

typically quite away from the center of mass, at a distance much larger than the cluster hydrodynamical radius.

2. A special strategy for generating random initial positions for a new primary particle, far upstream. This issue is non-trivial due to cluster rotation and fractal-like shapes. The upstream collision circle (where the particle emerges with uniform probability density) is found by backward integration of many relative trajectories from the cluster vicinity, using the far-field approximation to the particle-cluster hydrodynamical interaction; it is crucial here to include the cluster rotation. It is shown that the so-defined interception area is always an upper bound and, hence, includes all possible collisions when the full hydrodynamical problem is solved. Still, on the average, for moderate van der Waals attraction, 45 trajectories have to be generated until one collision is found to add a new particle to a large cluster. Simulating 80 growth realizations with N up to 100 spheres required a total of 300 000 particle-cluster relative trajectories.

3. A highly efficient and precise multipole algorithm to solve the full Stokes flow problem for cluster-particle interaction, including the case of near-contact. This algorithm, vastly different from other, approximate schemes for multiparticle problems, incorporates high-order multipoles (up to order 100) with rotational transformations of spherical harmonics by Wigner functions [1], economical truncation of multipole expansions [1,2] and the idea of geometry perturbation [3] to effectively include lubrication. The Stokes problem for an isolated cluster is a much simpler, particular case, and does not require high-order multipoles.

It is shown that the aggregate settling velocity, relative to the primary particle velocity, when ensemble-averaged for every N over many growth realizations, quickly attains a fractal scaling and is well approximated by $1.12\ N^{0.48}$ for all N>20 and "moderate" van der Waals attraction. The exponent 0.48 is found to be independent of the strength of the molecular force (provided that it remains short-range), while the prefactor shows only a few percent variation when the non-dimensional Hamaker constant is varied in a broad range. For our aggregates, the average gyration radius, a usual measure to characterize fractal clusters, is extremely close to the average hydrodynamical radius for all N >4. The fractal dimension of rigid monodisperse aggregates formed in sedimentation is predicted to be 1.92 as the only possible value for the growth process with sequential particle addition and realistic hydrodynamics; in contrast, in DLCA and RLCA Monte Carlo *ad hoc* models, the fractal dimension is an arbitrary parameter. It is also found for our clusters that, on the average, the outer cluster radius is twice as large as the hydrodynamical radius. In a cluster of size N>>1, most particles are connected to one or two neighbors (33% and 43% of particles, respectively), although there are also particles with 3 or 4 neighbors (19% and 5%, respectively).

[1] Zinchenko A.Z. and Davis R.H. 2000 An efficient algorithm for hydrodynamical interaction of many deformable drops. J. Comput. Phys., vol. 157, pp.539-587.

[2] Zinchenko A.Z., Rother M.A. and Davis R.H. 2011 Gravity-induced collisions of spherical drops covered with compressible surfactant. J. Fluid Mech., vol. 667, p.369-402.

[3] Zinchenko A.Z. 1998 Effective conductivity of loaded granular materials by numerical simulation. Phil. Trans. Roy. Soc. Lond. A, vol. 356, pp. 2953-2998.

(440e) Thermal Rupture Force Distributions in Colloidal Gels

Furst, E. M., University of Delaware
Whitaker, K., Rowan University
Swan, J. W., California Institute of Technology
Shindel, M., University of Delaware
Originally presented on: 11/6/2013 9:30:00 - 9:45:00

Colloidal gels yield under imposed or gravitational stresses. Such yielding can be an inherently useful property, for instance in controling the flow of consumer care products, or reflect limits in the stability against sedimentation, which ultimately affects product shelf life. Microscopically, the "bonds" between particles in the gel network are broken during yielding. In thermally fluctuating colloidal systems, there is not a single force required to achieve rupture, but a spectrum, as thermal forces can both augment and inhibit the bond breaking. We demonstrate measurement and interpretation of the distribution of rupture forces between pairs of colloidal particles bonded via the van der Waals attraction. The otherwise irreversible bond is broken by pulling the particles apart with optical tweezers. We show that an ensemble of the particle trajectories before, during and after the rupture event may be used to produce a high fidelity description of the distribution of rupture forces. Bond breakage measurements are then extended to depletion interactions to understand the rheology of depletion gels.

(440f) Shear-Induced Structure and Migration of Colloidal Particles in Concentrated Polymer Solutions

Breedveld, V., Georgia Institute of Technology
Peterson, E. C., Georgia Institute of Technology
Originally presented on: 11/6/2013 9:45:00 - 10:00:00

Suspensions with a viscoelastic continuous phase can exhibit interesting and unexpected microstructural features. One of these is the formation of small, vorticity-aligned particle clusters in simple shear, which has occasionally been observed in the past [e.g., DeGroot *et al.* (J. Colloid Interface Sci., 1994) and Pasquino *et al.* (Rheol. Acta, 2010)], without fully uncovering the underlying mechanism. We used rheo-SALS (*in situ* small-angle light scattering on a rheometer) to probe the dynamics of this phenomenon for a suspension of colloidal silica particles in shear-thinning entangled polymer solutions. It was found that shear rates in the shear-thinning regime resulted in particle aggregation and orientation of aggregates along the vorticity axis. In contrast, application of strain in the Newtonian, low-shear regime did not yield clusters, and even led to restoration of a uniform particle distribution in an already ordered, pre-sheared suspension. Comparison of the SALS patterns with simultaneously measured viscosity data provided interesting insight into the microstructural evolution in these suspensions under shear, revealing the decoupled dynamics of aggregation and flow orientation. Additional experiments with a Boger fluid that exhibits less shear-thinning, but enhanced normal stresses showed that these effects were predominantly caused by shear-thinning.

In addition to the simple shear flow experiments, these dilute colloidal suspensions were also subjected to channel flow through rectangular glass capillaries. Concentration profiles were determined via confocal microscopy. Depending on shear rate, cross-channel migration of particles was observed in different directions: toward the center of the channel at low shear rates and toward the walls at high shear rates.

Controlling and optimizing the migration, aggregation and alignment of colloidal particles in concentrated polymer solutions is highly relevant for the processing of "mixed matrix membranes", which combine a continuous polymer phase with embedded inorganic particulates that enhance the membrane properties.

(440g) Suspension of Rigid Particles Settling in Density Stratified Fluids

Doostmohammadi, A., University of Notre Dame
Ardekani, A. M., University of Notre Dame
Originally presented on: 11/6/2013 10:00:00 - 10:15:00

Particulate flows comprise important environmental processes in atmosphere and ocean where the physics of the fluid is markedly affected by variation of the fluid density through temperature and/or salinity variation. Our recent study shows that the settling dynamics of a particle in a linearly stratified fluid can be significantly influenced by the presence of density gradients. Moreover, we show that the interaction between a pair of particles in a stratified fluid can substantially differ from a homogeneous fluid and stratification leads to attraction between the particles. The results suggest that the presence of background density gradient can have a direct consequence on the microstructure of particulate flows of stratified fluids. The suspension of spherical particles is investigated in periodic cells and three-dimensional full non-linear solutions are provided to establish the fundamental effects of stratification on many-particle systems, a challenge that has to date gone unmatched. Direct comparison between the microstructure and pseudo turbulent properties of homogeneous and stratified fluids is provided.

(440h) Phase Diagram for a 2-D Colloidal System With An Isotropic Potential

Du, D., Rice University
Doxastakis, M., University of Houston
Biswal, S. L., Rice University
Originally presented on: 11/6/2013 10:15:00 - 10:30:00

We demonstrate how a two-dimensional paramagnetic colloidal system under the influence of a rotational magnetic field (PURM) can be used to induce an isotropic potential between the particles. The interaction potential well depth can be precisely tuned from 5kT to 40kT. We make use of this experimental system and complement our experiments with a Monte Carlo simulation to show how this system can be used to probe the phase behavior of systems that exhibit long-range attractive potentials, such as a monoatomic system.

A highly tunable long-range attractive interaction is induced by a rotating magnetic field generated by two pairs of coaxial air-core solenoids connecting with multi-frequency AC power supply. The solenoids are calibrated to be orthogonal to each other. The repulsive component to the interaction potential is the short-range electrostatic interactions. The potential well formed by summing the two interaction potentials and is tuned by changing the input voltage of power supply. The phase diagram can be mapped out by using samples of different particle concentration and different field strength.

In the Monte Carlo simulation, a density-dependent pair potential is used due to the many body effect shown by our previous study. We start from solving the Laplace equation that describes the magnetic field distribution of paramagnetic colloid system under external field. Then Maxwell stress tensor is used to calculate the magnetic force experienced on each particle and compare with superposed force from pair interaction. The density-dependent pair potential is obtained by the comparison above.

A grand canonical ensemble is used in the Monte Carlo simulation. A fixed chemical potential allows us to simulate phase transition from one phase to another without coexistence. Pressures and pair distribution function near transition have been plotted out to locate the phase point for transition for different densities. The temperature-density phase diagram is thus scanned, mapped out, and is in good agreement with that obtained from experiments. This phase diagram shows good qualitative agreement with the one obtained from experiment as well as the one of a typical atomic system.

(440i) Transient Rheology of a Clay-Polymer Suspension During Accelearated Nanocomposite Formation

Winter, H. H., National Science Foundation
Originally presented on: 11/6/2013 10:30:00 - 10:45:00

A novel exfoliation mechanism for clay/polymer suspensions reduces the energy that is required to release clay sheets from stacks (clay particles) and accelerates the formation of a nanocomposite network structure. The clay is montmorillonite in which the natural counterions were replaced by alkane counter-ions. The polymer is an endfunctionalized polybutadiene. Time-resolved rheometry on freshly mixed samples shows the evolving connectivity of the clay-polymer structure. Sample preparation is simple and fully reproducible since no shearing or sonication is required to facilitate exfoliation. We propose an explanation for the novel exfoliation process and provide an experimental test of the hypothesized mechanism.

(440j) Segregation of Particles Undergoing Shear Flow in a Counter Rotating Plate Geometry

Praharaj, S., University of Notre Dame
Leighton, D. T. Jr., University of Notre Dame
Originally presented on: 11/6/2013 10:45:00 - 11:00:00

Recently there has been much interest in the use of inertial secondary currents driven by centrifugal forces to achieve particle separations. A key application is the separation between deformable and non deformable particles (eg. circulating tumor cells in blood). In a microfluidic geometry, for example, separation of tumor cells has been demonstrated in a gently curved large aspect ratio rectangular channel [1,2]. These separations rely on a complex interplay between inertial secondary currents, size exclusion, inertial lift, deformation-induced lift, sedimentation velocity and shear induced diffusion giving rise to cross-streamline segregation. This particle distribution then interacts with the inertial secondary current to yield the separation. In this study we focus on a related problem: particle migration in the classic von Karman geometry of counter rotating parallel plates. As is well known, when the plates are at precise counter rotation ($s = -1$) the radial secondary current is such that the flow is outward at both the rotating plates and inwards at the center [3]. For particles in this geometry, the mean shear causes both an inertial lift [4] and deformation induced lift [5] forcing particles towards the center of the gap. This migration is balanced by shear induced diffusion and, for non-neutrally buoyant particles, by sedimentation. For very dilute suspensions, the shear induced dispersivity is small, and the balance is primarily between the lift force and the sedimentation velocity of the particles. Because the lift scales with r^2, this balance gives rise to an equilibrium radial position at which point the radial convective flux is zero. Although weak, shear induced diffusion causes a radial spread in the particle distribution about the equilibrium location. In this paper we experimentally demonstrate the existence of such a band and show that the width is quantitatively consistent with shear induced dispersion. In addition, the location of this band is qualitatively consistent with the inertial lift mechanism of Ho & Leal [4]. At higher concentrations, a band is also shown to exist, however the situation is complicated by modification of the inertial secondary current due to viscosity variation arising from the non uniform concentration profile between the plates. In this higher concentration region of stronger shear induced dispersion, the phenomenon may be modeled in terms of an average radial flux and a shear-induced Taylor dispersivity.

References:

[1] Hou, H.W. *et al.* Isolation and retrieval of circulating tumor cells using centrifugal forces. *Sci. Rep.* 3, 1259; DOI:10.1038/srep01259
(2013).

[2] Bhagat, A. A. S. *et al.* Continuous particle separation in spiral microchannels using dean flows and differential migration. *Lab Chip* 8, 1906–1914 (2008)

[3] J.F. Brady and L. Durlofsky. On rotating disc flow. J. Fluid Mech. vol. 175, pp. 363-394, (1987).

[4] B. P. Ho and L. G. Leal. Inertial migration of rigid spheres in two-dimensional unidirectional flows. *J. Fluid Mech.*, 65(2):365-400, (1974).

[5] P.C. Chan and L.G. Leal. The motion of a deformable drop in a second-order fluid, J. Fluid Mech., 92(1):131-170, (1979).

(449a) Reactive Mixing of Hydrocarbons and Water At Near-Critical Or Supercritical Conditions

Wu, G., MIT
Timko, M. T., Worcester Polytechnic Institute
Raghavan, A., Massachusetts Institute of Technology
Ghoniem, A. F., Massachusetts Institute of Technology
Originally presented on: 11/6/2013 8:30:00 - 8:50:00

The 1-D reactive mixing of an n-hexadecane droplet doped with a small fraction of hexyl sulfide in a pool of finite volume of supercritical water (SCW) is modeled. First, low temperature non-reactive cases are considered to study the phase separation behavior between hydrocarbons and water. Next, the temperatures are selected such that no phase separation is expected. Hexadecane and hexyl sulfide cracking, and coking are approximated using overall one-step chemical kinetics. This simplified reaction network provides, at least qualitatively, accurate prediction when compared with experimental results. Thermodynamics and transport sub-models are used to estimate the relevant physical properties. The current study provides microscale information, such as the coupling of mixing and chemical reaction processes, and insight into large scale mixing and reactions, such as how the system can be optimized by adjusting the temperature and water-to-hydrocarbon mass ratio, for various supercritical water applications. By considering reactions of hydrocarbon cracking, desulfurization, and coking, the current model is most relevant to the application of oil upgrading and desulfurization in supercritical water.

(449b) Molecular and Microkinetic Modeling of Catalytic Supercritical-Phase Fischer-Tropsch Synthesis

Asiaee, A.,
Alley, B.,
Benjamin, K. M., South Dakota School of Mines and Technology
Originally presented on: 11/6/2013 8:50:00 - 9:10:00

The use and advantage of supercritical fluids (SCFs) as solvents in heterogeneous catalysis has been explored in several systems, including supercritical water oxidation, green chemistry, hydrothermal biomass conversion (gasification and pyrolysis), and catalytic Fischer-Tropsch (FT) synthesis in supercritical hexane. In the particular case of FT catalytic chemistry, the use of supercritical hexane has shown improved CO conversion and an improved product slate. While there has been much experimental work in the area of heterogeneous catalysis in supercritical fluids, and in the development of SCF FT technology in particular, there is no current molecular-level understanding of the role(s) of the SCF during catalysis.

This work involves molecular and microkinetic modeling of catalytic SCF FT synthesis. Density functional theory (DFT) calculations have been applied to model the kinetics and thermodynamics of Fischer-Tropsch synthesis reactions on a model Co(0 0 1) catalyst in presence of supercritical hexane. In this regard, by utilization of the DFT-GGA method, the important elementary reactions have been determined to find out the probable mechanisms and investigate the effect of pressure and supercritical solvent on these catalytic reaction pathways. An existing microkinetic model in the literature has been modified to include these pressure and solvent effects of supercritical hexane on catalytic thermodynamics (chemisorption) and kinetics, using a combination of the aforementioned DFT calculations and equation of state calculations. Finally, reactor product yields have been compared with experimental results reported in literature, and net rate and sensitivity analysis results from the model are presented and discussed.

(449c) Low Temperature Non-Catalytic Transfer Hydrogenation of Coal in Supercritical CO2

Mondal, K., Southern Illinois University

Sims, A., Southern Illinois University

Elhussein, H.,

Soti, M., Southern Illinois University

Originally presented on: 11/6/2013 9:10:00 - 9:30:00

The paper describes the findings from experimental investigations related to the development and characterization of a low temperature (<150 °C) non-catalytic process using a hydrogen transfer agent (instead of molecular hydrogen) for coal dissolution in supercritical CO_2. The key idea is that one hydrogen atom from water (W) and one hydrogen atom from the hydrogen transfer agent (HTA) will be used to hydrogenate the coal. Since the products of coal dissolution are expected to be non-polar and polar while the supercritical CO_2, which is expected to promote disruption the coal structure, enhance the rates of hydrogenation and dissolve the non-polar molecules and remove them from the reaction site, is non polar, a polar modifier (PM) modifier for CO_2 is added to aid in the dissolution and removal of the polar components. In addition, to allow a seamless transport of the ions (HTA) and byproduct (reduced HTA) between the aqueous and organic phases, a phase transfer agent (PTA), is added.

$COAL + HTA + H_2O \rightarrow COAL.H_2 + reducedHTA$

The attractive feature of this process is that the liquid chemicals are obtained without requiring to first convert coal to CO and H_2 units. The results show that the process can extract chemicals from coal at temperatures ranging between 80 - 140 °C. The results also show that straight chain hydrocarbons for transportation fuels as well as chemicals such as aromatics with one two or three benzene rings that are needed as chemical precursors for commercial organics, can be extracted from coal by this process. The amount of liquid produced was found to be enhanced if a coal structure disruption step is carried out before the coal hydrogenation step. This can be achieved by the sudden expansion of the supercritical CO_2. Incidentally the use of supercritical CO_2 not only provides high dissolution strength but also allows for easy separation of the chemicals from the solvent (normally gaseous under room conditions). The effect of the mixture composition of coal/W/HTA/PM/PTA as well as the effect of temperature and pressure will be presented. It should be noted that this process can be used for the extraction of chemicals from biomass, presumably even from lignocellulosic materials.

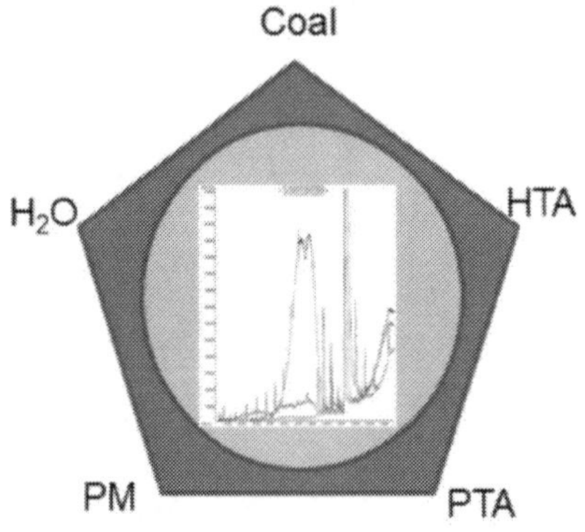

(449d) Evaluating and Improving a Kinetic Model for the Hydrothermal Liquefaction of Microalgae

Valdez, P. J., University of Michigan
Savage, P. E., University of Michigan
Originally presented on: 11/6/2013 9:30:00 - 9:50:00

Hydrothermal liquefaction in hot compressed water converts wet algal biomass into an energy-dense biocrude. Hydrothermal treatment of biomass to produce biofuels circumvents energy and resource intensive steps of drying and solvent extraction. We have developed a reaction network and kinetic model to describe the hydrothermal liquefaction of microalgae slurries [1]. Assuming lumped product fractions and pseudo first-order kinetics, we simplified the modeling of hydrothermal liquefaction while maintaining accuracy to describe a complex system of reactions. We will report the estimated Arrhenius parameters, rates, and sensitivity of each reaction pathway in the reaction network. We calculated activation energies of the gas formation pathways during liquefaction to be 66 and 80 kJ/mol, which are within range to a previously reported value of 71 kJ/mol [2]. We verified the validity of the reaction network and kinetic model with additional liquefaction experiments and data from other researchers using the same microalga. By incorporating the biochemical content of the algae into the model we are able to unify results between microalgae with varying composition of lipids, proteins, and carbohydrates. We determined that the optimized conditions for maximizing biocrude yield are at >300 °C and <20 min.

[1] PJ Valdez, PE Savage, Reaction Network for the Hydrothermal Liquefaction of *Nannochloropsis* sp., Manuscript Submitted, Algal Res., March 2013.

[2] Q Guan, PE Savage, C Wei, Gasification of alga *Nannochloropsis* sp. in supercritical water, J. Supercrit. Fluids 61 (2012) 139-145.

(449e) Influence of Water Content On Reaction Rates of Near and Supercritical Methanol Transesterification of Bio-Oil

Tschannen, R. E., Ohio University
Gonzales, A. C., Ohio University
Lee, S., Ohio University
Originally presented on: 11/6/2013 9:50:00 - 10:10:00

Biodiesel is a renewable, environmentally friendly fuel that is produced from the transesterification of triglycerides with methanol. This process can be carried out in two different ways: catalytically at low temperatures and pressures or non-catalytically in near and supercritical methanol. One important advantage of the supercritical production of biodiesel is the ability to process triglycerides with higher concentrations of water and free fatty acids without undergoing undesirable side reactions or greatly decreasing yield. At the conditions necessary for non-catalytic transesterification, water's ability to dissociate into ions is enhanced. This study examines the influence of water in varying concentrations to elucidate any effects on reaction rates due to catalyzing effects from ions formed from the dissociation of water. In these experiments, the process temperature was varied from 240°C to 300°C and the weight fraction of water in methanol was varied from 0 to 1.

(449f) Hydrothermal Deoxygenation of Fatty Acid

Miao, C., Washington State University
Chakraborty, M., Washington State University
Marin Flores, O. G., Washington State University
Dong, T., Washington State Univeristy
Wang, Y., Washington State University
Chen, S., Washington State University
Originally presented on: 11/6/2013 10:10:00 - 10:30:00

Fatty acid alkyl esters are renewable liquid biodiesel fuel, but the high oxygen content limits their application. As a better alternate, low-oxygen content hydrocarbon has received increased interest. The deoxygenation of fatty acid via hydrodeoxygenation however, requires a large amount of H_2, which significantly limits the commercialization of this process. Decarbonylation is another pathway to remove oxygen with the formation of CO. Water-gas shift reaction can be a potential way to produce H_2 from H_2O and CO. Therefore, we proposed to combine the decarbonylation with water-gas shift reaction to produce in-situ H_2, which will accelerate the deoxygenation of fatty acid.

Our lab has been working on employing hydrothermal deoxygenation reaction in aqueous media for this application. We aimed at removing oxygen in the form of DeCOx without or with a little amount usage of H_2. We use fatty acid as model compounds to screen several transition metal catalysts on different support and found Ni/ZrO_2 to be promising. The catalyst showed great selectivity for converting palmitic acid to pentadecane in aqueous phase. Without hydrogen usage, 50% conversion of palmitic acid was achieved at 300C and the selectivity to linear paraffin was 70%. Under hydrogen and palmitic acid molar ratio at 1:2, 82% conversion was achieved with the selectivity to linear paraffin at 50%. Three other fatty acids, stearic, oleic, and linoleic acid were also compared to illustrate the effect of fatty acid saturation on the conversion rate and products selectivity. Saturated fatty acid was observed to be more effectively deoxygenated to hydrocarbon than unsaturated fatty acid. The main organic product was the paraffin with one carbon less than the precursor fatty acid. The results provide another possible pathway: to conduct deoxygenation of fatty acid in subcritical water phase. The lower H_2 consumption in this process will reduce the cost for the production of oxygen-free bio-fuels.

(449g) Synthesis of Renewable Diesel By Catalytic Deoxygenation of Soybean Oil in a Supercritical Fluid-Assisted Fixed Bed System

Kim, S. K., Seoul National University
Kim, J., Korea Institute of Science and Technology
Originally presented on: 11/6/2013 10:30:00 - 10:50:00

The catalytic dexoygenation of natural triglyceride by hydrotreating has received considerable attention as a promising technology to obtain more useful renewable transportation fuel because such diesel-like hydrocarbons has properties similar to those of petroleum diesel, i.e., zero content of oxygen without metallic or acidic impurities. Hydrotreating of natural triglyceride is the gas-liquid-solid reaction which is controlled by the availability of hydrogen at the catalyst surface. However, reactant-catalyst contact is limited by both, the low solubility of hydrogen and the high mass transfer resistance of the liquid phase. Consequently, the catalyst efficiency is lowered and the hydrotreating of triglyceride generally required a large amount of heat (300 ~ 450 °C of reaction temperature) and a large quantity of hydrogen (> 50 MPa of H_2 pressure) to increase the hydrogen solubility. The use of adequate supercritical fluid can bring the reactive mixture into a homogeneous condition, increasing the rate to gas-like reaction rates. In the present study, various supercritical fluids, including carbon dioxide, propane, and hexane, were introduced with hydrogen and soybean oil to a fixed bed reactor containing pre-activated $CoMoS_x/\gamma\text{-}Al_2O_3$ and the effect of supercritical fluid-addition on the production of renewable diesel was investigate. Among those supercritical fluids, the addition of supercritical propane was found to be able to reduce the reaction temperature as much as 50 °C to obtain similar yield compared to those obtained by the reaction without supercritical fluid as a result of facilitated heat transfer. Hydrocarbon distribution of liquid product and analysis on the gas product composition indicated that the addition of supercritical propane during the hydrotreating of soybean oil promoted specific deoxygenation pathways, decarbonylation and decarboxylation, consuming small amount of hydrogen compared to hydrodeoxygenation pathway. As a result, a quantity of hydrogen required to obtain diesel-like hydrocarbons was reduced to 70 % compared to the reaction without supercritical fluid. Since the decarboxylation and decarbonylation are mildly endothermic reaction, reduced heat transfer resistance drove deoxygenation reaction to thermodynamically favorable pathways.

(452a) Interactions At Heterogeneous Surfaces: Regrouping Colloidal Interactions to Achieve Biomimetic Behavior

Santore, M. M., University of Massachusetts Amherst

Originally presented on: 11/6/2013 8:30:00 - 9:00:00

Textbooks on colloidal phenomena teach us to describe the electrostatic and dispersion interactions at interfaces using single parameters (Hamaker constants, surface potentials or charge densities), an approach which treats surface as uniform. Real surfaces, be they mineral, polymeric, or biological, present heterogeneous surface chemistry that complicates their interactions. It is recognized that clustered rather than uniform presentation of attractive chemistries, for instance peptide sequences, enhances biological interactions. Likewise, heterogeneity in charge and acid-base interactive groups has been invoked to explain colloidal instabilities and bacterial adhesion (that DLVO theory fails to predict). A lack of understanding the energy landscape of real surfaces prevents their description by convolved versions of DLVO and related approaches. Our lab, has, however, developed a series of model surfaces with well-characterized electrostatic heterogeneity and, through systematic variations in the interface, demonstrated how heterogeneity alters the surface forces. This talk addresses the fundamental aspects of heterogeneous surface interactions and then translates the interactions to static and dynamic particle adhesion. In particular, the influence of the energy and distribution of the heterogeneity will be addressed. The talk will then demonstrate how interactions at heterogeneous interfaces are particularly sensitive to physical interfacial features such as curvature and mechanical softness, and how this sensitivity can be exploited for separation and microfluidic sensing schemes. Finally, the talk will address the role of heterogeneity in dynamic adhesion, from capture –hold-release protocols to continuous behaviors such as rolling. These will be demonstrated with particles and spherical bacteria. Throughout the talk, parallels between synthetic and biological interfaces will be drawn, probing the extent to which charge and other chemical clusters can be treated conceptually as biological receptors.

(452b) Formation of Submicron/Nanoscale Features in Polymeric Or Colloidal Monolayers

Russel, W. B., Princeton University
Lau, C. Y. C.,
Originally presented on: 11/6/2013 9:00:00 - 9:15:00

The potential applications of two-dimensional order in submicron features have stimulated the development of cheap and fast fabrication techniques in recent years. We have shown that electrohydrodynamic patterning of polymer films can produce features as small as 350 nm quickly and economically but dielectric breakdown imposes a fundamental limit at that scale. To overcome that limit we turned to colloidal assembly via flow coating with nanoparticles of 50-100 nm diameters. Control of the deposition to achieve large areas of a uniform film is straightforward, but creating extensive crystalline domains in a close-packed monolayer is not. That motivated deposition of crystalline dispersions created by deionizing the solutions. The deposition followed faithfully a process model accounting for a power-law viscosity, but high shear rates disrupted the initial order producing domains similar to those achieved at dilute concentrations. At low shear rates order was partially preserved but the domain sizes were still not extensive. The most successful effort to create long-range order was achieved via the Langmuir-Blodgett process by adding excess free ligands to dispersions of gold nanoparticles. The excess ligands, oleylamine, not only improve the order upon deposition, but also anneal a poorly ordered array to form an ordered monolayer of 20,000 particles [C.Y. Lau, *et al. Langmuir* **27** 3355 (2011)].

(452c) Single Molecule Studies of Ion Specific Effects On the Hydrophobic Interaction At the Solid-Liquid Interface

Nelson, N., University of Colorado at Boulder
Schwartz, D. K., University of Colorado Boulder
Originally presented on: 11/6/2013 9:15:00 - 9:30:00

Understanding the effects of salts on the dynamics of single molecules is important for many biologically and industrially relevant systems. Ions are present in many industrial processes and in all *in vivo*environments and have long been known to impact molecular properties such as water solubility and surfactant self-assembly. A particularly interesting category of salts is the Hofmeister series. In this series, salts such as sodium fluoride increase the strength of the hydrophobic effect, whereas salts on the other end of the spectrum, such as sodium thiocyanate, decrease the strength of hydrophobic interactions.

While there are a large number of studies detailing the macroscopic effects of ions in solution, there is a dearth of experimental data describing the effect of these salts on the molecular level. Data presented here represent molecular level insights into the impact of salts on the behavior of molecules at the solid-liquid interface. These data were obtained using single molecule fluorescence techniques to determine the effect various ions have on the adsorption rate, surface diffusion, and surface residence time of fluorescently labeled fatty acids. Sodium fluoride greatly increased the adsorption rate of the fatty acid compared to sodium thiocyanate, but probe dynamics after adsorption remained relatively unchanged. This indicates that an ionic species in solution will have a significant impact on the behavior of that molecule while in the bulk, but has limited effect on individual molecules once they are adsorbed to the solid-liquid interface.

(452d) Understanding the Temperature Dependence of Solid-Liquid Interfacial Properties

Rane, K. S., University at Buffalo, The State University of New York
Errington, J. R., University at Buffalo, The State University of New York
Originally presented on: 11/6/2013 9:30:00 - 9:45:00

Knowledge of the temperature dependence of solid-liquid interfacial properties is necessary to design interfacial systems with a desired thermal response. In our recent work, we have studied how various wetting properties depend on temperature and the strength of the substrate-fluid interaction.[1] We are now interested in understanding the observed temperature dependence in terms of the nature of the fluid and solid substrate. We are particularly interested in conditions for which solid-fluid interfacial properties show negligible dependence on temperature. In this presentation, we show how applying Gibbs formalism to express the interfacial tensions in terms of their energetic and entropic components provides a clearer picture on the above mentioned phenomena. These components help us in quantitatively relating the observed temperature dependence of interfacial properties to solid-fluid or fluid-fluid interactions and the arrangement of fluid molecules near the solid. We then explain how these energetic and entropic components are calculated from Monte Carlo simulations using free energy based approach developed in our group. Results are the provided, wherein this strategy is used to explain the temperature dependence of wetting properties of simple models of ionic liquids on solid substrates.

(1) Rane, K. S.; Kumar, V.; Errington, J. R. *Journal of Chemical Physics* **2011**, *135*, 234102.

(452e) Hydrophobic and Hydrophilic Interactions in Aqueous Mixtures of Alcohols At a Hydrophobic Surface

Ballal, D., Rice University
Chapman, W. G., Rice University
Originally presented on: 11/6/2013 9:45:00 - 10:00:00

Aqueous solutions of alcohols are important because of applications in separation processes and biological systems. They are also interesting because of the strong concentration dependence of interfacial properties as well as their anomalous properties such as maxima in heat capacity[1] and viscosity [2]. It is believed that the molecular structuring of water and alcohol around each other in solution is responsible for the anomalous behavior[3, 4]. The structure and thermodynamics of these mixtures have been extensively studied in the bulk. But less work has been done towards studying water-alcohol mixtures at interfaces. The interfacial structuring and properties are significant for application in alcohol purification processes and biomolecular structure. Here we study aqueous mixtures of short alcohols (methanol, ethanol, 1- and 2-propanol) at a hydrophobic surface using interfacial Statistical Associating Fluid Theory (iSAFT) [5, 6] which is a perturbation density functional theory. We find that the addition of a small amount of alcohol to water decreases its interfacial tension drastically. This trend in interfacial tension can be explained by the structure of water and alcohol. The presence of a hydrophobic surface disrupts the hydrogen bonding network of water. The hydrophobic group of an added alcohol preferentially goes to the surface preserving the structure of water in the bulk. The different alcohols structure in different ways with 1-propanol being the most effective in packing the surface since the hydrophilic segment in 1-propanol is an end segment as compared to 2-propanol where the hydrophilic segment is in the middle. 1-propanol also has a longer hydrophobic tail than ethanol and methanol making it more effective. We also consider the structuring of aqueous solutions of a mixture of alcohols.

References

1. Noskov, S.Y., M. Kiselev, and A. Kolker, *Molecular dynamic study of the anomalous behavior of heat capacity in a methanol-water mixture.* Journal of structural chemistry, 1999. **40**(2): p. 253-261.

2. Franks, F. and D.J.G. Ives, *The structural properties of alcohol-water mixtures.* Quarterly Reviews, Chemical Society, 1966. **20**(1): p. 1-44.

3. Dixit, S., et al., *Molecular segregation observed in a concentrated alcohol-water solution.* Nature, 2002. **416**(6883): p. 829-832.

4. Davis, J.G., et al., *Water structural transformation at molecular hydrophobic interfaces.* Nature, 2012. **491**(7425): p. 582-585.

5. Jain, S., A. Dominik, and W.G. Chapman, The Journal of Chemical Physics, 2007. **127**(24): p. 244904-12.

6. Bymaster, A. and W.G. Chapman, The Journal of Physical Chemistry B, 2010. **114**(38): p. 12298-12307.

(452g) A Thermodynamic Model for the Wetting Characteristics of Hierarchical Physically-Patterned Surfaces

Bell, M. S., Pennsylvania State University
Shahraz, A., Pennsylvania State University
Fichthorn, K., The Pennsylvania State University
Borhan, A., The Pennsylvania State University
Originally presented on: 11/6/2013 10:15:00 - 10:30:00

Recent advances in nanotechnology have allowed for the creation of biologically-inspired superhydrophobic surfaces, which have a variety of applications including self-cleaning glass, low-drag surfaces and microfluidics. These surfaces are often made by incorporating roughness on micro and nano length scales, though a solid understanding of the role of hierarchical roughness in causing superhydrophobicity remains elusive. We present a two-dimensional thermodynamic model of the wetting of a hierarchically-grooved surface for droplets with negligible Bond number. By investigating wetting phase diagrams over the microscale parameter space, we find that surfaces with single-scale roughness are superhydrophobic only in small regions of the phase space—most notably in regions where the surface features are much smaller than the size of the droplet—while the addition of nanoscale roughness greatly extends the region of phase space in which the surface is superhydrophobic, so that even droplets with sizes on the order of those of the microscale features display high contact angles.

(452h) Energetics of the Cassie/Wenzel Transition On Parallel Cylinders: Applications in Characterization of Bird Feathers and in Designing Robust Friction Reducing Meshes

Srinivasan, S.,
Chhatre, S.,
Guardado, J.,
McKinley, G. H., Massachusetts Institute of Technology
Cohen, R. E., Massachusetts Institute of Technology
Originally presented on: 11/6/2013 10:30:00 - 10:45:00

In this work, we present a unifed model to evaluate the external pressure at which a wetting transition occurs on a liquid-solid composite interface that rests on an array of parallel cylinders. We extend previous work in characterizing three possible modes of Cassie/Wenzel transition (i.e, depinning, meniscus sagging and coalescence) on an array of parallel cylinders, and develop analytical expressions to determine the magnitude of the energy barrier at the onset of each transition in terms of the radius of the cylinder R, a dimensionless parameter D* that characterizes the geometric spacing, and the equilibirum contact angle θ_e. We then apply this analytical framework to explain two wetting phenomena: (i) Characterization of the hierarchical topography of bird feathers using goniometric measurements in order to explain the diving and wing spreading behavior of underwater diving birds using the thermodynamic analysis developed above. We perform contact angle measurements using a set of polar and non-poplar probing liquids on dip-coated feathers of 6 bird species, and apply our model to determine the dimensionless spacing ratio ($1 < D^* < 2$) and an effective radius of the quasi-heirarchial structure R_{eff}. (ii) Towards engineering optimal design geometries of superhydrophobic mesh surfaces that can maximize giant liquid slip in laminar flow while simultaneously providing robustness against the Cassie-Wenzel transition.

(452i) Wetting Transition of Ethanol-Water Droplet On Smooth and Pillared Surfaces: A Molecular Dynamic Study

Metya, A., IIT Kanpur
Khan, S., Indian Institute of Technology Kanpur
Singh, J. K., Indian Institute of Technology, Kanpur
Originally presented on: 11/6/2013 10:45:00 - 11:00:00

Wetting phenomena of a liquid to solid are ubiquitous in nature [1]. It has technological relevance, as large number of industrial, manufacturing, biological applications, require uniform and rapid wetting; whereas, others demand poor wetting. It is necessary to achieve such spreading, possibility by adding surface active agents such as surfactants or alcohols which modifies the interfacial tension [2]. Ethanol in water, as a mixture, is used in numerous applications for medical, paints, perfumes, and deodorants products. Though some works on nano-confined ethanol-water mixtures have been conducted, wetting transition of such mixture on on rough surfaces is limited in the literature.

In the present work, the effect of ethanol concentration and the spreading mechanism of ethanol-water droplets on smooth and textured surfaces are investigated using molecular dynamic simulations. The macroscopic property such as contact angle is found to decreases with increasing ethanol concentration in the droplet. We rigorously study the system size effect on the contact angle of water-ethanol drop on smooth graphite surface as a function of ethanol concentration. It has been found that the ethanol molecules accumulate in the three phase contact line and the vapor-liquid interface. However, wetting behavior of the droplet on rough surfaces is quite different from that on the smooth surface. We carefully examine the different wetting states as a function of surface fraction and pillar height. Ethanol molecules are found to accumulate in the grooves rather than on the top of the pillar. For pillared surfaces, the ethanol-water drop is found in the partial Wenzel state at lower ethanol concentrations. On the other hand, at higher concentration it is in the complete Wenzel state which is in good agreement with the experimental results[3]. We also attempt to quantify the behavior of the mixture in the Cassie-Baxter and Wenzel states in terms of continuum relations seen recently [4].

References

[1] A. Otten and S. Herminghaus, Langmuir **20**, 2405 (2004).

[2] M. Lundgren, N. L. Allan, and T. Cosgrove, Langmuir **18**, 10462 (2002).

[3] J. Fidler and P.M. Rodger, J. Phys.Chem. **103**, 7695 (1999).

[4] F. Leroy and F. M. Plathe, J. Chem. Theo. and Comp. 8, 3724-3732(2012).

(456a) The CO2-Solubility and Viscosity Enhancing Potential of CO2-Philes Functionalized With Aromatic Groups

Lee, J. J.,
Cummings, S., University of Bristol
Perry, R. J., Global Research, General Electric
Beckman, E. J., University of Pittsburgh
Enick, R., University of Pittsburgh
Originally presented on: 11/6/2013 8:30:00 - 8:55:00

The low viscosity of CO_2 at typical enhanced oil recovery (EOR) conditions is responsible for a poor mobility ratio that causes viscous fingering and poor sweep efficiency, leading to reduced efficiency and yield. To overcome this problem, there is a need to develop CO_2-soluble additives that will increase the effective viscosity of CO_2 without the use of a co-solvent. The only known polymeric direct thickener poly(fluoroacrylate-co-styrene) (polyFAST) has been shown to significantly increase the viscosity measured by falling cylinder viscometry and Berea sandstone core mobility experiments. The proposed mechanism of aggregation responsible for this viscosity enhancement is believed to be π-π stacking between aromatic rings. However, high molecular weight polymers like polyFAST require significant amounts of expensive fluorinated moieties in order to impart solubility. In addition to their cost, fluorinated compounds have undesirable environmental impacts.

Therefore, this work focuses on the development of less expensive and safer hydrocarbon-based CO_2 thickeners. Because our prior studies have demonstrated that unacceptably high pressures are required to dissolve non-fluorous polymers, this study entails the use of small, associating compounds to create a thermodynamic illusion of a polymer. Each CO_2 thickening candidate consists of at least one non-fluorous CO_2-philic segments and at least two slightly CO_2-phobic functionalities that promote intermolecular aggregation (e.g. aromatic groups). Solubilities of a range of newly synthesized molecules in CO_2 and their associated viscosity enhancing abilities will be presented. Preliminary results indicate that it is possible to design such novel CO_2-thickening candidates that are up to 1wt% soluble in CO_2 at pressures commensurate with EOR.

(456b) Solute Solubility Parameter Determination Using Supercritical Fluids

King, J. W., GENESIS-Engineering Research Center

Originally presented on: 11/6/2013 8:55:00 - 9:20:00

Solubility parameters can be determined or estimated by several methods, namely: calculation from P-V-T-ρ data and equation of state, measuring a solute's solubility in a number of liquids of varying solubility parameter, estimated using group contribution methods, Hansen 3-D solubility spheres, and inverse chromatography. By using appropriate equations for calculating solubility parameters of a sub- or supercritical fluid combined with corresponding states theory, and solute solubility data in supercritical carbon dioxide, ethane, fluoroform, chlorotrifluoromethane, 1,1 – difluoroethane, sulfur hexafluoride, and nitrous oxide – the solubility parameters for a diverse array of solutes have been determined using these dense fluids at pressures up to 2000 bar and temperatures from 40 – 120 C. This approach which uses the varying solubility parameter as a function of pressure/temperature for one fluid will be compared with solubility parameter data determined using a variety of liquid solvents. Solubility data from several literature resources (1-3) and unpublished data on complex lipid mixtures will be presented and include solutes such as synthetic antioxidants, beta-carotene, adamantine, dyestuffs, vegetable oils, low molecular weight polystyrenes and polyethylene glycols, polynuclear aromatic hydrocarbons, and synthetic esters. Comparison of the "supercritical" solvent-derived solubility parameters with those from the literature and calculated by group contribution methods is favorable suggesting this approach can be used to determine solubility parameters for optimizing supercritical fluid extraction and fractionation. Finally, the application of solubility parameter will be shown to be applicable to rationalizing extraction conditions used in ultrahigh pressure SFE.

(1) R.B. Gupta and J-J Shin, "Solubility in Supercritical Carbon Dioxide", CRC Press, Boca Raton, FL (2007)

(2) T. Kraska, K. O. Leonhard, D. Tuma and G. M. Schneider, Journal of Supercritical Fluids 23 (2002) 209–224

(3) Lyle Bowman, "Dense Gas Chromatographic Studies", Ph.D. Thesis, University of Utah, 1976

(456c) High-Pressure Phase Behavior of Synthesis Gas in Hydroformylation Systems: Experimental Data and Prediction

Vogelpohl, C., Technische Universit t Dortmund
Brandenbusch, C., Technische Universit t Dortmund
Sadowski, G., Technische Universit t Dortmund
Originally presented on: 11/6/2013 9:20:00 - 9:45:00

Synthesis gas (CO/H_2) is one of the key components in the hydroformylation process where it reacts with olefins to form aldehydes. Classical production approaches such as the Ruhrchemie/Rhone-Poulenc process are well understood, but fail for the processing of long-chain olefins. The arising challenge is the homogeneously-catalyzed hydroformylation of these long-chain olefins which is currently investigated both in academia and industry.

One promising approach for the hydroformylation of long-chain olefins to aldehydes is the application of thermomorphic multicomponent solvent (TMS) systems, which consist of at least a polar and an apolar solvent [1]. For kinetic studies as well as for process design and optimization the gas-liquid phase-behavior is of high interest. To be able to describe the gas solubility in these complex systems, the gas solubility in all applied solvents as well as the hydroformylation educt (olefin) and product (aldehyde) have to be taken into account.

The aim of this work is to predict the influence of temperature, pressure, as well as of the composition of the multicomponent solvent system, on the synthesis gas solubility using the PC-SAFT equation of state [2]. Therefore the solubility of synthesis gas in the pure solvents as well as in binary and quaternary solvent systems containing components of the TMS and the hydroformylation educt / product was taken into account. Measurements were performed at different temperatures and solvent compositions.

Experimental high-pressure gas-solubility data was gained in a high-pressure variable-volume view cell applying both, the visual synthetic method for gas solubility in homogeneous systems and a newly developed and validated non-visual synthetic method for solubility measurements in immiscible solvents like PC/n-decane. Synthesis gas as well as CO solubilities were determined in pure components like n-decane, PC, DMF, 1-dodecene, n-dodecanal, as well as in binary and quaternary mixtures of the liquids for a temperature range of 303 to 363 K and at pressures up to 15 MPa. Binary gas solubilities and pure-component data were used to fit pure-component parameters and the binary interaction parameter for the PC-SAFT. The experimental data of ternary and higher gas-liquid systems was used to validate the prediction by PC-SAFT without fitting any parameters to this data.

It could be shown, that the temperature as well as the composition of the solvent system has a decisive influence on the high-pressure gas solubility. For all systems investigated, an increase in the gas solubility with increasing temperature was observed. Additionally, higher amounts of the polar solvent (DMF or PC) led to lower solubility of synthesis gas and CO. It could be shown that PC-SAFT is able to predict the synthesis-gas solubility in complex systems (hydroformylation educt, product and TMS) with very good accuracy. Besides the gas solubility itself, also the temperature dependency of the gas solubility is predicted correctly without the need of temperature dependent binary interaction parameters.

The results now enable a further specification of the process window and optimization of the reaction performance for hydroformylation reaction in TMS systems.

[1] Y. Brunsch, A. Behr, Angew. Chem. 125 (2013) 1627.

[2] J. Gross, G. Sadowski, Ind. & Eng. Chem. Res., 40 (2001) 1244-1260.

(456d) Molecular Modeling of Retrograde Vaporization in Enhanced Oil and Gas Recovery

Cort s-Morales, A. D., University of Minnesota
Diamantonis, N. I., National Center for Scientific Research Demokritos
Michalis, V., The Petroleum Institute
Peters, C. J., Petroleum Institute
Economou, I. G., National Center for Scientific Research Demokritos
Siepmann, J. I., University of Minnesota
Originally presented on: 11/6/2013 9:45:00 - 10:10:00

Liquid dropout due to retrograde condensation during production of rich natural gas leaves valuable condensate fluids in the reservoirs, leading to significant loss of well productivity and potentially unfeasible recoveries. Thus, accurate modeling of fluid properties across a wide range of conditions is vital for the design and optimization of enhanced oil and gas recovery processes. In this work, Gibbs-ensemble Monte Carlo simulations using transferable force fields, the Peng-Robinson equation of state, and the perturbed-chain statistical association fluid (PC-SAFT) theory are applied to investigate double retrograde vaporization in vapor-liquid phase equilibria of fluid mixtures near the critical point of the lower-boiling compound. Multicomponent mixtures representative of rich natural gas and enhanced recovery fluids are discussed. The molecular simulations allow one to provide molecular-level understanding of the solution behavior of these mixtures at relevant reservoir operation conditions, whereas the equations of state serve as supporting data and guide for the set-up of Gibbs ensemble simulations (i.e., by providing information on appropriate overall composition and volume/pressure to be used). The agreement obtained between these approaches and with limited experimental data demonstrates the versatility of molecular simulations in advanced *PVT* modeling.

(456e) Analysis of Nonuniformity of Methanol-Water Mixture in Sub- and Supercritical State Via MD Simulation

Ono, T., Tohoku University

Ota, M., Tohoku University

Sato, Y., Tohoku University

Inomata, H., Tohoku University

Originally presented on: 11/6/2013 10:10:00 - 10:35:00

Analysis of Nonuniformity of Methanol-Water Mixture in Sub- and Supercritical State via MD Simulation

Takumi Ono, Masaki Ota, Yoshiyuki Sato and Hiroshi Inomata

Research Center of Supercritical Fluid Technology, Tohoku University, Sendai, Japan

Abstract

Alcohol-water mixtures exhibit interesting anomalies in their macroscopic properties.These features arise from inhomogeneous mixture caused by hydrogen bonding and this factor is influenced by temperature and pressure effects. We carried out molecular dynamics simulation to evaluate the nonuniformity from the macroscopic point of view. The remarkable nonuniformities of mixture were characterized by the distribution of number density of molecules in local domain at low methanol composition at 300 °C or less.

Introduction

Alcohol-water mixtures are scientifically intriguing and important for many fields ranging from basic molecular research to widespread industrial. Simple alcohol such as methanol or ethanol and water mix incompletely, which is a phenomenon described in terms of negative excess entropy [1, 2]. This well-known effect has been attributed to hydrophobic headgroups creating ice-like or clathrate-like structures in the surrounding water. Related to the inhomogeneous mixing, furthermore, distinct solvation regimes as a function of alcohol concentration have been experimentally reported [3]. This indicates that alcohol-water mixtures have nonuniformity at the molecular level. The fact that even sub- and supercritical water the hydrogen bonding still exists is reported [4, 5]. It is expected that inhomogeneous structure attributed by the hydrogen bonding is also influenced by temperature and pressure effects. For the purpose of solution structure analysis of methanol-water mixture and effect of temperature and pressure, some physical properties of alcohol-water mixture were measured at high temperature and pressure including sub- and supercritical condition [6-8]. Since the hydrogen bonding structures affect the density, our group has measured the density of methanol-water mixture and performed molecular dynamics simulation for understanding of the relationship between the macro and micro properties at 300 ~ 400 °C, high pressure condition [9]. However, still no consensus exits on the precise microscopic picture at the molecular level. In particular, the study of degree of nonuniformity in mixture is scarce. In this work, we calculated the distribution of molecular arrangement of methanol-water mixtures in ambient, sub- and supercritical states using molecular dynamics simulation to discuss the nonuniformity at the molecular level.

Computational details

The molecular models used in this study were the 3-sites flexible models proposed by Liew et al. [10] for water and by Honma et al. [11] for methanol. Both models can represent the *PVT* and saturated properties of the fluid in the liquid and supercritical state. The three sites for methanol are oxygen (O), hydrogen (H) and methyl group (-CH3). Both flexible models were formulated as a sum of intra- and intermolecular potential by using the angular form of the Toukan-Rahman (TR) potential [12]. The intermolecular potential used a Lennard-Jones (LJ) 12-6 potential and a Coulombic potential. Unlike interactions were computed using the Lorentz-Berthelot combination rule.

$$\sigma_{ij} = (\sigma_{ii} + \sigma_{jj})/2 \qquad (1)$$

$$\varepsilon_{ij} = \sqrt{\varepsilon_{ii}\varepsilon_{jj}} \qquad (2)$$

The LJ and Coulombic potential parameters for water and methanol were those given in the original studies [10, 11]. All simulations were performed with *NVT*-ensembles for several methanol mole fractions. The conditions in the simulation were set in the 25, 300, 400 °C and the number density N / V was set to the experimentally determined densities (0.1, 25MPa) [9, 13] to allow isobaric analysis at any given temperature. The total number of molecules was 2048. The temperature was initially (~ 60 ps) controlled by momentum scaling and afterwards (60 ps ~) by the Nosè-Hoover thermostat for computational stability. The equation of motion was integrated using the velocity Verlet technique. Time steps were 1 fs for the intermolecular motion and 0.2 fs for the intramolecular motion. Statistical sampling in the simulation was done for 600 ps after 100 ps equilibration.

Definitions of Nonuniformity

We divided into the computational cell to cubes of X Å and counted the number density n of methanol and water molecules of each divided cells to evaluate the nonuniformity of mixture. Figure 1 shows the schematic diagram of the evaluation of nonuniformity for methanol-water mixture. We performed above sampling at 50 fs intervals and calculated the standard deviation of the n of the molecules in the divided cells to evaluate the nonuniformity of the methanol-water mixtures.

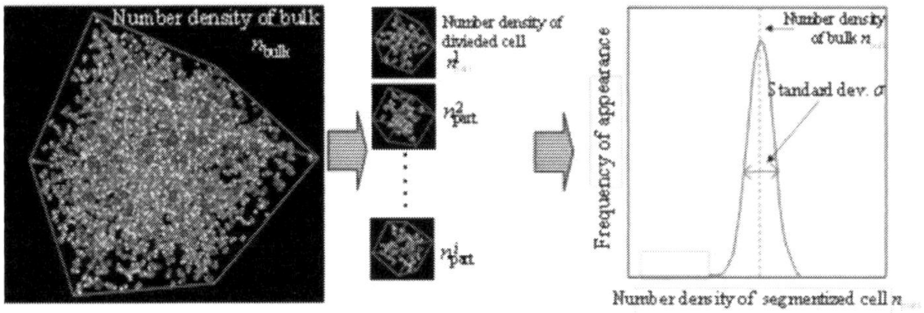

Figure 1 Schematic diagram of the evaluation of nonuniformity for methanol-water mixture.

Results and discussions

Figure 2 and 3 show n for each molecules in bulk and standard deviation of n for each molecules as a function of methanol composition in a 8 Å divided cubic cell (X = 8Å). The standard deviations of each component at 25 °C - 0.1 MPa and water molecule at

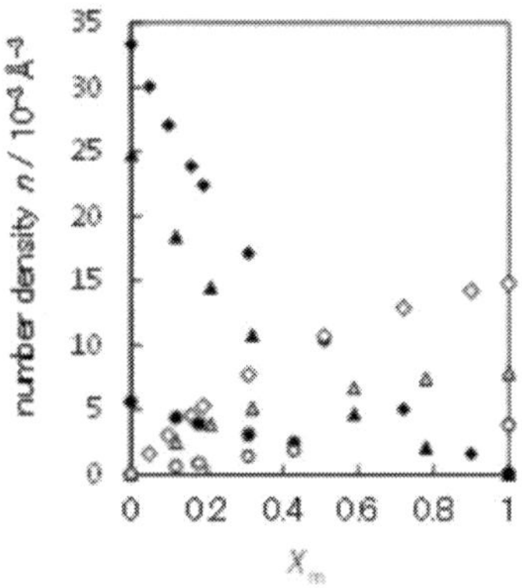

Figure 2 X_m dependence of the number density n of for methanol (open symbols) and water (filled symbols) in bulk: 25 °C-0.1MPa (diamonds), 300°C-25MPa (triangles), 400 °C-25MPa (circles).

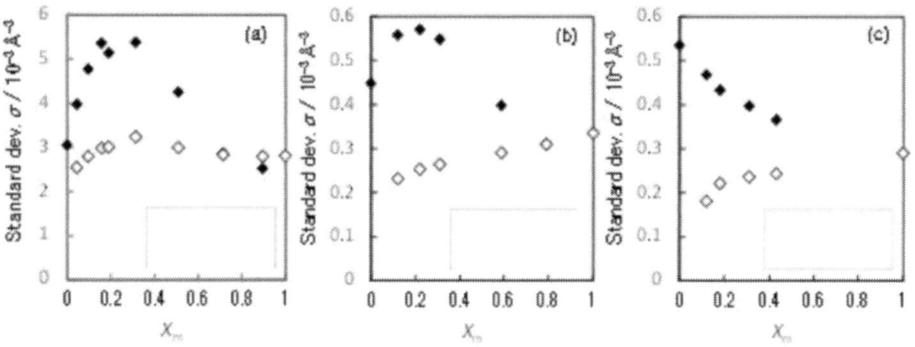

Figure 3 X_m dependence of the standard deviation s of number density n distribution for water and methanol at (a) 25 °C-0.1MPa, (b) 300°C-25MPa, (c) 400 °C-25MPa: methanol (open diamond) and water (filled diamonds).

300 °C - 25 MPa have maximum around methanol composition $X_m = 0.2 \sim 0.3$ although n value in bulk monotonously varied with composition. However this maximum was not observed in methanol at 300 °C - 25 MPa and both molecules at 400 °C - 25 MPa. The fact that effects of hydrophobic hydration, so-called "ice-like" or "clathrate-like", are remarkable in low X_m has been reported [14], hence this maximum suggests that inhomogeneous spatial distributions of the molecules caused by the inhomogeneous solution structure exist from the microscopic point of view. In addition, even under high temperature and pressure, such as 300 °C - 25 MPa, the inhomogeneous local structures proved to be remaining. On the other hand, we assumed that these non-uniform local structures cannot be maintained at 400 °C - 25 MPa because of intense thermal motion of molecules.

References

[1]
F. Franks, D.J.G. Ives, Quarterly Reviews, 20 (1966) 1-&.

[2]
F. Franks, J.E. Desnoyers, Water Science Reviews, 1 (1985) 171-232.

[3]
T. Sato, A. Chiba, R. Nozaki, J. Chem. Phys., 112 (2000) 2924-2932.

[4]
Y. Ikushima, K. Hatakeda, N. Saito, M. Arai, J. Chem. Phys., 108 (1998) 5855-5860.

[5]
M.M. Hoffmann, M.S. Conradi, J. Am. Chem. Soc., 119 (1997) 3811-3817.

[6]
E. Bulemela, P. Tremaine, S. Kawa, Fluid Phase Equilib., 245 (2006) 125-133.

[7]
C.B. Xiao, H. Bianchi, P.R. Tremaine, Journal of Chemical Thermodynamics, 29 (1997) 261-286.

[8]
D.T. Safarov, A.N. Shakhverdiev, High Temp., 39 (2001) 395-400.

[9]
T. Ono, S. Kobayashi, T. Hoshina, Y. Sato, H. Inomata, Fluid Phase Equilib., 302 (2011) 55-59.

[10]
T. Honma, C.C. Liew, H. Inomata, K. Arai, J. Phys. Chem. A, 107 (2003) 3960-3965.

[11]
T. Honma, S. Kuzuhara, C.C. Liew, H. Inomata, Fluid Phase Equilib., 194 (2002) 271-280.

[12]
K. Toukan, A. Rahman, Physical Review B, 31 (1985) 2643-2648.

[13]
M. Iglesias, B. Orge, J. Tojo, Fluid Phase Equilib., 126 (1996) 203-223.

[14]
M. Kiselev, S. Noskov, Y. Puthovski, T. Kerdcharoen, S. Hannongbua, Journal of Molecular Graphics & Modelling, 19 (2001) 412-416.

(457a) The Use of Free Energy Methods to Study Morphological Transitions in Vesicles and Cells Induced By Membrane Remodeling Proteins

Natesan, R., University of Pennsylvania
Tourdot, R., University of Minnesota
Radhakrishnan, R., University of Pennsylvania
Originally presented on: 11/6/2013 8:30:00 - 8:48:00

We investigate the interplay between cell membrane curvature induced at the atomic scale, due to specialized peripheral membrane proteins, and the resulting membrane morphologies, of varying complexity, observed at the mesoscale. The biological membrane, in our approach, is represented by a dynamically triangulated surface while the proteins are modeled as curvature fields on the membrane, that are either isotropic or anisotropic. In order to compare with experiments, we have focused on the ENTH domain containing EPSIN whose curvature field is modeled as isotropic, and on the BAR, Exo70 and ESCRT family proteins whose curvature fields are determined to be anisotropic, both in experiments and in molecular simulations. Thermal undulations in the membrane and cooperativity in the curvature fields, due to the stabilization of a nematic phase, collectively drive the membrane into different morphological states (*buds, tubules, etc.*) that resemble those in cellular experiments *in vivo* and vesicle experiments *in vitro*. These complex shapes are emergent in our phenomenological model, that extends the Helfrich Hamiltonian to include the role of anisotropic elasticity and orientational order.

We determine the relative stability of the above mentioned shapes based on the free energy of these membrane configurations, determined using the methods of Thermodynamic Integration (TI) and Bennett Acceptance (BA). Based on the absolute free energy computed using TI and the relative free energies computed using BA, we predict the critical protein density and critical protein induced curvature required for tubulation of invaginations and exvaginations in spherical membranes. Results are shown for the case of Exo70 protein and N-BAR domains and also compared against those determined from experiments.

The membrane model has been extended to study how a functionalized nanocarrier binds to a fluctuating membrane surface mediated by multivalent receptor-ligand interactions. Our studies show that thermal undulations tend to enhance the multivalency of binding, leading to a stronger binding affinity, when the free energy change due to a bond formation is large compared to thermal energy. Further, formation of multiple bonds leads to considerable membrane deformation around the binding sites, which we hypothesize to be a precursor to activation of the various protein machineries behind the internalization of nanocarrier by cells.

[1] N. Ramakrishnan, P. B. Sunil Kumar and Ravi Radhakrishnan, Physics Reports (article in review)

(457b) Molecular Simulation Study of Ionic Association in Lidocaine Docusate, An Active Pharmaceutical Ingredient Ionic Liquid and Its Derivative

Shah, J. K., University of Notre Dame
Smith, D. J.,
Maginn, E. J., University of Notre Dame
Originally presented on: 11/6/2013 8:48:00 - 9:06:00

Until very recently, the design of so-called first and second generations of ionic liquids focused on tailoring their physical and chemical properties for a specific application. The third generation of ionic liquids is designed to exhibit enhanced biological activity arising from their constituent ions. For example, lidocaine hydrogen chloride is an analgesic while sodium docusate is an emollient. Lidocaine ion when paired with the anion docusate produces a colorless ionic liquid lidocaine docusate that has been shown to prolong analgesic effects. However, it is not clear what the microscopic state of the ions is when dissolved in water. Such questions are vital to understanding how ion pairing might affect biological activity of such an ionic liquid.

In this talk, we elucidate the nature of ionic interactions in water by calculating the potential of mean force from molecular dynamics simulation. First, force field parameters are developed for the ions lidocaine and docusate. Adaptive biasing force methodology is then applied to determine the potential of mean force as a function of the distance between the two ions. The propensity for ionic association is then compared with that for the two derivatives: lidocaine chloride and sodium docusate. In addition, we report the association constants for these ionic liquids and discuss organization of the ions and water in terms of radial distribution functions.

(457c) Coarse-Grained Description of Phase Behavior in 38-Member Clusters of Lennard-Jones Particles

Sehgal, R. M., University of Massachusetts Amherst
Ford, D., University of Massachusetts Amherst
Maroudas, D., University of Massachusetts Amherst
Originally presented on: 11/6/2013 9:06:00 - 9:24:00

Small clusters of Lennard-Jones particles are a well studied system with interesting phase behavior. The behavior found in these clusters has common features with the phase behavior, thermodynamics, and kinetics governing protein folding, glass formation, and other intriguing phenomena in complex systems. The 38-member Lennard-Jones particle cluster (LJ_{38}) is especially interesting, as it has octahedral as well as different icosahedral minimum-energy configurations, leading to both solid-solid and solid-fluid phase coexistence.

In this presentation, we report results of a systematic investigation of the phase behavior of the LJ_{38} cluster. To study this phase behavior, we conducted Monte Carlo (MC) simulations and employed windowed Monte Carlo-umbrella sampling (MC-US) to generate free-energy landscapes (FELs). We constructed these FELs over a broad temperature range to develop a comprehensive picture of the thermodynamically stable configurations as well as the transitions observed between them. The FELs were generated within a coarse-grained framework, which employs diffusion mapping data mining techniques and is able to capture the phase behavior of interest over the temperature range examined. The reduced dimensionality of the coarse-grained description and the choice of coarse variables (order parameters) used in this coarse graining are addressed in detail. By analyzing the resulting FELs, we predict the conditions for the stability of different crystalline configurations and the coexistence between these crystalline configurations and the fluid-like states. This FEL analysis yields phase-diagram information, which can describe not only the bulk-like phase behavior (i.e., coexistence at a single point) but also the complex phase behavior arising from the thermodynamic system smallness inherent to these LJ_{38} clusters.

(457d) Effect of the Nano-Confinement of Cross-Linked Epoxy By Carbon Nanotubes On the Thermo-Mechanical Properties of Their Nanocomposites

Khare, K. S., Texas Tech University
Khare, R., Texas Tech University
Originally presented on: 11/6/2013 9:24:00 - 9:42:00

In this work, we use atomistically detailed molecular simulations to study the thermo-mechanical properties of nanocomposites formed by cross-linked epoxy and carbon nanotubes (CNTs). We use simulations to focus on the interphase, i.e. a part of the polymer matrix, which is strongly affected by the specific interfacial interactions between the matrix and the filler. We will present results on the volumetric, structural, dynamic, and mechanical properties of nanocomposites that have varying amounts of the interphase (as a result of controlling the extent of dispersion of CNTs), and varying properties of the interphase (due to the use of either pristine or functionalized CNTs). Our observations for the effect of interfacial interactions on the glass transition behavior lend support to the conceptual analogy between this interphase and the nanoconfinement of polymers by solid substrates. Since the interphase is responsible for the load transfer between the matrix and the filler, the altered polymer physics at the interphase has an impact on the overall mechanical properties of these nanocomposites.

(457e) Thermodynamics in Nano-Porous Electrodes

Kiyohara, K., National Institute of Advanced Industrial Science and Technology (AIST)
Asaka, K., National Institute of Advanced Industrial Science and Technology (AIST)
Originally presented on: 11/6/2013 9:42:00 - 10:00:00

Thermodynamics of fluids in porous media is qualitatively different from that in the bulk due to the finite-size effects, varying dimensionality, and the surface forces that are introduced by the porosity. Thermodynamics of gas molecules in porous media has been extensively studied [1]. On the other hand, it is relatively recently that thermodynamics of electrolytes in porous electrodes has been drawing particular attention [2].

For the thermodynamics in porous electrodes, voltage plays an important role as a thermodynamic field, in addition to chemical potential and temperature which are the thermodynamic fields also necessary for describing adsorption of gas molecules. Having this in mind, we have developed a Monte Carlo simulation method for studying thermodynamics of porous electrodes in the constant-voltage grand-canonical ensemble [3]. With this method, the anomalous pore size dependence of capacitance [4] and phase transition in porous electrodes [5] have been studied.

We extended the method for two-component electrolytes and applied it to a two component primitive model, for which the diameters of the ions are different for the two components. We found some thermodynamic behaviors that are similar to those of one component electrolytes such as the pore size dependence on the capacitance. Moreover, we also found other behaviors that are specific to multi-component systems. For example, it was found that the ratio of the content of the two components forming the electrical double layers changes as a function of thermodynamic parameters such as voltage. Those findings are discussed in relation to practical problems such as development of electrochemical capacitors or ion-selective adsorbents.

This work was supported by JSPS KAKENHI Grant Number 24550169. Part of the computation in this work was performed using the supercomputers of Research Institute for Information Technology of Kyushu University and those of Cybermedia Center of Osaka University.

[1] L. D .Gelb, K. E. Gubbins, R. Radhakrishnan, and M. Sliwinska-Bartkowiak, Rep. Prog. Phys., 62, 1573 (1999).

[2] J. Chmiola, G. Yushin, Y. Gogotsi, C. Portet, P. Simon, and P. L. Taberna, Science, 313, 1760 (2006).

[3] K. Kiyohara and K. Asaka, J. Chem. Phys., 126, 214704 (2007); 2007 AIChE Annual Meeting.

[4] K. Kiyohara T. Sugino, and K. Asaka, J. Chem. Phys., 132, 144705 (2010).

[5] K. Kiyohara, T. Sugino, and K. Asaka, 2010 AIChE Annual Meeting; J. Chem. Phys., 134, 154710 (2011); J. Chem. Phys., 136, 094701 (2012).

(457f) Coarse Grain Model of Nanostructured Alkaline Exchange Membranes: Phase Behavior and Transport Property Predictions

Mao, S., Stanford University
He, S., Stanford University
Coletta, E., Stanford University
Toney, M., SLAC National Accelerator Laboratory
Frank, C. W., Stanford University
Spakowitz, A. J., Stanford University
Originally presented on: 11/6/2013 10:00:00 - 10:18:00

Polymers with randomly grafted side chains can nanoscale phase separate due to the incompatibility between different polymer segments. This type of nanostructured material is used as high hydroxide conductivity alkaline exchange membranes (AEM) in fuel cell applications, where nanoscaled water channels enhance anion transport. In this work, we explore the thermodynamics and transport properties of the material using coarse grained polymer models. We utilize analytical theory and simulation techniques to predict the phase behavior and characterize transport properties of AEM. Field-based Monte Carlo simulations are performed with a coarse-grained Gaussian chain model with Flory-Huggins parameter and melt compressibility. Local order-disorder transitions (ODT) are identified with competition between interfacial energy and the entropic penalty of stretching the random-copolymer chains. Our theoretical formalism has been developed to characterize the diffusive property of the nanostructured media. The theory and model provide a framework that is amendable to direct comparison with small-angle X-ray scattering (SAXS) and conductivity measurements. The results are also compared and contrasted with predictions from self-consistent field theory (SCFT) of block copolymers. Further phase behavior predictions of copolymers will give insights to wider applications in designing materials with nanoscale phase segregation.

(457g) Theoretical Study of Nanostructured Alkaline Exchange Membrane Transport Property

Mao, S., Stanford University
Spakowitz, A., Stanford University
Coletta, E., Stanford University
He, S., Stanford University
Toney, M., SLAC National Accelerator Laboratory
Frank, C. W., Stanford University
Originally presented on: 11/6/2013 10:18:00 - 10:36:00

Polymers with randomly grafted side chains at low enough temperature micro-phase separate due to the incompatibility between different polymer segments. This type of nanostructured material is used in high hydroxide conductivity alkaline exchange membrane (AEM) design for fuel cells, where nanoscaled cylindrical water channels enhance anion transport. The thermodynamics and physical properties of the material have not been explored thoroughly. We have developed theories and simulation techniques to predict the phase behavior and characterize transport properties of AEM. Monte Carlo simulations are performed with coarse-grained Gaussian chain model with Flory-Huggins parameter and melt compressibility. Local order-disorder transitions (ODT) are identified with competition between interfacial energy and the entropic penalty of stretching polymers. Theoretical formalism has been developed to characterize the diffusive property of the micro-structured media. The theory and model provide a framework that is amendable to direct comparison with the small-angle X-ray scattering (SAXS) structural characterization and the conductivity performance measurements. The results are also compared and contrasted with predictions from self-consistent field theory (SCFT) of block copolymers. Further phase behavior predictions of copolymers will give insights to wider applications in designing materials with nanoscale structures.

(458a) Direct Numerical Simulation of Inertial Particles in Isotropic Turbulence At High Reynolds Numbers

Ireland, P. J., Cornell University
Collins, L. R., Cornell University
Originally presented on: 11/6/2013 8:30:00 - 8:45:00

We use direct numerical simulations (DNS) to study the motion of inertial particles in isotropic turbulence. The primary motivation is to advance our understanding of the role turbulence plays in the evolution of atmospheric clouds. It is known that clouds develop faster than microphysical models based on gravitational settling alone can predict, and it is our hypothesis that turbulence-driven collisions are responsible for that discrepancy. An open question is how to extrapolate DNS at Reynolds numbers (based on the Taylor microscale) in the low hundreds to atmospheric conditions, with Reynolds numbers of order 10,000. While the computational demands of DNS make the atmospheric condition unachievable on today's computers, a recent advance in our simulation code has enabled us to perform DNS on a 2048 cubed lattice and achieve a Reynolds number of about 600, the highest value to date for a multiphase flow simulation. We will present results from those simulations, with a focus on the statistics that impact the collision kernel, namely the radial distribution function, which quantifies the effect of particle clustering, and the relative velocity probability density function, which controls the number of intersecting trajectories per unit time. The results reveal interesting trends with Reynolds number that are informing a theory under development in our group, and that will enhance our ability to extrapolate to atmospheric conditions.

(458b) Flow Induced Particle Separation in Turbulent Channel Flow

Nguyen, Q. T., The University of Oklahoma
Papavassiliou, D. V., The University of Oklahoma
Srinivasan, C., University of Oklahoma
Originally presented on: 11/6/2013 8:45:00 - 9:00:00

Particle separation has been an issue in many engineering disciplines and applications, due to purity requirements in pre-treatment processes, or in post-reacting flows. Several methods have been invented, with use of additional equipment, by employing effects of electric field[1, 2], centrifugal force[3, 4] etc. Our study finds that turbulence, which is usually pursued to provide excellent mixing efficiency[5-7], could separate particles with different Schmidt numbers, Sc, with no use of outside force or devices. Direct numerical simulation is used with tracking of scalar markers (Lagrangian scalar tracking, LST) to obtain particle positions as they travel in a flow. Since particles with different Sc have different values of dispersion, mixed clouds of particles, after a long enough time, will result in separate clouds of particles with different Sc. The efficiency of separation is affected by the distance from the wall at which particles are released into the flow and by the difference in Sc between particles. The mechanism for this flow-induced separation will be discussed, and its relation to the development process of instantaneous puffs of particles. Criteria for a successful separation will also be discussed.

References

[1] X.-B. Wang *et al.*, Anal. Chem. (Wash.) **72** (2000).

[2] O. Gaal, *Electrophoresis in the Separation of Biological Macromolecules* (John Wiley & Sons Ltd, 1980).

[3] L. J. Gimbert *et al.*, Environ. Sci. Technol. **39**, 5 (2005).

[4] S. Juen *et al.*, Superlattices Microstruct. **11** (1992).

[5] L. Valino, C. Dopazo, and J. Ros, Phys. Rev. Lett. **72** (1994).

[6] B. I. Shraiman and E. D. Siggia, Phys. Rev. Lett. **77** (1996).

[7] J. Schumacher and K. R. Sreenivasan, Phys. Rev. Lett. **91** (2003).

(458c) Modeling Dilute Gas-Solid Turbulent Flows Using Moment Methods

Dunn, D. M.,
Squires, K. D., Arizona State University
Originally presented on: 11/6/2013 9:00:00 - 9:15:00

Numerical simulation offers a powerful tool for modeling particle-fluid interactions in turbulent two-phase flows. Although Lagrangian particle-tracking methods are a plausibly accurate approach, these models are often limited to dilute flows and can be inaccurate in regions of locally large particle concentrations where inter-particle interactions and effects of two-way coupling can be significant. These and other considerations motivate the current effort aimed at implementing Eulerian-based approaches that treat the particle phase as a continuum. The specific focus of the current effort is on modeling dilute particle-laden turbulence in which the gas-phase carrier flow is populated with a second phase of small, dispersed solid particles possessing material densities much larger than that of the carrier flow, and consequently large particle Stokes numbers. The approach adopted in this work is derived from the quadrature-based method of moments (QMOM) developed originally by McGraw (1997), and is known as the conditional QMOM method (CQMOM) developed by Yuan and Fox (2011).

Simulations are conducted of a particle-laden turbulent boundary layer, both with and without a coherent vortex periodically introduced into the flow that distorts the boundary layer. The gas-phase carrier flow is computed using Direct Numerical Simulation of the incompressible Navier-Stokes equations. The dispersed solid phase is coupled to the fluid via a drag force acting on the particles and modeled using CQMOM. The boundary layer develops spatially from a turbulent inflow condition; in cases in which the coherent vortex is used to distort the flow, the vortex impinges into the boundary layer around 20 initial boundary layer thicknesses downstream of the inlet. The simulations show that the carrier flow drives the particulate phase via the drag force and with resulting structural interactions, e.g., preferential concentration of particles, similar to those observed in Lagrangian particle tracking simulations. Further comparisons are made against simulations using Lagrangian particle tracking of the dispersed phase and demonstrate the utility of the Eulerian approach, e.g., with statistical descriptors in reasonable agreement between the two methods. Instantaneous and time-averaged statistics are presented, as well as phase-averages for the cases involving time-periodic impinging vortices.

(458d) Turbulence Modeling of Collisional Gas-Particle Flows

Capecelatro, J. S., Cornell University
Desjardins, O., Cornell University
Fox, R. O., Iowa State University
Originally presented on: 11/6/2013 9:15:00 - 9:30:00

Starting from the kinetic theory (KT) model for granular flow, the exact Reynolds-average (RA) equations have been derived for the particle phase in a collisional gas-particle flow. The corresponding equations for a constant-density gas phase have been derived from a model that includes drag and buoyancy coupling with the particle phase. The fully coupled hydrodynamic model is written in terms of the particle-phase volume fraction, the particle-phase velocity and the granular temperature (or total granular energy). In contrast, the RA model solves for the RA particle-phase volume fraction, the phase-average (PA) particle-phase velocity, the PA granular temperature, and the PA turbulent kinetic energy of the particle phase. Thus, unlike in most previous derivations of turbulence models for moderately dense gas-particle flows, a clear distinction is made between the PA granular temperature, which appears in the KT constitutive relations, and the particle-phase turbulent kinetic energy, which appears in the turbulent transport coefficients. The exact RA equations contain unclosed terms due to nonlinearities in the hydrodynamic model. We introduce closures for some of these terms and validate them using mesoscale direct-numerical simulation results for moderately dense turbulent riser flow. Of particular interest are the unclosed source terms due to phase iteractions such as drag coupling for mean momentum and turbulent kinetic energy, and turbulence production due to the mean velocity difference.

(458e) High-Resolution Simulations of One- and Two-Phase Homogeneous Isotropic Turbulent Flows

Komrakova, A. E., University of Alberta
Eskin, D., Schlumberger
Derksen, J., University of Alberta
Originally presented on: 11/6/2013 9:30:00 - 9:45:00

A way to produce a liquid-liquid dispersion or emulsion is to agitate two immiscible liquids in a stirred tank at sufficiently large Reynolds number. Since the dispersion is primarily formed as a result of drops interacting with turbulent eddies leading to drop breakup, it is necessary to know the energy input into the system that can provide the required intensity of the flow. To obtain this information, experiments are usually performed. With the rapid development of computational facilities, numerical simulations become a valuable addition to experimental studies.

In the present study, direct numerical simulations of a turbulently agitated liquid-liquid dispersion have been performed using a free energy lattice Boltzmann method. Large parallel computations were carried out in a three-dimensional, fully-periodic, cubic domain with an edge size of L=500 lattice units [lu] resulting in 125 million cells. Throughout this domain, statistically stationary homogeneous isotropic turbulence was generated by means of linear forcing. It sustains a constant turbulence source during the entire simulation. Also it produces realistic energy input at integral length scales (~L) that eventually dissipates at small scales (Kolmogorov length scale η_k). Kolmogorov scales of 1, 5 and 10 lattice units were considered. The possibility to generate turbulence with these η_k in a domain of 500^3 was examined. The viscous sub-range of the turbulent energy spectrum was always reproduced. The reproduction of the inertial sub-range depended on the L/η_k ratio. High-resolution simulations with η_k=10 [lu] revealed the smallest (~η_k) turbulent structures, while the large domain size allowed to capture integral eddies. The velocity magnitude fields in a cross-section corresponding to different Kolmogorov length scales are shown in Fig.1. The influence of numerical parameters on turbulence generation was also investigated.

With the adopted turbulence generation method the energy input into the system ε is known a priory. Thus, when liquid-liquid systems are considered, it is possible to determine if the specified ε is sufficient to obtain a dispersion effect desired. During the two-phase flow simulations the second phase was instantaneously injected into the fully-developed turbulent flow. The entire dispersion formation process was visualized capturing the drop/eddy interactions. A dispersed phase field is shown in Fig.2 for the case of η_k=1 [lu], the dispersed phase volume fraction of 0.008 at a time instant of $332\tau_k$ (a Kolmogorov time scale). The final drop size distributions (DSDs) for different energy input values were determined. Since the final emulsion properties, such as apparent viscosity, rheology, interfacial area, depend on the DSD, the developed numerical tool allows estimating these properties.

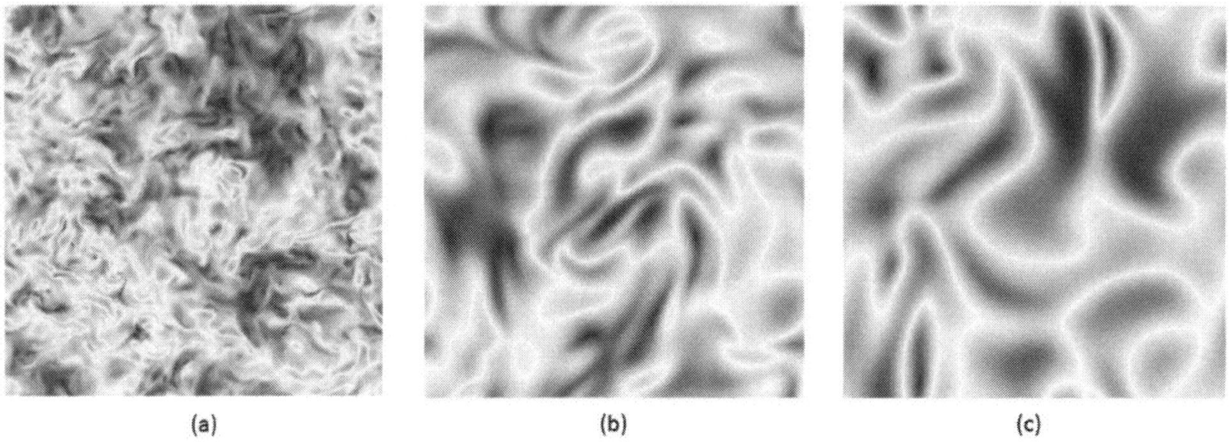

Fig 1. The velocity magnitude fields in the cross-sections (500x500 [lu]); (a) η_k=1 [lu]; (b) η_k=5 [lu]; (c) η_k=10 [lu].

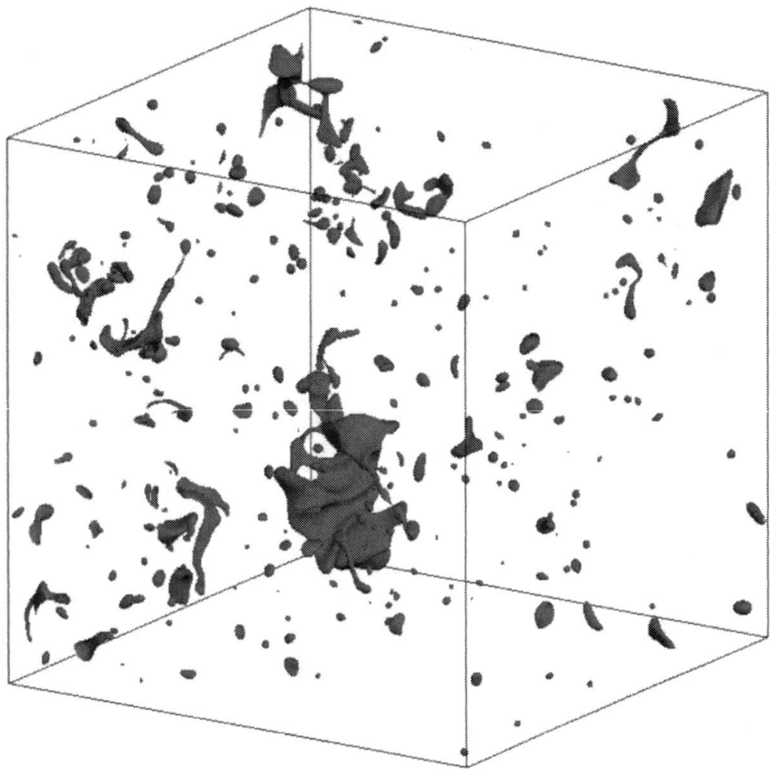

Fig.2. A dispersed phase field at time instant $332\tau_k$, η_k=1 [lu], the dispersed phase volume fraction is 0.008.

(458f) Experimental Validation of High Reynolds Number CFD Simulations of Heat Transfer in a Pilot-Scale Fixed Bed Tube

Dixon, A. G., Worcester Polytechnic Institute
Walls, G., East Coast Metrology
Stanness, H., Johnson Matthey Catalysts
Nijemeisland, M., Johnson Matthey Catalysts
Stitt, E. H., Johnson Matthey Catalysts
Originally presented on: 11/6/2013 9:45:00 - 10:00:00

The simulation of detailed flow and transport inside the interstitial space of packed tubes using computational fluid dynamics (CFD) is becoming a standard approach to obtaining insight into fixed bed transport phenomena. At the high flow rates of industrial practice, flow and heat transfer are highly turbulent and complex, and comparisons to experiment are necessary. In this study, CFD simulations of heat transfer in fixed beds of spheres were validated by comparison to experimental measurements in a pilot-scale rig. The comparisons were made for particle Reynolds numbers in the range $2200 < Re < 27000$ for tube-to-particle diameter ratio $N = 5.45$, and in the range $1600 < Re < 5600$ for $N = 7.44$. Radial temperature profiles were obtained at four axial positions in the heated bed spaced 0.16 m apart.

CFD models of a 0.72 m long tube containing 1000 spheres ($N = 5.45$) and a 0.35 m long tube of 1250 spheres ($N = 7.44$) were solved to obtain well-developed flow fields. These provided inlet velocity profiles to reduced models of 0.20 m heated packed length, consisting of 304 spheres for $N = 5.45$ and 722 spheres for $N = 7.44$. The measured temperature profiles were used as inlet boundary conditions and profiles 0.16 m downstream were computed. Verification with a medium and a fine mesh showed negligible difference between the profiles.

Contact points were treated by one of three methods: global reduction of sphere size resulting in gaps, or local insertion of bridges at particle-wall contact points with either surface flattening or bridges at the particle-particle contact points. The gaps method gave slightly poorer results; the two bridges methods were better and indistinguishable from each other. CFD simulations compared well to the experimental data: trends with Re, N and bed depth were captured, and quantitative agreement of temperature profiles was reasonable.

(458g) Meso-Scale Direct Numerical Simulations of Mono-Disperse Bubbly Flows

Panicker, N. S., Iowa State University
Passalacqua, A., Iowa State University
Fox, R. O., Iowa State University
Originally presented on: 11/6/2013 10:00:00 - 10:15:00

Disperse gas-liquid flows involved in industrial application such as bubble columns are typically modeled using continuum hydrodynamic models, supplemented with Reynolds-average turbulence models to describe the effects of the fluctuations in the phase velocity on the mean flow properties. These turbulence models are derived assuming that the hydrodynamic model constitutes a reliable mesoscopic description of the multiphase system, and applying a phase averaging procedure. Numerous terms originate from the averaging procedure, such as, for example, phase turbulent kinetic energy production and dissipation, phase fraction – velocity correlation, pressure gradient – phase fraction covariance, which require closure. In this work we perform mesoscale direct numerical simulations of mono-disperse bubbly flows in the inhomogeneous regime, in order to establish the relevance and the behavior of each of the terms originated from the phase averaging procedure of the mesoscopic model as a function of the dispersed phase fraction. The computational domain under consideration is constituted by a three-dimensional column with squared cross-section. Periodic boundary conditions are applied to the sides of the column to avoid wall effects, while air is fed at the bottom, and pressure is fixed at the top. Simulations are performed for a range of inlet gas volume fraction between 1% and 50%, and for an inlet gas velocity between 0.01 and 0.03 m/s. The magnitude and the trends of the turbulent quantities and of the terms that originate from the phase averaging procedure are extracted from the mesoscale DNS and reported.

(458h) Large Eddy Simulation of a Precipitate Flow With Qmom

Abboud, A. W., Institute for Clean and Secure Energy, University of Utah
Saad, T., Institute for Clean and Secure Energy, University of Utah
Thornock, J., University of Utah
Smith, S. T., Institute for Clean and Secure Energy, University of Utah
Originally presented on: 11/6/2013 10:15:00 - 10:30:00

A precipitation reaction with four competing solid phases was studied using our LES code in two different geometries. One simulation uses intersecting pipes flowing across a static mixer, while the other uses a confined impinging jet reactor (CIJR). The Reynolds number in the simulations is approximately 27,000 based on the inlet conditions. The complex geometries contain short inlet sections, which does not allow for the CFD code to develop the expected turbulence. In order to develop this turbulence, a digital filter method has been used to generate artificial turbulence which accurately matches prescribed first and second order fluid statistics. To account for the subfilter mixing scale, the scalar variance and scalar dissipation rate are calculated using a two equation model, which transports the second moment of the mixture fraction and the squared mixture fraction gradient.

The properties of the solid precipitate phase are calculated with the general population balance equation (GPBE). To obtain numerical closure to the problem the quadrature method of moments (QMOM) is utilized. The GPBE contains a full set of particle source terms including birth, growth & aggregation kernels, which are highly dependent on the local chemistry of the fluid phase. The physics for this problem have a timescale separation, such that the aqueous phase ions are considered to be in equilibrium. These equilibrium concentrations have been pre-tabulated with an external software package to reduce computational cost. The GPBE also utilizes a subgrid multi-environment approach consisting of three delta functions, where the mixing of these environments is determined by the local scalar variance value.

(458i) Coriolis-Induced Redistribution of Turbulent Kinetic Energy and Atmospheric Scintillations

Petty, C. A., Michigan State University
Koppula, K. S., Rochester Institute of Technology
Benard, A., Michigan State University
Originally presented on: 11/6/2013 10:30:00 - 10:45:00

In spanwise fully-developed rotating channel flows, the coupling between the Coriolis acceleration and the fluctuating velocity causes the primary and the secondary normal Reynolds momentum flux differences to flip signs in the outer region on the high pressure side of the symmetry plane. This redistribution of energy by the Coriolis acceleration occurs in a region where the intrinsic mean vorticity is zero.

In this presentation, the dispersion of a passive additive within the zero intrinsic vorticity region will be discussed. For low rotation numbers, the theory shows that the transverse component of the passive additive flux is mitigated by a coupling between the shear component of the Reynolds stress and the longitudinal component of the mean passive additive flux. At high rotation numbers, the dispersion coefficient in the transverse (cross flow) direction is four times larger than the dispersion coefficient in the spanwise direction. Surprisingly, the dispersion coefficient in the longitudinal direction is relatively small.

It is noteworthy that a Coriolis-induced anisotropy in the index-of-refraction of the atmosphere may provide a practical means to determine the relative latitude and the relative longitude between two fixed points on the surface of the Earth. Thus, atmospheric scintillation phenomenon may be an essential component in the mechanics of bird orientation and navigation.

Koppula, K.S., S. Muthu, A. Benard, and C. A. Petty 2013, "The URAPS closure for the normalized Reynolds stress", Physica Scripta, to appear June 2013.

Koppula, K.S., A. Benard, and C. A. Petty, 2011, "Turbulent Energy Redistribution in Spanwise Rotating Channel Flows", Ind. Eng. Chem. Res., 50 (15), 8905-8916.

Koppula, K. S., A. Benard, and C. A. Petty, 2009, "Realizable Algebraic Reynolds Stress Closure", Chem. Eng. Sci., 64, 4611-4624.

(458j) Transition and Turbulence in a Wall-Bounded Channel Flow At High Mach Number :

Pradhan, S., Indian Institute of Science
Kumaran, V., Indian Institute of Science
Originally presented on: 11/6/2013 10:45:00 - 11:00:00

Title : Transition and turbulence in a wall-bounded channel flow at high Mach number

Author(s) : S. Pradhan and V. Kumaran

Affiliation : Department of Chemical Engineering, Indian Institute of Science, Bangalore-560 012, India

Abstract :

The turbulence in the viscous, compressible flow in a wall-bounded channel is studied at high Mach $M = U_m / sqrt(\gamma k_B T_w/m)$, and Reynolds numbers $Re = (\rho_m U_m H)/\mu_w$ using direct simulation Monte Carlo (DSMC) simulations. Here, H is the channel half-width, U_m is the mean velocity, ρ_m is the mean density, T_w is the wall temperature, m is the molecular mass, μ_w is the molecular viscosity based on the temperature at the isothermal wall, and k_B is the Boltzmann constant. The laminar-turbulent transition is accompanied by a discontinuous change in the friction factor even at high Mach number. The transition Reynolds number increases faster than linearly with Mach number, and the Knudsen number at transition (also proportional to the ratio of Mach and Reynolds numbers) passes through a maximum as the Mach number is increased. This maximum value is small, less than 0.009, indicating that transition is a continuum phenomenon even at high Mach numbers. The transition Reynolds number predicted by the linear stability analysis is significantly higher than that observed in simulations, though its variation with Mach number is qualitatively similar.

In the turbulent channel flow, there is slip in both the mean and fluctuating velocities at the wall; there is also significant temperature slip. These have to be incorporated in continuum DNS models to get accurate results. The variation of the ratio of the mean free path and Kolmogorov scale show the same power law behavior as those in a lid-driven cavity. In the inertial sub-layer, the Van-Driest transformed velocity is found to accurately follow the logarithmic law of the wall.

We have also find that the smallest length scale for the velocity gradients is comparable to, or smaller than, the mean free path. Though this appears unusual, it should be noted that the smallest length for the gradients is the distance between molecules, and not the mean free path. In our simulations, the inter-molecular distance turns out to be much smaller than the mean free path. In fact, the intermolecular distance is smaller than the cell size, whereas the Kolmogorov scale and the mean free path are larger than the cell size.

Even though the distance between molecules is the smallest length scale for gradients, the mean free path is the length scale for molecular transport, since the kinematic viscosity and thermal conductivity are proportional to the product of the mean free path and the fluctuating velocity. The present results suggest that the smallest scale for transport could be much larger than the smallest scale for gradients, thereby suggesting non-local transport in high Mach number turbulent flows at the smallest scales.

A modification of the linear velocity profile in the viscous sub-layer near the wall, which takes into account temperature and density variations, is derived. The power law variation of the velocity and temperature is predicted under the assumption that the increase in temperature across the viscous sub-layer is larger than the wall temperature. It is found that the scaling laws do depend on the molecular model, through the dependence of viscosity and thermal conductivity on the temperature. The predicted power law, are found to be in good agreement with simulations, for two different molecular models, the hard-sphere and the variable hard-sphere.

(464a) The Role of Anisotropy for the Assembly of Hard Colloidal Nanoparticles

Damasceno, P. F., University of Michigan
Engel, M., University of Michigan
Glotzer, S. C., University of Michigan
Originally presented on: 11/6/2013 12:30:00 - 12:50:00

The exploration of new anisotropic building blocks have received much attention as a way of increasing the variability and complexity of materials possible to be achieved by self-assembly. Here we present the results of a systematic and extensive computational study of hard polyhedral particles [1,2] and their subsequent assembly into a diverse range of complex structures. We show that by utilizing more complex, anisotropically designed building blocks new structures can be self-organized purely from entropy maximization principles. Moreover, we find that already in the liquid, the coordination number for each particle correlates to the type of crystal structure (Liquid Crystal, Crystal or Plastic Crystal) that will be assembled. Those results suggest the existence of a rationalizing principle, equivalent to Linus Pauling's valency bond theory, governing the possible assembly of anisotropic nanoparticles into crystalline structures.

[1] Pablo F. Damasceno, Michael Engel & Sharon C. Glotzer. ACS NANO (2012).

[2] Pablo F. Damasceno, Michael Engel & Sharon C. Glotzer. SCIENCE (2012)

(464b) Assembly of Patchy Colloidal Rods

Shah, A. A., University of Michigan
Schultz, B., University of Michigan, Ann Arbor
Glotzer, S. C., University of Michigan
Solomon, M. J., University of Michigan
Originally presented on: 11/6/2013 12:50:00 - 13:10:00

Assembly of Patchy Colloidal Rods

Aayush A Shah[1], Benjamin Schultz[2], Sharon C. Glotzer[1,2,3], and Michael J. Solomon[1, 3]

[1] Macromolecular Science and Engineering, University of Michigan, Ann Arbor, MI

[2] Physics, University of Michigan, Ann Arbor, Ann Arbor, MI

[3] Chemical Engineering, University of Michigan, Ann Arbor, Ann Arbor, MI

e-mail: aayushs@umich.edu

We report the synthesis and assembly of patchy rods that combine both shape and interaction anisotropy. These particles are synthesized by sequentially combining evaporative deposition of chrome and gold with the uniaxial deformation of polymeric colloidal particles. We assemble these patchy particles at various salt concentrations and demonstrate patch functionality using newly developed two-channel confocal imaging and image processing. We further direct the assembly of our shape anisotropic patchy particles using AC electric fields. The metallic patch has a significantly higher polarity in AC fields as compared to the polymer component of the particle and this leads to formation of unique chain-like structural motifs. Finally, we achieve reversible reconfiguration in our assemblies by employing AC electric fields on self-assembled particle structures.

(464c) Entropically Patchy Particles

van Anders, G., University of Michigan
Ahmed, N. K., University of Michigan
Smith, R., University of Michigan
Engel, M., University of Michigan
Glotzer, S. C., University of Michigan
Originally presented on: 11/6/2013 13:10:00 - 13:30:00

Directional entropic forces that cause particle alignment have recently been proposed in systems of hard polyhedra that order into crystals. Here we provide a means of quantifying these forces and find them to be of the order of a few kT in monodisperse systems of hard particles. Based on the similarity of the anisotropy of these directional entropic forces to the directional enthalpic forces present in systems of enthalpically patchy particles, we propose and show that these forces originate from "entropic patches," which are features of particle shape that promote local dense packing. Using the notion of entropic patches, we engineer particle shape to target the assembly of specific crystal structures. We show that this procedure can be generalized as anisotropy dimensions, similar to those exploited for enthalpically patchy particles, thereby mapping out a large design space for colloidal self assembly with entropy as the sole mechanism underlying ordering. Our techniques also allow the quantification of the role of shape in systems where both enthalpic and directional entropic forces are at play.

(464d) Competition of Shape and Interaction Patchiness for Self-Assembling Nanoplates

Ye, X., University of Pennsylvania
Chen, J., University of Pennsylvania
Engel, M., University of Michigan
Millan, A. J., University of Michigan
Li, W., Massachusetts Institute of Technology
Qi, L., Massachusetts Institute of Technology
Xing, G., University of Pennsylvania
Collins, J. E., Intelligent Material Solutions, Inc.
Kagan, C. R., University of Pennsylvania
Li, J., Massachusetts Institute of Technology
Glotzer, S. C., University of Michigan
Murray, C. B., University of Pennsylvania
Originally presented on: 11/6/2013 13:30:00 - 13:50:00

Progress in nanocrystal synthesis and self-assembly enables the formation of highly ordered superlattices. Recent studies focus on spherical particles with tunable attraction and polyhedral particles with anisotropic shape and excluded volume repulsion, but the interplay between shape and particle interaction is only starting to be exploited. Here we present a joint experimental-computational, multi-scale investigation of a class of highly faceted planar lanthanide fluoride (LnF_3) nanocrystals (nanoplates, nanoplatelets). The nanoplates self-assemble in a hexane wetting layer at the liquid-air interface into long-range ordered tilings. Using Monte Carlo simulation, we demonstrate that the nanoplate assembly can be understood from maximization of packing density only in first approximation. Explaining the full phase behavior requires an interaction specificity of nanoplate edges, which originates from the atomic structure as confirmed by density functional theory calculations. Despite the apparent simplicity in particle geometry, the combination of shape-induced entropic and edge-specific energetic effects directs the formation and stabilization of unconventional long-range ordered assemblies not attainable otherwise.

Reference: Nature Chemistry, in press, doi: 10.1038/nchem.1651 (2013)

(464e) Disk-Shaped Colloids: the Synthesis and Self-Assembly of Layered Crystals

Cheng, Z., Texas A&M University
Shuai, M., Texas A&M University
Mejia, A. F., Texas A&M University
Originally presented on: 11/6/2013 13:50:00 - 14:10:00

Colloidal suspensions of disks are encountered in natural and industrial processes with particles ranging from clays to discotic liquid crystals to red blood cells. However, compared with calamitic systems (rods), research on liquid crystals of colloidal disks is relatively scarce due to the lack of model systems. Here, we fabricate disks via the exfoliation of layered crystals. We have performed systematic experiments on the synthesis of layered crystal α-ZrP and revealed the control of size, aspect ratio and size polydispersity of disk-shaped crystals. The growth of the disks is mediated by oriented attachment, taking place continuously throughout the hydrothermal treatment between various sized disks. We observed I-N transition at as low as 0.2% in volume fraction Φ when thickness to lateral size ratio ξ reached about 0.001. More importantly, the width of the I-N transition was observed to sensitive to the lateral size polydispersity. As our results indicate, *highly anisotropic platelets*, with relatively low size polydispersity, *open a large range* in particle concentration or volume fraction *f for the formation of various phases of discotic liquid crystals*. Therefore, our capability to fabricate disks/nanoplates with controlled aspect ratios, especially disks/plates with large anisotropic shape, is essential for the fully understanding of the behavior of discotic materials, and facilitating their applications.

(464f) Capillary Alignment of Long Cylinders On Curved Interfaces

Yao, L., University of Pennsylvania
Cavallaro, M. Jr., University of Pennsylvania
Bleier, B., University of Pennsylvania
Stebe, K. J., University of Pennsylvania
Originally presented on: 11/6/2013 14:10:00 - 14:30:00

When a cylindrical microparticle attaches to a fluid interface, it creates a distortion owing to its wetting boundary conditions which changes the area of the interface. If the interface is curved, the interface area, and hence the capillary energy of the system, depends on the particle alignment and location. In prior work, our group has shown that microcylinders with moderate aspect ratio (L/2R~3; L and R denoting the particle length and radius, respectively) align with their major axes along one of the principle axes of the interface and migrate along paths defined by the curvature gradients to regions of high curvature. This alignment and migration minimizes the capillary energy. Here, we show that the preferred orientation of microcylinders changes with aspect ratio; long cylinders prefer to orient perpendicularly to short cylinders and migrate to high curvature zones. More complex behavior is also observed: long cylinders often deviate from simple paths traced by curvature gradients, and can align their major axes at some angle to the principal axes of the interface. We discuss these observations in the context of the excess area of the distortion field around the particle.

(464g) Assembly of Microparticles On Thin Smectic Films

Gharbi, M. A., University of Pennsylvania
Yang, S., University of Pennsylvania
Kamien, R. D., University of Pennsylvania
Stebe, K. J., University of Pennsylvania
Originally presented on: 11/6/2013 14:30:00 - 14:50:00

Colloidal particles organize spontaneously at fluid interfaces owing to a variety of interactions to form well organized structures that can be exploited to synthesize advanced materials. While the physics of colloidal assembly at isotropic interfaces is well understood, the mechanisms that govern interactions between particles at liquid crystal interfaces are not yet clearly established. In particular, smectic liquid crystal films offer important degrees of freedom that can be used to direct particles into new structures. In this work, we report on the behavior of micrometric silica particles with homeotropic anchoring confined within or at interfaces of smectic films. We study the interactions and self-assembly of these particles as a function of film thickness in both supported and in free standing films. When particles are captured in thin membranes, they induce distortions of the smectic interface to satisfy wetting properties at particle boundaries, leading to capillary interactions between the particles. These capillary interactions compete with elastic interactions owing to particle-induced distortions in the smectic layers. The resulting potential drives assembly of the spheres into different structures ranging from 1D chains to 2D aggregates. By increasing the thickness of the smectic, we control the formation of focal conic domains (FCDs) and their organization. The FCDs interact with particles at the interface and can be used to direct the formation of complex particle structures. By organizing the defects, we direct the location of these complex structures. Recent progress in understanding the process of particle self-organization is presented.

(472a) Methanol Synthesis From CO2 - Experiment and Theory

Studt, F., SLAC National Accelerator Laboratory
Abild-Pedersen, F., Technical University of Denmark
N rskov, J. K., Technical University of Denmark
Originally presented on: 11/6/2013 12:30:00 - 12:50:00

The active site of the industrial methanol synthesis catalyst $Cu/ZnO/Al_2O_3$ is identified with a combined experimental and theoretical study. We investigate the reaction mechanism of both, CO and CO_2 hydrogenation, on this site using density functional theory calculations. In combination with microkinetic modeling, the kinetics of CO and CO_2 hydrogenation are described and we are able to shed some light on some of the open questions of methanol synthesis. We extend the analysis over a range of different transition metals and simplify adsorption and transition state energies using scaling relations between different adsorbates on surfaces. This enables us to describe CO_2 hydrogenation using simple descriptors and allows for the construction of activity volcanos for these two hydrogenation processes. Based on the activity volcanos we identified new catalyst leads based on Ni and Ga for the CO_2 hydrogenation to methanol at ambient pressure. Experiments showed that these catalysts are indeed selective and active for this process.

(472b) The Adsorption of CO_2 On Anatase TiO_2(101) Surfaces in the Presence of Subnano Ag and Pt Clusters: Implications for Promising Photocatalysis

Yang, C. T., University of South Florida
Bhethanabotla, V. R., University of South Florida
Joseph, B., Clean Energy Research Center, University of South Florida
Originally presented on: 11/6/2013 12:50:00 - 13:10:00

The broad objective of this research is to address two significant challenges facing the planet: reducing Green House Gas emission and the production of sustainable fuels by harnessing solar energy. Photocatalysis is an possible way to use solar energy to convert CO2 to fuels[1]. CO2 to form CO2 anion is commonly considered as the first and key step of the CO2 photoreduction mechanism, followed by interactions of CO2 anion and H · to form hydrocarbons[1, 2]. Due to the highest reduction potential (vs. NHE) of this step compared to the others[1], the study of it benefit the feasibility of successive steps, improving the generation of clean energy from sunlight.

The electronic structure of the catalyst surface governs the easiness of the formation of the CO2 anion from CO2; this formation relies on the adsorbed forms of CO2 on the catalyst surface. The angle of the bend form of the adsorbed CO2 determines the barrier to the transfer of photoexcited electrons to CO2 to form CO2 anion [3-5]. This is attributed to the changing of CO2 LUMO energy as O-C-O bond angle varies[6]. Many researchers have studied the adsorbed forms of CO2 on different TiO2 surfaces and other semiconductor along with the anticipated electron transfers [3, 5, 7, 8]. In recent years, due to the experimental capability, subnano clusters of several atoms are reported to possess dynamic structural fluxionality[9] and a large fraction of under-coordinated surface atoms[10], which grants the reaction enhanced reactivity. To the best of our knowledge, the first principle study of photoreduction considering the incorporation of subnano metal clusters on TiO2 surfaces is lacking. Therefore, one goal of this work is to study the effects of ultrasmall clusters of Ag, Pt deposited on TiO2 on the first and key step of the CO2 photoreduction. Another goal is the understanding of the sub-bandgap introduced by these clusters, which make possible the efficient utilization of the sunlight[11]. Perfect anatase TiO2(101) surface as well as the existence of the oxygen vacancy are studied along with the Ag and Pt octamers. Vienna Ab Initio Simulation package (VASP) is used to investigate the various adsorbed forms of CO2 on the model surfaces. In comparison of with the literature results, the electronic understandings of CO2 at the interface composed of subnano clusters and the conventional photocatalyst, TiO2, are believed to contribute the CO2 photoreduction to shed some light on these new class photocatalysts in the CO2 photoreduction.

The interplay of the electronic structures of the subnano clusters deposited on the anatase TiO2(101) and the CO2 is the focus of this study. We demonstrate how the presence of the subnano clusters on the anatase TiO2(101) influences the CO2 adsorption as well as the density of states of such model. The altered properties of the adsorbed CO2, and new CO2 adsorption sites on the surface of clusters and at interfaces are expected to be found. These findings give valuable information for the design of more efficient photocatalysts for CO2 photoreduction to generate light hydrocarbon fuels.

References

1. Indrakanti, V.P., J.D. Kubicki, and H.H. Schobert, *Photoinduced activation of CO2 on Ti-based heterogeneous catalysts: Current state, chemical physics-based insights and outlook.* Energy & Environmental Science, 2009. **2**.

2. Usubharatana, P., et al., *Photocatalytic Process for CO2 Emission Reduction from Industrial Flue Gas Streams.* Industrial & Engineering Chemistry Research, 2006. **45**(8): p. 2558-2568.

3. He, H., P. Zapol, and L.A. Curtiss, *A Theoretical Study of CO2 Anions on Anatase (101) Surface.* The Journal of Physical Chemistry C, 2010. **114**(49): p. 21474-21481.

4. Indrakanti, V.P., H.H. Schobert, and J.D. Kubicki, *Quantum Mechanical Modeling of CO2 Interactions with Irradiated Stoichiometric and Oxygen-Deficient Anatase TiO2 Surfaces: Implications for the Photocatalytic Reduction of CO2.* Energy & Fuels, 2009. **23**(10): p. 5247-5256.

5. Rodriguez, M.M., et al., *A Density Functional Theory and Experimental Study of CO2 Interaction with Brookite TiO2.* Journal of Physical Chemistry C, 2012. **116**(37): p. 19755-19764.

6. Freund, H.J. and M.W. Roberts, *Surface chemistry of carbon dioxide.* Surface Science Reports, 1996. **25**(8): p. 225-273.

7. Indrakanti, V.P., J.D. Kubicki, and H.H. Schobert, *Quantum Chemical Modeling of Ground States of CO 2 Chemisorbed on Anatase (001), (101), and (010) TiO 2 Surfaces.* Energy & Fuels, 2008. **22**(4): p. 2611-2618.

8. Liu, L., et al., *Surface Dependence of CO2 Adsorption on Zn2GeO4.* Langmuir, 2012. **28**(28): p. 10415-10424.

9. Hakkinen, H., et al., *Structural, electronic, and impurity-doping effects in nanoscale chemistry: Supported gold nanoclusters.* Angewandte Chemie-International Edition, 2003. **42**(11): p. 1297-1300.

10. Vajda, S., et al., *Subnanometre platinum clusters as highly active and selective catalysts for the oxidative dehydrogenation of propane.* Nat Mater, 2009. **8**(3): p. 213-216.

11. Mazheika, A.S., et al., *Theoretical Study of Adsorption of Ag Clusters on the Anatase TiO2(100) Surface.* The Journal of Physical Chemistry C, 2011. **115**(35): p. 17368-17377.

(472c) Microkinetic Modeling of CO2 Hydrolysis Over Zn-(1,4,7,10-tetrazacyclododecane) Catalyst Based On First Principles: Revelation of Rate-Determining Step

Ma, R., Northwestern University
Broadbelt, L. J., Northwestern University
Originally presented on: 11/6/2013 13:10:00 - 13:30:00

The emission of greenhouse gases, including carbon dioxide, has been put forth as the main cause of global warming.[1,2] Carbon capture and sequestration are regarded as a potential means of mitigating this warming effect.[2] One possible remedy is to convert carbon dioxide catalytically to HCO_3^-. In nature, HCA II, a member of the human carbonic anhydrase (CA) family of enzymes, is one of the fastest known enzymes with a k_{cat} for CO_2 hydrolysis at a value of 10^6 s^{-1}, which approaches the diffusion limit.[3] However, due to the high cost of isolation and purification, as well as the unstable nature of enzymes when removed from their natural environment, biomimetic catalysts are a more desirable choice whenever they are available. 1,4,7,10-tetraazacylododecane ([12]aneN_4 or cyclen) complex of zinc, to the best of our knowledge, has the highest activities among all biomimetic catalysts to date, with a rate constant of 3300 $M^{-1}s^{-1}$.[4] In this study, density functional theory (DFT) calculations were used to unravel the reaction mechanism and evaluate thermodynamic parameters. We mapped the entire catalytic cycle for CO_2 hydration by the zinc-cyclen complex and built a microkinetic model based on first principles kinetics and thermodynamics. The dependence of the observed reaction rate constant on the pH has a sigmoidal shape, which is similar to HCA II.[5] The inflection point is identical to the pK_a value of the zinc-cyclen complex, indicating that the rate-limiting step involves Zn-cyclen-OH^+. The reaction rate constant is calculated to be 3650.5 $M^{-1}s^{-1}$, while the experimental value has been reported to be 3300 $M^{-1}s^{-1}$ or 2691.5 $M^{-1}s^{-1}$.[4,6] Since our microkinetic model captured the experimental data well, we then analyzed each elementary step to identify the rate-determining step of the catalytic cycle through calculating the ratio of the net rate to the forward rate of reaction. Our model indicated that adsorption of CO_2 is the rate-determining step. In addition, we analyzed the initial CO_2 adsorption rate, which corresponds to the reaction: Zn-cyclen-OH^- + CO_2 → Zn-cyclen-HCO_3^-. The rate constant for this reaction is calculated to be 1.82×10^5 $M^{-1}s^{-1}$, which is on the same order of magnitude as the CO_2 adsorption rate constant of a novel biomimetic catalyst reported by Huang and coworkers.[7] Ultimately, these results provide a basis for the design of new biomimetic catalysts for CO_2 hydrolysis.

(1) Montzka, S. A.; Dlugokencky, E. J.; Butler, J. H. *Nature* **2011**, *476*, 43.

(2) Chu, S. *Science* **2009**, *325*, 1599.

(3) Khalifah, R. G. *J Biol Chem* **1971**, *246*, 2561.

(4) Zhang, X. P.; Vaneldik, R. *Inorganic Chemistry* **1995**, *34*, 5606.

(5) Silverman, D. N.; Lindskog, S. *Accounts Chem Res* **1988**, *21*, 30.

(6) Koziol, L.; Valdez, C. A.; Baker, S. E.; Lau, E. Y.; Floyd, W. C.; Wong, S. E.; Satcher, J. H.; Lightstone, F. C.; Aines, R. D. *Inorganic Chemistry* **2012**, *51*, 6803.

(7) Huang, D. G.; Makhlynets, O. V.; Tan, L. L.; Lee, S. C.; Rybak-Akimova, E. V.; Holm, R. H. *Proc. Natl. Acad. Sci. U. S. A.* **2011**, *108*, 1222.

(472d) DFT Studies of Electrochemical Reduction of CO2 On Cu Electrodes

Nie, X., The Ohio State University
Asthagiri, A., University of Florida
Janik, M., Penn State University
Originally presented on: 11/6/2013 13:30:00 - 13:50:00

Electrocatalytic reduction of CO_2 to alcohols would provide a synthetic sustainable way to convert renewable energy sources to transport fuels, however, large overpotentials and inability to control product selectivity are significant barriers to application. To date only Cu electrodes have shown significant hydrocarbon products but at relatively high overpotentials. Interestingly, methane and ethylene are the dominant products instead of methanol. Furthermore, the product distribution is sensitive to the Cu facet and morphology. Understanding these selectivity features and identifying the key reaction steps for CO_2 reduction on Cu electrodes will be an important step in designing improved catalysts. We will report the results of a systematic density functional theory (DFT) study to evaluate the potential-dependent barriers of elementary steps for CO_2electroreduction on the Cu(111) surface that incorporates the role of water solvation in a computationally tractable manner.

We find that O–H bond formation reactions occur through water-assisted H-shuttling, whereas C–H bond formation occurs via direct bonding with adsorbed H on the Cu(111) surface with negligible H_2O involvement. Furthermore, CO reduction to COH or CHO intermediate is the key selectivity step for production of methane/ethylene over methanol. Methane formation goes through a COH intermediate, which eventually reduces to CH_x species that can produce both methane and ethylene as observed experimentally. We will also present ongoing work on the impact of additional solvation, other copper facets and step edges, coverage effects, as well as exploring other potential materials for CO_2 electrocatalytic reduction.

(472e) First Principles Insight Into CO2 Hydrogenation to Methanol On Ceria Catalysts

Cheng, Z., Washington University in St. Louis
Lo, C. S., Washington University in St. Louis
Originally presented on: 11/6/2013 13:50:00 - 14:10:00

CO_2 catalytic hydrogenation to Methanol is of considerable importance in the energy industries and greenhouse gas mitigation. Although extensive experimental and theoretical efforts have been carried out based on metal catalysts such as Cu and Pd in the past decades, the metal oxide catalysts have not been well studied yet. The most fundamental questions on CO_2 hydrogenation on metal oxide, such as the reaction mechanisms, the key reaction intermediates and rate-determining steps are still in debate. Ceria (CeO_2) is commonly used for oxidation and reduction reactions, since Ce ably and reversibly converts between Ce4+ and Ce3+ upon release and storage of oxygen. Our previous study has shown that CO_2 molecule can be activated at oxygen vacancies of reduced ceria (110) by transferring a net charge of $-0.923|e|$, following which the surface is partially re-oxidized. Thus, reduced ceria (110) represents a promising catalyst for CO_2 hydrogenation. In this present work, a comprehensive reaction network for CO_2 hydrogenation to methanol on ceria (110) is studied using periodic density functional theory calculations and first principles kinetic Monte Carlo simulation. In all possible intermediates, our calculation results show that carbine diol, formic acid and methynol are not feasible due to the high formation energy. Also, direct formyl hydrogenation to formaldehyde (H_2CO), the key intermediate for methanol synthesis, is not feasible due to the high activation barrier. Instead, we find that H-formalin to formaldehyde is kinetically more favorable. The formaldehyde then converts into methoxy (H_3CO) rather than H_2COH, followed by a consecutive hydrogenation step to form methanol. Thus, two reaction channels to methanol are identified (i) COOH pathway via a carboxyl intermediate and (ii) HCOO pathway via a formate intermediate. The rate-limiting step of each pathway and the theoretical selectivity is determined. This study of CO_2 hydrogenation to methanol on Ceria (110) provides new insights into methanol synthesis chemistry and ceria catalyst application.

(472f) Activity Descriptors for CO2 Reduction On Transition Metal Surfaces

Shi, C., Stanford University
Hansen, H. A., Stanford University
Norskov, J. K., Technical University of Denmark
Originally presented on: 11/6/2013 14:10:00 - 14:30:00

The reversible electrochemical reduction of CO_2 into fuels would allow for the sustainable use of hydrocarbon fuels and production of commodity chemicals all while maintaining current fossil fuel-based infrastructure. A great deal of work has been done both experimentally and (in the last decade) theoretically on elucidating the mechanism of CO_2 reduction on transition metal surfaces. So far, however, the best catalyst, both heterogeneous and homogenous, found to reduce CO_2 to hydrocarbons such as methane and ethylene is metallic copper and why this is the case has remained somewhat of a mystery until recently. Unfortunately, copper performs this process at a reasonable (> 0.2 mA/cm^2) current efficiency only at extremely high overpotentials (> 1V).

The search for a better catalyst through usage of the Computational Hydrogen Electrode (CHE) using DFT calculations in previous works has shown that the reason why Cu is so good at reducing CO_2 to methane is that it sits at the top of the CO_2 reduction limiting potential volcano when the (211) stepped Cu surface is compared to the (211) surfaces of other transition metals primarily due to the slope of the *CO to *CHO step. Our work extends this work on limiting potentials through the CHE model to first include the effects of higher *CO coverage on the strong-CO binding metal surfaces, Ni, Pt, Pd, and Rh, where we expect steady state *CO coverage in electrochemical reducing conditions to be significant. Next, we also include limiting potential volcanoes for the (111) and (100) surfaces of FCC transition metals. There, we find that for the strong-CO binding metals on the terrace surfaces it is the *CO to *COH step that is potential limiting, and that this limiting potential is actually lower than that of *CO to *CHO on Cu(211). Finally, we compare the activity the hydrogen evolution reaction (HER) to CO_2 reduction to methane on those surfaces, and find that all of the strong-CO binding surfaces are much better relative hydrogen evolution catalysts than the weak-CO binding surfaces, Cu, Ag, and Au, which helps explains their experimentally observed low activity for CO_2 reduction.

(472g) Mechanistic Studies of the Catalytic Reduction of CO2 to Methanol

Lim, C. H., University of Colorado Boulder
Holder, A., University of Colorado at Boulder
Musgrave, C. B., University of Colorado, Boulder
Originally presented on: 11/6/2013 14:30:00 - 14:50:00

We employ quantum chemical calculations to investigate the mechanism of homogeneous CO_2 reduction by pyridine (Py) in the Py/p-GaP system. We find that CO_2 reduction by Py commences with $PyCOOH^0$ formation where: a) protonated Py (PyH^+) is reduced to PyH^0, b) PyH^0 then reduces CO_2 by one electron transfer (ET) via nucleophilic attack by its N lone pair on the C of CO_2 and finally c) proton transfer (PT) from PyH^0 to CO_2 produces $PyCOOH^0$. The predicted enthalpic barrier for this proton coupled ET (PCET) reaction is 45.7 kcal/mol for direct PT from PyH^0 to CO_2. However, when PT is mediated by one to three water molecules acting as a proton relay the barrier decreases to 29.5, 20.4 and 18.5 kcal/mol, respectively. The water proton relay reduces strain in the transition state (TS) and facilitates more complete ET. For PT mediated by a three water molecule proton relay, adding water molecules to explicitly solvate the core reaction system reduces the barrier to 13.6 - 16.5 kcal/mol, depending on the number and configuration of the solvating waters. This agrees with the experimentally determined barrier of 16.5 ± 2.4 kcal/mol. We calculate a pKa for PyH^0 of 31 indicating that PT preceding ET is highly unfavorable. Moreover, we demonstrate that ET precedes PT in $PyCOOH^0$ formation, confirming PyH^0's pKa as irrelevant for predicting PT from PyH^0 to CO_2. Furthermore, we calculate adiabatic electron affinities in aqueous solvent for CO_2, Py and Py•CO_2 of 47.4, 37.9, 66.3 kcal/mol respectively, indicating that the anionic complex $PyCOO^-$ stabilizes the anionic radicals CO_2^- and Py^- to facilitate low barrier ET. As the reduction of CO_2 proceeds through ET and then PT, the pyridine ring becomes aromatic and thus Py catalyzes CO_2 reduction by stabilizing the PCET TS and the $PyCOOH^0$ product through aromatic resonance stabilization. Our results suggest that Py catalyzes the homogeneous reductions of formic acid, and formaldehyde en route to formation of CH_3OH through a series of one-electron reductions analogous to the PCET reduction of CO_2 examined here, where the electrode only acts to reduce PyH^+ to PyH^0.

(474a) A Molecular Dynamics Simulation Study On Active Diffusiophoresis of Swimming Colloids

Sharifi-Mood, N., Department of Chemical Engineering, Levich institute
Koplik, J., City College of the City University of New York
Maldarelli, C., City College of New York
Originally presented on: 11/6/2013 12:30:00 - 12:45:00

Microbots and nanobots are locomoting objects which are designed to transverse liquids in a programmed way along small scale landscapes in highly imaginative applications such as the targeted delivery of drugs to individual cells, roving sensors for chemical detection or deployed agents for capturing targeted molecules, engines for micropumping, shuttles for moving cargo through microfluidic cells, and transporters for directly assembling supramolecular structures from molecular building parts. Phoretic mechanism has long been considered as a top-down approach to accomplish colloid locomotion, however the attendant miniaturizations in the force fields through the small scale venues envisioned for these locomotors are challenging and therefore the idea of self-propulsion has been thrived as an alternative approach for autonomous motion of colloids. We study theoretically one potential transduction mechanism, diffusiophoresis, which derives from unbalanced van der Waals forces which are exerted on the molecules of the colloid by low molecular weight solute molecules (less than one nanometer in diameter) distributed asymmetrically around the colloid. The van der Waals interaction of a solute molecule in solution near a colloid with all the molecules comprising the colloid is confined to an interaction zone around the particle with thickness L of the order of $1-10$ nm (a few to several molecular diameters). Within this zone, the concentration gradient required for the colloid motion can be generated and sustained by an irriversible surface chemical reaction of solutes only on one face of the colloid. We utilized the molecular dynamics (MD) simulation to investigate the colloid self-propulsion due to a surface chemical reaction. We focus on a simple system in which only three species (solvent, solute and molecules belong to colloid) exist in our simulation and the pair interactions were depicted via modulated Lennard-Jones interaction potential. We furthermore assigned a greater strength of attractive interaction for the particular pair (solute vs. colloid molecules) to actuate the motion which result in an inward flux of solute into the colloid and ultimately a specifically adsorbed layer of solutes due to a balance of attractive vs repulsive forces. Within this layer, we adopt the reaction on the active cites simply by converting solute to solvent. The colloid trajectories clearly demonstrated the enhanced motility of the swimming colloid compared to the controlled ones (colloid in the solvent bath undergoes only brownian motion) and with the aid of MD simulation results, we quantified the force as well as the directed velocity of the swimming colloid at nanoscale and lastly compared them with the continuum description.

(474b) Structure and Dynamics of Swimming Microorganisms in External Fluid Flows

Chilukuri, S., Rensselaer Polytechnic Institute
Collins, C. H., Rensselaer Polytechnic Institute
Underhill, P. T., Rensselaer Polytechnic Institute
Originally presented on: 11/6/2013 12:45:00 - 13:00:00

Suspensions of swimming microorganisms are one example of active colloidal systems. The activity keeps the system always far from equilibrium and leads to unique structure and dynamics. Many researchers have studied how interactions between active colloids alter the collective structures formed. Researchers have also examined interactions with confining walls that leads to a non-uniform density of cells (they concentrate near the walls). In this talk we will discuss computer simulations of these systems in external fluid flows. The simulations use the Brownian dynamics methodology including hydrodynamic interactions of the organisms with the walls.

We find that the external flow dramatically alters the distribution of the organisms across the channel. We examine the role of the strength of the external flow and the flow type (Couette versus Poiseuille) which alters the cell distribution. The change in distribution can be understood using the orientation distribution of the cells, and how the flow alters that distribution. We will also discuss the dynamics of the cells, and how those dynamics could impact other processes such as biofilm formation.

(474c) Margination and Segregation of Self-Propelled Particles in Blood Flow

Kumar, A., University of Wisconsin-Madison
Graham, M. D., University of Wisconsin-Madison
Originally presented on: 11/6/2013 13:00:00 - 13:15:00

Self-propelled particles like swimming bacteria and microbots have been envisioned for various applications in the human body including drug delivery and minimally invasive surgery. The efficacy of these particles in their respective applications will depend critically on their flow behavior in blood vessels. A property of particular interest will the distribution of the particles in the wall normal direction, i.e., whether the particles flow close to the walls (marginate) or close to the vessel center (antimarginate). In this work, we numerically examine the flow behavior of self-propelled particles in a suspension of deformable capsules, the latter being a model for red-blood-cells. This binary suspension is subjected to simple shear flow in a slit geometry, i.e., between two parallel walls. The effect of a variety of parameters is characterized including the particle shape, aspect ratio, size, swimming speed and the mode of propulsion, i.e., whether the particles are pushers or pullers. Our studies indicate that dead swimmers (zero swimming speed) accumulate in the near wall region at the edge of the capsule depleted layer, i.e., the swimmers marginate. At sufficiently high volume fractions of capsules, both the pushers and pullers demarginate, i.e., they escape from the near wall region to the channel interior. The degree of demargination increases with increasing swimming speeds. In contrast, at low capsule volume fractions, the pushers are found to flow close to the wall irrespective of their swimming speed, while the pullers demarginate with increasing swimming speeds, just as in the case of high volume fraction suspensions. The demargination of particles is found to occur due to their collision with capsules, which may instantaneously orient them in the wall normal direction, thereby leading to their escape to the interior due to their self-propulsion provided sufficient clearance is present. The difference in demargination behavior of pushers and pullers at low volume fractions arise because the pushers tend to get attracted towards the wall, while pullers tend to get pushed away from the wall. This, coupled with the fact that at low volume fractions the capsule depleted layer thickness can be larger than the particle size, leads to the aforementioned difference in the behavior of pushers and pullers. Additional insights on the effect of size and shape of the particles will also be provided. The results discussed in this study is expected to provide useful guidelines for the design of self-propelled particles for drug delivery and other applications in blood flow.

(474d) Aggregation in a Suspension of Self-Propelled Magnetic Colloidal Particles Studied Via Brownian Dynamics Simulations

Rivera, L. Y.,
C rdova-Figueroa, U. M., University Of Puerto Rico - Mayag ez
Originally presented on: 11/6/2013 13:15:00 - 13:30:00

Ferrocolloids interacting via dipole-dipole interactions subject to an external magnetic field form aggregates of various sizes at different time scales. Inserting self-propulsion ability to these particles promises significant benefits in the development of novel materials. It is well understood that by enhancing diffusion, in this case by self-propulsion, the time scale for particle interactions is reduced to an extent of forming aggregates much faster. It is unclear, however, the effect of self-propulsion of these particles, particularly at very high propulsion speeds, to cluster size formation in relation to the applied magnetic field and interparticle interactions. In this work, we propose a simple colloidal model consisting of a monodisperse suspension of self-propelled magnetic particles immersed in a Newtonian fluid subject to a constant external magnetic field. Using Brownian dynamics simulation, we interrogate the suspension over time by measuring cluster properties, such as mean squared radius of gyration, mean cluster mass, number of clusters and cluster diffusivities. All these properties are then compared for various particle speeds, magnetic field strengths, and dipolar coupling constants.

(474e) Brownian Dynamic Simulation of Soft Magnetic Ellipsoidal Mesoparticle Suspensions

Torres-Diaz, I., University of Puerto Rico at Mayaguez
Rinaldi, C., University of Puerto Rico
Originally presented on: 11/6/2013 13:30:00 - 13:45:00

Suspensions of sub-micron magnetic mesoparticles composed of soft magnetic materials are recently being considered for various applications, such as in cancer treatment. In these suspensions the static and dynamic behavior of the particles depends on both the magnitude of the applied magnetic field and the relative orientation of the particles with the applied magnetic field, in addition to hydrodynamic and thermal effects. Here, the rotational motion of magnetically-soft tri-axial ellipsoidal particles suspended in a Newtonian fluid was studied using rotational Brownian dynamics simulations. The approach considers the stochastic angular momentum equation (Langevin equation) in an orientation space described by quaternion parameters. The algorithm quantifies the magnetization of a monodisperse particle suspension in dilute limit conditions under applied steady and time-varying magnetic fields. The algorithm also takes into account the field-dependent relative permeability of the particle's bulk material. We have considered the equilibrium magnetization at different dimensionless field amplitudes and dynamic susceptibility in response to an oscillating magnetic field at different dimensionless frequencies for ellipsoidal mesoparticle suspensions with different geometric aspect ratios and different particle sizes.

(474f) On Aggregation Phenomena in a Disk-Like Hematite Particle Suspension in a Simple Shear Flow By Means of Brownian Dynamics Method

Satoh, A., Akita Prefectural University

Originally presented on: 11/6/2013 13:45:00 - 14:00:00

In addition to the magnetic spherical particles, magnetic particles with a variety of shapes and magnetic properties can be synthesized in the form of rod-like, disk-like and cubic-like particles [1-3]. Magnetic suspensions of these particles may be expected to be developed as a functional fluid that exhibits functional characteristics that will be useful for application in specific situations. Representative possible applications are high-density recording materials [4], optical units [5], actuators and dampers [6] and surface modifying technology [7].

In contrast to ordinary ferromagnetic particles, disk-like hematite particles have a characteristic feature in that it is magnetized in the oblate surface direction, normal to the particle axis direction, and it exhibits a weaker magnetization in comparison with the magnetite particles. Hence, characteristic features of orientational distribution and rheology are expected to be significantly different from those of the ferromagnetic spherical or rod-like particles. This has stimulated the motivation for the present study by means of the Brownian dynamics simulations [8,9] that takes into account the spin Brownian motion about the particle axis in addition to the ordinary rotational Brownian motion about a line normal to the particle axis.

From this background, in the present study, we have investigated aggregation phenomena in a suspension composed of disk-like hematite particles by means of Brownian dynamics simulations. The magnetic moment of the hematite particles lies normal to the particle axis direction, and therefore the present Brownian dynamics method takes into account the spin Brownian motion about the particle axis in addition to the ordinary translational and rotational Brownian motion. These particles are assumed to conduct translational and rotational Brownian motion in a simple shear flow with addition of a uniform magnetic field applied in the direction normal to the shearing plane. First, the validity of the present simulation method has been clarified by comparing with results obtained by the fully-established Monte Carlo method. Then, from performing the simulations, we have investigated the influence of the magnetic particle-field and the particle-particle interactions, shear rate and the volumetric fraction of particles on particle aggregation phenomena. Snapshots of aggregate structures are used for a qualitative discussion, and the cluster size distribution, radial distribution function and the orientational correlation functions of the direction of particle axis and magnetic moment are the focus for a quantitative discussion. The main results are summarized as follows. The significant column-like clusters are formed at a magnetic particle-particle interaction much larger than in the case of a magnetic spherical particle suspension. This is because the rotational Brownian motion has a significant influence on the formation of clusters for a suspension composed of disk-like particles with large aspect ratio. An applied magnetic field is found to decrease the formation of column-like clusters. A shear flow is found not to have a significant influence on the internal structures of clusters, but it does influence the cluster size distribution of the column-like clusters.

[1] E. Matijevic and P. Scheiner, J. Colloid Interface Sci. **63**, 509 (1978).

[2] M. Ozaki, S. Kratohvil and E. Matijevic, J. Colloid Interface Sci. **102**, 146 (1984).

[3] M. Ozaki, M. Senna, M. Koishi and H. Honda, *Particles of specific functions, in Fine Particles* (edited by T. Sugimoto), (pp. 662-682, Marcel Dekker, 2000).

[4] J.W. Harrell, S. Kang, Z. Jia, D.E. Nikles, R.W. Chantrell, and A. Satoh, Appl. Phys. Lett. **87**, 202508 (2005).

[5] S. Furumi, Y. Sakka and J. Nanosci, J. Nanosci. Nanotech. **6**, 1819 (2006).

[6] W.A. Bullough (Ed.), *Electro-Rheological Fluids, Magneto-Rheological Suspensions and Associated Technology* (World Scientific, Singapore, 1996).

[7] A. Satoh and Y. Sakuda, Molec. Phys. **108**, 2105 (2010).

[8] A. Satoh, *Introduction to Practice of Molecular Simulation: Molecular Dynamics, Monte Carlo, Brownian Dynamics, Lattice Boltzmann and Dissipative Particle Dynamics* (Elsevier Insights, 2010).

[9] A. Satoh, *Introduction to Molecular-Microsimulation of Colloidal Dispersions* (Elsevier, Amsterdam, 2003).

(474g) Hydrodynamic Behavior of Length Sorted Carbon Nanotubes

Silvera Batista, C., National Institute of Standards and Technology
Khripin, C., National Institute of Standards and Technology
Zheng, M., National Institute of Standards and Technology
Fagan, J. A., National Institute of Standards and Technology
Originally presented on: 11/6/2013 14:00:00 - 14:15:00

Anisotropic nanoparticles such as single-walled carbon nanotubes (SWCNTs) can have exceptional physical properties, and there is much interest in developing them through processing techniques while in aqueous dispersion. However, current understanding of the particle behavior in solution is not sufficient to enable rigorous measurement of critical aspects of a dispersion such as the particle size distribution. Our goal is to validate the use of hydrodynamic characterization for rod-like nanoparticles such as SWCNTs that have very large ranges of aspect ratios, but constant diameters.

Theoretical approximations that describe the hydrodynamic effects of shape have been developed since the 1930s for 1-D non-spherical particles such as ellipsoids or cylinders. Experimental verification of such approximations, however, has lagged the theoretical description due to a lack of appropriate particle systems both small enough to satisfy assumptions for the flow fields and with sufficient variability in aspect ratio. In general, experimental works that have evaluated the friction coefficient of rodlike particles have been limited to low aspect ratios due to these constraints, whether for variable particles such as gold nanorods or DNA, or monomodal rod-like particles such as TMV or Fd virus.

In this work, SWCNTs were used as model rod-like particles to experimentally validate expressions for the friction coefficients that are the result of the slender body theory or numerical methods. SWCNTs are an ideal model particle for this purpose. Different length populations of constant buoyancy SWCNTs were generated via centrifugation sorting followed by size exclusion chromatography. A total of 19 populations of SWCNTs with average aspect ratios varying from approximately 30 to 400 were generated from two different diameter SWCNT samples. Characterization of the SWCNT sedimentation in an analytical ultracentrifuge was used to compare to the hydrodynamic theory. The experimental data agrees well with predictions from numerical methods such as in [1] and slender body theory as reported by Batchelor [2]. The theory was then used to invert the sedimentation coefficient distributions for comparison to independently measured distributions via atomic force microscopy. Results of the measurements, comparison to the theoretical predictions and implications will be presented.

[1] M. L. Mansfield and J. F. Douglas, Macromolecules 41, 5422 (2008).

[2] G. K. Batchelor, J. Fluid Mech. 44, 419 (1970).

(474h) Dispersion of a Suspension Plug in a Periodic Shear Flow

Azadi, G., Brown University
Maxey, M., Brown University
Tripathi, A., Brown University
Originally presented on: 11/6/2013 14:15:00 - 14:30:00

The dispersion of a suspension plug, as a non-fully populated system has not previously been studied experimentally. We investigate the shear-induced dispersion of a suspension plug in an oscillating shear flow at low Reynolds number in a circular capillary. Paramagnetic particles with d= 100 μm are used to form a plug inside a capillary filled with particle-free fluid with ID= 1.16 mm, using magnets. The variables studied are volume fraction, amplitude and number of oscillations. We demonstrate that the irreversibility of the suspension system is a function of strain amplitude and volume fraction. Experimental results, including the dispersion coefficient are compared with the theoretical results of simulation studies.

(474i) Simulation of Macromolecules Immersed in Stretching Flow By Dissipative Particle Dynamics; Interpretation of Response Modes By Proper Orthogonal Decomposition

Caswell, B., Brown University
Deng, M., Brown University
Grinberg, L., Brown University
Karniadakis, G. E., Brown University
Originally presented on: 11/6/2013 14:30:00 - 14:45:00

We employ Dissipative Particle Dynamics (DPD) to create a stretching flow in a domain with periodic boundary conditions. The flow features a stagnation point flow at the center of the simulation box. Test bodies, such as macromolecules, colloids, cells or vesicles, are initially placed at the stagnation point, and time records of their kinematics are recorded. Post processing is carried out with Proper Orthogonal Decomposition (POD), a tool commonly employed in the analysis of simulated turbulence, and adapted for particle systems (L. Grinberg, Proper orthogonal decomposition of atomistic flow simulations, Journal of Computational Physics 231 (2012), 5542). In this work POD is employed to separate the basic modes of deformation of the test body from the thermal fluctuations. The probability distribution function (PDF) of the fluctuations in the vicinity of the stagnation point is obtained for both the stretching and compressing regions of the base flow. The method will be illustrated with the modes of an inextensible filament, modeled as a bead-spring chain, over the whole range of bending resistance.

(474j) Predicting the Yield Stress of Solvent-Free Nanoparticle Fluids Resulting From Configurational Entropy of Tethered Space-Filling Oligomers

Roy, A., Cornell University
Koch, D. L., Cornell University
Originally presented on: 11/6/2013 14:45:00 - 15:00:00

The yield stress is determined for a solvent-free nanoparticle fluid system known as nanoparticle-organic hybrid materials (NOHMs). These materials consist of inorganic spherical nanoparticle cores (with diameters of about 10 nm) with dense brushes of tethered oligomers and no unattached molecular solvent. Despite their solvent-free nature, these materials have been observed to flow when subject to stresses exceeding a yield stress. In this paper we will derive the yield stress resulting from the oligomer mediated many-core interactions. The oligomers are modeled as bead-soft-springs tethered to the surfaces of cores which form an incompressible liquid filling the interstices between the cores. The configuration of the oligomers for a given core configuration is determined by minimizing the configurational free energy of the oligomers subject to the constraint that the number of oligomer beads per unit volume be nearly constant. For simplicity we consider a periodic array of cores with an initial face centered cubic configuration which is then subjected to affine core motion in a simple shearing motion. The yield stress can be determined as the maximum slope of a plot of the free energy per volume with respect to strain. The predicted yield stress is high on the order many kPa and is intermediate between nk_BT and nMk_BT where n is the number of cores per unit volume and M is the number of oligomers per core and is thus intermediate between the entropic free energy of the particles and that of the oligomers. The theory predicts that the yield stress decreases with increasing core volume fraction and with increasing molecular weight. These surprising trends arise because oligomers can more easily fill the interstitial space when the interparticle spacing is smaller and when oligomer's radius of gyration is larger. The predicted yield stress is compared with estimates based on oscillatory shear rheology by Agarwal et al. (Nano Lett. 2010, 10, 111-115) for core volume fraction of 0.111 and oligomer radii of gyration of 2.71 and 3.63 nm.

(477a) Remediation of Diesel-Contaminated Soil With a Biostimulation and Bioaugmentation Joint Process

Li, X.,
Lu, D., Tsinghua University
Zheng, L., Tsinghua University
Originally presented on: 11/6/2013 12:30:00 - 12:44:00

Remediation of diesel-contaminated soil with a biostimulation and bioaugmentation joint process

Xiaosen Li, Diannan Lu and Zheng Liu*

Department of Chemical Engineering, Tsinghua University, Beijing 100084, China

lixs09@mails.tsinghua.edu.cnludiannan@mail.tsinghua.edu.cn

*Corresponding author: Phone: 86-10-6277-9876

Email: liuzheng@mail.tsinghua.edu.cn

A new remediation process that integrates biostimulation and bioaugmentation was proposed for diesel-contaminated soil. In addition to the determination of the residue diesel in soil, the total dehydrogenase activity, the microbial community changes in terms of composition and quantity of both microbial community and functional genes were probed by molecular biological techniques. In this research, five bacterial strains capable of degrading diesel were isolated from contaminated sites, in which Rem-E gave the highest degradation efficiency compared to the others and thus used as the additive strain for the bioaugmentation. As for biostimulation, monopotassium phosphate and ammonium sulfate were added to achieve an optimal C:N:P being determined as 100:1.25:1. After 10 day biosimulation, the content of diesel reduced from 2.7% to 1.6%. The joint process by biostimulation and bioaugmentation, however, gave a reduction of disel content from 2.7% to 0.6% in 40 days. Pyrosequencing for the total DNA extracted from the soil showed the evolution of the microbial community and qPCR revealed the change of the functional genes during the remediation process. The result revealed that coupling degrading strain and source of phosphorus and nitrogen led to the highest dehydrogenase activity and degradation rate, which also recover almost all the microbial community in the soil after the remediation. This work showed the great potential of the coupling technology composed of bioaugmentation and biostimulation for the remediation of the oil-contaminated soil. It also offers a viable route for the application of molecular biological techniques as enabling tools in the monitor, design, implementation and assessment of bioremediation practice.

(477b) Electrically Enhanced Contaminant Degradation in Sediment Caps

Zhang, R., Tianjin University of Technology

Originally presented on: 11/6/2013 12:44:00 - 12:58:00

In-situ capping is known as an attractive and cost-effective method for remediation of contaminated sediments. Sediment caps with bioremediation will ultimately significantly improve it performance, but reports about methods to enhance contaminant degradation in sediment caps still very little. Electrically enhanced bioactive barrier □BIOLANCE□is a new process for in situ remediation reducible compounds in soil or sediments. Chlorinated benzenes are significant contaminants to environment.

Many chlorinated organic compounds are persistent organic pollutants due to their resistance to biodegradation and chemical degradation. The objective of this study was to determine if electrodes placed in sediment and poised at several volts could create a redox gradient and provide electron acceptor/donor to stimulate chlorinated contaminant degradation, which will prepare for sediment caps application. The results demonstrated that choose graphite as electrodes can lead to sustainable evolution of hydrogen. Hydrogen evolution displayed zero-order kinetics with the time in the initial stages in different voltage. Even high voltage can produce more hydrogen, but it can not keep longer time because the overpotentials on electrode surfaces will prevent its function. In our experiments, 4V is the best choice. At this condition, the redox and pH change were detected. It is possible that reductive dechlorination could have been achieved with lower redox potentials in the presence of enough hydrogen electron donors through electrodes. Just while sulfate reducing or the other favorable electron acceptor may have reducing priority than chlorinated compounds. The experiments of T-cell capping biodechlorination use electrodes will be further studied.

(477c) pH-Dependent Transport of Energy By-Products in the Subsurface

Prigiobbe, V., University of Texas at Austin
Hesse, M., University of Texas at Austin
Bryant, S., University of Texas at Austin
Originally presented on: 11/6/2013 12:58:00 - 13:12:00

pH-dependent transport of energy by-products in the subsurface

Valentina Prigiobbe[a], Marc A. Hesse[b,c], and Steven L. Bryant[a,c]
[a]*Department of Petroleum and Geosystems Engineering*, [b]*Department of Geological Sciences*, [c]*Institute for Computational Engineering and Sciences, The University of Texas at Austin, Austin, Texas, U.S.A.*

Abstract

The transport of solutes through reactive porous media can be enhanced by the e ffect of pH-dependent adsorption/desorption processes. This eff ect can be particularly critical when the solutes, such as toxic metals and radionuclides, have hazardous impact on the environment and the health.
To study the e ffect of pH-dependent adsorption/desorption on the toxic metal and radionuclide transport, we have developed an analytical and a numerical model coupling transport and surface complexation for a single-phase incompressible fluid. We have analyzed a one-dimensional (1D) system of conservation laws to account of the pH-dependence and longitudinal dispersion and to gain a physical intuition of the phenomenon as well as a 2D system to include the eff ect of transverse dispersion and to investigate more realistic scenarios.
An important feature of these reactive transport problems is that one of the conservation laws is for the total hydrogen ($c_{ht} = c_h - c_{oh}$), also called acidity, which gives rise to an additional nonlinearity to the systems distinguishing them from other competitive adsorption/desorption problems where hydrogen is not involved.

Here, we present the results for two cases. A first one of pH-dependent transport of naturally-occurring toxic metals in shallow aquifers mobilized through a brine acidi

ed by carbon dioxide (CO_2) leaking from deep geological storage [1,2]. A second case of pH-dependent transport of adsorbing radionuclides accidentally discharged into the subsurface from temporally storage pools [3,4]. In both cases, we have analyzed the transport through a reactive porous medium composed of hydrophilic iron-oxide.
In the fi

rst case, the 1D solution of the intrusion of a CO_2-acidi

fied brine into a shallow aquifer has two retarded fronts separated by a spike of concentration of desorbed metal with a maximum much larger than the initial value. This is a classical behavior of pH-dependent adsorption, which occurs when the adsorption of the proton is more favorable than of the metal. In a 2D system, the structure of the plume resembles the 1D solution which, migrating into the aquifer with a maximum concentration much larger than the initial one, may present a health hazard.
In the second case, the 1D solution of the injection of an alkaline brine containing radionuclides into an aquifer consists of a retarded classical front and an unretarded non-classical pulse traveling at the average fluid velocity. The suitable initial and boundary conditions associated with the interaction of pH-dependent adsorption and the longitudinal dispersion are responsible for this phenomenon. In a 2D system in addition to the retarded front and the fast pulse, a fast spreading plume forms due to mixing induced by transverse dispersion. The unretarded pulse is a new transport phenomenon due to the interaction of pH-dependent adsorption and hydrodynamic dispersion, it leads to the migration of a contaminant with a velocity much faster than expected and it should be taken into account in the design of contaminated site remediation.

References

[1] Apps, J.A., Zheng, L., Zhang, Y., Xu, T., Birkholzer, J.T. (2010), Evaluation of potential changes in groundwater quality in response to CO_2 leakage from deep geologic storage. Transp Porous Med 82, 215-246.
[2] Kharaka, Y.K., Cole, D.R., Hovorka, S.D., Gunter, W.D., Knauss, K.G., and Freifeld, B.M. (2010), Gas-water-rock interactions in Frio formation following CO_2 injection: Implications for the storage of greenhouse gases in sedimentary basins. Geology 34, 577-580.
[3] Prigiobbe, V., Hesse, M. A., and Bryant, S. L. (2012a), Fast strontium transport induced by hydrodynamic dispersion and pH-dependent sorption. Geophys Res Lett 39 L18401.
[4] Prigiobbe, V., Hesse, M. A., and Bryant, S. L. (2012b), Anomalous reactive transport in the framework of the theory of chromatography. Transp Porous Med 91, 127-145.

(477d) Experimental Investigation Using a Contaminant-Monitoring Sytem in Storm Drains: A Dichotomous Preliminary Risk Assessment

Lackey, D., University of Alabama at Birmingham
Peters, R. W., University of Alabama at Birmingham
Originally presented on: 11/6/2013 13:12:00 - 13:26:00

As humans have built cities using concrete, asphalt, and large areas of impervious materials, they have potentially disturbed the normal hydrologic circle, and have learned that they must build storm drains to convey excess rainwater and prevent flooding. But these storm drains have a drawback in that they can also lead to pollution of natural environments if they carry contaminants such as antifreeze, fertilizers, paints, used motor oil, tossed pet waste, cigarette butts, and litter. Everything other than pure water is a potential contaminant that degrades water quality. Pollution sources entering storm drains include: automobile emissions, residential wastes, agricultural runoff, urban activities, construction runoff, and businesses [Hyperspective Studios, 2002].

In the summer of 2010, in conjunction with the UAB Occupational Health and Safety (OHS) Department, Dr. Robert W. Peters, and two of his graduate students conducted a project involving a risk assessment of contamination of runoff water flowing into storm drains on the UAB campus [Lackey *et al.*, 2010]. The project was successful enough to spin off into further studies but it had one key shortcoming: it considered only the effects of point sources on storm drains and largely ignored the effects of non-point sources. The details of that study, as well as the steps that will be taken to address that shortcoming, and what further studies the results will lead to, are listed below.

To begin the study, the UAB OH&S Department gave the civil engineering graduate research assistants a series of overhead maps of the UAB campus that have marked storm drain inlet locations from a previous study. The study took place over two main strategic areas on the UAB campus, one that contained the majority of storm drains on the western academic side of campus, and the other that contained the majority of the storm drains on the eastern side of campus which is more focused on medical research. The students were helped by the inlet locations provided on the maps by O&HS, but there were several differences between the locations given on each map and the actual locations of the inlets. Therefore, the project involved an update on the actual locations of each inlet as well as a risk assessment of the likelihood of dumping unwanted chemicals.

In the study, the students identified four main types of inlets:

1. Standard Covered Curbside: The type that contain a concrete or iron cover and engulf large volumes of water from adjacent roadways and parking lots during a rainstorm.
2. Uncovered Grate: These are often found in the middle of parking lots or on pedestrian

walkways. Occasionally, these are even found in grassy areas where irrigation is common.

3. Combination Curbside Grate: Essentially a combination of Types A and B, these are found in the same areas as Type A.
4. Elongated Grate: Found in the same types of areas as Type B, and extremely common at entrances of parking decks.

In addition to the types, each inlet was also assigned a risk assessment value of 0 to 3 based on the priority of its labeling. A color was assigned to each as well for easier readability:

0 (blue) – Marker already present on drain inlet

1 (green) – Low dumping risk

2 (yellow) – Moderate dumping risk

3 (red) – High dumping risk

Some of the criteria for determination of the risk as 1, 2, or 3 are as follows:

1. Default condition; generally assigned unless the inlet was near a probable pollution source such as a trash receptacle, dumpster, laboratory entrance, or parking space.
2. Assigned if the inlet was near a potential moderate pollution source, such as a laboratory entrance or trash can, or if the inlet was on or adjacent to a parking space.
3. Assigned if the inlet was next to a dumpster or other probable major pollution source, or two conditions or more that would result in "2" applied.

The major results of the study, in terms of number of inlets of each type and risk factor, are displayed below in Table 1 and Table 2.

Table 1. Summary of type and risk factor of inlets in Strategic Area 1 (Main Campus).

Area 1 – Main Campus

Risk Factor	**Type A**	**Type B**	**Type C**	**Type D**	**TOTAL**
Level 1	68	64	3	7	**142**
Level 2	11	20	1	0	**32**
Level 3	0	2	1	0	**3**
ALL	**79**	**86**	**5**	**7**	**177**

Table 2. Summary of type and risk factor of inlets in Strategic Area 2 (Medical Campus).

Area 2 – Medical Campus

Risk Factor	**Type A**	**Type B**	**Type C**	**Type D**	**TOTAL**
Level 1	112	49	0	20	**181**
Level 2	7	6	0	2	**15**

Level 3	1	4	0	3	**8**
ALL	**120**	**59**	**0**	**25**	**204**

In late 2012, the O&HS department decided that they not only want to install warning stencils on the UAB campus, but also sensors that would detect whenever harmful chemicals are present. The idea was expanded on to not only call for "artificial nose" chemical sensors but also a contaminant monitoring system that would map out the data from each sensor in real-time and send alerts to concerned parties whenever harmful chemicals are detected. Despite an initiative to cut costs as much as possible, the budget only calls for 50 of these "artifical noses". Therefore, priority needs to be assigned to the drains most in need of monitoring. When determining this, it turns out that anthropogenic point sources that result from dumping, while being the best criteria for whether or not to use a warning stencil, are not neccessarily the best criteria to decide whether a storm drain is a good candidate for a contaminant monitoring system, or at least not the only criteria. In order to complete the picture, non-point sources such as emissions from traffic must also be considered. O&HS has determined that 11th Avenue and University Boulevard are the roads on the UAB campus that have the highest traffic, and therefore the most at risk for contamination by traffic emissions. Some other adjacent roads have also been determined to have significant traffic.

Therefore, it would be best to expand the study to a two-phase, or dichotomous, risk assessment. The scale will be expanded from 1-5, and storm drains which are adjacent to high-traffic roads will have their risk factor raised by two, while storm drains which are adjacent to medium-traffic roads will have their risk factor raised by one. The new scale will become:

0: marked storm drain not on a high-or-medium-traffic road.

1: unmarked low point-source risk storm drain not on a high-or-medium-traffic road or marked storm drain on a medium-traffic road.

2: unmarked medium point-source risk storm drain not on a high-or-medium-traffic road, unmarked low point-source risk storm drain on a medium-traffic road, or marked storm drain on a high-traffic road.

3: unmarked high point-source risk storm drain not on a high-or-medium-traffic road, unmarked medium point-source risk storm drain on a medium-traffic road, or unmarked low point-source risk storm drain on a high-traffic road.

4: unmarked high point-source risk storm drain on a medium-traffic road, or unmarked medium point-source risk storm drain on a high-traffic road.

5: unmarked high point-source risk storm drain on a high-traffic road.

Storm drains which receive overall risk values of 5 will be the first to be monitored, followe.d by those with values of 4, and then those with values of 3, either until all 50 sensors have been placed or those of the 50 not meant to monitor "blank" base cases will be used. A "blank" would be one assigned a risk value of 0 or 1 that is used to determine if labeling or proximity to either point or non-point sources really does have an effect on contamination. And once the target drains are determined, the next phase of the project, which involves installing the sensors into the storm drains and beginning to analyze data generated by the contaminant-monitoring system, may proceed.

References Cited:

Hyperspective Studios, 2002. “Storm Drains”, website: http://protectingwater,com/storm-darins.html;

website accessed: October 16, 2012.

Lackey, D.A., A.N. Kajale, R.W. Peters, and J.D. Hagan, 2010. "The Effects of Labeling Storm Drains on the University of Alabama at Birmingham Campus", Poster paper presented at the Alabama Water Resources Conference, Orange Beach, Alabama (September 8-10).

(477e) Innovative Polymer Quantification and Management At a Large Scale Remediation Project

Grieco, S., O'Brien & Gere Engineers, Inc.
Redding, A.,
Lim, N.,
Originally presented on: 11/6/2013 13:26:00 - 13:40:00

Contaminated sediment remediation projects typically require a physical/chemical sediment pore-water treatment plant for metals and organics. In many cases, the sediments are thickened and dewatered for cost effective management and disposal. The polymer used for thickening can be problematic for downstream pore-water treatment. Specifically, this paper reports on a treatment system that experienced excessive "sticky" solids build-up in the clarifier and difficultly removing these solids through normal operations. The polymer also increased backwashing of Granular Activated Carbon (GAC), and decreased adsorption performance of Granular Activated Carbon (GAC).

In order to optimize thickening polymer management, this program identified the threshold of residual polymer that adversely impacted the pore-water treatment plant chemical precipitation & gravity settling unit operation. A mathematical relationship was established between a field-utilized China Clay Procedure and laboratory particle charge density (PSD) to establish an acceptable real-time simple procedure to validate the operating range and upper threshold for polymer dosing.

This program also quantified the impacts of residual polymer on the GAC. Equilibrium isotherms and rapid small scale column tests (RSSCTs) were conducted to confirm that the excess polymer was causing the observations in the full-scale GAC operations. Laboratory testing was used to quantify reduction of the bed volumes processed before both excess pressure drop and benzene breakthrough occurred. The program evaluated chemical types and dosages to dislodge and/or oxidize the excessive polymer that adhered to the carbon in the GAC beds.

The results of this testing were used in the field to optimize polymer dosing, which lowered operating costs and reduced system maintenance. The results also identified the impacts of GAC cleaning (both forward maintenance dosing as well as during backwash) on both pressure drop and adsorption capacity recovery.

(477f) Agile Multiple Water Network Designs for Soluble Contaminant Transport Identification and Site Remediation

Cai, T., Lamar University
Xu, Q., Lamar University
Originally presented on: 11/6/2013 13:40:00 - 13:54:00

Water network design (WND) is widely studied and employed in many industrial processes to minimize freshwater consumption and wastewater generation. Previous studies have been focusing almost exclusively on identifying one optimal WND for a studied process system. Actually, the quest for agile manufacturing by some multi-purpose industrial processes needs a comprehensive WND having the maximum operational flexibility to deal with various production requirements. Thus, the most desirable way for such agile operation is to couple various WNDs into one super design, and then through an effective control strategy to accomplish different production requests by switching among these WNDs. However, this has added the difficulty in contaminant transport identification once the network has been exposure to soluble hazards. In this paper, a novel methodology for the design of such a flexible super WND has been developed. It simultaneously considers the integration of multiple WNDs, network control switch based on different purposes and soluble contaminant transport identification in order to target timely site remediation. The objective of this optimization problem is a super WND coupled with all the recipe-based WNDs; the number of regular and three-way valves needed for the super WND; the optimal design of all of the valve placements in terms of the minimum total valve-switching times (TVST) to swap one pair of WNDs at a time; the control strategy (valve opening/closing strategy) to switch among WNDs in order to identify contaminant transport and reduce/terminate its further impact. The developed systematic methodology is general and thus applicable for many potential network integration problems. Case studies have demonstrated the efficacy of the development and its significance in the context of agile operation.

(477g) Modeling Electrokinetic-Soil Cleaning By Spatial Averaging

Tijaro-Rojas, R.,
Arce, P. E., Tennessee Technological University
Originally presented on: 11/6/2013 13:54:00 - 14:08:00

To predict the transport of ionic species into soils, such as in the acid mine drainages, a model will be implemented. An anisotropic porous medium and application of external forces as electric potential will be used in a multi-scale modeling approach. A minimum of key simplifications will be assumed, in order to capture the fundamental system behavior in model soils that mimic actual ones. Geometry of the control volume, medium properties, and intensity of electric potential constitute important variables to be considered. Since, the mass transport of the ionic species in soils takes place in a domain that displays different scales (molecular, microscopic, macroscopic and mesoscale), a multi-dimensional (spatial) averaging approach will be applied. Once, the spatial-averaged model become available, computational approaches will be used to illustrate the behavior of fundamental aspects. As a complementary part of this research, an experimental approach is being developed to guide and validate the predicted results.

(477h) Dynamics, Source Diagnosis, and Risk Assessment of Persistent Organic Pollutants in Water and Sediments of the Urban Aquatic Bodies

Kumar, R.,

Originally presented on: 11/6/2013 14:08:00 - 14:22:00

Abstract:

Organic compounds which are persistent, posses toxic characteristics, liable to bioaccumulate, prone to long range atmospheric transport and can cause adverse environmental and human health effects at locations near and far from their sources can be defined as persistent organic pollutants (POPs). Polycyclic aromatic hydrocarbons (PAHs) and organochlorine pesticides (OCPs) are the two important categories of POPs which are highly toxic to human beings, animals and plants. Polycyclic aromatic hydrocarbons are the product of any combustion process involving materials containing carbon and hydrogen namely coal, oil, petrol, wood etc. Emissions from the anthropogenic activities predominate, but some PAHs in the environment arise from natural combustion such as forest fires and volcanoes. PAHs occur naturally in crude oil and form a significant component of the petroleum products such as paints, creosote (used in wood preservation) and asphalt (used for road paving). There are some minor biogenic sources like plants, algae/phytoplankton and microorganisms and small amounts are formed by diagenesis in the lake sediments.

Organochlorine pesticides comprise a group of structurally diverse compounds with a large number of chlorine atoms. These pesticides have a tendency to bioaccumulate in the food chain and pose a great threat to the human health and the global environment. In spite of the ban on their use they are still being used in some of the developing countries, especially in the tropical regions, for health, agriculture and preservation purposes. Pesticides that are not bound in soils or taken up by the plants and animals can drain into rivers and lakes and move into the aquatic food chain.

The lakes and the rivers are two important types of surface water systems containing only 0.3% of the total global fresh water which is generally used for drinking purpose. These surface water resources, therefore, are precious and extremely sensitive and once degraded it would take immense efforts to revive them. In view of the above water and sediment samples were collected from two different types of urban surface water bodies of India. One of the water bodies is a lake system of Bhopal (Madhya Pradesh-a central state of India) and the other body is the stretch of river Yamuna at Delhi-the capital city of India. The aquatic systems under investigation are the typical urban aquatic systems and exist in the densely populated areas. Both of the systems are in different geographical locations. These surface water bodies being urban are expected to suffer with some similar types of anthropogenic activities. The various types of industrial, agricultural and domestic activities in the catchment area create different level of anthropogenic stress.

For the extraction of PAHs from water and sediment samples the method followed by Chen et al. [1] was adopted. The mobile phase was a gradient of acetonitrile and degassed water (50% acetonitrile held for 7 min; linear gradient to 80% acetonitrile in 7 to 20 min; 80% acetonitrile held till 25 min and linear gradient to 95% acetonitrile from 25 to 30 min) at a flow rate of 1.5 mL min^{-1}. The analysis was carried out at 254 nm and the individual PAH was identified and quantified by comparing the retention time and peak area with those of the standards. The procedure for extraction and cleanup of OCPs from both water and sediments is the same as outlined by Wan et al. [2]. The mobile phase was acetonitrile and water (70:30) with isocratic mode and a flow rate of 1.3 mL min^{-1}. The analysis of aldrin, endrin, pp'-DDT and α-endosulfan was performed at the wavelength 220, 228, 238 and 214 nm, respectively. The individual pesticide was identified and quantified on the basis of retention time and peak area. PAHs and OCPs were analyzed with HPLC equipped with a UV-detector (Agilent 1100, USA) and recorder. An octadecyl endcapped RP-C_{18} column (4.6 mm×250 mm, 5 μm particle size) was used.

The total PAHs (∑PAH) concentration in the water of the lakes and the river ranged from 4.85 to 18.2 ng/mL and 0.29 to 46.2 ng/mL, respectively. The profile of individual PAH in the water of both the aquatic systems is similar but different in sediments. The total PAHs concentration in sediments ranged from 952 to 5353 ng/g and 649 to 7910 ng/g for the lakes and river, respectively. The maximum pesticide concentrations in water and sediments of the lakes were 4.12 (aldrin), 2.26 (endrin), 6.17 (pp'-DDT), 5.98 ng/mL (α-endosulfan) and 167 (aldrin), 124 (endrin), 513 (pp'-DDT), 349 ng/g (α-endosulfan), respectively. The aldrin and endrin were absent at all the sites of the Yamuna. The maximum concentration of pp'-DDT and α-endosulfan in river water was 3.37 ng/mL and 3.55 ng/mL and in sediments 525 ng/g and 194 ng/g, respectively.

A perusal of the data points out that certain areas of the two aquatic systems are severely contaminated with PAHs and pesticides. The ∑PAH concentrations in water at some points of the lake system are above the lethal concentration and may cause acute toxicity to the exposed organisms. The data in the case of river indicate that both water and sediments are highly contaminated with PAHs. In both the cases the source of PAHs is mainly pyrogenic in nature. The potential toxicity due to total PAH concentrations will be rare in the lake system. The scenario is totally different in the case of river sediments where at most of the sites the toxic effects will be occasional. The apparent reasons for the deterioration of the said water bodies due to PAHs are dumping of untreated waste and a common practice of open burning of a variety of solid waste on the banks of the aquatic bodies. Apart from this, localized activity like plying of diesel operated motor boats is also contributing to the PAH pollution in the lake system. The major contributing activities responsible for PAHs contamination to the river through surface runoff and atmospheric transport are thermal power plants, heavy vehicular traffic load and domestic and industrial waste. A comparison of the two aquatic systems under study indicates that the ∑PAH concentrations in the river are about two times higher than those detected in the lakes. The PAH profiles in the water of the two systems are similar but the profiles in the sediments are different.

As far as the contamination due to pesticides is concerned aldrin is detected only at a few sites of the lake system and the other pesticides are observed only at sites surrounded by agricultural land. Endrin, pp'-DDT and α-endosulfan, wherever detected in water, are generally above the prescribed limit of acute toxicity. Aldrin and endrin are not detected in water and sediments of the river. Both the water and the sediments of the river are heavily contaminated with α-endosulfan and pp'-DDT. The biological impairments due endrin in the lakes and pp'-DDT in the river will be frequent as their levels in the sediments are above PEL. The accumulation of pesticides in the sediments may affect the natural bacterial composition and can lead to anaerobic conditions which may result in more toxic byproducts such as sulfide and ammonia.

References:

[1]. Chen, B., Xiaodong, X., Lizhong, Z., Jing, W., Yanzheng, G., Kun, Y., Xueyou, S. and

Baofeng, L., "Distributions of polycyclic aromatic hydrocarbons in surface waters, sediments and soils of Hangzhou City, China", Water Res. 38, 3558 (2004).

[2]. Wan, M.T., Kuo, J. and Pasternak, J., "Residues of endosulfan and other selected organochlorine pesticides in farm areas of the lower Fraser valley, British Columbia, Canada", J. Environ. Qual. 34, 1186 (2005).

(477i) Mass Transfer and Reaction Kinetics Limitations in Hazardous Waste (Organic and Metal) Treatment

Chawla, R., Howard University
Ighere, J., Howard University
Mlusu, J., Howard University
Farrell, A.,
Originally presented on: 11/6/2013 14:22:00 - 14:36:00

Improper storage and disposal practices have rendered hundreds of sites contaminated with hazardous substances. Some of the commonly found compounds include chlorinated organics such as trichloroethylene (TCE) and metals such as hexavalent chromium, Cr(VI), which occur simultaneously. This paper presents the feasibility of simultaneously treating TCE and Cr (VI), while addressing the problems related to (i) mass transfer limitations in chemical oxidation of TCE in soils, (ii) Effect of TCE oxidation on desired reduction of Cr (VI) to Cr (III), and (iii) removal of Cr(VI) using selected chelating agents.

Results of using alcohol cosolvents and household detergents for enhancing TCE solubility in water, thereby reducing mass transfer resistance in aqueous and soil systems will be presented. Redox reactions were studied using oxidizing agents, e.g. permanganate (MnO_4^-) for the remediation of TCE in the presence of Cr (VI) and using ferrous ions, Fe (II) for reduction of Cr (VI) to Cr(III) in the presence of TCE. In the redox reactions, TCE is oxidized to chloride, hydrogen, and carbon dioxide while permanganate is reduced to manganese dioxide (MnO_2). Similarly, Cr(VI) is reduced to Cr(III) and Fe(II) is oxidized to Fe(III). Batch experiments were performed to determine the reaction kinetics and extent of redox reactions of TCE and permanganate in the presence and absence of hexavalent chromium in aqueous systems. The reaction was determined to be second order overall, first order with respect to both TCE and MnO_4^- in the aqueous system. The TCE-Cr(VI) reaction was determined to be second-order overall, first order with respect to both Cr(VI) and TCE. The TCE-$KMnO_4$-Cr(VI) reaction at stoichiometric molar ratios (1:2:1), based on TCE reactions yielded 100% TCE oxidation. Cr(VI) was found to have a significant positive effect on the oxidation of TCE as it acted as an additional oxidant.

In the chelation study, batch experiments were performed to determine the degree and rate of chelation of Cr(VI) for three chelating agents- disodium dihydrogen ethylenediamine tetraacetate (Na_2H_2EDTA), ethylenediamine tetraacetic acid (H_4EDTA) and Nitrilotriacetic acid (NTA) Kinetic analysis indicated that the order of reaction was 1.32 with respect to Cr(VI) for Na_2H_2EDTA, 1.17 in the case of H_4EDTA and 2.38 for NTA. It was also found that the rate constants for the individual complexation reactions were 6.89×10^{-2} $[\text{mg Cr(VI)/g}]^{1-n}\text{hr}^{-1}$, 4.98×10^{-2} $[\text{mg Cr(VI)/g}]^{1-n}\text{hr}^{-1}$ and 7.25×10^{-2} $[\text{mg Cr(VI)/g}]^{1-n}\text{hr}^{-1}$ respectively.

(477j) A Simulation Of Weathering Of Colombian Crude Oils In The Colombian Caribbean Sea

Ramírez, J., Universidad Nacional de Colombia – Sede Medellín, Facultad de Minas, Bioprocesos y Flujos reactivos

Merlano, A., Universidad Nacional de Colombia – Sede Medellín, Facultad de Minas, Bioprocesos y Flujos reactivos

Guerrero-Gallego, J., Universidad Nacional de Colombia – Sede Medellín, Facultad de Minas, Grupo de investigación en Oceanografía e Ingeniería Costera - OCEANICOS

Peláez, D.,

Hernandez, J., Universidad Nacional de Colombia – Sede Medellín, Facultad de Minas, Grupo de investigación en Oceanografía e Ingeniería Costera - OCEANICOS

Osorio, A., Universidad Nacional de Colombia – Sede Medellín, Facultad de Minas, Grupo de investigación en Oceanografía e Ingeniería Costera - OCEANICOS

Molina, A., Universidad Nacional de Colombia – Sede Medellín, Facultad de Minas, Bioprocesos y Flujos reactivos

Originally presented on: 11/6/2013 14:36:00 - 14:50:00

A SIMULATION OF WEATHERING OF COLOMBIAN CRUDE OILS IN THE COLOMBIAN CARIBBEAN SEA

Juan Ramírez[1], Aura Merlano[1], Juan Guerrero-Gallego[2], Daniel Peláez[2], Juan Hernández[2], Andrés Osorio[2], Alejandro Molina[1,*]

[1]*Universidad Nacional de Colombia ? Sede Medellín, Facultad de Minas, Bioprocesos y Flujos reactivos, Medellín, Colombia*

[2]*Universidad Nacional de Colombia ? Sede Medellín, Facultad de Minas, Grupo de investigación en Oceanografía e Ingeniería Costera - OCEANICOS, Medellín, Colombia*

[*]*Corresponding author: amolinao@unal.edu.co*

A module, MEUN (*Módulo de Envejecimiento Universidad Nacional*), that describes the processes that occur due to the interaction, also known as weathering, of the crude in an oil spill with the atmosphere and the ocean, was developed. This module, programmed in FORTRAN, couples individual sub-models available in the literature to describe the evaporation, emulsification, dispersion and spreading processes that characterize the first week after an oil spill in the ocean. MEUN predicts as well the variation in density and viscosity over time as a result of the weathering processes.

In MEUN the evaporation submodel considered the pseudocomponent approach proposed by Jones, (1997). Knowledge of the True Boiling Point (TBP) curve and API gravity of the crude oil of interest, allowed the estimation, with Aspen Plus V7.2, of the thermodynamic properties (vapor pressure, mole fraction and mole volume) for each pseudocomponent. The weathering module includes the first-order kinetic expression proposed by Mackay et al., (1980) to model the rate of the emulsification process. The kind of emulsions formed (stable, mesostable, unstable or entrained water), the maximum water content of the emulsion and the evaporative thresholds to have a change in the type of the emulsion formed were interpolated from the experimental work developed by Fingas and Fieldhouse, (2004). MEUN couples the dispersion expression, developed by Delvigne and Sweeney, (1988), that takes into account the significant wave height and wave period to predict the amount of crude oil dispersed in the water column by the wave action. To predict the spreading rate of the oil slick, MEUN uses the equation developed by Lehr et al., (1984) based on observations in controlled oil spills. This expression considers the imbalance of forces acting on the oil slick and the effect of wind velocity to predict the spreading rate. The models developed by Lehr et al., (2002) and by Mackay et al., (1980), included in MEUN, predict the increase in density and viscosity, respectively, that results from the evaporation and emulsification processes.

All sub-model's constants were adapted to the specific requirements of Colombian crudes, particularly to Cusiana (°API 44.1) and Vasconia (°API 24.5). While the first one represents light crudes, the second one is an example of heavy oil. These two crudes have high production and require marine transport. The simulations results of the atmospheric model WRF (Weather Research and Forecasting Model), the wave model SWAN (Simulating WAve Nearshore) and the ocean model ROMS (Regional Ocean Modeling System) gave a typical ocean and atmospheric data of *Golfo de Morrosquillo* (latitude 09°29' N and longitude 75°46' W). Two scenarios, for high (4 m/s - 12 m/s) and low (0 m/s - 4 m/s) wind velocity, and the effect of time-averaging wind velocity, significant wave height and sea surface temperature were studied.

As an example of MEUN's predictions, Figure 1 shows the variation of the evaporated volume fraction with time after the spill for Cusiana and Vasconia oils. As expected, the lighter oil yields a higher evaporation fraction. Figure 1 also suggests that, while an average wind (discontinuous lines in Figure 1) yields reasonable predictions for a high wind condition, it significantly overestimate evaporation when the wind velocity (continuous lines in Figure 1) is low.

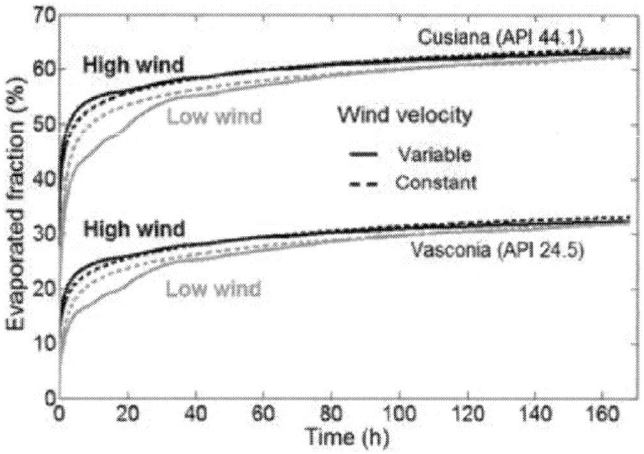

Figure 1. Predicted variation of the evaporated fraction for Cusiana and Vasconia crude oils after a spill in the Caribbean Sea under high and low wind velocity scenarios and variable and constant wind velocity.

Figure 2 summarizes MEUN's predictions as a plot of the volume fraction evaporated, dispersed and remaining on the sea surface. Because of differences in the physicochemical properties of the Colombian crude oils, Cusiana is more affected by evaporation and dispersion than Vasconia. According to MEUN, under high-wind conditions, a combination of evaporation and dispersion causes in Cusiana a decrease in the fraction of crude oil remaining on the ocean surface that is more than twice the decrease for Vasconia crude oil. While the results in Figure 1 and 2, as well as the fact that MEUN predicts an increase in viscosity of two to three orders of magnitude as the slick evaporates and the water makes an emulsion with the oil, agree with previous observations of weathering of oils in the ocean, a more complete validation of MEUN with experiments is currently under way using a wind tunnel and a rotating cylinder for evaporation and emulsion, respectively, at conditions typical of the Colombian Caribbean Sea.

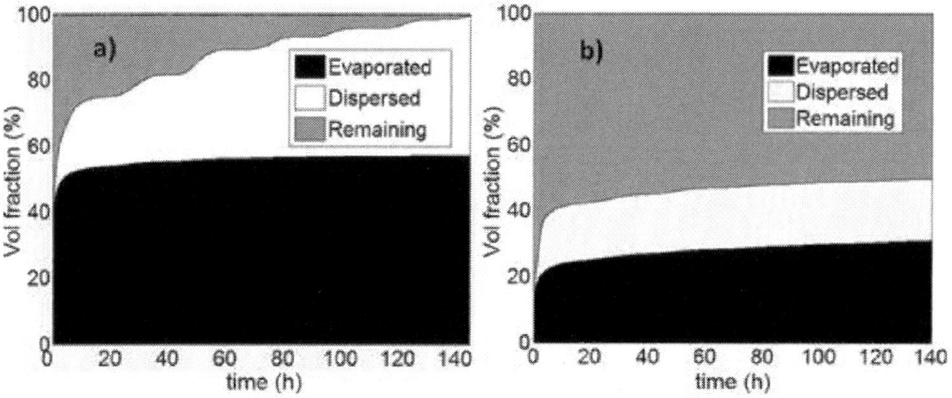

Figure 2. Distribution of the spilled crude oil between evaporated, dispersed and remaining on the ocean fractions for a) Cusiana and b) Vasconia oils. Predictions according to *MEUN*.

REFERENCES

Delvigne, G. A. L., & Sweeney, C. E. (1988). Natural dispersion of oil. *Oil and Chemical Pollution, 4*(4), 281?310.

Fingas, M., & Fieldhouse, B. (2004). Formation of water-in-oil emulsions and application to oil spill modelling. *Journal of hazardous materials, 107*(1-2), 37?50.

Jones, R. K. (1997). A simplified pseudocomponent oil evaporation model. *20th Arctic Marine Oilspill Program Technical Seminar.* (pp. 43?61). Ottawa, Ontario: Environment Canada.

Lehr, W. J., Fraga, R. J., Belen, M. S., & Cekirge, H. M. (1984). A new technique to estimate initial spill size using a modified fay-type spreading formula. *Marine Pollution Bulletin, 15*(9), 326?329.

Lehr, W., Jones, R., Evans, M., Simecek-Beatty, D., & Overstreet, R. (2002). Revisions of the ADIOS oil spill model. *Environmental Modelling & Software, 17*(2), 189?197.

Mackay, D., Buist, I., Mascarenhas, R., & Patterson, S. (1980). Oil Spill Processes and Models. Report No. EE/8. *Environment Canada* (p. 192). Ottawa, Ontario.

(477k) Increasing Strength of Thickened Oil Sand Tailings with Geopolymers

Xu, Y., Natural Resources Canada
Karimi, S.,
Originally presented on: 11/6/2013 14:50:00 - 15:04:00

The government of Alberta recently imposed new regulations requiring that oil sand tailings deposits must reach strength of 5 kPa one year after deposition. However, due to the limitations of gravitational settling and thickener operation, fine oil sand tailings can be thickened only to certain degree, and the strength may not meet the regulatory requirement. CanmetENERGY and Total E&P Canada (TEPCA) have conducted an investigation on the feasibility of using geopolymer additives to improve the strength of the deposit.

In this project, fly ash (type C) and cement were used as additives. Mature fine tailings (MFT) containing 35 wt% solids and thickened tailings (TT) with solids contents of 50 wt% were used as tailings material. Geopolymer was mixed with tailings and the treated tailings were distributed into 250-mL jars, some of which were closed, while others were left open. The solids content in the closed jars remained constant because no water evaporated, but increased in the open jars as the tailings dried in air at ambient conditions. The tailings strength was measured in each jar after a specified time.

These experiment results indicate that the addition of small amount of geopolymer increased the tailings strength significantly in both open and closed jars under the test conditions. Addition of about 5 wt% fly ash to TT increased the yield stress of the tailings to over 5 kPa after the solids content increased to around 60 wt% under natural drying. Similarly, adding around 5 wt% fly ash to the MFT increased the yield stress to over 5 kPa when the tailings were dried to about 55 wt% solids. Adding 3 wt% Portland cement also increased the yield stress of MFT or TT to more than 5 kPa when the solids concentration increased to 55 wt % for MFT and 65 wt% for TT. By comparison, untreated MFT or TT had to be dried to over 70 wt% and 85 wt%, respectively, in order for the yield stress to reach 5 kPa.

(486a) How Do Water and Protons Move Through Ionomers?

Benziger, J., Princeton University

Originally presented on: 11/6/2013 12:33:00 - 12:57:00

Ionomer membranes are the heart of Polymer Electrolyte Membrane Fuel Cells. The ionomer membrane transports protons from the anode to the cathode. Proton conductivity is a strong function of water activity. In Nafion, the archetypal PEM for fuel cells the proton conductivity increases by more than six orders of magnitude as water activity increases from 0 (0% humidity) to 1 (saturated vapor). Water transport in Nafion also increases by more than two orders of magnitude as water activity increases.

We have examined in detail the transport and properties of Nafion, sulfonated polyether etherketone (SPEEK) and sulfonated polystyrene (SPS) as functions of temperature and water activity. Ionomers separate into hydrophilic domains of sulfonic acid and water dispersed in a hydrophobic matrix. Water and protons are transported through the hydrophilic domains. The volume fraction of the hydrophilic phase increases with water activity, which also changes the structure of the hydrophilic domains. A percolation threshold of 0.1 vol fraction for proton and water transport is found for 1100 EW Nafion; above the percolation threshold water diffusion and proton mobility increase quadratically with water volume fraction. The percolation thresholds for SPEEK and SPS are larger, indicating different shapes of hydrophilic domains for different ionomers. We will show how ionomer structure is related to the shape and percolation of hydrophilic domains.

Water transport in PEMs involves the coupling of interfacial mass transport, diffusion and polymer swelling. We will show that interfacial transport resistance at the vapor/membrane interface is the limiting transport resistance at high water activity. The interfacial transport resistance is a key reason why Nafion is such a good PEM for fuel cells. We will show how the coupling of internal diffusional resistance and external interfacial transport resistances are couple in ionomers and how it affects the coupled water and proton transport.

(486b) Membranes With Sulfonated Perfluorocyclobutane (PFCB) Ionomer for Fuel Cells

Jiang, R.,
Fuller, T., General Motors Company
Gittleman, C., General Motors
Originally presented on: 11/6/2013 12:57:00 - 13:21:00

Membranes with Sulfonated Perfluorocyclobutane (PFCB) Ionomer for Fuel Cells

Ruichun Jiang, Timothy Fuller, Craig Gittleman

Global Fuel Cell Activities, General Motors Company

895 Joslyn Ave, Pontiac, MI 48340, USA

Ruichun.Jiang@GM.com

Polymer electrolyte fuel cells (PEFC) have been regarded as one of the most promising power sources for various applications. As the heart of PEFC, the proton exchange membrane, or PEM, plays a critical role in proton transport, electrical isolation and reactant gas separation between anode and cathode components of a fuel cell. Perfluorosulfonic Acid (PFSA) ionomers, such as NafionÂ® from Dupont, are the most commonly used materials for PEMs because of their favourable proton conductivity, and mechanical and chemical durability characteristics. There are, however, drawbacks associated with the use of PFSA membranes. These include high cost and high gas permeability which contributes to reductions in both fuel cell efficiency and durability. To overcome these deficiencies, hydrocarbon membranes with aromatic backbones have received considerable attention as PEM materials due to their lower gas permeability and decreased production costs.

A new class of hydrocarbon block copolymer membrane based on perfluorocyclobutane (PFCB) chemistry was developed at General Motors for fuel cell PEM applications. The PEMs with sulfonated PFCB ionomer demonstrate benefits of low production costs, good proton conductivity, and lower gas (H_2, O_2, N_2) permeability than PEM with PFSA ionomer. One of the challenges for sulfonated PFCB based PEMs is enhancing their mechanical stability. Several strategies, including blending PFCB ionomer with a poly(vinylidene fluoride) (PVDF) elastomer, and introducing reinforcement layer containing expanded polytetrafluoroethylene (ePTFE) have been investigated to develop composite PFCB PEMs with improved mechanical properties.

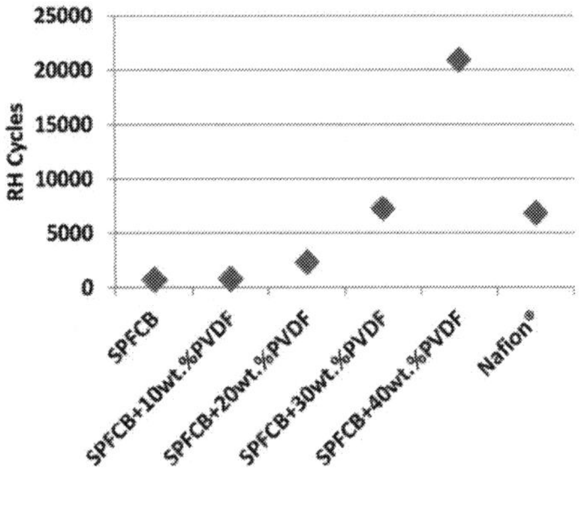

Performance and durability of fuel cell membranes based on PFCB ionomer are greatly influenced by the content of PVDF, as well as by the properties of ePTFE reinforcement layer. The figure on the right shows that the mechanical durability in RH cycling tests of the sulfonated PFCB/PVDF blend membranes were improved with increasing level of PVDF content. More performance and durability information of PFCB/PVDF and PFCB/PVDF-ePTFE types of fuel cell membranes will be reported and discussed.

(486c) Increased Hydrophobic Content Multiblock Copolymers for Use As Proton Exchange Membranes in Direct Methanol Fuel Cells (DMFC)s

Rowlett, J., Virginia Tech
Shaver, A., Virginia Tech
Kim, Y. S., Los Alamos National Laboratory
Zelenay, P., Los Alamos National Laboratory
McGrath, J. E., Virginia Tech
Originally presented on: 11/6/2013 13:21:00 - 13:45:00

Two series of increasing hydrophobic poly(arylene ether nitrile) and poly(arylene ether ketone) oligomers were successfully synthesized. These poly(arylene ether) hydrophobic oligomers consistent of different amounts of methyl groups on the bisphenol A co-monomer (bisphenol A, dimethyl bisphenol A, and tetramethyl bisphenol A), and were combined with either difluorobenzonitrile or difluorobenzophenone to from the hydrophobic segments. In addition to further increase the hydrophobicity of the moieties, small mole percent's of the bisphenol A monomer series (0-25%) were replaced by the fluorinated bisphenol A analog 4,4'-hexafluoroisopropylidenediphenol. Only up to twenty-five percent of the fluorinated monomer was used in the hydrophobic copolymers due to the concern over methanol cross-over, which has been shown to greatly increase with higher fluorine content.

Both series of hydrophobic oligomers were then reacted with a poly(arylene ether sulfone) hydrophilic oligomer (BPS100) via a low temperature coupling reaction to produce the hydrophilic-hydrophobic alternating multiblock copolymers. The resultant block copolymers formed a tough, durable, and transparent membrane when cast from solution, with structure confirmation and composition done using ^{1}H and ^{13}C NMR spectroscopy. Mechanical properties were examined using tensile testing, and used in combination with size exclusion chromatography (SEC) to confirm high molecular weight was achieved. In addition the conductivity, swelling ratio, and fuel cell performance was measured and compared against that of the nafion membrane. Membrane electrode assembly (MEA) and fuel cell performance was carried out by Los Almas National Laboratory (LANL) in collaboration with this work. The synthesis and characterization of these multiblock copolymer membranes will be discussed.

(486d) Controlling Mechanical and Transport Properties of Layer-By-Layer Electrospun Mat Composite Membranes for Fuel Cell Applications

Liu, D. S., Massachusetts Institute of Technology
Rutledge, G. C., Massachusetts Institute of Technology
Hammond, P. T., Massachusetts Institute of Technology
Originally presented on: 11/6/2013 13:45:00 - 14:09:00

The continued development of thin-film solid polymer electrolytes with improved mechanical and ion transport properties is critical for the further advancement of electrochemical energy devices. For hydrogen and methanol fuel cells, the proton exchange membrane (PEM) is the electrolyte and has to have high protonic conductivity, low fuel crossover, and be mechanically and chemically stable. More recently, there has been considerable interest in improving the mechanical properties and durability of the proton exchange membranes (PEMs) without sacrificing its transport properties. For methanol fuel cells in particular, there is a need for a cheap, mechanically robust PEM that also has high protonic conductivity and low methanol permeability.

LbL assembly allows for the controlled deposition of alternating polyelectrolytes at the nanometer scale. An LbL system composted of poly(diallyl dimethyl ammonium chloride) (PDAC) and sulfonated poly(2,6-dimethyl 1,4-phenylene oxide) (sPPO) has shown to have proton conductivity comparable to that of Nafion®, the industry standard, but with methanol permeability two orders of magnitude less than Nafion's. While this system by itself is mechanically deficient when hydrated, the mechanical properties can be greatly improved if the film is spray-coated on an electrospun fiber mat to form a composite membrane. With vacuum assisted spray-LbL assembly, the fibers can be individually coated, and the mat filled; mats as thin as 15 μm can be coated. Without vacuum, the spray-LbL assembly can form a continuous fuel-blocking layer with properties similar to the free-standing LbL film by itself. Combined, composite membranes with methanol permeability twenty times lower than Nafion and through-plane proton selectivity five times greater than Nafion could be fabricated.

As anticipated, the mechanical properties of the underlying electrospun mat were passed on to the composite membrane. The composite membranes were further strengthened by annealing the electrospun mat prior to coating. In addition, the swelling of the composite membranes in water was significantly lower than the LbL-only system and over a magnitude lower than Nafion. The mechanical hysteresis was also lower than Nafion under hydrated conditions. The composite membranes were tested in an operational direct methanol fuel cell with results showing a higher open circuit voltage and comparable cell resistances when compared to Nafion.

(486e) First-Principles Modeling of Proton Conductivity in Perfluorosulfonate Fuel Cell Membranes

Zhang, B., University of Tennessee
Edwards, B., The University of Tennessee
Originally presented on: 11/6/2013 14:09:00 - 14:33:00

Proton conductivity through a perfluorosulfonate polymer electrolyte membrane was investigated using an extension of a recent model developed for the purpose of quantitatively describing transport of charged species through the ionomer interface of a typical membrane [Kumar et al., JCP, 138, 064903, 2013]. In this study, the membrane was modeled as a collection of cylindrical pores with the anionic groups (i.e., $-SO_3^-$) assumed to be fixed at a prescribed periodic spacing along the pore wall, as determined by the material's equivalent weight and surface charge density. Microscopic potentials for the interactions between the various chemical species present within the pore were scaled up to a mesoscopic level, where transport equations were derived via the generalized bracket formulation of nonequilibrium thermodynamics in terms of the density and velocity of each species (hydronium ions and water) within the pore. These coupled transport equations were solved based on reasonable boundary conditions, and an expression of conductivity was deduced in terms of the following quantities: water content, equivalent weight, temperature, density, and the monomeric structure of the perfluorosulfonate chain. Theoretical predictions of this model were compared against experimental data of conductivity in three different membranes, Nafion 117 (EW = 1100), a 3M membrane (EW = 900), and a Dow membrane (EW = 800), for different values of water content and temperature. Theoretical predictions of the model matched quantitatively the experimental data without the necessity of using empirical fitting parameters.

(486f) Functionalized Inorganic Particle-Doped Composite Proton Exchange Membranes for Enhancement in Fuel-Rejecting and Proton-Conducting Property

Wu, H., Tianjin University
Zhao, Y., Tianjin University
Jiang, Z., Tianjin University
Originally presented on: 11/6/2013 14:33:00 - 14:57:00

Organic-inorganic hybrid proton exchange membranes have gained tremendous attention for use in direct methanol fuel cells due to their combined advantages originated from both the inorganic and organic components. A large variety of polymers have been blended with various inorganic particles by either physical blending or sol-gel process to generate inorganic phase in polymer matrix. The addition of carefully selected inorganic components leads to inhibited methanol crossover and improved stability. The inorganic dopants are usually functionalized to improve the organic-inorganic interfacial compatibility and to introduce additional pathways for proton transfer at the interface, thus increasing proton conductivity of the resultant hybrid membranes. Inorganic dopants such as titania and silica were applied widely for the ease of availability and functionalization. In our previous work various functional groups including phosphonic acid groups, sulfonic acid groups, carboxylic acid groups, amine groups and amino acid groups by means of chelation absorption of these functional groups onto titania. In this study, two kinds of phosphorylated silica submicrospheres are synthesized and incorporated into sulfonated poly(ether ether ketone) (SPEEK) to fabricate hybrid membranes. Phosphorylation is carried out by the reaction of phosphorus oxychloride and the epoxy groups on the silica surface introduced by glycidyl-silane (SiP-I) or poly(glycidyl methacrylate) (SiP-II). The length of the chains and the amount of phosphoric acid groups grafted on the silica surface are tuned by the above two different phosphorylation methods. The dispersion of SiP-II, anti-swelling and methanol resistant property of the SPEEK/SiP-II hybrid membranes are enhanced. And the proton conductivity is increased by doping with both kinds of phosphorylated silica particles, while, interestingly, the SPEEK/SiP-I hybrid membranes with less amounts of phosphoric acid groups show unexpectedly much higher proton conductivity than the SPEEK/SiP-II hybrid one at the same doping content. A proton conductivity of 0.335 S cm^{-1} at 60 ^{o}C and 100%RH is obtained. The result is tentatively discussed in terms of effective proton-conducting groups/sites instead of the absolute amount of acidic groups.

(488a) NO Reduction By H2 On Platinum Catalysts: Fundamental Mechanistic Studies

Farberow, C. A., University of Wisconsin - Madison
Bai, Y., University of Wisconsin - Madison
Dumesic, J. A., University of Wisconsin-Madison
Mavrikakis, M., University of Wisconsin - Madison
Originally presented on: 11/6/2013 12:30:00 - 12:50:00

The catalytic conversion of nitric oxide, an unwanted byproduct of the combustion of hydrocarbons in air, to inert N_2 is of fundamental and practical importance.[1] Platinum catalysts, when utilized for the low temperature reduction of NO by H_2, demonstrate high conversion of NO but low selectivity toward N_2 and unfavorable production of side-products, namely N_2O and NH_3.[2] In this study, we combine experimental reaction kinetic studies, periodic, self-consistent Density Functional Theory calculations, and microkinetic modeling to understand the origin of the activity and selectivity of Pt in NO reduction by H_2. Multiple Pt facets are considered to elucidate the reaction structure sensitivity. The detailed reaction mechanism, including the role of hydrogen in NO activation, is described. Due to the strong adsorption of NO on Pt, the effect of NO coverage on the surface chemistry is accounted for in the DFT calculations, and accordingly, in the microkinetic model.

1. Heck, R. M.; Farrauto, S. T.; Gulati, S. T., *Catalytic Air Pollution Control: Commercial Technology* Wiley, New Jersey **2009.**

2. H. G. Stenger, J. S. Hepburn, *Energy & Fuels* **1987**, *1*, 412.

(488b) The Role of Surface Oxides in NOx Storage Reduction Catalysts: A DFT and kMC Study

Ni, Z.,
Aich, P., University of Illinois at Chicago
Hoffmann, M., Technische UniversitÃ¤t MÃ¼nchen
Reuter, K., Fritz-Haber-Institut der Max-Planck-Gesellschaft
Meyer, R. J., University of Illinois - Chicago
Originally presented on: 11/6/2013 12:50:00 - 13:10:00

NOx storage reduction (NSR) catalysts have drawn great attention due to their ability to accommodate NOx reduction under the lean burn conditions demanded by more fuel efficient engines [1]. When using Pd as the active metal, a monolayer PdO surface is suggested to be the thermodynamically favored surface during the oxidation catalysis process [2,3]. In this work, the first-principles Monte Carlo methods are used to explore the role of surface oxide during catalysis cycle in NSR catalysts. A reconstructed ($\sqrt{5}\times\sqrt{5}$) R27°PdO(101)-Pd(100) which was used as a model surface in previous examinations of CO oxidation [2] has been used as our catalyst model. Density function theory (DFT) calculations have been employed to examine thermodynamics and kinetics of the atomic-level elementary steps for NSR process. The use of H_2 as a reductant has also been explored. A phase diagram has been constructed to determine the thermodynamic stability of the surface under the fuel-rich mode conditions of NSR. Based on the results from the DFT calculations, first-principles kinetic Monte Carlo (kMC) simulations have been used to investigate the stability range of different surface structures under both the oxidation (O_2 rich) and reduction (fuel rich) portions of the NSR process. kMC results show that surface oxide is stable in fuel-lean mode, though phase-diagram shows bulk PdO is the most phase under that particular thermodynamic condition[3]. kMC results indicate there is pressure difference of NO partial pressure for decomposition of surface oxide and that for forming surface oxide from prime metal at the same O_2 partial pressure. H-O phase-diagram produced by DFT linked to thermodynamic methods indicates under standard NSR operating condition, monolayer PdO is stable while preliminary kMC results show that thin film oxide starts decomposing around partial pressure of H_2 at 10^{-3} atm with O_2 partial pressure of 1atm.

[1] N. Takahashi, H. Shinjoh, T. Iijima,et al, Catal. Today 27, 63 (1996)

[2] J. Rogal, K. Reuter, and M. Scheffler, Phys. Rev. B 75, 205433 (2007).

[3] J. Jelic and R. J. Meyer, Phys. Rev. B 79, 125410 (2009)

(488c) Surface Chemistry of Oxygenates On Pt(111) Modified By Mo and Re

Robinson, A.,
Medlin, J. W., University of Colorado
Originally presented on: 11/6/2013 13:10:00 - 13:30:00

Biomass-derived compounds are highly functionalized, making them difficult to selectively process into a desired product because they can interact with the surface in multiple, often uncontrolled ways. Addition of metals including rhenium and molybdenum to supported catalysts has been shown to facilitate C-O bond activation and influence selectivity for a number of deoxygenation reactions. However, the mechanism for the improvements is not known, hampering efforts to design better catalysts. Surface science techniques using single crystals can provide a fundamental understanding of the role of oxophilic metals (Re, Mo) as promoters to Pt group catalysts for deoxygenation reactions.

Our initial investigations have focused on how oxophilic modifiers interact with water and related surface species such as O, H, and CO. The majority of deoxygenation reactions using biomass-derived feedstocks are performed in the aqueous phase; it has been proposed that oxophilic modifiers allow water to dissociate and create acidic sites that alter the catalyst's behavior. Temperature programmed desorption (TPD) and Auger electron spectroscopy (AES) have been used to investigate this chemistry on unmodified and Mo-coated Pt(111) surfaces. The Mo-coated catalysts form surface oxides following even low-temperature exposure to oxygen. Preliminary results suggest that these surface oxides primarily influence reactivity of small molecules (water, CO) by blocking surface sites. On the reduced surface, Mo appears to have a more pronounced effect on surface adsorption energies. Possible mechanisms by which Mo influences reactions of small oxygenates on Pt will be discussed.

(488d) First-Principles Study of Methanol Steam Reforming On Cobalt and Cobalt Oxide Surfaces

Asthagiri, A., University of Florida
Luo, W., The Ohio State University
Originally presented on: 11/6/2013 13:30:00 - 13:50:00

Steam reforming of bio-derived alcohols or dimethyl ether for the production of hydrogen is a potential substitute to fossil fuels for both energy storage and automotive fuel. Co-based catalysts have been shown to be promising catalysts for ethanol steam reforming, but there is still much debate about the nature of the active site, the role of the oxide support, and reaction mechanisms. *In operando* XAFS studies indicate a mixture of cobalt metal and oxide phases under reaction conditions but the exact role of the two phases is still not clear. We will report density functional theory (DFT) based studies of methanol steam reforming (MSR) on both the metallic Co and Co oxide surfaces. The MSR reaction is chosen since it provides a simpler set of reaction pathways to explore than ethanol or butanol. We examined the adsorption of reactants, products, and intermediates that are along the MSR reaction pathways suggested by previous studies. For each species we determined the most stable structure and based on this information we have constructed a free energy diagram for the reaction pathways for reforming methanol on Co(0001) surface. We have also used the nudged elastic band (NEB) method to determine the transition states and energy barriers for important elementary steps to study the kinetics, especially for the competing steps that effect product selectivity. We will discuss the implications of our calculations to experimental results and other DFT studies of MSR on other metal surfaces. We will also present on going studies of the same reaction steps for MSR on CoO(100) and CoO(111) surfaces.

(488e) Initiation of Autocatalytic Surface Explosion Reactions

Gellman, A. J., Carnegie Mellon University
Mhatre, B. S., Carnegie Mellon University
Pushkarev, V., Carnegie Mellon University
Holsclaw, B., Carnegie Mellon university
Lawton, T., Tufts University
Sykes, C., Tufts University
Originally presented on: 11/6/2013 13:50:00 - 14:10:00

Autocatalytic reaction mechanisms are observed in a range of important chemical processes including catalysis, radical-mediated explosions, and biosynthesis. Because of their complexity, the microscopic details of autocatalytic reaction mechanisms have been difficult to study on surfaces and heterogeneous catalysts. On surfaces, these types of reactions were originally discovered during studies of formic acid decomposition on Ni(110) and the mechanism has often been described as a vacancy-mediated surface explosion. Autocatalytic decomposition reactions of tartaric acid (TA) adsorbed on Cu(110) offer molecular-level insight into aspects of these processes, which until now, were largely a matter of speculation. The decomposition of TA/Cu(110) is initiated by a slow, irreversible process that forms vacancies in the adsorbed TA layer, followed by a vacancy-mediated, explosive decomposition process that yields CO_2 and small hydrocarbon products. Initiation of the explosive decomposition of TA/Cu(110) has been studied by measurement of the reaction kinetics, time-resolved low energy electron diffraction (LEED), and time-resolved scanning tunneling microscopy (STM). Initiation results in a decrease in the local coverage of TA and a concomitant increase in the areal vacancy concentration. Observations of explosive TA decomposition on the $Cu(651)^S$ surface suggest that initiation does not occur at structural defects in the surface, as has been suggested in the past. Once the vacancy concentration reaches a critical value, the explosive, autocatalytic decomposition step dominates the TA decomposition rate. The onset of the explosive decomposition of TA on Cu(110) is accompanied by the extraction of Cu atoms from the surface to form a (6,7;-2,1) overlayer that is readily observable using LEED and STM. The explosive decomposition step is second-order in vacancy concentration and accelerates with increasing extent of reaction.

(488f) Component Surface Concentrations of Gas Mixtures By Langmuir Theory and Ideal Adsorbed Solution Theory (IAST)

Yasukazu, K., Tsinghua University

Dezheng, W.,

Originally presented on: 11/6/2013 14:10:00 - 14:30:00

[Introduction] Langmuir theory is used in reaction kinetics to calculate the surface concentration or coverage of each gas component on solid catalysts that are exposed to gas mixtures. However, this theory is only consistent with thermodynamics when the monolayer capacities are the same for all species [1]. In zeolites, there can be spaces a small molecule can occupy but which are too constrained for a bigger species or there are acidic sites where one species can adsorb but which are too weak for a less basic species, which would cause the monolayer capacity to differ for different species. Krishna and Baur [2] have shown using molecular simulation that the different monolayer capacities of different hexane isomers in a MFI zeolite led to erroneous reaction kinetics when Langmuir adsorption isotherms were used, and that ideal adsorbed solution theory (IAST) should be used for surface concentration calculations in place of Langmuir theory. In this study, we used experiments with binary component mixtures to examine whether Langmuir theory or IAST is better for surface concentration calculations. We compared the estimated surface concentrations by Langmuir and IAS theories of binary mixtures of methanol and dimethyl ether (DME) and C_3H_8 and DME on a SAPO-34 catalyst with experimental measurements. Since the results are of interest for the methanol-to-olefins (MTO) reaction, we also compared the estimated surface concentrations by Langmuir and IAS theories of multicomponent mixtures at 350°C.

[Experiment] Adsorption isotherms of pure gases and binary component gas mixtures on a SAPO-34 zeolite at room temperature to 100°C were measured. Together with the adsorption isotherms, the differential heats of adsorption were also simultaneously measured with a heat flow microcalorimeter in the volumetric manometry apparatus. The total dead volume of the system was 40 ml, which allowed the composition of the gas phase to be sampled by mass spectrometry (MS) after equilibrium was achieved for each dosage. The MS-measured gas phase compositions were used to experimentally verify the calculation of the individual surface concentrations on the sample. The composition of the binary gas mixtures used were methanol/DME = 63/37 [mole%] and C_3H_8/DME = 68/32 [mol%]. The sample used was a SAPO-34 catalyst powder synthesized using $SiO_2/Al_2O_3 = 0.24$, which had an acid site density of 1.0 mmol/g. The sample was first pretreated at 400°C for 5 hours under a vacuum of 10^{-5} Pa. After that, sample was kept at the adsorption temperature of 25, 60, 80 or 100°C.

[Results and Discussion] First, pure component adsorption isotherms of methanol, DME, C_3H_6, and C_3H_8 were measured at 25, 60 and 100°C. The saturated surface concentration of methanol is quite different from those of the other adsorbates which differ slightly from one another. The saturation coverage of methanol is approximately twice that of the other gases, which may be due to the formation of methanol clusters stabilized by hydrogen-bonded networks of the type reported for HZSM-5 [3]. These isotherms were fitted by the single site Langmuir equation or, when it did not give a good fit, the dual site Langmuir equation. The parameters obtained were used to calculate surface concentrations of binary gas mixtures in the same temperature range using the Langmuir and IAST theories to compare these with experimental measurements. These parameters were also extrapolated to the MTO reaction temperature of 350°C to predict individual surface concentrations at 350°C by both Langmuir theory and IAST. In the methanol and DME adsorption experiments, some irreversible adsorption were observed, and a second measurement after evacuation was performed to get an isotherm without irreversibly adsorbed molecules for curve fitting. The component surface concentrations from the methanol-DME mixture by the Langmuir theory and IAST showed significant deviations, esp. at high pressures. The isotherm and differential adsorption heat curve from methanol-DME binary mixture adsorption at 80°C are shown in Fig. 1, together with the corresponding curves for pure methanol and DME. The pure gas isotherms showed that the methanol saturation coverage is much larger than that of DME. In the calorimetric curves, the initial heat of adsorption was 60 kJ/mole, which decreased to 45 kJ/mole when 0.5 mmole/g had been adsorbed for all the gases used. After that, there was a plateau in the calorimetric curves, which indicated the homogeneity of the adsorption sites in SAPO-34. With methanol, DME and their mixture, falloffs in the heat of adsorption were then observed at 6.5, 27 and 6.5 kPa, respectively, corresponding to surface concentrations of 2.1, 1.0 and 1.5 mmole/g. The falloff in the heat of adsorption indicated the saturation coverage at which regular chemisorption sites were all occupied, and the remaining amount adsorbed after the falloff began was attributed to adsorption on weak sites. This calorimetric trend indicated that the dual site Langmuir equation should be used for fitting the pure component isotherms. The fits are illustrated in Fig. 1a. Figure 2 shows the individual surface concentrations of methanol and DME measured in the mixture adsorption experiment, and the calculated amounts by Langmuir theory and IAST. During adsorption from the gas mixture, there were constant increases in methanol and DME surface concentrations up to 6.5 kPa where the calorimetric curve began its falloff. At higher adsorption pressures, a DME surface concentration *decrease* with increasing pressure was observed, i.e., there was desorption of previously adsorbed DME. A similar surface concentration *decrease* with increasing adsorption pressure was simulated by Krishna and Baur [2] who attributed it to entropy effects from the adsorption of molecules with different saturation capacities. Fig. 2 shows that IAST gave good calculated surface concentrations that also included the desorption with increasing pressure, while there were significant deviations in the calculated surface concentrations by Langmuir theory, esp. at higher pressures where it failed to predict the desorption. This result may suggest that Langmuir theory should not be used in the reaction kinetics, which was the conclusion of Krishna and Baur. However, our results also showed that deviations in the calculated surface concentrations by Langmuir theory were significant only after the regular chemisorpion sites were saturated, that is, the deviations appeared when adsorption on the weak adsorption sites became significant. Further work is required to understand if these sites are involved in the reactions.

[Conclusion] Component surface concentrations from gas mixtures calculated using Langmuir theory and IAST that were compared with experimental values showed that the calculated values by Langmuir theory were wrong when the saturation coverages of the individual gases were different and the total adsorbed amount from the mixture exceeded the number of regular adsorption sites.

[References] (1) D.D. Do, Adsorption Analysis: Equilibria and Kinetics, Imperial College Press, London, 1998. (2) R. Krishna, R. Baur, *Chem. Eng. Sci.* 60 (2005) 1155-1166. (3) C.-C. Lee, R. J. Gorte, W. E. Farneth, *J. Phys. Chem. B* 101 (1997) 3811-3817.

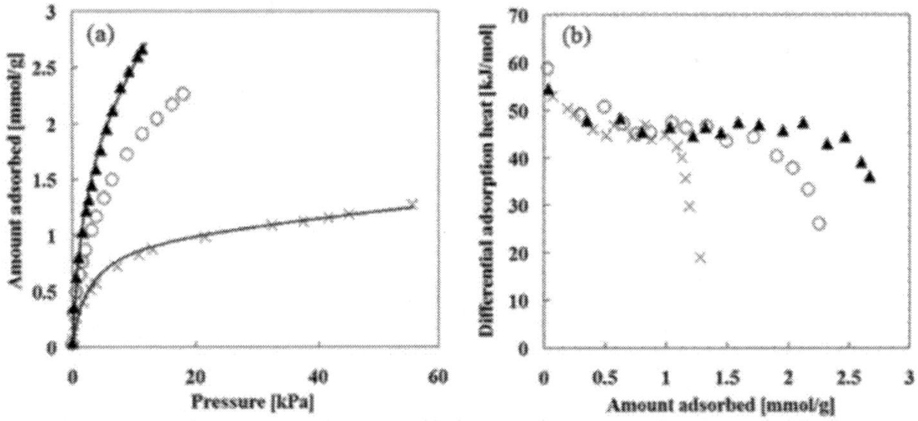

Fig. 1 (a) Adsorption isotherms and (b) differential adsorption heat curves of methanol (▲), DME (×) and methanol-DME mixture (○). The solid lines in (a) were fitted with the dual site Langmuir equation.

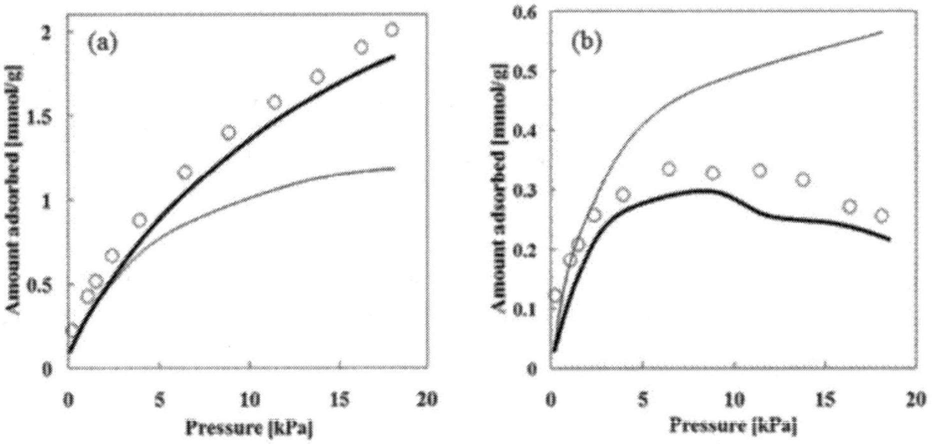

Fig. 2 Component surface concentrations of (a) methanol and (b) DME in the gas mixture adsorption experiment. Experimental measurements (○), and calculated by IAST (bold lines) and Langmuir theory (thin lines)

(488g) Active Metal Surface Area: PULSE Chemisorption, Temerature Programe Technique and Cyclic Voltammetry

Namdeo, A., IIT Bombay
D, M., IIT Bombay
Mahajani, S. M., Indian Institute of Technology
Suresh, A. K., IIT Bombay
Sarkar, A., IIT Bombay
Originally presented on: 11/6/2013 14:30:00 - 14:50:00

In heterogeneous catalytic reactions, a comparison, hence selection, of catalysts is to be based on intrinsic rates, expressed for example as Turnover Frequencies (TOF). A good measurement of the dispersed active metal surface area (degree of dispersion, Active metal surface area) is important to calculate the turn over frequency (TOF).TOF can be defined as: number of moles of given product/reactant formed/consumed per catalytic site per unit time for the given set of reaction conditions. Values of TOF express intrinsic catalytic ability and hence offer a straightforward means to compare different catalysts and for comparisons of data for the same catalyst prepared by different procedures or characterized in different laboratories. Our aim in this work is to find and compare alternative methods to determine the active metal surface area (AMSA) of noble metal catalysts. For evaluating the active metal surface area of the catalyst, we have used pulse-chemisorption, Temperature program techniques (TPD) and cyclic voltammetry technique (CV). Active surface has been calculated from cyclic voltammtry (CV) using adsorption of H^+ and Cu^{2+} onto the surface from solutions. The values are compared to each other and also compared with the obtained pulse chemisorptions and TPD values.

(500a) Alkyl Ethoxylate Surfactants With High Hydrophilic-Lipophilic Balance for Smart CO2-Water Foam in Enhanced Oil Recovery

Chen, Y., the University of Texas at Austin
Elhag, A. S., UT Austin
Reddy, P. P., UT Austin
Ou, A. M., UT Austin
Hirasaki, G., Rice University
Nguyen, Q. P., University of Texas
Johnston, K. P., University of Texas at Austin
Originally presented on: 11/6/2013 12:30:00 - 12:55:00

The sweep efficiency for CO_2 enhanced oil recovery can be improved by forming CO_2/water (C/W) foams with nonionic alkyl ethoxylate surfactants with high Hydrophilic-Lipophilic Balance (HLB). With long ethylene oxide (EO) chains as hydrophilic heads, these alkyl ethoxylate surfactants have high cloud points (70 °C or higher) in the presence of brines with NaCl concentrations up to 120 g/L, which is of interest for foam stabilization in reservoirs with high temperatures and salinities. Due to their high HLB value, these alkyl ethoxylate surfactants partition strongly to low salinity brine phase versus a dodecane phase at moderate temperatures (24-40 °C). Moreover, they reduce the interfacial tension between CO_2 and low salinity brine (~1% Total dissolved solids (TDS)) at moderate tempeature. Thus, they form "smart" shear-thinning CO_2/water foams in sand packs with different permeabilities without stabilizing dodecane/brine emulsions, which is of interest in mobility control of C/W foam for CO_2 enhanced oil recovery.

(500b) Phase Behavior of Spherical Polyelectrolyte Brush Solution in Compressed Carbon Dioxide

Liu, L., East China University of Science and Technology
Zhang, R., East China University of Science and Technology
Cang, Y., East China University of Science and Technology
Zhang, J., East China University of Science and Technology
Li, T., East China University of Science and Technology
Guo, X., East China University of Science and Technology
Originally presented on: 11/6/2013 12:55:00 - 13:20:00

Phase Behavior of Spherical Polyelectrolyte Brush Solution in Compressed Carbon Dioxide

Lixiao Liu, Rui Zhang*, Yu Cang, Jianchao Zhang, Tao Li, Xuhong Guo*

State Key Laboratory of Chemical Engineering, East China University of Science and Technology, 130 Meilong Road, Shanghai 200237, China.

**Corresponding authors. Tel.: +86-21-64253491; E-mail:* r.zhang@ecust.edu.cn guoxuhong@ecust.edu.cn

Abstract: Spherical polyelectrolyte brushes (SPB), consisting of polystyrene cores with linear macromolecules grafted onto their surfaces, were widely used as nano-reactors for synthesis of metal nanoparticles, adsorption of proteins and ideal carrier for high-efficiency catalyst, et. al. Phase equilibrium of polyelectrolyte brushes-solvent-carbon dioxide (CO_2) ternary systems were first investigated experimentally. Polystyrene-poly acrylic acid (PS-PAA) was selected as model solute, while ethanol was chosen as the organic solvent. The cloud-point pressure is the critical pressure which the one phase system changes to two-phases. It has been measured under different temperatures, the solid contents and the sizes of the brushes. The results were shown in Fig. 1. Compared with the cloud-point of different size of PS-PAA (135nm and 450nm), we know that the pressure for phase-separation of the system decreases with increasing size under same temperature. The reason maybe the bigger size of PS-PAA has longer brush chain with same core and that means it's more sensitive to the solvability of CO_2 because CO_2 is easier to enter the circle of brush chain. When the temperature increases from 273.15K to 333.15K, the pressure for cloud-point increases from 3.8MPa to 6.7MPa at PS-PAA(135nm) 1% in ethanol. It can be explained that the solubility of CO_2 in organic solvent decreases with increasing of temperature, so the higher pressure is needed to modulate the solvability of the mixed solvent. The cloud-point pressure exhibits an increasing trend with the increasing of PS-PAA solid content in ethanol. It indicates that the SPB play an important role in the phase equilibrium. It will provide us a potential method about separate and recover polyelectrolyte brush both scientific research and industry application.

Keywords: phase behavior, spherical polyelectrolyte brush, compressed CO_2;

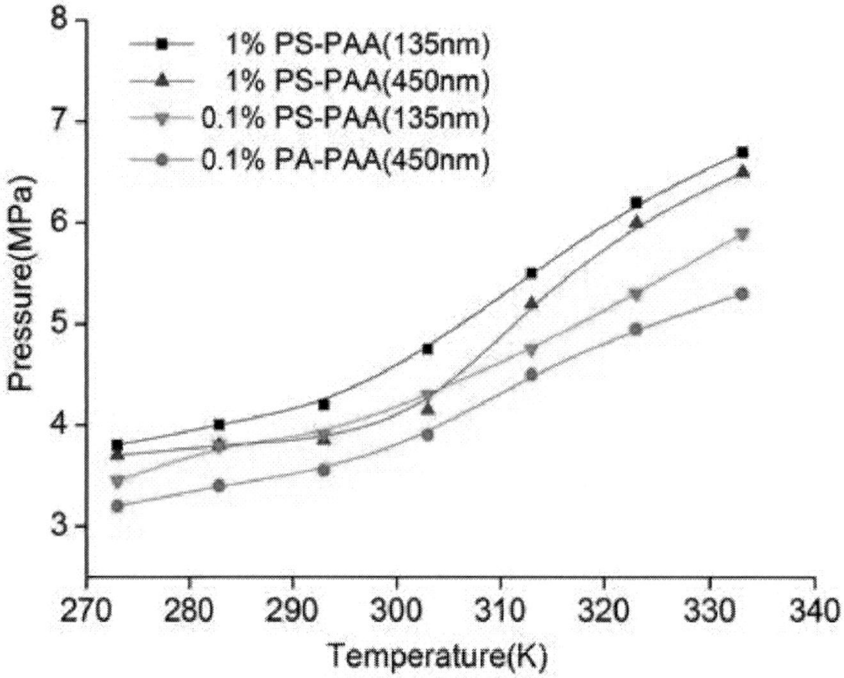

Figure 1. Phase diagram in PS-PAA/ethanol /CO_2 system

(500c) Effect of Compressed CO2 On the Solubility of Inorganic Salt in Reverse Micelle

Zhang, J., East China University of Science and Technology
Zhang, R., East China University of Science and Technology
Cang, Y., East China University of Science and Technology
Liu, L., East China University of Science and Technology
Guo, X., East China University of Science and Technology
Li, T., East China University of Science and Technology
Originally presented on: 11/6/2013 13:20:00 - 13:45:00

Effect of Compressed CO_2 on the Solubility of Inorganic Salt in Reverse Micelle

Jianchao Zhang, Rui Zhang*, Yu Cang, Lixiao Liu, Tao Li, Xuhong Guo*

State Key Laboratory of Chemical Engineering, East China University of Science and Technology, 130 Meilong Road, Shanghai 200237, China.

**Corresponding authors. Tel.: +86-21-64253491; E-mail: r.zhang@ecust.edu.cn*

guoxuhong@ecust.edu.cn

Abstract: Reverse micelles are thermodynamically stable surfactant aggregates with polar cores, which are formed spontaneously as certain types of surfactants dissolve in nonpolar solvents. It's well known that a small amount of salt can reduce critical micelle concentration (CMC) of surfactant at normal pressure. However, the work about the compressed CO_2 can enhance the solubility of inorganic salt obviously in reverse micelles has not been reported before.

In this work, the phase behavior of compressed CO_2 on sodium bis(2-ethylhexyl) sulfosuccinate (AOT) in octane with different concentration of inorganic salt was studied. It was found that the suitable pressure which can form one phase region increases with the decreasing of salt concentration at high salt concentration range (Figure 1). Different inorganic salts with valence state have been used to examine the effect of metal ion on the AOT-octane-inorganic salt-water-CO_2-quintuple system. It was found that compressed CO_2 can dissolve higher concentration of NaCl than that of $CaCl_2$ at the same W0, which is defined by water to AOT molar ratio. For example, thanks to compressed CO_2, the maximum concentration value of NaCl and $CaCl_2$ aqueous solution with the W_0 at 30 which can be dissolved is 5.0mol/L and 2.5 mol/L, respectively. The main reason is probably the divalent ion can reduce the thickness of electric double layer more effective than monovalent ion do. The clear point pressure increases with the W0 increasing at the same salt concentration. In other words, the area of one phase of quintuple system turns to smaller with the increasing of W_0. This work will provide the valuable suggestion for the application of surfactant under the inorganic salt condition.

Keywords: compressed CO_2, solubility, inorganic salt, reverse micelle

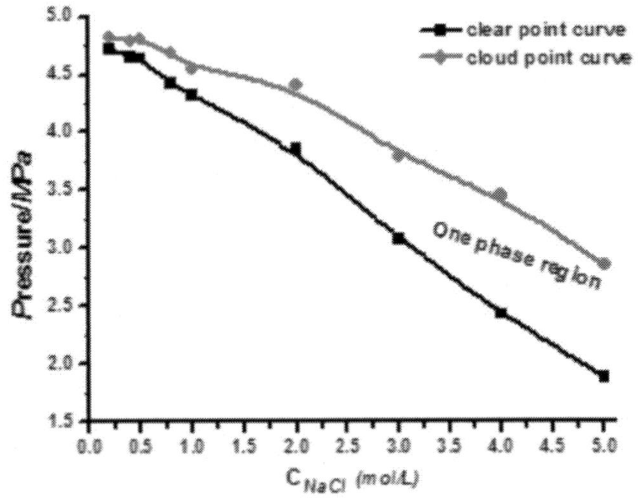

Figure 1. The phase diagram of AOT-octane-inorganic salt-water-CO_2 quintuple system at 303.2K.

(500d) Rational Design of Hydrofluoroolefin-Philic Moieties: A Combined Computational and Experimental Approach

Yang, L., Wayne State University
Borges, G., Universidade Tiradentes
Cabral, V., State University of Maring
Franceschi, E., Universidade Tiradentes
Dariva, C., Universidade Tiradentes
da Rocha, S. R. P., Wayne State University
Originally presented on: 11/6/2013 13:45:00 - 14:10:00

Hydrofluoroalkanes (HFAs) are currently the propellants of choice in many industrial applications, including refrigeration and in medical aerosols, as they have zero ozone depleting potential (ODP). However, HFAs have a relatively high global warming potential (GWP), at 1430 for HFA-134a and 3200 for HFA-227. As regulations become more stringent, new alternatives to HFAs are being sought. Hydrofluoroolefins (HFOs), with their zero ODP and very low GWP - 6 for HFO-1234ze and 4 for HFO-1234yf, are promising next generation working fluids. In many applications involving propellants, the dispersion of low solubility species is of essence. Because of the dual hydrophobic and lipophobic nature of HFOs, solute solubility is expected to be of major concern for this class of solvents. The design of HFO-philic moieties and surface active agents for HFOs is thus of great relevance in furthering the applicability of such environmentally friendly propellants.

In this work we use a combination of computational and experimental tools, including *ab initio* calculations, molecular dynamics (MD) simulations, and high pressure tensiometry to address HFO-philicity. Pair interaction binding energy from *ab initio* calculations was used to determine an HFO-philicity scale for the alkane, ether, and ester moieties, with the ether and ester moieties having the strongest interaction with HFO. The potential of such chemistries as HFO-philic tails in non-ionic surfactants was tested by determining their impact on the tension of the water-HFO interface, and the surfactant balance using both MD simulations and high-pressure tensiometry. The interfacial tension of HFO|H_2O interface from MD and tensiometry was found to be in excellent agreement. The PO-based surfactant was capable of reducing the interfacial tension of the binary interface by 30.9 mN/m. The optimum balance for that surfactant class was determined to be at 60EO% by MD simulations and 50EO% by tensiometry. This study represents the first attempt to systematically design dispersing agents for HFOs. Reverse aggregates of water in HFO may find applications in the delivery of water-soluble or water-dispersible therapeutics using medical sprays and in other industrial applications.

(500e) Numerical Simulation of Dripping and Jetting in Supercritical Fluids / Liquid Micro Coflows

Guillaument, R., I2M - TREFLE - IPB - University of Bordeaux
Erriguible, A., I2M - TREFLE - IPB - University of Bordeaux
Aymonier, C., Institute of condensed Matter Chemistry of Bordeaux
Marre, S., Institute of Condensed Matter Chemistry of Bordeaux
Subra-Paternault, P., Laboratoire TREFLE (Transfert, Ecoulements Fluides Energetiques)
Originally presented on: 11/6/2013 14:10:00 - 14:35:00

In this work, a two dimensional simulation of segmented micro coflows of CO_2 and water in microcapillaries (20 < T (°C) < 50 and 8 < p (MPa) < 16.5) was carried out using a combination of the one-fluid model and the volume of fluid (VOF) method to describe the two-phase flow and a penalty method to account for the wetting property of the capillary walls. The computational work was validated by comparing numerical and experimental results in both the dripping and jetting regimes (Figure 1). The agreement of the calculated pressure difference across the droplet or jet interface with the Laplace-Young's law was assessed as supplementary criteria. The effects of CO_2 / Water interfacial tension (5 < s ($mN.m^{-1}$) < 35) and wall wettability (contact angle CO_2 / wall varying from 0 to 180°) on the segmented water – supercritical CO_2 microflows were specially described. It was shown that switching the wall surface from hydrophilic to hydrophobic by tuning the contact angle allows for changing the droplet curvature so that the continuous water phase eventually undergoes a phase inversion resulting in water droplets/slugs formation in a continuous CO_2 phase.

(514a) Hydrophobic, Electrostatic, and Time-Dependent Polymer Bridging Forces At Surfactant-Modified Silicone Surfaces

Rapp, M., University of California, Santa Barbara
Donaldson, S. H. Jr., University of California, Santa Barbara
Gebbie, M. A., University of California, Santa Barbara
Das, S., University of California, Santa Barbara
Kaufman, Y., Ben Gurion University
Israelachvili, J. N., University of California, Santa Barbara
Originally presented on: 11/6/2013 12:35:00 - 12:55:00

We have synthesized covalently grafted polydimethylsiloxane (PDMS) thin films on extended gold surfaces that behave as collapsed polymer brush surfaces in solution, with uniform thickness, surface coverage, and surface chemistry. The Surface Forces Apparatus (SFA) was used to measure the surface forces due to surfactant adsorption and self-assembly at these hydrophobic interfaces. We show that polymeric surfactants —containing PDMS anchoring domains and quarternary ammonium cationic headgroups—assemble at the PDMS surface, leading to both an effective charge reversal at the hydrophobic interface and strong adhesion energies to mica surfaces. Molecular structure and charge are shown to govern the assembly of surfactants on these surfaces: PDMS bola-surfactants assemble to form fluctuating surface aggregates that extend into solution, leading to long-ranged and time-dependent attraction to mica surfaces, while other surfactants behave as an adsorbed layer of smeared-out charges and their interaction with mica can be described by the electric double layer theory of asymmetric surfaces. Additionally, specific Coulombic interactions between the surfactant headgroups and mica result in time-dependent polymer bridging forces as surfactants are pulled from the PDMS interface. Thus, surfactant self-assembly is presented as a method to tailor the complex and dynamic surface forces at hydrophobic interfaces, and can promote strong wet adhesion between hydrophobic and mineral surfaces.

(514b) Integration of Gold Nanoparticles Into Bilayer Structures Via Adaptive Surface Chemistry

Lee, H. Y., Penn State University
Shin, S. H. R., The Pennsylvania State University
Abezgauz, L. L.,
Danino, D., Technion
Bishop, K. J. M., Penn State University
Originally presented on: 11/6/2013 12:55:00 - 13:15:00

We describe the spontaneous incorporation of amphiphilic gold nanoparticles (Au NPs) into the walls of surfactant vesicles. Au NPs were functionalized with mixed monolayers of hydrophilic (deprotonated mercaptoundecanoic acid, MUA) and hydrophobic (octadecanethiol, ODT) ligands, which are known to redistribute dynamically on the NP surface in response to changes in the local environment. When Au NPs are mixed with preformed surfactant vesicles, the hydrophobic ODT ligands on the NP surface interact favorably with the hydrophobic core of the bilayer structure and guide the incorporation of NPs into the vesicle walls. Unlike previous strategies based on small hydrophobic NPs, the present approach allows for the incorporation of water-soluble particles even when the size of the particles greatly exceeds the bilayer thickness. The strategy described here based on inorganic NPs functionalized with two labile ligands should in principle be applicable to other nanoparticle materials and bilayer structures.

(514c) Tuning Protein Nanoclusters With Molecular Crowding Agents (Depletants)

Hung, J., University of Texas at Austin
Dinin, A. K., University of Texas at Austin
Borwankar, A. U., University of Texas at Austin
Dear, B. J., University of Texas at Austin
Wilson, B. K., University of Texas at Austin
Maynard, J. A., University of Minnesota
Truskett, T. M., The University of Texas at Austin
Johnston, K. P., University of Texas at Austin
Originally presented on: 11/6/2013 13:15:00 - 13:35:00

Stabilizing proteins at high concentration is of broad interest in drug delivery for treatment of cancer and many other diseases. Proteins have a tendency to undergo irreversible aggregation, gelation or precipitation at high concentrations due to unfolding caused by specific short ranged forces. We create highly concentrated antibody dispersions comprising dense equilibrium nanoclusters of therapeutic protein molecules, which upon dilution *in vitro* dissociate to individual protein molecules. The key is to control nanocluster size and interparticle spacing within the clusters by manipulation of the particle colloidal forces. The interaction potential, which depends upon the electrostatic, Van der Waals, steric and depletion forces, governs thermodynamic self-assembly and particle morphology.

A hierarchy of intracluster and intercluster interactions has been controlled to provide colloidal stability, low viscosities, and stable protein for drug delivery. The clusters are formed by balancing depletion attraction between protein molecules against electrostatic repulsion. Our previous studies have focused primarily on trehalose as the depletant or crowding agent. In the current study we show substantial additional control over nanocluster morphology and the viscosity of the protein dispersions by varying the size and polarity of the crowding agent. We also report new types of results for the effect of depletant on the nanocluster sizes as a function of the depletant properties.

(514d) Controlling Viscoelasticity of Nanoemulsions Using Polymer-Surfactant Self-Assembly

Kim, J., University of California, Santa Barbara
Gao, Y., University of California, Santa Barbara
Hebebrand, C., University of California, Santa Barbara
Peirtsegaele, E.,
Helgeson, M. E., Massachusetts Institute of Technology
Originally presented on: 11/6/2013 13:35:00 - 13:55:00

Nanoemulsions have been considered for a range of advanced applications where fine control over their rheological properties is desired, independent of their colloidal stability. To address this issue, we have developed a scheme for imparting thermoreversible viscoelasticity to model oil-in-water nanoemulsions. Specifically, we have shown that polymer-surfactant interactions in the aqueous phase cause thermoreversible gel-like viscoelasticity at low temperatures, distinct from their emulsion-free counterparts. Interestingly, these gels exhibit Maxwell-like viscoelasticity that follows time-temperature superposition (TTS) over a wide temperature range - the first time such behavior has been observed in emulsions. Small angle neutron scattering (SANS) shows that the low-temperature nanoemulsion gels exhibit temperature-insensitive microstructure, confirming that the gel-like viscoelasticity arises entirely from the dynamics of the system. We hypothesize that this low-temperature gelation arises from a temporary polymer network mediated by polymer-surfactant interactions at the droplet interface. To test this hypothesis, we use TTS to probe the relation of viscoelastic properties and material parameters and compare them with temporary network theory. The results show that the viscoelasticity is critically linked to the energetics of polymer-surfactant complexation, as confirmed by isothermal titration calorimetry (ITC). We have also studied the steady shear rheology of these nanoemulsion gels, which exhibit characteristics of extreme shear thinning and shear banding. Rheo-SANS measurements indicate the formation of shear-induced structure of the droplet phase, which we believe is linked to the macroscopic rheology by non-linear deformation of the gel network. Overall, these studies demonstrate polymer-surfactant complexation as a highly generic mechanism to control both the linear and non-linear rheology of nanoemulsions, which could be exploited to design new materials with dynamically controlled rheology.

(514e) Photodirected Assembly of Custom-Designed Polyelectrolyte Complexes

Okoye, N. H.,
de Silva, U. K.,
Wengatz, J. A.,
Lapitsky, Y., University of Toledo
Originally presented on: 11/6/2013 13:55:00 - 14:15:00

When oppositely charged polyelectrolytes are mixed in aqueous solutions they self-assemble into insoluble complexes ranging from micro- and nanoparticles, to thin films, to macroscopic fibers and capsules. This phenomenon has an electrostatic driving force and finds use in numerous applications (e.g., controlled release, separation and sensor technologies), many of which require polyelectrolyte complex (PEC) structures with well-defined geometries. To expand the range of their structural motifs, we have developed a simple strategy for directing PEC assembly between two commonly-used polyelectrolytes – poly(allylamine) (PAH) and poly(styrene sulfonate) (PSS) – using light. Deprotonated PAH (which was neutral) was mixed with anionic PSS and a photoacid generator (diphenyliodonium nitrate), and irradiated with long-wave ultraviolet light ($\lambda = 365$ nm). The photoactivated acid generation led to the protonation of PAH amines at the photoirradiation site and (because the neutral PAH became cationic) resulted in the rapid PAH/PSS association into either colloidal or macroscopic PECs. The shape of these PECs was easily tuned using a photomask, which controlled the photoirradiation pattern, and thus the shape of the resulting PECs. The formation and structure of these PECs were characterized through phase studies, microscopy and isothermal titration calorimetry (ITC), and suggest photodirected assembly as an effective tool for preparing PECs with custom-designed, anisotropic geometries.

(514f) Self-Assembly of Super-Stiff Adhesive Gels From Common Polyelectrolytes

Huang, Y., University of Toledo
Lawrence, P., University of Toledo
Lapitsky, Y., University of Toledo
Originally presented on: 11/6/2013 14:15:00 - 14:35:00

Ionic crosslinking of polyelectrolytes with multivalent counterions can yield a wide array of self-assembled complexes. These complexes range from liquid coacervates to solid gels and attract widespread interest in fields such as controlled release, antibacterial coatings and water treatment. Here, we will show that certain mixtures of commonly-used polyelectrolytes and multivalent counterions can self-assemble into extremely stiff hydrogels (G' ~ 10^6 Pa). Such stiff gels, to our knowledge, have hitherto only been prepared through dense covalent crosslinking or via saloplastic processing of polyelectrolyte complexes, but never through spontaneous polymeric self-assembly. In addition to their high moduli, these gels are very adhesive and stick to a wide variety of substrates (including Teflon®, skin and glass). The analysis of their formation via isothermal titration calorimetry (ITC), phase studies and rheology revealed that they typically form when the polymers and multivalent counterions are mixed at near-stoichiometric charge ratios and that their formation and dissolution can be triggered by changes in pH and ionic strength. The effects of polymer structure and polyelectrolyte/counterion affinity on the gel properties have also been examined and will be discussed in this talk.

(514g) Incorporation of Additives in Self-Assembled Block Copolymers: From Molecular Solvents to Nanoparticles

Sarkar, B., University at Buffalo - The State University of New York (SUNY)
Alexandridis, P., University at Buffalo - The State University of New York (SUNY)
Originally presented on: 11/6/2013 14:35:00 - 14:55:00

Block copolymers provide self-assembled systems with great versatility as to the types of nanostructures formed and the conditions under which they are stable. Interesting hybrid nanocomposite materials can result from the arrangement of nanoparticles in block copolymer matrices. We are addressing the interplay between block copolymer organization and nanoparticle properties as affected by the solvents present, and are exploring the continuum between molecular solvents and nanoparticles. In particular, we discuss here lyotropic liquid crystalline systems formed by hydrated poly(ethylene oxide)-poly(propylene oxide) block copolymers (Pluronics), and their ability to accommodate within the same structure polar organic solvents, polyhedral oligomeric silsesquioxanes (POSS), or silica nanoparticles. We relate the solvent/particle degree of loading and location in the block copolymer structure to enthalpic and entropic considerations that emanate from the solvent/particle size and chemical functionality.

Sarkar, B.; Lakshmichand, J.; Alexandridis, P. Self-Assembly of Amphiphilic Block Copolymers in Ternary Solvent Mixtures: Lyotropic Liquid Crystalline Phase Behavior and Structure. *Macromol. Chem. Phys.* **2012**, *213* (23), 2514-2528. DOI: 10.1002/macp.201200438

Ayandele, E.; Sarkar, B.; Alexandridis, P. Polyhedral Oligomeric Silsesquioxane (POSS)-Containing Polymer Nanocomposites. *Nanomaterials* **2012**, *2* (4), 445-475. DOI: 10.3390/nano2040445

Sarkar, B.; Alexandridis, P. Self-Assembled Block Copolymer - Nanoparticle Hybrids: Interplay between Enthalpy and Entropy. *Langmuir* **2012**, *28* (45), 15975-15986. DOI: 10.1021/la303568e

(515a) Substrate Constraints Modify the Rayleigh Spectrum of Vibrating Sessile Drops

Chang, C. T., Cornell University
Bostwick, J. B.,
Steen, P. H., Cornell University
Daniel, S., Cornell University
Originally presented on: 11/6/2013 12:30:00 - 12:45:00

In this work, we study the resonance behavior of mechanically-oscillated, sessile water drops. By mechanically oscillating sessile drops vertically and within prescribed ranges of frequencies and amplitudes, a rich collection of resonance modes are observed and their dynamics subsequently investigated. We first present our method of identifying each mode uniquely, according to their geometric patterns and through association with spherical harmonics. Next, we compare our estimated resonance frequencies of drops to theoretical predictions using both the classical theory of Lord Rayleigh for free, oscillating drops, and new theory by Bostwick/Steen that explicitly considers the effect of the solid surface on drop dynamics. Finally, we report observations and analysis of drop mode mixing, or the simultaneous coexistence of multiple mode shapes within the resonating sessile drop driven by one sinusoidal signal of a single frequency. During mode-mixing, distinct modes within a single drop are demonstrated to coexist, while retaining their respective individual mode dynamics. The dynamic response of a deformable liquid drop constrained by the substrate it is in contact with is of interest in a number of applications, such as drop atomization and ink jet printing, switchable electronically controlled capillary adhesion, optical microlens devices, and digital microfluidic applications where control of droplet motion is induced by means of a harmonically driven substrate.

(515b) Superhydrophobic Surfaces Through Micromolding Surface-Initiated Polymerization

Jennings, G. K., Vanderbilt University
Escobar, C. A., Vanderbilt University
Ristau, J.,
Cooksey, T., Vanderbilt University
Mahfuz, N. N., Vanderbilt University
Spellings, M., Vanderbilt University
Originally presented on: 11/6/2013 12:45:00 - 13:00:00

Nature provides a vast and highly optimized array of surface architectures that have given plants and animals unique advantages in their environments. In particular, hundreds of plant species are equipped with superhydrophobic leaves, which minimize disease threat and effectively channel rainwater near the roots for rapid uptake. This research presentation introduces a new method to prepare superhydrophobic coatings that replicate the architectures of superhydrphobic leaves but have diverse surface and polymer chemistries that are not confined by nature's building blocks. Termed micromolding surface-initiated polymerization, we first form the mold of a superhydrophobic leaf, fill that mold with a liquid monomer, and press the monomer-filled mold against a surface that has been initiated for polymerization. The growing polymer chains are confined by the mold and the surface features become positive replicas of the original leaf. We will demonstrate the use of this method to replicate both natural and artificial surfaces with a partially fluorinated polymer that exhibits ultralow critical surface tensions and is prepared by surface-initiated ring-opening metathesis polymerization to exhibit robust attachment to the underlying surface. The strengths and limitations of this technique to prepare superhydrophobic coatings from source plants such as *Aristolochia esperanzae* and *Trifolium repens* will be described, as well as the relation of surface topography to the wetting and barrier properties of the coating. The presentation will also describe strategies to boost the yield and thickness of the polymerization so that larger scale topographies can be accurately reproduced.

(515c) Effect of the Chemical Structure of the Nanometer-Thick Polymer Coatings On the Simultaneously Oleophobic/Hydrophilic Behavior

Li, L.,
Wang, Y.,
Originally presented on: 11/6/2013 13:00:00 - 13:15:00

Effect of the chemical structure of the nanometer-thick polymer coatings on the simultaneously oleophobic/hydrophilic behavior

Lei Li and Yongjin Wang

Department of Chemical and Petroleum Engineering, University of Pittsburgh, 3700 O'Hara Street, Pittsburgh, PA 15261

Simultaneously hydrophilic/oleophobic surfaces are highly desirable in many important applications, such as anti-fogging. However, such wetting behavior is not expected for traditional surfaces since previous experimental results showed that, in general, a solid surface is more wettable to oil than to water. Interestingly, a couple of recent studies showed that it is possible to achieve simultaneous oleophobicity/hydrophilicity by applying some nanometer-thick polymer on some solid substrate. However, the underlying mechanism remains unclear to date. In the current study, we investigate the effect of chemical structure of the nanometer-thick polymer on the simultaneous hydrophilic/oleophobic behavior. Three nanometer-thick perfluoropolyethers with different end-groups were dip-coated on the silicon wafer (with native oxides). Both water contact angle (WCA) and hexadecane contact angle (HCA) were measured on these polymer-coated surfaces. The results show that simultaneous oleophobicity/hydrophilicity only occurs to some polymers but not the others. It was also found that the HCA changes significantly with time only for the polymer showing simultaneous oleophobicity/hydrophilicity. These findings further support a mechanism previously proposed by us [1]. Anti-fogging experiments also indicated that there is a clear correlation between simultaneous oleophobicity/hydrophilicity and the anti-fogging performance.

Acknowledgement: We thank NSF (CMMI 1233161) for the financial support!

[1] Li L.; Wang Y.; Gallaschun C.; Risch T. and Sun J. "Why can a nanometer-thick polymer coated surface be more wettable to water than to oil?", JOURNAL OF MATERIALS CHEMISTRY, 2012, 22(33), 16719-16722

(515d) Improved Stability of Functional Low-Density Self Assembled Monolayers On Gold: A Generalized Approach

Safazadeh Haghighi, L., University of Kentucky
Stevens, C., University of Kentucky
Berron, B. J., University of Colorado
Originally presented on: 11/6/2013 13:15:00 - 13:30:00

Self-assembled monolayers are a convenient method for the alteration of surface chemistry and architecture. The development of low density monolayers (LDMs) has enabled an additional level of control over interfacial chemistry, where the lateral spacing of the surface bound functional groups are altered. This packing density directly influences the conformational freedom of surface tethered molecules, where surface interactions with solvents, analytes, and reactants are all modulated. Critically, LDMs on gold are currently limited in both functionality and stability. We report a generalized approach where "click" thiol-yne chemistry enables the formation of click-LDMs with arbitrary tail group chemistry.

We demonstrate click-LDMs through the preparation of carboxyl-terminated LDMs on gold by adsorption of 1,10 decanedithiol on gold and subsequent grafting reaction of 10-undecynoic acid in presence of a photoinitiatior and light. The resulting LDMs are exceptionally stable with respect to time, temperature and solvents. The click-LDMs showed a high charge-transfer resistance suggesting stable structuring of the monolayer's base, while both symmetric and asymmetric methylene stretching FTIR modes are red shifted from that of densely packed monolayers supporting a decreased ordering of the alkyl chains compared to traditional SAM. These analyses are consistent with the expected densely-packed, stable region of 1,10 decanedithiol near the gold surface and a covalently region of non-crystalline region which interfaces the environment. Contact angle goniometry is consistent with the expected intermediate surface energy of exposed carboxyl and methylene functionalities. The specificity of the click chemistry enables the examination a large library of functionalities.

(515e) Effects of Organic Additives On Hydration At Silicate and Aluminate Particle Surfaces

Sangodkar, R. P., University of California, Santa Barbara
Smith, B. J., University of California, Santa Barbara
Roberts, L. R., Roberts Consulting Group
Funkhouser, G. P., Halliburton Inc.
Doherty, M. F., University of California, Santa Barbara
Chmelka, B. F., University of California
Originally presented on: 11/6/2013 13:30:00 - 13:45:00

Hydration and crystallization processes are often influenced by the presence of low concentrations of additives or contaminants. In many cases, this involves competitive adsorption of additive species and water at inorganic oxide surfaces, which can have significant effects on surface hydration that drastically affect the mechanical and rheological properties of these complicated mixtures. A technologically important example is the use of dilute concentrations (<1 wt%) of organic additives that adsorb at silicate and aluminate particle surfaces and thereby inhibit hydration reactions in cement-water slurries. This maintains the fluidity of the cement-water mixtures by preventing the formation of cross-linked silicate and aluminate species that are central to the development of overall mechanical strength in cements. To understand the effects of organic additives on the hydration of silicate and aluminate particles, it is important to establish the compositions and interactions of the additives and water with inorganic surface species. Such understanding can be obtained by using nuclear magnetic resonance (NMR) spectroscopy in conjunction with electron microscopy, X-ray diffraction, and X-ray fluorescence, which allow hydration processes to be monitored over different length scales. Specifically, molecular-level interactions of industrially relevant concentrations (~0.1 wt%) of organic additives, such as saccharides and phosphonates, adsorbed at inorganic oxide surfaces are determined by using state-of-the-art NMR techniques. The results establish that sucrose molecules adsorb non-selectively at both anhydrous and hydrated silicate sites, but selectively at anhydrous aluminate sites. By comparison, phosphonates adsorb non-selectively at anhydrous and hydrated silicate and aluminate sites. The molecular-level compositions, structures and site-specific interactions in hydrating cements correlate with their macroscopic properties, such as, compressive strength and slurry viscosity. The analyses shed insights on hydration processes in cement-water mixtures, their constituent inorganic oxide species, and the influence of industrially relevant additive concentrations on cement setting.

(515f) Effect of Lateral Roughness On the Adsorption of Surfactants At Various Temperatures

Hamon, J.,
Grady, B. P., University of Oklahoma
Originally presented on: 11/6/2013 13:45:00 - 14:00:00

Previous work by our group has shown that the effect of roughness on the adsorption of surfactants at room temperature increases the rate of adsorption at low times, and decreases the equilibrium adsorption and finally makes adsorption more sensitive total impurities near the critical micelle concentration. In this study, we varied the temperature with both rough and flat surfaces in quartz crystal microbalance experiments for cetyl trimethyl ammonium bromide and found that the adsorption decreased for the flat surfaces. Results will be presented on the same experiments for rough surfaces. We have also found that a common "high-purity" alcohol ethoxylate shows characteristics of poly(ethylene oxide) contamination, and we are currently developing a procedure to remove this component so we can study this molecule as well.

(515g) Intermittent Desorption-Mediated Surface Diffusion

Skaug, M. J., University of Colorado Boulder
Mabry, J., University of Colorado Boulder
Schwartz, D. K., University of Colorado Boulder
Originally presented on: 11/6/2013 14:00:00 - 14:15:00

The mobility of molecules on a solid surface plays a key role in diverse phenomena such as friction and self-assembly and in surface-based technologies like heterogeneous catalysis and molecular targeting. To understand and control these surface processes, a universally applicable model of surface transport at solid-liquid interfaces is needed. However, unlike diffusion at a solid-gas interface, little is known about the mechanisms of diffusion at a solid-liquid interface. Using single molecule tracking at a solid-liquid interface, we found that a diverse set of molecules, including polymers and proteins, underwent intermittent random walks with non-Gaussian displacements. This contrasts with the normal random walk and Gaussian statistics that are commonly assumed for molecular surface diffusion. The molecules became temporarily immobilized for random waiting times between surface displacements produced by excursions through the bulk fluid. We propose that intermittent hopping is universal to molecular surface diffusion at a solid-liquid interface.

(515h) Microfluidic Platform for Isolating Nucleic Acid Targets Using Sequence Specific Hybridization Near Solid Liquid Interface

Wang, J., Brown University
Morabito, K., Brown University
Tang, J.,
Tripathi, A., Brown University
Originally presented on: 11/6/2013 14:15:00 - 14:30:00

The separation of target nucleic acid sequences from biological samples has emerged as a significant process in today's diagnostics and detection strategies. In addition to the possible clinical applications, the fundamental understanding of target and sequence specific hybridization on surface modified magnetic beads is of high value. In this paper, we describe a novel microfluidic platform that utilizes a mobile magnetic field in static microfluidic channels, where single stranded DNA (ssDNA) molecules are isolated via nucleic acid hybridization. We first established efficient isolation of biotinylated capture probe (BP) using streptavidin-coated magnetic bead. We find that BP binding is restricted to roughly one probe molecule per streptavidin molecule on bead surface. Subsequently, we investigated the hybridization of target ssDNA with BP bound to bead and explained these hybridization kinetics using a dual-species kinetic model. The number of hybridized target ssDNA molecules was determined to be about 6.5 times less than that of BP on the bead surface, due to steric hindrance effects. The hybridization of target ssDNA with non-complementary BP bound to bead was also examined, and non-specific hybridization was found to be insignificant. Lastly, we demonstrated highly efficient capture and isolation of target ssDNA in presence of non-target ssDNA, where as low as 1% target ssDNA can be detected from mixture. The microfluidic method described in this paper is significantly relevant and is broadly applicable, especially towards point-of-care biological diagnostic platforms that require binding and separation of known target biomolecules, such as RNA, ssDNA or protein.

(515i) Transient Hydrate Formation/Dissociation of Methane Hydrates in Porous Media At Hydrate Stable Conditions

Babu, P., National University of Singapore
Linga, P., National University of Singapore
Palmer, A.,
Khoo, B. C., National university of singapore
Tan, T. S., Singapore Institute of Technology
Rangsunvigit, P., The Petroleum and Petrochemical College, Chulalongkorn University, Bangkok, Thailand
Originally presented on: 11/6/2013 14:30:00 - 14:45:00

Natural gas hydrates are crystalline compounds formed from mixtures of water and methane or other gases at sufficiently high pressures and low temperatures. They are considered a potential unconventional energy resource. It is estimated that the liberation and subsequent production of just 15% of the trapped gas in hydrates would provide the world with energy for 200 years at the current level of energy consumption. Understanding the fundamental behavior of hydrate formation and dissociation in porous media is necessary in order to avoid the potential hazards during production from the hydrate reservoirs in the marine sediments and permafrost regions. In this study, methane hydrate formation and dissociation in porous media was observed. Experiments were carried out at 8.0 MPa, 274.15 K and different water saturation levels (50 and 100%). Two types of porous media (silica sand and activated carbon) were employed. In silica sand bed, hydrate crystals were observed to form in the inter-particle pore space and a stable and clear hydrate front movement across the crystallizer was observed. On the other hand, nucleation of hydrate crystals on the surface of the activated carbon grain and then dissociation in the hydrate stable region was observed in activated carbon bed. This occurrence of hydrate formation and dissociation processes within the hydrate stability region is defined as transient hydrate formation/dissociation. Based on our results, we postulated that particle size, pore space and water saturation level might play a role in transient hydrate formation/dissociation. When a smaller sized activated carbon grains were used, a clear and stable hydrate front was observed. Our results suggests that pore space and its interconnectivity plays an important role in hydrate formation in porous media.

(515j) Impact of Tablet Solid Fraction On Disintegration Rate

Hilden, J., Eli Lilly and Company
Kramer, T., Eli Lilly and Company
Originally presented on: 11/6/2013 14:45:00 - 15:00:00

The absorption rate or extent of absorption of an active pharmaceutical ingredient (API) in patients can be significantly affected by the time required for a tablet to mechanically disintegrate in the stomach. Depending upon the dosage form, tablet disintegration times can be designed to range from tens of seconds for orally disintegrating tablets to several hours for extended release tablets. Tablet composition is critical to controlling the disintegration rate. However, even for a fixed composition, the disintegration rate can be significantly impacted by the size and porosity of the tablet. In this work, a mathematical model is derived to relate tablet disintegration time to tablet thickness and porosity. Experimental data are queried from electronic lab notebooks using proprietary internal software and used to assess the validity of the disintegration rate model. Tablet compaction profiles are analyzed and results are presented from over 400 batches of tablets.

(517a) Invited Talk: Thermoplastic Elastomers and Thermosets Derived From Vegetable Oils

Robertson, M., University of Houston
Wang, S., University of Houston
Yang, G., University of Houston
Rohde, B., University of Houston
Originally presented on: 11/6/2013 12:30:00 - 13:00:00

A great challenge to overcome is the replacement of traditional petroleum-based polymers with materials derived from sustainable, alternative resources. Vegetable oils are an attractive substitute for petroleum sources for the derivation of polymers, due to their abundance, low cost, and lack of toxicity. We are developing two classes of polymers derived from vegetable oils: thermoplastic elastomers and thermosets.

Of particular interest is the design of nanostructured polymers from sustainable sources with well-defined molecular characteristics and competitive properties to petroleum-derived materials. We are developing triblock copolymers with hard, glassy outer blocks and a soft, rubbery midblock which can act as thermoplastic elastomers and pressure sensitive adhesives. We have utilized reversible addition fragmentation chain transfer polymerization for polymerizing fatty acrylates, for the production of triblock copolymers. The thermal and mechanical properties of the polymers are readily tuned by varying the composition of fatty acids used as the raw material source for the polymers.

Epoxy resins have been synthesized containing epoxidized soybean oil (ESO) and traditional petroleum-derived epoxy resin components (the diglycidyl ether of bisphenol A, DGEBA, and methylene dianiline). Significantly higher curing temperatures are required for ESO as compared to DGEBA. Increasing the concentration of ESO in the resin significantly decreased the glass transition temperature. However, epoxy resins with high ESO content (60-100 wt% relative to the total epoxide-bearing molecules in the mixture) exhibited a second, higher temperature peak in the derivative of the thermogravimetric analysis weight loss curve. Significant differences were observed in the hydrolytic degradation characteristics of the polymers at 80 °C in sodium hydroxide (NaOH) solutions. The presence of degradable cleavage points throughout the network structure may provide a valuable route for recycling these materials after their useful lifetime.

(517b) Non-Classical Sphere Forming in A-B-A-C Tetrablock Copolymers

Zhang, J., University of Minnesota
Sides, S., Tech-X Research Corporation
Bates, F. S., University of Minnesota
Originally presented on: 11/6/2013 13:00:00 - 13:20:00

AB diblock and ABC triblock copolymers have been studied thoroughly. ABAC tetrablock copolymers, representing the simplest variation from ABC triblock by breaking the molecular symmetry via inserting some of the A block in between B and C blocks, have been studied systematically in this research. The model system is poly(styrene-b-isoprene-b-styrene-b-ethylene oxide) (SISO) tetrablock terpolymers and the resulting morphologies were characterized by nuclear magnetic resonance, gel permeation chromatography, small-angle X-ray scattering, transmission electron microscopy, differential scanning calorimetry and dynamic mechanical spectroscopy. Three novel phases are first discovered in a single component block copolymer melts: hexagonally ordered spherical phase, dodecagonal quasicrystalline (QC) phase and Pm3n cubic phase. In particular, the discovery of QC phase bridges the world of soft matters to that of metals. These unusual sets of morphologies will be discussed in the context of segregation under the constraints associated with the tetrablock molecular architecture. Theoretical calculations based on the assumption of Gaussian chain statistics provide valuable insights into the molecular configurations associated with these morphologies.

(517c) Evaluating the Structure and Properties of Polyimide-Poly (Ethylene Glycol) Materials for Fuel Cell Membrane Applications

Coletta, E., Stanford University
Toney, M., SLAC National Accelerator Laboratory
Frank, C. W., Stanford University
Originally presented on: 11/6/2013 13:20:00 - 13:40:00

Polymer electrolyte fuel cells have the potential to provide cleaner and more efficient energy. Often, current polymer electrolyte membranes for these fuel cells are limited in terms of their operating conditions, especially with regard to humidity and temperature. By using different polymers than have been traditionally used, more versatile operating conditions and better efficiencies may be achieved. The structural details of current polymer electrolyte membranes is often not studied in great detail. By performing a fundamental and in depth study of the structure of potential polymer membrane materials, more insight can be gained in terms of achieving good performance through material design. A multi-polymer system has been created and analyzed in terms of the relationship between ionic conduction properties and nanometer scale structure.

In the current work, aromatic polyimide and poly (ethylene glycol) (PEG) composite membranes have been synthesized for fuel cell polymer electrolyte membrane applications. Because different properties are needed for an effective polymer electrolyte membrane, multiple polymers often need to be used. Also, these multi-polymer systems often lead to phase separation and self-assembly that can greatly impact material properties. Aromatic polyimides are known to be very stable, both thermally and mechanically. Poly (ethylene glycol) (PEG) has been known to provide proton conduction properties under certain conditions. The goal is to exploit the drastically different properties of the polymers and understand how the structure of these materials relates to the conductivity in this hybrid system. The polyamic acid precursors to these polymer systems were synthesized by a classical, random, one pot poly-condensation method that has been shown to work for many different types of polyamic acids. The poly (ethylene glycol) (Mn 990 to 3400) was used in pre-polymerized form and incorporated in the polyamic acids through chemical bonding to achieve structural self-assembly on the nanometer length scale. These polymer systems were converted to polyimide-PEG systems through thermal imidization. This thermal processing can achieve a variety of film properties easily if varied. Composition and morphology of these hybrid systems was varied to create a family of materials.

The polyamic acid precursors and polyimides were characterized using Fourier transform infrared spectroscopy. Thermal properties of the films were analyzed using thermal gravimetric analysis. Small angle x-ray scattering was used to determine the polymer structure on a length scale between 0.5 and 80 nanometers, because it has been shown that features on this length scale can often lend themselves to achieving good conduction properties. Oftentimes conductivity of materials requires control of the material structure on a very small length scale and understanding how structure can improve ion transport is important in fuel cell membranes. Since liquid electrolytes are often needed to provide any significant conduction properties for polymers, inorganic acid or ionic liquids were incorporated into the films by soaking. The films were then analyzed for proton conductivity properties using electrochemical impedance spectroscopy in conjunction with cyclic voltammetry. Effects of temperature and humidity were studied with regards to the conductivity. These composite systems were analyzed in order to determine composition-structure-property relationships for fuel cell membrane applications. Composite or hybrid systems have been studied for structure, or properties, but rarely is the relationship between structure and properties well understood.

(517d) Molecular Influence On Segmental Dynamics of Select Model Poly(urethane urea) Elastomers

Hsieh, A. J., U.S. Army Research Laboratory
Chantawansri, T. L., University of California, Santa Barbara and Army Research Laboratory (AMSRD-ARL-WM-MA)
Hu, W., University of Massachusetts, Amherst
Originally presented on: 11/6/2013 13:40:00 - 14:00:00

Segmental dynamics in a series of 4,4'-dicyclohexylmethane diisocyanate- diethyltoulenediamine-poly(tetramethylene oxide) based poly(urethane urea) (PUU) elastomers have been investigated. Scale-bridging of segmental dynamics determination was accomplished through a well-coupled multiscale characterization and multiscale modelling approach. New experimental observations obtained by dielectric analysis and solid-state nuclear magnetic resonance (NMR) not only clearly elucidate the key material attributes that are critical to a material's response at the moment of target interaction with either accelerating projectile or other stress wave excitation, but also serve to provide validation with respect to high strain-rate deformation results from simulation. Solid-state NMR results further reveal the molecular dynamics associated with respective hard segment and soft segment domains, as well as with various phase mixed regions, thus allowing for better corroboration with microstructure details discerned from atomic force microscopy analysis as well as for enabling synthesis of hierarchical PUU elastomers with desired dynamic deformation response to provide enhanced protection under extreme dynamic environments.

(517e) Dynamic Stiffening of Liquid Crystal Elastomers

Agrawal, A.,
Patra, P.,
Chapman, W. G., Rice University
Ajayan, P. M., Rice University
Verduzco, R., Rice University

Originally presented on: 11/6/2013 14:00:00 - 14:20:00

Dynamic Stiffening of Liquid Crystal Elastomers

Aditya Agrawal[1], Prabir K. Patra[4], Pulickel M. Ajayan[2], Walter G. Chapman[1], and Rafael Verduzco[1]

[1]Rice University, Department of Chemical and Biomolecular Engineering, Houston, TX

[2]Rice University, Department of Mechanical Engineering and Materials Science, Houston, TX

[3]Rice University, Department of Biochemistry and Cell Biology, Houston, TX

[4]University of Bridgeport, Departments of Biomedical Engineering and Mechanical Engineering, Bridgeport, CT

Liquid crystal elastomers (LCEs) are rubbery, crosslinked polysiloxane networks with liquid crystal side-groups. Due to the presence of liquid crystalline order, LCEs respond to heat, light, and magnetic and electric fields. Monodomain LCEs, which have uniform ordering of the nematic director, exhibit reversible shape changes to external stimuli. Here, we explore the rheological properties of LCEs in response to low-amplitude, repetitive deformations. While conventional elastomers respond reversibly, nematic LCEs exhibit a dramatic increase in stiffness ? up to 90 % - when subjected to a low-amplitude, repetitive compression. By studying a systematic series of materials, we demonstrate that the stiffness increase is directly influenced by the liquid crystal content of the elastomers, the presence of a nematic liquid crystal phase and the use of a dynamic as opposed to static deformation. Through a combination of rheological measurements, polarizing optical microscopy and 2-D X-ray diffraction, we demonstrate that self-stiffening arises due to rotations of the nematic director in response to dynamic compression, and show that the behavior is consistent with the theory for nematic rubber elasticity. We explore the response of nematic LCEs to uniaxial extension and to varying frequencies and amplitudes of deformation. We also investigate the use of dynamic deoformations to align higher order LCEs, including chiral nematic and smectic networks. The use of a low-amplitude, dynamic deformation represents a potentially straightforward route to achieving uniform alignment in LCEs.Previous work with liquid crystal elastomers has focused primarily on ?soft elastic' deformations at large strains, but our findings indicate rich behavior at previously overlooked low-strain, dynamic deformations.

Keywords: Liquid crystal elastomers, stiffening, dynamic mechanical analysis.

(517f) Modification of Linear Prepolymers to Control Polymerization-Induced Phase Separation in a Free-Radical Photo-Polymerization

Szczepanski, C.,
Stansbury, J. W., University of Colorado-Denver , School of Dental Medicine
Originally presented on: 11/6/2013 14:20:00 - 14:40:00

Multi-phase, heterogeneous polymer networks have the potential to alleviate the stress experienced during a polymerization reaction[1-3]. One approach to network heterogeneity is through polymerization induced phase separation (PIPS). Here we present a method to adjust and expand the limits at which PIPS leads to effective stress reduction by varying the molecular weight of the prepolymer added to induce phase separation during a free-radical polymerization.

Polymerization stress leads to both internal and interfacial defects of a network, which manifest as cracks within a network or delamination of the material from the substrate to which it has been applied. To improve the utility and longevity of polymeric materials, our research has focused on understanding and optimizing both materials-based and processing-based techniques to reduce and eliminate polymerization shrinkage and stress, while maintaining critical properties such as mechanical strength, overall conversion and appearance.

Through PIPS, phase separation is favored as the entropy of mixing decreases during reaction, causing a multi-phase network to be formed from the polymerization of an initially homogeneous multi-component monomer/prepolymer mixture. Previously, we have reported a method to introduce network heterogeneity via PIPS in a one-step, purely photo-initiated, free radical polymerization[4]. Polymer networks formed via photo-initiation have the advantage of spatial and temporal control of the reaction - which is favored, and often necessary for application in stereo-lithography, coatings and biomaterials. In our approach, a bulk dimethacrylate polymerization (TEGDMA-based) was modified by the addition of linear non-reactive prepolymers: poly-methyl, ethyl, and butyl methacrylates (PMMA, PEMA, and PBMA) to induce phase separation shortly after photo-irradiation. The use of photo-initiation and ease of implementation of this technique broadens the scope of applications for PIPS, especially those where biological compatibility is essential.

One advantage of PIPS is that the phase separation process and the resulting phase structure can be modified through tuning of the kinetics and thermodynamics of the polymerization reaction. In a photo-initiated system, simple kinetic tuning can be achieved by varying the rate of initiation through irradiation intensity. Thermodynamics of the reaction can be modified through the design of prepolymers and selection of the monomer(s). Unfortunately, the kinetic and thermodynamic factors cannot be probed and tested separately, since they are co-dependent. The degree and extent of phase separation that occurs during PIPS is a direct result of the dynamic competition between kinetic formation and thermodynamic instability in the network.

In our studies utilizing PMMA, PEMA, and PBMA to modify TEGDMA-based polymerizations, we observed differences in resulting phase structure, domain size, efficiency at stress reduction, and kinetics of phase separation based on the prepolymer utilized and the loading level thereof. Many of these differences arise from the varying molecular weight and glass transition temperature as well as the composition of the prepolymers. In this work, we present systematic studies of prepolymer physical properties, specifically molecular weight, and how it can be used to manipulate the balance between kinetics and thermodynamics during PIPS, the resulting phase structure, and stress reduction efficiency.

Three different PMMA prepolymers were synthesized, with increasing molecular weights (1.5K, 11K, and 120K Da). As the molecular weight decreases, the minimum loading level at which two distinct phases exist in the final polymer network increase (1.5K – 10wt%, 11K – 5wt%, 120K – 1wt%), as indicated by the tan delta profile post cure. This indicates that as molecular weight decreases, the prepolymer behaves more like filler within the network. Only at sufficiently high loading levels will thermodynamic instability be encountered during the polymerization, thus promoting PIPS.

Using real-time tensometry, we measured the efficiency of polymerization stress reduction as a function of loading level, molecular weight of the prepolymer, and irradiation light intensity. By increasing the irradiation light intensity, the rate of polymer network formation increases, and the time for diffusion of incompatible phases decreases. This results in a decrease in the extent of phase separation, and can lessen the amount of stress reduction. For PMMA with a molecular weight of 1.5 KDa, the kinetic network formation dominates and eliminates stress reduction via PIPS at irradiation intensities equal to or greater than 300μW/cm^2. However, this threshold occurs at 5mW/cm^2 for 11 KDa prepolymer, and 20mW/cm^2 for 120 KDa prepolymer, indicating that by increasing the molecular weight of prepolymer, stress reduction via PIPS can still occur in high intensity photo-polymerizations. This is a favorable result for applications such as dental restorative materials where minimum reaction time and maximum overall conversion is favored. Our future work will investigate approaches to alter the order of polymerization of incompatible phases, and whether this can be achieved through prepolymer modification. Additionally, we will apply this understanding to work in co-polymer based systems appropriate for dental restorative applications.

Works Cited

1. Velazquez R, Sanchez F, Yanez R, and Castano VM.Journal of Applied Polymer Science 2000;78(3):586-591.

2. Velazquez R, Ceja I, Guzman J, and Castano VM.Journal of Applied Polymer Science 2004;91(2):1254-1260.

3. Li W and Lee LJ.Polymer 2000;41(2):685-696.

4. Szczepanski CR, Pfeifer CS, and Stansbury JW. Polymer 2012;53(21):4694-4701.

(517g) Control of Thermal Degradation of Poly(lactic acid) Using Poly(ã-glycidyloxypropylsiloxane) Microspheres As Chain Extenders

Han, T., East China University of Science and Technology
Xin, Z., State Key Laboratory of Chemical Engineering, East China University of Science and Technology, School of Chemical Engineering
Shi, Y.,
Originally presented on: 11/6/2013 14:40:00 - 15:00:00

Control of thermal degradation of poly(lactic acid) using poly(γ-glycidyloxypropylsiloxane) microspheres as chain extenders

Ting Han, Zhong Xin*, Yaoqi Shi

State Key Laboratory of Chemical Engineering, School of Chemical Engineering, East China University of Science and Technology, P.O. Box 545, Meilong Road 130, Shanghai 200237,People's Republic of China

Corresponding author. Tel.: +86 2164252972; fax: +86 2164240862.

E-mail address: xzh@ecust.edu.cn (Z. Xin).

Poly(lactic acid) (PLA) is a linear, aliphatic thermoplastic polyester that can either be semi-crystalline or amorphous depending on the stereopurity of the polymer backbone. PLA can be synthesized from 100% renewable resources such as fermentation products of corn and sugar beets. Due to its good biodegradability, biocompatibility, high mechanical strength, and excellent shaping and moulding properties,PLA should have been widely applied and is on the expectation to substitute petroleumbased polymers.

However, the processing instability of PLA (i.e.,thermal, oxidative, and hydrolytic degradations), hinders its development in practice, These degradations will cause the cleavage of polymer chains and a decrease in molecular weight, resulting in a deterioration of the rheological properties. Moreover, the low melt viscosity of PLA can also destroy its performance in the blow molding and foaming process. To overcome such shortcomings, attempts to control the melt rheology of PLA by means of increasing molecular weight to compensate for the molecular weight decrease caused by processing degradation, have motivated considerable research efforts. Copolymerization, blending and filling techniques can be used, among of which, the incorporation of chain extension agents into PLA to produce extended PLA is more attractive due to its lower cost .

Chain extension reactions not only offer the opportunity for enhancement of physical and chemical properties by increasing the molecular weight but also introduce new functional groups onto the PLA backbone paving the way for preparation of composites, laminates, coated items, and blends/alloys with improved properties and cost effectiveness. In the meantime they are economically advantageous because they can be carried out in the melt, with only low amounts of chain extending agents, and separate purification steps are not required. In industry, chain extension reactions have been developed to generate branching structures to increase shrinkage and extension of melt-spun PLA fibers.

For the purpose to eliminate degeration during the processing of PLA, the functional polysilsesquioxane microspheres(PSQ), synthesized by one-step process with high content of epoxy groups, were introduced as a novel chain extension agent to graft onto the commercial PLA with relatively controlled structures via functional group reactions through melt processing. The characterization of the product was, investigated by rheometer, FTIR, SEC and melt index meter.

Firstly, the effects of reaction time on the properties of branched poly(lactic acid) grafted by Poly(γ-glycidyloxypropylsiloxane) microspheres (PSQ-g-PLA) was fully investigated. With the increase of reaction time, there was only a slight change in molecular weight, between 200.0kg·mol^{-1} and 220.0kg·mol^{-1}(PDI=1.7), which was indicative of inhibited degeration of PLA. During the process of thermal degradation, PSQ could work as a chain extender to covalently bond the segments of PLA through functional group reaction, which dynamically balanced the molecular weight. With the increase of reaction time, melt index of PSQ-g-PLA showed a increasing trend, but the change is not obvious. The melt index showed a minimum of 3.793g·10min^{-1} at the reaction time of 5min. The rheological properties of PSQ-g-PLA, dynamic shear viscosity, storage modulus, loss modulus were greatly improved compared with the neat PLA. From the plot of viscosity (η) vs frenquecy (γ) in low frequency,the existence of branched structure was confirmed and 10min-40min was enough for increasing molecular weight compensate for the molecular weight decrease caused by processing degradation and increase the melt viscosity. With the increase of reaction time, the mechanical properties showed the same trend with rheological properties. The stiffness and toughness of PSQ-g-PLA were simultaneously enhanced when the processing time was 10min and 20min. The impact strength reached to a maximum value of 60 J m^{-1} at the reaction of 30min, which was two times more than that of the neat PLA.

Secondly, the effects of processing temperature to PSQ-g-PLA have been studied. With the increase of the reaction temperature, the molecular weight showed the maximum of 250.4kg·mol^{-1} at 200 Â°C, which can be attribute to the balance of the reaction rate and the thermal degradation of polylactic acid. The melt index showed the minimum of 3.938g·10min^{-1} at 190 Â°C. The rheological properties showed the same trend as the molecular weight, and the dynamic shear viscosity showed the maximum value at 190 Â°C,which indicated the optimized temperature for higher degree of branching.

Lastly, the influence of the additive concentration to the grafting reaction have been researched. With the content of PSQ increased, the molecular weight showed a maximum of 196.7kg·mol^{-1} at 2%, which provided us convincing evidence to hypothesize that a small amount or excess PSQ will aggravate the degradation of polylactic acid, while melt index showed a minimum of 3.624g·10min^{-1} at 4%. Compared with neat PLA, shear viscosity of each PSQ-g-PLA sample increased which implied the formation of branched structure. A positive correlation between viscosity and additive concentration was found below the content of 2%, while a plain was found after 2%.

According to the values of shear viscosity, the reaction time, temperature and the additve concentration were optimized to 10mins, 190 Â □?C and 2%, respectively. The coresponding shear viscosity of PSQ-g-PLA was 4500 Pa.s, approximately six times that of neat PLA, while the increment by peroxide initiated branching was just 25%. The M_w of PSQ-g-PLA was 250.4kg mol^{-1}, about three times more than that of neat PLA, and the melt index was 3.624 g 10min^{-1}, reduced to almost the half of the neat PLA. More importantly, the stiffness and toughness of PSQ-g-PLA were simultaneously enhanced with the addition of functional polysilsesquioxane microspheres. the impact strength was increased by more than two-fold, up to 60 J m^{-1}. Finally, this structural information was correlated with its foaming ability, with guidance on its potential applications in polymeric foam or film.

(520a) Long-Range Angular Correlations in Liquid Water

Liu, Y., University of California, Riverside
Wu, J., UC Riverside
Originally presented on: 11/6/2013 12:30:00 - 12:50:00

The microscopic structure of liquid water is of fundamental importance to our understanding of aqueous systems including hydrophobic phenomena and vast biological processes at the microscopic scales. In contrast to that for simple fluids, the local structure of liquid water reflects not only the molecular excluded volume effects but also a partially tetrahedral network of hydrogen bonds among water molecules. However, it is commonly believed that, near ambient conditions, the structure of liquid water lacks long-range correlations as shown in the structure factor or the pair distribution functions obtained from by x-ray or neutron diffraction techniques or from molecular simulations. In this work, we demonstrated, by using a molecular density functional theory, the existence of long-range orientational correlations in liquid water at a length scale of about 40 Å. The angular correlation function attenuates in an oscillatory manner consistent with recent experimental studies. We also examined the angle dependence of density profile and found that the long-range correlations are mainly in the dipolar directions and little influenced by the rotational angles.

(520b) Direct Numerical Test of the Statistical Mechanical Theory of Hydrophobic Interactions

Chaudhari, M., Tulane University
Pratt, L. R., Tulane University
Originally presented on: 11/6/2013 12:50:00 - 13:10:00

This work tests the statistical mechanical theory of hydrophobic interactions, isolates consequences of excluded volume interactions, and obtains B2 for those purposes. Cavity methods that are particularly appropriate for study of hydrophobic interactions between atomic-size hard spheres in liquid water are developed and applied to test aspects of the Pratt-Chandler (PC) theory that have not been tested. Contact hydrophobic interactions between Ar-size hard-spheres in water are significantly more attractive than predicted by the PC theory. The corresponding results for the osmotic second virial coefficient are attractive (B2 <0), and more attractive with increasing temperature (Delta B2/Delta T < 0) in the temperature range 300K < T < 360K. This information has not been available previously, but is essential for development of the molecular-scale statistical mechanical theory of hydrophobic interactions, particularly for better definition of the role of attractive intermolecular interactions associated with the solutes.

(520c) Dissecting Hydrophobic Hydration and Association

Remsing, R. C., University of Maryland
Weeks, J. D., University of Maryland

Originally presented on: 11/6/2013 13:10:00 - 13:30:00

We use appropriately defined short ranged reference models of liquid water to clarify the different roles local hydrogen bonding, van der Waals attractions, and long ranged electrostatic interactions play in determining the structure and thermodynamics of the solvation and association of apolar solutes in water. While local hydrogen bonding interactions dominate hydrophobic effects involving small solutes, longer ranged electrostatic and dispersion interactions are found to be increasingly important in the description of interfacial structure around large solutes. The hydrogen bond network sets the solute length scale at which a crossover in solvation behavior between these small and large length scale regimes is observed. This network is also found to be crucial for the existence of the phenomena of entropy convergence. Unbalanced long ranged forces acting on interfacial water molecules are also important in hydrophobic association, as will be illustrated here through analysis of the association of model methane and buckminsterfullerene solutes.

(520d) Not a Tiny Bit of Iceberg Near a Hydrophobic Molecule

Kim, J., University of California, Riverside
Tian, Y., UC Riverside
Wu, J., UC Riverside
Originally presented on: 11/6/2013 13:30:00 - 13:50:00

In a lengthy article published almost 70 years ago, Frank and Evans suggested that a rare gas atom or a non-polar molecule dissolved in liquid water result in freezing of the surrounding water molecules even at room temperature. While these authors rightfully cautioned with a footnote that the word freezing or iceberg does not imply that the water structure near the solute is exactly ice-like, the behavior of water molecules at the hydrophobic surface, more specifically, the iceberg picture, has been a topic of perennial debate in physical chemistry. By an integrated description of hydrophobic hydration based on thermodynamics, molecular simulation and the scaled-particle theory (SPT), we analyzed the effect of several hydrophobic solutes on hydrogen bonding among water molecules over a broad range of temperatures. A careful analysis of different contributions of the intermolecular forces to the total solvation energy provides unequivocal evidence that a hydrophobic solute reduces the degree of hydrogen bonding in the neighboring water molecules in comparing to that in the bulk liquid at all temperatures.

(520e) Solid-Water Interfaces: Fundamental Observations to Enhance Practical Devices

Striolo, A., University of Oklahoma

Originally presented on: 11/6/2013 13:50:00 - 14:10:00

A number of experimental and simulation results consistently show that a solid surface perturbs interfacial liquids for up to 1-2 monolayers. Our group is interested in understanding how it is possible that such short-ranged perturbations result in macroscopic observations that, in the case of aqueous interfaces include hydrodynamic slip, salt-specific effects relevant for protein folding and precipitation, self assembly of lipids and nanoparticles, and even the preservation of artistic heritage. (For example, Giotto's frescos in the Scrovegni's chapel, Padua, Italy, and the Mogao caves, in Dunhuang, China, a lasting but frail testimony to the Buddhist influences on modern Chinese culture, are endangered by the humidity caused by the breathing visitors.) Our phylosophy consists in emploing molecular simulations to provide the details needed for interpreting, and when possible predicting experimental observations. I will present the results for structure and dynamics of water on a few solid-liquid interfaces, and how such predictions have been tested by AFM. I will then discuss how such observations could be extended to understand the partition of aqueous electrolytes (e.g., NaCl vs. $CaCl_2$) within narrow carbon pores, and how such effects could lead to the design of enhanced electrodes for water desalination and energy-storage devices.

(520f) Nanoscale Behavior of DNA Oligomers Near Hydrophobic and Hydrophilic Functionalized Surfaces

Elder, R. M., University of Colorado at Boulder
Jayaraman, A., University of Colorado, Boulder
Originally presented on: 11/6/2013 14:10:00 - 14:30:00

A fundamental understanding of the interactions of DNA with surfaces, and the behavior of DNA near surfaces, is important for emerging biomaterials technologies, such as DNA origami and the self-assembly of DNA-grafted nanoparticles. We use atomistic molecular dynamics and biased sampling techniques to study single-stranded DNA (ssDNA) oligomers on model hydrophilic and hydrophobic functionalized surfaces to elucidate the molecular interactions between the DNA, the surface, and water molecules. ssDNA oligomers are amphiphilic and adsorb strongly to both hydrophilic and hydrophobic surfaces, but different forces govern adsorption to the two surfaces. When the ssDNA is in the vicinity of the hydrophilic surface, it forms attractive hydrogen bonds with the surface species, while facing a repulsive force from the dense, strongly adsorbed water molecules at the surface. When the ssDNA is in the vicinity of the hydrophobic surface, it experiences weak hydrophobic attraction to the surface and a surprising repulsive force associated with a low-density region of water near the hydrophobic surface.

(520g) Multi-Scale Concepts in the Molecular Theory of Electrolyte Solutions: Mcmillan-Mayer and Quasi-Chemical Theory

Zhang, W., Tulane University
You, X., Tulane University
Pratt, L. R., Tulane University
Originally presented on: 11/6/2013 14:30:00 - 14:50:00

Multi-scale concepts in the molecular theory of electrolyte solutions: McMillan-Mayer and Quasi-Chemical Theory

W. Zhang, X. You, and L. R. Pratt

Department of Chemical and Biomolecular Engineering

Tulane University, New Orleans, LA 70118

The role of statistical mechanical theory in understanding electrolyte solutions at a molecular level has shifted substantially over the past few decades. The principal reason is that molecular simulation has steadily improved in precision and scope. By now, simulation calculations bypass dominating obstacles of non-simulation statistical mechanical theory without comment or resolution. Nevertheless, some of those theoretical obstacles reflect our conceptual understanding, and deserve resolution even if alternative brute force computations are widely available. Multi-scale concepts are needed to enable difficult chemical-scale *ab initio* molecular dynamics (AIMD) calculations to be applied to obtain the thermodynamics of complex electrolyte solution. For example, charges are transferred from charge donor to charge acceptor when one ion approaches another counter-ion. Development of statistical mechanical theory can assist with a better understanding of such chemical effects as charge transfer between ions.

Here we address one such conceptual point, namely the coarse-graining that eliminates the solvent molecules from direct consideration when treating electrolyte solutions. It is common to consider ions in solutions with the solvent replaced by a uniform dielectric medium. In a primitive case, for example, hard-spherical ions are considered, the solvent is not actively present, and, when the ion hard spheres do not overlap, the interactions between ions is assumed to be $q_iq_j/4Â^1\varepsilon r$ where q_i is the formal charge on the ith ion, $(\varepsilon/\varepsilon_0)$ is the relative dielectric constant of the solvent, and r is the separation of the ionic charge. Typically called "primitive models," such models are not justified by direct molecular-scale observation of the solvent, *i.e.*, the solvent is not actually a dielectric continuum. The required justification is the target of statistical mechanical theory, and clearly must be more subtle.

Proposed justifications are often intuitive,[1] and sometimes solely empirical as in fitting modeled thermodynamic properties to data.[2] But the theory for elimination of the solvent has been conclusively considered: it is the McMillan- Mayer (MM) theory,[3],[4],[5] and it is the pinnacle of the theory of coarse-graining for the statistical mechanics of solutions. "Primitive model" is then synonymous[5],[6] with "McMillan-Mayer model."

The foremost feature of MM theory is that the solvent coordinates are fully integrated out. The statistical mechanical problem that results from MM analysis treats the solute species (ions) only, but with effective interactions that are formally fully specified. Those effective interactions typically are complicated.[7],[8] Though it can be argued that no sacrifice of molecular realism is implied by MM theory, cataloging the multi-body potentials implied by a literal MM approach is prohibitively difficult.[9] Use of MM theory to construct a specific primitive model for a system of experimental interest has been limited.[7,8]

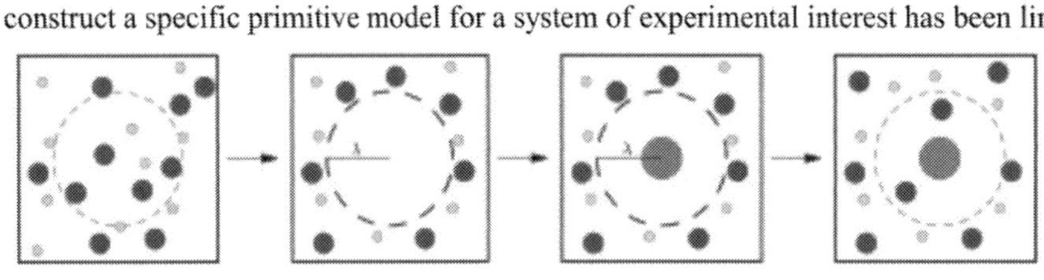

Figure SEQ Figure * ARABIC 1 Evaluation of the excess chemical potential of the interesting ion (red

(520h) Model-Free Test of Local Density Mean-Field Behavior in Electric Double Layers

Giera, B., University of California Santa Barbara
Kober, E., Los Alamos National Laboratory
Squires, T. M., University of California, Santa Barbara
Shell, M. S., University of California Santa Barbara
Originally presented on: 11/6/2013 14:50:00 - 15:10:00

Electric double layers (EDLs) form at all interfaces between charged surfaces (including electrodes, colloids, proteins, and cell membranes) and electrolytes or ionic liquids. EDL structure governs the differential capacitance, electrokinetic flow, surface conductivity, capacitive desalination, and rational design of supercapacitors, for example. The standard starting point to modeling EDLs are "local density approximations" (LDAs) that assume uncorrelated, bulk-like ion interactions to predict the spatial ion density profiles. LDAs are ubiquitous in many fields, e.g., models of ionic liquids, ion distributions near liquid-liquid interfaces, and electrochemical cells, because they are simple yet physically insightful. Despite their widespread use, numerous computations and experiments measure deviations from LDAs in a variety of electrochemical systems.

We present a general and powerful formalism to identify and elucidate LDA behavior in EDLs that follows from a remarkably simple observation: in any LDA, an ion does not know where the interface is located. Without assumptions, our approach enables one to analyze planar EDLs and determine directly from contact value expressions whether any LDA can possibly succeed, without considering any specific LDA models. If an LDA will work, measured profiles of the free charge density for many different electrolyte systems will collapse onto a single, master curve as a function of an intrinsic, derivable similarity variable. Where measured EDL profiles do not collapse, the test reveals the boundaries of LDA feasibility space, beyond which more sophisticated treatments capable of addressing correlation effects are needed.

We demonstrate this model-free test using extensive molecular dynamics simulations of EDLs with detailed ions and interactions in which LDAs are by no means guaranteed to work. Our model-free approach is broadly applicable to any planar EDL system and will be valuable in assessing potential LDA behavior in more general electrolytes, e.g. where solvation effects, emergent electrostatic correlations, and/or discrete interfacial charges play an important role.

(525a) Patchy Particles Made Easy: Developing General Colloidal Approaches to Produce Anisotropic Building Blocks for Novel Functional Materials

Klupp Taylor, R. N., University of Erlangen-Nuremberg
Bao, H., University of Erlangen-Nuremberg
Sadafi, F. Z.,
Originally presented on: 11/6/2013 15:15:00 - 15:35:00

Conventional applications of small particles generally make use of isotropic interactions to form dense, closely packed layers or dilute dispersions in a matrix. There is however, a growing interest in the exploration of particles with anisotropic properties. On the one hand, anisotropy enables particles to behave as "intelligent" building blocks, assembling according to asymmetrical interactions to form hierarchical structures with novel characteristics. On the other hand, exciting technological possibilities to improve existing materials and processes will be developed. Metallic Janus and patchy particles have drawn much attention in the past few years. They have many potential applications in the field of optics, sensors, drug delivery, catalysis or self-assembled hierarchical materials. However, versatile and scalable methods for forming metallic patches on dielectric nanoparticles are still rather scarce. One common approach is to vacuum deposit noble metal such as gold onto a monolayer of silica or polystyrene particles. Another is to introduce particles at the boundary between two phases with the patch-forming chemical reaction selectively taking place in one phase. In our group a method of fabrication of silver and silver/gold patches on silica nanospheres via heterogeneous nucleation and surface growth has been established. Compared to the abovementioned approaches, our method is based on facile liquid phase colloidal techniques without the need for chemical functionalization or templating and allows the morphology and properties of the nanostructures to be influenced by process variables. Recently we have expanded this work considerably to explore the potential generality of the technique.

In this contribution we will report on the use of different core particles, including silica and anionic and cationic polystyrene nanospheres, which permit modulation of the interactions between core, metal precursor and the growing patch. Our systematic experiments show that while heterogeneous nucleation is promoted by electrostatic driven accumulation of the patch precursor, the interfacial chemistry of the core particle has a critical influence on the mode of growth of gold or silver patches. For instance, for increasing ratios of ascorbic acid (reductant) to the chloroauric ion (gold precursor) the gold patch morphology on cationic polystyrene spheres changes significantly from small nanospheres to finely structured dendritic-like patches. The activation of such surface diffusion-limited growth is considered in the framework of accepted models, like diffusion limited aggregation. We will also report on attempts to control the location of heterogeneous nucleation, a critical step in developing a reliable process for producing patchy particles in large quantities with uniform morphology.

(525b) Modeling of Tunable Structural Re-Configuration of Janus Colloidal Particles

Beltran-Villegas, D. J., University of Michigan
Larson, R. G., University of Michigan

Originally presented on: 11/6/2013 15:35:00 - 15:55:00

Colloidal particles can assemble into a myriad of structures by virtue of the many interaction forces available to them. Variable range attraction and repulsion and the recently explored non-isotropic character, exemplified by Janus particles, are examples of the versatility of colloidal particles as building blocks. A systematic approach to understand the assembly of Janus colloids, as a function of Janus balance and particle concentration is not yet available. In this work we study the phase behavior of Janus particles as a function of the strength of interaction, Janus balance and volume fraction of spherical particles. A secondary goal of this work is the assessment of re-configurability of the structures found. Our results show the range of stability of several structures, including a fluid phase of small clusters, bilayers and worm-like aggregates. We find the bilayer structures are very stable over a range of phase space and provide a good pre-cursor to hexagonally close-packed structures. These findings enable the understanding of the assembly process of Janus building blocks and provide a framework with which to study the kinetics of structure change.

(525c) Design of Multipolar Colloidal Rods Using Discontinuous Molecular Dynamics

Rutkowski, D., North Carolina State University
Velev, O. D., North Carolina State University
Hall, C. K., North Carolina State University
Originally presented on: 11/6/2013 15:55:00 - 16:15:00

Colloids with anisotropic charge distributions hold promise for creating a number of useful new materials including optical materials with novel symmetries, electrical materials for information storage, and dampers for controlling vibrations in structures. Because the experimental characterization of the many possible types of multipolar colloidal particles is currently infeasible, the search for novel colloidal materials can be enhanced and guided by simulations of colloidal system assembly. We have simulated a system of dipolar rods with an aspect ratio of 2 to 1 using discontinuous molecular dynamics (DMD). Each dipolar rod was modeled as three overlapping spheres held in a rod shape to represent excluded volume and two smaller, embedded spheres to represent the charges that make up the dipole. From NVT simulations of these particles we have discovered the existence of fluid, string-fluid, and "gel" phases at low volume fractions and nematic phases at high volume fractions. A phase diagram has been constructed using a variety of methods to delineate the boundaries between the various phases. Our nematic phase boundary at high volume fraction and temperature is consistent with results from Frenkel and Bolhuis on hard spherocylinders. Additionally, we have found evidence for temperature-dependent hexagonal and square ordering for the rods at high volume fractions. We have also simulated dipolar rods of various lengths to learn how this modifies the phase diagram.

(525d) Stop Flow Lithography in Perfluoropolyether (PFPE) Microfluidic Channels

Bong, K. W., Harvard Medical School, Massachusetts General Hospital, and Department of Chemical Engineering in Massachusetts Institute of Technology
Lee, J., MIT
Doyle, P. S., Massachusetts Institute of Technology
Originally presented on: 11/6/2013 16:15:00 - 16:35:00

Stop Flow Lithography (SFL) is a microfluidic-based particle synthesis method for creating anisotropic multifunctional particles with applications that range from MEMs to biomedical engineering. Polydimethylsiloxane (PDMS) has been typically used to construct SFL devices as the material enables rapid prototyping of channels with complex geometries, optical transparency, and large gas permeability. However, PDMS is not compatible with organic solvents which limit the current range of materials which can be synthesized with SFL. Here, we demonstrate a fluorinated elastomer, called perfluoropolyether (PFPE), can be an alternative oxygen permeable elastomer for SFL microfluidic flow channels. We fabricate PFPE microfluidic devices with soft lithography and synthesize anisotropic multifunctional particles in the devices via the SFL process - this is the first demonstration of SFL with oxygen lubrication layers in a non-PDMS channel. We benchmark the SFL performance of the PFPE devices by comparing them to PDMS. As PFPE has three times lower elastic modulus than PDMS, the PFPE channels need longer relaxation time that leads to the decrease of the particle production rates. However, the PFPE devices can exhibit similar SFL capabilities with PDMS devices except the particle throughput. We synthesized particles in both PFPE and PDMS devices under the same SFL conditions and found the difference of particle dimensions was sub-microns. PFPE devices can greatly expand the range of precursor materials which can be processed in SFL because the fluorinated devices are chemically resistant to organic solvents, an inaccessible class of reagents in PDMS-based devices. As one demonstrative application, we synthesize electrically conducting particles from organic precursors with stable dispersion of graphene.

(525e) Formation of "Colloidal Molecules" Induced By AC Electric Fields

Ma, F., Colorado School of Mines
Wang, S., Colorado School of Mines
Wu, D. T., Colorado School Of Mines
Wu, N., Colorado School of Mines
Originally presented on: 11/6/2013 16:35:00 - 16:55:00

Colloidal analogues of real molecules with controlled architecture are important building blocks for making mesocopic materials with exotic properties. Existing methods of making colloidal molecules mostly rely on particles that are anisotropic in terms of geometry or interfacial properties. However, those anisotropic particles themselves are challenging to make. In this talk, we demonstrate a simple and versatile method to build colloidal molecules based on isotropic particles that are commercially available or easy to synthesize. By applying an external AC electric field, we show that the apparently isotropic particles experience anisotropic interactions, which lead to the formation of colloidal oligomers. Moreover, those colloidal oligomers can "react" with others to form macro-colloidal molecules or two-dimensional non-close packed structures. We also illustrate the striking similarity between oligomers assembly and chemical reactions of real molecules. Our method can be used to generate complex patterns or to build complex building blocks, all based on simple spherical particles with isotropic properties. Because our method is physical in nature, its potential application to a broad range of particles with different chemical properties is envisioned.

(525f) Using Shape Anisotropy to Toughen Disordered Particle Assemblies

Lee, D., University of Pennsylvania
Zhang, L.,
Brugarolas, T., University of Pennsylvania
Feng, G.,
Originally presented on: 11/6/2013 16:55:00 - 17:15:00

Disordered solids constitute an important class of materials that find wide-ranging applications from energy devices to bone replacement. A major roadblock limiting the widespread utilization of these amorphous materials is their tendency to undergo mechanical failure under small load. The fundamental understanding on their mechanical behavior is imperative for the development of new strategies to enhance their mechanical properties (*hardness*, *modulus*, and *fracture toughness*). In our work, we study the effect of particle shape anisotropy on the mechanical behavior of disordered nanoparticle packings. We study the mechanical response of disordered nanoparticle packings made of TiO_2 prolate ellipsoids with various aspect ratios using nanoindentation. We observe striking similarities in the deformation mechanism of disordered particle assemblies to that of metallic glasses, which are random packings of metallic atoms. It is demonstrated that anisotropic particles greatly suppress shear band formation and toughen particle packings without sacrificing their strength, which implies that tuning constituent-anisotropy may be a new strategy to enhance toughness in disordered solids.

(525g) Rheology of Janus Particle Suspensions Via Brownian Dynamics Simulation

De La Cruz, R. A.,
Cordova-Figueroa, U. M., University Of Puerto Rico - Mayag ez
Originally presented on: 11/6/2013 17:15:00 - 17:35:00

Recent experimental and theoretical studies on the suspension of patchy or Janus colloidal particles have examined collective structures, phase diagrams, and self-assembly at different times scales and particle concentrations. However, only a few efforts have focused on understanding their potential rheological advantages in comparison to their isotropic counterparts. In order to make some progress in this field, we propose a simple model consisting of Janus particles immersed in a Newtonian fluid interacting via a combination of square-well (SW) and hard-sphere (HS) surface potentials and in the absence of hydrodynamic interactions. Using Brownian dynamics simulation, the osmotic pressure, the effective viscosity, and normal stress differences are obtained at equilibrium and away from it, by varying the PÃ©clet number -- the ratio between shear and Brownian forces -- and particle volume fraction. To provide a better description of the microstructure of the suspension and explain the time-dependent rheological properties, we characterize at each time interval the number of clusters, the number of particles in each cluster, the radius of gyration, and the cluster effective size. The results are compared to isotropic suspensions interacting with only HS or SW potentials.

(532a) The Effect of Precursor Salt, Reduction Promoter, and Drying Rate On the Performance of Iron Fischer-Tropsch Catalysts Prepared By Solvent Deficient Precipitation (SDP)

Hecker, W. C., Brigham Young University
Schaffers, W.,
Albretsen, M.,
Argyle, M. D., Brigham Young University
Originally presented on: 11/6/2013 15:15:00 - 15:35:00

The Fischer-Tropsch (FT) process has been a valuable means of producing liquid hydrocarbons and fuels since its invention over 90 years ago. This process continues to draw significant interest in light of current environmental and energy concerns. The development of better - more active, selective, and stable - FT catalysts is an area of ongoing research.

The work presented in this paper will discuss three important variables that have an impact on the performance of iron FT catalysts prepared by solvent deficient precipitation (SDP), a novel dry-mix preparation method that produces Fe FT catalysts in a significantly shorter time than it takes to prepare standard precipitated catalysts. The FT performance (activity/selectivity, physical and chemical properties) of unsupported Fe-Cu-K-SiO_2 catalysts made from 4 different iron precursor salts: 1) iron nitrate, 2) iron citrate, 3) iron oxylate, and 4) iron sulfate – will be compared and discussed. Secondly, we examine the use of manganese instead of copper as a promoter for enhancing reduction of the Fe during preparation. Thirdly, we look at the effect of drying temperature on the performance of a series of six catalysts made from the nitrate salt. Each catalyst was first dried at 60°C, and then ramped to 80,100,115, 130, or 150°C. It was found that the catalyst dried at 60/130°C had significantly higher activity, lower methane make, and better overall performance than the other catalysts.

(532b) Selective Adsorption of Manganese Onto Rhodium for Optimized Mn/Rh/SiO2 Alcohol Synthesis Catalysts

LIU, J.,
Tao, R.,
Zhao, G.,
Klie, R. F., University of Illinois at Chicago
Miller, J. T., Argonne National Laboratory
Marshall, C. L., Argonne National Laboratory
Regalbuto, J. R., University of South Carolina
Meyer, R. J., University of Illinois - Chicago
Originally presented on: 11/6/2013 15:35:00 - 15:55:00

Using supported rhodium-based catalysts to produce alcohols from syngas provides an alternative route to conventional fermentation methods. When left unpromoted, Rh catalysts have a strong selectivity towards methane. However, promotion with early transition metal elements has been shown to be effective to increase alcohol selectivity. Therefore, a key design objective is to increase the promoter-metal interaction to maximize their effectiveness. This can be achieved by the use of the Strong Electrostatic Adsorption (SEA) synthesis method, which utilizes pH control to steer the promoter precursor (in this case MnO_4^-) onto Rh oxide supported on SiO_2. Mn promoted catalysts were created using both SEA and traditional dry impregnation (DI) and subsequently characterized by STEM and EXAFS. The reduction behavior of the catalysts was also obtained by XANES and TPR. Reactivity results for CO hydrogenation showed significant improvement of ethanol selectivity is achieved when the promoter was prepared by SEA in comparison of promoter prepared by DI, suggesting that intimate interaction between the promoter and the active metal is critical for high selectivity.

(532c) Investigating the Effect of Mass Transfer in Supercritical Phase Higher Alcohol Synthesis Over a Cu-Co Based Catalyst

Xu, R., Auburn University
Roberts, C. B., Auburn University
Originally presented on: 11/6/2013 15:55:00 - 16:15:00

There is sustained interest in the synthesis of higher alcohols from synthesis gas (syngas) based on the potential for producing high octane value fuels and fuel additives. Higher alcohols are also valuable in view of their applicability as either raw materials or solvents for producing high value-added chemicals. The abundance of diverse carbonaceous sources for producing syngas, including coal, natural gas and biomass, makes the catalytic synthesis of higher alcohols from syngas appealing as an alternative to petroleum resources. The Cu-Co catalyst is one of the most common catalyst types used in the synthesis of higher alcohols (HAS) from syngas. It is believed that the nature of the interaction between the Cu and Co species in these catalysts significantly impacts the CO conversion and the selectivity towards higher alcohols.

Since the HAS process is so highly exothermic in nature, a supercritical fluid (SCF) can be introduced as a reaction medium in order to better manage this exothermic process while also providing other benefits. SCFs are attractive media for various chemical reactions because of their unique properties, including enhanced heat transfer and mass transfer operations. Introducing a supercritical solvent into HAS can facilitate single phase operation while simultaneously providing gas-like viscosity and liquid-like diffusivity. As such, introduction of a SCF in this heterogeneous catalytic process can enhance the in situ extraction of low volatility products from catalysts pores, diminish transport limitations, and integrate desorption and separation processes. In addition, higher alcohol synthesis from syngas involves a significant reduction in volume due to the stoichiometry of the reaction. As such, higher selectivity towards $C_{2+}OH$ can be expected when applying a properly selected SCF medium into the process based on the application of elevated operating pressures while simultaneously managing the significant exotherm.

In this work, a Cu-Co based catalyst was investigated for the synthesis of higher alcohols from syngas under both gas phase and supercritical phase conditions. The catalyst was prepared by coprecipitation methods and tested in a fixed bed reactor using supercritical hexanes as the reaction medium. The influence of catalyst pellet size on the catalytic performance, the mass transfer of reactants, and the alcohol product distribution has been evaluated under both gas phase and supercritical hexanes phase conditions. The formation of CH_4 was significantly reduced under supercritical hexanes reaction conditions and the effect of various operating conditions on product yield and selectivities have been systematically determined. The results of these catalytic investigations, both with and without the supercritical hexanes medium, demonstrate that the supercritical medium has a significant effect on the selectivity and the productivity towards higher alcohols.

(532d) A Step Towards Scaling Up the Design of a Non-Conventional Fischer Tropsch Reactor: Validation of Simulation Studies On Heat and Mass Transfer Behavior Inside the Reactor Bed

Elbashir, N. O., Texas A&M University - Qatar
Blank, J. H., Texas A&M University - Qatar
Hussain, R.,
Bani Nassr, L., Texas A&M University at Qatar
Kasht, A.,
Afzal, S.,
Originally presented on: 11/6/2013 16:15:00 - 16:35:00

The Fischer Tropsch synthesis is the focal process in which the synthesis gas (a CO and H_2 mixture) is converted to ultra-clean liquid based fuels and value-added chemicals. Many factors impede the current commercial FTS reactor technologies which can mainly be attributed to transport and thermal limitations due to the complex nature of the reaction. Our efforts have been mainly focusing on scaling up reactor technology that facilitate running the reaction in the supercritical fluid (SCF) media to leverage several advantages over the conventional technologies [1]. We started with a series of macro and micro scale investigations aiming at better understanding the phase behavior of the non-ideal reaction mixture while simultaneously identifying its influence on the reaction kinetics [2] and the *in-situ* diffusion behavior [3] of the reactants and products inside catalyst pores. On the macro scale investigations we conducted series of process integration and optimization studies coupled with the development of sophisticated dynamic control systems [4]. Furthermore, we identified the most applicable solvent(s) for the SCF-FTS using different but interrelated approaches that include phase behavior, techno-economic and safety analysis for the proposed process [5]. These investigations supported our efforts on building a bench-scale reactor in order to verify several simulation studies, which is the focus of this communication.

Our modeling efforts to scale up this reactor initially focused on establishing sufficient tools and parameters to differentiate between operating the fixed-bed reactor on either conventional gas phase or in SCF-FTS media. We utilized different diffusion rates into the catalyst particle throughout the catalyst bed as well as different assumptions on the amount of wax and liquids present in capillary pores and the bulk phase [2, 3]. We utilized different diffusion rates into the catalyst particle throughout the catalyst bed as well as different assumptions on the amount of wax and liquids present in capillary pores and the bulk phase.

The reactor bed behavior is normally combined with a severe pressure/temperature gradient depending on the reaction media. In essence, the catalyst bed has been modeled as various sections whereby different regimes exist and local conditions apply. By generating diffusion profiles for each regime we were able to determine a catalyst effectiveness profile over the catalyst bed for the specified rate of reaction expression. Our efforts continued by simultaneously calculating the effect of heat and mass diffusion in and out of the particles through different media such as wax and supercritical hexane. The uniqueness of our model is that it is capable of simultaneously evaluating the catalyst effectiveness profiles that relates in and out catalyst pores while at the same time measures the temperature profiles inside catalyst pores The results presented in Figure 1a show an example of the diffusion of CO from the catalyst boundary inwards while the composition of the fluid inside the catalyst pore is changing as a result of CO consumption. In one case we have considered the catalyst pores as filled with supercritical hexane (equivalent to SCF-FTS), and in the second case we have considered the pores to be filled with wax (equivalent to gas-phase FTS). These reaction mixtures have different effective diffusivities, resulting in different levels of penetration of the reactant CO into the catalyst capillary pores. For this reason, the level of H_2 remains near constant and high throughout the catalyst particle. Figure 1b shows the temperature distribution changes when modeling for SCF-hexane as the dominating reaction mixture whereby the obtained profile shows almost isothermal behavior. This is in an excellent agreement with our previous experimental investigations of the lab scale reaction behavior as reported for the SCF-FTS [4]. Nevertheless, scaling up the SCF-FTS reactor technology requires experimental validation of the capabilities of these simulations models to predict the reaction behavior over a wide range of operating conditions and catalyst bed sizes, which will be the focus of the current study utilizing our sophisticated bench scale reactor bed.

Figure 1a, and b: (a) Left: intraparticle CO diffusion into the particle. (b) Right: intraparticle temperature distribution as heat is generated by the product synthesis inside the catalyst particle.

Related Literature:

1. Blank J., Bani Naser L., Hussain, R., Elbashir, N. O. (2013) "Modeling of Heat and Mass Transfer Limitations in Fischer Tropsch's Fixed Bed Reactor: Comparisons between Supercritical Phase and Gas Phase" 10th Natural Gas Conversion Symposium (NGCS) Extended Abstract, Synfuels Session, Conference Proceedings, January 2013, Doha, Qatar.
2. Mogalicherla A., Elbashir N. O. (2011) "Development of a Kinetic Model for Supercritical Fluids Fischer Tropsch Synthesis" Energy & Fuels; 25(3); 878-889.
3. Mogalicherla A., Elmalik E; Elbashir N. O. (2012) "Enhancement in the Intraparticle Diffusion in the

Supercritical Phase Fischer-Tropsch Synthesis" Chemical Engineering and Processing: Process Intensification, 62; 59-68.
4. Elbashir N. O., Bukur D. B., Durham E., Roberts C. B. (2010) "Advancement in Fischer-Tropsch Synthesis via Utilization of Supercritical Fluids as Reaction Media" *AIChE Journal*; 56 (4) 997-1015
5. Elmalik E., Tora, E., El-Halwagi M. M., Elbashir N. O. (2011) "Solvent Selection for Commercial Supercritical Fischer-Tropsch Synthesis Process" Fuel Processing Technology, 92; 1525-1530.

(532e) Highly Active and Stable Silica-Doped, Alumina-Supported Fe Fischer-Tropsch Catalyst

Keyvanloo, K., Brigham Young University

Hecker, W. C., Brigham Young University

Bartholomew, C. H., Brigham Young University

Originally presented on: 11/6/2013 16:35:00 - 16:55:00

Typical commercial iron Fischer-Tropsch (FT) catalysts are unsupported and consist of precipitated iron promoted with copper, potassium, and SiO_2. However, most precipitated iron catalysts are structurally too weak to be used in slurry-bubble column reactors (SBCR), the most thermally efficient and economical reactors for FTS [1]. The use of a carrier or support has the potential to greatly increase the strength and longevity of these catalysts; however, previous research on silica- and alumina-supported Fe has not been promising as it has shown activity and productivity values ½ and ¼, respectively, of those for unsupported iron [2].

To overcome the low activity and attrition resistant problems, in this study we explored using 4 different alumina supports, including an original silica-doped g-alumina (AlSi). We found that AlSi after thermal treatment at 1100°C still maintained excellent pore properties (200 m^2/g, 1.06 cm^3/g, 24 nm) which allowed higher iron loadings (40%) without pore blocking or iron aggregation. The 4 catalysts were prepared by co-impregnation with non-aqueous solution, were tested in a fixed bed reactor, and were compared to unsupported Fe catalysts from the literature. A St. Gobain alumina-based catalyst and the AlSi-based catalyst showed the best activity. Both were nearly three-fold more active than the silica-supported catalyst reported by Bukur [2], and 6-fold more active than his alumina-supported catalyst. More importantly, the activity of our catalysts at 260°C was nearly 20% higher than Bukur's most active *unsupported* catalyst. In addition the AlSi based catalyst was tested for 700 h and showed excellent stability. In fact its activity after 700 h is about 15% *higher* than its initial activity.

The results of the present work indicate that the preparation protocol used here produces supported iron catalysts of high extent of reduction, high dispersion, very high FTS activity, moderate hydrocarbon selectivity, and excellent stability. In addition it was shown that our high pore volume SiO_2 doped alumina (AlSi) support served as a superior support for supported Fe FT catalysts.

References:

1. Bukur, D.B. and X. Lang, Highly Active and Stable Iron Fischer-Tropsch Catalyst for Synthesis Gas Conversion to Liquid Fuels. *Ind. Eng. Chem. Res.*, 1999. 38(9): p. 3270?3275.

2. Bukur, D.B. and C. Sivaraj, Supported iron catalysts for slurry phase Fischer?Tropsch synthesis. *Applied Catalysis A: General*, 2002. 231(1-2): p. 201-214.

(532f) Fischer Tropsch Catalysis: From Fundamentals to Engineering Practice
Li, Y., Chinese Academy of Sciences
Originally presented on: 11/6/2013 16:55:00 - 17:15:00

The Fischer-Tropsch catalysis and theoretical studies behind the recent breakthroughs in the 4000 bbl/d scale coal-to-liquids plants in China using Synfuels China's Fischer-Trospch (F-T) synthesis technology is summarized. DFT calculations for adsorption and reaction mechanisms on different facets of iron carbide species in F-T catalysts provide guidance to the development of an industrial F-T catalyst. In addition, kinetics can be very important tool for F-T reactor and process development. However, the F-T catalytic kinetics is very complicated and the intrinsic kinetics has not been obtained. The detailed kinetics has been the major tool to validate the slurry reactor design for productivity and for providing a controllable chemical environment for the catalyst used. The difficulties are mainly due to that no kinetics model could take the surface changing of the catalyst into account.

It is concluded that the F-T chemistry and catalysis need to be fully understood where a combination of experimental and theoretical tools is imperative. The knowldge gained from fundamental study should be combined with engineering devleopment during FT process scaleup.

(532g) Combined Aging Processes for SNG Production By Bio-Syngas Methanation

Li, H., Laboratoire de Catalyse et Spectrochimie, CNRS-ENSICAEN, University of Caen
Travert, A.,
Maug , F.,
Dreibine, L., Universit Lyon 1, CNRS, UMR 5256, IRCELYON, Institut de recherches sur la catalyse et l'environnement de Lyon
Schuurman, Y., Institut de Recherches sur la Catalyse et l'Environnement de Lyon (IRCELYON, UMR 5256, CNRS; Universit Claude Bernard Lyon 1
Mirodatos, C., Universit Lyon 1, CNRS, UMR 5256, IRCELYON, Institut de recherches sur la catalyse et l'environnement de Lyon
Meunier, F. C., Universit Lyon 1, CNRS, UMR 5256, IRCELYON, Institut de recherches sur la catalyse et l'environnement de Lyon
Originally presented on: 11/6/2013 17:15:00 - 17:35:00

Combined Aging Processes for SNG Production By Bio-Syngas Methanation

H. Li[1], A. Travert[1], F.MaugÃ©[1], L. Dreibine[2], Y. Schuurman[2], C. Mirodatos[*2], F.C. Meunier[1,2]

[1] Laboratoire de Catalyse et Spectrochimie, CNRS-ENSICAEN, University of Caen, 14050 Caen, France

[2] UniversitÃ© Lyon 1, CNRS, UMR 5256, IRCELYON, Institut de recherches sur la catalyse et l'environnement de Lyon, 2 avenue Albert Einstein, F-69626 Villeurbanne, France.

Synthesis gas obtained from biomass gasification (biosyngas, a mixture of CO, CO_2, H_2 and impurities) is pertaining to the wider societal agendas not only as a sustainable source of fuels and chemicals, but also as a mean for dealing with global warming challenge. A major barrier in the commercialization of biomass gasification is the presence of bio-impurities such as ammonia, HCl, H_2S and tars in the gas products that are detrimental to downstream processes. Two important downstream processes are the conversion of syngas to (i) methane and light hydrocarbons (Substitute Natural Gas, SNG) and (ii) higher hydrocarbons (Fischer-Tropsch synthesis, FT). The exact role and effects of biomass-derived impurities on the catalysts used for these reactions are poorly known. Furthermore the way that such aging factors can combine with structural aging processes such as metal sintering under operating conditions is not clearly documented either. In the present study, we focused on the SNG process, using Ni based catalysts (alumina supported) in order to evaluate the kinetic sensitivity of various aging factors, combining the effect of model poisons such as ammonia, acetonitrile, heptanes, toluene and benzene together with metal particle re-structuring by sintering under SNG process conditions.

To perform this study, various operando techniques were used, essentially DRIFT spectrometry [1], completed by magnetic measurements and SSITKA analysis.

To sum up the main effects observed with time on stream and added poisons concentrations, the following features can be outlined:

(i) The CO carbonyl IR signal appeared to be correlated with the loss of activity, in the absence of added poisons. This irreversible ageing process was assigned from previous studies [2] to metal sintering and restructuring, as a consequence of metal carbonylation leading to Ni transfer from small to larger particles.

(ii) The deactivating effect of N containing poisons like acetonitrile, NO and ammonia, superimposed with the previous irreversible Ni sintering, was totally or partially reversible. Though acetonitrile and NO were at least partially converted into ammonia under the methanation conditions, the degree of deactivation was ranked as acetonitrile> ammonia> nitric oxide, suggesting that various poisoning modes (e.g. competitive adsorption, coking, electronic effects) might be present.

iii) The effect of adding hydrocarbon compounds representative of tar by-products coming from the bio-mass gasification revealed that ethylene, heptanes and benzene did not show obvious poisoning effects. However, toluene did show somewhat irreversible poisoning effect on accelerating the deactivation rate comparing with the blank reference test.

iv) Accompany with the loss of carbonyl, there were usually an increase of the carboxylate species, which became normally saturated after few hours deactivation during methanation, independant of the increasing concentration of any poison introduction or any continues increasing carbonyl loss. Therefore, the carboxylate can be considered only as in function of time, and can hardly be correlated with carbonyl loss, neither with deactivation process alone.

A tentative rationale of these complex aging processes will be proposed on the basis of a previous kinetic study of the methanation reaction, based the changes in carbonyl and CH_x active intermediates concentrations, directly measured by SSITKA (switch from ^{12}CO to ^{13}CO and reverse) [3].

For example, by observing that the reversible poisoning effect of N containing species hardly changes the nature and concentration of carbonyl species over the catalyst surface, it is concluded that it affects essentially surface controlling steps such as hydrogen adsorption and/or hydrogenation of CH_x reacting intermediates deriving from carbonyls decomposition. Electronic effects via N interaction with Ni or reactive adspecies combined with Ni surface restructuring might originate the observed decays. For the case the added hydrocarbons, the absence of electronic effects would explain their negligible toxicity, unless surface coking may lead to diffusion limitation for the reacting syngas feedstock.

Strategies for improving the resistance of methanation catalysts to this type of poisoning can be proposed from this mechanistic and kinetic re-investigation of one of the oldest catalytic reaction studied till now.

REFERENCES:

[1] Effective bulk and surface temperatures of the catalyst bed of FT-IR cells used for in situ and operando studies. H. Li, M. Rivallan, F. Thibault-Starzyk, A. Travert, F.C. Meunier, *PCCP*, **2013**, DOI: 10.1039/c3cp50442e

[2] CO hydrogenation on a nickel catalyst. Part I : Kinetics and modeling of a low temperature sintering

process.

M. Agnelli, M. Kolb, and C. Mirodatos, *J. Catal.,* **1994**, 148, 9-21.

[3] CO Hydrogenation on a Nickel Catalyst. II. A Mechanistic Study by Transient Kinetics and Infrared Spectroscopy. M. Agnelli, H.M. Swaan, C. Marquez-Alvarez, G.A. Martin, C. Mirodatos, *J. Catal.* **1998**, 175, 117.

(533a) Non-Equilibrium Distributions and Hydrodynamic Coupling Distort the Measurement of Nano-Scale Forces Near Interfaces

Swan, J. W., California Institute of Technology
Furst, E. M., University of Delaware
Originally presented on: 11/6/2013 15:15:00 - 15:30:00

We calculate the displacement of a single spherical particle from the minimum of a harmonic well positioned near a plane wall and immersed in a uniform flow. A failure to account for the fluctuations in particle position orthogonal to the plane (leading to fluctuations in hydrodynamic drag) results in large discrepancies with the naive displacement calculated by assuming no fluctuations in the balance of forces. The chief criterion for neglecting such fluctuations is that the stiffness of the harmonic potential exceeds the thermal stresses on the particle by at least two orders of magnitude. For micrometer diameter particles typically employed in force spectroscopy of DNA, macromolecules and molecular motors, this can lead to errors of up 100% in the measured properties. The supplemental material to the article provides an implementation of this model intended to fit experimental measurements for the stiffness of the harmonic potential constraining the particle.

(533b) Coarsening Colloidal Gels: Micro-Mechanics and Rheology

Zia, R. N., Princeton University
Landrum, B.,
Russel, W. B., Princeton University
Originally presented on: 11/6/2013 15:30:00 - 15:45:00

We study the evolving structure and time-dependent rheological properties of an aging colloidal gel, with a focus on understanding the non-equilibrium forces that drive late-age coarsening. The gel is formed from a dispersion of Brownian spheres that interact via a hard-sphere repulsion and short-range attraction, as would occur in the presence of a polymer depletant, for example. The O(kT) strength of attractions leads to an arrested phase separation, and the resulting structure is a bi-continuous, space-spanning network that exhibits elastic and viscous behaviors: the gel may sustain its weight under gravity, or flow under shear. With O(kT) attractions the colloid-colloid bonds are reversible, giving rise to a continuous breakage/formation process as the gel ages. This balance favors coarsening over time, accompanied by an increase in feature size and heterogeneity in the gel. We show here that anisotropic surface migration leads to heterogeneous coarsening, and that this migration is driven by gradients in particle-phase stress.

(533c) Heterogeneity and Large-Scale Dynamics in Nanoemulsion Colloidal Gels

Helgeson, M. E., Massachusetts Institute of Technology
Gao, Y., University of California, Santa Barbara
Originally presented on: 11/6/2013 15:45:00 - 16:00:00

Nanoemulsions are known to form "organohydrogels", solid-like phases with rheology similar to structurally arrested states in attractive near hard sphere suspensions. In this work, we use comprehensive experimental studies to reveal the mechanism of gelation in these systems, providing insight into the mechanics of both soft and hard-sphere colloidal gels at high volume fraction. The use of materials with thermosensitive interdroplet attractions allows for careful examination of the gel transition by combining a number of techniques, including bulk rheology, scattering, and optical microscopy. In particular, we employ newly developed dynamic microscopy techniques to track both the structure, dynamics, and topology of large-scale density fluctuations in the gelling fluid across a wide range of length and time scales *in situ*.

We find that the nanoemulsion gels exhibit strong structural and dynamical heterogeneity at length scales significantly larger than the characteristic size of interdroplet clusters. At the onset of gelation, strong cluster-cluster correlations appear, indicating the formation of colloid-rich and colloid-poor regions in the fluid. As gelation proceeds, the growth and relaxation of these correlations reveal that gelation proceeds by arrested phase separation vis-à-vis spinodal decomposition. Specifically, the dynamics of cluster-cluster correlations are ballistic above a characteristic length scale, which grows linearly with time until complete structural arrest is observed. We track the dependence of arrest as a function of both the colloid volume fraction and degree of quenching into the spinodal region, and find that arrest approximately corresponds with the point at which the colloid-rich region approaches the attractive jamming transition. These results suggest rules for controlling both the porosity and mechanical properties of a large class of particulate gels formed in polymer-colloid mixtures.

(533d) Fluid Flow Through Networks in the Collapse of Colloidal Gels
Mertz, A., University of Colorado - Denver
Graham, A. L., Los Alamos National Laboratory
Feng, S., Los Alamos National Lab
Redondo, A., Los Alamos National Lab
Ingber, M., University of New Mexico
Originally presented on: 11/6/2013 16:00:00 - 16:15:00

In unstable colloidal gels, the viscous flow through the non-neutrally buoyant colloidal network determines the initial rate of collapse[1]. The resistance to this flow through the network is characterized by the permeability. The permeability is generally accepted to be a power-law function of volume fraction, where the power is dependent on the fractal dimension of the network. To test this with direct numerical simulations, randomly generated diffusion-limited-cluster-aggregated networks of spheres and other particle are generated along with ordered lattices that span the containing structure or periodic cells. These networks will be characterized using the fractal dimension by measuring the length of the network strands with different sized measurement scales[2].

The permeability is then determined using finite element solutions of the Navier-Stokes equations of pressure-driven flow of Newtonian fluids through the networks. With this method, we calculate the permeability's dependence on volume fraction, particle size and shape, and fractal dimension. In addition, a comparison of periodic boundary conditions to networks in finite size containers is used to test the effect of particle pressure on the dynamics in these systems[3].

[1] S. Manley, J. M. Skotheim, L. Mahadevan, and D. A. Weitz, "Gravitational Collapse of Colloidal Gel," PRL **94**, 218302 (2005).

[2] B. Mandelbrot (1967). "How Long Is the Coast of Britain? Statistical Self-Similarity and Fractional Dimension", Science **156** 3775, 636(May 5, 1967).

[3] Y. Yurkovetsky and J. F. Morris, "Particle Pressure in Sheared Brownian Suspensions," J. Rheol. **52**, 141 (2008).

(533e) Interparticle Force Measurements, Microscopic Dynamics, and the Linear Rheology of Colloidal Depletion Gels

Hsiao, L. C., University of Michigan
Whitaker, K. A., Rowan University
Furst, E. M., University of Delaware
Solomon, M. J., University of Michigan
Originally presented on: 11/6/2013 16:15:00 - 16:30:00

We report an experimental study that relates two-particle force measurement by optical trapping and microscopic dynamics from confocal microscopy to the linear viscoelasticity of a colloidal depletion gel ($\varphi = 0.20$). Fluorescent poly(methyl methacrylate) spherical particles are dispersed in a mixture of 75 vol% cyclohexane:25 vol% cyclohexyl bromide ($\Delta n = 0.03$, $\Delta\rho/\rho = 0.12$). Non-adsorbing polymer is used as the depleting agent at various concentrations (MW = 900,000 g/mol, $c = 5.3$ to 15.9 mg/ml) to induce an attractive potential. The optical tweezer measurements show a linear increase in the average rupture force as the attraction strength increases, with possible long-range interactions present at high polymer concentrations. The gels exhibit a linear elastic modulus that increases as a function of the non-adsorbing polymer concentration *c*. Particle tracking measurements of the single particle mean-squared displacement quantify the localization of particles in the gel. We discuss the correlation between the interparticle force measurements, the microscopic dynamics, and the linear rheology in light of gelation theory, including results from mode coupling theory.

(533f) Microstructure of Colloids in Sheared Boger Fluids

Perera, M. T., Lehigh University
Gilchrist, J. F., Lehigh University
Originally presented on: 11/6/2013 16:30:00 - 16:45:00

Microstructures of high volume fraction colloidal suspensions in Boger Fluids are measured using high speed confocal microscopy. Unlike pure hydrodynamic interactions, wall effects have a large effect on the kinematics of the flow and the microstructure. What is unclear is why low weight fraction high molecular weight polymers have a significant effect on structure even when the polymer is localized. Through fluorescence imaging, the polymer phase is seen to remain coiled and isotropic in low shear regions of the flow and extended in the flow direction in high shear. Results demonstrating the effect of weight fraction, molecular weight, and relaxation time of the polymer phase will be shown.

(533g) Microrheology of Soft Particle Pastes: Forced Motion of a Tagged Particle in a Jammed Suspension

Mohan, L.,
Cloitre, M.,
Bonnecaze, R. T., The University of Texas at Austin
Originally presented on: 11/6/2013 16:45:00 - 17:00:00

Soft particle pastes like microgels and compressed emulsions are densely packed, disordered suspensions of soft and deformable particles. The particles are jammed beyond random close packing (φ >0.64). They behave like weak elastic solids at rest and low stresses but flow like liquids above the yield stress and this unique feature makes them useful as advanced rheological additives to process materials including high-performance coatings, textured food and personal care products. To relate the particle scale behavior to the macroscopic rheology we study the forced motion of a tagged particle through jammed suspensions with the help of particle scale simulations. A random particle is chosen and a known constant external force is applied on it and its motion in response is tracked. At small forces the particles are trapped in cages which resists the external force whereas at larger forces the tagged particle breaks free of the cage. This provides the microscopic evidence of yielding behavior in these materials. The average velocities of these particles in response to the applied forces are computed and the microscopic mobilities of these particles are extracted from them. The maximum distance through which the effect of the disturbance caused by this single tagged particle travels is also computed to determine the effectiveness of the particle-particle facet network present in these materials in transmission of localized microscopic forces. The simulations are performed with different packing fractions to understand the effect of volume fraction which gives information about the "microscopic cage strength" at different packing densities. The simulations thus provide means to probe the mirco and nano scale microstructure and properties of these materials including particle scale mobilities, transmission of localized disturbances and cage strength.

(533h) Rheology and Microstructure of Concentrated, Near Hard-Sphere Colloidal Dispersions Under Steady Shear and Large Amplitude Oscillatory Shear in All Three Planes of Shear

Gurnon, A. K., University of Delaware
Porcar, L., Institut Laue-Langevin
Wagner, N. J., University of Delaware
Originally presented on: 11/6/2013 17:00:00 - 17:15:00

We present the first experimental measurement of the complete 3-D microstructure of a shearing, near hard-sphere colloidal dispersion that is made possible by the development of novel instrumentation implemented on the small angle neutron scattering (SANS) beamlines of the ILL and the NIST Center for Neutron Research. This work is motivated by the intriguing rheology of Shear Thickening Fluids (STFs), which are concentrated colloidal suspensions that exhibit increases in viscosity under high shear rates of deformation. STFs aptly respond to an imposed impact stress and therefore, these novel materials have the potential to be implemented in any number of impact protection applications. STFs have been critically studied for their steady shear properties [N.J. Wagner and J. F. Brady, *Physics Today*. **2009.** 62 (10): 27-32]. However, little is understood about the nonlinear dynamic properties of STFs. To address this, we utilize large amplitude oscillatory shear (LAOS) to investigate the dynamic rheological properties of a model STF suspension [Kalman, D.P. and N.J. Wagner, *Rheologica Acta.* **2009**, 48: 897-908]. We develop a new flow-SANS sample environment for time and spatially resolved measurements in the plane of flow (1-2 plane) and combine this with rheo-SANS in the radial direction (1-3 plane) and tangential direction (2-3 plane) to elucidate the connection between the microstructure and the measured shear rheology. For the first time, the microstructure of a colloidal suspension undergoing steady shear is measured in all three planes of shear. The flow-induced microstructure under steady and oscillatory shear flow is compared to the rheological behavior via stress-SANS laws that separate the thermodynamic and hydrodynamic components of the stress that drive shear thinning, shear thickening, and first normal stress differences. A new non-equilibrium microstructure is discovered during a LAOS that proves to be important for understanding the rheology of concentrated colloidal suspensions. This first, complete measurement of rheological and microstructural properties is invaluable for testing constitutive models for colloidal suspension rheology.

(533i) Microrheology of Drug Particulates in Drying Polymeric Films

Huynh, P., Rutgers University
Takhistov, P., Rutgers, The State University of New Jersey
Originally presented on: 11/6/2013 17:15:00 - 17:30:00

The introduction of nano- and micro-scale particles into polymer blends creates a large internal surface between the solid particles and the polymer, changing substantially the thermodynamic and mechanical properties of the polymer matrix Adsorption of polymers from solutions of their mixtures on solids plays an important role in the formation of API composite structures with evenly distributed i.e. non-aggregated particulate materials.

The effect of the particles on the phase separation in forming polymeric networks can be related to the selectivity of the component adsorption at the interface with a solid phase, leading to the redistribution of the system components between the bulk and the surface layers. We have studied the filmgel-formation process of several drug loadedhydroxypropyl methylcellulose formulations using a new optical technology based on diffusing-wave spectroscopy. This unique and simple technique allows a non-destructive real-time monitoring of the drying process providing unique information on rheological properties at the particle lenghth scale. Measured kinetics of film formation provide a new vision of the successive steps of the mechanisms taking place (evaporation, packing, etc.). Obtained data are in good corellation with the data on the phase behavior of the polymeric solution.We also present experimental data and theoretical models that predict transport properties and stability of this colloidal system.

(533j) Brownian Dynamics Simulations of Cubic Nanoparticle Suspensions

Mallavajula, R., Cornell University
Koch, D. L., Cornell University
Archer, L. A., Cornell University
Originally presented on: 11/6/2013 17:30:00 - 17:45:00

The rheological properties of suspensions of Brownian cube-shaped particles are interesting because of the greater increase in the translational freedom caused by layering relative to suspensions of Brownian spheres.

At very low volume fractions, the sharp edges and corners affect flow profile around the cube. In the finite element method, the sharpness of the edges and corners can be varied using the grid spacing on the surface of the cube. The stress acting on the particle was found to increase with the sharpness of the cube. A theoretical scaling for this increase was predicted using the 2D flow profile around the corners. The results for the stress were fitted with this scaling to obtain the results for the cubes with sharp edges and corners. At moderate volume fractions, Brownian dynamics simulations were carried out to study the structure and flow behavior of suspensions. Simulations were performed over wide range of volume fractions and Pectlet numbers. Under shear, cubic particles were observed to behave like spheres circumscribed over it for volume fractions less than 0.25. Interesting differences in the rheological properties between spheres and cubes are analyzed based on the structure, order parameters, rotational and translational diffusivities of the particles in the suspension.

(541a) Facile Anhydrous Proton Transport On Hydroxyl Functionalized Graphene

Choudhury, P., University of Pittsburgh
Saidi, W. A., University of Pittsburgh
Johnson, J. K., University of Pittsburgh
Originally presented on: 11/6/2013 15:18:00 - 15:42:00

Proton transport is a critical important property of proton exchange membrane (PEM) fuel cells. We propose a nano-structured material capable of fast water-free proton transport. These materials can be constructed from functionalized surfaces (e.g. graphene, SWNT, Si(001), MOFs, etc) with different functional groups (e.g. -OH, -COOH, -SO3H, -NH, etc.) that form hydrogen bonded networks. In this work, we predict that hydroxyl functionalized graphene will conduct protons in the complete absence of water due to a self-assembled hydrogen bond network of OH groups on the surface.Â Activation energy of hydroxyl functionalized graphene is similar to that of Nafion, a standard polyelectrolyte proton exchange membrane. The single-file proton diffusion and proton mobility on the surface of both 1-D and 2-D hydrogen bonded graphane membranes will be discussed.

Acknowledgement

We gratefully acknowledge support from DTRA under Contract No. HDTRA1-09-1-0008 and from NSF TeraGrid (XSEDE) resources under allocation numbers [TG-DMR100097], [TG-SEE090006], and [TG-DMR110091]. We thank K. D. Jordan , M. E. Tuckerman and J. T. Yates, Jr. for helpful discussions. We thank the Center for Simulation and Modeling at the University of Pittsburgh for providing computational support.

(541b) Polysulfone-Based Anion Exchange Membrane Fuel Cell: Molecular Dynamics Simulation Approach

Han, K. W., Georgia Institute of Technology
Ko, K. H., Georgia Institute of Technology
Choi, J. I., Georgia Institute of Technology
Jang, S. S., Georgia Institute of Technology
Originally presented on: 11/6/2013 15:42:00 - 16:06:00

In this study, the nanophase-segregated structures and transport properties of quaternary ammonium grafted polysulfone membranes are investigated using molecular dynamics (MD) simulation method. For this, we develop a new force field from a reference density functional theory modeling with B3LYP and 6-31G** in order to describe the hydroxide anion. The bond stretching force constant is determined to reproduce the quantum mechanical vibrational frequency. The atomic charges are determined by Mulliken population analysis. Through the annealing procedure, the nanophase-segregated structure is developed as a function of water contents such as 10 and 20 wt %. The extent of nanophase-segregation is evaluated by the structure factor analysis, which can be compared with the experimental small angle scattering data. Once the equilibrium structures are obtained, another MD simulation is run for data collection from which the diffusion of water and hydroxide are analyzed using the mean-square displacement analysis by assuming that the diffusion is a Gaussian diffusion. The nanophase-segregated structures and the transport properties are compared to the proton exchange membrane consisting of the same polymer backbone except for the acidic functional group.

(541c) Permethylcobaltocenium Based Hydroxide Exchange Membranes With High Stability

Gu, S., University of Delaware
Wang, J.,
Fang, Q., University of Delaware
Kaspar, R. B., University of Delaware
Zhang, B., University of Delaware
Yan, Y., University of Delaware
Originally presented on: 11/6/2013 16:06:00 - 16:30:00

Permethylcobaltocenium Based Hydroxide Exchange Membranes with High Stability

Shuang Gu, Junhua Wang, Qianrong Fang, Robert B. Kaspar, Bingzi Zhang and Yushan Yan*

Department of Chemical & Biomolecular Engineering, University of Delaware

5 Innovation Way, Newark, DE 19716, USA

yanys@udel.edu

Hydroxide (OH^-) exchange membranes (HEMs) have become the research focus in energy conversion electrochemical devices (fuel cells, electrolyzers, and solar hydrogen generators) during the past decade owing to their intrinsic ability to work with non-precious-metal catalysts as well as their desirable tolerance to CO_2 contamination[1]. However, unlike their counterpart—proton (H^+) exchange membranes (PEMs) that are based on the highly stable sulfonic-acid functional group ($-SO_3H$) and thus have excellent chemical and thermal stability, HEMs suffer from limited chemical stability due to their currently used ammonium hydroxide cationic functional group ($-NR_3^+OH^-$). The sensitive degradation pathways (such as nucleophilic substitution, Hoffman elimination, and ylide formation) of ammonium hydroxide not only limit the working temperatures for HEM applications but also raise the strong concern for long-term HEM stability. Alternative cationic functional groups based on the same nitrogen element (such as pyridinium, guadinium, and imidazolium) and other elements (phosphorus: phosphonium, sulfur: sulfonium, and ruthenium: bis(terpyridine)ruthenium complex[2]) have also been introduced, bringing new features (improved solubility, improved thermal stability, and/or increased charge number) for HEMs but the practice suggests their chemical stability still remain limited. The limited chemical stability of ammonium and other cations may be rationalized as the result of their flexible valence electron structures, allowing for many degradation pathways possible.

Alkali metal cations (typically, Li^+, Na^+, and K^+) have excellent chemical stability, because of their non-flexible closed-shell valence electron configurations ([$1s^2$], [$2s^2 2p^6$], [$3s^2 3p^6$], respectively) that are the same as those of noble gases, but they are very challenging to immobilize for HEM applications. Bearing a unit of positive charge, biscyclopentadienyl cobalt sandwich cation (e.g., $(C_5H_5)_2Co(III)^+$, or cobaltocenium) satisfies the 18-electron rule for transition metal complexes, in which Co(III) also has closed-shell valence electron configuration. The cobaltocenium hydroxide is considered as a strong base, and in fact the cobaltocenium is the most stable metallocenium in alkaline media[3]. Equally importantly, cobaltocenium can also be immobilized to polymers in both main-chain and side-chain.

Permethylcobaltocenium, i.e., decamethylated cobaltocenium, is expected to have even higher structure stability than the normal cobaltocenium, due to the charge delocalization caused by strong electron donation from the ten methyl groups[4]. This is evidenced by over 600 mV negative shift of formal potential observed [−1.24 of $(C_5Me_5)_2Co(III)^+/(C_5Me_5)_2Co(II)$ vs. −0.63 V of $(C_5H_5)_2Co(III)^+/(C_5H_5)_2Co(II)$ referring the SHE, the same solvent of CH_2Cl_2[5]]. Here we present, for the first time, the permethylcobaltocenium-functionalized polymer HEMs that show very high thermal and chemical stability, promising to applications in durable and robust electrochemical devices.

References

[1] J. R. Varcoe, R. C. T. Slade, *Fuel Cells* **2005**, *5*, 187-200.

[2] Y. P. Zha, M. L. Disabb-Miller, Z. D. Johnson, M. A. Hickner, G. N. Tew, *J Am Chem Soc* **2012**, *134*, 4493-4496.

[3] G. Wilkinson, F. A. Cotton, *Prog Inorg Chem* **1959**, *1*, 1-124.

[4] N. G. Connelly, W. E. Geiger, *Chem Rev* **1996**, *96*, 877-910.

[5] U. Koelle, F. Khouzami, *Angewandte Chemie-International Edition in English* **1980**, *19*, 640-641.

Key words: Cobaltocenium, Hydroxide exchange membrane, Stability, Fuel cells, Electrochemical devices.

(541d) Durability and Performance Of An Anion Exchange Membrane Separator For Electrochemical Energy Storage and Conversion Applications

Yun, S.,
Parrondo, J., Illinois Institute of Technology
Ramani, V., Illinois Institute of Technology
Originally presented on: 11/6/2013 16:30:00 - 16:54:00

All-vanadium RFBs (VRFB) are very interesting from an operational point of view because the intermixing of positive and negative solutions does not cause irreversible damage. In this work, we synthesized a novel AEM based on cardo-polyetherketone (PEK-C), and evaluated its stability under RFB working conditions. The membrane stability was evaluated by measuring the changes in the chemical structure and mechanical properties after immersion in vanadium (V) solutions (strongly oxidizing) for 1 month. Post-mortem analysis of the membrane using 2D NMR techniques confirmed that there was no degradation in chemical structure after such immersion. The vanadium (IV) membrane permeability of quaternized PEK-C/TMA^+ membrane ($8.2\pm0.2 \times 10^{-9}$ cm^2 s^{-1}) was 35 times lower than for Nafion® membranes ($2.9\pm0.2 \times 10^{-7}$ cm^2 s^{-1}) confirming the suitability of anion exchange membranes for RFB applications. Water absorption isotherms (VTI-SA, TA instruments) showed moderate water uptakes (30wt%) - this is a necessary requisite for adequate mechanical integrity during long term RFB operation. No changes were observed in the ultimate tensile strength of the membranes after 1 month immersion in the vanadium (V) solution (19 ± 2 and 21 ± 1 MPa respectively). Redox flow battery (5 cm^2 single cell) testing was carried out and the charge/discharge of the battery was performed at a current density of 100mA cm^{-2} over 20 cycles initially. The coulombic efficiency was close to 100%, and the voltage and energy efficiencies were around 80% for every cycle. The stability of the anion exchange membrane under RFB operating conditions is a promising result that will facilitate widespread use of inexpensive AEMs as separators in RFBs.

Likewise, the same membrane chemistries can also serve as separators in anion exchange membrane fuel cells. Studies on this front are currently ongoing; the presentation will also discuss the efficacy and stability of the above membranes in alkaline fuel cell applications.

(541e) Polymer/Carbon Nano-Tube Composites for Alkaline Fuel Cell Applications

Lue, J.,

Originally presented on: 11/6/2013 16:54:00 - 17:18:00

Polybenzimidazole (PBI)-functionalized multi-walled carbon nano-tubes (CNTs) are obtained via an ozone-mediation route. The PBI nano-composites containing small amounts (<1%) of CNTs are prepared using solution casting methods. These thermally stable PBI and PBI/CNT composites are doped with potassium hydroxide (KOH) solution to prepare hydroxide-conducting electrolytes for alkaline direct methanol fuel cell (ADMFC) applications. The CNTs-containing composites exhibit higher fractional free volumes and higher water diffusivities. The CNTs also promote ionic conductivity of electrolytes and improves the fuel cell performance. Gas diffusion electrodes (GDEs) without polytetrafluoroethene (PTFE) treatment give superior cell power density compared to commercial E-tek GDEs, which contain hydrophobic PTFE layers. When the fuel cell is fed with 2 M methanol in 6 M KOH (as the anode fuel) and humidified oxygen (as the cathode oxidant), the system achieves a maximum power density of 104.7 mW cm^{-2} at 90°C. These KOH-doped PBI/CNT composites have the potential to be used in high temperature alkaline fuel cell applications.

(541f) Di-Quaternized DABCO-Based PEEK Hydroxide Exchange Membranes With Improved Ionic Conductivity

Wu, X., Dalian University of Technology
He, G., Dalian University of Technology
Wang, J., Dalian University of Technology
Yan, X., Dalian University of Technology
Originally presented on: 11/6/2013 17:18:00 - 17:42:00

Di-quaternized DABCO-based PEEK hydroxide exchange membranes with improved ionic conductivity

Xuemei Wu, Gaohong He*, Junjun Wang, Xiaoming Yan

State Key Laboratory of Fine Chemicals, Research and Development Center of Membrane Science and Technology, Dalian University of Technology, Dalian 116024, PR China.

Abstract

Polymer Electrolyte Membrane Fuel cells are considered to be clean and efficient power sources. Among the several kinds, hydroxide exchange membrane fuel cells (HEMFCs) operate in alkaline media with faster kinetics of the oxygen reduction, lower methanol permeability and easier water management, therefore superior to proton exchange membrane fuel cells (PEMFCs) in some aspects such as cost and application on direct methanol fuel cells (DMFCs). Hydroxide exchange membrane (HEM) is a key component of HEMFC, which conducts hydroxide ions and prevents crossover of reactants. Significant research works have been done in recent years to develop new HEMs. However, there are still two main challenges facing with HEMs, i.e. low ionic conductivity and poor stability of basic groups. In this study, di-quaternized 1, 4-diazabicyclo [2,2,2] octane (DABCO)-based PEEK (di-QDPEEK) hydroxide exchange membranes are prepared via the two nitrogen atoms of DABCO to improve ionic conductivity. And the strain free conformation of DABCO provides good resistance to both Hoffman degradation and nucleophilic substitution by OH^-.

We design both mono- and di-quaternized DABCO-based PEEK membranes with similar ionic exchange capacity (IEC) to investigate the influence of distribution of quaternary ammonium functional groups on the microstructures and electrochemical properties of the HEMs. For low IEC around 0.8, di-QDPEEK membrane exhibits much higher ionic conductivity (2.9 times) and swelling ratio (1.7 times) as compared with the mono-QDPEEK membrane. However, for high IEC around 1.2, di-QDPEEK membrane exhibits much higher ionic conductivity (2.1 times) and much lower swelling ratio

(0.6 fold) as compared with the di-QDPEEK membranes. Based on our data, a schematic model with scattered or concentrated distributions of the quaternary ammonium groups in the mono- and di-quaternized membranes is proposed. In the di-quaternized membranes, quaternary ammonium functional groups concentrate in a local scale, which enhance the interactions between ionic groups, leading to well hydrophilic-hydrophobic micro-phase separation and better interconnected pathways for ions through the hydrophilic domains. As a result, di-quaternized membranes exhibit much higher ionic conductivity and special swelling behavior as compared with the mono-QDPEEK membranes of similar IEC. More details about the microstructures-properties correlations, such as TEM image, glass transition temperature and mechanical strength of the mono- and di-QDPEEK membranes will be present in this work. For di-QDPEEK of IEC about 1.69, high ionic conductivity of around 35.5 mS cm^{-1} at 20 °C is achieved, which is within the top values reported in the literatures.

Keywords: Separations, Hydroxide exchange membranes, Fuel cells, DABCO, PEEK, Ionic conductivity

Acknowledgements

The authors thank Program for National Science Fund for Distinguished Young Scholars of China (Grant no.21125628), NSFC (Grant no. 20976027 and 21176044) for supports of this work.

* **Corresponding author.** Tel: +86 411 84707892 E-mail address: hgaohong@dlut.edu.cn

(545a) Effects of Self-Compression On Metal Clusters

Li, L., Stanford University
Larsen, A., Universidad del PaÃs Vasco
Romero, N. A.,
Abild-Pedersen, F., Technical University of Denmark
Greeley, J. P., Purdue University
N rskov, J. K., Technical University of Denmark
Originally presented on: 11/6/2013 15:15:00 - 15:35:00

Traditionally, periodic unit cells are used to model the surface properties of metal catalysts, even those that exist as nanoparticles. We are interested in finding more realistic ways of modeling these particles, in order to determine whether the traditional approach is valid at this length scale. We have used density functional theory to study the effects of relaxation on gold and platinum clusters, ranging from 13 to 923 atoms, as well as on their island analogs. Calculations of this size have been made possible by leadership computing resources, containing up to 40,960 nodes, at Argonne National Laboratory. We find that self-compression leads freestanding clusters to converge to weaker oxygen adsorption energies than the values calculated from traditional surface science slabs. The island systems proposed were able to capture this compression effect and gave better predictions of adsorption energies on clusters. However, the agreement between the two systems appeared only after quantum-size-effects ended. We find that below the critical size of 561 atoms on gold, and 147 on platinum, electronic features on freestanding clusters cannot be represented by continuous systems.

(545b) Adsorption of Heterobifunctional 4-Nitrophenol On the Ge(100) Surface

Shong, B., Stanford University
Hellstern, T. R., Stanford University
Bent, S. F., Stanford University
Originally presented on: 11/6/2013 15:35:00 - 15:55:00

Group IV semiconductor materials play a crucial role in integrated circuits. Recently, high carrier mobility has drawn attention to Ge as an alternative channel material. However its surface requires modification to be utilized in devices, since the native oxide of germanium is much less stable compared to SiO_2. The (100) surface of Ge under vacuum reconstructs to a 2 × 1 pattern after proper preparation. Each dimer possesses partial double bond character so that alkene-like reactions occur, and also charge separation within each dimer allows acid/base-like reactions. The ordered dimers that form rows separated by trenches serve as the local reaction sites. On the other hand, organic functionalization of surfaces could be interesting since that way the versatility of the organic chemistry could be coupled with the current semiconductor technologies. Especially, adsorption of bifunctional molecules renders possibilities of having one terminus of the molecule available for further chemical modification of the surface. Most studies on the adsorption of bifunctional molecules conducted so far have focused on homobifunctional molecules, where the two functionalities are identical; consequently, dual reaction of a single molecule with the surface was found in many cases, reducing the number of available reaction sites. A heterobifunctional molecule may be useful by undergoing an asymmetric reaction where one terminus would preferentially react with the surface, keeping the other unreacted.

In this study, adsorption of a heterobifunctional molecule, 4-nitrophenol, on the Ge(100) surface was explored using multiple internal reflection Fourier transform infrared (MIR-FTIR) spectroscopy and X-ray photoelectron spectroscopy (XPS) experiments combined with density functional theory (DFT) calculations. Since the two functional groups are both reactive with Ge(100), there are three possible adsorbate structures: (1) singly bound through the O-H dissociation of the hydroxyl group, (2) singly bound through the cycloaddition of the nitro group, and (3) dually bound with both functionalities. DFT calculations show all these products have large enough adsorption energies to be found at room temperature as final products, and that conversion to (3) from (1) would be feasible while (2) is unlikely to be transformed to (3) because of a high activation barrier. The Ge(100) surface was exposed to increasing amounts of nitrophenol to test the coverage dependence of adsorption experimentally. The IR spectrum of the physisorbed molecules shows that the presence of the free nitro group could be identified with its strong characteristic absorption bands. While the $-NO_2$ bands are clearly discernible in saturated chemisorption spectra at room temperature, they are not observed at low surface coverage. Therefore structure (1) can be understood to appear only at higher coverage. Also, multiple peaks overlapping at the ring stretch region even at low coverage suggest that a mixture of products is formed, most likely by structures (2) and (3). In the XPS spectra, the existence of both unbound $-NO_2$ and unbound –OH are suggested. The free nitro group is most apparent close to saturation, confirming the findings from IR. Moreover, through XPS we could quantify the fraction of each adsorbate at saturation. In conclusion, we show that 4-nitrophenol adsorbs on Ge(100) through both of its functionalities at low coverage, but forms singly adsorbed products through either the $-NO_2$ or –OH group as saturation coverage is approached.

(545c) Determining the Effects of Oxygen On Multifunctional Alcohol Decomposition On Pd(111)

Williams, R.,
Medlin, J. W., University of Colorado Boulder
Originally presented on: 11/6/2013 15:55:00 - 16:15:00

To aid in developing more selective catalysts for biorefining applications, we have studied the surface chemistry of multifunctional alcohols using surface science techniques. In this contribution, we describe the effects of adsorbed oxygen on the decomposition pathways of a variety of alcohols, including benzyl alcohol and furfuryl alcohol, on Pd(111) via temperature programmed desorption (TPD) and high resolution electron energy loss spectroscopy (HREELS) experiments. On clean Pd(111), furfuryl alcohol and benzyl alcohol proceed through competing reaction pathways of deoxygenation or decarbonylation. It was found that deoxygenation is favored at high coverage, with a bimolecular reaction between adsorbed multifunctional alcohols accelerating the deoxygenation pathway. The reaction mechanisms for decomposition of a variety of alcohols has also been elucidated in the presence of oxygen, utilizing carefully selected isotopes to track particular C-H and O-H bonds, with TPD and HREELS. Our results show that pre-adsorbed oxygen strongly influences decomposition activation barriers, the selectivity of C-C and C-O scission, desorption products, and the overall extent of alcohol decomposition. The relative coverages of the alcohol and oxygen play a significant role in the reaction chemistry. The results indicate that interactions between adsorbed oxygen and alcohol-derived intermediates play a major role in the selective oxidation of functional alcohols on Pt-group catalysts.

(545d) High-Throughput Characterization of Composition, Structure and Reactivity of PdxCu1-X Alloys

Gumuslu, G., Carnegie Mellon University
Kondratyuk, P., Carnegie Mellon University
Yin, C., US Department of Energy
Miller, J. B., Carnegie Mellon University
Gellman, A. J., Carnegie Mellon University
Originally presented on: 11/6/2013 16:15:00 - 16:35:00

Alloy catalysts often display better activity, selectivity, and stability than their pure components. Most experimental studies of alloy catalysts try to correlate changes in catalytic activity with electronic and/or structural properties using a limited number of single-composition alloy samples. In this work, we use high-throughput techniques, developed in our laboratory, to discern relationships among catalytic activity, microstructure and electronic structure across continuous composition space.

Alloy libraries were prepared as Pd–based binary Composition Spread Alloy Films (CSAFs) which include all possible compositions of the alloy in the form of a thin film deposited on a substrate. Pd_xCu_{1-x} CSAFs were prepared by evaporative co-deposition of the components onto Mo substrates, followed by annealing at 800 K. To evaluate alloy activity for H_2 dissociation as a function of alloy composition, H_2-D_2 exchange ($H_2 + D_2 \rightarrow 2HD$) kinetics were measured at 100 discrete locations (alloy compositions) across the CSAF surface using a unique multichannel microreactor array. Microstructure and electronic structure were characterized as functions of composition across the CSAF surface by spatially resolved X-ray photoemission spectroscopy (XPS), UV photoemission spectroscopy (UPS), and electron backscatter diffraction (EBSD).

Binary Pd_xCu_{1-x} CSAFs, spanning the composition range x= 0.05→0.95, possess the features of the bulk PdCu phase diagram at the annealing temperature (800 K). H_2-D_2 exchange experiments, performed at atmospheric pressure and over a temperature range from 300 to 600 K, reveal maximum exchange activity at $x \sim 0.70$. Activation barriers to H_2 dissociative adsorption and recombinative desorption, estimated from microkinetic analysis of the reaction data, correlate with valence band electronic structure, characterized by UPS, across composition space. For example, the barrier to dissociative adsorption decreases as average d-band energy increases at Pd-rich (high x) compositions. This work illustrates use of a high-throughput methodology to construct composition-structure-activity relationships across broad composition and structure space for optimization and fundamental understanding of multicomponent catalytic materials.

(545e) Inhibiting Role of Dimer Surface Species in the Dehydration of Alcohols Over γ-Alumina

DeWilde, J. F., University of Minnesota
Kang, M., University of Minnesota
Chiang, H., University of Minnesota
Ho, C. R., University of Minnesota
Hickman, D. A., The Dow Chemical Company
Bhan, A., University of Minnesota
Originally presented on: 11/6/2013 16:35:00 - 16:55:00

Steady state kinetic measurements verify that surface dimer species composed of alcohol, water, and alcohol-water co-adsorbed surface species inhibit the synthesis of olefins and ethers from the unimolecular and bimolecular dehydration of ethanol and n-propanol on gamma-alumina at 488K. Exposure of the catalyst surface to water prior to reaction lowered the synthesis rates for olefins and ethers, demonstrating that water irreversibly poisons a fraction of the active sites for alcohol dehydration over gamma-alumina. Kinetic isotope studies demonstrated that the rate-limiting step for the formation involves the either cleavage of the C_β-H bond or desorption of water after formation of the olefin molecule (KIE = 2.4), while the synthesis of ethers are limited by cleavage of either the C-O bond of the alcohol molecule or an O-Al bond of a surface-bound reaction intermediate. Different degrees of inhibition of the rates of olefin and ether synthesis by pyridine indicate that the active sites for olefin and ether formation are both acidic and non-equivalent. In-situ titrations of the active sites for ether synthesis identify an active site density of 0.1 sites nm^{-2} at 488K. The selectivity for olefin formation is a factor of ten larger for 1-propanol dehydration than for ethanol dehydration, indicating a carbocation-like transition state for the unimolecular dehydration of alcohols on gamma-alumina. Dimer-inhibited mechanisms for the unimolecular and bimolecular dehydration pathways of ethanol and 1-propanol on gamma-alumina consistent with measured steady state and isotopic kinetic measurements and titrations are presented and evaluated.

(545f) Nanosize-Induced Resonant Photocatalysis On Metal Clusters

Kale, M., University of California, Riverside
Avanesian, T., University of California, Riverside
Christopher, P., University of California, Riverside
Originally presented on: 11/6/2013 16:55:00 - 17:15:00

Photon-induced electron transfer between solid-state materials and molecules is of fundamental importance for the design of many solar energy conversion processes including dye sensitized solar cells and photocatalysis. Heterogeneous photocatalysis is typically assumed to be a two-step process where formation of electron-hole pairs occurs in the solid-state photocatalyst (semiconductors or large metal particles/single crystals) followed by a subsequent charge transfer to adsorbates. This process of energetic charge carrier generation and transfer results in wavelength dependent quantum efficiencies that strictly follow the absorption spectrum of the solid state photocatalysts.

In this work, we show evidence for direct photon-induced metal-to-adsorbate electron transitions that induce efficient photocatalytic processes on Pt nanoclusters. This process is fundamentally different than what occurs on large metal particles or semiconductors as it is a single step process. As a result of the single step nature of the process the wavelength dependence of photocatalytic CO oxidation on small Pt clusters show a resonant behavior, unique to this excitation mechanism. Particle size dependent and *in-situ* diffuse reflectance UV-Vis Spectroscopy experiments are utilized to examine the nanosize-induced direct electron transfer process. We also compare the results for photocatalytic carbon monoxide, ammonia and hydrogen oxidation reactions on Pt clusters to develop a general understanding of how the nature of the chemical transformation dictates the excitation mechanism (two-step vs. direct process). Our studies show that the adsorbate-induced electronic structure of the metal-adsorbate complex is an indicator of which excitation mechanisms will be prominent under visible light illumination on small metal nanoclusters. Our results will be discussed in the context of utilizing solar energy to enhance important chemical reaction rates efficiently and with state-selectivity.

(545g) The Role of Geometric and Electronic Effects in Creating Active Sites On the Surface of Lead-Doped Bismuth Molybdates

Licht, R. B., University of California Berkeley
Getsoian, A., University of California - Berkeley
Bell, A. T., University of California, Berkeley
Originally presented on: 11/6/2013 17:15:00 - 17:35:00

Multicomponent bismuth molybdenum oxide catalysts have been used industrially for over 60 years for the oxidation and ammoxidation of light olefins. Despite their long history, little is understood about the role of each element, or how favorable interactions between elements lead to a better performing catalyst. The most heavily studied reaction in the literature, and that which we continue to investigate, is the oxidation of propene to acrolein. This reaction proceeds via a Mars and van Krevelen mechanism, where the rate-determining step is the initial abstraction of a hydrogen radical from propene[1].

Many of the industrially used catalysts are multi-phase and contain on the order of 10 different elements, making them difficult to model computationally. However, there are a number of multicomponent single-phase systems based on the scheelite structure, the parent structure of α-bismuth molybdate, that have been studied experimentally. For example, Sleight *et al.* have studied a range of $Pb_{1-3x}\varphi_xBi_{2x}MoO_4$ for the oxidation of propene to acrolein[2]. They found that while pure scheelite $PbMoO_4$ is not active, doping with a small amount of bismuth and introduction of charge compensating defects (x~0.06) result in a catalyst with a higher activity than that of pure $Bi_2(MoO_4)_3$.

Recent advances have identified the active site for alpha hydrogen abstraction of propene on $Bi_2(MoO_4)_3$ to consist of a molybdenyl oxygen perturbed by a neighboring bismuth atom at ~3Å distance[3]. The reaction barrier was found to be the spin crossing from the singlet to the triplet state, in which a molybdenum is reduced by one electron and the associated oxygen is activated in order to later abstract a hydrogen radical from the propene. The nearby bismuth both destabilizes the singlet state via repulsion with the lone pair and stabilizes the triplet via interaction of the Bi 6p and the Mo=O π^* orbital, allowing this reaction to occur with a reasonable barrier.

The present work extends previous investigations in this area by further examining the effect of the neighboring element and the molybdate ion environment on the reaction barrier. Q-Chem calculations were performed on a variety of relevant clusters that mimic the environment in various scheelite structures. We find that both the distance and directionality of the neighboring element with respect to the active molybdenyl oxygen play key roles in its ability to perform the initial hydrogen abstraction. This finding is used to explain reaction barriers calculated in VASP using DFT+U density functional theory on a range of

$Pb_{1-3x}\varphi_x Bi_{2x}MoO_4$ structures. The experimental inactivity of $PbMoO_4$ to propene oxidation is reproduced in VASP, and is due to the suboptimal geometry of the pure scheelite structure. However, the introduction of defects and bismuth into the scheelite $PbMoO_4$ distorts the structure to an improved geometrical configuration at the surface, resulting in a catalytically active material.

References:

[1] Grasselli R.K.; Burrington J.D. *Industrial & Engineering Chemistry Product Research and Development* **1984**, *23*, 393-404

[2] Sleight A.W.; Linn W. J *Annals of the New York Academy of Sciences* **1976**, *272*, 22-44

[3] Getsoian A.; Shapovalov V.; Bell A.T. *Journal of Physical Chemistry C* **2013**, *117*, 7123-7137

(549a) On the Frequency of Large Waves in Vertical Gas-Liquid Annular Flows

Zadrazil, I., Imperial College London
Matar, O. K., Imperial College London
Markides, C. N., Imperial College London
Originally presented on: 11/6/2013 15:15:00 - 15:30:00

A flow of gas in the core of a vertical pipe and of liquid in the annulus between the pipe wall and the gas phase is referred to as an annular flow. Annular flows can be found in both upwards (when both phases travel against gravity) and downwards pipe orientations. Both upwards and downwards annular flows are of high industrial importance; they can be found in the oil-and-gas industry (e.g. raisers, transfer pipelines and gas-liquid oil wells) or various industrial units, such as condensers, evaporators, distillation columns and chemical reactors.

One of the most important features of annular flow is the presence of large waves (often referred to as disturbance waves in the literature). These do not only carry a high fraction of the liquid phase being transported, but are also considered crucial for the process of liquid entrainment into the gas phase. These waves are usually characterised in terms of their amplitude (relative to mean film thickness), specific wave length (which is the distance between two successive crossings of the mean film thickness) and interfacial frequency.

In this paper results will be presented from an extensive effort to characterise the statistics of the interfacial frequency of these large waves, in both upwards and downwards annular flows using conductivity-probe and Laser-Induced Fluorescence (LIF) techniques. The two different measurement methods are used for the quantification of the interfacial wave frequency: (i) from direct counting of the waves having a thickness of 2-5x the mean film thickness; and (ii) by carrying out a power spectral density analysis from which the frequency at the maximum power is identified. Importantly, our results indicate that the large amplitude waves are best identified by direct wave counting, as opposed to the analysis of the power spectral density that, traditionally in the literature, is the most widely accepted method for this purpose.

(549b) Transport of Materials Into-and-Out of Dead End Pores

Kar, A., The Pennsylvania State University
Chiang, T. Y., Penn State University
Rivera, I. O.,
Sen, A., The Pennsylvania State University
Velegol, D., Penn State University
Originally presented on: 11/6/2013 15:30:00 - 15:45:00

Achieving transport inside dead end pores is not currently possible with pressure driven mechanisms, which limits our ability to obtain material from these pores, or to put material into them. Here we show that the phenomenon of diffusioosmosis can be used to drive flows in dead end pores, and the flows can be used to sweep otherwise-trapped materials from the pore into a sink, or vice versa. These flows arise along the walls of the channel due to an imposed or self-generated salt gradient. We observe that tracer particles have a finite axial speed of roughly 10 μm/s, as well as a finite crosswise transport due to mixing inside these pores caused from the transient nature of the diffusioosmotoc flows. Using electrokinetic modeling along with independently measured system parameters, we find solid agreement between predicted speeds and our experimental data. The temporal nature of the salt gradients may explain some of the other phenomena like the LoSal waterflooding process and DNA translocation in solid-state nanopores.

(549c) Experimental Investigations of Transport Processes and Interfacial Phenomena in Micellar Liquid/Liquid Systems

Paul, N., Technische Universit t Berlin
Kraume, M., Technische Universit t Berlin
Originally presented on: 11/6/2013 15:45:00 - 16:00:00

One of the main goals of "green chemistry" is: using water as a solvent. Many substances are hardly soluble in water; hence multiphase reactions do not proceed. By using surfactants the water solubility of such substances is increased which results in an increase of the reaction rate. Due to their amphiphilic structure surfactants adsorb at interfaces. This is the place where the important transport processes occur. Therefore, an interference of the transport processes by surfactants is obvious. In previous works a reduction of the mass transfer in the presence of surfactants was attributed to two mechanisms: the physicochemical effect and the change in the fluid dynamics. Due to the adsorption at the interface an additional mass transfer resistance is created. Furthermore, the fluid dynamics will change. Quantifying the influences of surfactants on the mass transfer is the goal of this work. Therefore, the mass transfer is observed at the smallest transfer unit: at single droplets.

For the experimental investigations the system water (continuous phase), octan-1-ol (dispersed phase), Triton X-100 (surfactant) and the azo dye Pyridine-2-methylaniline (transferred component) were observed. The liquid/liquid mass transfer and the fluid dynamics were measured during a single drop-rise in special test cells. Furthermore, experiments to quantify the occurring interfacial phenomena were carried out. The determination of the interfacial tension was carried out by two different methods (pendant drop, spinning drop). A special method to determine the rigidity of liquid/liquid interfaces was developed by applying colloidal probe atomic force microscope measurements. Furthermore, oscillating drop measurements were carried out to determine the interfacial rheology.

The results show: An increase of surfactant concentration causes a reduction of liquid/liquid mass transfer. The mechanisms mentioned above are responsible for this reduction, but the results of this work show that only regarding these mechanisms is not enough. For high surfactant concentrations the liquid/liquid mass transfer is reduced dramatically — even more than predicted by the known mechanisms in the literature. Therefore, other interfacial phenomena must be taken into account.

Besides the experimental investigations of mass transfer processes the focus of this works lies on the determination of interfacial phenomena in micellar liquid/liquid systems. The characteristics of the liquid/liquid interface were determined by applying the oscillating drop method as well as colloidal probe atomic force microscope measurements at the liquid/liquid interface. Both measurement techniques show that with an increase of surfactant concentration the rigidity of the liquid/liquid interface increases; although the interfacial tension decreases. This is caused by a change in the phase behaviour at the liquid/liquid interface, due to the high local surfactant concentration. The appearance of high viscous microemulsion layers, or even liquid crystalline layers, ends up in an additional mass transfer resistance. For an integrated description of liquid/liquid mass transfer in micellar systems this interfacial phenomenon of the change in phase behaviour must be regarded, otherwise the liquid/liquid mass transfer is overestimated.

(549d) Transport Across Fluid Interfaces and Lipid Membranes: Role of Interfacial Fluctuations

Kopelevich, D. I., University of Florida

Originally presented on: 11/6/2013 16:00:00 - 16:15:00

Transport of molecules and nanoparticles across liquid-liquid interfaces and flexible membranes of nanometer thickness plays an important role in various technological applications as well as biological processes. The rate of this transport is determined in part by the height of the energy barrier, which depends on the solute affinity for liquids separated by the interface as well as the membrane composition and density. Another factor affecting the solute transport is the local solute diffusivity associated with the Brownian forces acting on the solute. In a homogeneous solvent these forces are created by collisions between the solute and the solvent molecules occurring on the timescale significantly faster than the timescale of the solute motion. This allows one to approximate the Brownian forces as white noise and model the solute motion by a Langevin equation for the translational degrees of freedom of the solute.

Interfacial systems possess an additional source of the Brownian forces, namely forces caused by relatively slow fluctuations of the interfaces. The timescale of these fluctuations is comparable with the timescale of the solute transport. Therefore, these Brownian forces cannot be modeled as white noise and one needs to explicitly include them into the Langevin equation. The solute transport then corresponds to motion on a multi-dimensional energy landscape parametrized both by the solute and the interfacial degrees of freedom.

In this talk we present analysis of transport of a hydrophobic nanoparticle across a lipid bilayer. The system is investigated using coarse-grained molecular dynamics (MD) simulations and the free energy landscape of the system is reconstructed using constrained MD simulations. A simple model for the energy landscape of the solute-membrane system is proposed and validated. This model explicitly accounts for effects of the membrane undulations on the solute transport. The developed model is quite general and is expected to be applicable to a wide range of interfacial systems.

(549e) Effective Properties of Real Biofilm Shell

Guibert, R., Institut de M canique des Fluides de Toulouse
Davit, Y., Institut de M canique des Fluides de Toulouse
Quintard, M., Institut de M canique des Fluides
Debenest, G., Institut de M canique des Fluides de Toulouse
Originally presented on: 11/6/2013 16:15:00 - 16:30:00

Biofilms are involved in several natural or engineered systems. Their presence within porous media can substantially affect the hydrodynamic and the transport process. In this work, considering a simple porous medium including a real biofilm structure, we describe a complete work-flow to numerically evaluate effective properties of the medium.

We analyze the influence of a multi-species basal biofilm developed over a plastic bead with a diameter of three millimeters. The biofilm thickness is close to one millimeter. The biofilm shell used for this study is generated under control in laboratory conditions and imaged using x-ray computed micro-tomography with a resolution of nine millimeters cube per voxel.

Several important preprocessing steps are necessary to achieve direct numerical simulations, including image processing and mesh generation. In order to evaluate the effectives properties (permeability and effective diffusion) of this porous medium, we consider first biofilm as impermeable (solving Stokes equations), and as a porous medium in a second time (using a fictitious domain approach).

Theses different approaches are compared and also permit simple geometric optimization.

(549f) Probing Dilational Interfacial Stresses of Complex Interfaces Using a Microscale Spherical Bubble

Kotula, A. P., Carnegie Mellon University
Anna, S. L., Carnegie Mellon University
Originally presented on: 11/6/2013 16:30:00 - 16:45:00

The dilational properties of complex fluid-fluid interfaces are often correlated with the stability and bulk rheology of emulsions and foams. Probing the dilational response of an interface is challenging since it is often difficult to isolate dilational, shear, and other deformation modes. In this talk, we generate purely dilational deformation of a spherical bubble pinned at the tip of a capillary tens of micrometers in diameter. We oscillate the pressure jump across the interface at small amplitude while optically measuring the time-dependent radius of the spherical interface. The resulting dilational modulus can contain components arising from both interfacial rheological properties and transport to the interface from the bulk solution. Each of these phenomena can depend on radius of curvature of the bubble and applied frequency. Through careful analysis of the force balance on the interface in conjunction with several existing constitutive models, we show that radius of curvature and frequency can be used together to separate thermodynamic and dynamic effects. Various interfacial phenomena can be separated by examining the real and imaginary parts of the complex dilational modulus, as well as the radius-dependent crossover frequency at which the two moduli are equivalent. We validate these findings experimentally using two common nonionic surfactant systems: $C_{12}E_8$ and Tween 80 at air-water interfaces. In the case of $C_{12}E_8$, which adsorbs reversibly, the dilational modulus is solely a result of diffusion-limited transport to the interface. Tween 80, which is known to form an irreversibly adsorbed monolayer, exhibits a Kelvin-Voigt-type viscoelastic response. Finally, we apply this analysis to an interface stabilized by a mixture of colloidal silica and the cationic surfactant CTAB that is used in the generation of particle-stabilized emulsions and foams.

(549g) Parametric Study of Drop Motion Through a Narrow Orifice

Bordoloi, A., University of Minnesota
Longmire, E., University of Minnesota
Originally presented on: 11/6/2013 16:45:00 - 17:00:00

In many natural as well as industrial applications, such as geological CO_2 sequestration, underground oil recovery, and other porous media flows, drops move through a solid constriction by displacing a continuous liquid. Both the constriction geometry and the wettability of the surface influence the drop motion in such flows. In this study, we simplify the constricting structure to a thin plate with a circular orifice with orifice-to-drop diameter ratio $d/D<1$. Drops of water/glycerin with Bond number (Bo) 2-10, released above and axisymmetric to the orifice, encounter the plate after reaching terminal speed. We examine the effects of Bo, orifice-to-drop diameter ratio (d/D), orifice edge geometry (rounded/sharp) and surface wettability (hydrophobic/hydrophilic) on the passage of the drop using high-speed imaging. It is observed that the rounded edge prevents direct contact of drops with the orifice surface, and the subsequent dynamics are unaffected by the surface wettability. By contrast, sharp-edged orifices yield contact between drop and orifice, and the surface wettability influences the subsequent dynamics. We generate regime maps for different d/D and Bo delineating domains of drop capture, release, and release with breakup. In sharp-edged orifices, pinning of contact lines at the edge leads to an overall slowing of drop motion. This leads to some additional captures of drop compared to round-edged case. For sharp-edged hydrophilic case, a fraction of drop is always captured at the orifice due to strong wettability of the surface.

(549h) Droplet Traffic in a Microfluidic Loop

Wang, W. S., Texas Tech University
Maddala, J., Texas Tech University
Pate, B.,
Huynh, D. T.,
Rengaswamy, R., Texas Tech University
Vanapalli, S. A., Texas Tech University
Originally presented on: 11/6/2013 17:00:00 - 17:15:00

We study the dynamics of a train of confined droplets passing through a fluidic loop, where a main channel bifurcates into two branches and recombines. Prior works have shown that even though droplets enter the loop at uniform time intervals, they can exit the loop at either periodic or aperiodic time intervals. Despite a growing body of literature on this problem, there have been little direct comparisons between experiments and models, and the mechanisms that cause transitions between periodic and aperiodic behaviors remain unknown.

We conduct experiments in an easily assembled millifluidic loop device and record the dynamics of several hundreds of drops as a function of droplet spacing. In the given duration of our experiments, we observe three distinct dynamics – sustained periodic behaviors, intermittent behaviors (i.e., system switches between a finite number of periodic states) and aperiodic behaviors (i.e, the system shows a large number of periods). We compare the experimental data with simulations in which drops are modeled as mobile fluidic resistors. Using Poincare maps and period lengths as determined by a spacing quantization rule, we find good agreement with experimental data for certain inlet droplet spacings, particularly when period 2 behaviors and other stable periodic behaviors are exhibited. In some cases, however, our simulations match experiments only after adjusting certain parameters (e.g. droplet resistance and inlet drop spacing) from our predicted values. We will discuss possible experimental non-idealities that need to be implemented in simulation models to better capture the experimentally observed behaviors.

Furthermore, using both simulations and experiments, we find that simultaneous droplet entrance and exit events are the main source of transitions from periodic to intermittent and aperiodic behaviors. We also observe that the more distant droplet entrance and exit events are from each other, the more stable and resilient the resulting behavior will be to fluctuations in inlet drop spacing. Our results impact the design of microfluidic devices in which fluidic networks are used to autonomously control droplets for applications in lab-on-chip technologies.

(549i) Modeling of Diffusion in Stratified Epithelia Using Smoothed Particle Hydrodynamics

Nitsche, L. C., University of Illinois at Chicago
Nitsche, J. M., University at Buffalo, The State University at New York
Kasting, G. B., University of Cincinnati Academic Health Center

Originally presented on: 11/6/2013 17:15:00 - 17:30:00

Detailed microscopic diffusion models of solute transport through stratified epithelia serve as valuable tools to understand solute flux pathways, make reasonable predictions of transport rates and tissue concentrations, and quantify transport mechanisms underlying biological function. Thus, for instance, brick-and-mortar models of the stratum (SC, barrier) layer of skin[1] can provide the quantitative basis to predict permeation of topically applied and transdermally delivered drugs, and assess risks associated with chemical exposures. Such models have evolved to incorporate two-dimensional cross-sectional geometries of cells and surrounding lipid membranes based on actual bitmapped microscope images,[2] and idealized three-dimensional polygonal geometries,[3-7] which are usually approached by finite element and finite volume methods.[2-7] Smoothed particle hydrodynamics (SPH) is a general, mesh-free method[8-10] for solving partial differential equations in more intricate and realistic domain geometries that can overtax these grid-based numerical techniques. As yet, it has seen essentially no application to biological diffusion problems. This talk develops the extensions of SPH needed to realistically address such problems, and applies it to derive significant mechanistic conclusions from test problems describing dermal absorption.

Traditional SPH is based upon grid-free, inter-node interpolation of the solute concentration field using a suitable window or kernel function. We use a non-interpolant variant of SPH that corresponds to tracking mesoscale pseudo-particles of solute that interact via the kernels.[11-13] Kernel sums over boundary particles represent an effective "chemical potential" of the walls and allow efficient treatment of arbitrary domain shapes.[14-16] For the resultant non-interpolant boundary-sum (NIBS) SPH method, a chaining mesh reduces all sums to O(N) operations. Three phenomena prevalent in biological systems define the extensions needed for biological applications of SPH, namely: (i) partitioning equilibria at the interfaces between aqueous and lipid phases; (ii) strong diffusional anisotropy of fluid-phase phospholipid-cholesterol bilayers in viable cell layers, and ordered gel-phase arrays of ceramides, cholesterol and long-chain fatty acids in keratinized epithelia; and (iii) binding to albumin and other binding/transporter proteins in both cytoplasm and extracellular fluid of viable cells, as well as keratin in cornified cells. We show in detail how these three phenomena can be incorporated into the NIBS-SPH method, including kinetic rates of exchange between mobile and bound populations of particles in the case of binding.

Solute concentration profiles, effective permeability coefficients for perpendicular penetration, and effective lateral diffusion coefficients quantifying in-plane solute mobility are calculated for realistic microscopic representations of sections of (a) human stratum corneum, and (b) underlying viable epidermis, based on realistic physicochemical parameterizations for the model compounds water, hydrocortisone and glucose. It is shown that the diffusional anisotropy of intercellular lipid phases has a dramatically affect on concentration profiles and fluxes, strongly controverting numerical results based on the assumption of an isotropic lipid phase that has commonly been made to date.[2-7]

1. Mitragotri S, Anissimov YG, Bunge AL, Frasch HF, Guy RH, Hadgraft J, Kasting GB, Lane ME, Roberts MS 2011. Mathematical models of skin permeability: an overview. *Int J Pharm* **418**:115-129.

2. Barbero AM, Frasch HF 2005. Modeling of diffusion with partitioning in stratum corneum using a finite element model. Annals of Biomedical Engineering **33**:1281-1292.

3. Rim JE, Pinsky PM, van Osdol WW 2008. Using the method of homogenization to calculate the effective diffusivity of the stratum corneum with permeable corneocytes. *J Biomech* **41**:788-796.

4. Rim JE, Pinsky PM, van Osdol WW 2009. Multiscale modeling framework of transdermal drug delivery. *Ann Biomed Eng* **37**:1217-1229.

5. Naegel A, Hansen S, Neumann D, Lehr CM, Schaefer UF, Wittum G, Heisig M 2008. In-silico model of skin penetration based on experimentally determined input parameters. Part II: Mathematical modelling of in-vitro diffusion experiments. Identification of critical input parameters. *Eur J Pharm Biopharm* **68**:368-379.

6. Naegel A, Heisig M, Wittum G 2009. A comparison of two- and three-dimensional models for the simulation of the permeability of human stratum corneum. *Eur J Pharm Biopharm* **72**:332-338.

7. Muha I, Naegel A, Stichel S, Grillo A, Heisig M, Wittum G 2011. Effective diffusivity in membranes with tetrakaidekahedral cells and implications for the permeability of human stratum corneum. *J Membr Sci* **368**:18-25.

8. Gingold RA, Monaghan JJ 1977. Smoothed particle hydrodynamics – theory and application to nonspherical stars. *Monthly Notices R Astron Soc* **181**:375-389.

9. Lucy LB 1977. Numerical approach to testing of fission hypothesis. *Astron J* **82**:1013-1024.

10. Monaghan JJ 2005. Smoothed particle hydrodynamics. *Reports Progr Phys* **68**:1703-1759.

11. Chaubal CV, Leal LG 1999. Effects of flexibility on liquid crystalline polymer behavior: The nematic broken rod. *J Polym Sci B: Polym Phys* **37**:281-300.

12. Chaubal CV, Leal LG 1999. Smoothed particle hydrodynamics techniques for the solution of kinetic theory problems Part 2. The effect of flow perturbations on the simple shear behavior of LCPs. *J Non-Newtonian Fluid Mech* **82**:25-55.

13. Chaubal CV, Srinivasan A, Eğecioğlu Ö, Leal LG 1997. Smoothed particle hydrodynamics techniques for the solution of kinetic theory problems.1. Method. *J Non-Newtonian Fluid Mech* **70**:125-154.

14. Nitsche LC, Zhang W 2002. Atomistic SPH and a link between diffusion and interfacial tension. *AIChE J* **48**:201-211.

15. Nitsche LC, Zhang WD, Wedgewood LE 2006. Asymptotic basis of the L closure for finitely extensible

dumbbells in suddenly started uniaxial extension. *J Non-Newtonian Fluid Mech* **133**:14-27.

16. Nitsche LC 2006. Accurate asymptotic formulas for the transient PDF of a FENE dumbbell in suddenly started uniaxial extension followed by relaxation. *J Non-Newtonian Fluid Mech* **135**:109-116.

(549j) Effect of Protein Adsorption On Dewetting of Silicone Hydrogel Contact Lens Materials

Bhamla, M. S.,
Fuller, G. G., Stanford University
Originally presented on: 11/6/2013 17:30:00 - 17:45:00

Contact lens related discomfort and dryness is a major complaint amongst soft contact lens wearers. Previous studies have shown that over time, proteins from the tear film accumulate on the surface of a contact lens, leading to reduced wettability of the lens and ultimate dry eyes. In order to better understand this phenomenon, the Fuller lab has developed an apparatus to measure the dewetting of pure water films on a contact lens material. Lenses are stretched flat on an elevation stage built on a miniature Langmuir trough. By raising the stage, the lens surface captures a sessile droplet. Liquid is slowly drained and dewetting dynamics captured using a CCD camera. Film thickness is verified using laser interferometry.

We discuss the consequences of the adsorbed proteins on the dewetting dynamics of pure water on different contact lens hydrogel materials. Our findings indicate that adsorbed lysozyme modifies the surface heterogeneity and hydrophobicity of the hydrogel surfaces leading to increased dewetting.

(570a) Designing Dissipative Self-Assembly

Osorio Vivanco, A. F., University of Michigan
Szleifer, I., Northwestern University
Olvera de la Cruz, M., Northwestern University
Glotzer, S. C., University of Michigan

Originally presented on: 11/6/2013 15:20:00 - 15:40:00

With recent developments in experimental techniques, we can begin to consider the synthesis and fabrication of switchable building blocks that can dynamically switch between two or more states introducing new non-equilibrium, dissipative dynamics into the self-assembly process. Particles that can toggle their shape or interactions based on some external stimulus are examples of switchable particles. In this work, we present some of the new possibilities that arise from introducing switchability and non-equilibrium dynamics into the self-assembly process. These possibilities include the stabilization of novel steady-state structures, the enhancement of self-assembly speed and propensity, the ability to capture, or dynamically arrest, a pattern that was previously only available as a transient structure as the system evolved towards equilibrium, and the ability to dynamically tune the phase and length scale of the self-assembled systems by adjusting an external non-thermodynamic control signal.

This material is based in part upon work supported as part of the Non-Equilibrium Energy Research Center (NERC), an Energy Frontier Research Center funded by the U.S. Department of Energy, Office of Science, Office of Basic Energy Sciences under Award Number DE-SC0000989.

(570b) Effect of Pressure On Micelle Aggregation and Solubilization

Meng, B.,
Ashbaugh, H. S., Tulane University
Originally presented on: 11/6/2013 15:40:00 - 16:00:00

In deep sea application of dispersants, pressure can have a significant effect on their self-assembly properties and oil solubilization. We have undertaken a molecular simulation study to piece apart the molecular-level contributions to pressure induced micellar aggregation. The effect of pressure on the partial molar volumes in aqueous solution were investigated for three different surfactants including the anionic surfactant sodium decyl sulfate, the cationic surfactant decyl trimethyl ammonium bromide, and non-ionic surfactant penta ethylene glycol monodecyl ether. The partial molar volumes of surfactants in micellar and monomeric dispersed states have been simulated in the pressure range of 1-2500 bar. The partial molar volume change on micelle formation transforms from a positive value to a negative value as the pressure increases, which corresponds to the observed maximum value in the critical micelle concentration vs pressure curve. The tail group of the surfactant contributes most to the partial molar volume change has been demonstrated using Kirkwood-Buff theory. The effects of pressure on micelle solubilization of non polar gases is also discussed.

(570c) pH-Responsive Behavior of Aqueous CTAB/NaSal Solutions

Umeasiegbu, C., University of Houston
Krishnamoorti, R., University of Houston
Balakotaiah, V., University of Houston
Originally presented on: 11/6/2013 16:00:00 - 16:20:00

pH-responsive behavior of aqueous CTAB/NaSal solutions

Chinedu Umeasiegbu[1], Ramanan Krishnamoorti[1,2], and Vemuri Balakotaiah[1,2]

Department of Chemical and Biomolecular Engineering[1] and Petroleum Engineering Program[2], University of Houston, Houston, TX 77204

Certain self-assembly systems exhibit pH response becoming viscoelastic upon an increase in pH and this can be utilized in ensuring effective stimulation in heterogeneous carbonate reservoirs. In this study, we investigate the pH-responsive behavior of an aqueous solution of cetyltrimethylammonium bromide (CTAB) in the presence of sodium salicylate (NaSal). The fluid can be switched between gel-like and a fluid-like within a narrow pH range. Dynamic light scattering (DLS) results revealed that the pH-sensitive rheological behavior was due to a transition in microstructure between wormlike and spherical micelles and that the pH-sensitivity originated from the changing hydrophobic interactions between the salicylate anion and CTAB amphiphile as pH varies. Using the bubble recoil test, a phase diagram showing the pH at which different concentrations of aqueous CTAB (C_d) and NaSal (C_s) solutions transition from viscoelastic to nonviscoelastic was obtained. These transition pH values were also investigated using microrheology data extracted from particle tracking of polystyrene probes dispersed in these solutions. It was revealed that the transition pH of these solutions and the associated change in rheology and viscosity of the system upon transition was dependent on the NaSal to CTAB ratio (C_s/C_d) and on the concentration of CTAB (C_d).

(570d) Design of a Biocompatible Oil Dispersant for Deep Water Applications Using Discontinuous Molecular Dynamics

Benner, S.,
Hall, C. K., North Carolina State University
Genzer, J., North Carolina State University
Originally presented on: 11/6/2013 16:20:00 - 16:40:00

Oil spills have caused major environmental incidents several times over the past 50 years, and similar occurrences are likely to happen in the future. Dispersants are commonly used to clean up oil spills, however current dispersants show mild to moderate toxicity to aquatic wildlife. The goal of this research is to develop a biocompatible oil dispersant, that effectively stabilizes the oil/water interface, allowing ocean bacteria to naturally degrade hydrocarbons. The backbone of the dispersants is chitosan, a naturally occurring polysaccharide found in the exoskeletons of ocean crustaceans, with hydrocarbon groups attached at selective locations on the backbone to achieve amphiphilic behavior. We will describe our efforts based in discontinuous molecular dynamics (DMD) simulations to explore ways to selectively attach the hydrocarbon groups such that the resulting hydrophobically modified chitosan will tend to wrap around oil molecules. Our first DMD simulations are being performed on a system of 200 aldehyde molecules (each containing 12 carbons) in water to observe the formation of micelles. The aldehyde molecules micellize in water to hide the hydrophobic alkane chains and expose the reactive hydroxide group. Our next simulations will add 50 chitosan molecules to the system, which surround the micelles creating hydrophically modified chitosan. After the amine groups of the chitosan bond to the hydroxide groups of the aldehyde, the chitosan will cause the micelle to open, exposing the hydrocarbon chains that are bonded to the chitosan. Finally, 150 hydrocarbon molecules (each containing 15 carbons) are added to the system representing oil droplets. The hydrocarbon molecules bonded to the chitosan backbone penetrate the oil droplets, causing the hydrophobically modified chitosan molecule to completely surround the oil droplet.

We are investigating the micelle's ability to remember its original micelle structure when it is exposed to hydrocarbon molecules. The use of DMD over traditional molecular dynamics (MD) simulations drastically reduces the simulation time, and allows for the study of larger systems over longer periods of real time. This research could reveal an effective, environmentally friendly, alternative to current oil dispersants.

(570e) pH-Responsive Amphiphiles in Enhanced Oil Recovery

Chen, I. C., Texas A&M
Yegin, C.,
Akbulut, M., Texas A&M University
Originally presented on: 11/6/2013 16:40:00 - 17:00:00

In this talk, a novel approach based on the complexation and supramolecular assembly of an amino-amide and maleic acid to control viscosity of aqueous displacement fluids will be described. The addition of only 2 wt % of adaptable amphiphile/maleic acid into water increases the viscosity of water by 4.5×10^5 times. This superior viscosity behavior is ascribed to the formation and entanglements of layered cylindrical supramolecular assemblies having diameters of several hundred nanometers. Furthermore, the viscosity of the amphiphile solution can be increased 12 times by changing pH from 4 to 8 in a reversible manner. Such a property can be very beneficial for oil recovery applications when the injectivity becomes a limitation. In addition, the use of switchable viscosity can reduce the energy cost associated with pumping large volumes of viscous displacement fluids. We also demonstrate the proof-of-concept for the use of adaptable amphiphile solutions in EOR as oil displacement fluids through column experiments. Overall, this presentation describes that pH-switchable supramolecular assemblies have very intriguing properties that can significantly impact enhanced oil recovery technologies.

(570f) The Cluster Fluid Phase of Systems With Competing Short and Long-Ranged Interactions

Sweatman, M. B., University of Edinburgh
Lue, L., University of Strathclyde
Originally presented on: 11/6/2013 17:00:00 - 17:20:00

We investigate the behavior of low density systems of spherical particles that interact with a short-ranged attractive and long-ranged repulsive (SALR) force. Using a multi-scale statistical thermodynamic model that goes beyond the mean-field level, we find that this system can form a uniform fluid of well defined clusters of particles, which are large, but finite in size. The clusters form within a narrow range of model parameters, and the transition from a uniform unaggregated fluid to a uniform dispersion of clusters (a cluster fluid) occurs at a critical cluster concentration, or CCC, in analogy to the critical micelle concentration in aqueous surfactant solutions. Within this model, the clusters are described as liquid droplets dispersed within a vapor of unaggregated particles. Minimizing the Helmholtz free energy functional yields the equilibrium cluster size and density, as well as the liquid and vapor densities for a given bulk density. The CCC is less than the bulk binodal vapor density, which is metastable. A formula for the average cluster size is determined, which essentially minimizes the cluster self-energy; this might find use in estimating effective solute-solute interactions from experimental cluster size data. We also find that the cluster size distribution is relatively narrow for the SALR system investigated here. As the strength of attractive interactions is increased, we find the cluster size increases until it diverges, although in experiments macroscopic crystals might be formed before this limit is reached. At bulk densities higher than the CCC, the cluster density increases, yet the cluster size remains nearly constant, and we predict the location of a phase transition to a non-uniform, ordered dispersion (a cluster solid) based on the packing fraction of clusters. The CCC increases with decreasing strength of attractive interactions, and we find a new first-order phase transition, a cluster vapor to cluster liquid transition, within a very narrow range of model parameters. We suggest the theoretical framework developed here could be applied to many physical systems of interest that exhibit strong effective attractive interactions competing with screened Coulomb interactions, such as colloid-polymer mixtures, many kinds of biomolecular dispersion, and perhaps even simple liquid mixtures that exhibit hydrogen bonding and de-protonation. The theoretical framework outlined here for the disordered cluster phase can be extended to ordered phases, and more complex cluster-forming systems, such as micellar solutions.

(570g) The Location and Interaction Between Propyl Gallate and Nonionic Micelles

Liang, X.,
Liu, Q., Institute of Process Engineering, Chinese Academy of Sciences
Originally presented on: 11/6/2013 17:20:00 - 17:40:00

Propyl gallate (PG) is widely used in foods, cosmetics and pharmaceuticals. The solubilization of PG in nonionic micelles is few reported. The solubilization of PG in nonionic micelles formed from Genapol® X-080 (GX-080) and Triton™ X-100 (TX-100) was investigated by NMR methods. Moreover, association constants (K_a) of PG and the surfactants were calculated by using the first-order reversible fast exchange model. In the presence of PG, the critical micelle concentration (CMC) values of the surfactants decrease. The ^{1}H NMR and nuclear Overhauser effect (NOE) experiments indicated that PG is solubilized in the shell region of TX-100 micelles, while in GX-080 micelles the aromatic protons of PG interact with the hydrophilic EO chain and more hydrophobic parts penetrate into the micelle core. The values of K_a for GX-100 and TX-080 binding to PG were found to be 700 ± 65 M^{-1} and 352 ± 30 M^{-1}, respectively. The studies on the location and the interaction between antioxidant and surfactants will facilitate the understanding of the mechanism of antioxidant effect and the application of antioxidant in pharmaceutical and cosmetic industry.

(576a) Introductory Comments for Thermodynamics Needs of Chemical Industry

Elliott, J. R., The University of Akron

Originally presented on: 11/6/2013 15:15:00 - 15:30:00

Background – A lack of industrial- academic interactions in the area of thermodynamics has been noticed by several academic researchers as well as industrial practitioners. This could have a long term impact on the future of thermodynamics research and lead to a lack of improved models and good quality data in the future. Europe is seeing a lot of industrial-academic interactions through the Working Party on Thermodynamics and Transport Properties (WP-TTP). This forum seeks to fill this interactions gap for North America based thermodynamics researchers and is open to others as well.

Purpose – Provide a forum for interactions between academic researchers and industrial practitioners so as to

(a) help increase the academic understanding of industrial practices and applications,

(b) help industrial practitioners understand the current activities and future trends in thermodynamics research,

(c) increase common basis of thinking, synergy, and possible technical collaborations between these two groups

(d) help increase industrial applicability of research activities in applied thermodynamics,

(e) provide a forum for industrial practitioners to share examples of important applications of thermodynamics for chemical processes, interesting cases of property behavior, and important classes of pure components and mixtures etc.

(f) encourage research in an important traditional area of chemical engineering.

(g) help identify and disseminate research gaps related to thermodynamics.

(h) help bring important areas of research to the notice of funding agencies.

These introductory comments are intended to put the upcoming presentations into perspective as the audience prepares for a panel discussion designed to identify industrial research topics in thermodynamics that require intensified academic research.

(576b) European Federation of Chemical Engineering - Working Party On Thermodynamics and Transport Properties

Economou, I. G., National Center for Scientific Research Demokritos

Originally presented on: 11/6/2013 15:30:00 - 15:50:00

The European Federation of Chemical Engineering Working Party on Thermodynamics and Transport Properties (WP-TTP) was found in 2007 with the following objectives:

- To promote thermodynamic, transport and physical property measurements, correlations, predictions, and simulations in the academic and industrial chemical engineering communities, the national societies, as well as the national and international funding agencies,
- To promote concerted education activities to Chemical Engineers in the areas of thermodynamics and transport properties,
- To offer a suitable forum to Chemical Engineers for exchange of ideas, the development of new basic and applied research projects and collaborations through the organization of conferences, workshops and other activities,
- To coordinate research activities at the European level by identifying chemical systems and conditions that are important for novel chemical processes and for which data and / or correlations are non-existent or inaccurate so that further work is required,
- To make available recent developments in the area through an electronic newsletter that will be published every three months.

Today, the WP-TTP has 38 full members representing 24 European countries and 6 guest members, including a US member. 8 members are from industry while the remaining are from academia, national labs and government.

In this presentation, we will discuss recent activities of the WP emphasizing those that aim to promote academia – industry collaboration in research and development. In particular, we will present an outline of a recent publication related to the needs for accurate data and models for physical properties by the industry [1]. This publication was based on an extensive survey to industry in Europe, North America and Asia. An additional WP activity is the establishment of a series of workshops under the general title "Industrial Use of Thermodynamics". Two workshops were organized in 2010 in Würzburg, Germany and in 2012 in Lyon, France with more than 100 participants each time from industry and academia. The third one will take place in Eindhoven, The Netherlands in the summer of 2014.

References

[1] E. Hendriks, G.M. Kontogeorgis, R. Dohrn, J.-C. de Hemptinne, I.G. Economou, L. Fele ?ilnik and V. Vesovic, "Industrial Requirements for Thermodynamics and Transport Properties", *Ind. Eng. Chem. Res.*, **49**(22), 11131 – 11141 (2010).

(576c) Capec Consortium: Effective Academic-Industrial Interactions Strengthened By Surveys and Analyses of Gaps in Properties Modeling

O'Connell, J. P., University of Virginia
Gani, R., Technical University of Denmark
Originally presented on: 11/6/2013 15:50:00 - 16:10:00

J.P. O'Connell and R.A. Gani : The CAPEC Consortium has operated for many years. Starting with UNIFAC correlations for activity coefficients, the research has branched out to many other properties like critical properties, heats of formation, boiling points, heats of vaporization, and many others. One key to CAPEC's success has been the establishment of an industrial advisory committee at an early stage. This committee meets annually to provide guidance about which properties are of top priority for the coming year(s). This process will be reviewed along with the resulting compilation of properties of interest and its evolution with time. One key to CAPEC's success has been the establishment of an industrial advisory committee at an early stage. This committee meets annually to provide guidance about which properties are of top priority for the coming year(s). This process will be reviewed along with the resulting compilation of properties of interest and its evolution with time.

(576d) Ab Initio Prediction of Thermodynamic Equlibria of Pharmaceutically Relevant Solutes in Organic Solvents : An Application of COSMO-SAC Formalism

Tabora, J. E., Bristol-Myers Squibb
Cohen, B., Bristol Myers Squibb
Xiong, R., University of Delaware
Sandler, S. I., University of Delaware
Originally presented on: 11/6/2013 16:10:00 - 16:30:00

This is a temporary abstract.

This work will describe the structure or an industry-academia collaboration with the goal of improving ab intio models to estimate thermodynamic properties of pharmaceutically relevant entities in organic solvents.

Results from simulations will be presented from case studies and will be contextualized to the goals of process development

Additional information will be provided shortly.

(576e) Towards Meeting Industry Needs: A Review of SAFT Applications

Adidharma, H., University of Wyoming

Originally presented on: 11/6/2013 16:30:00 - 16:50:00

Since the inception of SAFT more than two decades ago, SAFT has been theoretically improved and applied to wider practical applications. Although SAFT, as expected, will continue to face new challenges in the future, it has become one of the most important thermodynamic models available to chemical industries.

Various systems, including small, branched, polar, and associating molecules, polymers, electrolyte solutions, and ionic liquids have been modeled using SAFT. In addition to the well-known SAFT capability of describing common thermodynamic properties, such as density, vapor pressure, and phase equilibria, recent applications demonstrate that SAFT is also able to describe and predict thermal properties, critical micelle concentrations, and even the properties of biomaterials and solid phases. Coupled with other theories, such as the friction or free volume theory, the density functional theory, or the van der Waals and Platteeuw theory, SAFT becomes even more powerful in solving practical problems. These various applications will be sampled and presented.

(576f) Modeling Electrolyte Systems: Recent Progress and Current Challenges

Anderko, A., OLI Systems, Inc.

Originally presented on: 11/6/2013 16:50:00 - 17:10:00

Recent advances in modeling electrolyte systems will be reviewed in the context of well-established and emerging needs in the chemical process industry. Development of models for systems containing aqueous, mixed-solvent, and non-aqueous electrolytes remains an active area of research. A large fraction of simulation needs can be met by using excess Gibbs energy models that account for ionic interactions on a semi-empirical basis. Here, one of the main challenges lies in correctly predicting ionic reaction equilibria, which usually play an important role in systems containing electrolytes. For this purpose, it is essential to account for both the standard-state properties of species as a function of temperature, pressure, and solvent environment and for their interactions at finite concentrations. In addition to thermodynamic properties and phase equilibria, it is increasingly important to account for transport properties (e.g., electrical conductivity) and surface properties (i.e., surface and interfacial tension) in practical applications involving electrolytes. Equations of state for electrolytes present an alternative to the excess Gibbs energy models and become particularly attractive for high-temperature applications. At such conditions, predicting the changes of ionization as a function of density and temperature remains an important challenge. Various applications that are providing a stimulus for electrolyte theory, models, and experimental techniques will be discussed. Such applications include problems associated with flow assurance (in particular, mineral scaling) and CO_2 capture and sequestration including the interactions between carbon dioxide and minerals. Of particular interest is the role of electrolyte properties in energy storage and hydrogen production. There is increasing interest in the properties of electrolytes in the context of materials synthesis and processing high-value inorganic materials. Furthermore, properties of electrolytes are essential for understanding and mitigating electrochemical corrosion. Examples of such applications will be discussed and analyzed to identify the gaps in electrolyte data and models.

(576g) Panel Discussion On Thermodynamics Needs of Chemical Industry

Mathias, P. M., Fluor Corporation
Gupta, S.,
Elliott, R., University of Akron
Lira, C. T., Michigan State University

Originally presented on: 11/6/2013 17:10:00 - 17:50:00

Background – A lack of industrial- academic interactions in the area of thermodynamics has been noticed by several academic researchers as well as industrial practitioners. This could have a long term impact on the future of thermodynamics research and lead to a lack of improved models and good quality data in the future.

Through this panel discussion, we hope to form teams to develop two documents: (a) Thermophysical Property Needs for Various Unit Operations and (b) Measurement and Modeling Gaps for Various Thermophysical Properties. Future possibilities include Thermophysical Properties for Chemical Products, Properties for Bioseparations, … Volunteers for these teams will be sought at the meeting.

(577a) Invited: Polymer Stabilization of Peptide Helicies

Ashbaugh, H., Tulane University

Originally presented on: 11/6/2013 15:15:00 - 15:37:00

Helical peptide sequences generally exhibit a greater helical propensity in a protein context than when expressed independently. Mixing helical peptides with synthetic polymers offers the potential for increasing secondary structural stability. In this talk I present our simulation studies of polymer stabilization of helical sequences. In the first half of the talk, I present a preliminary bond fluctuation model simulation study of helix stabilization by polymeric crowding agents. Fitting our simulation results to a Zimm-Bragg-like model for the helix/coil transition, we find the helicies are slightly destabilized by the hydrogen bonding enthalpy despite the fact the crowding polymers exhibit no specific interactions with the peptide. Nevertheless the helix is ultimately entropically stabilized by the reduction of the conformational freedom of the peptide in the coil state.

Rather than using polymer concentration as a tool to stabilize peptide structure, polymer oligomers can be covalently attached to peptides to impart biological activity to synthetic materials. Recently, new amphipathic peptide-poly(ethylene glycol) conjugates were introduced (Shu, J., et al. Biomacromolecules 2008, 9, 2011), which displayed enhanced peptide helicity upon polymer functionalization while retaining tertiary coiled-coil associations. In the second half of this talk, we report simulations of helix stabilization by conjugation with poly(ethylene glycol). The polymer oxygens are shown to favorably interact with the cationic lysine side chains, providing an alternate binding site that protects against disruption of the peptide hydrogen bonds that stabilize the helical conformation. When the peptide lysine charges are neutralized or poly(ethylene glycol) is conjugated with polyalanine, the polymer exhibits a negligible effect on the secondary structure. We also observe the interactions of poly(ethylene glycol) with the amphipathic peptide lysines tends to segregate the polymer away from the nonpolar face of the helix, suggesting no disruption of the interactions that drive tertiary contacts between helicies.

(577b) Thermodynamics and Simulation of the Negative Pressure Folding and Unfolding of Trp-Cage and GB1 Beta-Hairpin Miniproteins

Hatch, H. W., Princeton University
Stillinger, F. H., Princeton University
Debenedetti, P. G., Princeton University
Originally presented on: 11/6/2013 15:37:00 - 15:52:00

Although hot, cold and high pressure denaturation are well characterized, the possibility of low and negative pressure denaturation remains virtually ignored. Proteins under negative pressure, however, are important in applications such as the e ect of medical ultrasounds on proteins in the body, and the survival of proteins in the xylem and the adjacent parenchyma cells of vascular plants. In addition, low and negative pressure unfolding is fundamentally important in obtaining a complete understanding of protein stability, and naturally complements previous studies of high pressure denaturation. We use extensive REMD simulations totaling 255 microseconds to obtain folding and unfolding equilibrium for the Trp-cage miniprotien and GB1 β-hairpin in explicit water. While the Trp-cage miniprotein destabilizes at negative pressure, GB1 β-hairpin stabilizes at negative pressure. The ultimate goal is to elucidate the molecular mechanistic e ect of negative pressure on protein folding and unfolding by structural analysis of the proteins and hydration water.

(577c) Atomistic Simulation of Polyglutamine in Solution

Fluitt, A. M., University of Nebraska-Lincoln
de Pablo, J. J., University of Wisconsin-Madison
Originally presented on: 11/6/2013 15:52:00 - 16:07:00

Peptides containing a long repeat of glutamine residues, called polyglutamine or polyQ, are widely studied, both as a useful model system for protein misfolding and aggregation and in relation to their own role in the molecular-level pathogenesis of neurodegenerative disease. The early-stage aggregates, or oligomers, of polyQ-containing peptides are believed to be especially cytotoxic, but their characterization is frustrated by their small size and transient nature. A combination of atomistic molecular dynamics simulations and nonlinear spectroscopies possesses the structural and temporal resolution necessary for characterizing polyQ oligomers. The reliability of simulation results depends in part on the accuracy of the potential energy functions which describe intermolecular interactions. In this work, we systematically analyze the performance of several atomistic models of biomolecules, or "force fields," in reproducing two essential properties of simple polyQ peptides in dilute aqueous solution: its lack of regular secondary structure and its poor solvation by water that drives collapse. We observe substantial differences in the predicted properties depending on the force field used. We identify the force fields that achieve good qualitative and quantitative agreement with available experimental data, and we discuss the implications of our results as they relate to the study of polyQ and intrinsically disordered peptides.

(577d) Molecular Dynamics Simulations of Self-Assembly of Nonviral Gene Delivery Complexes By Pegylated Peptides and siRNA Molecules

Fu, I. W., University of California, Irvine
Wong, S., Unversity of California, Irvine
Edson, J., University of California, Irvine
Markegard, C. B., University of California, Irvine
Kwon, Y. J., University of California, Irvine
Nguyen, H. D., University of California, Irvine
Originally presented on: 11/6/2013 16:07:00 - 16:22:00

The design of peptide-based polyelectrolyte complexes as viable non-viral vector is of great interest in gene delivery systems due to its ability to overcome significant challenges such as biocompatibility, stability, and degradation. Molecular interactions between polyplexes formed from cationic peptide-based polymers and siRNA have been shown to largely influence the effectiveness of the therapeutic system. Moreover, structural characteristics such as the charge density, molecular weight and changes in the physiological conditions can contribute and dictate the resulting physiochemical behavior of the siRNA/polymer polyplexes. Our studies examine molecular-level kinetic mechanisms of self-assembly by PEGylated-poly(Lysine) and siRNA molecules to form polyplexes via molecular dynamics simulations using multiscale models. We also investigate the role of external biological stimuli (e.g. pH, salt concentration, PEGylated-poly(Lysine)/SiRNA concentrations, or temperature) on stability and size distribution of the assembled complexes. Additional study is also on an analogous system of PEGylated-poly(ketalized-Serine) that has been recently shown to enhance gene delivery efficiency compared to poly(Lysine). Comparison between the two systems will enable a more comprehensive understanding of the relationship between the structural design and the biological functionality. The findings of this research will aid experimentalists by identifying optimal assembly and disassembly conditions to improve the design of polyplexes for gene therapy.

(577e) Elucidating the Molecular Recognition of CXCR4 By a Dual Tropic HIV-1 gp120 V3 Loop Using Free Energy Calculations and Molecular Dynamics Simulations

Tamamis, P., Princeton University
Floudas, C. A., Princeton University
Originally presented on: 11/6/2013 16:22:00 - 16:37:00

The primary step of the human immunodeficiency virus type 1 (HIV-1) cell entry is the interaction of the viral envelope glycoprotein (comprising subunits gp41 and gp120) with the host leukocyte glycoprotein receptor, CD4, and the two chemokine receptors CXCR4 / CCR5 on the surface of host cells. The molecular recognition of CXCR4 or CCR5 by gp120 is mediated through the V3 loop, a fragment of gp120. The binding of the V3 loop to CXCR4 or CCR5 determines the cell tropism of HIV-1 and constitutes a key step prior to HIV-1 cell entry. Thus, elucidating the molecular recognition of CXCR4 by the V3 loop is important for understanding HIV-1 viral infectivity, tropism and designing HIV-1 inhibitors. We employed a comprehensive set of computational tools, predominantly based on free energy calculations and molecular dynamics (MD) simulations to investigate the molecular recognition of CXCR4 by a dual tropic V3 loop. We report the first V3 loop : CXCR4 complex structure, and our results are in remarkable agreement with experiments, as well as, reveal an abundance of specific intermolecular interactions contributing to the stability of the binding conformation. This work sheds light into the functional role of V3 loop and CXCR4 residues associated with HIV-1 gp120 activity, and therefore, it provides gateways for understanding HIV-1 viral infectivity, tropism, and designing novel anti-HIV-1 inhibitors of protein gp120.

(577f) Invited: Probing Surface Effects On the Orientation and Conformation of Adsorbed Proteins With Multiscale Simulations

Pfaendtner, J., University of Washington
Originally presented on: 11/6/2013 16:37:00 - 16:59:00

Adsorption of biomolecules plays an important role in far range processes from biocatalysis to surface fouling. Understanding and controlling adsorption remains elusive due to persistent difficulties in quantitatively coupling relevant thermodynamic and kinetic driving forces to the atomic scale features of these systems. In principle, molecular dynamics or Monte Carlo simulations could be used to provide some of these insights. However, the time and length scales accessible by classical simulations are normally insufficient. This talk highlights recent examples from our group in the development and application of new methods to study protein adsorption, orientation and conformation on engineered and natural surfaces. Three examples are presented. The methods are explained and developed using an experimentally characterized model system (LK peptides adsorbing to self-assembled monolayers). Following this the role of hydrophobicity and surface charge is explored on by studying the folding of a small helical peptide as a function of its distance from the surface. The final example studies the role of the surface in controlling the structure and orientation of the protein statherin. Throughout the talk we also focus on methods by which the biased (non-equilibrium) metadynamics simulations can be reweighed to recover unbiased estimates of experiment-relevant observables.

(577g) Modeling Protein Conformational Stability On Chromatography Media With a Gō-Like Model

Zhong, E. D., University of Virginia
Shirts, M. R., University of Virginia
Originally presented on: 11/6/2013 16:59:00 - 17:14:00

In protein-based drug development, hydrophobic interaction chromatography (HIC) is a vital commercial method used to separate desired proteins from complex mixtures. However, interactions between the protein and the chromatographic surface used for separation may lead to unfolding or misfolding of the protein. We test how well a Gō-like model simulated with Hybrid Monte Carlo (HMC) can describe folding and unfolding of proteins on chromatographic surfaces as well as explore the trends connecting surface unfolding to temperature and the level of surface attraction.

The Gō-like model implemented in this study uses a residue-level coarse grain representation of the protein as a linear chain of beads and requires the classification of all possible contacts as either native or nonnative. The surface is modeled as a flat plane of point interaction sites, which are treated as native contacts to protein hydrophobic residues. With this level of approximation, we can guarantee convergence of the folding curves while still preserving residue-level details of the protein stability.

In this study, we compare the efficiency of HMC to traditional MD and pure MC simulation protocols for simulating coarse-grained models of biopolymers. We also quantitatively characterize the thermal stability of the B1 domain of protein G, on surfaces of varying attractive strength in terms of the PMF with respect to surface distance, PMF with respect to fraction native contacts, and entropic and enthalpic contributions to folding and unfolding by the surface. Finally, we compare the residue-level stability from simulations to experimental (HXMS) hydrogen exchange mass spectroscopy both in solution and in the adhered phase in actual HIC operation.

(577h) The Influence of Non-Ideal Effects On DNA Hybridization On Surfaces

Welling, R. C., Brigham Young University
Pratt, K. E., Brigham Young University
Knotts, T. A. IV, Brigham Young University
Originally presented on: 11/6/2013 17:14:00 - 17:29:00

The hybridization of single-stranded DNA in solution (the target molecule) to single-stranded DNA tethered to a surface (the probe molecule) is a key phenomenon involved in many technologies such as microarrays, DNA origami, self-assembly, templating, and surface biofunctionalization. DNA is useful in these applications precisely because the hybridization process can be exploited to yield information and direct surface immobilization. However, hybridization in these technologies occurs in an environment far removed from the cellular context, and the implications of these differences are not completely understood.

One of the difficulties of studying hybridization on a surface is that no experimental technique exists to precisely determine the effects of the inhomogeneous environment on the process. As such, molecular simulation has emerged as the primary method to investigate DNA-surface interactions. In this role, simulation has provided a variety of insights into the process, but the vast majority of these studies have been done with relatively idealized systems. The idealities include tethering the DNA to the surface at only one location and simulating only one probe hybridizing with only one target. Studies to date that have used multiple surface probes have either not studied the hybridization process or started with a hybridized duplex and focused only on melting.

This presentation will summarize several efforts to move away from the "ideal" cases and model the system in a manner that more closely approximates the real system. Results will be presented on two major themes: tethering the probe to the surface at multiple locations and hybridization of a target to a group of probes. The former is important because fabrication techniques do not always result in the probe extending perpendicular to the surface. The latter is important because each "dot" of DNA on the surface consists of many probe molecules of the same sequence.

In each case, replica exchange molecular dynamics and/or umbrella sampling were used to simulate hybridization, and the results offer surprising insights that are contrary to initial theories. For example, it was hypothesized that tethering the probe to the surface at multiple locations would inhibit the ability of the target to hybridize and decrease duplex stability, but the opposite occurred. It was also theorized that the presence of multiple probes on the surface—and the accompanying entanglements—would also destabilize the duplex, but the data showed an increase in stability. These and other results that will be presented give a more complete picture of surface hybridization than previously understood and offer hope that future design improvements of DNA-surface technologies are possible.

(577i) Tensile Mechanics of Helical Polypeptides

Torabi, K., Northwestern University
Schatz, G. C., Northwestern University
Originally presented on: 11/6/2013 17:29:00 - 17:44:00

Helices are amongst the most abundant secondary structures found in proteins. Here we report a thermodynamical model for helix-coil transition of polypeptides under mechanical tension. The proposed algorithm incorporates the empirical-based helix-coil algorithm of AGADIR [Viguera et al. J. Mol. Biol. 284, 173-191 (1998)] with the "partially freely rotating chain model" of random coil peptides [Hanke et al. EPL 92, 53001 (2010)]. Our model estimates the equilibrium value of overall and residue based helicity of a given polypeptide sequence under a given tensile force. Due to good agreement of the helix-coil algorithm AGADIR with numerous experimental data, this model is capable of making quantitative predictions of single molecule manipulations of helical peptides. The potential of mean force generated by the model elucidates the highly elastic characteristic of helical polypeptides as opposed to energy dissipative forced-unfolding of the tertiary structure of the proteins. We explains why nature chooses helical structures in certain domains of biological molecular machines where highly elastic feature of these structures avoids a large amount of energy dissipations and lead to higher energy efficiencies.

(595a) Pore Formation in 1,2-Dimyristoyl-Sn-Glycero-3-Phosphocholine Bilayers and Listeria Moncytogenes By Low Concentrations of Antimicrobial Peptide Melittin

Zhou, L., Purdue University
Wu, X., Purdue University
Wu, X., Purdue University
Singh, A., Purdue University
Bhunia, A. K., Purdue University
Narsimhan, G., Purdue University
Originally presented on: 11/7/2013 8:35:00 - 8:50:00

Antimicrobial peptides belong to a large and varied class of relatively short peptides family, which usually are largely positively charged, and constitute host-defense system in many living organisms, capable of killing inter alia, Gram negative and Gram positive bacteria by permeabilzation of cell membrane through pore formation. In the first part of this study, pore formation in DMPC bilayers by antimicrobial peptide melittin was monitored by leakage of fluorescent dyes (calcein, FD4 and FD20) of different molecular weights encapsulated within liposomes which indicated a lagtime which was larger for smaller melittin/liposome ratio and larger molecular weight of dye. Pore formation was also demonstrated through direct observation of dye leakage using confocal microscopy. A mathematical model for the prediction of time of pore formation was proposed accounting for (i) adsorption of melittin onto bilayer (ii) surface aggregation and (iii) energy barrier for formation of pore by melittin aggregates that may be interspersed with phospholipid heads. Model predictions were compared with experimental data. In the second part of this study, the antimicrobial activity of native melittin as well as its mutant (26 amino acids: GIGAVLKVLTTGLPALKSWIKRKRQQ in which Ile-17 residue of melittin was replaced by a Lys residue in order to decrease its hydrophobicity with more charge) against different strains of Gram positive bacteria *listeria monocytogenes* was characterized by absorbance and plate count which also indicated a lagtime for deactivation that was concentration dependent. Because of decreased interaction with mammalian cell, the cytotoxicity of mutant melittin was found to be less than that for native.

(595b) Molecular Dynamics Study of the Cytolysin A Pore Forming Toxin Using Atomistic and Coarse Grained Simulations

Desikan, R., Indian Institute of Science
Patra, S. M., R V College of Engineering
Maiti, P. K., Indian Institute of Science
Ayappa, K. G., Indian Institute of science
Originally presented on: 11/7/2013 8:50:00 - 9:05:00

Pore forming toxins are a unique class of proteins that assemble in biological membranes to form stable pores, which ultimately lead to cell lysis. Understanding the self-assembly and molecular mechanism of pore formation by these toxins, and their effect on the surrounding membrane has implications in disease control, cancer and gene therapies. Cytolysin A (ClyA) is a membrane-associated cytolytic protein expressed by Escherichia coli and other enterobacteria. It is secreted in a water soluble form, and the hypothesized mechanism is that it inserts into the membrane, undergoes a conformational change, and oligomerizes on the surface of the membrane. Using a recently reported surfactant crystallized structure of the dodecameric pore-complex formed by ClyA, we carried out all-atom (CHARMM 27), united-atom (GROMOS 54A7) and coarse-grained (MARTINI) molecular dynamics simulations of the assembled pore complex in a POPC membrane. These multi-scale molecular dynamics simulations enable us to address various aspects of the pore-complex and its effect on the lipid membrane surrounding it.

In the all atom simulations, the pore structure is found to be stable over a 140 ns simulation, independent of the salt content. The all-atom simulations indicate a two fold increase in the overall number of salt bridges in the membrane pore complex when compared with the crystal structure data, suggesting increased stabilization of the pore complex in the membrane. The MARTINI simulations were carried out for a time duration of 5 microseconds to further verify the stability of the pore complex in the membrane. The united-atom simulations were run for a total of 120 ns. Using these relatively long time scale simulations we analyzed the lipid dynamics around the pore complex and observed the effect of the pore-complex on the lipids in its vicinity. United-atom and MARTINI simulations of the intermediate oligomers shed light on the pore forming mechanism. In the absence of crystal structures of these intermediate oligomers, these models can be used to obtain valuable insights into the collective lipid movement and dynamics around these complexes. The coarse-grained simulations show rapid evacuation of the lipids from the interior of the octamer and higher level intermediates to the bulk membrane, thus suggesting that lipid evacuation occurs well before the final stage of the dodecamer oligomerization.

(595c) Protein Adsorption At the Electrified Air-Water Interface: Effects On Foam Stability

Engelhardt, K., University of Erlangen-Nuremberg
Peukert, W., University of Erlangen-Nuremberg
Braunschweig, B., University of Erlangen-Nuremberg
Originally presented on: 11/7/2013 9:05:00 - 9:20:00

Protein adsorption at the electrified air-water interface: Effects on foam stability

Kathrin Engelhardt, Wolfgang Peukert, Björn Braunschweig

Institute of Particle Technology

University Erlangen-Nuremberg,Germany

The surface chemistry of ions, water molecules and proteins as well as their ability to form stable networks in foams can influence and control macroscopic properties such as the stability of a macroscopic foam. Despite the great importance of protein adsorption at liquid interfaces, a molecular level understanding of proteins and their interactions at aqueous interfaces has been elusive.

Therefore, we have addressed the adsorption of the proteins bovine serum albumin (BSA) and β-lactoglobulin (bLG) at the air–water interface with vibrational sum-frequency generation (SFG) and ellipsometry. SFG provides specific information on the composition and average orientation of interfacial molecules, while complementary information on the adsorbate thickness can be obtained with ellipsometry.

Adsorption of charged proteins at the water surface leads to an electrified interface, pH depending charging and electric field-induced polar ordering of interfacial H_2O and proteins. Varying the bulk pH of protein dilutions changes the intensities of the protein vibrational bands substantially, while dramatic changes in OH stretching bands of interfacial H_2O are simultaneously observed.

These observations are related to the pH dependent charging of BSA and bLG and have allowed us to determine the isoelectric point of BSA and bLG directly at the air-water interface. At a pH near the isoelectric point proteins form an amorphous network of possibly agglomerated proteins, while highly charged proteins form monolayers with strong repulsions. Finally, we provide a direct correlation of the molecular structure of the investigated protein interfaces with foam stability and rheology[1].

[1] Engelhardt, K., Rumpel, A., Walter, J., Dombrowski, J., Kulozik, U., Braunschweig, B., Peukert, W., Protein adsorption at the electrified air–water interface: implications on foam stability, Langmuir 2012, 28, 7780–7787

(595d) Partitioning of Surface-Active Peptides Into Lipid-Based Microbubbles

Badami, J. V., The City College of New York
Tu, R., The City College of City University of New York
Originally presented on: 11/7/2013 9:20:00 - 9:35:00

Over the past decade, the ability to design assemblies that are capable of providing a means for site-specific targeting, loading/administration of therapeutics, and visualization by clinical diagnostic imaging modalities has become an attractive avenue for localized drug delivery. Microbubbles have emerged as a particular subset of such "theranostic" materials due to their intrinsic contrast-enhancing properties in the presence of ultrasonic pressure waves, along with the potential to decorate the surface with various functional molecular entities. Consisting of a hydrophobic gas core encapsulated by a monolayer of surface-active materials, the lateral interactions of molecules at the gas/liquid interface dictate the overall system properties. While early formulations of microbubble systems have been composed of mainly lipid and polymer permutations, the advancement of these particles relies on the successful conjugation of targeting/therapeutic species to the surface. One particular class of bioactive molecules that fulfill a wide degree of functions including, binding, ion/membrane transport, cellular signaling, and catalysis are proteins. Proteins however, are highly complex biomolecules consisting of numerous intramolecular interactions that promote explicit structural conformations. This work aims to circumvent the higher order complexity of such structural configurations by designing surface-active peptides that adopt well-defined, alpha-helical secondary structures and are capable of partitioning into the surface of novel lipid/polymer microbubbles. We have designed amphiphilic peptides that display differing charge distribution down the cylindrical axis of the helix and aim to quantify the influence of charge separation on the lateral phase behavior of the monolayer shell, bubble mechanics, and interfacial ordering.

(595e) A Charged Surface Induces Oriented Alignment of Like-Charged Peptides and Reduces Their Aggregation

Zhang, L., Tianjin University
Dong, X., Tianjin University
Jiang, S., University of Washington
Sun, Y., Tianjin University
Originally presented on: 11/7/2013 9:35:00 - 9:50:00

Aggregation of protein folding intermediates is a critical issue encountered in protein refolding for the production of recombinant proteins, so inhibiting protein aggregation is essential for enhancing *in vitro* protein refolding yield. Recently, it was found that like-charged ion-exchange particles were highly efficient for suppressing the aggregation, leading to a significant increase in protein refolding yields (G.-Z. Wang, X.-Y. Dong, Y. Sun, Biotechnol. Bioeng. 2011, 108, 1068). The working mechanism of like-charged resin was proposed to be the formation of spatially oriented alignment of protein molecules induced by the repulsion of protein dipoles by the like-charged surfaces. This maximizes the electrostatic repulsion between protein molecules and leads to the inhibition of protein aggregation. This work was designed to examine this hypothesis. All-atom (AA) models of positively charged and neutral surfaces were constructed, while a dipeptide, KL, with a net charge of +1e (2 positive charges and 1 negative charge) at pH 7, was utilized to mimic a protein molecule that has an uneven surface charge distribution with an electric dipole. Two peptide molecules were equilibrated in an aqueous solution, and then placed over a surface to investigate their molecular behaviors (steric alignment and distribution) and molecular interactions by molecular dynamics simulation. It is found that both the peptide molecules are excluded from the like-charged surface within a few nanoseconds. During this process, oriented alignment forms immediately after the beginning of the simulation. Namely, the electric dipoles become perpendicular to the charged surface with the negatively charged end facing the surface. Moreover, inter-molecular interaction energy reveals the increase of electrostatic repulsion between the two orientedly aligned peptide molecules. By contrast, neither oriented alignment nor increase of electrostatic repulsion between two dipeptides is observed near the neutral surface. This work has thus demonstrated that charged surface can induce oriented alignment of like-charged peptide molecules, leading to the increase of electrostatic repulsion between peptide molecules, which would inhibit protein aggregation in a protein folding process as observed in experiments.

(595f) Single Molecule Observations of Fibronectin Conformation and Its Interaction With Model Integrins

Kastantin, M. J., University of Colorado
McLoughlin, S., University of Colorado Boulder
Kaar, J. L., University of Colorado Boulder
Schwartz, D. K., University of Colorado Boulder
Originally presented on: 11/7/2013 9:50:00 - 10:05:00

Recognition of fibronectin by integrins is an important step in cell adhesion that can mediate cellular response to a protein-coated surface. This work observes the fibronectin-integrin interaction at the single-molecule level using total internal reflection fluorescence microscopy (TIRFM) and resonance energy transfer (RET) to simultaneously identify fibronectin conformation as well as its integrin-binding ability.

We use a mutant protein that contains fibronectin domains 8-10 (Fn_{8-10}) as well as the integrin-binding peptide sequence, RGD, and synergy sequence, PHSRN. Fn_{8-10} is mutated at two amino acid positions to incorporate fluorescence donor and acceptor dyes that report on protein conformation by exploiting the phenomenon of resonance energy transfer (RET). Unlike many protein labeling strategies that nonspecifically label cysteine or lysine residues, our approach allows bottom-up experimental design with a unique RET signal that is sensitive to Fn conformational changes at the single molecule level. In order to test integrin binding ability in Fn_{8-10}, we use an integrin-mimicking peptide labeled with a second RET acceptor dye. Unique spectral signals allow us to distinguish intramolecular RET, which reports on the conformation of Fn_{8-10}, from intermolecular RET, which simultaneously identifies the binding of the integrin mimic. The link between fibronectin conformation and integrin-binding ability is observed on several different chemically modified glass substrates.

(595g) Examination of the Interaction Between the Alzheimer's Disease Related Tau Protein and Model Lipid Membranes

Jones, E. M., University of New Mexico
Wiater, M., (2) Worcester Polytechnic Institute
Biernat, J., Max Planck Unit for Structural Biology
Mandelkow, E., Max Planck Unit for Structural Biology
Chi, E. Y., University of New Mexico
Originally presented on: 11/7/2013 10:05:00 - 10:20:00

In this study, we characterize in depth the interaction of the Alzheimer's disease (AD) related tau protein and model lipid membranes. The misfolding and aggregation of the intrinsically disordered, microtubule-associated tau protein into neurofibrillary tangles is implicated in the pathogenesis of AD. However, the mechanism by which tau aggregation causes neuronal dysfunction remains unclear. Prior lines of research suggest that tau's interaction with the plasma membrane may modulate tau aggregation. Our previous studies show that tau's strong interaction with anionic lipids induces disruptions to membrane lipid packing, morphology, and structural integrity. Binding of tau protein to lipids of the plasma membrane may play an important role in the misfolding, aggregation, and cytotoxicity of tau. Membrane lipids may provide a favorable local environment which facilitates misfolding and aggregation of tau protein. Membrane permeabilization may also serve as a pathway by which protein aggregates exert toxicity through cell membrane disruption, resulting in the alteration of ion homeostasis and disregulation of neuronal signal transduction. We investigate tau's propensity to interact with membranes to elucidate the disruptive effects these interactions induce in the membrane. To examine this interaction, we use Surface Plasmon Resonance (SPR) to study the binding of tau protein constructs to lipid bilayers of differing charge composition. SPR allows for both qualitative and quantitative data on interactions of proteins with lipid membranes. We obtain and compare equilibrium and kinetic binding parameters for tau construct ligands interacting with both pure zwitterionic and anionic lipid containing systems. To further investigate we also perform vesicle dye leakage experiments using tau constructs and vesicles of varying lipid composition. Using these results, we comprehensively study the role of lipid bilayer composition on tau-lipid membrane binding and membrane disruption.

(595h) Dynamics of Monolayer Collapse and Folding

Choi, S. Q., University of Minnesota
Zasadzinski, J. A., University of Minnesota
Originally presented on: 11/7/2013 10:20:00 - 10:35:00

When subjected to a large compression, a two-dimensional thin film collapses. To date, several collapse mechanisms are known for lipid monolayers: fracture, solubilization, and folds. We present a novel method that enables us to distinguish "folds" from other mechanisms and, furthermore, quantify folds. We simply measure fluctuations of the surface tension when the collapse occurs, and compute the mean squared fluctuation, which provides various properties of folds (e.g. a size and time that folds remain). We validate this technique using mixed lipid monolayers of DPPC and POPG, which exhibit the different mechanisms of collapses by varying POPG fraction. Moreover, visualization of such monolayers with the measurements reveals that formation of folds occurs randomly, and depends strongly on compression rates and a composition of POPG, showing that collapse is governed by a kinetic, rather than an equilibrium process.

By understanding the nature of monolayer collapse, we can begin to design optimized replacement lung surfactants for the treatment of neonatal respiratory distress syndrome (nRDS). Monolayer collapse determines the minimum surface tension at an air-water interface, which must be 10 mN/m for an effective lung surfactant. We expect that the various lipid and protein components of lung surfactant act to modify the kinetics of collapse by modifying the activation energy of the kinetic process of monolayer folding. We also know that saturated dipalmitoylphosphatidylcholine (DPPC) is necessary for a low surface tension monolayer; we also know that DPPC adsorbs slowly to the interface without assistance from the other lipid and protein species of lung surfactant. By studying the relationship between the fluctuations and the chemical composition of the surfactant, we hope to determine the roles of the various components of lung surfactant on our way to creating an entirely synthetic lung surfactant for treatment of nRDS.

(595i) Tug of War At the Air/Water Interface: The Role of Cholesterol and SP-B On Controlling the Line Tension of Model Lung Surfactants

Dhar, P., University of Kansas
Waring, A.,
Kaviratna, A., University of Kansas
Originally presented on: 11/7/2013 10:35:00 - 10:50:00

Lung surfactant is a complex mixture of lipids and proteins that work to regulate the interfacial tension in the lung, minimizing the work of breathing as well as insuring uniform lung inflation. A lack or dysfunction of lung surfactant can result in various respiratory diseases, such as Neonatal Respiratory Distress Syndrome (NRDS), and Acute Respiratory Distress Syndrome (ARDS). In this study, we will present our research findings on (a) how small and elevated amounts of cholesterol control the adsorption and surface tension lowering properties of synthetic surfactant mixtures (b) how cholesterol-surfactant protein SP-B alter the morphology and energy of a model lung surfactant film. A langmuir trough coupled with a Wilhelmy plate pressure sensor has been used as an *in vitro* model to record the surface activity of our model surfactant mixtures at the air/water interface in the alveoli. Using fluorescence microscopy, alterations in the domain morphology of lipid molecules is recorded as a function of cholesterol and SP-B content. The size distribution is then fitted with a theoretical model recently developed by us, that allows us to simultaneously monitor changes in the line tension l and dipole density difference m . Using this technique, we report that (a) individually *Cholesterol and surfactant protein SP-B show opposing effects on the lateral organization of domains, (b) SP-B plays a more dominating role than cholesterol when modulating the line tension at the interface.* This may well be because of electrostatic interactions between the negatively charged unsaturated lipids in our model surfactant and the positively charged protein.

(603a) Impact of Water On the Mechanism of Fischer-Tropsch Synthesis On Ru Catalysts

Hibbitts, D. D., University of California at Berkeley

Loveless, B.,

Neurock, M., University of Virginia

Iglesia, E., University of California, Berkeley

Originally presented on: 11/7/2013 8:30:00 - 8:50:00

Impact of Water on the Mechanism of Fischer-Tropsch Synthesis on Ru Catalysts

David D. Hibbitts[1], Brett T. Loveless[1], Matthew Neurock[2*] and Enrique Iglesia[1*]

[1]Department of Chemical and Biomolecular Engineering, University of California at Berkeley, Berkeley, California 94720, United States

[2]Departments of Chemical Engineering and Chemistry, University of Virginia, Charlottesville, Virginia 22904, United States

*iglesia@berkeley.edu, mn4n@virginia.edu

Fischer-Tropsch synthesis (FTS) forms large hydrocarbons from synthesis gas (H_2-CO) derived from biomass, coal, or natural gas. The specific mechanism of C-O cleavage and C-C bond formation steps in FTS remains the subject of active debate, even after many experimental and theoretical inquiries. Rigorous kinetic and isotopic data and theoretical calculations have shown that at the saturation CO* coverages prevalent during FTS [1-5], C-O activation occurs predominantly with assistance by chemisorbed H-atoms (H*) on Ru and Co surfaces, instead of via direct CO dissociation on vicinal vacant sites [2,3]. Previous studies have also shown that H_2O increases FTS turnover rates and product molecular weight on Co catalysts, [4,5] but without definitive mechanistic interpretations. The effects of water on turnover rates and selectivities on Ru-based catalysts are reported here and interpreted mechanistically using density functional theory (DFT) calculations.

CO turnover rates are proportional to H_2 pressure on Ru/SiO_2 and obey a rate equation (Eq. 1) similar to that reported on Co catalysts [2,3,4] at low CO conversions and H_2O pressures.

$$r_{CO} = \frac{\alpha(CO)(H_2)}{[1+K_{CO}(CO)]^2} \qquad (1)$$

DFT calculations indicate that a sequence of elementary steps involving H-addition to CO* to form formyl (HCO*) species followed by a kinetically-relevant H-addition to form hydroxymethylene (*HCOH*), a species that decomposes irreversibly to form methylidyne and hydroxyl (CH* + OH*) is consistent with Eq. 1. The kinetically-relevant transition state has an energy of 193 kJ mol^{-1} compared to a CO*-covered surface and H_2 in the gas phase, the resting state of the catalyst.

CO turnover rates on Ru/SiO_2 increase monotonically with H_2O pressure (up to 0.3 MPa) before reaching nearly constant values (Figure 1). This suggests that H_2O may increase the rate of kinetically-relevant CO activation steps and that a H_2O-derived intermediate may block active sites on the catalyst surface at higher H_2O pressures. DFT calculations were employed to determine how H_2O could alter the kinetically-relevant step or increase its rate constant. Results show that H_2O can mediate the formation of COH* via a H-shuttling mechanism in which H* transfers to a nearby H_2O molecule to form a short-lived H_3O^+ intermediate which can then protonate the O of CO*. Following this, COH* undergoes H-addition at the C to form *HCOH* which then dissociates with H_2O nearby as a solvent in the kinetically-relevant step. The transition state for this step has an energy of 122 kJ mol^{-1} compared to a CO*-covered surface with H_2 and H_2O in the gas phase.

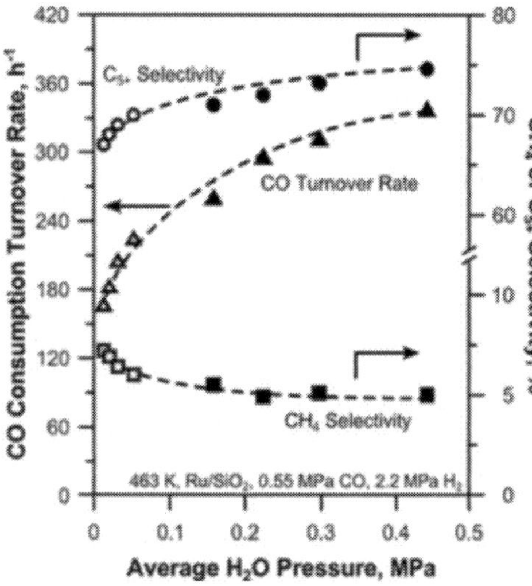

Figure 1. CO consumption rate (r or p), selectivity to CH_4 (™ or ˜), and selectivity to C_{5+} products (£ or ¢) as a function of the average H_2O partial pressure on a 5 wt% Ru/SiO_2 catalyst (463 K, 2.9 MPa, H_2/CO = 4.5). (Open symbols) Space velocity runs; (Closed symbols) H_2O-addition runs.

The presence of a H_2O-mediated path requires the addition of a H_2O-dependent term to the rate equation (Eq. 1). Furthermore, the weaker effects of H_2O observed above 0.3 MPa indicate the presence of one or more H_2O-derived intermediates blocking active sites on the catalyst surface, leading to a revised rate law in Eq. 2.

$$r_{CO} = \frac{\alpha(CO)(H_2)+\beta(CO)(H_2)(H_2O)}{[1+K_{CO}(CO)+K_{H_2O}(H_2O)+K_{OH}(H_2O)(H_2)^{-0.5}+K_O(H_2O)(H_2)^{-1}]^2} \qquad (2)$$

References

1. Mims, C., McCandlish, L., *J. Phys. Chem.* **91**, *929* (**1987**).
2. Ojeda, M., Nabar, R., Nilekar, A., Ishikawa, A., Mavrikakis, M., Iglesia, E., *J. Catal.* **272**, *287* (**2010**).
3. Loveless, B. T., Buda, C., Neurock, M., Iglesia, E., *J. Am. Chem. Soc.* **135**, *6107* (**2013**).
4. Iglesia, E., Reyes, S., Madon, R., Soled, S., *Adv. Catal.* **39**, *221* (**1993**).
5. Krishnamoorthy, S., Tu, M., Ojeda, M.P., Pinna, D., and Iglesia, E., *J. Catal.* **211**, *422* (**2002**).

(603b) Effects of Coverage Dependent Adsorbate-Adsorbate Interactions for CO Methanation On Transition Metal Surfaces

Lausche, A. C., University of Michigan

Medford, A., Stanford University

Norskov, J. K., Stanford University

Studt, F., SLAC National Accelerator Laboratory

Originally presented on: 11/7/2013 8:50:00 - 9:10:00

Recent advances in electronic structure theory have enabled researchers to understand the kinetics of surface reactions on the molecular level. Using microkinetic-modeling techniques, trends in the reaction rates and selectivities of heterogeneous catalysts can be predicted for a variety of chemical reactions.

These strategies, however, are based on a series of assumptions that may not be appropriate under all reaction conditions. The mean field model, which describes reaction kinetics with low coverages of reaction intermediates, may represent one such example. It has been reported previously that the dissociation of strongly bonded molecules (such as N_2, O_2, and CO) typically takes place under conditions where the surface coverages of intermediates are relatively high. In such cases, the trends in predicted rates and selectivities may be influenced by interactions between surface adsorbed reaction intermediates. Research described in this paper investigated the effect of these interactions on the adsorption energies (and hence reaction rates) of transition metal surfaces for the CO methanation reaction.

The DFT-calculated adsorption and transition state energies of the relevant intermediates were calculated for the 211 surfaces of Ag, Cu, Pd, Pt, and Rh. As described in the literature, the binding energies for these intermediates on a given surface can be scaled with the binding energies of carbon and oxygen. Likewise, the energies of a given transition state species can be scaled with the energies of the reaction products. By combining these scaling relations with a microkinetic modeling technique, we calculated the reaction turnover frequencies for CO methanation (both with and without interaction effects) for a range of carbon and oxygen binding energies (Figure 1).

In general, trends in the predicted methanation rates were unaltered by the inclusion of adsorbate-adsorbate interactions. The highest methanation rates were predicted for Ru, Co, and Ni, which is consistent with earlier experimental and theoretical studies. Inclusion of adsorbate interactions resulted in an increase in the predicted rates for materials with the strongest binding energies of C and/or O (e.g., Fe and Re). These increases were primarily due to an increase in the concentration of active sites via a reduction in surface poisoning due to strongly bound intermediates.

In general, the results of this analysis support the use of the mean field model for determining qualitative trends in simple reactions like CO methanation. Nevertheless, the inclusion of interaction effects is suggested to be increasingly important in cases where reaction selectivities play an important role.

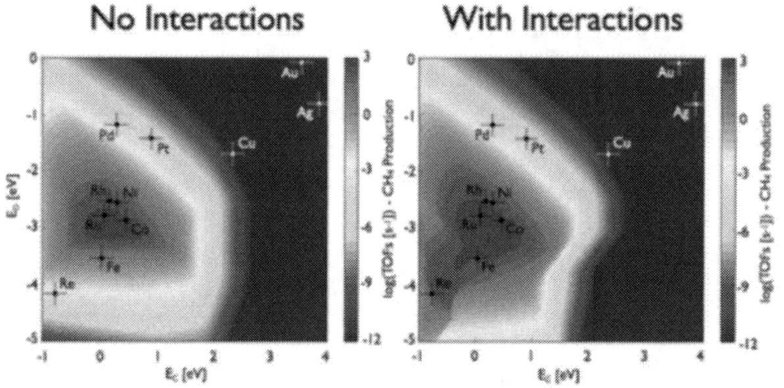

Figure 1 *- Theoretical turnover frequencies as a function of the carbon (ÆE$_C$) and oxygen (ÆE$_O$) binding energies with and without interaction effects (T = 523 K, P = 1 bar, 1% CO in H_2). Carbon and oxygen binding energies for the (211) surfaces of selected transition metals are depicted.*

(603c) Manganese Promotion of Silica-Supported Cobalt Fischer-Tropsch Synthesis Catalysts

Johnson, G., University of California, Berkeley
Werner, S.,
Bell, A. T., University of California - Berkeley
Originally presented on: 11/7/2013 9:10:00 - 9:30:00

Fischer-Tropsch synthesis (FTS) has long been investigated as a chemical process for producing transportation fuels using non-petroleum feedstocks such as natural gas, coal, or biomass. While modern cobalt FTS catalysts are highly selective toward long-chain hydrocarbons used in liquid fuels, methane is an unwanted byproduct from these catalysts [1]. Consequently, a large body of contemporary FTS research concerns methods for shifting the product distribution in favor of higher molecular weight hydrocarbons. Certain non-catalytically active elements, referred to as promoters, are known to improve catalyst performance when added to FTS catalysts in small amounts. Promoting cobalt FTS catalysts with manganese has repeatedly been found to shift the product distribution to heavier, more olefinic products and enhance catalyst activity in some circumstances [2, 3].

This presentation will discuss our ongoing investigations into the role of manganese promoters in cobalt FTS catalysts supported on silica. The catalysts were characterized by XRD, TEM, and STEM-EELS to assess particle size distribution and composition. The STEM-EELS elemental maps showed a preferential association between cobalt and manganese. Kinetics experiments in a fixed bed micro-reactor were conducted to assess apparent activation energies and partial order dependencies on reactants. Activity and selectivity data at different space velocities and long-term stability data were collected.

Using a direct hydrogenation synthesis procedure, we observed strong sensitivity of methane selectivity reduction on manganese loading for Mn/Co atomic ratios of less than 0.05. Variation of catalyst synthesis procedure and manganese loading were used to identify an optimal manganese loading for activity and selectivity enhancement. The possibility of promotion effects arising from cobalt-manganese surface alloy formation will be discussed.

References

[1] Zhang, Q. et al. ChemCatChem. 2 (2010) 1030.

[2] Bezemer, G.L. et al. J. Catal. 237 (2006) 152.

[3] Dinse, A. et al. J. Catal. 288 (2012) 104.

(603d) Higher Alcohols Synthesis From Syngas Over Three-Dimentionally Ordered Macroporous Cu-Fe Catalysts: The Active Site

Lu, Y., Mississippi State University
Yu, F., Mississippi State University
Hu, J., Mississippi State University
Zhou, P., Mississippi State University
Originally presented on: 11/7/2013 9:30:00 - 9:50:00

Higher alcohols (C_2+OH) synthesis is the challenging area in the field of C1 chemistry. Higher alcohols have higher water tolerance, reduced fuel volatility, lower vapor lock tendency and higher volumetric heating values. Higher alcohols synthesis from syngas or syngas derived from biomass, coal, or natural gas reforming is an important process for the production of oxygenates fuels, fuels additives and other intermediates for value-added chemical feedstock such as medicine, cosmetic, lubricants, detergents, and polyester. Cu-based catalysts containing metal active toward Fischer–Tropsch synthesis such as Cu-Fe based catalyst, is considered as the most promising catalysts for higher alcohols synthesis. Recently, three-dimensionally ordered macroporous (3DOM) materials with pores sized in the submicrometer had got much attention. Connected macropores with high porosity in 3DOM catalysts can permit facile transport of guest molecules, and increase the catalytic activity in potential catalysis.

In the paper, 3DOM Cu-Fe catalyst was synthesized by using poly (methyl methacrylate) (PMMA) colloidal crystal template (CCT) method. 3DOM Cu-Fe catalyst was characterized by SEM, HRTEM, XRD, XPS, Mössbauer spectroscopy. The catalytic performance of 3DOM Cu-Fe based catalyst was evaluated under model syngas in a half-inch fixed-bed reactor system. The results showed that 3DOM Cu-Fe catalysts have high selectivity of C_2^{+}OH. Cu^0 species is the active site for alcohols synthesis. χ-Fe_5C_2 is the active site for olefin and paraffin synthesis and it also plays a very important role for chain growth. There is a synergetic effect between Cu^0 and χ-Fe_5C_2 for higher alcohols synthesis. High activity and selectivity was generated by two factors. First, the presence of steps at the Cu surface is required, which can be stabilized by bulk defects like stacking faults or twin boundaries terminating at the surface. The second requirement is the presence of χ-Fe_5C_2 around the defective (stepped) Cu surface, which in the high-performance catalyst is a result of a synergetic effect leading to partial coverage of the Cu particles with χ-Fe_5C_2.

(603e) Selective Synthesis of C2+ Terminal Alcohols By CO Hydrogenation Over Cocu-Based Catalysts

Xiang, Y., Universit Libre de Bruxelles
Kruse, N., Universit Libre de Bruxelles
Originally presented on: 11/7/2013 9:50:00 - 10:10:00

While short-chain C_2-C_6 mixed alcohols play an increasingly important role as fuel additives long-chain C_8-C_{14} 1-alcohols form the feedstock for plasticizers, detergents and lubricants. Generally, 1-alcohols can be obtained from natural feedstocks such as fatty acid esters or synthezied from petrochemical sources through hydroformylation and subsequent hydrogenation of aldehydes. More recently, alternative procedures for long chain 1-alcohols production have been developed as, for example, anti-Markovnikov 1-olefin hydration and one-pot hydroformylation-hydrogenation using homogeneous catalysis. On the other hand, given the continuing problems to comply with the Markovnikov rule and the noble metal recovery the question may be asked if a heterogeneous process based on the Fischer Tropsch technology can be developed to produce strictly 1-alcohols. We delt with this question in the recent past and provide here an account of our successful attempts to boost the selectivity of short-chain and long-chain 1-alcohols synthesis.

Our approach is based on tuning the composition of CoCu-based catalysts. In a first step, our efforts concentrated on increasing the mixibility of both metals (only a maximum of 9% Cu can be dissolved in Co metal) by adding a third metal. Moreover, we chose to co-precipitate all metals as oxalates and to decompose these precursors by hydrogen-assisted TPDec (temperature programmed decomposition). This procedure allowed us to produce ternary systems such as "CoCuMn". It should be noted that these systems do not contain a classical support material. 3D Atom-Probe-Tomography demonstrated single nanosized "CoCuMn" particles to have a core-shell structure, with Co forming the core. Most interestingly, atom mapping using the same technique revealed the thin outer shell to contain all metals but to be otherwise dominated by Cu atoms. It should also be noted that some of the co-precipitated Mn-oxalate decomposed to Mn-oxides (mainly Mn_5O_8) which was catalytically inactive in CO hydrogenation but acted as a dispersant and thus conferred the catalyst a high BET specific surface area.

In this paper, we compare the catalytic performance of "CoCuMn" with that of "CoCuMo" and "CoCuNb". Under conditions of low CO conversion (<10%), catalyst compositions of $Co_1Cu_1Mn_1$, $Co_1Cu_1Mo_1$ and $Co_1Cu_1Nb_{0.4}$ showed 1-alcohols selectivities of more than 50 wt% and sometimes reach more than 90%. The α-ASF chain lengthening probability for alcohols over these catalysts was quite different, however. While a unique α value was obtained for "CoCuMn" (an optimum C_8-C_{14} 1-alcohol slate with $\alpha\approx0.65$ and a CO conversion close to 20% was achieved at T=240 °C and a total pressure of 60 bar (H_2/CO=2)) both $Co_1Cu_1Mo_1$ and $Co_1Cu_1Nb_{0.4}$ ASF plots showed two chain lengthening probabilities depending on the chain length. For $Co_1Cu_1Mo_1$ we found a behavior very close to that of pure CoCu ($\alpha\approx0.28$) in the range of C_1 to C_7. For C_{7+}, however, the α-values usually exceeded those of $Co_1Cu_1Mn_1$. Just the opposite ASF chain lengthening behavior was observed for $Co_1Cu_1Nb_{0.4}$. In that case, the α-values started high up to C_4 and then remained close to those of pure CoCu.

Our research strategy thus proved to be viable for tuning the synthesis to long-chain 1-alcohols for plasticizers, surfactants or detergents production, or to short-chain C_1-C_5 alcohols with a possible use as fuel additives.

Keywords: CO hydrogenation, long- and short-chain 1-alcohols, CoCu-based catalysts, Mn, Nb and Mo promotion.

(603f) Insights Into Mechanism of Syngas Reaction On Mo2C Catalysts --- Identifying "Descriptor" Elementary Steps for Complex Reaction Systems

Li, L., Georgia Institute of Technology
Sholl, D. S., Georgia Institute of Technology
Originally presented on: 11/7/2013 10:10:00 - 10:30:00

With the limited fossil fuel resources, extensive research efforts have been devoted to find alternative building blocks in the chemical industry. Syngas ($CO + H_2$) is potentially promising because it can be derived from both conventional and renewable energy sources and produce a range of different downstream products.

As a syngas reaction catalyst, molybdenum carbides have been reported in applications towards various products such as Fischer-Tropsch synthesis, water gas shift, and higher alcohol synthesis. It was also found experimentally that selectivity could be further modified by the addition of alkali promoter. In principle, finding the "descriptor" controlling product selectivity could eventually help us designing catalyst selective to specific product. However, little progress has been made so far due to the complexity of the syngas reaction mechanism.

In this work, we seek to provide a broad perspective of syngas reaction network on Mo_2C and determine individual elementary steps most sensitive to product selectivity. We first built a kinetic model containing steps relevant in Fischer-Tropsch synthesis, methanol synthesis, ethanol synthesis and water gas shift reaction. Reaction energies were then calculated from Density Functional Theory while activation energies were approximated from Bronsted-Evans-Polanyi (BEP) relations. With these energies as inputs, reaction in the kinetic model was simulated. Our computed selectivity was consistent with experimental selectivity without any data fitting. Sensitivity analysis was applied to determine steps sensitivity to the overall model. Our results suggested CH_x intermediates on the surface were mostly produced through H-assisted CO dissociation although it might not be the rate determining step due to surface coverage effect. We hope this work could give insights into the reaction mechanism of syngas reaction systems and serve as a guideline of applying a complex kinetic model to determine "descriptor" elementary steps.

(603g) Mechanistic Study of Fischer-Tropsch Synthesis: C-O Scission and Initial CH Formation Pathways for Realistic Surface Coverages

Banerjee, A., National University of Singapore
Gunasooriya, G. T. K. K., National University of Singapore
Saeys, M., National University of Singapore
Originally presented on: 11/7/2013 10:30:00 - 10:50:00

Fischer-Tropsch synthesis transforms synthesis gas, a mixture of CO and H_2, to long-chain hydrocarbons and water. In recent years, interest in Fischer Tropsch synthesis as an attractive route to convert natural gas, coal and biomass to clean transportation fuels, has surged. Supported cobalt catalysts are often preferred due to their high activity, selectivity towards long chain hydrocarbons, and low CO_2 selectivity. The detailed reaction mechanism of this complex reaction however remains intensely debated. C-O scission is a key step in the FT mechanism, and occurs either by direct cleavage of CO at defect sites, via a hydrogen-assisted pathway, or after formation of a C-C bond via CO insertion. Our kinetic study [1] shows that the propagation mechanism for the CO insertion mechanism is consistent with available experimental kinetic and selectivity data [2,3]. However, the initiation and termination steps for the CO insertion mechanism remain to be elucidated. In this presentation, we analyze various reaction paths for the formation of the chain initiating CH species on a CO saturated Co surface [1] using density functional theory. These CH species either hydrogenate to methane or initiate chain growth to long chain hydrocarbons. Next, a detailed microkinetic model is developed by combining the initiation reactions with chain growth and termination pathways.

References

1. Zhuo, M.; Borgna A.; Saeys, M.; J. Catal., 290, 217 (2013)
2. Schweicher J.; Bundhoo A.; Kruse N.; J. Am. Chem. Soc., 134, 39, 16135 (2012)
3. Tuxen, A.; Carenco, S.; Chintapalli, M.; Chuang, C; Escudero, C.; Pach, E.; Jiang, P.; Borondics, F.; Beberwyck, B.; Alivisatos, A. P.; Thornton, G.; Pong, W.; Guo, J.; Perez, R.; Besenbacher, F.; Salmeron, M.; J. Am. Chem. Soc., 135, 6, 2273 (2013)

(604a) Silica Nanoparticle - Amphiphilic Polymer Interactions in Aqueous Dispersions

Alexandridis, P., University at Buffalo - The State University of New York (SUNY)
Sarkar, B., University at Buffalo - The State University of New York (SUNY)
Bodratti, A. M., University at Buffalo - The State University of New York (SUNY)
Tsianou, M., University at Buffalo - The State University of New York (SUNY)
Originally presented on: 11/7/2013 8:35:00 - 8:50:00

We consider colloidal dispersions of silica nanoparticles in aqueous media containing poly(ethylene oxide)-poly(propylene oxide)-poly(ethylene oxide) (PEO-PPO-PEO) amphiphilic block copolymers (Pluronics), with a specific interest on how solution conditions such as pH and temperature, and the presence of salts, polar organic solvents, or PEO homopolymers, affect the (i) organization of the amphiphiles on the nanoparticle surface and in the bulk solution, (ii) ensuing interactions between nanoparticles, and (iii) macroscopic properties of the dispersions. The fundamental knowledge thus gained guides the design of formulations incorporating nanoparticles, and the spatial organization of nanoparticles in block copolymer matrices.

Sarkar, B.; Venugopal, V.; Tsianou, M.; Alexandridis, P. Adsorption of Pluronic Block Copolymers on Silica Nanoparticles. *Colloids Surfaces A: Physicochem. Eng. Aspects* **2013**, *422*, 155-164. DOI: 10.1016/j.colsurfa.2013.01.010

Sarkar, B.; Venugopal, V.; Bodratti, A. M.; Tsianou, M.; Alexandridis, P. Nanoparticle Surface Modification by Amphiphilic Block Copolymers in Aqueous Media: Role of Polar Organic Solvents. *J. Colloid Interface Sci.* **2013**, *397*, 1-8. DOI: 10.1016/j.jcis.2013.01.034

(604b) Diffusion and Transport of Nanoparticles in Arrays of Nanoposts

He, K., University of Houston
Babaye Khorasani, F., University of Houston
Conrad, J. C., University of Houston
Krishnamoorti, R., University of Houston
Originally presented on: 11/7/2013 8:50:00 - 9:05:00

We study the diffusive dynamics and convective transport of nanoparticles of diameter 200—400 nm in microfabricated arrays of cylindrical nanoposts using differential dynamic microscopy and single particle tracking. Posts of diameter 500 nm and height 10 μm were spaced by 1.2—10 μm on a square lattice. First, we measured dynamics of nanoparticles in quiescent dilute dispersions. As the spacing between posts was decreased, the diffusive dynamics slowed and on all length scales were best represented by a stretched exponential function. Both the relative diffusivity and the stretching exponent decreased linearly with increased confinement and, equivalently, with decreased void volume. Second, we measured the dispersion of nanoparticles as they flowed through the nanopost arrays. The distributions of particle velocity in the longitudinal and transverse directions exhibited distinct scaling behaviors with increasing confinement. Moreover, with increasing confinement the transverse dispersion became progressively non-Gaussian and increasingly important relative to the longitudinal dispersion. The appearance of non-Gaussian and stretched dynamics suggests that the transport properties of nanoparticles become increasingly cooperative when strongly confined.

(604c) Covalent Grafting of Phenylphosphonic Acid On Calcium Niobate Platelets

Shori, S., The Pennsylvania State University
Ploehn, H. J., University of South Carolina
Originally presented on: 11/7/2013 9:05:00 - 9:20:00

Inorganic platelets (such as exfoliated montmorillonite) are often treated with alkylammonium cations to improve their compatibility with more hydrophobic polymers. As an alternate route, we are exploring covalent grafting of organic groups onto the face surfaces of synthetic inorganic platelets. In this work, we investigate covalent grafting of phenylphosphonic acid (PPA) onto the face surfaces of calcium niobate (CN), a Dion-Jacobsen layered perovskite. CN can be fully exfoliated in aqueous solutions of tetrabutylammonium hydroxide (TBA), as confirmed by AFM; we denote the product as TBA-CN. We envisioned that PPA could be easily and fully grafted onto the face surfaces of exfoliated TBA-CN, which has reactive apical oxygen atoms. This hypothesis has been explored through studies of PPA grafting onto TBA-CN (and unexfoliated CN) in pure water, aqueous TBA solution, and various alcohols. Characterization results (NMR, FT-IR, XPS, XRD, and TGA) are used to quantify the extent of PPA grafting and different grafting modes (mono-, bi- and tridentate). The results are rationalized in terms of a more realistic picture of the grafting process, representing the first step towards the synthesis of polymer-grafted platelets (PGPs) prepared by surface-initiated polymerization.

(604d) Particle Dispersion and Particle Interactions in Ionic Liquids

Wagner, N. J., University of Delaware
Gao, J., Univesity of Delaware
Ndong, R.,
Shiflett, M. B., DuPont Company
Originally presented on: 11/7/2013 9:20:00 - 9:35:00

Shear thickening dispersions of colloidal particles in ionic liquids are being developed for possible use to improve the ballistic, puncture and abrasion resistance of space suits and micrometeorite and orbital debris (MMOD) shielding for spacecraft. Ionic liquids are proposed as the solvent phase of STFs formulation for space application because of their stability over the broad range of temperatures and low volatility. However, there is little known about dispersing particles into ionic liquids or the rheology of such novel dispersions. Suspensions of concentrated amorphous hydrophilic silica in ionic liquid 1-Butyl-3-methylimidazolium tetrafluoroborate ([Bmim][BF_4]) are reported to show some shear thickening behavior, but the successful dispersion of the particles is questionable. Simulations and experiments have been conducted to study the interparticle interactions in ionic liquids, showing that the high ionic concentration in ionic liquids effectively screens the electrostatic repulsion between silica particles, leading to significant particle aggregation. Therefore, to successfully disperse nanoparticles into the ionic liquid [Bmim][BF_4], we have modified the surface of silica nanoparticles with fluorocarbon 1H, 1H, 9H-hexadecafluoro-1-nonanol. This steric layer is thought to both make the silica compatible with the ionic liquid and to provide a steric stabilization layer.

A combination of rheology, dynamic light scattering, and small angle neutron scattering are applied to test an unproven hypothesis in literature, namely that stabilization of particles in this ionic liquid is a consequence of a solvation layer induced by hydrogen bonds between ionic liquid [Bmim][BF_4] and surface groups on the particles. Our rheological results indicate that the coated silica particles disperse in [Bmim][BF_4] at room temperature and aggregate and gel at high temperature, which may be due to the disruption of the solvation layer at higher temperatures by weakening the hydrogen bonding between the steric layer and the ionic liquid. SANS data enable determining the interparticle interactions and suspension microstructure in the ionic liquid under different temperatures and concentration conditions. The SANS data (especially in the low q regimes) indicates that the suspension microstructure changes significantly with temperature and concentration. Modeling the structure factor using a core-shell sticky hard sphere model allows us to directly determine the stabilizing effect of the solvation layer in the ionic liquid. This work also has potentially important implications for environmental and energy engineering as ionic liquids are leading candidates for remediation, separation, and recycling of nuclear wastes.

(604e) Angle-Resolved Second-Harmonic Light Scattering of Colloidal Interfaces

Sauerbeck, C., University of Erlangen-Nuremberg
Braunschweig, B., University of Erlangen-Nuremberg
Peukert, W., University of Erlangen-Nuremberg
Originally presented on: 11/7/2013 9:45:00 - 10:00:00

In many colloids macroscopic properties such as stability are controlled by the inherent interfaces and interactions of the particle surface with the solvent. It is, therefore, of great importance to gain a detailed understanding of the interfacial molecular structure. This task is, however, highly challenging since with many experimental techniques, interfacial molecules cannot be unambiguously separated from molecules in the bulk. Nonlinear optical techniques in particular second-harmonic generation (SHG), are inherently surface sensitive and can be powerful tools in interface science. For colloidal particles, second-harmonic scattering (SHS) patterns exhibit distinct angular features that are strongly dependent on the particle size as well as on the polarization and orientation of molecules at the particle interface[1,2]. The SHS intensity is determined by the molecular hyper-polarizabilites of the interfacial molecular structure and thus contains information about molecular species, number densities and orientations of surface adsorbed molecules. For charged particles, the interfacial electric field causes polar ordering of water molecules within the particles electrical double layer and gives rise to additional second-harmonic light scattering. The field strength of this contribution is linearly dependent on the interfacial electric field and can, thus, be used to determine the electric potential of the particle surface via the so-called Χ$^{(3)}$method.

In our study we have used second-harmonic light scattering to investigate the interfacial molecular structure and the surface charge of amphoteric particles. The surface potential of these particles changes its polarity when the isoelectric point is being crossed. The total second-harmonic field generated by the charged particles is composed of contributions directly from the interface and of contributions whose origin is the surface charge. Both contributions interfere depending on the sign of the surface potential. Hence the second-harmonic field strength depends not only on the absolute value of the surface charge, but also on the direction of the resulting static electric field. The observed interference in the second-harmonic field allows us to determine the net orientation of interfacial water molecules. Additional charge screening experiments at conditions where the particles carry a strongly positive, negative or no net charge corroborate our conclusions and have allowed us to determine the surface charge densities of the particles by applying the Χ$^{(3)}$method.

[1] Schürer, B., Wunderlich, S., Sauerbeck, C., Peschel, U., Peukert, W.,Phys. Rev. B, 82 (2010) 241404

[2] Schürer, B., Hoffmann, M., Wunderlich, S., Harnau, L., Peschel, U., Ballauff, M., Peukert, W., J. Phys. Chem. C, 115 (2011) 18302–18309

(604f) Modulating Homogeneous Crystal Nucleation in DNA-Coated Colloids

Crocker, J. C., University of Pennsylvania
McGinley, J. T. III, University of Pennsylvania
Jenkins, I., University of Pennsylvania
Sinno, T., University of Pennsylvania
Originally presented on: 11/7/2013 10:00:00 - 10:15:00

While DNA-coated particles have been reported to form a variety of different colloidal crystalline structures, the factors determining nucleation (or its absence) remain poorly understood. As with other colloidal crystals, we find that homogeneous nucleation rates depends on the ratio of the particle radius to the range of the DNA-induced attractive interactions. Unexpectedly, we find that particle-to-particle variations in DNA content, which give rise to comparable heterogeneity in interaction strength, significantly accelerate homogeneous nucleation rates in some crystals. This is demonstrated in detailed Langevin Dynamics simulations and experiment. The physical determinants of such DNA content variations, and how they can be engineered to modulate nucleation, will be discussed.

(604g) Reconfigurable Self-Assembly of DNA-Coated Colloids Through Competing Hybridization Reactions

Rogers, W. B., University of Pennsylvania
Manoharan, V. N., Harvard University
Originally presented on: 11/7/2013 10:15:00 - 10:30:00

Due to its chemical specificity and combinatorial sequence space, DNA is emerging as an ideal material to guide the self-assembly of colloidal materials. The cooperative interplay of hydrogen bonding, pi-stacking, electrostatic, and hydrophobic interactions between tethered DNA sequences induces attractive pair interactions, which can direct particles to form clusters, ordered crystal lattices, or other interesting structures. In the simplest cases, DNA-mediated binding is a monotonic and near exponential function of temperature, resulting in a single, narrow temperature window for equilibrium assembly, which has frustrated efforts to make multicomponent or reconfigurable structures. Here we show that DNA-coated colloids that interact with competing DNA strands in solution, by a process called toehold exchange hybridization, exhibit temperature dependences that need not be exponential nor monotonic. Using this additional control over the assembly pathways, we design and experimentally demonstrate reconfigurable colloidal systems, ranging from colloidal molecules that undergo reversible, structural isomerization to binary superlattices with temperature-dependent composition. These unique interactions will be useful in the future design of multicomponent, hierarchical, and programmable self-assembling materials.

(604h) Instantaneous Formation and Aging Behavior of Charged Vesicle Gels

Seth, M., University of California, Santa Barbara
Leal, L. G., University of California, Santa Barbara
Originally presented on: 11/7/2013 10:30:00 - 10:45:00

Colloidal dispersions of cationic bilayer vesicles serve as base materials for commercial household products. During the manufacturing process, these dispersions are subject to a number of expansion/contraction flows. The long-term stability of such dispersions can depend, amongst other factors, on their processing history. Here, we study dispersions of charged multilamellar vesicles (MLVs) made from the cationic surfactant diethylesterdimethyl ammonium chloride (DEEDMAC), whose bilayers are in the crystalline (solid) state at room temperature. Typically under the application of shear, MLV dispersions display strong shear-thinning behavior. However when subject to a contraction flow through an extruder, above the main phase transition temperature of the bilayers, we find that such vesicle dispersions instantaneously transform into a 'jammed', glassy-state at the extruder outlet. The glassy dispersions behave as stiff gels, having visco-elastic moduli that are several orders of magnitude higher than their un-extruded counterparts. We use tools such as rheology and cryo TEM imaging to probe mechanisms that lead to the formation of such vesicle gels. Further, we closely examine the subsequent aging or breakdown of their microstructure; a process which occurs over time scales that depend on the dispersion composition as well as the thermotropic phase behavior of the surfactant.

(604i) Aggregation Rate of Colloidal Particles Under Reaction-Limited Conditions At High Concentrations

Lattuada, M., ETH Zurich

Originally presented on: 11/7/2013 10:45:00 - 11:00:00

Aggregation of colloidal particles plays a key role in a variety of applications. A very common situation is where colloidal particles aggregate in the presence of residual repulsive interaction, a situation commonly known as reaction-limited conditions. In the thirties, Fuchs extended Smoluchowski's equation, developed for the rate of aggregation of colloidal particles under diffusion-limited conditions, to the case of reaction-limited conditions. Nevertheless, its use is limited to sufficiently dilute conditions. In this work we investigate by means of Brownian-Dynamic simulations the rate of aggregation of colloidal particles at high volume fractions, up to 30%. In a previous work, we have developed a correction to Smoluchowski's equation for the rate of aggregation of colloidal particles under diffusion-limited conditions at high concentrations within the framework of trapping theory. This corrected aggregation kernel, accounting for concentration dependence effects, was implemented in a population-balance equations scheme and used to model the aggregation kinetics of colloidal particles undergoing diffusion-limited aggregation under concentrated conditions. It was found that classical Smoluchowski's theory substantially underpredicts the rate of aggregation as well as the onset of gelation, with deviations becoming progressively more severe as the particle volume fraction increases. Similarly, we observed that even under reaction-limited conditions the rate of aggregation of particles under concentrated conditions is substantially higher than that expected from Fuchs-Smoluchowski's theory. An attempt of developing a kernel that corrects for the effects of high concentrations has been made.

(605a) Reconfigurable Assemblies With Shape Shifting Particles

Nguyen, T. D., Oak Ridge National Laboratory
Glotzer, S. C., University of Michigan
Originally presented on: 11/7/2013 8:30:00 - 8:47:00

Utilizing nanoscale building blocks that can change shapes in response to environmental changes proves to be a promising approach to engineering bio-inspired nanomaterials[1,2]. We propose a generic scheme for shifting particle shapes to induce the assembly and reconfigurability of various ordered structures in two- and three-dimensional spaces. Starting from ordered structures made of spherical and rod-like particles, we alter the particle shape at different rates into other shapes such as square, triangle or zigzag (in 2D)[2] or tetrahedron and octopod (in 3D), and observe the change in the morphology of the structure as a function of the initial and target shapes and shifting rate. Using molecular dynamics simulation, we demonstrate that more efficient assembly, higher packing fractions and ordered structures with fewer defects are among the advantages that shifting the particle shapes can offer. We also discuss the challenges the approach should face compared to the traditional assembly method, i.e. without shape shifting.

References

1. T. D. Nguyen and S. C. Glotzer, ACS Nano, 4, 2585-2595, 2010.

2. T. D. Nguyen, E. Jankowski and S. C. Glotzer, ACS Nano, 5, 8892-8903, 2011.

(605b) Systematic Inference of Self-Assembly Pathways By Nonlinear Machine Learning

Long, A. W., University of Illinois
Ferguson, A. L., University of Illinois
Originally presented on: 11/7/2013 8:47:00 - 9:04:00

Controllable self-assembly has opened a new frontier for the fabrication of sophisticated functional aggregates, such as self-healing membranes for biomedical devices, nanoscale optoelectronics for photonic computing, and "smart" drug delivery vessels. Computer simulations provide a means to probe the structural mechanisms and interactions driving micro and nanoscale self-assembly, but it remains a challenge to identify the dynamic structural rearrangements driving these phenomena. The absence of good order parameters describing assembly has frustrated the establishment of a quantitative theoretical framework explicitly accounting for its thermodynamic and kinetic features, and hindered the rational design of programmable building blocks.

We have confronted this challenge by coupling nonlinear data mining algorithms to molecular simulation trajectories of self-assembly. Our approach enables us to systematically extract kinetically meaningful order parameters for assembly, infer assembly pathways and mechanisms, and quantify fluxes through competing assembly routes. In brief, we map the assembly trajectory to a dynamically updating interaction graph, in which nodes are monomers and edges bonds, enabling us to compute the structural similarity of the partially assembled clusters through graph matching. Using the quality of the match as a proxy measure for the dynamic proximity of pairs of clusters, we use the diffusion map nonlinear machine learning algorithm to extract the underlying self-assembly pathways describing the sequence of structural steps by which the system evolves into its terminal assembled state. The order parameters computed by the diffusion map correspond to the collective modes governing the long time evolution of the system, furnishing a kinetically meaningful low-dimensional subspace in which to infer assembly pathways and construct dynamically meaningful projections of the underlying free energy landscape for assembly.

We have applied our approach to the self-assembly of patchy colloids into icosahedral aggregates as a coarse-grained biomimetic model for viral capsid formation. In good agreement with previous work employing cluster size as a heuristic order parameter, our approach systematically identifies two qualitatively different assembly pathways corresponding to 1) monomeric addition, and 2) budding of aggregates from a liquid-like phase, and we recover kinetically meaningful collective order parameters describing the dynamic structural rearrangements along each path. Further, we quantify the flux through each pathway to compute a quantitative measure of the assembly propensity through each route as a function of temperature and chemistry of the colloidal monomers. Finally, we discuss how "assembly landscapes" may be used to guide the design of building blocks programmed to assemble desired aggregates that are both thermodynamically stable and kinetically accessible, by rational manipulation of the underlying free energy landscape.

(605c) A General Mechanism for Emergent Chirality in Achiral Hard Particle Assembly

Carmichael, S., University of California Santa Barbara
Shell, M. S., University of California Santa Barbara
Originally presented on: 11/7/2013 9:04:00 - 9:21:00

For centuries, chirality has been appreciated as a key component in understanding how matter orders. While intuitively chiral particles can self-assemble into chiral superstructures, it is less clear how achiral particles are driven to do the same. Here we show that there is a general mechanism that explains why many simple achiral particles can assemble into chiral materials. Namely, we use simulations of hard, regular polygons to show that surprisingly subtle particle shape modifications, such as corner rounding, can induce chiral symmetry breaking by deforming the underlying close-packed lattice. Not only does this mechanism quantitatively explain recent experimental results reporting chirality in the hard triangle system, but it offers an intuitive and general way by which chirality might emerge in nature. Moreover, because effective corner rounding can be realized by modulating repulsive interactions, this simple mechanism suggests tremendous potential for creating dynamically tunable chiral surfaces with a variety of applications.

(605d) Quantitative Modeling of DNA Grafted Nanoparticle Self-Assembly

Vo, T.,
Srinivasan, B., Columbia University
Kumar, S., Columbia University
Gang, O.,
Venkatasubramanian, V., Columbia University
Originally presented on: 11/7/2013 9:21:00 - 9:38:00

Nanoparticle super-lattice engineering, which seeks to design DNA-grafted NPs that self-assemble into desired structures, is a growing field of research because of its relevance to optical band gap materials, sensors etc. In this protocol, single-stranded DNA sequences are grafted to NPs. This motif corresponds to a central NP core grafted with chains with "sticky" ends. DNA "sticky" ends across two particles with complementary strands can hybridize into double-stranded DNA, thus creating an emergent network of interconnected NPs, which under certain conditions crystallizes into a given lattice structure. A simple, but popular model for the prediction of DNA-grafted NP crystal structure was devised by Macfarlane. This Complementary Contact Model views DNA-NP complexes as "fuzzy spheres" where contacts between NPs lead to permanent DNA hybridization. However, despite its success, the model is limited in its prediction abilities to four broad classes of structure and cannot differentiate between different crystals within those classes. Here, we propose an extension of the model to allow for grafting of two different types of complementary linkers on each particle and incorporate hybridization between particles for both similar and different particles. These modifications enabled us to predict grafting density of DNA linkers on each particle that will result in a transition from a body-centered cubic structure to a face-centered cubic structure formation. Furthermore, the model can now differentiate between different types of crystal structure from the same class such as a disordered FCC, CuAu, and regular FCC structure – in a manner that is apparently in good agreement with the recent experiments of Gang and his coworkers. These new developments allow for a wider selection of design parameters of DNA-NP mixtures and thus can increase the selectivity of the final structure of the system.

(605e) Fundamental Thermodynamic and Kinetic Models for Self-Assembly of Small Clusters of Colloidal Particles

Thyagarajan, R., Indian Institute of Technology Madras
Sehgal, R. M., University of Massachusetts Amherst
Maroudas, D., University of Massachusetts Amherst
Bevan, M. A., Johns Hopkins University
Ford, D., University of Massachusetts Amherst
Originally presented on: 11/7/2013 9:38:00 - 9:55:00

The self-assembly of finite clusters of colloidal particles into crystalline objects is a topic of technological interest, as a route to produce photonic crystals and other meta-materials. Such assembly problems are also of fundamental scientific interest because they involve thermodynamically small systems, with a number of particles (10 to 1000) that is far below the bulk limit. One assembly method involves the use of thermodynamic variables (e.g. temperature, depletant concentration) or external fields as actuators, to modify the level of interparticle attraction. Robust methods for assembling defect-free target structures will ultimately require reduced-dimension process models that link the particle-level dynamics of the colloids to the actuator states. We have developed a three-part strategy for developing such process models. First, we employ diffusion maps (DMaps) on raw trajectory data to identify slow, low-dimensional manifolds in the system dynamics. Second, we identify convenient observables, or order parameters (OPs), that strongly correlate with the low-dimensional DMap coordinates; this step may involve a feedback loop with the DMap process itself. Third, we use a Fokker-Planck or Smoluchowski formalism to build free energy and diffusivity landscapes in the OPs, which serve as our reduced-dimension process models. We demonstrate our strategy on two colloidal systems that have been synthesized and studied experimentally.

(605f) Patterned Substrates to Direct Self-Assembly of Particle Monolayers

Ferraro, M. E., The University of Texas at Austin
Truskett, T. M., The University of Texas at Austin
Bonnecaze, R. T., The University of Texas at Austin
Originally presented on: 11/7/2013 9:55:00 - 10:12:00

As current lithographic techniques approach practical engineering limits for resolution, directed self-assembly of nanoparticles becomes an attractive scalable nanomanufacturing process for creating ordered arrays of particles at a variety of length scales that could be used both as patterning agents and functional materials. However the roles of interparticle forces and external fields on directed self-assembly of particles is not well understood. In this presentation density functional theory (DFT) and molecular dynamics (MD) simulations are used to explore the use of larger scale patterned substrates to drive smaller scale directed self-assembly of particle monolayers. Square patterned substrates with varying energy barriers at length scales N-fold the final desired particle pitch are considered (N>1). Ranges of N, bulk density and patterned substrate field strength are identified that disrupt the entropically favored hexagonally close-packed lattice and promote square lattice formation for hard-spheres. Event-driven MD simulations are then employed to verify the predictions from DFT and to analyze the dynamics of the self-assembly process. The effects of interparticle forces and other geometries of patterned substrates are considered. These DFT and MD results are used to discuss and define the energetically and kinetically accessible spaces for non-hexagonal lattice formation.

(605g) Molecular Dynamics Simulations of Polymer-Surfactant Interactions

Surampudi, L. N.,
Ashbaugh, H. S., Tulane University
Originally presented on: 11/7/2013 10:12:00 - 10:29:00

The self-assembly of surfactants in the aqueous media is an area of active research. The presence of the polymer molecules gives rise to changes in the solution and the interfacial properties of the surfactant compared to the corresponding polymer-free systems. In this work, we have analyzed the polymer-surfactant aggregation phenomenon by looking at the interactions between the anionic sodium dodecyl sulfate (SDS) surfactant and two non-ionic polymers – poly-(ethylene oxide) (PEO) and poly-(vinylpyrolidone) (PVP) using molecular dynamics (MD) simulations in water medium. From the simulation results, it is observed that the polymer chains adsorb on the hydrophobic regions of the SDS micelle surface in a pattern similar to the beaded necklace model as it was observed in experimental studies. The association of hydrophobic regions in the polymer and micelle suggest that the hydrophobic effect is the driving force for association. As hydrophobic effect is a function of temperature, our research is concentrated on studying the polymer-micelle interactions at various temperatures in the range of 280-360K. In all the temperature studies, the nonionic polymer displayed interaction with the anionic surfactant by adsorbing on the surface region of the micelle. The Surface Accessible Surface Area (SASA) of the micelle adsorbed by polymer (PEO and PVP) was small compared to the surface area of the polymer-free micelle, proving that the polymer shields the micelle from water promoting hydrophobic interactions between the polymer and micelle. The Potential of Mean Force (PMF) between the polymer and the surfactant assembly was also evaluated at various temperatures by pulling short chains of PEO and PVP from the surface of the micelle. The strength of interactions between micelle and polymer per unit surface area of the polymer was analyzed as a function of temperature. These results provide valuable insight into the temperature dependent hydrophobic interactions between nonpolar polymers with the anionic surfactants.

(605h) Nucleation, Growth, and Transformations in DNA-Linked Binary Colloidal Superlattices

Jenkins, I., University of Pennsylvania
Sinno, T., University of Pennsylvania
McGinley, J. T. III, University of Pennsylvania
Crocker, J. C., University of Pennsylvania
Originally presented on: 11/7/2013 10:29:00 - 10:46:00

A promising approach for realizing versatile colloidal assembly is to graft single-stranded DNA oligomer brushes onto the surfaces of particles in order to create tunable attractive interactions between them. Using this concept, micro- and nanoscale particles have now been successfully assembled into bulk crystalline phases, including a variety of ordered, binary superlattice structures. However, there remain open questions regarding the prediction of nucleation and growth behavior in such structures as a function of DNA brush characteristics. Moreover, recent observations of diffusionless transformations in binary colloidal crystallites further motivate a more mechanistic understanding of this class of materials.

Here, we apply a combination of Monte Carlo simulations, Brownian dynamics simulations, and free energy calculations to generate a detailed picture for the assembly of binary colloidal superlattice crystals. We first describe a pseudo-phase diagram that includes both thermodynamic and kinetic influences on homogeneously nucleated colloidal crystallites and demonstrate how kinetic limitations during crystal growth play a direct role in establishing the observed phase. Next, we also present a detailed analysis of diffusionless transformations between body-centered and face-centered cubic configurations of binary colloidal crystals and demonstrate that hydrodynamic factors play an important role in establishing the final structure.

(605i) Role of Symmetry in the Self Assembly of Space-Filling Honeycombs

Schultz, B., University of Michigan, Ann Arbor
Damasceno, P. F., University of Michigan
Engel, M., University of Michigan
Glotzer, S. C., University of Michigan
Originally presented on: 11/7/2013 10:46:00 - 11:03:00

The self-assembly of hard polyhedral particles is driven by the maximization of entropy. As particles are forced into increasingly dense configurations, the system seeks states with maximal entropy, which are often ordered crystalline structures. The notion of directional entropic forces - statistical forces that promote particles to bind or align in certain directions - have been developed to support observations that facets on particles can mimic covalent bonds in atomic crystals. Within this framework, we set out to design particles that would self-assemble into honeycombs, or space-filling arrangements of polyhedra, that are isostructural to atomic crystals. We use the Voronoi tessellation to generate hard, space-filling polyhedra, or "Voronoi particles," that are guaranteed to stabilize the target structure at infinite pressure, analogous to designer potentials that stabilize arbitrary crystals structures with complex radially symmetric potentials. We characterize the thermodynamics, kinetics and self-assembly behavior of these polyhedra and demonstrate how an understanding of the role of symmetry and particle shape can guide design to improve assembly behavior.

(611a) Free Energy Barriers in Wetting-Mode Transitions of Droplets On Grooved Surfaces

Shahraz, A., Pennsylvania State University
Borhan, A., The Pennsylvania State University
Fichthorn, K., The Pennsylvania State University
Originally presented on: 11/7/2013 8:30:00 - 8:52:00

Wetting behavior of a liquid droplet on a rough or physically-patterned surface is typically characterized by either the Cassie mode, in which the droplet resides on top of the surface features, or the Wenzel mode, in which the droplet penetrates into the surface features. For fixed surface topology and droplet size, only one of these modes represents the global free-energy minimum. The other state can be metastable and sufficiently long-lived to be observed experimentally, due to a free-energy barrier that hinders the transition to the global free-energy minimum. In this work, we examine the wetting behavior of a droplet on a grooved surface using both a nano-scale molecular dynamics (MD) approach and a macro-scale model based on free energy minimization. We observe the metastable states in our molecular dynamics (MD) simulations, and use forward flux sampling to study the kinetics of the Cassie-Wenzel transition. We find that the global-minimum wetting states emerging from our MD simulations are consistent with those predicted by our macroscopic model. We also find that the free-energy barrier for transition to the global free-energy minimum depends on both droplet size and surface topology, with a transition-state ensemble consisting of droplets that are on the verge of initiating/breaking contact with the substrate at the bottom of the grooves.

(611b) Particle Pinning At Liquid-Liquid Interfaces

Razavi, S., City College of New York
Colosqui, C. E., Levich Institute of Physico-Chemical Hydrodynamics
Koplik, J., City College of the City University of New York
Kretzschmar, I., The City College of New York, The City University of New York
Originally presented on: 11/7/2013 8:52:00 - 9:14:00

Physical and chemical heterogeneity are known to influence the interaction of colloidal particles with liquid interfaces. Such heterogeneities can produce undesirable or desirable effects depending on the specific application. Recent breakthroughs in particle synthesis have enabled the fabrication of particles with complex geometries and anisotropic surface properties. Hence, a fundamental understanding of the effects of surface features can be exploited in novel industrial processes. In this work, via molecular dynamics simulations we compute the free energy landscape for Brownian particles of different size and surface morphology that straddle the interface between two immiscible liquids. Our results demonstrate that nano-scale surface features can produce sufficiently large energy barriers that effectively "lock" the particle at specific angular orientations with respect to the liquid-liquid interface. These results provide new insights on the rotational dynamics of Brownian particles at interfaces. Our findings are relevant to various applications (e.g., reactive emulsions, peptide assembly) where is desirable that particles remain for long time periods in a specific angular orientation at the fluid interface.

(611c) Quantitative Measurement of Fluorescent Molecule Adsorption Via Confocal Microscopy

Shieh, I. C., University of California
Zasadzinski, J. A., University of Minnesota
Originally presented on: 11/7/2013 9:14:00 - 9:36:00

Contrast in confocal microscope images of phase-separated monolayers at the air-water interface can be generated by the selective adsorption of water-soluble fluorescent dyes from the subphase. The optical sectioning capability of confocal imaging minimizes the fluorescence signal from the subphase as well as allowing for quantitative measurement of the dye concentration in the monolayer. The partitioning of the water-soluble dyes depends strongly on the surface pressure or area/molecule measured from the isotherms. The extent of adsorption is consistent with a free area model previously used to model lateral diffusion and surface viscosity. The rates of adsorption and desorption are governed by Langmuir-type kinetics. Unlike fluorescent lipids, the water-soluble dye is in exchange equilibrium between the interface and the subphase so that the dye is not concentrated as the proportions of the two phases change at coexistence. This approach may be extended to study the adsorption of a variety of subphase-soluble molecules and interfacial films, as well as allow multi-color imaging of the monolayers during dynamic processes.

General acceptance of complex monolayer and bilayer phase behavior, especially phase coexistence and critical phenomena, has relied on the visualization of fluorescently tagged lipids preferentially segregated between phases. This segregation, which is due to differences in local molecular organization corresponding to the different phases, provides the necessary contrast to visualize even very subtle differences in packing density, molecular tilt, and short and long-range ordering. Since the technique was introduced, fluorescence microscopy has settled numerous debates over the molecular scale organization of monolayers and bilayers. However, fluorescent lipids, like many of the other lipids in monolayers and bilayers, are effectively insoluble in the surrounding aqueous subphase and are trapped in the monolayer or bilayer. As a result, expelling the fluorescent lipid from one phase means concentrating the fluorescent lipid in another phase, which can result in fluorescence quenching, alterations in the structures of both the less ordered (dye-rich) and more ordered (dye-poor) domains, or perturbations of the mechanical properties of the monolayer. Improvements in the sensitivity of cameras and other recording devices, and improved quantum efficiencies of fluorescent dyes, have led to a gradual decrease in the overall mole fraction of fluorescent lipid required for good imaging, from 2 mol% in early experiments to 0.05 - .1 mol% in recent work. An alternative to fluorescently labeled lipids is Brewster Angle Microscopy (BAM), in which the contrast in the image is developed by small variations in refractive index between different areas in the monolayer, due to different phases, molecular density, or molecular tilt. Resolution limitations and ease of use have limited the general acceptance of BAM.

Here we show that the selective adsorption of a fluorescent water-soluble dye from the subphase to the monolayer provides similar contrast in confocal images of monolayer phase separation and critical phenomena to insoluble lipid dyes. However, unlike insoluble lipid dyes, the water-soluble dye remains at a concentration at a given surface pressure or area per molecule that is set by a dynamic equilibrium between the monolayer and subphase. The optical sectioning of the confocal microscope allows the interfacial fluorescence of the monolayer-adsorbed dye to be visualized and rejects the fluorescence from the subphase, which is not possible with conventional wide-field microscopy. By proper de-convolution of the confocal image, the total adsorption and the adsorption/desorption rates of the dye can be quantified. Dye adsorption can modeled using a simple, free area model previously used to describe diffusion in condensed phases. The amount of dye adsorption depends non-linearly on the free area which is the difference between the area per molecule at a given surface pressure measured by the isotherm, and the minimum, close packed area of the lipids making up the monolayer. The rates of dye adsorption and desorption can be described with classic Langmuir adsorption kinetics. These methods might be extended to quantify monolayer adsorption of a variety of labeled proteins, polymers, and other materials of interest.

(611d) Oscillation of Particles On An Interface Due to Surface Tension Gradients

Shardt, O., University of Alberta

Lee, G., University of Alberta

Derksen, J., University of Alberta

Mitra, S. K., University of Alberta

Originally presented on: 11/7/2013 9:36:00 - 9:58:00

In the 19th century, Lord Rayleigh studied the erratic motion of camphor scrapings on water in air [P. R. Soc. London 47 (1889) 364-367]. More recently, Nakata and Hayashima [J. Chem. Soc. Faraday Trans. 94 (1998) 3655-3658] identified two modes in the motion of isolated camphor scrapings. The motion of camphor scraping occurs due to surface tension gradients around the particles. These gradients are maintained by the sublimation camphor from the water surface into the air above. We describe a qualitatively similar system in which the part exhibit a sustained oscillation in contrast to the erratic motion of camphor scrapings. In our system, the particle calcium propionate powder, which is less dense than water and floats on the air-water interface while dissolvin hypothesize that the synchronized motion is due to surface tension gradients along the interface and the interac between nearby particles.

The oscillating system was characterized by measurements of the particle motion and the physical properties of calcium propionate solutions. High speed imaging (211 fps) was used to record the oscillation of the powder on surface of solutions with varying initial calcium propionate concentrations. The three frames shown below illus the motion of the particles. The image sequences were analyzed using particle image velocimetry (PIV) to quan the temporal and spatial structure of the oscillation; two sample velocity fields are provided below. A typical frequency of the oscillation in the velocity at a fixed point on the surface is 2 Hz. The surface tensions of calciu propionate solutions with varying concentations were determined with the pendant drop method. The surface tension decreases with increasing calcium propionate concentration.

We also compare the experimental results with a one-dimensional model to gain insight into the observed beha The predicted dependence of the oscillation frequency on the initial solute concentration is compared with the experiments. The model predicts fast dissolution at low initial concentrations, oscillation at moderate concentrations, and no motion at high concentrations, three modes of behaviour that are also seen in the experiments.

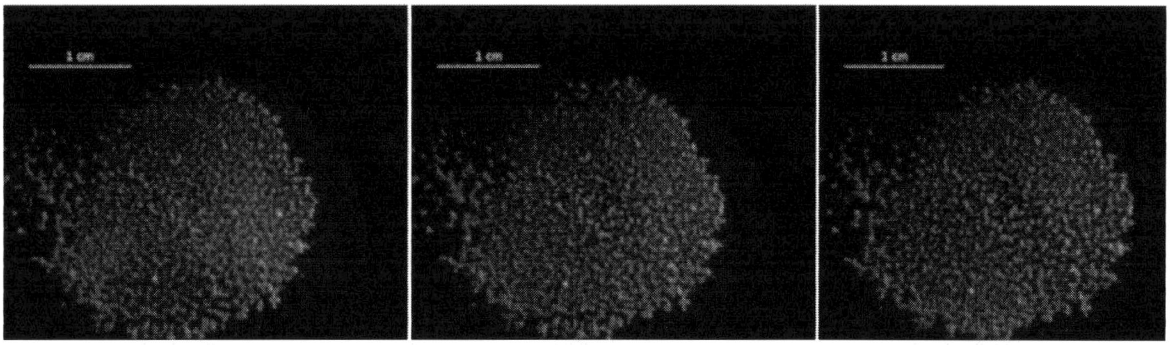

Sample frames 95 ms apart showing particle motion. Note the lower particle coverage near the bottom of the fi frame.

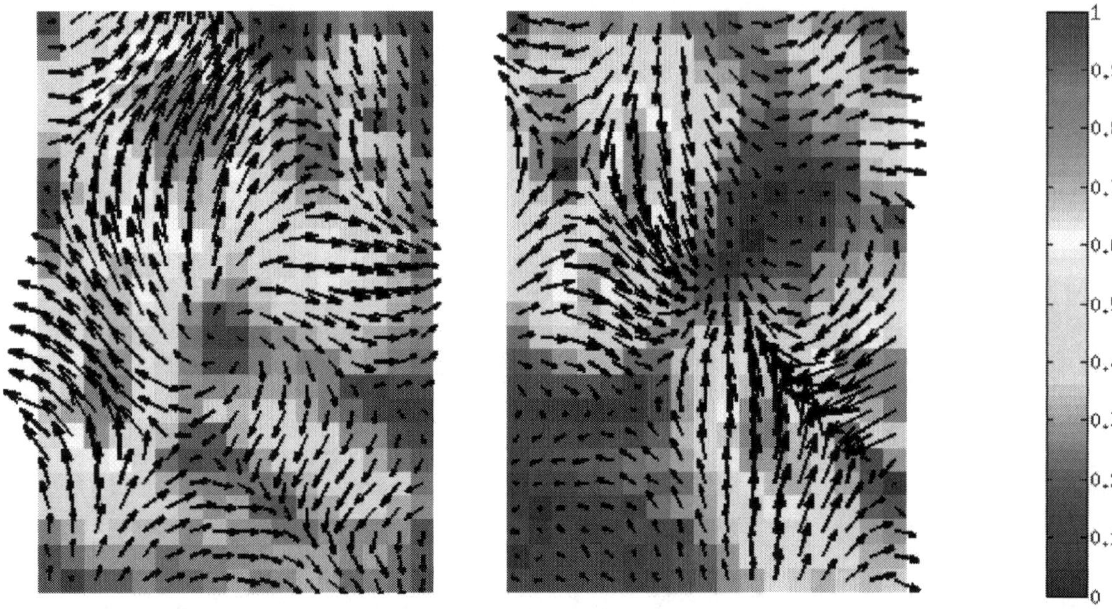

Sample velocity fields after PIV analysis in a 1.7 cm wide by 2.4 cm high region of the powder-covered surface Colours and arrow lengths represent velocity magnitude. Motion is towards the centre in the first field and awa; from the centre in the second (140 ms later).

(611e) Engine-Payload Self-Propelling Particles for Efficient Disruption and Cleanup of Oil Films On Water

Sharma, R., North Carolina State University
D'Costa, L., North Carolina State University
Velev, O. D., North Carolina State University
Originally presented on: 11/7/2013 9:58:00 - 10:20:00

A self-propelling particle uses energy (such as chemical, thermal or electrical) for propulsion by generating local gradients of chemical or other potential. We will first present a class of gel based self-propelling particles moving in a programmed oscillatory mode at water/air interfaces propelled by Marangoni effect. These gel boats floating on the water surface are driven by a hydrogel reservoir releasing an ethanol flux that is periodically disrupted by the bulk flows around the particles. The pulse interval and the distance propelled in a pulse by these gel floaters were interpreted based on the rate of ethanol mass-transfer. The model allowed us to design particles that perform "dances" of repeatable periodic sequences of forward/backward, or rotary steps, and travel in complex pre-programmed trajectories on the liquid surface (*Langmuir* **28**, 10128, 2012). We will then report how on this basis we designed and tested new types of functionalized self-propelling floaters driven by surfactant release that may serve as a new platform for environmental remediation. Previous research on oil collection deals mostly with the synthesis of new immobile adsorbent materials. We demonstrated that self-propelling adsorbent particles can achieve highly efficient oil recovery by rapidly moving over the water surface. The "engine-payload" particles act as motile collectors that can function with a broad range of adsorbent materials. In addition, the Marangoni propulsion leads to oil film disruption, which depends on the type and solubility of the surfactant. In more general plan, this technique offers a radical departure from the present practice of engineering processes where particles adsorb and process a certain component passively, by diffusion of the component towards the stationary (or carried with the fluid) particles. The use of self-propelling adsorbing, catalytic and other particles may allow redefining and drastically intensifying the mass-transport processes in fluid-particle systems.

(611f) Instability-Driven Streak Formation During Convective Deposition

Weldon, A. L., Lehigh University
Gilchrist, J. F., Lehigh University
Originally presented on: 11/7/2013 10:20:00 - 10:42:00

The formation of streaks, often unwanted localized multilayer particle deposition oriented in the direction of deposition, is studied across a robust variety of process conditions. In systems where streaks form, many streaks are initiated and then coarsen and merge as the deposition progresses. Image analysis of streaks give a detailed account of their formation and development. Insights into the formation of this stochastic process will be given and suggest ways to suppress the formation of streaks.

(611g) Investigation of Interfacial Processes Leading to Anisotropic Silver Nanostructures On 2D Monolayers of Silica Spheres

Klupp Taylor, R. N., University of Erlangen-Nuremberg
Spies, I.,
Hanisch, M.,
Originally presented on: 11/7/2013 10:42:00 - 11:04:00

One-dimensional nanostructures comprising noble metals are promising materials for a range of electronic, photonic, biomedical and catalysis applications. In particular, their anisotropic morphology opens up possibilities for charge or photonic transport as well as providing a means for the self- or directed-orientation of the building blocks during fabrication of a functional structure e.g. using electrical fields or fluid forces. In order to enable widespread exploitation of one-dimensional nanostructures, there needs to be more effort to develop and understand simple and scalable routes to achieve them. In this contribution we describe how one-dimensional silver nanoparticle assemblies can be formed at a solid-air interface in the absence of explicitly added reducing and templating agents. We show that in this extremely simple but highly sensitive process, dynamic and strongly anisotropic effects occur during aging of a 2D monolayer of silica nanospheres that had been treated with the ammoniacal silver complex ($Ag(NH_3)_2^+$). Electron microscopy reveals that 5 – 10 nm silver nanoparticles are arranged in necklace-like assemblies on the silica spheres and up to a certain distance from them on the substrate. We investigate the formation of the nanoparticle necklaces by varying simple parameters of the liquid phase and aging treatments. We also describe attempts to scale-up the process from a few mm^2 to cm^2, particularly focussing on the formation of the nanostructures on 2D silica nanoparticle monolayers deposited by dip-coating on silicon wafers. We show that together with the simple chemical treatments, dynamic effects at the silica particle monolayer-air interface can lead to a number of interesting morphologies being produced, included the nanoparticle necklaces seen in the smaller scale work. These findings, as well as studies of the interfacial properties of silica help us to narrow down our mechanistic understanding of the nanostructure formation by the synergistic chemical behavior of the used materials, namely $Ag(NH_3)_2^+$ and amorphous silica particles. In particular, the presence of dissolved silicates, a result of the high pH of the ammoniacal silver complex appear to be critical.

Besides potential uses in transparent electrical conductive coatings, substrates for surface-enhanced Raman spectroscopy (SERS) or for electromagnetic shielding the 1D nanostructures produced by our process could be particularly interesting for possible novel optical properties of the structures. In the latter case we believe that mesh-coated silica particles formed by their further metallization may display resonances in their magnetization at optical frequencies and thus meets requirements for the formation of, for example, negative index metamaterials.

(614a) Selective Oxidation of 1,2-Propanediol to Pyruvate Or Lactate in Alkaline Media Over Supported Au and Pt Nanoparticles in Anion Exchange Membrane Fuel Cell (AEMFC) and Electrocatalytic Flow Reactors

Chadderdon, D., Michigan Technological University
Xin, L., Michigan Technological University
Qi, J., Michigan Technological University
Zhang, Z., Michigan Technological University
Brady, B.,
Janik, M., Pennsylvania State University-University Park
Li, W., Michigan Technological University

Originally presented on: 11/7/2013 8:30:00 - 8:50:00

Renewable biomass has received great interest for its potential to replace petroleum as the primary feedstock for liquid fuels, chemicals, and polymeric materials. Biorenewable 1,2-propanediol (PDO) may be produced from the catalytic hydrogenolysis of glycerol, a major co-product of biodiesel production. PDO is a 3-carbon (C3) diol featuring vicinal primary and secondary alcohol groups which may be selectively oxidized to form lactate, which has potential as a feedstock in biodegradable polymer production, or pyruvate which is an important building-block in chemical, pharmaceutical, and agrochemical industries. In this talk, I will present the selective oxidation of PDO to pyruvate or lactate in an anion exchange membrane fuel cell (AEMFC) and a custom-made AEM-based electrocatalytic flow reactor, both using carbon supported Pt and Au nanoparticles (Pt/C or Au/C) on carbon cloth as the anode catalyst layer. The main oxidation product on Pt/C was lactate (85-90%), while Au/C displayed potential-controlled selectivity to pyruvate (20-55%) with lactate as the other main product. At sufficient anode potential, lactate was further oxidized to pyruvate on Au/C while C-C bond breaking was preferred on Pt/C. Based on the observed oxidation products and additional three-electrode voltammetry study of product intermediates, a reaction pathway was proposed to map the unique catalytic properties of Au/C and Pt/C. Density-functional theory (DFT) simulations were performed to calculate reaction energies, and the theoretically favorable pathway agreed with the experimentally proposed pathway. This work unveils new insight to the role of supported nanoparticle electrocatalysts for production of important chemicals from biorenewable sources.

(614b) Catalytic Activity of Platinum Nanoparticles for H2 Formation

Skulason, E., University of Iceland
A. Faraj, A.,
Kristinsd ttir, L.,
Hussain, J.,
Garden, A. L., University of Iceland
Jonsson, H., University of Iceland
Originally presented on: 11/7/2013 8:50:00 - 9:10:00

Theoretical calculations based on density functional theory (DFT) and transition state theory have led to a deeper understanding of catalytic activity and thus helped develop improved heterogeneous catalysts, a task that is vital for the design of processes with higher energy and atom efficiency in the chemical industry. By providing information about the factors controlling the reactivity, such as the energetics of molecule-surface interaction and the kinetics of elementary processes at solid surfaces, trends in the catalytic activity as the chemical composition of the catalyst is varied have been established and predictions made for new and improved catalysts.

The interaction of hydrogen with the surface of platinum metal is of fundamental importance to a wide range of technologies including electrolysis of water and hydrogen fuel cells. The metal is typically dispersed as small particles embedded in a matrix. The active sites of such particles can be under-coordinated atoms at edges where crystal facets meet, or at corners where edges meet. A 3 nm particle, which is at the lower end of the size range used in fuel cells today, contains on the order of 1000 atoms. DFT calculations of the catalytic activity of a whole nanoparticle of this size generally require a prohibitively large computational effort. An alternative approach is to construct a periodic array of the sites of interest, such as edges between micro-facets. The missing-row reconstructed Pt(110)-(1x2) surface can be considered a very small model of edges between (111) facets and has been studied both experimentally and theoretically [1,2].

In the current work [3], the associative desorption of hydrogen at edges and facets on Pt nanoparticles was studied using DFT. The goal was to identify catalytically active sites on Pt nanoparticles for the hydrogen evolution reaction as well as the hydrogen oxidation reaction. The adsorption sites were modeled by periodic, face centered cubic slabs representing an array of edges between two (111) micro-facets or edges between (111) and (100) micro-facets. The width of the facets in the periodic representations was systematically increased to reach converged results for binding and activation energy. Under typical electrochemical conditions (for small applied voltage vs. SHE), edges between (111) micro-facets were found to be several orders of magnitude more active than edges between (100) and (111) micro-facets or at terraces.

References:

[1] S. Gudmundsdóttir, E. Skúlason, and H. Jónsson, Phys. Rev. Lett, **108**, 156101 (2012).

[2] S. Gudmundsdóttir, E. Skúlason, K.-J. Weststrate, L. Juurlink, and H. Jónsson, Phys. Chem. Chem. Phys. **15**, 6323 (2013).

[3] E. Skúlason, A. A. Faraj, L. Kristinsdóttir, J. Hussain, A. L. Garden and H. Jónsson, Top. Catal. (submitted).

(614c) Electrode Potential Dependent Activation Barriers Approximated With Density Functional Theory

Esopi, M., Pennsylvania State University
Regula, M.,
Nie, X., The Ohio State University
Asthagiri, A., University of Florida
Janik, M. J., Pennsylvania State University
Originally presented on: 11/7/2013 9:10:00 - 9:30:00

Density functional theory (DFT) studies of electrocatalytic systems are widespread, but the calculation of activation barriers for elementary steps involving electron and ion transfer remains challenging. A simple and transferable DFT approach to estimate these barriers for inner sphere electrochemical reactions, focusing on approximating the electron transfer coefficient, will be presented. Our approach follows from the assumption that an electrocatalytic elementary step is inherently an inner-sphere reaction, for which attainment of the transition state requires nuclei rearrangement local to an adsorbed species, and electron transfer is rapid once the transition state is attained. The Arrhenius concept (transition state theory) of the reaction rate then holds, and the Born-Oppenheimer approximation allows for evaluation of reaction rates without explicit consideration of electron conduction. We replace the need to locate a transition state for an electrochemical reaction (reduction of an adsorbed species by a proton/electron pair, for example) with the location of a transition state for an analogous non-electrochemical reaction (hydrogenation). It is assumed that the transition states for these two reactions are identical at one specific electrode potential. Once the transition state is located, the activation free energy, G_{act}^{0}, may be assigned to the potential, U^{0}, at which the chemical potential of the H* species is equivalent to the true $H^{+} + e^{-}$ reactant. The activation barrier can be extrapolated as a function of electrode potential using Butler-Volmer theory, with the symmetry factor estimated by examination of the transition state and reaction coordinate.

The talk will further detail this approach and demonstrate that the method produces activation barriers consistent with experimental electrokinetics, and also yields values of the electron transfer coefficient that make sense and fit well with the other data. The first elementary step of CO_2 reduction will be used as a detailed example case, with application to other reactions discussed.

(614d) Bifunctional Alkaline Electrocatalysts for HOR and HER: A First-Principles Study

Zeng, Z.,
Greeley, J., Purdue University
Originally presented on: 11/7/2013 9:30:00 - 9:50:00

Alkaline fuel cells and electrolyzers may offer significantly improved performance over devices using acid electrolytes. Among other advantages, non-precious metal catalysts have sufficient stability to be used in alkaline media, leading to lower system costs. However, alkaline systems have often suffered from lower activity than corresponding acid-based systems. Recently, it has been shown that bi-functional Ni-(hydr)oxide/Pt(111) electrocatalysts manifest activities that approach those of acidic fuel cells. In spite of the promising properties of these and related systems, however, the atomic-scale details underlying their operation remain largely unknown, and the present study focuses on using atomistic computational techniques to elucidate these atomic-scale phenomena.

In this talk, we describe a combined thermodynamic and Density Functional Theory (DFT) analysis of oxidation/reduction of transition metal (hydr)oxides deposited on precious metal electrocatalyst surfaces. We show how this formalism can be used to develop Pourbaix diagrams and surface phase diagrams of thin (1-2 ML) Ni (hydr)oxide films/islands deposited on close-packed Pt substrates. We then extend these models to describe three-phase boundaries between the films, the Pt substrate, and the surrounding electrolyte, and we study the hydrogen oxidation reaction (HOR) and hydrogen evolution reaction (HER) in alkaline media at these interfaces. We find that a few partially hydrated-Ni/Pt edges are the most likely candidates for the active phases. In the hydrogen oxidation region, there is a high coverage of OH adsorbed at the edges, while in the hydrogen evolution region there are enough vacancies to accommodate H_2O adsorption. Kinetic analysis confirms that OH-H association and H_2O dissociation are promoted at these interfaces, in comparison to mono-functional Pt catalysts. We close with a discussion of the potential for further improvement of this and related bi-functional electrocatalysts in alkaline environments.

References

1. "Enhancing Hydrogen Evolution Activity in Water Splitting by Tailoring Li+-$Ni(OH)_2$-Pt Interfaces" R.Subbaraman, D.Tripkovic, D.Strmcnik, KC Chang, M.Uchimura, A. P. Paulikas, V.Stamenkovic, N, M. Markovic, *Science* 334 (2011) 1256

2. "Trends in activity for the water electrolyser reactions on 3d M(Ni,Co,Fe,Mn) hydr(oxy)oxide catalysts" R. Subbaraman, D. Tripkovic, K. Chang, D. Strmcnik, A.Paulikas, P. Hirunsit, M. Chan, J. Greeley, V. Stamenkovic, and N. Markovic. *Nature Materials* **11**(2012) 550.

(614e) Electrocatalytic Selective Oxidation of Glycerol to Tartronate and Electricity Generation Using Au, Pd Mono and Bimetallic Nanoparticles Supported On Carbon Nanotube As Anode Catalyst in Anion Exchange Membrane Fuel Cell

Qi, J., Michigan Technological University
Xin, L., Michigan Technological University
Chadderdon, D., Michigan Technological University
Qiu, Y., Michigan Technological University
Jiang, Y., Michigan Technological University
Zhang, Z., Michigan Technological University
Li, W., Michigan Technological University
Originally presented on: 11/7/2013 9:50:00 - 10:10:00

Electrocatalytic selective oxidation of glycerol to tartronate with electricity generation using Au, Pd mono and bimetallic nanoparticles supported on carbon nanotube as anode catalyst in anion exchange membrane fuel cell

Ji Qi[1], Le Xin[1], David Chadderdon[1], Yang Qiu[1], Yibo Jiang[2], Zhiyong Zhang[1], Wenzhen Li[1]

[1]*Department of Chemical Engineering, Michigan Technological University*

[2]*Department of Civil & Environmental Engineering, Michigan Technological University*

As a renewable biomass feedstock, glycerol can be valorized to a series of chemical products, among which tartronic acid is one of the most expensive chemicals. Since glycerol oxidation reaction (GOR) is spontaneous reaction with negative Gibbs free energy change, it's suitable to investigate GOR in fuel cell to simultaneously obtain high yield of value-added product and high electricity output. Studying the GOR in fuel cell not only have profitable application value, but also provides deep scientific insights into the difference between electrocatalytic reaction process and heterogeneous catalytic reaction process.

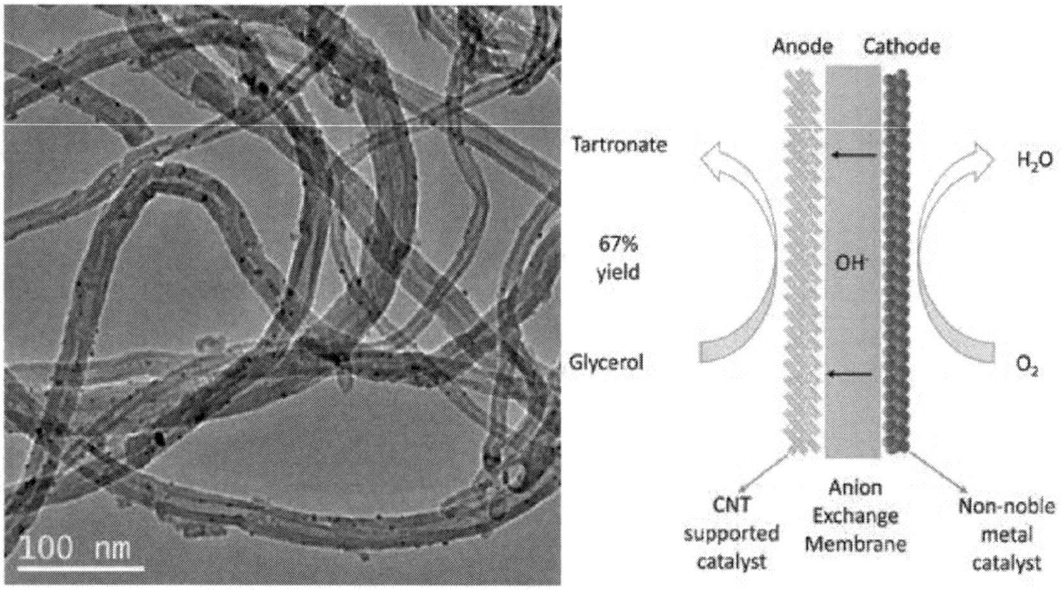

In the present work, Au, Pd mono and bimetallic nanoparticles supported on carbon nanotube (CNT) were prepared and used for glycerol oxidation in anion exchange membrane fuel cell. In the past, fuel cell anode potential window is very narrow. Therefore, high selectivity of tartronate can only be obtained by consuming electricity in electrolysis cell which has a relatively large anode potential window. We successfully designed the electrode structure of fuel cell to lower down the anode potential in the fuel cell and thus prevent overoxidation of tartronate to mesoxalate and C-C bond cleavage. Also, the reaction rate is increased by alloying PdAu, leading to high instantaneous electricity output. During the 8 h reaction process, stable 50 mW cm^{-2} to 5 mW cm^{-2} can be generated with 67% yield of tartronate finally achieved using PdAu alloyed nanoparticles supported on CNT as anode catalyst. Future work includes optimizing the catalyst synthesis method and other operating parameters to further increase yield of different kinds of target products.

(614f) Understanding ORR Kinetics Through Microkinetic Modeling: Revisiting the Tafel Slope

Holewinski, A., University of Michigan
Linic, S., University of Michigan
Originally presented on: 11/7/2013 10:10:00 - 10:30:00

The oxygen reduction reaction (ORR), $O_2 + 4H^+ + 4e^- = 2H_2O$, on Pt electrodes is known to limit performance in low-temperature hydrogen fuel cells. A mechanistic understanding of the reaction is thus critical to the rational design of improved electrode catalysts. One often-debated phenomenon is the deviation of the observed ORR kinetic behavior on Pt from the linear characteristic predicted by the electro-kinetic Tafel equation: [delta]E=a+b *log*(i). An ORR Tafel plot generally displays a sharp change in slope.

Herein, we utilize microkinetic modeling to demonstrate analytically that rapid changes in the kinetic behavior (Tafel slope and reaction orders) of electrochemical reactions are an inherent property of electrode kinetics that involve multiple elementary steps and adsorbed intermediates, irrespective of commonly proposed causes such as changes in rate-limiting step or interactions between adsorbates. For the ORR on Pt, our analysis reveals the measured Tafel slope, including its characteristic shift, is consistent with a rate-limiting initial electron transfer. We show that surface oxygen species (primarily hydroxyl) impede the rate at small overpotentials through site blocking even when their removal involves fast, quasi-equilibrated steps. Thus, while higher ORR turnover rates would intuitively be expected when increasing reactivity toward oxygen, a more reactive surface will also push the equilibrium-driven decomposition of H_2O to OH to lower potentials, poisoning the surface for relevant operating conditions. A major breakthrough in ORR catalysis will likely require materials that can decouple the binding energies of oxygen and hydroxyl groups, which generally scale together.

(614g) In-Situ XAS Study of Self-Switching Electrocatalyst for Oer and ORR

Jiao, F., University of Delaware

Originally presented on: 11/7/2013 10:30:00 - 10:50:00

Oxygen evolution and reduction through heterogeneous catalysis have attracted much attention because they play important roles in renewable and sustainable energy applications. Traditional battery technology is based on intercalation mechanism and no catalytic process is involved. Recent breaking-through in rechargeable lithium oxygen batteries introduces OER and ORR into battery system for the first time. After intensive screening active catalysts in the past few years, α-MnO_2 have been identified as the most active electrocatalyst for both reactions, while the fundamentals of how this catalyst is involved in OER/ORR during electrochemical charging/discharging haven't yet been explored.

Our previous studies in photocatalytic oxygen evolution from water suggest that Mn^{3+} exhibits much higher activity in oxygen evolution than Mn^{4+} and it is known that Mn^{4+} is only active for oxygen reduction reaction from the literature. Our hypothesis is that α-MnO_2 in the composite oxygen cathode switching its oxidation state during charging and discharging due to lithium intercalation/de-intercalation. At initial discharging, Mn is +4 in α-MnO_2 and it acts as ORR electrocatalyst. At the end of discharging, the oxidation state of Mn is close to +3 because lithium intercalation into α-MnO_2 significantly reduces the oxidation state of Mn. When the battery is on charging mode, Mn^{3+} in Li_x-α-MnO_2 acts as oxygen evolution site to decomposition of lithium peroxide to lithium and oxygen. During deep charging to 4.5V, the lithium ions intercalated into α-MnO_2 structure are removed and Mn atoms are back to Mn^{4+} again (see the figure). Such switching mechanism greatly promotes the electrocatalytic activity of α-MnO_2 nanowires in both ORR and OER during cycling at the same system.

Because the catalyst in the electrochemical cell is in a complicated environment (gas, liquid and solid, three phases), in-situ techniques are appreciated. In this presentation, we will discuss our recent progress in using in-situ X-ray absorption techniques to monitor the oxidation state of Mn during the electrocatalytic processes at real time and to elucidate the electrocatlytic mechanism of ORR/OER on the surface of α-MnO_2. We have successfully fabricated a custom electrochemical cell, which allows us to monitor the electrocatalyst structure evolution in real time when subjected to high flux synchrotron X-ray source. This represents the first attempt to utilize in-situ XAS techniques to study the mechanism of OER and ORR in an electrochemical cell at real time. The XAS data clearly show the oxidation state of Mn shifted to 3+ during discharging and returned to 4+ when the cell was charged. This in-situ technique will help us reveal the detailed mechanism in electrocatalytic OER/ORR and advance our understanding of the fundamentals.

(615a) Transparent Polyaniline Nanocomposite Films for Electrochromism and Electrochemical Energy Storage Applications

Wei, H., Lamar University
Zhu, J., Lamar University
Wu, S., Agilent Technologies, Inc, Chandler
Wei, S., Lamar University
Guo, Z., Lamar University
Originally presented on: 11/7/2013 8:30:00 - 8:55:00

In the paper, transparent polyaniline (PANI) nanocomposite films for electrochromism and electrochemical energy storage applications were investigated. Tungsten oxide (WO_3) and graphite oxide (GO) were selected as the inorganic nanoparticles. The nanocomposite films were fabricated via direct electropolymerization of the aniline monomers on to the WO_3 or GO coated-indium tin oxide (ITO) glass slides prepared by spin coating technique. The chemical structure and morphology as well as the crystalline structure of the composite films were studied using Fourier transform infrared spectroscopy (FT-IR), atomic force microscopy (AFM), and X-ray diffraction (XRD). The results confirmed the chemical interactions between the PANI matrix and the nanoparticles. The optical properties and electrochemical capacitive behaviors of the composite films for electrochromic (EC) and energy storage devices applications were investigated using spectroelectrochemistry (SEC), cyclic voltammetry (CV) and galvanostatic charge-discharge measurements. Multi-color electrochromism at different potentials were observed in the nanocomposite films. The nanocomposite thin films also exhibited higher coloration efficiency and comparable areal capacitances as compared to those of the pure PANI. Significantly enhanced cycling stability was found in the nanocomposite thin films, which is believed to arise from the interactions between the polymer matrix and nanoparticles.

(615b) Electrodeposition of Cu-Sn-In Alloys From Citrate Electrolytes

Zahmi, S., Northeastern University
Kim, H., Northeastern University
Podlaha, E.,
Originally presented on: 11/7/2013 8:55:00 - 9:20:00

The electrodeposition of CuSnIn alloys and different combinations of these alloys from a citrate electrolyte are compared and their complexation with citrate species simulated, with a comprehensive mass balance model to assess the different complexed species with variable combinations of the metal ions. The partial current densities were determined for different combinations of metals (*i.e.,* Sn, In, InSn, CuSn, CuIn, CuInSn) using a three electrode configuration with a constant applied potential deposition at room temperature. Experimental findings included the following: (1) the reaction rate of Sn was enhanced when codeposited with Cu and CuIn, (2) the surface morphology of copper containing alloys were rougher than the elemental deposits or alloys without Cu, and (3) the ratio of Sn:In in the deposit was altered when Cu was present. The enhancement was associated with an increase in surface area and a kinetic effect, but not due to changes in the metal-citrate species.

(615c) Solution Speciation Effects On Electrodeposited Cuprous Oxide Films for Solution-Processed Photovoltaics

Panzer, M. J., Tufts University
Originally presented on: 11/7/2013 9:20:00 - 9:45:00

Electrodeposited cuprous oxide (Cu_2O) is a p-type semiconductor that holds promise for constructing future photovoltaic devices using solution-based processes that require low energy input, feature benign chemistry, and contain earth abundant materials. Recently, we have successfully demonstrated the electrodeposition of single-phase, polycrystalline Cu_2O thin films from acidic lactate/Cu^{2+} solutions. A distinctive dendritic, flowerlike Cu_2O morphology is observed when films are grown from acidic lactate-based deposition baths; this stands in stark contrast to the faceted, dense films obtained from traditionally used alkaline solutions. Numerical simulation and UV-Vis absorption spectroscopy have been employed to demonstrate key differences in the relative distributions of various metal complex ions present under different growth conditions. Hot point probe measurements confirm the p-type conductivity of all Cu_2O films obtained in this work. Importantly, a short circuit current density of 23 uA/cm^2 under monochromatic green LED illumination (0.583 mW/cm^2) was measured for a Cu_2O film grown at pH 5.3, which was substantially larger than that of a Cu_2O film deposited from basic lactate solution. These findings suggest that gaining a deeper understanding of the effects of metal complex ion speciation on the ultimate optoelectronic properties of electrodeposited Cu_2O films can lead to improved performance for realizing efficient photovoltaic devices with this material.

(615d) Optical Properties of Titanium Dioxide Nanotube Arrays

Sokoloff, J., Northeastern University
Abdelmoula, M.,
Lu, W. T., Northeastern University
Menon, L., Northeastern University
Richter, C., Yale University
Originally presented on: 11/7/2013 9:45:00 - 10:10:00

Nanostructured metal oxides are promising substrates for several ‘next generation’ photovoltaic and photo-electrochemical cell architectures like DSSC’s, organic-inorganic solar cells and all inorganic solar cells. Metal oxides generally have advantages like long term stability, low cost and in materials like ZnO or TiO_2, relatively long carrier lifetimes and carrier mobility. When designing solar cells or photonic devices that incorporate nanostructured metal oxides it is important to have good models for how light is reflected, scattered, transmitted and absorbed in these structures.

We fabricated titanium dioxide nanotube arrays with electrochemical anodization and studied the reflection and transmission of light from the UV to the NIR through these arrays. Anodic nanotube arrays consist of open tubes (with diameters ~ 100 nm) on one end and capped or close tubes on the other end. Our measurements revealed that the total reflectance from the open nanotube array end is significantly higher (15% at 400 nm) than the reflectance from the closed nanotube end for short wavelengths (blue and UV light). This anomalous behavior is not easily explained by simple classical optical models. Attempts to explain the data by several models including theoretical models based on the eikonel approximation, photonic crystal band theory and disorder induced scattering will be presented. Disorder induced scattering appears to be the primary factor causing the anomalous behavior at short wavelengths.

(615e) Novel Synthesis Route for Symmetric and Asymmetric Diarylethene Molecules By Modifying Pendant Groups

Zhang, S., The Key Lab of Organic Optoelectronics and Molecular Engineering of Ministry of Education
Originally presented on: 11/7/2013 10:10:00 - 10:35:00

Diarylethene(DTE) molecules have two forms, open form and close form, which have different properties in many aspects, such as absorption spectrum, refractive index, permittivity and chemical properties. The two forms can switch when irritating by UV radiation and visible light. DTEs are one of the key materials for volume holographic storage. But the synthesis process for DTEs are always too complicated for practical applications. A new route for synthesizing symmetric and asymmetric diarylethene molecules bearing different functional groups was introduced. Treated by $AlCl_3$, 2-chloro -5- methyl thiophene converted to 1,5-bis(5-chloro-2-methylthiophen-3-yl)pentane-1,5-dione. After cyclization and reduction, (1,2-bis(5-chloro-2-methylthiophen-3-yl)cyclopent-1-ene, BCM-DTE) was obtained. DTEs with different pendant groups were obtained by Suzuki reaction using different boric acids. And symmetric and asymmetric DTEs can be synthesized at the same time. The whole process takes only 3 steps which are all simple and easily controlled. Massive DTEs can be produced by this method in commercial use. Also these DTEs have typical photochromic properties and could have promising applications in volume holographic storage.

(618a) Selective Oxidation of Propene On Bismuth Molybdenum Based Oxide Catalysts

Zhai, Z., University of California - Berkeley
Wang, X., University of California, Berkeley
Getsoian, A., University of California - Berkeley
Bell, A. T., University of California - Berkeley
Originally presented on: 11/7/2013 8:30:00 - 8:50:00

The selective catalytic oxidation of lower olefins to produce α,β-unsaturated aldehydes is one of the essential processes for preparing monomers and intermediates in industry. Among the most successful of these reactions is the selective oxidation of propene to acrolein [1], a commodity chemical produced at the level of 10^8kg/year [2]. The most effective catalysts for the oxidation of propene are multi-component oxide systems based on bismuth molybdate. However, it is still difficult to explain what the role of the additional elements is in the multicomponent bismuth molybdate catalysts.

In this work, we aim to understand the effects of catalyst composition on catalyst activity and selectivity. To do this, we first prepared and investigated a series of scheelite-type multi-component bismuth molybdate catalysts doped with a third element (iron, gallium or vanadium, etc). Several characterizations and oxidation of propene to acrolein reaction were then performed on these catalysts. We conducted. XRD data showed that all the catalysts have similar scheelite-type structures. X-ray absorption near edge spectroscopy (XANES) was used to characterize metal oxidation state before and after exposure of the catalyst to propene. The diffuse reflectance UV-VIS-NIR spectra were measured to calculate band gap based on the absorption edge. Measurements of reaction rates and product distributions were performed on these catalysts. All these measurements illustrate that the production of the main product of these catalysts, acrolein, follows first order dependence of C_3H_6 and zero dependence of O_2 under operation conditions for most catalysts except gallium doped bismuth molybdate. For gallium doped catalysts, our measurements show they have the highest activation energy, while vanadium doped catalysts provide the lowest one. These activation energies have positive correlations with the band gaps for different catalysts with different additives. Generally, the structures of the catalyst, as well as the electronic properties, are influenced by the properties of additional elements, i.e. iron, gallium or vanadium. How these properties further influence acrolein production is discussed.

[1] G.W. Keulks, L.D. Krenzke, T. M. Notermann, Adv. Catal. 27 (1978) 183-225.

[2] G. Ertl, H. Knozinger, F. Schuth, J. Weitkamp (Eds.) Handbook of Heterogeneous Catalysis, Wiley

(618b) Kinetic Analysis of the Reactivity of Lattice Oxygen in Molybdenum Dioxide Based Catalysts Under Pulse-Type Experiments

Cuba Torres, C. M., Washington State University
Norton, M. G., Washington State University
Ha, S., Washington State University
Originally presented on: 11/7/2013 8:50:00 - 9:10:00

Molybdenum dioxide (MoO_2) is a distorted rutile oxide which displays metallic character, high oxygen anion conductivity and recently has been reported to exhibit catalytic activity towards a partial oxidation of a variety of hydrocarbons to produce syngas with a high coke resistance. The coke resistance is believed to be possible due to the high mobility of its lattice oxygen that facilitates the oxidation of the adsorbed carbon species. However, when MoO_2 is exposed to reducing conditions, such as low O_2/C ratio in the partial oxidation or is employed as an anode material for a solid oxide fuel cell (SOFC), MoO_2 will undergo a thermodynamic phase transformation to Mo_2C, and thus, catalytic activity and selectivity would be modified. In order to better understand this phenomenon, a fundamental study is presented to analyze and correlate the role of lattice oxygen driven force with its oxidation activity by using a pulse technique approach. This assessment was done under different operating temperatures where repeated pulses of a reducing gas were sent to the sample to titrate the lattice oxygen while product gases were analyzed using a Real Time Gas Analyzer (RTGA). Changes in the stoichiometric formula and phase composition were obtained using Thermogravimetric Analysis (TGA) and X – Ray diffraction (XRD).

(618c) Monitoring the Catalytic Conversion of Methane to Benzene By Supported MoOx/ZSM-5 Catalysts With In Situ and Operando Molecular Spectroscopy

Tang, Y., Lehigh University
Jehng, J. M.,
Wachs, I. E., Lehigh University
Originally presented on: 11/7/2013 9:10:00 - 9:30:00

The catalytic conversion of methane to benzene represents a highly desirable route for conversion of natural gas into liquid fuels.[1-3] The supported MoO_x/ZSM-5 catalyst system has been reported to be the most promising catalyst system for methane dehydroaromatization (DHA) to aromatics (mostly benzene).[4-6] The structure of the Mo species present in the supported MoO_x/ZSM-5 catalysts and its catalytic properties for methane DHA have been the focus of many publications, but the nature of the supported Mo species and their anchoring sites on the ZSM-5 support (Si sites or Al sites) are still not resolved.[1-6] For the first time, this catalytic system has been investigated with *in situ* and *operando*molecular spectroscopy during methane conversion to allow establishment of molecular structure-activity relationships that can guide the design of improved catalysts.

In the present study, supported Mo/ZSM-5 catalysts (Mo=0.67-3.33wt % and Si/Al=15-140) were examined with *in situ* FTIR (ZSM-5 surface hydroxyls), UV-vis (domain size of surface MoOx) and Raman (molecular structure) spectroscopy under dehydrated conditions and by *operando* Raman-MS spectroscopy during methane DHA. *In situ* FTIR indicates that the supported MoO_x species preferentially anchor at the surface Brønsted acid sites (3610 cm^{-1}) of the ZSM-5 support. *In situ* UV-vis and Raman reveal that the supported MoO_x phase is 100% dispersed on the ZSM-5 support as isolated dioxo $(O=)_2MoO_2$ and mono-oxo $O=MoO_4$ surface species. *Operando* Raman-MS result suggests that reduction of MoO_x is required but not the rate determining step for the reaction. The nature of reduced Mo species will be examined by *in situ* EXAFS and XANES. The methane DHA reaction pathway by the supported Mo/ZSM-5 catalysts proceeds by initial reduction of the surface MoO_x species that forms CO/CO_2 and transformation to carburized molybdenum species (Mo_2C, Mo_2C_{1-x} or MoO_xC_y). The intermediate $[CH_2]$ species initially oligomerize to C_2H_x products that further oligomerize to form aromatic compounds. Carbonaceous deposits, however, also form during methane DHA that lead to catalyst deactivation by coking. These new fundamental insights provide strategies for the molecular design of improved supported MO_x/ZSM-5 catalysts for the methane DHA reaction.

(1) Choudhary, T. V.; Aksoylu, E.; Goodman, D. W., *Catal. Rev.* **2003,**45, 151-203

(2) Burns, S.; Hargreaves, J. S. J.; Pal, P.; Parida, K. M.; Parija, S., *J. Mol. Catal. A: Chem* **2006,**245, 141-146

(3) Ismagilov, Z. R.; Matus, E. V.; Tsikoza, L. T., *Energy Environ. Sci.* **2008**, 1, 526–541

(4) Borry, R. W.; Kim, Y. H.; Huffsmith, A.; Reimer, J. A.; Iglesia, E., *J. Phys. Chem. B* **1999**, 103, 5787–5796

(5) Zheng, H.; Ma, D.; Bao, X.; Hu, J. Z.; Kwak, J. H.; Wang, Y.; Peden, C. H. *J. Am. Chem. Soc.* **2008**, 130, 3722-3723

(6) Kung, H. H.; Kung, M. C., *Adv. Catal.* **1985,** 33, 159-19

(618d) The Microstructure and Catalytic Behavior of M1 and M2 Phases With Bulk Mo-V-M-O Compositions (M= combination of Sb, Nb, Te and Ta) in Propane Ammoxidation to Acrylonitrile

Woo, J., University of Cincinnati
Borisevich, A., Oak Ridge National Laboratory
Guliants, V., University of Cincinnati
Originally presented on: 11/7/2013 9:30:00 - 9:50:00

Abstract

Acrylonitrile, ACN, is a top 50 chemical produced globally. The bulk mixed metal Mo-V-M oxides (M= combination of Nb, Te, Sb, and Ta) consisting of so-called "M1" and "M2" phases are highly promising catalysts for a one-step propane ammoxidation to ACN. The M1 phase has been proposed to be responsible for propane activation and its selective transformation to ACN. The *ab* planes [001] in M1 phase have been proposed to contain the active and selective sites for propane ammoxidation. However, the structure-directing (in a sense of the Mo/V distribution in surface *ab* planes) and catalytic roles of these M substituents in the *ab* planes are poorly understood. Specifically, there is a lack of the fundamental understanding of the Sb functions in the Mo-V-Sb-O, Mo-V-Sb-Nb-O, and Mo-V-Sb-Ta-O catalysts. Moreover, the M2 phase has not been studied as extensively as the M1 phase, whereas this phase was found to play a role in the ammoxidation of the propylene intermediate. The synergy between the Mo-V-Te-Nb-O M1 and M2 phases was reported previously. However, very little is known about synergistic effects for the Mo-V-Sb-O, Mo-V-Sb-Nb-O, and Mo-V-Sb-Ta-O compositions, which are also active and selective for propane ammoxidation to ACN.

In this study, the M1 and M2 phases possessing the Mo-V-Sb-Ta-O, Mo-V-Te-Ta-O, Mo-V-Sb-Nb-O, Mo-V-Te-Nb-O, and Mo-V-Sb-O compositions were prepared by hydrothermal (HT), microwave (MW), and slurry-evaporation (SE) syntheses, which usually resulted in M1/M2 phase mixtures. These mixed phases were treated by H_2O_2 to selectively remove the M2 phase and obtain pure M1 phases.

The M1 phases for STEM characterization were prepared by sectioning epoxy-embedded M1 crystals producing slices of uniform thickness (ca. 50 nm) which eliminated significant uncertainty in the atomic contrast of various metal lattice sites encountered in previous studies of mechanically crushed M1 crystals. The HAADF STEM images obtained were quantified and analyzed statistically in terms of atomic contrast of metal lattice sites, atomic column by column, using Gatan Digital Micrograph (DM) software running DM scripts. A large number of HAADF-STEM images were analyzed with high accuracy using DM scripts. The statistically analyzed intensity of atomic columns suggested that the Mo-V-M-O catalysts show different Mo/V distribution in the *ab* planes. The atomic contrasts of various metal lattice sites confirmed the location of Ta in the center of pentagonal site and Sb in the hexagonal channel. The STEM-based partial occupancy data, especially the Mo/V distribution in *ab* planes from Mo-V-M catalysts, were correlated to their experimentally observed catalytic performance in propane ammoxidation. Additionally, mixtures of M1 and M2 and pure M2 phases with various Mo-V-M-O compositions were also investigated in propane and propylene ammoxidation to understand the catalytic behavior of the M2 phases and the M1/M2 synergy in this selective ammoxidation reaction.

(618e) Improving the C2 Selectivity for the Oxidative Coupling of Methane Reaction By Doping Rare-Earth Oxide Based Catalysts With Lithium

Elkins, T., University of Florida
Hagelin-Weaver, H., University of Florida
Originally presented on: 11/7/2013 9:50:00 - 10:10:00

The oxidative coupling of methane (OCM) has been extensively studied over the past few decades. Ethylene (one of the two C_2 hydrocarbon products in OCM), as the shortest chain hydrocarbon, is an extremely valuable feedstock material for synthesizing longer chain hydrocarbons. Early work in the OCM field discovered doping a magnesium oxide support with lithium produces an active catalyst under OCM conditions with a relatively high C_2 product selectivity. This catalyst system has since been a reference system in many studies due to its activity and simplicity. However, lithium is volatile under OCM reaction conditions. Within the first few hours of reaction, the majority of the lithium will leave the surface of the catalyst resulting in a sharp decrease in activity. In this study we have doped lithium onto supported samarium oxide and terbium oxide catalysts to observe the stability of these catalysts by monitoring the changes in activity during a time-on-stream study. Samarium oxide, and other select rare-earth oxides (REOs), have been established as some of the highest performing single component OCM catalysts. Surface sensitive characterization techniques, such as XPS and FTIR-DRIFTS, were utilized to identify surface species that could be important for enhancing the C_2 selectivity for this reaction.

(618f) Novel UV-Visible/Temperature-Programmed Reduction Methodology for in-Depth Structural Characterization of Cu in Mixed Metal Oxides

Bravo-Suarez, J. J., The University of Kansas
Subramaniam, B., University of Kansas
Chaudhari, R. V., The University of Kansas
Originally presented on: 11/7/2013 10:10:00 - 10:30:00

Novel UV-visible/Temperature-programmed Reduction Methodology for in-Depth Structural Characterization of Cu in Mixed Metal Oxides

Cu containing mixed metal oxides (MMOs) are an important class of catalytic materials. Examples of these materials include the well-known $CuO/ZnO/Al_2O_3$ industrial catalyst for the low-pressure conversion of syngas to methanol ($CO+2H_2 \leftrightharpoons CH_3OH$) and the low-temperature water gas shift reaction ($CO+H_2O \leftrightharpoons CO_2+H_2$), Cu chromite, and Cu/ZnO for (de)hydrogenation and oxidation reactions. Many characterization techniques are available for the structural characterization of Cu in these catalysts such as X-ray photoelectron spectroscopy (XPS), extended X-ray absorption fine structure (EXAFS), and X-ray diffraction (XRD)/pair distribution function (PDF). In this presentation, we will describe a simple methodology for the structural characterization of Cu in $CuMgAlO_x$ MMOs based on a combination of commonly available UV-visible spectroscopy and temperature programmed reduction methods of untreated and N_2O-passivated MMO. The use of $CuMgAlO_x$ MMOs is particularly advantageous because: (1) highly dispersed Cu containing MMOs can be prepared over a wide range of Cu compositions from well-defined precursor structures, namely, $CuMgAlCO_3$ layered double hydroxides; (2) the CuO UV-visible absorption bands in $CuMgAlO_x$ do not overlap with those of MgO or Al_2O_3; and (3) CuO is the only reducible species in the MMO. These conditions allowed the unambiguous determination of Cu composition (by H_2-TPR), domain size (from CuO optical absorption edge energies as measured by UV-visible spectroscopy), and relative fractions of oligomeric CuO species in the bulk and on the surface of the MMOs (by UV-visible+N_2O/H_2-TPR). As a result, by increasing the Cu content in the $CuMgAlO_x$ MMOs from 4 to 21 at. %, it was possible to determine the relative concentrations of isolated and oligomeric CuO species, displaying average numbers of nearest CuO neighbors between about 2 to 4. Additionally, the distinct reducibility patterns of untreated and N_2O-passivated $CuMgAlO_x$ MMOs could be correlated with the corresponding CuO domain sizes. An example will be presented in which well-dispersed Cu species in reduced $CuMgAlO_x$ are found to be highly active for C-O coupling reactions of alcohols and aldehydes to the corresponding esters. The developed methodology could be extended to the study of other Cu-containing MMOs and supported catalysts and can provide useful structural information of heterogeneous catalysts for which highly isolated and/or oligomeric surface metallic or metal oxides may be required.

(618g) Characterization and Performance of Alumina Supported Rare Earth Oxides for Hot Gas Desulfurization of Syngas

Feist, B. J.,
Hill, J. M.,
Originally presented on: 11/7/2013 10:30:00 - 10:50:00

During the production of syngas from coal gasification, a number of gas phase contaminants are produced, such as particulates, tars, hydrogen sulfide and metallic vapors. In order to prevent fouling of downstream processes, these contaminants must be removed from the syngas prior to its ultimate use. The current project is focused on the removal of H_2S, which is in concentrations of 1700 – 14,000 ppmv for syngas produced by coal gasification [1,2]. The permissible limit of H_2S is dependent upon the ultimate application but typically lies in the low ppmv range. Current technologies exist for the removal of H_2S from syngas, but a low temperature adsorption process is utilized, which incurs an efficiency penalty for any process requiring high temperature syngas[3]. For an integrated gasification combined cycle plant, approximately a 10% increase in overall efficiency can be obtained through the use of hot gas desulfurization[3,4]. Previous studies of sorbents for high-temperature desulfurization have focused on Cu-, Fe-, Mn- and Zn-based regenerable sorbents, with studies of rare-earth based sorbents becoming more prevalent in recent years [1,5]. Due to the vapor pressure of metallic Zn and the equilibrium limitations of metallic Cu, Mn oxide, and Fe oxide, these materials are unsuitable for desulfurization above 600°C to produce a high purity syngas. Rare-earth oxide sorbents are stable in reducing atmospheres and have favorable equilibrium constants for desulfurization. In addition, it has been shown that alumina supported rare-earth oxides form a highly dispersed rare-earth aluminate when the loading is kept below a critical limit[6]. This high dispersion of the rare earth elements minimizes the required loading of these costly materials.

Alumina-supported rare earth oxides, prepared at different weight loadings (1-10%) using impregnation, have been characterized with volumetric chemisorption, Temperature Programmed Desorption (TPD), and Diffuse Reflectance Infrared Fourier Transform Spectroscopy (DRIFTS) to investigate the binding between the adsorbate and the oxide. Due to the similarities in the binding sites between CO_2 and H_2S, CO_2 has been used as a probe molecule. DRIFTS data suggests that the addition of lanthanum to alumina results in formation of a La_2O_3 domain that retains the adsorption characteristics of unsupported La_2O_3. Using this information, the CO_2adsorption behaviour has been related to the hot gas desulfurization performance. In addition, the characterization and performance of other supported rare earth oxides including cerium oxide, samarium oxide and europium oxide will be discussed.

[1] Meng, X., de Jong, W., Pal, R., & Verkooijen, A. H. Fuel Process. Technol., 2010. 91(8), 964.

[2] Meng, X.M., De Jong, W., Verkooijen A.H.M. Environ. Prog. Sustainable Energy. 2009. 28(3), 360 [3] Gangwal, S., Portzer, J., Roberts, G., & Kozup, S. Engineering Evaluation of Hot-Gas Desulfurization with Sulfur Recovery. Tech. rep., Research Triangle Institute, 1998.

[4] Gasper-Galvin, L. D., Atimtay, A. T., & Gupta, R. P. Ind. Eng. Chem. Res., 1998. 37(10), 4157.

[5] Cheah, S., Carpenter, D.L., and Magrini-Bair, K.A. Energy Fuels, 2009. 23. 5291.

[6] Bettman, M., Chase, R. E., Otto, K., & Weber, W. H. J. Catal., 1989. 117(2), 447.

(639a) Supercritical Water Upgrading of Heavy Oils: Historical Perspective

Timko, M. T., Worcester Polytechnic Institute
Wu, G., MIT
Tester, J. W., Cornell University
Green, W. H. Jr., Massachusetts Institute of Technology
Ghoniem, A. F., Massachusetts Institute of Technology
Originally presented on: 11/7/2013 8:32:00 - 8:41:00

Continuous increases in the sulfur content and average molecular weight of remaining petroleum reserves are straining refinery capacity using current technologies. New technologies are required for cleaner, energy efficient utilization of sulfur rich heavy oils. Treating heavy oils and bitumen in supercritical water has been known since the 1980s to reduce their molecular weight distributions and sulfur contents. Crucially, supercritical water treatment does not promote coke formation, does not require catalysts, and reduces heavy oil asphaltene content. However, commercialization has been held back by contradictory literature data and a lack of fundamental understanding of the chemical reactivity, thermodynamic phase behavior, and transport phenomena at supercritical conditions. This paper will present an overview of the historical development of supercritical water upgrading of heavy crude oils, with a focus on identifying knowledge gaps in several key areas: 1) the role of water, 2) the role of phase behavior and mixing, 3) the chemical mechanism, 4) the use of catalysts, and 5) the fate of sulfur bearing compounds.

(639b) Supercritical Water Process to Upgrade Heavy Oil

Abdulhadi, A., Saudi Aramco
Garhoush, M., Saudi Aramco
Alali, F.,
Jishi, A., Saudi Aramco
Al-Mulla, F., Saudi Aramco
Albahar, M., Saudi Aramco
Al-Otaibi, B.,
Alshareef, A. H., Saudi Aramco
Shafei, E., Saudi Aramco
Punetha, A. K., Saudi Aramco
Lee, J. H., Saudi Aramco
Choi, K. H., Saudi Aramco
Qarzouh, M.,
Al-Dossary, M.,
Originally presented on: 11/7/2013 8:41:00 - 8:56:00

Supercritical water, as an environmentally friendly solvent, found potentially a new application area: heavy oil upgrading. Although water is a very common species in exploration and production of petroleum, direct usage in upgrading of heavy oil has not been applied industrially. Basic chemistry governing upgrading reaction under supercritical water is believed to resemble that of conventional thermal cracking such as visbreaking and coking. However, presence of supercritical water has capability to organize random radical reactions to suppress coking and gas generation. Major huddles to the scale-up of such a novel process are high energy cost required to heat-up large quantities of water, heavy-duty equipment to handle harsh operation conditions, and plugging of process lines by coke or sludge. These technical challenges can be resolved based on the fundamental understanding of the process reactions and fluids. This paper will present the recent advances in supercritical water technology for heavy oil upgrading as well as the proper approach to facilitate its industrial application.

(639c) Evaluation of Reactivities of Various Sulfur Groups in Arabian Heavy Crude During SCW Treatment Using GCxGC-FID/SCD

Kida, Y., MIT
Class, C. A., MIT
Carr, A. G., MIT
Green, W. H. Jr., Massachusetts Institute of Technology
Originally presented on: 11/7/2013 8:56:00 - 9:14:00

Arabian heavy crude oil is treated with supercritical water (SCW) in a batch reactor. Detailed analysis of the crude feed and product is made possible by a two dimensional GC (GCxGC) with two types of detectors: FID and SCD, a sulfur specific detector. GCxGC's separate compounds by volatility (first column) and polarity (secondary column), allowing complex mixtures such as crude oil to be separated into individual compounds. An FID and SCD detector working in series following the GCxGC separation allows the sulfur compounds to be seen separately from the non-sulfur containing hydrocarbons, which are the overwhelming majority of the compounds in crude. SCW desulfurization proceeds via interactions between the sulfur-containing compounds and the non-sulfur containing compounds. The capability of separating individual compounds and seeing sulfur compounds and hydrocarbons separately gives us insight into how SCW treatment upgrades crude oils.

(639d) Hydrothermal Liquefaction of Food Waste and Model Food Waste Compounds

Zastrow, D. J., Mainstream Engineering Corporation
Jennings, P. A., Florida Institute of Technology

Originally presented on: 11/7/2013 9:24:00 - 9:39:00

Solid waste is generated at a rate of more than 4 lb per person per day. A significant portion of the non-recyclable waste is food waste, with high water content and low energy density. While most thermochemical conversion processes require a relatively dry feedstock, hydrothermal liquefaction uses subcritical water as the reaction medium and is therefore compatible with high-moisture feedstocks. We have studied the viability of converting high-moisture food waste into a liquid fuel. A surrogate food waste stream was formulated and the yield and bio-oil quality were measured at various process conditions. Yields as high as 45% on a dry mass basis were observed at a reactor temperature of 315 °C.

To further understand hydrothermal liquefaction of food waste, three model food compounds were studied—starch, soybean oil, and casein. The reaction mechanism, product distribution, and product yields for each compound were investigated at reactor temperatures ranging from 250–315 °C, using a sodium carbonate catalyst and a syngas head gas. The pure-component data were compared to data collected for the surrogate food waste under the same operating conditions. A lumped kinetic model describing the hydrothermal liquefaction of starch is also proposed.

(639e) Microfluidics Studies of Hydrocarbon-Water Systems At High Pressure and Temperature

Ates, A., Cumhuriyet University
Liu, N., Institute of Condensed Matter Chemistry of Bordeaux
Timko, M. T., Worcester Polytechnic Institute
Aymonier, C., Institute of condensed Matter Chemistry of Bordeaux
Green, W. H. Jr., Massachusetts Institute of Technology
Marre, S., Institute of Condensed Matter Chemistry of Bordeaux
Originally presented on: 11/7/2013 9:39:00 - 9:54:00

Supercritical water upgrading of heavy crude oils is a combined physical/chemical process involving chemical reaction, phase behavior, and mixing. Fundamental understanding of each of these major phenomena is lacking. The purpose of this study is to obtain experimental data on phase behavior and mixing for heavy oil model compounds and water near its critical point. Lab-on-chip microfluidic devices are ideal for mixing and phase behavior studies because of their excellent optical and spectroscopic access. In this study, the dissolution rate measurements of hydrocarbons in the water at conditions near the mixture critical point were studied through a droplet-based approach and the findings were compared with the microscopic mixing models described in Guang Wu's 2012 papers [*J. Supercrit. Fluids, 67, p:29 and 72, p:150*].

Toluene and n-dodecane were chosen as a model compounds. The miscibility of a single component hydrocarbon droplet (toluene or n-dodecane) or a mixture of toluene and n-dodecane at varying ratios (1:3, 1:1 and 3:1) in the water was investigated over the temperature range 50 - 400°C at p = 100 bar in a Silicon-Pyrex microreactor. The phase behavior of the hydrocarbon droplet during "stop-flow" conditions as a function of temperature was observed with an Olympus microscope equipped with a digital camera. The composition of each phase was measured using a confocal Raman spectrometer (Thermo Scientific DXR Raman) with a 532 nm wavelength excitation laser operating at 10 mW output power. Measurable water solubility in the hydrocarbon starts above 150°C at which temperature the hydrocarbon droplet swells due to thermal expansion and water absorption. When the water concentration in the droplet reaches the solubility limit, a third phase consisting of hydrocarbon and water mixture was observed. On the other hand, the Raman analysis of the water phase showed negligible hydrocarbon solubility in water.

Studies were performed with "small" (define) and "large" hydrocarbon droplets. Small toluene droplets swell with increasing temperature up to 200°C, shrinks above 200°C, and dissolves completely at 310°C. These temperatures are shifted higher for large toluene droplets, presumably a consequence of lower mass transfer rates. For larger toluene droplets, a third phase was observed at 300°C; the droplet shrinks and disappears at 340°C. The behavior of dodecane droplet in water is similar to toluene, but the solubility of dodecane in water is lower. Namely, small dodecane droplets swell with increasing temperature up to 350°C, then shrink and disappear at 390°C. Large dodecane droplets remain stable up to 400°C, the limit of the current method. The addition of dodecane into toluene increases the miscibility temperature from 320°C to 400°C, depending on ratio of toluene to dodecane.

(639f) The Effects of Process Conditions On Fast Hydrothermal Liquefaction of Microalgae

Faeth, J. L., University of Michigan
Savage, P. E., University of Michigan
Originally presented on: 11/7/2013 9:54:00 - 10:12:00

Transportation fuels are dominated by high energy-density liquids, a vast majority of which are petroleum-based. Petroleum is a finite resource, and the combustion of petroleum fuels releases carbon that had previously been stored underground, contributing to the greenhouse gas effect and global climate change. The production of biofuels from biomass can reduce dependence on petroleum and reduce greenhouse gas emissions. Algae are a particularly desirable biomass feedstock due to their fast growth rates, high oil content, and tolerance for a variety of cultivation conditions. However, algae are inherently very wet, and many biomass conversion processes, like pyrolysis, require dry feedstocks. Hydrothermal liquefaction is an established biomass conversion process that avoids energy intensive drying steps and produces biocrude, a biofuel precursor. Recent research in this field led to the development of fast hydrothermal liquefaction (FHTL), a novel process reducing the amount of time necessary to produce biocrude to perhaps to just 60 seconds. This decrease in reaction time reduces the size of reactor vessels required for continuous biomass conversion, subsequently cutting capital costs associated with such a process. Since FHTL was just recently discovered, there remains a large gap in the literature surrounding FHTL.

This presentation elucidates the effects of time, temperature, heating rate, biomass loading and other process variables on the yields and compositions of biocrudes and various byproducts produced via FHTL. The information gained through this research will identify preferred process conditions for FHTL as the technology is examined and considered for scale-up. Further, characterization of the various byproducts of FHTL will facilitate the environmental decisions that are associated with large-scale processes. Taken together, this information will provide a comprehensive idea of the sustainability of FHTL as a biomass conversion process for the production of liquid transportation fuel precursors.

(639g) Hydrothermal Liquefaction of Wastewater Derived Green Alga

Roberts, G. W., University of Kansas
Fortier, M. O. P., University of Kansas
Sturm, B. S. M., University of Kansas
Stagg-Williams, S. M., University of Kansas
Originally presented on: 11/7/2013 10:12:00 - 10:30:00

Hydrothermal liquefaction (HTL) of algae has gained fast popularity within the research community because if it's superior pairing with algae as a conversion process to fuels and chemicals. HTL is a whole cell conversion process not limited to high-lipid content algae for high biocrude yields; typically 10-30% higher than initial lipid content, and uses subcritical water chemistry as the chemical driving force reducing the need for extensive dewater and solvent extraction as with pyrolysis and biodiesel production. The carbon rich biocrude product from HTL is most similar to that of petroleum crude than any other bio-oil reported to date. It has recently been reported that combining HTL conversion and wastewater cultivation relieves many of the sustainability concerns as identified by the National Research Council.[1, 2] Wastewater derived algae gives very different product profiles compared to fertilized controlled-growth cultivated algae, mainly in terms of biocrude molecular and elemental content with a great increase in solid biochar production. This paper reports on the HTL of various green algae and evaluates the feedstock and products in terms of its yield and characterized by GC-MS, calorimetry, DSC/TGA, FT-IR, ICP-OES, elemental content, and distillate fractions.

References:

1) National Research Council of the National Academies. Committee on the Sustainable Development of Algal Biofuels. *Sustainable Development of Algal Biofuels in the United States*; National Academy Press: Washington, D.C., 2012.

2) Griffin W. Roberts, Marie-Odile P. Fortier, Belinda S. M. Sturm, and Susan M. Stagg-Williams, Promising Pathway for Algal Biofuels Through Wastewater Cultivation and Hydrothermal Conversion, *Energy and Fuels*, 27 (2), 857–867 (**2013**).

(645a) Experiment and Theory for the Dew Point of Water in Compressed CO2

Meyer, C. W., NIST
Harvey, A. H., National Institute of Standards and Technology
Originally presented on: 11/7/2013 8:30:00 - 8:52:00

When compressing and transporting CO_2 for carbon capture and sequestration, it is important to know the water dew point as a function of temperature, pressure, and water content; this determines the degree to which it is necessary to dry the CO_2 in order to avoid condensation that can lead to corrosion in pipelines. This dew-point relationship is sensitive to the vapor-phase nonideality, on which there is limited theoretical information and only a few scattered experimental data at conditions relevant for pipeline transport. Dew-point data can also inform thermodynamic modeling of other parts of the process where H_2O and CO_2 are present.

To provide needed data, we have constructed an apparatus to measure the water mole fraction at saturation in a compressed gas. The apparatus consists of a saturator and the NIST gravimetric hygrometer (an instrument originally designed for work in humidity standards). The saturator humidifies the gas by bringing it into equilibrium with liquid water at a precisely controlled temperature and pressure. Then, the gravimetric hygrometer measures the water mole fraction of the humid gas with a relative uncertainty better than 0.1%. For CO_2, the apparatus can operate at pressures up to 5 MPa and at temperatures from approximately 1 °C to 85 °C.

We have measured dew-point data for several isotherms both above and below room temperature. Preliminary analysis indicates that these data are consistent with what would be expected based on recent theoretical predictions by Wheatley and Harvey, based on an *ab initio* potential-energy surface, of the interaction second virial coefficient between water and carbon dioxide.

(645b) Analytical and Pore-Network Calculations of Methane Saturation Resulting From Hydrate Dissociation in Porous Media

Tsimpanogiannis, I. N., National Center for Scientific Research
Lichtner, P., Los Alamos National Laboratory
Originally presented on: 11/7/2013 8:52:00 - 9:14:00

The capability of clathrate hydrates to store large amounts of gases has attracted significant attention from the scientific and industrial communities. Practical applications such as the storage and transportation of "energy-carrier" gases (methane, hydrogen), or "green-house" gases (carbon dioxide), mixture separation, and desalination, are among those of interest. Flow assurance, ocean-floor stability, and hydrates as a major source of well-drilling hazards are issues of major concern to the oil/gas industry [1].

The discovery of large amounts of natural gas-hydrate deposits in oceanic or permafrost sediments has rendered hydrates as a possible future energy source. Of interest to this study is the case of methane hydrate in porous media, as a source of methane for energy applications. For this purpose the hydrate needs to dissociate in order for the gas phase to be released, and subsequently to be produced through production wells. Methods that are considered in order to achieve the particular goal of hydrate dissociation include pressure reduction, thermal stimulation, hydrate-inhibitor injection or combinations of the above methods. The gas saturation that results from the dissociation of methane hydrate inside aqueous-saturated porous domains is a crucial parameter that determines if the resulting methane gas phase can be produced. To this purpose the gas phase needs either to form a percolating gas cluster that connects to the producing well, or reach the well, as a result of gas bubble mobilization due to viscous or buoyancy forces.

In this study we develop predictive tools for methane gas saturation that results from hydrate dissociation in porous media at different length scales, ranging from the single-pore scale, up to the pore-network scale [2]. Porous-medium overall, as well as, transversely-averaged, spatially specific, gas saturations are calculated. The analytical calculations are compared with the results obtained from detailed simulations using pore networks [3] and both results are found to be in very good agreement. Pore networks have been used successfully to capture the randomness of real porous media, and simulate processes involving flow, heat/mass transfer, and reactions in porous media.

References

[1] Sloan, E. D.; Koh, C. A. Clathrate Hydrates of Natural Gases, 3rd edition, Taylor & Francis. CRC Press, 2008.

[2] Tsimpanogiannis, I. N.; Lichtner, P. C. *J. Phys. Chem. C* **2013**, In press.

[3] Tsimpanogiannis, I. N.; Lichtner, P. C. *Phys. Rev. E* **2006**, *74*, 056303.

(645c) Exploring the Effect of Pore Morphology On Model Electrochemical Double-Layer Capacitor Systems

Phillips, K. A., North Carolina State University
Gubbins, K. E., North Carolina State University
Originally presented on: 11/7/2013 9:14:00 - 9:36:00

Electrochemical double-layer capacitors (EDLCs) have drawn much attention as energy storage devices in recent years due to their long cycle life as well as their ability to uptake and discharge energy very rapidly and to store more energy than typical capacitors. Molecular simulations are employed to study electrochemical properties of model EDLC systems. A variety of simple pore sizes, shapes, and configurations are studied, as well as model carbons, which reproduce structural features of disorder carbons used in EDLC experiments. Surface charge densities of 0 C/m^2, 5 C/m^2, 15 C/m^2, and 25 C/m^2 are applied to the carbon structure, which are representative of typical values for EDLCs. Solutions of 1M NaCl, NaF, and $NaBF_4$ are used as electrolytes in each system. We study the effect of surface charge on the electrochemical and structural properties of the electrolyte solution. Of particular interest are: capacitance of the model EDLC system, selectivity of ions to reside in various pore sizes, charge density of the electrolyte solution across the electrode system, and the solvation structure of ions in various confining geometries compared to the bulk solution between electrodes.

(645d) Thermodynamic Energy and Exergy Efficiency Limits of Photosynthesis

Silva, C., University of Pennsylvania
Seider, W. D., University of Pennsylvania
Lior, N.,

Originally presented on: 11/7/2013 9:36:00 - 9:58:00

With recent concerns about sustainable resources and environmental protection, growing attention has been focused upon biological sources of chemicals and fuels; however, to analyze such bioprocesses for commercial viability, the theoretical efficiencies must be known as an upper-bound on performance. Since almost all exergy contained in biomass originated from solar radiation, photosynthesis is the gateway to sustainable bioprocess development. Yet, despite the mechanism for photosynthesis being known for decades, there is still no well-established definition for its efficiency. The two types of studies that exist in the literature consider either the thermodynamic/physical effects of the process (evaporation, carbon dioxide sequestration, temperature changes) [1], [2], [3] and ignore the complex biochemical reaction mechanism, or the converse [4] [5]. The disjunction is due to the dichotomy that exists between biology and traditional thermodynamics. This study dissects the complex bio-processes involved in photosynthesis and analyzes their energy and exergy effects, bridging the gap and creating a clearer picture that is acceptable to both thermodynamicists and biologists. For the light reactions, reduction potentials were used to analyze the flow of excited, high-energy electrons through photosystem II and photosystem I, resulting in an exergy efficiency of 25%. For the dark reactions, free energies of formation were taken from the literature and used to calculate the Gibbs free energy and exergy changes; exergy changes due to concentration changes were assumed to be negligible for both the light and dark reactions. The resulting efficiency of the dark reactions was 89.5%. An exergy balance was formulated for water flows throughout the system to account for exergy destruction due to transpiration. Similarly, an exergy balance was constructed to account for losses due to ATPsynthase as the proton-motive force is used to synthesize ATP. Heuristic values for photorespiration, photo-degradation, and absorbance were taken from the literature and combined with the theoretical analysis above, yielding an overall photosynthetic efficiency of 1.9%. The overall efficiency shows good agreement with recent publications in the literature [1], [2]. Ultimately, the largest losses were due to poor absorbance and electron transfer through the photosystems.

References:

[1] Bisio G, Bisio A. Some Thermodynamic Remarks on Photosynthetic Energy Conversion. Energy Conversion Management 1998; 39(8):741-748.

[2] Petela R. An approach to the exergy analysis of photosynthesis. Solar Energy 2008; 82: 311-328.

[3] Reis HA, Miguel AF. Analysis of the exergy balance of green leaves. Int. J. Exergy 2006; 3(3): 231-237.

[4] Lems S, van der Kooi HJ, de Swaan Arons J. Exergy analyses of the biochemical processes of photosynthesis. Int. J. Exergy 2010; 7(3): 333-351.

[5] Chain RK, Arnon DI. Quantum Efficiency of Photosynthetic Energy Conversion, Proceeding of the National Academy of Science 1977; 74(8): 3377-3381.

(645e) Energetics and Mechanism of Desorption of Small Molecules From Cellulose Crystal Surface: Atomic Force Microscopy (AFM) and Molecular Simulation Study

Muthukumar, L., Texas Tech University
Liao, Y. C.,
Weeks, B. L., Texas Tech University
Khare, R., Texas Tech University
Originally presented on: 11/7/2013 9:58:00 - 10:20:00

The cello-oligosaccharide intermediates formed during enzymatic hydrolysis of cellulose are adsorbed on the cellulose crystal surface thus hindering further action of the enzyme. In this work, we have used a combination of AFM experiments and molecular modeling to decipher both the energetics and mechanism of the process of desorption of cello-oligosaccharide molecules from the cellulose crystal surface. For this purpose, molecules of model compounds containing different functional groups are attached to the AFM tip and the adhesion force between these molecules and the cellulose crystal surface is determined using AFM experiments. The steered molecular dynamics (SMD) technique is used to determine the adhesion force as well as the mechanism of desorption of these same molecules from the cellulose crystal surface. Simulation results obtained for different crystal surfaces as well as for varying number of adsorbed molecules are directly compared with AFM experiment results.

(645f) Modelling Super-Equilibrium in Biomass Gasification With the Constrained Gibbs Energy Method

Kangas, P., VTT Technical Research Centre of Finland
Koukkari, P., VTT Technical Research Centre of Finland
Hannula, I.,
Originally presented on: 11/7/2013 10:20:00 - 10:42:00

Background

Biomass gasification can be seen as a viable option for generating electricity from biomass with high efficiency or for the production of liquid biofuels and chemicals. Gasification is a thermo-chemical conversion process that takes place at high temperature and reductive environment. For such conditions thermodynamic equilibrium is often assumed, but not reached in practice: during biomass gasification light hydrocarbons (e.g.: CH_4, C_2H_2, C_2H_4, C2H6, C_3H_8, C_6H_6), ammonia, tars and char are also formed. These phenomena cannot usually be modelled with equilibrium assumptions.

In order to increase the accuracy of gasification models, both empirical and mechanistic models are often applied alone or combined with thermodynamic equilibrium calculations [1]. This leads to a 'dual' approach where different parts of the process and reactions are modelled independently and results are merged in post-processing.

This study presents a unified solution for simultaneous calculation of the super-equilibrium reactions and related reaction enthalpies in the gasification process. The solution is based on the Constrained Free Energy (CFE) method where equilibrium computation is extended with additional immaterial constraints for solving the local or partial constrained equilibrium instead of the global thermodynamic equilibrium.

Methodology

First applications of the constrained thermodynamic equilibrium method in Gibbs energy minimization were the conservation of immaterial properties [2] and inclusion of reaction kinetics into the Gibbs'ian multicomponent models [3,4]. In the course of time the method has found new applications, ranging from modelling sorption in pulp suspensions to determining surface tension of liquid mixtures [5]. Quite recently CFE has been used to estimate the super-equilibria of alkali metals and sulphur in black liquor combustion [6]. Today constrained free energy method is seen as a versatile tool for various fields of application.

In this study the CFE methodology is applied for describing the super-equilibrium occurring in biomass gasification. Light hydrocarbons, ammonia, tars and char tend to decompose if thermodynamic equilibrium calculation is performed for high temperatures, and thus constraints are needed for the modelling of their presence in the super-equilibrium conditions.

This a computational study with all experimental data obtained from literature. The empirical model of super-equilibrium applied is from [1]. Thermodynamic data is from [7]. ChemSheet [8] is used as the modelling tool as it allows extending the thermodynamic system and thus enables the calculations of constrained free energy models.

Results

In the present work different approaches for the modelling of global or local equilibrium of biomass gasification have been assessed. First (i) thermodynamic equilibrium without any constraints is computed and used as reference. Later additional constraints are introduced to system, such as amount of (ii) char, (iii) tars, (iv) ammonia. The light hydrocarbons are defined by (v) the amount of carbon and by (vi) amount of hydrogen in hydrocarbons. In addition constraints such as (vii) amount of double bonds in hydrocarbons or (viii) methane amount are evaluated. Finally a fully constrained system, where all species are defined according to [1] is reported.

The super-equilibrium of major syngas components (CO, CO_2, H_2, H_2O and CH_4) can be predicted with the proposed constrained free energy method. Accuracy of the model increases when an additional constraint for hydrogen in hydrocarbons is defined. At the same time the partial equilibrium moves further from the global equilibrium.

While the adapted constraints were successful in evaluating the super-equilibrium of major species, the respective predictions for minor hydrocarbon components (C_2H_2, C_2H_4, C_2H_6, C_3H_8, C_6H_6) were not as satisfying. The selection of modelled light hydrocarbon components seems to be vital, if more accuracy is required. In addition by defining all species with separate constraints, it is possible to increase the accuracy of results, but with the cost of decreasing the degrees of freedom of the thermodynamic system.

Summary

Constrained free energy method was successfully applied for the calculation of the super-equilibrium of light hydrocarbons, ammonia, tars and char in biomass gasification. Gasification can be solved as restricted partial equilibrium with a single calculation step. Thus, there is no need for post-processing corrections for e.g. volume or enthalpy flows. For a reasonable accuracy of the simulation model, it is not necessary to define constraints for all reactions, but a set of more general constraints (e.g. carbon in hydrocarbons) remain sufficient.

Distinct benefit of presented approach is that chemical reactions, gasification enthalpy and states of the system (such as volume) are estimated concurrently. This method is also resilient: the number of details describing constraints can vary from simpler experimental results and models to complicated mechanistic models. A promising application area for the presented methodology could be as part of large scale process simulation where detailed thermodynamic models can be implemented and effects in gasification chemistry could conduct changes in the entire process.

References

[1] Hannula, I. and Kurkela, E., "A parametric modelling study for pressurised steam/O2-blown fluidised-bed gasification of wood with catalytic reforming", Biomass and Bioenergy, 38:58–67 (2012).

[2] Alberty, R.A., "Thermodynamics of the formation of benzene series polycyclic aromatic

hydrocarbons in a benzene flame", The Journal of Physical Chemistry, 93(8):3299–3304 (1989).

[3] Keck, J.C., "Rate-controlled constrained-equilibrium theory of chemical reactions in complex systems", Progress in Energy and Combustion Science, 16(2):125–154 (1990).

[4] Koukkari, P., "A physico-chemical method to calculate time-dependent reaction mixtures", Computers & Chemical Engineering, 17(12):1157–1165 (1993).

[5] Koukkari, P. and Pajarre, R., "Calculation of constrained equilibria by Gibbs energy minimization", Calphad, 30(1):18–26 (2006).

[6] Kangas, P., Koukkari, P., Lindberg, D. and Hupa, M., "Modelling black liquor combustion with the constrained Gibbs energy method", Proceedings, 8th International Black Liquor Colloquium, Belo Horizonte, Brazil (2013).

[7] Roine, A., Lamberg, P., Mansikka-aho, J., Björklund, P., Kentala, J., Talonen, T., et al., HSC Chemistry 6.12., Outotec Research Oy, (2007).

[8] Koukkari, P., Penttilä, K., Hack, K. and Petersen, S., "CHEMSHEET – An Efficient Worksheet Tool for Thermodynamic Process Simulation", In: Microstructures, Mechanical Properties and Processes, Y. BrÃ©chet (Ed.), Wiley-VCH Verlag GmbH & Co. KGaA, Weinheim (2005).

(645g) Microscopic Insights Into the Electrochemical Behavior of Non-Aqueous Electrolytes in Electric Double-Layer Capacitors

Wu, J., UC Riverside
Jiang, D. E., Oak Ridge National Laboratory
Originally presented on: 11/7/2013 10:42:00 - 11:04:00

Electric double-layer capacitors (EDLC) are electrical devices that store energy by adsorption of ionic species at the inner surface of porous electrodes. Compared with aqueous electrolytes, ionic liquid and organic electrolytes have the advantage of larger potential windows, making them attractive for the next generation of EDLC with superior energy and power densities. The performance of both ionic liquid and organic electrolyte EDLC hinges on the judicious selection of the electrode pore size and the electrolyte composition that requires a comprehension of the charging behavior from a microscopic view. In this perspective, we discuss predictions from the classical density functional theory (CDFT) on the dependence of the capacitance on the pore size for ionic-liquid and organic-electrolyte EDLC. CDFT is applicable to electrodes with the pore size ranging from that below the ionic dimensionality to mesoscopic scales, thus unique for investigating the electrochemical behavior of the confined electrolytes for EDLC applications.

(646a) Combustion Dynamics of Mechanically Alloyed Al-Mg Powders in Turbulent Flames

Corcoran, A., New Jersey Institute of Technology
Aly, Y., New Jersey Institute of Technology
Dreizin, E. L., New Jersey Institute of Technology
Originally presented on: 11/7/2013 8:30:00 - 8:55:00

Metal powders are widely used as high-energy density additives to energetic formulations. Recent efforts focused on development of new metal-based materials, that are expected to ignite more readily and burn faster than pure metal powders, such as Al, used in many solid propellants, explosives, and pyrotechnics. Full scale tests of energetic systems with multiple new components are expensive and thus predictive models for particle ignition and combustion rates are required in order to establish whether the novel materials are compatible with the present system configurations. Most current models and descriptions are based on laboratory experiments performed in stationary or laminar combustion configurations and with relatively coarse powders of pure metals. Experimental data on various new materials, especially including fine powders burning in practically interesting turbulently mixing environments are lacking. In our recent work, an approach to characterize combustion of fine aluminum powders in turbulently mixing gas flows was discussed. In this paper, the approach is extended to characterize combustion of fine, mechanically alloyed Al·Mg powders. The powders with particle sizes roughly from 1 to 20 μm are prepared by ball-milling elemental aluminum and magnesium in a planetary mill. The Al/Mg ratios are varied to explore powders with different compositions. The experiments are aimed to measure burn times for the alloyed particles burning in environments with different levels of turbulence. For each material composition, the powders are injected using a nitrogen flow into an air-acetylene flame. Experiments are performed with both laminar flame and flames turbulized by auxiliary tangential jets of air. The air flows with adjustable flow rates are used to achieve different controlled levels of turbulence intensity. Optical emission of the burning particles was recorded using filtered photomultiplier tubes. Measured durations of individual particle emission pulses were assumed to represent their burn times; these data were classified into logarithmically spaced time bins and are correlated with the particle size distribution to obtain the burn time as a function of the particle size. It is observed that the effect of particle size on burn rate is stronger for the mechanically alloyed Al·Mg powders compared to that for pure aluminum. Experiments will be presented for powders of different compositions and burning in environments with different levels of turbulence. It is expected that these results will serve to develop a mechanistic model for burn rates of Al·Mg alloys in environments with varied turbulence intensities. The experimental methodology is expected to be useful for a broad range of reactive powders.

(646b) Formation of Reactive Composite Films

Doorenbos, Z., Innovative Materials and Processes, LLC
Leaming, S., South Dakota School of Mines & Technology
Haines, C., US Army RDECOM-ARDEC
Kapoor, D., US Army RDECOM-ARDEC
Puszynski, J., South Dakota School of Mines and Technology
Originally presented on: 11/7/2013 8:55:00 - 9:20:00

The use of metal/metal, metal/metal-oxide reactive or pyrophoric systems for defense, aerospace, commercial and electronic industries have been widely used for over fifty years. As new requirements were established more sophisticated methods for the formation of these reactive materials have been developed. To meet new requirements, specifically faster energy release, many of these technologies require the use of nanoscale materials. In the past decade, the use of nanoscale materials to tune the combustion characteristics of energetic materials has been investigated. The special attention was paid to pyrophoric and nanothermite systems.

This contribution will focus on the formation of thin pyrophoric metals and metal/metal oxide reactive substrates reinforced with carbon nanofibers, MWNTs and/or carbon veils. These reactive substrates are formed using tape casting or pressure filtration techniques. The combustion characteristics of such reactive substrates are determined using pyrometers along with high speed and thermal cameras. The effect of reactive substrate composition (i.e. metal, metal oxide, fiber and binder(s)) along with substrates thickness on the combustion characteristics will be presented. Impact, friction and ESD sensitivity thresholds will also be presented for the metal/metal oxide reactive substrates.

(646c) Mechanically-Alloyed Alâ^™Mg Powders With Customized Particle Sizes and Compositions

Aly, Y., New Jersey Institute of Technology
Schoenitz, M., New Jersey Institute of Technology
Dreizin, E. L., New Jersey Institute of Technology
Originally presented on: 11/7/2013 9:20:00 - 9:45:00

Adding aluminum to propellants, pyrotechnics, and explosives is a common way to boost their energy density. A number of approaches have been investigated that shorten aluminum ignition delay, increase burn rate, and decrease the tendency of aluminum droplets to agglomerate. Previous work showed that particles of mechanically alloyed Al·Mg powders burn faster than similarly sized particles of pure aluminum. However, prepared mechanically alloyed powders were coarser than fine aluminum commonly used in energetic formulations. This work is focused on preparation of mechanically alloyed, composite Al·Mg powders in which both internal structures and particle size distributions are adjusted. Powders with 50 - 90 at. % Al were prepared and characterized. Milling protocol is optimized to prepare equiaxial, micron-scale particles suitable for laboratory evaluations of their oxidation, ignition, and combustion characteristics. Particle size distributions are measured using low-angle laser light scattering. Electron microscopy and x-ray diffraction are used to examine particle morphology and phase makeup, respectively. Combustion of aerosolized powder clouds is studied using a constant volume explosion apparatus. Combustion of individual particles is studied using a laser ignition setup. Oxidation and decomposition are studied using thermo-analytical measurements. For all materials, ignition and combustion characteristics are compared to one another and to those of pure Al. Additionally, ignition and oxidation characteristics are compared to determine which events lead to particle ignition. Compositions with improved performance (i.e., shorter ignition delays, higher burn rates, and faster pressurization rates) compared to pure Al are identified.

(646d) Selection of Pyrotechnic Formulations Based On Data Generated By Thermochemical and Thermodynamic Software Packages (INVITED)

Redner, P., US Army, RDECOM-ARDEC
Oyler, K., US Army RDECOM-ARDEC
Csernica, C., US Army, RDECOM-ARDEC
Mehta, N., US Army RDECOM-ARDEC
Puszynski, J., South Dakota School of Mines and Technology
Originally presented on: 11/7/2013 9:45:00 - 10:10:00

Thermodynamic and thermochemical analyses are a very important aspect of the selection phase for the development of new energetic formulations as they can greatly enhance trial and error explorations in the lab. There are several commercially available software packages, including: Cheetah, PEP Code, NASA_Lewis code, FCT, and HSC Chemistry, but each of them has a number of limitations regarding adequate predictability of the fundamental properties of complex reactive mixtures. This presentation will focus on a comparative study of several binary, ternary and quaternary pyrotechnic mixtures with and without gas generation.

The specific attention will be on condensed phase systems consisting of Al, B, and Si as fuels and oxides, iodates and nitrates as oxidizers, in addition to some other materials of interest. The presentation will summarize the pros and cons of each software package for its potential use for the prediction of thermodynamic properties of condensed phase reactive systems in an array of pyrotechnic applications.

(646f) Extension of the Transferable Potentials for Phase Equilibria to Nitroaromatics

Bhatnagar, N., Wayne State University
Kamath, G., University of Missouri-Columbia
Potoff, J. J., Wayne State University
Originally presented on: 11/7/2013 10:35:00 - 11:00:00

United-atom force fields, based on the Transferable Potentials for Phase Equilibria (TraPPE), are developed for twelve nitro-aromatic compounds, which include 2,4-dinitrotoluene (2,4-DNT), 2,6-dinitrotoluene (2,6-DNT), 3-nitrotoluene (3-NT), 4-nitrotoluene (4-NT), 1,3-dinitrobenzene (1,3-DNB), 1,4-dinitrobenzene (1,4-DNB), 2,4-dinitroanisole (DNAN), 1,3,5-trinitrobenzene (TNB), 2,4,6-trinitrotoluene (TNT), 2-nitroanisole (2-NAN), 4-nitroanisole (4-NAN) and n-methyl-p-nitroaniline (MNA). 1-Octanol-water and air-water partition coefficients are predicted for the optimized TraPPE-UA force field with adaptive biasing force molecular dynamics simulations, and compared to available experimental data. Log Kow values are predicted with an average absolute deviation of 0.2 log units, while Henry's law constants are predicted to with an average absolute deviation of 0.5 log units. Two additional models are presented for energetic materials with five membered rings for which no experimental data are available in the open literature: 3,5-dinitropyrazole (DNP) and 3-nitro-1,2,4-triazole-5-one (NTO). Investigation of the local microstructure around each solute reveals that 1-octanol is able to form hydrogen bonded chains around the solute, while little organized microstructure was observed around the solutes in water.

(659a) DNA Microarrays On UV-Modified Surfaces for Differentiating Clostridium Species

Yang, K. L., National University of Singapore
Liu, Y.,
Originally presented on: 11/7/2013 12:30:00 - 12:45:00

DNA microarrays are powerful tools for the detection and identification of bacteria. However, commercially available DNA chips are very expensive and each chip often contains a very high density of information. In this study, we report a simple approach to activate inert hydrocarbon monolayers with UV light to fabricate DNA microarrays. Unlike like traditional UV microarrays which require reactive functional groups on the surface, our DNA microarray is built upon an inert layer of N,N-dimethyl-n-octadecyl(3-aminopropyl)trimethoxysilyl chloride silane (DMOAP). This layer is activated by UV (254 nm) just prior to the immobilization of oligonucleotide probes. Our XPS results show that new functional groups such as alcohol (C-O), aldehyde (C=O) and carboxylic acid (O-C=O) form on the surface after the UV exposure. Among them, aldehyde groups are responsible for the immobilization of amine-label oligonucleotide. By using this approach, we further optimize several factors affecting the immobilization of oligonucleotides to obtain a DNA microarray on which the density of oligonucleotide is up to 0.1molecular/nm^2.This DNA microarray is able to differentiate solventogenic *Clostridium sp.* such as *Clostridium acetobutylicum, Clostridium butylicum, and Clostridium beijerinki.*

(659b) Contact Time and pH Dependent Interactions of Chitosan in Aqueous Solutions

Lee, D. W., University of California, Santa Barbara
Lim, C., Pohang University of Science and Technology (POSTECH)
Israelachvili, J. N., University of California, Santa Barbara
Hwang, D. S., POSTECH (Pohang University of Science and Technology)
Originally presented on: 11/7/2013 12:45:00 - 13:00:00

Chitosan is a cationic biomacromolecule that is the structural component of insect and crustacean exoskeletons. Due to its biocompatibility and biodegradability, chitosan is one of the most widely used biopolymers for various applications such as fabrics, cosmetics, water treatment, and food processing. Here, we investigated the effects of contact time and pH on the interactions mediated by chitosan in aqueous solutions using a surface forces apparatus (SFA). The chitosan film was adhesive to mica surfaces at all pHs tested (3.0, 6.5, and 8.5) and showed maximum adhesion energy (W_{ad} ~4.8 mJ/m^2) at pH 3.0 after a contact time of 1 hr. Cohesion energy between chitosan films also showed a maximum value (W_{co} ~6.4mJ/m^2) at pH 3.0 after a contact time of 1 hr. However, at high pH, the solubility of chitosan decreased significantly, causing the chitosan-chitosan interaction to be repulsive at all contact times. The contact time and pH dependence of the adhesion/cohesion interaction indicates that the confinement induces molecular group reorientations and interdigitations. Our results and characterization of these interactions should be applicable to the use of chitosan in various biomedical applications.

(659c) Using Phospholipase A2 to Create Air-Stable Supported Lipid Bilayers

Han, C. T., National Taiwan University
Chao, L., National Taiwan University
Originally presented on: 11/7/2013 13:00:00 - 13:15:00

Supported lipid bilayer platform has a potential to be applied as anti-biofouling surface and biosensor interface because of the high bio-compatability and the ability to keep biomolecule's orientation for specific ligand-receptor interaction. However, conventional supported lipid bilayers may delaminate as they are exposed to air-water interface, causing the difficulty in their long-term storage and real-life application. Therefore, developing methods to maintain the integrity and fluidity of lipid bilayers after dried and rehydrated can significantly broaden the application. In this work, we observed that a type of pheripheral enzyme, phosholipase A_2 (PLA_2), can induce the formation of a branch-liked gel domain in the supported lipid bilayers. The branch-liked gel domains give aid to the preservation of lipid bilayers when it is exposed to air-water interface., The diffusivity of the rehydrated lipid bilayers measured by the fluorescence recovery after photobleaching technique was found to be comparable to the one before dried, indicating that the fluidity of the lipid bilayer is nearly unchanged. Our work suggests that the gel domain induced by PLA_2 may play an important role in the retention of supported lipid bilayers, which leads to a new methodology of air-stable lipid bilayers formation.

(659d) Specific and Nonspecific Oligonucleotide Interfacial Dynamics On Modified Silica Surfaces

Monserud, J., University of Colorado Boulder
Schwartz, D. K., University of Colorado Boulder
Originally presented on: 11/7/2013 13:15:00 - 13:30:00

Single-molecule total internal reflection fluorescence microscopy was used to observe the dynamic behavior of ssDNA oligonucleotides on modified silica surfaces. High throughput molecular tracking was used to determine >75,000 molecular trajectories. Nonspecific interactions occur where the oligonucleotides interact with a surface lacking the oligonucleotide's compliment strand covalently bound to the surface. These interactions were monitored using total internal reflection fluorescence microscopy (TIRF) and labeled probe oligonucleotides in solution. Specific oligonucleotide interactions occur when target molecules in solution hybridize to probe molecules attached to the surface. These interactions were monitored using labeled probe and target oligonucleotides, where the labels are a resonance energy transfer (RET) pair and two-channel TIRF. The use of RET provides a signal when a molecule transitions to a hybridized state. Using the molecular trajectories acquired through RET TIRF experiments, the rate constants of molecules transitioning between hybridized and melted states, as well as the fraction of molecules in a specific state that transition to each potential state, can be extracted.

(659e) Lipid Domain Pixelation Patterns Imposed By Homogeneous Nanosurfaces Fabricated By Electron Beam Lithography

Ogunyankin, M. O., University of California, Davis
Longo, M. L., University of California, Davis
Originally presented on: 11/7/2013 13:30:00 - 13:45:00

Lipid molecules self-assemble into a dual leaflet form of smectic liquid crystal with fascinating properties. With the exception of compositions near a critical point, coexisting lipid phases exist as microscopic domains owing to high line energy, and their size is non-uniform. Therefore, a significant challenge exists in forming uniformly sized lipid domains existing as single addressable elements that could be arrayed at high density and provide compatibility with current nanometer-scale device elements. Formation and arraying of these domain "pixels" in a high throughput manner could be used to create optical devices, biosensor arrays, drug delivery devices, and artificial organelles where the major advantage is that the positioning of the domain pixels could be dynamically controlled by applied forces, e.g. electric fields. This work describes a technique for forming nanometer-scale pixelated lipid domains that are self-organized into geometric patterns residing on a square lattice. In this process, a lipid multibilayer stack is deposited onto a silicon substrate patterned with a square lattice array of poly(methyl methacrilate) (PMMA) hemispherical features formed by electron beam lithography. Lipid domain patterns are shown to be confined to the flat grid between the hemispheres and comprised of connected and individual domain pixels. Analysis of lattices of varying sizes shows that domain pattern formation is driven by mechanical energy minimization and packing constraints. We demonstrate single lattice sizes and a gradient in lattice size varying from the micrometer to the 100 nm scale applicable to precise arraying, patterning, and transport of biomolecules that partition to lipid domains. We study the dynamic evolution of the pixelation patterns of the liquid-ordered (L_0) phase in coexistence with the liquid-disordered phase in the lipid multibilayers, and show that the L_0 phase exists transiently on the curvature-patterned sections of the substrate. This apparent metastability of the pattern derives from the high line energy of a pixelation pattern where a Boltzmann distribution shows near zero equilibrium partitioning of the L_0 phase in the patterned regions. In addition, we demonstrate that the dynamics of the pattern metastability can be harnessed by implementing some simple variations to the underlying curvature substrate. This work stands as a stepping stone toward a deeper understanding of biological membranes. The well-defined nature of the domain patterns provide proof-of-concept that curvature patterning of substrates can be used to direct bending and line energy to attain cell-like dynamics.

(659f) Predicting Protein Stability At Biotic/Abiotic Interfaces With a Coarse Grain Model

Wei, S., University of Michigan
Brooks, C. L. III, University of Michigan
Originally presented on: 11/7/2013 13:45:00 - 14:00:00

Protein stability and activity at a biotic/abiotic surfaces are key factors for reliable performance of bio-chip based techniques such as biosensors. However, a detailed structural understanding of the protein-surface interaction is limited when experimental techniques are employed due to their limitations of atomic resolution. All-atom simulations are capable of providing atomistic information of protein structure on surfaces, however, the enormously large time scale required to generate thermodynamic quantities comparable to experimental results significantly restricts its application in such systems. In this work, we develop a carefully parameterized coarse grain model to compare protein structure generated from fast simulations with protein stability information from Sum Frequency Generation (SFG) spectroscopy and protein activity tests. A maleimide functionalized Self-Assembled Monolayer surface (SAMs) is used in experiments, which is simulated as a moderately-hydrophilic surface. Two proteins beta-gal and NsfB are tethered with different tethering sites in either loop or helix regions with different orientations on the surface. Simulation using the same set-up parameters generated protein structures that are consistent with the experimental results of protein stability and activity. Further studies from both simulation and experiment on a the alpha-helical peptide cecropin P1 is also performed with tethering at either the N- or C-termini . The results are also consistent between simulation and experiment. tour findings suggest that one may predict the structure and stability of mutant sequences of the peptide to inform further experiments on these systems and nicely augment the existing experiments.

(659g) Understanding the Solubility of Triblock Copolymers With Supported Lipid Bilayers

Wang, J., Rice University
Segatori, L., Rice University
Biswal, S. L., Rice University
Originally presented on: 11/7/2013 14:00:00 - 14:15:00

The interactions of amphiphilic poly(ethylene oxide) - poly(propylene oxide) - poly(ethylene oxide) (PEO-PPO-PEO) triblock copolymers and lipid membranes have been characterized. The solubility of these triblock polymers in a lipid membrane can be changed with temperature leading to the ability to create a tunable insertion within the membrane. Here, we present our results when this system is changed from a free standing lipid membrane to a supported lipid membrane, in which one lipid leaflet is confined by a solid support. With the use of microcantilevers and fluorescence recovery after photobleaching (FRAP), we measure the solubility and interactions of triblock copolymers (F68 and F98) within supported lipid bilayers. Two factors are found to affect the lipid polymer interactions: temperature and number of repeating units in hydrophobic PPO block. For both polymers, the polymer solubility with the membrane was increased with higher temperature on microcantilevers. However, FRAP results show that the polymer with the longer PPO block, F98, significantly hinders the diffusion of lipids, suggesting a highly ordered and expanding structure of lipid bilayer. This conclusion is also confirmed by the copolymer induced surface stress change of lipid bilayer on microcantilevers, and further illustrates the various conformations that these polymers can take on when interacting a supported lipid membrane. These results are also in contrast with that found for free membranes, suggesting that the solid support affects the polymer-lipid interactions. A Langmuir isotherm model and a free area theory are applied to describe the lipid polymer interactions at the sensor surface, assess the binding affinity constants, and analyze the effect of triblock copolymers on lipid lateral diffusion.

(659h) Liquid Crystal Orientational Transformations Induced By Interfacial Biomolecular Interactions

Noonan, P. S., University of Colorado Boulder
Roberts, R. H., University of Colorado at Boulder
Schwartz, D. K., University of Colorado Boulder
Originally presented on: 11/7/2013 14:15:00 - 14:30:00

We have studied how biomolecular interactions at interfaces, in particular aptamer-ligand binding events, can induce orientational transformations in an adjacent bulk liquid crystal phase. Physical properties of liquid crystalline materials such as long-range orientational order, responsiveness to external stimuli, and optical anisotropy make them advantageous for use as transduction elements in sensing applications. Thus, we have demonstrated that aptamer-ligand binding events, involving small molecules, at a surfactant-laden aqueous/liquid crystal (LC) interface can trigger sensitive and selective LC orientational transformations. Mechanistic studies involving circular dichroism and resonance energy transfer measurements revealed that the LC reorientation was directly related to a conformational change of the aptamer upon ligand binding. Specifically, under conditions where aptamer-ligand binding induced a conformational change from a relaxed random coil to more intricate secondary structures that enclose hydrophobic nucelobases (eg. double helix, G-quadruplex); an LC orientational transformation from planar to homeotropic was observed. These observations improve our understanding of how relatively subtle changes in the hydrophobicity of biomolecules at interfaces can induce changes to the associated interfacial structure. Furthermore, we demonstrate the potential for a label-free LC-based detection system that can simultaneously detect the presence of both small molecules and nucleic acids.

(659i) Porous Nanofilm Biomaterials: Biomolecular Adsorption and Cellular Response

Wu, C., Yale University
Aslan, S., Yale University
Gand, A., Universite de Cergy-Pontoise
Wolenski, J., Yale University
Pauthe, E., Universite de Cergy-Pontoise
Van Tassel, P. R., Yale University
Originally presented on: 11/7/2013 14:30:00 - 14:45:00

Hydrogel-like biomaterials are often too soft to support robust cell adhesion, yet methods to increase mechanical rigidity (e.g. covalent cross-linking of the gel matrix) can compromise bioactivity by suppressing the accessibility / activity of embedded biomolecules. Nanoparticle templating is reported here as a strategy toward porous, layer-by-layer assembled, thin (ca. 100 nm) polyelectrolyte films i) of sufficient mechanical rigidity to promote initial cell adhesion, and ii) exhibiting significant adsorption of bioactive species within the pore space. Latex nanoparticles are incorporated during layer-by-layer assembly, and following EDC-NHS cross-linking of the polyelectrolyte film, are removed *via* exposure to tetrahydrofuran (THF). THF exposure results in only a partial reduction in film thickness (as observed by ellipsometry), suggesting the presence of internal pore space. The attachment, spreading, and metabolic activity of pre-osteoblastic MC3T3-E1 cells cultured on templated, cross-linked films are statistically similar to those on non-templated films, and much greater than those on non-cross-linked films. Laser scanning confocal microscopy and quartz crystal microgravimetry indicate a high capacity for bioactive species adsorption (ca. 10% of film mass) wthin nanoparticle templated films. Porous nanofilm biomaterials, formed *via* layer-by-layer assembly with nanoparticle templating, promote robust cell adhesion and exhibit high bioactive species loading – and thus appear to be excellent candidates for cell-contacting applications.

(659j) Inhibition of Bacterial Adhesion and Attachment Using Aerogels

Akbulut, M., Texas A&M University
Oh, J.,
Originally presented on: 11/7/2013 14:45:00 - 15:00:00

With outbreaks of foodborne illnesses occurring more frequently, it is becoming increasingly necessary to develop of transformative intervention strategies to prevent pathogen adherence to produce and to completely remove adhered pathogens from the surface of produce. In this presentation, I am going to talk about the ability of chlorotrimethylsilane (CTMS) functionalized silica aerogel to completely stop the adhesion of *Salmonella enterica* serovar Typhimurium and enterohaemorrhagic *Escherichia coli* (EHEC). Neither pristine nor CMTS-functionalized non-porous silica materials displayed similar levels of bacterial-attachment inhibition compared to the hydrophobic silica aerogel. To be specific, upon the EHEC exposure of 10^8 CFU /ml, the hydrophilic quartz surfaces had an EHEC density of $6.38\pm0.74\times10^4$ bacteria/cm^2 while the hydrophobic quartz had $7.34\pm0.44\times10^3$ bacteria/cm^2. The hydrophobic aerogel had no bacteria on them. The superior bacterial anti-adhesion behavior is ascribed to the negative Hamaker constant and repulsive van der Waals interactions for the bacteria-aerogel-water system. Overall, the realization of repulsive van der Waals interactions in a bacterial system is significant in the context of surface and interface science and coating design.

(665a) Colloidal Microstructure of Bacterial Biofilms By Direct Visualization With Confocal Microscopy

Stewart, E. J., University of Michigan
Younger, J. G., University of Michigan
Solomon, M. J., University of Michigan
Originally presented on: 11/7/2013 12:35:00 - 12:50:00

Microscale cellular clustering and intercellular separation of surface adherent biofilms of the opportunistic pathogen *Staphylococcus epidermidis* are found to be well-described by colloidal measures such as the cluster distribution and the radial distribution function. These measures are found to depend significantly on the environmental stress or antibiotic stress present as the biofilm grows under conditions of steady flow. Bacterial biofilms are bacterial cells encapsulated in an extracellular polymeric substance consisting of polysaccharides, DNA, and proteins. Since bacterial size is on the order of 1 μm, we view bacterial biofilms as a colloidal composite material with the bacterial cells as the particles and the extracellular matrix as a viscoelastic hydrogel. Here we probe the cellular microstructure of *S. epidermidis* biofilms at varying environmental stress conditions by combining high-resolution confocal microscopy with colloidal image analysis techniques that identify individual cells. The microstructural properties characterized are the local number density, the cluster distribution, the radial distribution function, and the fractal dimension of biofilm image volumes. Environmental stressors of high osmotic pressure (776 mM NaCl) and sub-lethal antibiotic dose (1.9 μg/mL vancomycin) decreased the average bacterial local number density 10-fold. Density phenotypes associated with low, medium, and high local number densities were identified in unstressed biofilms, while environmentally stressed biofilms only contained medium- and high-density phenotypes. All biofilms considered exhibited short-range clustering on length scales associated with cell division (~ 1 μm). However, on length scales of ~ 6 μm, biofilms with high-, and medium-density phenotypes contained longer range connectivity and structures characteristic of densely packed disordered colloidal suspensions, while low-density phenotype biofilms contained clusters of bacteria with fractal structure (d_f=1.7±0.1). This clustering and connectivity has implications for the mechanical properties of biofilms, similar to how the statistical configuration of particles determines the elasticity of colloidal gels.

(665b) Inverse Emulsion Stability in Shear Flow

Lazzari, S., ETH Zurich
Maggioni, G. M., ETH Zurich
Soos, M., ETH Zurich
Wu, H., ETH Zurich
Morbidelli, M., ETH Zurich
Originally presented on: 11/7/2013 12:50:00 - 13:05:00

Inverse emulsions (IE) are of paramount importance in several industrial areas, such as cosmetics, pharmaceutics and oil industry. Many studies focused on their preparation, stabilization and destabilization mechanisms[1]. Despite this, no investigation has been reported in the literature on the shear-induced aggregation of IEs. Therefore, the aim of this work is twofold: i) to explore the possibility of the shear-induced gelation in IEs and ii) to identify the key parameters that regulate the observed destabilization and gelation phenomena.

The employed IE systems consist of soft, water-swollen polymer-based particles at large particle contents, dispersed in an organic solvent and stabilized by a mixture of steric surfactants. It is found that at a fixed shear rate, the shear-viscosity of the IE systems first decreases with time (shear thinning), and then, after reaching a local minimum, it increases explosively. Cryo-SEM experiments reveal that along the shear history, the IE systems undergo competition between coalescence and aggregation, leading eventually to fractal gelation.

With respect to that of non-coalescing colloidal systems, the shear-induced gelation[2] of the IE systems results clearly from a different mechanism: coalescence slows down the gelation process by reducing the total occupied volume of the clusters. Only when the aggregation rate becomes faster than the coalescence rate, the gelation phenomenon becomes possible. Through the shear-induced gelation, we have identified the main parameters affecting the shear-stability of the IE systems, and their effects can be well interpreted by considering their role in altering the competition between coalescence and aggregation kinetics.

Literature:

1. I. Capek, Adv. Colloid Interface Sci., 2010, 156, 35-61

2. A. Zaccone, H. Wu, D. Gentili and M. Morbidelli, Physical Review E, 2009, 80

(665c) Driving Crystal Transformations and Assembling Colloidal Clusters Via Reprogrammable DNA Interactions

McGinley, J. T. III, University of Pennsylvania
Jenkins, I., University of Pennsylvania
Sinno, T., University of Pennsylvania
Crocker, J. C., University of Pennsylvania
Originally presented on: 11/7/2013 13:05:00 - 13:20:00

DNA-coated polymer microspheres have proven to be versatile building blocks for the study of directed self-assembly. DNA-directed self-assembly is governed by a multi-dimensional phase diagram, controlled by an *nxn* matrix of specific interactions between *n* species. A previous study demonstrated that the "as nucleated" crystal structure is not necessarily the minimum free-energy structure for the system. Transformations from the nucleated crystal structure to the ground state structure can occur if a suitable pathway exits between them. We show that the addition of competitive, soluble DNA strands to the system can *reprogram* the initial interaction matrix, and define a new minimum free energy configuration, thus driving a desired transformation. Using this concept, I have been able to reprogram both BCC and FCC crystal structures, by selectively destroying bonds, to produce high yields of colloidal clusters, consisting of a single particle surrounded by 8 or 12 particles, respectively. The clusters are initially formed as cubes (for BCC) and cuboctahedron (for FCC), but upon annealing they undergo transformations to the lower free energy structures of skew-cubes and icosahedron, respectively. We also employ reprogrammable interactions to drive reversible crystal transformations between FCC and BCC structures, with each transformation being accomplished simply by the addition of a soluble DNA strand.

(665d) The Importance of Non-Ideal Diffusion and Entropic Packing Effects in Brownian Aggregation of Hard Spheres

Kelkar, A. V., Purdue University
Franses, E. I., Purdue University
Corti, D. S., Purdue University
Originally presented on: 11/7/2013 13:20:00 - 13:35:00

Colloidal dispersions are ubiquitous. Hence, understanding their stability is an important scientific endeavor. Stability of colloidal dispersions is a key challenge in the manufacture of coatings, environmental pollution, formulation of food products, drug delivery, membrane fouling, inkjet printing, and flow assurance in oil and gas production. One of the destabilization mechanisms is perikinetic aggregation, where particle transport occurs by diffusion. This was first described by Smoluchowski who considered the steady-state aggregation of initially mono-disperse hard spheres with an initial uniform concentration profile. This model matches simulation results for aggregation rates only in dilute systems, i.e., for particle volume fractions, φ, less than 0.0005 [1]. However, for most practical applications, particle volume fractions can be as high as φ = 0.4. A limited number of theoretical [2] and semiempirical approaches [3] have been developed for perikinetic aggregation in concentrated systems. These models predict aggregation rates that are higher than the predictions of the Smoluchowski model by an order of magnitude. Nonetheless, understanding of the mechanism that leads to the enhancement of aggregation kinetics is still limited. Transient effects have been shown to increase the rates of perikinetic aggregation, because of the unsteady-state flux that drives aggregation at short times [1]. The aggregation rates predicted by this model, however, only match Brownian Dynamics Simulations (BDS) results for φ up to about 0.1. In addition to these effects, non-ideal particle diffusion and entropic packing effects are found to be important, especially at higher concentrations. The Smoluchowski model assumes that particle diffusion is driven by a concentration gradient. This is strictly valid only in the infinite dilution limit. The more general approach is to consider the non-ideal diffusion of particles driven by a chemical potential gradient. Additionally, the initial uniform concentration profile used in the Smoluchowski model becomes unrealistic at high volume fractions. Entropic packing effects lead to a local 'liquid-like' structure in the concentrated dispersions. We have developed a new model based on the `liquid-state' dynamical density-functional theory that accounts for transient, non-ideal particle diffusion, and packing effects in perikinetic aggregation. The predictions of the new model for hard-sphere dispersions are in excellent agreement with BDS results for the time dependence of monomer number density and the half-times of aggregation for φ up to 0.35, and with gelation times (at which the volume fraction reaches the limit of a freezing transition) reported by Lattuada [2]. Thus, transient, non-ideal particle diffusion, and packing effects result in enhancement of aggregation kinetics of hard spheres. The model is then extended to account for thermodynamic and hydrodynamic interactions whose effects are re-evaluated for concentrated dispersions. Predictions of the model with interactions also agree with BDS results. Additionally, the transition between perikinetic and shear-induced or orthokinetic aggregation is re-examined and found to depend not only on the Peclet number as thought earlier, but also on time and particle volume fraction.

[1] A. V. Kelkar, J. Dong, E. I. Franses, and D. S. Corti, J. Colloid Interface Sci., **389**, 188, (2013)

[2] M. Lattuada, J. Phys. Chem. B, **116**, 120, (2012)

[3] M. Heine and S. E. Pratsinis, Langmuir, **23**, 9882, (2007)

(665e) Electrostatic Assembly of Aqueous Colloids for Polymer Solar Cell Applications

Richards, J. J., University of Washington
Pozzo, D., University of Washington
Originally presented on: 11/7/2013 13:35:00 - 13:50:00

The recent development of aqueous dispersions of conjugated polymer/fullerene composite nanoparticles (CNPs) has enabled the use of electrostatic assembly (EA) of active layers for the production of organic photovoltaics. EA is the spontaneous deposition of a charged material onto an oppositely charge surface. CNPs are readily produced using a number of anionic and cationic surfactants (i.e. sodium dodecyl sulfate (SDS) and dodecyl trimethyl ammonium bromide (DTAB)). Therefore, their charge can be easily customized for electrostatic deposition on a wide range of charged surfaces. We have recently demonstrated the spontaneous EA of poly(3-hexylthiophene) (P3HT)/[6,6]-phenyl-C_{61}-butyric acid methyl ester (PCBM) CNPs onto poly(3,4-ethylenedioxythiophene):poly(styrenesulfonate) (PEDOT:PSS) coated indium tin oxide (ITO). PEDOT:PSS coated ITO substrates assume a negative charge when immersed in water. When these substrates are immersed in a solution containing P3HT/PCBM CNPs stabilized with DTAB, spontaneous EA of monolayers of P3HT/PCBM CNPs results. By tuning the concentration of DTAB in the bulk phase of P3HT/PCBM CNPs as to minimize competitive surfactant adsorption, the surface density of the nanoparticles can be tuned for a fixed dipping time. This process opens up the prospect of producing thin-film active layers composed of monolayers of fully optimized conjugated polymer/fullerene CNPs. Because photovoltaic modules are composed of number of other materials which also have pH and salt dependent charging behavior in water, layer-by-layer (LBL) assembly of entire photovoltaic modules is possible. LBL processing for photovoltaic modules would be an inexpensive and scalable processing technique that could be applied to substrates with complex form factors (e.g. woven fibers) at low temperatures and without the release of volatile organic solvents during the drying process.

(665f) Competitive Adsorption Of Sodium Dodecyl Sulfate and Hydroxypropyl Methylcellulose On Drug MICRO-Particles

Takhistov, P., Rutgers, The State University of New Jersey
Huynh, P., Rutgers University
Originally presented on: 11/7/2013 14:00:00 - 14:15:00

In this work we report experimental and theoretical study of the adsorption of sodium dodecyl sulfate (SDS) and hydroxypropyl methylcellulose (HPMC) on griseofulvin particles, a critical step in many pharmaceutical processes. As the concentration of SDS increased, ζ potential of griseofulvin particles decreased, indicating strong adsorption of SDS. We show that SDS adsorption occurs via a single monolayer mechanism well-described in terms of Langmuir and Dubinin-Radushkevish isotherms. Adsorption of HPMC on griseofulvin is determined by two processes that occur simultaneously: self-association of the polymer and polymer-drug adsorption. One of the processes will dominate, depending on polymer concentration. Presence of the adsorbed surfactant adds more complexity to the phenomenon due to complex formation between the surfactant and polymer. We linked observed changes in the adsorption behavior to the phase behavior of polymeric solution, using de Gennes-Teraoka scaling theory Adsorption energy was evaluated theoretically by means of atomistic level force field models, where single point energy calculations were done on the individual molecules and energy-minimized conformations of their complexes The results from molecular calculations are in good qualitative agreement with our experimental findings. The obtained data help to understand the adsorption mechanisms of surfactant and polymer on hydrophobic solid particles, allowing to optimize the formulation of aqueous suspensions of these particles to ensure their stability.

(665g) Development of Particle Packed Layer in the Drying Process of Latex Paints

Komoda, Y., Kobe University
Yamane, T., Kobe University
Suzuki, H., Kobe University
Hidema, R., Kobe University
Originally presented on: 11/7/2013 14:15:00 - 14:30:00

In order to minimize the amount of volatile organic compound (VOC) in the coating industry, latex paints are frequently used as the substitution for VOC based paints. In latex paints, submicron sized resin particles are dispersed in water and, the resin particles are packed during drying and finally formed a film over the coated material after drying. One of the important physical properties of latex paints is a glass transition temperature, Tg, of dispersed resin particles. That is to say, the resin particles will produce uniform film when drying at the temperature higher than Tg, while the film becomes brittle at a lower drying temperature. As for the packing process of dispersed resin particles, the resin particles are first accumulated in the region near the drying surface and the packing of particle layer developed in the depth direction from the surface. This process of the particle packing layer formation have been directly observed using CryoSEM technique by Scriven or Francis. On the contrary, it is well known in the drying process of particulate dispersion system that the boundary of highly accumulated wet region and porous particle packed region can be clearly observed as a drying front and the drying front proceeds from the outer surface to the internal part of the material to be dried. The transport phenomena in the porous particle packed layer significantly affects the drying process of wet material inside of the drying front. So, we have experimentally investigated the drying process of latex paints focusing on the development of dried particle packed layer.

The latex paints used in this study contain two kinds of dispersed resin particles having different Tg value. One is Tg = -11C°, which means the dispersed particles are flexible under a room temperature, producing uniform film after drying. The other is Tg = 108C°, which means solid like resin particles are dispersed in water, resulting in a brittle dried film. The concentration of resin particles in each latex paint was adjusted to 23vol%. We have carried out direct observation of one-dimensional drying process of latex paint in a one-side opened square cell composed of two glass plates with the separation of 80um. Three sides of a square thin glass plate (18mm x 18mm) were stuck onto the other glass plate using double-side tapes, the thickness of which is 80um. The latex paint inserted into the thin space having the dimensions of 15mm x 15mm surrounded by top and bottom glass plates and surrounded by the double side tapes will be dried only from the open side. Then, the drying process of the latex paint was considered as uni-directional drying. The test cell placed on a transparent glass heater and then the drying process of the latex paint was observed using digital microscope with dark field observation mode. The temperature of the heater was controlled as 25 and 40C°. In this observation, we could clearly observed that the drying front proceeded uniformly in the direction normal to the open side of the test cell and measured the thickness of the particle packed layer as a distance between the open side to the drying front. The moving speed of the drying front was also calculated from the time variation of particle layer thickness.

The moving speed of the drying front was initially very large and then gradually decreased with time. In the initial stage of drying process until the thickness of the particle layer became much larger than the glass plate separation, the drying process could not considered as uni-directional. Additionally, the air flow near the open side significantly enhanced the evaporation of water because of thin particle packed layer. As for the effect of Tg on the drying process, the moving speed of the drying front for the latex paint of Tg = 108C° was much larger than that of Tg = -11C°. We also have carried out rheological measurement of latex paint in order to elucidate the difference in particle aggregation. As a result, the latex paint of Tg = 108C° were well dispersed showing a small storage modulus in a linear viscoelastic regime and a lower viscosity, while more aggregated latex paint of Tg = -11C° indicated a high viscosity at relatively small shear rates and a larger storage modulus at an oscillatory shear application with small shear strains. Therefore, since the aggregated structure of resin particles hold water in the structure and deteriorates water evaporation from the latex paint, the moving speed of the drying front at Tg = -11C° became relatively slow. In the following drying stage, the gradual decrease in the moving speed of the drying front was reasonably described by the development of particle packing layer. That is to say, as increasing in the thickness of particle packed layer, the water evaporated at the drying front required long time to diffuse out of the particle layer.

Since the development process of particle packer layer was governed by the diffusion of water vapor through porous particle packed layer, the moving speed of the drying front should be described by the one-dimensional Fick's diffusion equation in a porous material. Additionally, the permeation of water vapor though a particle packed layer was modeled by Cozeny-Karman equation introducing void fraction of particle layer. Therefore, we have obtained the formulation where the product of the moving speed of drying front and the thickness of particle layer is constant, and the void fraction in the particle layer can be determined from the constant value. As a result, we found that the decreasing behavior in the moving speed of drying front have been successfully correlated by the model established for each latex paint at different drying temperatures. Also, the packing density of the particle packed layer at Tg = -11C° was much larger than that for the packing density of maximum packing layer of solid spheres, which indicates flexible resin particles are deformed and then packed in the drying process. The higher packing density at the higher drying temperature indicates the resin particles seemed to have been well deformed and more tightly packed. On the other hand, the packing density of solid-like resin particles of Tg = 108C° was much smaller than the maximum packing density of solid sphere compensating for large clack formation in a particle packed layer.

(665h) Effect of Polyacrylamide Polymers On Floc Size and Rheological Behaviour of Papermaking Suspensions

Nasser, M.,

Originally presented on: 11/7/2013 14:30:00 - 14:45:00

The influence of cationic and anionic Polyacrylamides (PAM) on the floc size, and rheological behaviour of papermaking suspensions has been investigated. Cationic C496 and C492 and anionic A130LMW PAMs were used in this study. At optimum PAM concentrations, the floc sizes were larger in the presence of the anionic PAM than the cationic PAMs due to the repulsive forces between the anionic PAM and particle surface, leading to the formation of open structure flocs of less effective density. The results also show that increasing the cationic surface charge from 10 % to 35 % reduces the floc size and increases the effective floc density.

On the other hand, the magnitude of yield stress, elastic and viscous moduli are strongly dependent upon the floc structure; with greater values being observed for the cationic PAM than for the anionic PAM flocculated papermaking suspensions. The difference in the shear sensitivity may be attributed to floc structure-related adsorption. It was found that the yield stress and viscoelastic behaviour are important factors that should be considered when selecting the type of polyelectrolyte for use in a particular solid liquid separation. This work has been very successful in establishing the link between greater particle attraction and higher yield stress, and elastic and viscous modulus.

(665i) Monte Carlo Simulations of Nanofluids

Lotfizadeh, S., The Pennsylvania State University
Matsoukas, T., The Pennsylvania State University
Originally presented on: 11/7/2013 14:45:00 - 15:00:00

Adding nanoparticles to a fluid results in considerable improvement of the fluid thermal properties but the magnitude of the enhancement and the physical mechanism by which this enhancement occurs are open to debate. Maxwell's mean field theory, the standard model for these systems, predicts two limiting bounds for this enhancement depending on whether the dispersed phase (nanoparticles) has higher or lower conductivity than the continuous phase (fluid). This also suggests that the enhancement depends on the configuration and degree of aggregation of the particles, though such effects are not captured by Maxwell's theory. In our past work we demonstrated that aggregated nanoparticles have higher conductivity compared to fully dispersed particles at the same volume fraction. Here we employ a computational model to explain the experimental results and quantify the effect of aggregation on thermal conductivity.

We generate model clusters in a base fluid and evaluate the thermal conductivity of the system using a Monte Carlo algorithm. Briefly, we discretize the aggregate and embed a large number of random walkers whose displacement is biased by the thermal conductivity in the vicinity of their location. The thermal diffusivity and heat conduction are then obtained from the average mean-squared displacement.We vary the degree of aggregation in a systematic way and find the conductivity increases up to 30% above the lower Maxwell bound. We attribute the increase the formation of a neck between neighboring particles and show that a pair of partially merged particles has 15% higher conductivity than either the full dispersed or fully merged particles. Further increases in conductivity are achieved by allowing the particles to be hollow such that fluid is contained in the core. Clusters formed by hollow particles exhibit even higher enhancements, though in all cases these remain below the upper Maxwell bound. The results suggest that manipulation of the microstructure of particle-fluid systems has the potential to produce enhanced thermal properties and provides an explanation for the "anomalous" enhancements often reported by experimentalists who do not control the state of aggregation in the suspension.

(667a) Effect of Langmuir Constants On Hydrate Equilibrium Calculations

Tsimpanogiannis, I. N., National Center for Scientific Research
Economou, I. G., National Center for Scientific Research Demokritos
Stubos, A. K., National Center for Scientific Research "Demokritos"
Originally presented on: 11/7/2013 12:30:00 - 12:47:00

Under appropriate conditions of pressure and temperature water molecules can self-assemble to form cavities that are stabilized by the presence of guest molecules that can fit within the cavities. These non-stoichiometric, solid (ice-like), inclusion materials are known as clathrate hydrates and have significant applications in science and engineering. Hydrate equilibrium at the continuum (i.e., macroscopic) level is described by a theory based on statistical mechanics that has been developed, initially by van der Waals and Platteuw [1], and subsequently modified by others (e.g., see the extended review by Sloan and Koh [2]). In this type of theory the cavity occupancies (i.e., the fraction of cavities occupied by the guest molecules) are described by Langmuir-type functions of the gas fugacity. The Langmuir constants that appear in these functions are temperature, guest, as well as, cavity-type-dependent, and affect the accuracy of the calculations. Therefore, it is essential to develop dependable and accurate methods for the calculation of the Langmuir constants. Traditionally, two approaches are followed in order to calculate the Langmuir constants:

(i) An intermolecular potential is chosen to describe the water-guest interactions (e.g., the Lennard-Jones, the square-well, or the Kihara potential), and the potential parameters are obtained by a fitting procedure using available experimental hydrate equilibrium data. Following this approach results in very good agreement with hydrate equilibrium predictions (within the range of development of the method), however, the values for the potential parameters are significantly different than those calculated when using experimental data of second virial coefficients or transport properties to obtain the potential parameters. Therefore, extrapolation to hydrate equilibrium predictions outside the range of development becomes highly problematic.

(ii) The interaction energies between the guest molecules and the water molecules that form the hydrate cavities are calculated using ab initio methods [3]. The ab initio calculation results, subsequently, are fitted to an empirical formula (e.g., the Lennard-Jones plus a Coulombic charge-charge formula) in order to obtain the potential parameters [4]. The accuracy of calculations of ab initio potential energy surfaces depends on the level (e.g., HF, MP2, MP3, MP4, etc.) of the simulations, and the basis sets used. Increasing the complexity of the calculation, results in increasing significantly the computational cost.

Alternatively, the following approach could be used. Hydrate formation can be simulated as a special case of a process of gas adsorption in a solid matrix. Such processes have been simulated with Monte Carlo techniques with great success. The hydrate cavity occupancies, thus, could be calculated from the Monte

Carlo simulations at various temperatures and pressures, and the results can be fitted to Langmuir-type functions [5]. Such temperature and pressure-dependent functions, once they are obtained, can be introduced to models that use the van der Waals and Platteuw theory for hydrate equilibrium predictions. However, the hydrate equilibrium predictions of such macroscopic models seem to be very sensitive to the values of the Langmuir constants used. As a result of such sensitivity, other constants of the macroscopic thermodynamic models need to be re-adjusted in order to increase the accuracy of the calculations. In the current study we perform a detailed analysis of the sensitivity of the hydrate equilibrium calculations on the values used for the Langmuir constants. We attempt to delineate the boundaries of the values of the Langmuir constants obtained from Monte Carlo simulations that can be used by van der Waals–Platteuw–type models without having to re-adjust any other of the model parameters.

References

[1] van der Waals, J. H.; Platteeuw, J. C. *Adv. Chem. Phys.* **1959**, *2*, 1-57.

[2] Sloan, E. D.; Koh, C. A. Clathrate Hydrates of Natural Gases, 3rd edition, Taylor & Francis. CRC Press, 2008.

[3] Klauda, J. B.; Sandler, S. I. *J. Phys. Chem. B* **2002**, *106*, 5722-5732.

[4] Sun, R.; Duan, Z. *Geochim. Cosmochim. Acta* **2005**, *69*, 4411-4424.

[5] Papadimitriou, N. I.; Tsimpanogiannis, I. N.; Papaioannou, A. Th.; Stubos, A. K. *J. Phys. Chem. B* **2008**, *112*, 10294-10302.

(667b) Self-Assembly of Aqueous Surfactants On Patterned Surfaces: Insights From Coarse-Grained Simulations

Suttipong, M., The University of Oklahoma
Grady, B. P., University of Oklahoma
Striolo, A., University of Oklahoma
Originally presented on: 11/7/2013 12:47:00 - 13:04:00

The adsorption and self-assembly of surfactants are ubiquitous processes in several technological applications. From a fundamental point of view, understanding such processes is also important because it can have profound implications in the manufacture of nano-structured materials using bottom-up strategies. Although a lot is known about the adsorption of surfactants on homogeneous flat surfaces, limited information is available, in quantifiable terms, regarding the adsorption of surfactants on heterogeneous surfaces. To complement and enrich our recent experimental observations [Langmuir 27 (2011) 6091], we report here our first results obtained using dissipative particle dynamics simulations for the adsorption of model surfactants on patterned flat surfaces. The results show that the surface pattern can be used to direct the morphology of the self-assembled aggregates (flat monolayers, hemi-cylinders, hemi-spheres, hollow hemi-cylinders, and irregular structures have all been observed using the same surfactant but different surface patterns). We will discuss how our predictions could be tested experimentally, and how, if reliable, they could be used for the design of new materials.

(667c) Equations of State for Hard Polyhedra

Irrgang, M. E., University of Michigan Ann Arbor
Engel, M., University of Michigan
Glotzer, S. C., University of Michigan
Originally presented on: 11/7/2013 13:04:00 - 13:21:00

The role of shape in entropy-driven self-assembly has recently been highlighted in computer simulations of hard anisotropic particles. A rich diversity of crystal and other solid-like phases has been demonstrated in particular for hard polyhedra [1-5]. Hard polyhedra and other simple shapes can be good models for nano crystalline colloids when interactions are dominated by geometry. Though advancements in the synthesis of such nanoparticles[7] are expanding our ability to create novel fluids and structured hierarchical materials, our understanding of the thermodynamics of anisotropic particles is still limited. Here we present equations of state for systems of hard polyhedra spanning the low-density fluid to high-density solid states, obtained numerically from equilibrium Monte Carlo simulations. We compare the results to various forms of equations of state from free volume and scaled particle theory. We discuss trends in the behavior for different shapes, and show some general features common to all systems.

[1] Haji-Akbari, A. et al., 2009. Disordered, quasicrystalline and crystalline phases of densely packed tetrahedra. Nature, 462(7274), pp.773–7.

[2] Agarwal, U. & Escobedo, F. A., 2011. Mesophase behaviour of polyhedral particles. Nature Materials, 10(3), pp.230–235.

[3] Ni, R. et al., 2012. Phase diagram of colloidal hard superballs: from cubes via spheres to octahedra. Soft Matter, 8(34).

[4] Damasceno, P.F., Engel, M. & Glotzer, S.C., 2012. Predictive Self-Assembly of Polyhedra into Complex Structures. Science, 337(6093), pp.453–457.

[5] Damasceno, P.F., Engel, M. & Glotzer, S.C., 2012. Crystalline assemblies and densest packings of a family of truncated tetrahedra and the role of directional entropic forces. ACS Nano, 6(1), pp.609–14.

[6] Henzie, J. et al., 2012. Self-assembly of uniform polyhedral silver nanocrystals into densest packings and exotic superlattices. Nature Materials, 11(2), pp.131–7.

(667d) Formation of Interconnected Morphologies of Symmetrical Block Copolymer/Nanorod Comoposites Under Cylindrical Confinement: A Coarse-Grained Molecular Dynamics Study

Park, J. H., Cornell University
Joo, Y. L., Cornell University
Originally presented on: 11/7/2013 13:21:00 - 13:38:00

Our recent experiment and simulation on confined self-assembly of block copolymer (BCP)/nanoparticle (NP) composites in electrospun nanofiber have proven to be a useful guidance for fabrication of the composite materials with ordered nanostructures. Specifically, we have examined how nanoparticle presence can influence the formation of very different asymmetrical BCP morphologies under cylindrical confinement when compared with that of pure asymmetrical BCP, both qualitatively and quantitatively. Using the same coarse-grained molecular dynamics tool, we have further extended our study to self-assembly of nanorods with BCP under cylindrical confinement. Unlike nanoparticles, nanorods can readily be interconnected with each other, and also induce asymmetrical BCP morphologies at much lower loading that nanoparticles do. Such interconnected nanorods, when incorporated within nanofibers, have great potential to provide highly conductive pathways for energy applications, such as battery electrodes and separators. Symmetric BCP under cylindrical confinement with nanorod aspect ratio of 1, 5, and 10 are examined with three different types of nanorod-BCP attraction: a) neutral nanorod, b) *A* (wall-attractive phase)-attractive nanorods, and c) *B* (wall-repulsive phase)-attractive nanorods. As one may have expected, much less loading of longer nanorods are required to observe formation of radially and axially interconnected nanorods within the cylinder. The interconnected nanorods often, but not always, induce asymmetrical BCP morphology as well. From this trend, we were able to extract a dimensionless parameter, D/N, or the polymer domain diameter over the nanorod aspect ratio, to be correlated to the onset of nanorod interconnectivity and formation of asymmetrical BCP morphology. By developing a phase diagram with respect to D/N, one can predict the formation of desired BCP morphology, nanorod placement, and it connectivity within cylindrical confinement.

(667e) Parameterization of Coarse-Grained Models for Simulation of Self-Assembly: the Role of Molecule Rigidity

Lee, M. T., Rutgers, The State University of New Jersey
Vishnyakov, A., Rutgers, The State University of New Jersey
Neimark, A. V., Rutgers, The State University of New Jersey
Originally presented on: 11/7/2013 13:38:00 - 13:55:00

Micellization of surfactant solutions is a ubiquitous phenomenon in natural systems and technological processes, and its theoretical description is a cornerstone problem in physical chemistry of colloidal systems. However, successful attempts of quantitative modeling confirmed by experimental data remain limited. We devised a novel approach to parameterization of coarse-grained models: binary interaction parameters are fitted to the infinite dilution activity coefficients of binary solutions formed by reference compounds that represent coarse-grained fragments of surfactant molecules. Using this protocol, we derived dissipative particle dynamics (DPD) models of several surfactants of different chemical structures and obtained a quantitative agreement with experimental critical micelle concentration (CMC) and aggregation number (AN).

Another factor determining micellization in surfactant solutions is the rigidity of the molecules. For the first time, we present a systematic study of chain rigidity effect on CMC and AN using DPD simulations. Molecule rigidity was controlled by second-neighbor (?1-3?) harmonic bonds, or the harmonic angle potential between the nearest neighbor bonds. Compared to flexible molecules with the nearest neighbor bonds being the only type of bonded interactions, rigid molecules exhibited a lower critical micelle concentration, larger and better-defined micelles. By varying the strength of head-tail repulsion and chain stiffness, we constructed two-dimensional diagrams presenting how the critical micelle concentration and aggregation number depend on these parameters. We found that solutions of flexible and rigid molecules that exhibited approximately the same critical micelle concentration could differ substantially in the micelle size and shape depending on the chain stiffness. With the increase of surfactant concentration, primary micelles of more rigid molecules were found less keen to agglomeration and formation of non-spherical aggregates characteristic to flexible molecules.

Overall, our parameterization methodology can be recommended for studies of self-assembly and dynamics in a wide range of soft matter systems. However, rigidity has to be accurately accounted for in coarse-grained modeling of self-assembly.

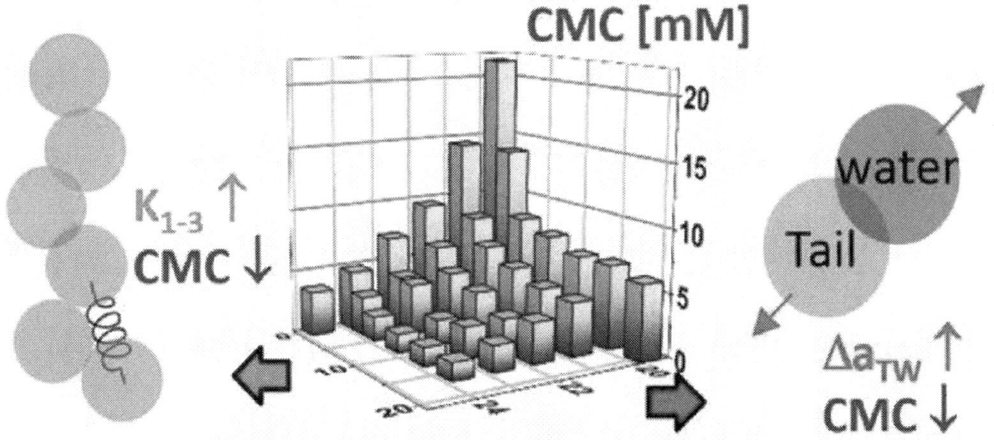

(667f) New Theory for Associating Fluids: Effects of Steric Hindrance, Ring Formation and Double Bonding

Marshall, B. D. II, Rice University
Chapman, W. G., Rice University
Originally presented on: 11/7/2013 13:55:00 - 14:12:00

Equations of state based on Wertheim's first order perturbation theory (TPT1) have proven to be a powerful tool in the modeling of associating fluids. In TPT1 the effect of bond angle is neglected such that the equation of state for associating molecules with a bond angle of 60 degrees is identical to that for associating fluids with bond angles of 180 degrees; that is, the effect of steric hindrance between association sites is neglected. Also neglected in TPT1 is the ability for molecules to associate into rings and double bond. For large bond angles these assumptions are valid; however, we show through molecular simulation that as bond angles are decreased steric hindrance between sites, association into rings and the possibility of double bonding must all be accounted for.

To model these small bond angle systems we develop a new molecular theory which captures each one of these small bond angle effects (steric hindrance, rings, double bonds). We explicitly include the effect of bond angle in each contribution to the free energy to develop an equation of state which is accurate over the full range of bond angles. We show by comparison to molecular simulation that the theory accurately predicts the pressure, internal energy and distributions of molecules in the various types of associated clusters (chains, double bonded, 3-mer rings, 4-mer rings, etc…). Further, we use the theory to demonstrate the effect of bond angle on the phase diagram of patchy colloid fluids.

(667g) Prediction of Phase Behavior of Strongly Associating Fluids By Explicit Consideration of Transitions in Local Fluid Structures

Chen, W. L., National Taiwan University
Hsu, C. C., National Taiwan University
Lin, S. T., National Taiwan University
Originally presented on: 11/7/2013 14:12:00 - 14:29:00

Accurate modeling of phase behavior of acetic acid has been of great interest both in academia and industry. The challenge is to accurately describe the various local structures (hydrogen bonding configurations) and the transition in the distribution of local structures associated with phase transitions. In this work we use the PR+COSMOSAC EOS to predict the phase behaviors of fluids containing acetic acid. The PR+COSMOSAC EOS does not use any species dependent parameter and does not require input of experimental data. However to accurately model strongly associating fluids, it is necessary to identify the important local fluid structures such as the monomer, cyclic dimer, chain fragment of acetic acid and the cross dimer of acetic acid and alcohol. The transformations among these structures are considered by chemical reactions. With the explicit inclusions of these associating structures the PR+COSMOSAC EOS is as accurate as the mod-UNIFAC+ Hayden-O'Connell EOS in describing the vapor liquid equilibrium of 19 binary mixtures (447 data points, temperature range from 293.15 to 426.39) is 6.49% (AARD-P%) in pressure and 2.22% (AAD-y%) in vapor composition, which is similar to those from the mod-UNIFAC+HOC 5.16% and 1.87%. The root-mean square error in predicting liquid composition in liquid-liquid equilibrium of 5 binary systems (51 data points, temperature range from 289.15K to 337K) is found to be 0.091, which is more accurate than that from mod-UNIFAC, 0.173. Our results show that the phase behavior of strongly associating systems can be predicted accurately by explicit consideration of the significant local fluid structures.

(667h) Molecular Insights Into Diphenylalanine Nanotube Assembly: All-Atom Simulations of Oligomerization

Jeon, J. H., University of California Santa Barbara
Mills, C. E., University of California Santa Barbara
Shell, M. S., University of California, Santa Barbara
Originally presented on: 11/7/2013 14:29:00 - 14:46:00

The simple diphenylalanine (FF) peptide self-assembles in aqueous solutions into nanotubes (FFNTs) with remarkable strength, thermal stability and other promising physical properties. FFNTs have found use in many applications, including as sacrificial templates and scaffolds for structuring inorganic materials, and as new hydrogels and biosensors. FFNTs can also form on substrates using vapor-deposition in vacuum, to create dense FF-nanoforests that have already found use in super-hydrophobic surfaces and high-performance supercapacitors. However, little is known about assembly mechanisms of FFNTs or the forces underlying their stability.

Here, we use detailed molecular simulations to study the early self-assembly stages of FF in water and in vacuum. First, we perform a variety of molecular dynamics (MD) simulations on small-oligomer formation in water and assess the balance of hydrogen bonds, electrostatic interactions, and side chain aromatic or hydrophobic forces, as well as the emergent structural motifs. We find that while electrostatic interactions steer FF peptides into more ordered dimers and trimers, the hydrophobic side chain interactions play a strong role in determining the structures of larger oligomers. However, capped (uncharged) versions of the peptide display dramatically different assembly behavior, and emphasize the importance of amphiphilicity to FF assembly. By comparing these results to simulations of the experimental FFNT X-ray crystal structure, we propose that the early structural templates to be formed are more amphiphili-like, rather than the salt-bridge-stabilized hexamer ring motif that has been proposed in the literature.

Second, we study the assembly of FF peptides in vacuum. During the vaporization process, the linear backbone of FF peptides is cyclized, and so its chemical structure is distinct from that in solution. We find that cyclo-FF molecules indeed have strong preferences to form vertically aligned ladder-like structures stitched together by hydrogen bonds. Such structures are reminiscent of the solution-phase oligomers, but appear much more stable. Importantly, the obtained ladder structures have some structural similarities with the crystal structure of cyclo-FF nanowires, although future work mush address how these ladders pack in a way to form nanotubes. In total, our results suggest novel pathways for assembly of FF in different environments and offer plausible explanations for the interactions that drive their unique properties.

(667i) Effect of Solvent Additives On the Self-Aggregating Properties of Phenyl-C61-Butyric Acid-Methyl-Ester (PCBM)

Tummala, N. R., The University of Oklahoma
Risko, C., Georgia Institute of Technology
Bredas, J. L., Georgia Institute of Technology
Originally presented on: 11/7/2013 14:46:00 - 15:03:00

In organic photovoltaics, the morphology at the interface between the electron-rich (donor) and electron-deficient (acceptor) materials in a bulk hetero-junction (BHJ) film directly impacts the critical processes of exciton dissociation and charge separation; on the other hand, aggregates of the individual components influence the charge-carrier mobilities in the bulk-like areas of the active layer. Solvent additives with high boiling points have been found to improve the efficiency of organic solar cells[1, 2] by aiding in the formation of improved morphological or topological[3-5] features within the active layer of the BHJ. Here, we aim to understand how solvent additives induce self-aggregation in solution, prior to spin-casting, through molecular dynamics simulations with force fields previously validated within our group. We evaluate properties like aggregation numbers, energies of aggregates of different sizes, cluster-size distributions in various solvents, and cluster shapes. By understanding the aggregation of PCBM clusters in chlorobenzene with and without solvent additives (*e.g.*, 1-chloronaphthalene and 1-8-diiodooctane), we determine the role of solvent additives in the self-aggregation processes and PCBM-PCBM orientations. We also evaluate the energetic landscape of a PCBM dimer in multiple solvent / solvent additive environments to investigate how local molecular packing affects the ordering of small-to-large clusters. These studies shed light on the morphological nature of the PCBM domains, an important feature that affects the power conversion efficiency of organic solar cells.

References

[1] J. K. Lee, W. L. Ma, C. J. Brabec, J. Yuen, J. S. Moon, J. Y. Kim, K. Lee, G. C. Bazan, A. J. Heeger, J. Am. Chem. Soc. 2008, 130, 3619.

[2] J. S. Moon, C. J. Takacs, S. Cho, R. C. Coffin, H. Kim, G. C. Bazan, A. J. Heeger, Nano Lett. 2010, 10, 4005.

[3] F. Etzold, I. A. Howard, N. Forler, D. M. Cho, M. Meister, H. Mangold, J. Shu, M. R. Hansen, K. Müllen, F. Laquai, J. Am. Chem. Soc. 2012, 134, 10569.

[4] L. Ye, S. Zhang, W. Ma, B. Fan, X. Guo, Y. Huang, H. Ade, J. Hou, Adv. Mater. 2012, 24, 6335.

[5] X. Liu, S. Huettner, Z. Rong, M. Sommer, R. H. Friend, Adv. Mater. 2012, 24, 669.

(673a) Electrocatalysis of Total Water Splitting On Perovskite Oxides and Oxynitrides

Montoya, J. H., University of South Carolina
Vojvodic, A., SLAC National Accelerator Laboratory
Norskov, J. K., Stanford University
Originally presented on: 11/7/2013 12:30:00 - 12:50:00

Perovskite oxides and oxynitrides have recently been the subject of multiple screening studies (Castelli et al, Persson et al) in order to determine which materials have a bandstructure conducive to overall water splitting. However, such screening studies neglect the effect of kinetic overpotentials for oxygen evolution (OER) and hydrogen evolution (HER). In the present work, we use density functional theory (DFT) calculations to describe the limiting potentials for both OER and HER on the 100 facet of various perovskite materials. Our results are found to be compatible with previous theoretical descriptions of oxide materials for total water splitting such as those proposed by Rossmeisl et al. More importantly, we demonstrate that perovskite photoelectrocatalysts with bandgaps well-suited for total water splitting may have kinetic limitations that make co-catalysts necessary. We also describe the theoretical activity of a variety of perovskite compounds for water splitting in order to identify active candidate materials for electrodes in photoelectrocatalytic or electrocatalytic water splitting devices.

(673b) Strontium Cobalt Ferrite Perovskite Materials As Anode Catalysts for Solid Oxide Fuel Cells: Effect of Ce Substitution At the A-Site

Fuller, A.,
Choi, H., The Ohio State University
Davis, J.,
Co, A.,
Ozkan, U. S., The Ohio State University
Originally presented on: 11/7/2013 12:50:00 - 13:10:00

Solid Oxide Fuel Cells (SOFCs) are direct electrochemical energy conversion devices considered to have very high prospects as an alternative source of power generation, due to their high efficiency and fuel flexibility. Their high operating temperature makes it possible to internally reform hydrocarbons for the production of hydrogen fuel, negating the need for expensive external reformers. The option of using the exhaust gases for cogeneration of power can also improve the efficiency of SOFCs to almost 80% [1]. However, the state-of-the-art Ni-YSZ (yttria-stabilized zirconia) anode is highly prone to coking in the presence of hydrocarbons and is readily poisoned by even ppm levels of sulfur in the fuel [2]. Perovskite-type oxides of the type ABO_3 have shown promise as anode electro-catalysts for SOFCs. Incorporation of a variety of elements into the A- and B-sites of these materials can be used to tailor their properties and make them active for oxidation reactions [3]. The present work is aimed at examining the effect of cerium substitution at the A-site of $SrCo_{0.2}Fe_{0.8}O_{3-\delta}$perovskites as sulfur-tolerant, anode catalysts for solid oxide fuel cells. The material's stability under typical anode conditions will be evaluated. Its catalytic activity for methane oxidation, its tolerance to sulfur poisoning along with its performance in an actual button cell will also be discussed.

References:

1. Yokokawa, H., et al., *Fundamental mechanisms limiting solid oxide fuel cell durability*. Journal of Power Sources. **182**(2): p. 400-412.

2. Matsuzaki, Y. and I. Yasuda, *The poisoning effect of sulfur-containing impurity gas on a SOFC anode: Part I. Dependence on temperature, time, and impurity concentration.* Solid State Ionics, 2000. **132**(3-4): p. 261-269.

3. Voorhoeve, R.J.H., et al., *Perovskite Oxides: Materials Science in Catalysis*. Science, 1977. **195**: p. 827-833.

(673c) Conversion Of Solar Into Chemical Energy On Plasmonic Metal Nanostructures

Andiappan, M., University of Michigan
Christopher, P., University of California, Riverside
Xin, H., University of Michigan
Linic, S., University of Michigan
Originally presented on: 11/7/2013 13:10:00 - 13:30:00

We will show that composite photo-catalysts combing plasmonic metallic nanoparticles of noble metals and semiconductor nanostructures exhibit improved photo-chemical activity compared to conventional photo-catalytic materials.[1,2] We will also show that plasmonic silver nanoparticles, optically excited with low intensity visible light, exhibit direct photo-catalytic activity. We will discuss underlying mechanisms associated with these phenomena.[2,3,4] We propose that this new family of photo-catalysts could prove useful for many heterogeneous catalytic processes that cannot be activated using conventional thermal processes on metals or photo-catalytic processes on semiconductors. I will show an example of such a process.[5]

1. D. B. Ingram, S. Linic, **JACS,** 133, 5202, 2011
2. Suljo Linic, Phillip Christopher and David B., **Nature Materials, 10**, 911, 2011.
3. Ingram P. Christopher, H. Xin, S. Linic, **Nature Chemistry,** 3, 467, 2011.
4. P. Christopher, H. Xin, M. Andiappan, S. Linic, **Nature Materials**, 11, 1044, 2012.

(673d) Molybdenum Sulfides Materials As Hydrogen Evolution Catalysts and Surface Protecting Layers for Highly Active and Stable Silicon-Based Water Splitting Photocathodes

Jaramillo, T. F., Stanford University
Benck, J. D., Stanford University
Kibsgaard, J., Stanford University
Chen, Z., Stanford University
Reinecke, B. N., Stanford University
Originally presented on: 11/7/2013 13:30:00 - 13:50:00

Photoelectrochemical (PEC) water splitting could provide a sustainable means of hydrogen fuel production.[1] Recent research in PEC water splitting has focused on developing materials suitable for application in a dual-absorber device configuration due to the high solar-to-hydrogen efficiencies tandem devices could enable.[2, 3]

Silicon is a promising candidate photocathode material for a tandem PEC device due to its abundance, relatively low cost, excellent charger carrier transport, and near-ideal band structure.[4-9] However, several challenges must be addressed to make silicon-based photocathodes efficient and economical. The surface of the silicon must be protected to prevent the formation of an insulating SiO_2 layer, which can degrade device performance.[8] The silicon must also be combined with an active catalyst to reduce the kinetic overpotential necessary to drive the hydrogen evolution reaction (HER) at the photocathode surface.[4-9]

Previous reports have demonstrated that high-quality SiO_2 tunneling layers or conductive TiO_2 layers can protect silicon photocathode surfaces.[8, 9] When combined with platinum catalysts, these photocathodes demonstrate excellent activity and stability on the order of days. Si photocathodes incorporating catalyst materials based on earth-abundant elements such as NiMo and $[Mo_3S_4]^{4+}$ cubanes have also been demonstrated, but significant improvements in the activity and stability of these precious-metal free devices are necessary to match the performance of Pt/Si structures.[4, 6]

Molybdenum sulfide materials offer a promising solution to these challenges. Nanostructured, crystalline MoS_2 HER catalysts possess excellent activity and stability.[10, 11] We show that MoS_2 can also confer these benefits of high activity and stability to silicon photocathodes. We fabricate conformal protective coatings of MoS_2 on silicon using a simple thermal synthesis. This technique yields silicon photocathodes that remain highly active after more than 24 hours of continuous operation. While the stability of this structure is among the best reported for silicon photocathodes, but the low density of HER-active MoS_2 edge sites at the electrode/electrolyte interface limits the voltage generated by this device. Recently we have developed a scalable HER catalyst with an increased number of active sites – supported thiomolybdate $[Mo_3S_{13}]^{2-}$ nanoclusters – which are particularly interesting as most sulfur atoms in the cluster exhibit a similar structural motif to those found at MoS_2 edges.[12] As illustrated in the figure below, incorporation of additional molybdenum sulfide HER catalysts, such as the $[Mo_3S_{13}]^{2-}$ clusters, to increase the density of active site densities improves the photocurrent onset voltage to within ~150 mV of the best reported Pt/Si photocathodes.

Our results demonstrate that molybdenum sulfides can be employed as multifunctional coatings for silicon photocathodes, serving as both catalysts and protecting layers. Based on our findings, we propose strategies for further improving the performance of molybdenum sulfide/silicon photocathodes.

1. M. G. Walter, et al., *Chem. Rev.*, **110**, 6446 (2010).

2. L. C. Seitz, et al., *in preparation* (2013).

3. M. F. Weber, et al., *J. Electrochem. Soc.*, **131**, 1258 (1984).

4. J. R. McKone, et al., *E&ES*, **4**, 3573 (2011).

5. S. W. Boettcher, et al., *JACS*, **133**, 1216 (2011).

6. Y. Hou, et al., *Nat. Mater.*, **10**, 434 (2011).

7. B. Seger, et al., *Angew. Chemie*, **124**, 9262 (2012).

8. B. Seger, et al., *JACS*, **135**, 1057 (2013).

9. D. V. Esposito, et al., *Nat. Mater.* (2013).

10. Z. Chen, et al., *Nano Letters*, **11**, 4168 (2011).

11. J. Kibsgaard, et al., *Nat. Mater.*, **11**, 963 (2012).

12. J. Kibsgaard, et al., *in preparation* (2013).

(673f) Study of Electrocatalytic Activity of Oxide-Based Materials in Li-Air Batteries: Towards Enhancing the Rates of Oxygen Evolution Reaction During Charging

Savithra, G. H. L., Wayne State University
Nacy, A., Wayne State University
Nikolla, E., Wayne State University
Originally presented on: 11/7/2013 14:10:00 - 14:30:00

Nonaqueous Li-air batteries are among the most promising energy storage technologies due to their high theoretical energy density and specific capacity. In practice, their performance is significantly limited by several factors including the high overpotential losses at the cathode. Among the most commonly used cathode materials for nonaqueous Li-air batteries are porous carbons due to their low cost, light weight, good storing capability for Li_2O_2 and excellent electron conductivity. In addition, porous carbon-based cathodes have also shown to possess oxygen reduction reaction (ORR) activity during discharge due to their surface functional groups and defect sites. One of the challenges with the carbon-based cathodes for Li-air batteries is the inability to electrocatalyze the oxygen evolution reaction (OER) during the charging process. Therefore, the oxidation of Li_2O_2 during the charging appears to be catalytically sensitive as shown by various research groups. In this contribution, we will discuss a promising class of oxide-based materials with mixed ionic and electronic conductivity as OER electrocatalysts for Li-air cathodes. We find that these oxides reduce the charging potentials significantly (from 4.2 V to 3.7 V) compared to the standard carbon based cathodes without an OER electrocatalyst. We have been able to control the size and shape of these oxide materials by using a reverse micelle synthesis method and study the effect on the electrocatalytic activity toward OER.

(673g) Mechanistic Insights Into the Electrocatalytic Lithium-O2 Reduction and Evolution Reactions

Dathar, G. K. P., Louisiana Tech University
Shelton, W. A. Jr., Oak Ridge National Laboratory
Xu, Y., Louisiana State University
Originally presented on: 11/7/2013 14:30:00 - 14:50:00

The lithium-air battery can far exceed the specific energy of lithium-ion batteries, but prototypes suffer from poor round-trip efficiency, low current density, and capacity fading. Unlike the O_2 reduction reaction in hydrogen fuel cells, which forms water that is the same as the solvent and can be vented to the atmosphere, the Li analog (Li-ORR) is usually carried out in aprotic organic solvents, resulting in two peculiarities: 1) The reduction products readily precipitate into solid phases; 2) organic solvents are susceptible to attack by reactive oxygen, making it difficult to assess the microscopic reaction mechanisms. There has been conflicting experimental evidence regarding the efficacy of electrocatalysts for this reaction. Nonetheless, the sluggishness of the Li-ORR/OER will need to be addressed before the rechargeable Li-air battery becomes a reality. We use DFT calculation-based modeling techniques to interrogate the intrinsic activity of metals, metal oxides, and carbon for the Li-ORR. We propose that on unreactive surfaces the reaction proceeds associatively through superoxide formation, whereas on reactive surfaces the initial reduction occurs between Li and dissociated O instead. The predicted Li-ORR activities form a volcano trend with respect to oxygen binding energy, in close agreement with Lu et al. The OER activities are evaluated in a similar framework, with the predictions broadly in line with experimental evidence. Our findings demonstrate that the intrinsic electrocatalytic activity toward Li-ORR/OER differs considerably for different substances and can be reasonably reflected theoretically.

(678a) The Acidity Of Aluminas and Silica-Aluminas: Catalytic Activity In The Conversion Of Ethanol

Phung, T. K., University of Genova
Lagazzo, A., University of Genova
Sanchez Escribano, V., Universidad de Salamanca
Rivero Crespo, M. A., University of Salamanca
Busca, G., University of Genova
Originally presented on: 11/7/2013 12:30:00 - 12:50:00

Abstract

The properties of a number of commercial aluminas and silica-alumina have been characterized by XRD, TEM, Surface area and porosity analysis, skeletal IR and Raman spectroscopies, IR studies of adsorbed probe molecules (hydroxy groups, pyridine, CO, CO_2). The catalytic activity of these materials in the dehydration of ethanol to ethylene and diethylether has been studied in a flow reactor as well as in an IR cell. The catalytic activity data have been discussed in relation to the surface chacarterization data. This allowed us to crytically discuss the most popular models for the surface sites of alumina and silica-alumina.

1. Introduction

Transitional aluminas are among of the most used materials in any field of technologies. However, the details of the crystal structure of the most common polymorph, γ-Al_2O_3, are still matter of controversy, being a defective non stoichiometric spinel [[1],[2]] or a different structure with the occupancy of non-spinel sites [[3]-[4],[5],[6]]. γ-Al_2O_3 starts to convert at ca 500 °C towards other transitional phases: however also the sequence of alumina phases obtained upon calcination is object of disagreement [[7]]; while some authors report the direct conversion of γ-Al_2O_3 into θ-Al_2O_3 near 650 °C, most studies find the formation of slightly different intermediate phases, closely related to that of spinel and of γ-Al_2O_3, formed continuously in the range 500-650 °C. Also in this case, different spinel tetragonal superstructures, such as those denoted as δ-Al_2O_3 [[8]] and γ'-Al_2O_3 [[9]], have been determined by different authors. Only above ca 650 °C the crystal chemistry of alumina becomes well established. In fact, at this temperature θ-Al_2O_3 is formed, whose "beta-gallia" structure is completely determined, that further transforms into α-Al_2O_3, corundum, completely above 1200 °C.

γ-Al_2O_3 and related structures are found in many industrial applications as catalysts and as catalyst supports. Most of reactions performed on alumina are performed in the 200-600 °C temperature range, and imply the use of materials whose real structure is not still established. Several techniques, such as theoretical calculation [1] and selective poisoning [2] are used to propose reliable models for its surface active sites, and, consequently, also detailed mechanisms for the reactions.

Aluminas are used today as the catalyst of the Claus process and for the synthesis of dimethyl ether and methyl chloride from methanol. They have been used in the sixties for producing ethylene from dehydration of bioethanol at temperature > 250 °C. This reaction can become useful again to produce bioethylene from bioethanol.

2 Experimental.

2.1. Catalysts

Four commercial alumina and one silica-alumina samples were investigated (see Table 1).

Table 1. Properties of investigated catalysts

Notation	Producer	Preparation	Surface area (m^2/g)	XRD phase	Na content (%)
V200	Versal UOP	From boehmite via precipitated $NaAlO_2$	200 ± 10	γ-Al_2O_3	< 0.04 *
P200	Puralox SBa Sasol	From boehmite via Al alkoxides	190 ± 10	γ-Al_2O_3	0.002 *
D100	Degussa/ Evonik	From flame hydrolysis of $AlCl_3$	95 ± 10	γ,δ-Al_2O_3	0,006
P90	Puralox SBa Sasol	Calcination of P200	87 ± 5	θ-Al_2O_3	0.002 *
SA330	Strem (13% Al_2O_3)		320 ± 10	amorphous	

* data from the producers

2.2. TEM

TEM analysis of samples, prepared in aqueous suspension in an ultrasonic bath, were recorded with a Zeiss EM 900 instrument.

2.3. XRD

The XRD analyses have been performed on a Siemens D-500 Diffractometer (Cu K radiation, Ni filter; 30 mA, 40 KV) equipped with the Diffract AT V3 software package.

2.4. Infrared spectroscopy (IR) experiments

2.4.1. Skeletal studies.

The characterization of catalysts was studies using Nicolet 380 FT-IR spectrometer. The samples were pressed into thin wafers with KBr and spectra where recorded in air.

2.4.2. Pyridine adsorption

The acidity measurements were determined by pyridine adsorption in an *in-situ* IR-cell using Nicolet 380 FT-IR Spectrometer. The pure powders were pressed into thin wafers and activated in the IR cell connected with a conventional gas-manipulation apparatus at 773 K for 1 h. After contact of activate catalyst samples with pyridine vapor (p_{Py}~ 1 torr) at room temperature during 15 min, the IR spectra of the surface species were collected in continuous evacuation at room temperature and upon increasing temperature.

2.4.3. CO adsorption

The CO adsorption/desorption process has been studied by Nicolet 380 FT-IR spectrometer. CO adsorption was performed at 130 K (real sample temperature measured by a thermocouple) by the introduction of a known dose of CO gas inside the low temperature infrared cell containing the previously activated wafers. The sample was saturated with CO using sufficiently high CO pressure (up to 20 Torr), until the intensity of the adsorbed species has raised a maximum. IR spectra were later collected evacuating at increasing temperatures between 130 and 273 K.

2.4.4. Ethanol adsorption

In order to study the mechanism of reaction, the adsorption/desorption process has been studied using Nicolet 380 FT-IR Spectrometer. Pressed disks of the pure catalyst powders were activated *in-situ* in the IR cell connected with a conventional gas-manipulation apparatus before any adsorption experiment. IR spectra of the surface species as well as of the gas phase were collected upon increasing temperature in static conditions (p_{EtOH}~ 4 torr).

2.5. Surface area and porosity

The BET surface areas and porosity have been measured with a Micromeritics Gemini 2380 instrument.

2.6. Catalytic experiments

Catalytic experiments have been performed at atmospheric pressure in a tubular flow reactor using 0.5 g catalyst (60-70 mesh sieved) and feeding 7.9% vol/vol ethanol in nitrogen with total flow rate 80 cc/min. The carrier gas (nitrogen) was passed through a bubbler containing ethanol (96%). The temperature in the experiment was varied stepwise from 423 K to 723 K.

Ethanol conversion was defined as usual:

$$X_{EtOH} = (n_{EtOH(in)} - n_{EtOH(out)})/n_{EtOH(in)}$$

While selectivity to product i is defined as follows:

$$S_i = n_i/(n_i(n_{EtOH(in)} - n_{EtOH(out)}))$$

where n_i is the moles number of compound *i*, and $v_i$is the ratio of stoichiometric reaction coefficients.

The outlet gases were analyzed by a gas chromatograph (GC) Agilent 4890 equipped with a Varian capillary column "Molsieve 5A/Porabond A Tandem" and TCD and FID detectors in series. In order to identify the compounds of the outlet gases, a gas chromatography coupled with mass spectroscopy (GC-MS) Thermo Scientific with TG-SQC column (15 m x 0.25 mm x 0.25 mm) was used.

Acknowledgements

TKP acknowledges funding by EMMA in the framework of the EU Erasmus Mundus Action 2.

References

[1]. B.C. Lippens, J.H. de Boer, Study of phase transformations during calcination of aluminum hydroxides by selected area electron diffraction, Acta Crystallogr., 17 (1964) 1312-1321.

[2]. A.R. Ferreira, M.J.F. Martins, E. Konstantinova, R.B. Capaz, W.F. Souza, S.S.X. Chiaro, A.A. Laitão, Direct comparison between two γ-alumina structural models by DFT calculations, J. Solid State Chem., 184 (2011) 1105-1111.

[3]. R.-S. Zhou, R.L. Snyder, Structures and transformation mechanisms of the η, γ and θ transition aluminas, Acta Crystallogr. B 47 (1991) 617-630.

[4]. M. Digne, P. Sautet, P. Raybaud, P. Euzen and H. Toulhoat, Use of DFT to achieve a rational understanding of acid–basic properties of γ-alumina surfaces, J. Catal., 226 (2004) 54–68.

[5]. G. Paglia, A.L. Rohl, C.E. Buckley, J.D. Gale, Determination of the structure of y-alumina from interatomic potential and first principles calculations: The requirement of significant numbers of nonspinel positions to achieve an accurate structural model, Phys. Rev. B 71 (2005) 224115-1-224115-16.

[6]. L. Smrcok, V. Langer, J. Krestan, γ-Alumina: a single-crystal X-ray diffraction study, Acta Crystallogr. C. 62 (2006) i83-i84

[7]. I. Levin, D. Brandon, Metastable Alumina Polymorphs: Crystal Structures and Transition Sequences, J. Am. Ceram. Soc., 81 (1998) 1995-2012

[8]. S.J. Wilson, J.D.C. McConnel, A kinetic study of the system γ-$AlOOH/Al_2O_3$, J. Solid State Chem., 34 (1980) 315-322.

[9]. G. Paglia, C.E. Buckley, A.L. Rohl, R.D. Hart, K. Winter, A.J. Studer, B.A. Hunter, J.V. Hanna, Boehmite derived [gamma]-alumina system. 1. Structural evolution with temperature, with the identification and structural determination of a new transition phase [gamma]'-alumina, Chem. Mater. 16 (2004) 220-236.

(678b) Hydrogen-Deuterium Exchange Over Ceria-Zirconia Mixed Metal Oxides

McNew, S., Georgia Institute of Technology
Lamont, O., Georgia Institute of Technology
D'Amico, A., Georgia Institute of Technology
Sievers, C., Georgia Institute of Technology
Originally presented on: 11/7/2013 12:50:00 - 13:10:00

Ceria based supports are commonly used in supported metal catalysts. In many cases these supports are of interest because their redox properties facilitate reactions that are commonly attributed to the metal particle. However, ceria catalysts also have activity for reactions that are typically metal catalyzed, such as the hydrogenation of acetylene.

The present work shows that the ability of ceria to activate hydrogen is enhanced for ceria-zirconia mixed metal oxide catalysts. H_2/D_2 isotopic exchange is used to probe this ability. It is shown that there are different types of active sites with different activation energies for the exchange reaction. In-situ IR spectroscopy is used to identify these sites. Detailed physico-chemical characterization is performed to obtain structure-property relationships. Most importantly, it is shown that the H_2/D_2 exchange activity increases with decreases crystallite size of ceria-zirconia. In addition, the presence of defect sites from surface reduction of ceria-zirconia has a strong impact on the H_2/D_2 exchange reaction.

Correlations of H_2/D_2 exchange activity with catalytic performance in other reactions will be discussed.

(678c) Synthesis-Structure-Function Relations for Supported Oxides, Including Quantification of Accessible Metal Oxide Active Sites On Ti-SiO2 Systems

Eaton, T., Northwestern University
Gray, K. A., Northwestern University
Notestein, J. M., Northwestern University
Originally presented on: 11/7/2013 13:10:00 - 13:30:00

Quantitative determination of the number of active sites in supported catalysts is required for full understanding. Characterizing site dispersion in supported metals is typically carried out through techniques such as CO chemisorption, direct particle imaging, or fitting to EXAFS. [1-3] While strong acid sites or the reducible fraction of a supported oxide are also straightforward, [4, 5] quantification of weaker Lewis sites is not routine. Specifically, many supported oxide catalysts undergo significant structural changes as surface densities increase from statistically isolated sites to crystallites, but there is not necessarily parallel quantification of the number of surface-accessible sites.

Here, we provide synthesis-structure-function relationships for supported oxides that include the effect of several grafting precursors on electronic structure (DRUV-vis, XAS, XRD) and reactivity in H_2O_2 activation, photo-oxidation, and CO_2 reduction for Fe, Ti, and other oxide domains supported on SiO_2. Borrowing from literature on self-assembled monolayers, we use the specific adsorption of phosphonic acid on the active oxide and not SiO_2 to estimate the number of surface accessible atoms. For example, the photocatalytic oxidation of benzyl alcohol is shown to have a rate of ~0.15 h^{-1} over SiO_2-supported titanium oxides of different precursor, loadings, electronic structure and crystallinity, when counting only the sites which phosphonic acid can titrate. Time permitting, other supported oxides and reactions will be discussed, with an emphasis on systems known to exhibit strong structure sensitivity.

[1] J.W. Niemantsverdriet, Spectroscopy in Catalysis, 2nd ed., Wiley-VCH, 2000.

[2] W.D. Williams, M. Shekhar, W.-S. Lee, V. Kispersky, W.N. Delgass, F.H. Ribeiro, S.M. Kim, E.A. Stach, J.T. Miller, L.F. Allard, Journal of the American Chemical Society 132 (2010) 14018-14020.

[3] J.M. Thomas, W.J. Thomas, Principles of Heterogeneous Catalysis, 1st ed., Wiley-VCH, 1997.

[4] J. Macht, R.T. Carr, E. Iglesia, Journal of Catalysis 264 (2009) 54-66.

[5] D. Prieto-Centurion, J.M. Notestein, Journal of Catalysis 279 (2011) 103-110.

(678d) Composition Function Relations for Methanol Dehydration and Oxidation On Acid-Redox Bifunctional Polyoxometalate Clusters

Deshlahra, P., University of Notre Dame
Carr, R. T., University of California at Berkeley
Phadke, N., University of Wisconsin, Madison
Iglesia, E., University of California, Berkeley
Originally presented on: 11/7/2013 13:30:00 - 13:50:00

Keggin polyoxometalate (POM) clusters are oxides with known structure and diverse compositions, which makes them suitable for developing rigorous composition-function relations via theoretical calculation of relevant descriptors of redox and acid properties. Here, we use bifunctional acid-redox Mo and W-based POM clusters with different central atoms ($H_3\mathbf{P}Mo_{12}O_{40}$, $H_4\mathbf{Si}Mo_{12}O_{40}$) and addenda atoms ($H_{3+n}P\mathbf{V}_{\mathbf{n=1-3}}\mathbf{Mo}_{\mathbf{12-n}}O_{40}$, $H_4P\mathbf{V}_{\mathbf{1}}\mathbf{W}_{\mathbf{11}}O_{40}$), and their salts, supported on silica, to explore the effect of local redox properties, measured as H-atom addition energies (HAE), on methanol oxidative dehydrogenation (ODH) rates and to extend previous studies of the effect of acid strength of charge-balancing protons, measured as deprotonation energy (DPE), on methanol dehydration turnover rates [1] to Mo-POM systems.

Methanol oxidation involves Mars van Krevelen (MVK) cycles and kinetic coupling of elementary steps that reduce and oxidize metal centers in oxide catalysts with H-abstraction by lattice oxygen from the substrate as the commonly observed rate limiting step. Measured H-abstraction rate constants were slightly lower for HSiMo than HPMo, but were insensitive to V-substitution of addenda atoms on acid form Mo-POM. These results are inconsistent with about 0.3 eV lower UV-vis edge energies measured on $HPV_{n=1-3}Mo_{12-n}$ than HPMo, and contradict the general expectation that V-susbtitution would lead to improved redox activity, but are consistent with previous results for unsupported and silica supported POM [2]. These findings were further examined using DFT based exploration of redox pathways which show that direct H-abstraction form a molecularly adsorbed methanol is more favorable on POM than dissociative pathway. H-abstraction occurs vial late TS and their calculated kinetic isotope effects are consistent with the measured values. The calculated TS energy, as expected, changes linearly with the local HAE value of the abstracting POM O-atom, because HAE reflects redox properties of POM by introducing electrons at metal centers through lattice O-atom (O*) via processes similar to those required for rate limiting H-abstraction steps. Consistent with the measured ODH rates, the HAE values were insensitive to V-substitution. Spin-density maps of unpaired d-electron introduced due to reduction via H-addition shows that V atom in acid-form V-substituted POM is not reduced since the d-electron is located at a Mo atom. Similar calculations on anionic POM without the protons, however, show that V-atom is reduced and show more negative HAE values compared to HPMo. These results indicate that commonly observed improved redox properties of VO_x groups can largely disappear in acid-form POM because of the additional proton introduced along with V-substitution stays with the VO_x groups as a V^{5+} replacing the Mo^{6+} needs this additional proton to satisfy its surrounding O ligands and the electronic effects of this acidic proton diminishes redox capability of VO_x. Such effects are likely to be less significant in liquid-phase reactions with dissociated protons and anions or for salt-form POM which do not contain such protons. HAE values emerge as relevant and accurate descriptor of redox property.

Re-oxidation steps were not rate limiting under conditions used in this work and the corresponding rate constant could not be obtained directly through steady-state experiments. Relative differences in re-oxidation rate constants for different compositions were assessed via UV-vis measurements of extents of reduction of catalysts during catalysis, and via time-scales of transients in extent of reduction due to changes in reactant concentrations [3]. The observed differences in re-oxidation were examined theoretically by exploring molecular O_2 adsorption and plausible subsequent O_2 activation routes involving O-atom migration, direct re-oxidization of vacancy pair, and reactions of one oxygen atom in bound peroxide structures with the organic substrate.

First-order rate constant (k_{mono}) for methanol dehydration decreases exponentially with increasing DPE due to decrease in stability of ion-pair transition states (TS) involved in acid catalyzed reactions which recover only a fraction of the electrostatic part of DPE [1]. The measured values of k_{mono} for Mo-POM lie below the k_{mono} versus DPE correlation previously developed for W-POM because they undergo reduction under reaction conditions which increases DPE relative to the values obtained using DFT calculations of intact clusters and, because of a larger covalent part of DPE in Mo-POM that is not recovered in ion pair transition states. Contribution of these two effects to the difference between the trends for the two addenda atoms is estimated, by calculation DPE for reduced clusters, and by calculations of H^+POM^- and TS^+POM^- classical electrostatic interactions using integration over of DFT derived cation and anion charge distributions along with rigorous assessment of interaction energies within the framework of thermochemical cycle relating DPE to TS energy [1].

References

1. Carr R., Neurock M., Iglesia E., *J. Catal.* 278, 78 (2011).
2. Liu H., Iglesia E., *J Phys. Chem. B*, 107, 10840 (2003); Brückman K., Tatibouët, J.M., Che, M., Serwicka E., Haber J., J. Catal., 139, 455 (1993).
3. Argyle, M.D., Chen, K., Iglesia, E., Bell, A.T., *J. Phys. Chem. B*, 109, 2414 (2005).

(678e) Nanoscale Characterization of Heteroatom Distributions in Boro- and Alumino-Silicate Frameworks

Hsieh, M. F., University of California, Santa Barbara
Garaga, M.,
Nour, Z.,
Aronson, M., Univ. of California, Santa Barbara
Cadars, S., University of California Santa Barbara
Zones, S. I., Chevron Energy Technology Company
Chmelka, B. F., University of California
Originally presented on: 11/7/2013 13:50:00 - 14:10:00

The local environments and distributions of heteroatoms in boro- or alumino-silicate materials are known to have strong influences on their macroscopic adsorption and reaction properties. This is hypothesized to be due to the variations of Brønsted acidity caused by the molecularly distinct surroundings of boron or aluminum atoms incorporated into different framework sites. Experimental characterization of such heteroatom distributions at the molecular-level (sub-nm) has been challenging, due to the complicated local order and disorder of heteroatom-containing silicate frameworks to which diffraction and other conventional techniques are insensitive. Solid-state nuclear magnetic resonance (NMR) spectroscopy overcomes many of these challanges by being sensitive to the local environments and spatial and/or bonding interactions of heteroatoms. In particular, two-dimensional (2D) NMR methods allow interactions of specific ^{11}B-O-^{29}Si or ^{27}Al-O-^{29}Si framework species to be identified and resolved by detecting their *through-space* (dipole-dipole) or *through-bond* scalar (*J*) interactions. Here, we analyze the local structures and distributions of boron and aluminum atoms in semi-crystalline boro- and alumino-silicates by directly determining the molecular site connectivities of ^{11}B or ^{27}Al heteroatoms and adjacent ^{29}Si atoms in the framework. The 2D NMR results, together with density-functional-theory (DFT) calculations, indicate that boron heteroatoms can be preferentially incoporated into specific ^{29}Si sites in layered and zeolitic boro-silicate frameworks, whereas aluminum heteroatoms appear to be broadly distributed in otherwise identical layered alumino-silicates directed using the same structue-directing agents as layered borosilicates. The results provide new insights on local heteroatom environments and distributions in boro- and alumino-silicates that are expected to aid understanding of their adsorption and reaction properties.

(678f) Enantiospecificity of Thin Metal Films On Chiral SrTiO3 Surfaces

Asthagiri, A., University of Florida
Yuk, S., The Ohio State University
Originally presented on: 11/7/2013 14:10:00 - 14:30:00

The separation of racemic mixture is an important step in today's pharmaceutical fields since the physiological and toxicological properties of enantiomer can be significantly different to its mirror-image counterpart. One option for separating enantiomers is the use of chiral metal surfaces, but while these surfaces have been shown to be enantiospecific they are limited in the degree of enantiospecificity, cost, and low-surface area. One possible method to enhance the degree of enantiospecificity and reduce cost is to grow metal nanostructures or thin films on chiral oxide surfaces. Using density functional theory (DFT) we are examining Pt grown on the chiral $SrTiO_3$ substrate due to the lattice matching between these materials.

As a first effort to understand the effects of oxide substrate on the adsorption properties of Pt films, we studied CO adsorption on Pt thin films supported on the SrO- and TiO_2-terminated $SrTiO_3$(100). With one monolayer (ML) of Pt, the presence of $SrTiO_3$ substrate enhances the CO adsorption energy by ~ 40% and in addition results in large differences in the adsorption energy at various sites on the Pt layer due to the underlying atoms of the oxide. Pt atoms sitting on top of oxide O atoms show the strongest biding to the CO molecule. This degree of heterogeneity will be important in enhancing the enantioselectivity of the chiral Pt/$SrTiO_3$ surfaces. The effects of oxide substrate are negligible after 2 ML of Pt, indicating the importance of deposited Pt thickness for maintaining any potential oxide effect.

We also examined the adsorption behaviors of R- and S-FAM (fluro-amino-methoxy) on the Cu(874), Pd(874), Pt(874), and a ML of Pt on $SrTiO_3$(874) in order to understand the enantioselectivity of oxide-supported chiral metal structures. Our preliminary results show that the adsorption behaviors of FAM are dependent on the position of fluorine atom of FAM and the electronic nature of chiral surfaces. The presence of underlying $SrTiO_3$ also strongly modifies the energetics of R- and S-FAM adsorbed on the chiral Pt ML compared to the bare Pt (874). We will present results from ongoing studies of the enantioselectivity for FAM and other small chiral molecules on ML thick metal films on chiral $SrTiO_3$(874) and discuss any general guidelines observed for enhancing enantiospecificity.

(678g) Increasing Oxygen Reduction On TiO2 Via Non-Precious-Metal Surface Doping

Muhich, C. L., University of Colorado at Boulder
Westcott, J. Y. IV, University of Colorado at Boulder
Weimer, A. W., University of Colorado at Boulder
Musgrave, C. B., University of Colorado, Boulder
Originally presented on: 11/7/2013 14:30:00 - 14:50:00

The photocatalytic activity of TiO_2 is limited by its high band gap, ~3.2 eV which limits photoabsorption to the small percentage of UV light in the solar spectrum and slow oxidation reduction reaction rates. This is particularly true in photocatalytic degradation of aqueous organic pollutants (OP). Traditionally Pt is deposited on TiO_2 to increase the photocatalytic OP degradation. Recently this was shown to increase the TiO_2 photocatalytic activity because Pt acts as an excellent catalyst for O_2 reduction reaction, the OP degradation rate limiting step. Pt provides high energy electrons that simultainusly reduce O_2 and allow for the formation of O-surface bonds. However, Pt is expensive and if too much Pt is deposited on the TiO2 surface the photocatalytic activity is worse than if there was no surface Pt because Pt clusters completely bridges the TiO2 band-gap and therefore acts as an electron/hole pair. We have utilized density functional theory (DFT) to investigate the use of non-Pt surface dopants to increase the O_2 reduction reaction. Among the non-metals investigated, B improves O_2 reduction the best, enabling O_2 reduction independent of dopant location, followed by C, which enables O_2 reduction only at interstitial dopant sites, while N doping does not allow for increased O_2 reduction. V, Nb, and Mo act to increase O_2 adsorption while Cr is not as effective. These results are consistent with the hypthosysis that filled bands sitting high in TiO_2's band gap are needed for O2 adsorption and therefore reduction.

(682a) Nanomaterials As Interfacial Modifiers in Energy Applications

Vander Wal, R. L., Pennsylvania State University

Originally presented on: 11/7/2013 12:30:00 - 12:55:00

Introduction

Energy and nanomaterials are intimately related. Most energy processes occur at the interface. Nanomaterials, when used as interfacial modifiers have the potential to significantly alter the energy landscape. Although carbon nanomaterials have been explored for a suite of energy applications, to-date the approach has been based on replacing bulk materials, not as interfacial modifiers. The work presented here will highlight this latter approach by showing results for a suite of energy applications from the author's own work.

Experimental

Experimental details will be presented at the presentation with regards to the synthesis and utilization of the nanomaterials. Our applications include energy conversion, generation, storage, efficiency, conservation and control.

Results and Discussion

Nanoscale materials are redefining the relation between material composition, size and properties. Chemical properties (e.g. reactivity) and physical properties (e.g. surface area) become a strong function of size at the nanoscale. Applications illustrated include the following selected examples from the author's work: catalysts, composite materials, supercapacitors, batteries, sensors, nanofluids, and lubricants. The key concept is that the nanomaterials serve as interfacial modifiers. Since most energy related processes are dominated by interfacial reactions, nanomaterials have the potential to dramatically affect energy conversion rates and magnitudes. Examples follow.

With regards to overall system integration and control, sensors will play a prominent role [1]. With ultrahigh surface exposure relative to bulk material, nanoscale materials are exceedingly sensitive to gas adsorption. Exploitation of nanoscale properties will lead to new NO_x sensors and in-cylinder oxygen sensors for combustion control in power generation facilities. Examples include catalyst coated metal oxide semiconductors capable of low temperature operation amidst complex gas environments.

Lightweight materials, a prime route to energy conservation particularly in transportation will reduce weight significantly, yielding substantial benefits in fuel efficiency with reduced emissions. Composites with substantial gains in Young's modulus, tensile strength, and EM shielding may be realized in polymeric composites using carbon nanotubes as an interfacial modifier rather than bulk filler [2]. The purpose of the foil is to serve as a gas impermeable barrier layer within the polymer composite.

Catalysis is central to fuel cell reactions and CO cleanup from the feeding, fuel stream. Preferential CO oxidation, so-called "PROX" is necessary to prevent catalyst poisoning by CO chemisorption, deactivating the catalyst. Catalytic conversion at low temperature is advantageous for continuity of thermal management and can be accomplished by Au, supported on reducible oxides such as CeO_2, TiO_2 and Fe_2O_3 [3]. Other applications for these catalysts systems include volatile organic compounds (VOCs) and potentially exhaust hydrocarbons in present automotive transportation systems.

Energy generation is often synonymous with energy harvesting of sunlight, a renewable and "free" source of energy, though quite arguably this is yet an energy conversion process – photons to electrons. Whether by dye-sensitized solar cell or all semiconductor-based architecture, anodic capture and conversion of incoming photons is drastically increased by nanomaterials with the anode [4]. Crystalline nanowires offer vectorial electron transport with mobilities ~ 100-fold greater than polycrystalline counterpart. Moreover their high surface area facilitates charge carrier injection from anchored dye or from an exterior semiconductor in the form of quantum dots or cladding as in a core-shell construction.

Advances in energy storage include batteries, ultra-capacitors. These can support lighting, appliances, a starter, cooling fans, transmission and hydraulic systems, fuel and air handling systems and ultimately enable hybrid systems. Towards these goals, substantial gains in Li ion battery cathode and anode materials have been realized using carbon nanofibers, coating processes and including elements such as tin and silicon. Modifications of the CNT surfaces increases the Li ion capacity beyond the theoretical limit of normal graphite. As an illustration, CNTs were directly fabricated for this end-application use upon ultra-fine SS mesh. Significantly no harvesting, purification or processing (using binders) was required. The CNTs could be used directly as synthesized [5].

Thermal energy transfer will benefit from nanofluids. Nanofluids can increase thermal conductivity and reduce radiator and heat exchanger size in power generation and cooling operations. Carbon-based nanofluids using nano-onions and carbon nanotubes have increased water conductivity by ~ 20% [6]. Being nanoscaled, they will not sediment, and with graphitic carbon content, can provide lubricity rather than abrasion to seals, valves and gears within fluid systems.

Lubrication is critical to efficient energy transfer many engine components and powertrain systems. Nanolubricants can bridge the gap between fluid and solid materials [7]. As additives within liquids or greases, synergistic properties may be realized, particularly in boundary-phase lubrication by preventing rheological thinning of the lubricant. Improved coating formulations and properties can reduce or eliminate fretting and pitting. Results with single-walled carbon nanotubes show superior performance relative to graphite, diamond like carbon (DLC) and even Teflon.

Conclusions

Nanoscale materials are redefining the relation between material composition, size and properties. Their applications to energy processes include selected examples from the author's work. Selected technologies include catalysis, composite materials, energy storage, sensors, thermal management, and tribology. Interfacial modification serves to affect changes in energy conversion rates, and transfer magnitudes. This appears to be a rather universal concept and suggests R&D directions for nano-engineering of interfaces.

Acknowledgements

Funding through The Penn State Institutes for Energy and the Environment (PSIEE) and the Pennsylvania

Keystone Innovation Starter Kit (KISK) is gratefully acknowledged.

References

1. Hunter, G. W., Vander Wal, R. L., Liu, C. C., Xu, J. C., and Berger, G. M., Sensor and Actuators B ***2009***, 138, 113-119.

2. Vander Wal, R. L., and Hall, L. J., Advanced Engineering Materials **2004,** 6, 48-52.

3. R. L. Vander Wal, Nanomaterials and Energy Applications: Interfacial Engineering – A Survey. (*in preparation).*

4. *ibid.*

5. Luo, Y., Vander Wal, R. L., and Scherson, D. A., Electrochemical and Solid State Letters **2003,** 6, A56-A58.Vander Wal, R. L., Mozes, S. D., and Pushkarev, Nanotechnology ***2009***, 20, 105702-10.

6. Vander Wal, R. L., Mozes, S. D., and Pushkarev, Nanotechnology ***2009***, 20, 105702-10.

7. Street, K. W., Miyoshi, K., and Vander Wal, R. L., "Application of Carbon Based Nano-particles to Aeronautics and Space Lubrication", in *Superlubricity*, Chapter 19, pp 311-340. Edited by Drs. Jean-Michel Martin and Ali Erdemir, Elsevier, Amsterdam (2007). (also NASA/TM-2007-214473).

(682b) Covalent Bonding of CdSe Quantum Dots to Carbon Nanotubes Without Ligands for Solar Energy Harvesting

Azoz, S., Yale University
Pfefferle, L., Yale University
Keskar, G.,
Originally presented on: 11/7/2013 12:55:00 - 13:15:00

Hybrid nanomaterials have a wide range of applications in modern technology; different functionalization strategies are being intensely sought for preparing nanocomposites with tunable properties and structures. Multi-Walled Carbon Nanotube (MWNT)/ CdSe Quantum Dot (QD) and Single-Walled Carbon Nanotube (SWNT)/ CdSe Quantum Dot (QD) heterostructures serve as an important example for an active component of solar cells. The attachment mechanism of CdSe QDs and carbon nanotubes is known to strongly affect the charge transfer between them and consequently to alter the efficiency of solar cell devices. Previous authors [e.g. 1] have shown that short ligands improve charge transfer; we show that more direct bonding results in significant improvement.

In this study, we present a novel method that enables the exchange of some of the organic capping agents on the QDs with carboxyl functionalized MWNTs and SWNTs upon ultrasonication. This produces a ligand-free covalent attachment of the QDs to the carbon nanotubes. EXAFS characterization reveals direct bond formation between the Cd in the CdSe QDs and the nanotubes.

Additionally, we characterize the optical and structural properties of the QD-MWNT heterostructures and investigate how these properties are affected by the attachment. The steady state photoluminescence of QDs is quenched to the noise level of the instrument when the QDs are covalently bonded to an oxygen group on the nanotubes.

Time-resolved fluorescence spectroscopy was also used to reveal more information about the dynamics of charge carriers within the heterostructures probed. We monitored the decay of exciton emission of CdSe after mixing with MWNT or SWNT using magnetic stirring and after reacting with MWNT or SWNT using ultrasonication. The normalized luminescence emissions of the materials when they are excited at 390 nm shows that the lifetime is <<3nm (the laser pulse width) for the QDs covalently bonded to the MWNT or the SWNT as no tail in the PL signal is observed. We will also discuss photoconductivity data.

[1] X. H. Peng, J. Y. Chen, J. A. Misewich and S. S. Wong, Chem Soc Rev, 2009, 38, 1076-1098.

(682c) Butanol Partitioning At Model Bacterial Membrane/Water Interfaces

Bothun, G., University of Rhode Island
Kurniawan, Y., University of Rhode Island
Scholz, C., University of Alabama, Huntsville
Originally presented on: 11/7/2013 13:15:00 - 13:35:00

Bacteria adjust their membrane lipid composition to counteract the fluidizing effects of alcohol and to adapt to elevated alcohol concentrations during fermentation. Bacterial membranes are rich in anionic phosphatidylglycerols (PG), but little is known regarding alcohol partitioning into anionic membranes, particularly for n-butanol, which exhibits clear advantages over ethanol as an alternative biofuel. This work examines the effects of lipid charge on n-butanol partitioning into model anionic bacterial membrane vesicles composed of dipalmitoyl phosphatidylcholine (DPPC) and dipalmitoyl phosphatidylglycerol (DPPG) in the absence and presence of salt. Above a threshold n-butanol concentration (0.135 M), the membranes were interdigitated irrespective of DPPG or salt concentration, consistent with previous results for neutral membranes such as DPPC. Increasing salt concentration let to greater n-butanol partitioning in DPPC membranes and caused aggregation/fusion. However, aggregation/fusion was prevented with increasing DPPG concentration (i.e. increasing membrane charge) and small vesicles were observed. The results suggest that n-butanol partitioning, and subsequent changes in membrane and vesicle structure, was driven by a balance between the 'salting-out' of n-butanol, interlipid electrostatic interactions, and interfacial cation binding and hydration. This is the first study to the best of our knowledge to examine the effects of n-butanol partitioning on model bacterial membranes composed of negatively charged lipids in the presence of salts.

(682d) Effects of Interfacial Forces On the Hydrodynamic Behavior of An Anaerobic Sequencing Batch Reactor (ASBR)

Maurina, G. Z., University of Caxias do Sul
Rosa, L. M., State University of Campinas
Beal, L. L.,
Baldasso, C.,
Gimenez, J. R.,
Pederiva, L.,
Torres, A. P.,
Sousa, M. P.,
Originally presented on: 11/7/2013 13:35:00 - 13:55:00

Hydrogen is known as a clean and ideal energy source due to its renewable status and green effect on the global environment (Nurtono et al., 2012). It can be obtained through physicochemical and biological routes. The hydrogen production by microorganisms (biohydrogen) is affected simultaneously by biological, physical, and chemical factors (Ding et al., 2010; Wang et al., 2009; Wang et al., 2010). Although its production is complex, with numerous interactions between gas, liquids and solids, current research has focused primarily on the biological and chemical characteristics (Nath and Das, 2004). However, the physical characteristics of the reactors are seldom well described (Cao et al., 2010).

In this sense, CFD tools can be valuable to study the fluid dynamics in bioreactors. Simulations can provide values in any location inside the reactor, and the influence of each force can be analyzed independently. However, there are difficulties when simulating the phenomena occurring in the reactor. The interaction between the dispersed gas phase and the continuous liquid phase, for example, is affected by interfacial forces (drag, lift and virtual mass forces) and the turbulence in the reactor. Therefore, the correct modeling of these forces is of prime importance for capturing the physics correctly (Tabib et al., 2008).

The present work aims to use CFD techniques in order to analyze the influence of interfacial forces on the behavior of a gas-liquid flow in a pilot-scale anaerobic sequencing batch reactor. The reactor has a capacity of 1 m^3, operating with a mixture of wastewater and biomass for biohydrogen production.

Numerical simulations were carried out considering gas bubbles and liquid injected at the bottom of the reactor. The phases leave the reactor through different tubes, located at the top of the reactor. The interaction between the disperse bubbles and the liquid phase in the turbulent flow inside the reactor is influenced by the drag, lift and virtual mass forces. Thus, in the present work the sensitivity of these forces has been evaluated through simulations using an Eulerian-Eulerian framework. Each case was calculated to simulate 300 s of flow, with the last 100 s considered to obtain average values.

Results indicates that the interfacial forces have their greatest influence in the bottom part of the reactor. The lift force shifts the maximum of gas volume fraction and axial velocities to the center of the reactor, while the virtual mass force accentuate those peaks. In the middle and top section of the reactor, the lift force has lower influence on the flow, but the virtual mass force still causes higher maximum values in the evaluated variables. References

- Cao, G-L.; Ren, N-Q.; Wang, A-J.; Guo, W-Q.; Yao, J.; Feng, Y-J.; Zhao, Q-L. 2010. Statistical optimization of culture condition for enhanced hydrogen production by thermoanaerobacterium thermosaccharolyticum w16. Bioresource Technology 101, pp. 2053-2058.
- Ding, J.; Wang, X.; Zhou, X-F.; Ren, N-Q.; Guo, W-Q. 2010. CFD optimization of continuous stirred-tank (CSTR) reactor for biohydrogen production. Bioresource Technology 101, pp. 7005-7013.
- Nath, K.; Das, D. 2004. Improvement of fermentative hydrogen production various approaches. Appl. Microbiol. Biotechnol. 65, pp. 520–529.
- Nurtono, T.; Nirwana, W.O.C.; Anwar, N.; Kusdianto; Nia, S.M.; Widjaja, A.; Winardi, S. 2012. A computational fluid dynamics (CFD) study into a hydrodynamic factor that affects a bio-hydrogen production process in a stirred tank reactor. Procedia Engineering 50, pp. 232-245.
- Tabib, M.V.; Roy, S.A.; Joshi, J.B. 2008. CFD simulation of bubble column - An analysis of interphase forces and turbulence models.
- Wang, X.; Ding, J.; Guo, W-Q.; Ren N-Q. 2010. Scale-up and optimization of biohydrogen production reactor from laboratory-scale to industrial-scale on the basis of computational fluid dynamics simulation. International Journal of Hydrogen Energy 35, pp. 10960-10966.
- Wang, X.; Ding, J.; Ren, N-Q.; Liu, B-F.; Guo, W-Q. 2009. CFD simulation of an expanded granular sludge bed (EGSB) reactor for biohydrogen production. International Journal of Hydrogen Energy 34, pp. 9686-0695.

(682e) Enhancing the Perfomance of Supercapacitors By Engineering Electrodes and Electrolytes

Ho, T. A., The University of Oklahoma
Striolo, A., University of Oklahoma
Originally presented on: 11/7/2013 13:55:00 - 14:15:00

Securing availability of both energy and potable water is becoming the challenge chemical engineering has to overcome to secure humans propsperity. Electric double layer capacitors could serve both for storing electricity and for desalinating water (capacitive desalination has been pioneered in the mid 1950's). Current research is focused on understanding how to improve the amount of electrolytes stored within the electrodes for a given applied voltage. We report here our recent molecular dynamics results along this research direction. We considered aqueous electrolytes and graphene electrodes (both pristine and patterned ones). Our results suggest that patterning the electrodes can in some cases enhance the capacitance. We explain these results using atomic-level structural obseravtions regarding the orientation of water, and the packing of NaCl ions near the charged electrodes. Quantities that could be verified experimentally will also be presented. The results presented could be useful for advancing our understanding of electric double layer capacitors, but also for undersing the transport of water and electrolytes in sub-surface formations.

(682f) Energy Storage Mechanism of Porous Supercapacitors

Feng, G., Vanderbilt University
Qiao, R., Clemson University
Cummings, P. T., Oak Ridge National Laboratory
Originally presented on: 11/7/2013 14:15:00 - 14:35:00

Electrical double layer capacitors (EDLCs), also called supercapacitors, have attracted considerable attention in the electrical energy community. To address the limitation of supercapacitors related to their moderate energy density, many efforts have focused on the development of porous carbon-based materials as promising electrodes for supercapacitors to improve their energy density. Despite considerable work on porous carbon materials, the details of what happens under nano-confinement within the pores still require in-depth exploration.

Using molecular dynamics (MD) simulation, we investigated supercapacitors with porous carbon electrodes filled with two types of electrolytes: aqueous electrolytes and room-temperature ionic liquids (RTILs). For aqueous electrolytes, we analysed the ion and water distribution inside different-sized carbon pore and developed a sandwich capacitance model to describe the capacitance of EDLs formed inside carbon micropores. This model is capable of predicting the anomalous enhancement of capacitance experimentally observed at a specific pore size. We show that the free energy penalty for an ion getting into the pore with size of 0.82 nm is less than 14 kJ/mol for both sodium and chlorine ions, which provides strong impetus for further developing nanoporous electrodes featuring subnanometer pores. For RTIL electrolytes, we studied the pore-size dependence of capacitance over a broad range of pore size, by performing MD simulations of EDLs inside slit-shaped micropores with RTILs. We observed that the capacitance of porous supercapacitors exhibits oscillatory behavior as a function of electrode nanopore size and attributed this phenomenon to the interference of two electrical double layers originating from each wall of the carbon slit.

(682g) Simulation of Electrochemical Double-Layer Capacitors Based On Carbon Nanotube Forests

You, X., Tulane University

Zhu, P., Tulane University

Pratt, L. R., Tulane University

Originally presented on: 11/7/2013 14:35:00 - 14:55:00

Simulation of Electrochemical Double-layer Capacitors based on Carbon Nanotube Forests

Xinli You, Peixi Zhu, and Lawrence R. Pratt

Abstract Unique properties of nanomaterials, and specifically of Carbon nanotube (CNT) forests based electrochemical double-layer capacitors (EDLCs), arise from their large surface areas.[1] In recent studies, we exhaustively studied propylene carbonate as a solvent for EDLCs, reported experimental and molecular simulation results on PC interfaces. PC wets graphite with a contact angle of 31° at ambient conditions.[2] Molecular dynamics simulations agree with this contact angle after 40% reduction of the strength of graphite-C atom Lennard-Jones interactions with the solvent. This validation paves the way to direct simulation of dynamical filling and performance of CNT-based EDLCs. Here, we present all-atom molecular dynamics simulations of electrochemical double-layer capacitors based on CNT forests with propylene carbonate tetraethylammonium-tetrafluoroborate (PC:TEA-BF4) solutions. Preliminary results are showed in FIG. 1. We explicit characterizations of the charging and discharging process, including electrolyte composition in the pores, the rates that charging responses equilibrate, and the possibility of bubble formation. Also, we consider the pore size dependence of EDLC characteristics and the structural stability of CNT forests.

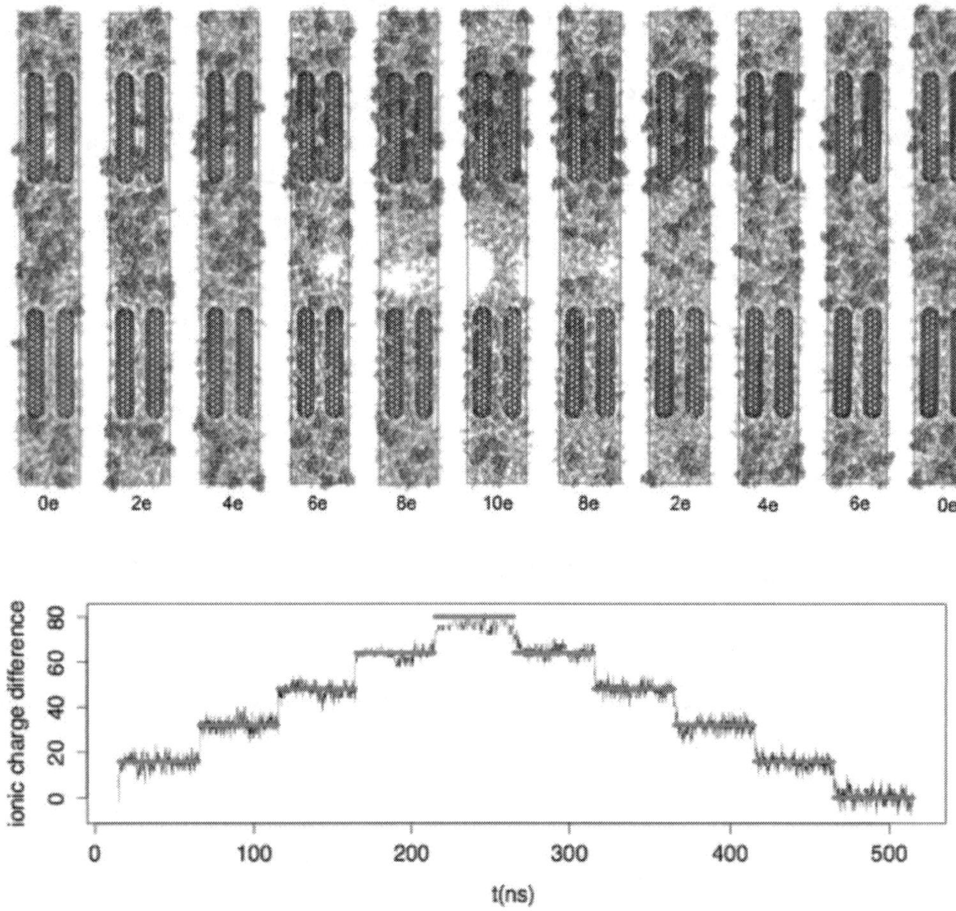

FIG. 1. Upper panel: snapshots, separated by 50 ns, of the molecular dynamical filling of CNT forests as the electronic charge on the nanotubes is incremented. The label below each frame is the electronic charge/CNT. Upper layer has negatively charged CNTs, and the lower layer has positively charged CNTs. Here there are four CNTs above and below, for a total of eight (8) tubes altogether. The solvent is propylene carbonate, and the electrolyte is tetraethylammonium tetrafluorate. T = 300K (established by a Nose-Hoover thermostat). Notice the demixing and formation of a bubble in the separating solution, as the forest becomes strongly charged in this constant volume calculation. Notice also the low occupancy of the pores by the ions at the lowest levels of charging. Lower panel: Difference in ionic charge (black) above and below the mid-plane of the upper panel. Notice the sub-nanosecond equilibration of the charge balance, in both filling and emptying.

[1] L. Yang, *et al., J. Am. Chem. Soc.* **131**, 12373-12376 (2009): "Molecular simulation of electric double-layer capacitors based on carbon nanotube forests."

[2] X. You, *et al., J. Chem. Phys.* **138**, 114708 (2013): "Interfaces of propylene carbonate."

(687a) Controllable Phase Transition of CuBTC Metal-Organic Framework and Its Application in CO2 Capture and Storage

Huang, L., North Carolina State University
Zhang, S.,
Gubbins, K. E., North Carolina State University
Lu, X., Nanjing university of technology
Originally presented on: 11/7/2013 12:30:00 - 12:47:00

As one of the most studied metal-organic frameworks, CuBTC has been considered promising in the fields of adsorption, separation and catalytic reaction. However, the expensive commercial price and the thermal/hydrostatic stability are two limiting factors for industrial utilization of CuBTC. Here we report a controllable and reversible 1D/3D phase transition of CuBTC structures, and demonstrate its application in CO_2 capture and storage. By *ab initio* density functional theory (DFT) calculation and reactive molecular dynamics simulation, we study at the electron and molecular level the stability of 1D and 3D CuBTC structures, the potential reversible phase transition, and the CO_2/CuBTC interactions. Together with the experimental evidence, we discuss the mechanism of the controllable 1D/3D phase transition, the role of hydrogen bond network, the ratio of 1D/3D strucutures and its experimental implementation. The fast and controllable characteristics of the 1D/3D phase transition make CuBTC material a promising candidate in CO_2 separation and storage.

(687b) Thermodynamic Stability of Hypothetical ZIF-8 Polymorphs Using Molecular Simulations

Gee, J., University of California Santa Barbara
Sholl, D. S., Georgia Institute of Technology

Originally presented on: 11/7/2013 12:47:00 - 13:04:00

Zeolitic imidazolate frameworks (ZIFs) are a class of metal-organic frameworks (MOFs) that are known to exhibit structural polymorphism. Despite the large number of reported ZIF polymorphs, the factors governing the stability of these materials remain poorly understood. In this work, we present a novel approach to investigate the thermodynamic stability of MOF polymorphs using molecular simulations. Our work employs a combination of thermodynamic integration (TI) and OFAST simulations to predict the free energy of solvation, ΔG_{solv}, and vibrational density of states (VDOS) calculations to determine the configurational energy of the framework. This approach is applied to a series of hypothetical ZIF-8 polymorphs under solvothermal conditions. Our simulations predict that the framework stability is governed by the configurational energy rather than the ΔG_{solv} of the framework. We find good agreement between OFAST and the more precise TI method for the prediction of the ΔG_{solv} of the framework. This methodology can be applied to study the effect of structural topology, solvent, and temperature on the stability of the framework for a wide range of nanoporous materials.

(687c) Adsorption of 5-Hydroxymethyl-Furfural (HMF) From Liquid Water/Dimethyl Sulfoxide (DMSO) Solutions Onto Hydrophobic Zeolites: An Experimental and Simulation Study

Xiong, R., University of Delaware
Leon, M., University of Delaware
Nikolakis, V., University of Delaware
Sandler, S. I., University of Delaware
Vlachos, D. G., University of Delaware
Originally presented on: 11/7/2013 13:04:00 - 13:21:00

The adsorption of 5-hydroxymethylfurfural (HMF), dimethyl sulfoxide (DMSO), and water from binary and ternary mixtures in hydrophobic silicalite-1 and dealuminated Y (DAY) zeolites at ambient conditions was studied via experiments and molecular modeling. HMF and DMSO adsorption isotherms were measured and compared to those calculated using a combination of grand canonical Monte Carlo and expanded ensemble (GCMC-EE) simulations. A method based on GCMC-EE simulation for dilute solutions combined with the Redlich-Kister (RK) expansion (GCMC-EE-RK) is introduced to calculate the isotherms over a wide range of concentrations. The simulations, using literature force fields, are in reasonable agreement with the experimental data. In HMF/water binary mixtures, large pore hydrophobic zeolites are much more effective for HMF adsorption but less selective because large pores allow water adsorption due to H2O-HMF attraction. In HMF/DMSO/water ternary mixtures, the HMF loading decreases with increasing DMSO fraction, rendering separation of HMF from water/DMSO mixtures by adsorption difficult. We show that the ratio of the energetic interaction in the zeolite to the solvation free energy is a key factor in controlling separation from liquid mixtures. Overall, our findings could have an impact on the separation and catalytic conversion of HMF and the rational design of nanoporous adsorbents for liquid-phase separations in biomass processing.

(687d) Preferential Adsorption From Liquid Ethanol-Water Mixtures in Alumina Pores

Phan, A. T., The University of Oklahoma
Striolo, A., University of Oklahoma
Cole, D. R., Oak Ridge National Laboratory
Originally presented on: 11/7/2013 13:21:00 - 13:38:00

Equilibrium MD simulations were employed to investigate interfacial liquid ethanol-water mixtures on several alumina slit-shaped pores. The results were quantified in terms of preferential adsorption from the bulk, density profiles, in-plane radial distribution functions, residence times at contact with the substrate, self-diffusion coefficients, orientational distribution functions, hydrogen bonds distributions, and reorientation autocorrelation functions. When pure ethanol is simulated, the results show that ethanol molecules in the first monolayer near a solid surface have a well-organized arrangement, while those in the second layer show bulk-like features. The predicted orientational distribution for ethanol molecules within the first layer is consistent with experimental results. When water is added to the system, our results show that water quickly replaces ethanol molecules from the first adsorbed layer, and preferentially partitions within narrow slit-shaped alumina pores. We will discuss how such results could be useful for the design of separation processes. The results could also be important for better understanding the behavior of fluids in sub-surface environments.

(687e) Computer Simulation of Water Sorption On Flexible Protein Matrices

Palmer, J. C., Princeton University
Debenedetti, P. G., Princeton University
Originally presented on: 11/7/2013 13:38:00 - 13:55:00

In the last two decades, protein therapeutics have emerged as the fastest-growing segment of the pharmaceutical industry, resulting in more than 200 new treatments for diseases such as AIDS, leukemia, melanoma, cystic fibrosis, and multiple sclerosis. Despite these advances, however, there remain a number of technical challenges in formulating and manufacturing new protein-based therapeutics. One of the most difficult tasks is to develop therapeutic formulations with long-term stability against the chemical degradation and protein aggregation processes that compromise the safety and efficacy of drugs that are stored in solution over long periods of time.

The standard approach for increasing the stability of protein therapeutics is to remove the biomolecules from their aqueous environment by incorporating them into vitrified solid materials using a process known as lyophilization or freeze-drying. In some cases, however, proteins often exhibit diminished therapeutic activity when they are reconstituted with water following storage in the glass state. To understand this behavior, we use computer simulation to explore how proteins are affected by dehydration and subsequent rehydration. We demonstrate a new computational approach we have developed to study water sorption on protein crystals and powders, and show that this technique allows us, for the first time, to make rigorous contact with experimental measurements designed to probe the protein hydration process. We also show that we are able to go beyond the current capabilities of experimental techniques and investigate how changes in hydration level can give rise to microscopic mechanical stresses that may alter the structure and biological activity of some proteins.

(687f) Adsorption of Polyols and Furfurals in All-Silica Zeolites

Bai, P., University of Minnesota
Tsapatsis, M., University of Minnesota
Siepmann, J. I., University of Minnesota
Originally presented on: 11/7/2013 13:55:00 - 14:11:00

Polyols and furfurals are important commodity compounds that can be produced through biomass conversion. To increase the efficiency and lower the energy cost of the separation of these molecules from reactor effluents, adsorptive separation using nanoporous materials has been proposed as a promising technique. In order to better understand the capacity and the selectivity of all-silica zeolites to polyols and furfurals, configurational-bias Monte Carlo simulations in the Gibbs ensemble were performed to compute adsorption isotherms from solution phases. The TraPPE force field is extended to furfural and its derivatives and intramolecular potentials are determined for some polyols; good agreement with experimental liquid density and vapor pressure data is found for the unary vapor-liquid equilibria of these compounds. The trends in the adsorption isotherms, the free energies of transfer, and selectivities of several selected polyol and furfural molecules are explained in terms of hydrogen-bond speciation and confinement effects in zeolite structures.

(687g) Self-Assembled Monolayer of Carboxylate On Calcite (10$\bar{1}$4) Surface: Molecular Dynamics Simulation Approach

Chun, B., Georgia Institute of Technology

Lee, S. G., Georgia Institute of Technology

Brunello, G. F., Georgia Institute of Technology

Choi, J. I., Georgia Institute of Technology

Jang, S. S., Georgia Institute of Technology

Originally presented on: 11/7/2013 14:11:00 - 14:28:00

Self-Assembled Monolayer of Carboxylate on Calcite $(10\bar{1}4)$ Surface:

Molecular Dynamics Simulation Approach

Byeong Jae Chun, Seung Geol Lee, Giuseppe F. Brunello, Ji Il Choi and Seung Soon Jang

Computational NanoBio Technology Laboratory, School of Materials Science and Engineering, Georgia Institute of Technology, 771 Ferst Drive NW, Atlanta, GA 30332-0245

To study the wettability alteration of calcite surface in oil reservoir, we investigate the molecular adsorption of carboxylate on the calcite $(CaCO_3)$ $(10\bar{1}4)$ surface using full-atomistic molecular dynamics (MD) simulation as well as density functional theory (DFT). There are three types of carboxylate adsorbate used for this study; benzoate, dodecanoate (C12), and octadecanoate (C18). Using DFT calculation, the molecular details of carboxylates are determined including the quantum charge and optimal configuration. Once the fundamental information of the adsorption models is attained, MD simulations are performed to obtain the equilibrium structures and evaluate the thermodynamic stability of the chemisorbed carboxylate and their self-assembled monolayer (SAM) on calcite. Our force field contains the newly developed parameters for the off-diagonal van der Waals interactions of carboxylate functional group with calcite surface. The given original force fields are employed for van der Waals interactions between the same type of atoms or molecules. Based on the thermodynamic stability of each adsorption case achieved by our MD simulations, we determine the optimal packing density of carboxylate SAM on the calcite surface for each case and characterize the corresponding structures. In order to determine the effect of solvation on the packing and structures of SAM on calcite surface, we simulate the same systems in the presence of solvent phase such as water and octane. Once equilibrated structures of carboxylate SAM are obtained, the free energy barrier for detachment of carboxylate from the surface is evaluated by applying potential of mean force (PMF) in order to assess the kinetics of wettability alteration as well as thermodynamic stability.

(687h) Elucidating the Details of Protein Transport in Polymer-Grafted Ion Exchangers Through Multi-Scale Modeling

Basconi, J., University of Virginia
Carta, G., University of Virginia
Shirts, M. R., University of Virginia
Originally presented on: 11/7/2013 14:28:00 - 14:45:00

In protein separations using ion-exchange chromatography, matrices with pores containing charged polymer grafts have exhibited increased binding capacity, faster adsorption kinetics, and a different diffusion mechanism as opposed to traditional macroporous matrices, though these enhancements depend strongly on the protein properties and concentration, as well as the buffer salt concentration. We use multi-scale modeling to study the factors which affect protein transport in polymer-grafted and macroporous materials, which in general depends on both the equilibrium adsorption and diffusion behavior of proteins in the different phases.

We have developed coarse-grained molecular models suitable for molecular dynamics (MD) simulations of proteins diffusing through representative pores of both polymer-grafted and macroporous materials. In simulations of lysozyme interacting with agarose-based matrices with charged dextran polymers, we obtain good agreement with experimental results for equilibrium binding capacities and effective transport rates, as well as their dependence on system parameters such as polymer graft density and surface topology.

The kinetic and equilibrium information from these MD simulations is also employed in a simple rate model to study the macroscopic protein concentration profiles in the different materials over time. With this model we observe faster adsorption kinetics and a different adsorption front for the polymer-grafted material as compared to the macroporous material, which is qualitatively consistent with experiments. We apply this multi-scale modeling approach to different types of proteins and adsorbed phases, as understanding how protein transport rates are affected by various protein and material properties should provide insight into the general mechanism by which charged polymers can enhance chromatographic performance.

(687i) Investigation of the Critical Point of Adsorption of Polymer Chains On Porous Adsorbents

Cimino, R. T., Rutgers, The State University of New Jersey
Neimark, A. V., Rutgers, The State University of New Jersey

Originally presented on: 11/7/2013 14:45:00 - 15:01:00

Liquid phase chromatography at critical conditions (LCCC) is a widely used technique for the separation of polymeric mixtures of similar molecular weight but different chemical properties. The critical point of adsorption (CPA) is defined as the point at which adsorption is independent of molecular weight, and is observed experimentally. While theory has been developed to evaluate the CPA for ideal chains, an adequate theoretical description of the CPA of real chains is still lacking. This work provides a theoretical basis for defining the CPA of a realistic polymer chain based on the free-energy and chemical potential of the chain, and provides model calculations of the CPA on a porous substrate using Gauge Cell Monte Carlo. In developing the model, several modes of adsorption are considered: adsorption approaching the external surface of an adsorbent particle, complete adsorption within a pore, adsorption of chains tethered to the external surface of the particle, and the adsorption of chains which are partially translocated – called "flower" configurations. The summation of the Boltzmann-weighted free energies of these configurations are used as the partition function for calculating system variables – notably the partition coefficient and chemical potential of the polymer chain.

(689a) Understanding Pickering Emulsions Using Multi-Scale Simulations

Striolo, A., University of Oklahoma

Originally presented on: 11/7/2013 12:40:00 - 13:00:00

Pickering emulsions are becoming extremely popular in the academic community. However, few applications benefit from such interesting systems. We argue that better understanding their fundamental properties, especially related to their stability, will enable a host of practical applications. The stability of Pickering emulsions is related to structural and dynamical properties of interfacial nanoparticles. Our group has investigated such properties by means of multi-scale simulations (from atomistic to coarse-grained, to phenomenological). We considered several nanoparticle types, including spherical and elliptical, Janus or homogeneous with different amounts of hydrophilic/hydrophobic surface groups. We focus on (a) the effect of particle properties on self-assembled aggregates structure, and (b) systems of different composition. For a given system composition, we investigate the effect of increasing nanoparticle density. We will present here a survey of our recent results, quantified in terms of interfacial tension reduction, radial distribution function, order parameters, contact angle, and self-diffusion coefficient, etc. The goal is to identify properties that could be tested experimentally.

(689b) Effects of Vapor Adsorption Isotherm On Adhesion, Friction, and Wear of Solid Interfaces

Kim, S. H., The Pennsylvania State University
Al-Azizi, A.,
Barthel, A.,
Originally presented on: 11/7/2013 13:00:00 - 13:20:00

In tribology, there are many examples where we expect certain adhesion, friction and wear behaviors based on mechanical properties of solids or structures of molecules involved; but find drastically different results. There are clear demonstrations that fundamental understandings of tribological systems at nano- or molecular scales are still far from realistic. For example, it is often expected that friction and wear behaviors of glass and silica are similar since both are based on amorphous Si-O networks and their moduli are similar. Thus, many scientists study silica surfaces as a model system for more complicated glasses. However, their wear behaviors are drastically different in humid environments. We expect the wear is always lower when the friction coefficient is lower; but diamond-like carbon (DLC) exhibits no wear in alcohol vapor environments even though its friction coefficient is much higher than the values measured in dry conditions in which hydrogenated DLC surface shows wear. These are a few examples where conventional solid contact mechanics or tribochemistry models cannot explain the interfacial processes. These observations can be explained only when we fully understand the mechanochemical interplay at nanoscales between the solid surfaces and the molecules adsorbed from the environment. This talk will address adsorption isotherms of vapor molecules and their effects on mechanical properties of solid interfaces.

(689c) Load-Induced Hydrodynamic Lubrication of Porous Polymer Films

Pesika, N. S., Tulane University

Originally presented on: 11/7/2013 13:20:00 - 13:40:00

In this study, we explore the mechanisms involved in the lubrication of porous polymer-based surfaces or coatings with ultra-low coefficients of friction arising from the partial hydrodynamic repulsive forces acting between shearing solid yet compliant surfaces. Through tribological experiments, using a universal materials tester, we show how it is possible to shift the lubrication regime from boundary lubrication to hydrodynamic lubrication even at relatively low shearing velocities. We hypothesize that the extruding liquid from the pores provides a repulsive hydrodynamic force, which aids in minimizing contact between the shearing surfaces. Conventional phololithography and molding techniques are used to fabricate the porous polydimethyl siloxane (PDMS) surfaces. Specifically, the effect of the pore density and pore size on the lubrication properties will be elucidated. In addition, the effect of the lubricating liquid (including water and an aqueous sodium dodecylsulfate solution) on the friction forces will be studied. Potential applications for these low friction biocompatible surfaces include their use in the fabrication of joint implants.

(689d) Direct Measurement of the Surface Tension of a Soft Elastic Hydrogel: Exploration of Elasto-Capillary Instability in Adhesion

Chakrabarti, A., Lehigh University

Chaudhury, M. K., Lehigh University

Originally presented on: 11/7/2013 13:40:00 - 14:00:00

Elasto-capillarity plays an important role in a variety of surface phenomena. In an attempt to investigate the role of elasto-capillarity in the adhesion of ultra-soft gels, it was found that an adhesively stressed thin film of a soft hydrogel confined between two rigid flat substrates auto-roughens with its dominant wavelength (λ) exhibiting pronounced dependence on the film thickness (H). A linear stability analysis confirmed that this long wavelength instability (λ ~$7H$) is due to an elasto-capillary effect, the implementation of which required direct measurements of the surface tension and the elasticity of the gel. A novel method is introduced to estimate the surface tension (γ) of the gel from the fundamental spherical harmonic of its hemispherical cap that is excited by an external noise. The resonance vibration frequency of the hemispherical gel cap's free surface was found to vary with its volume as $\omega \sim V^{-0.55}$, being very close to that of water in magnitude and character, i.e., $\omega \sim V^{-0.5}$. Analogous to the Stokes' experiment in liquids which is performed to measure the viscosity, a static Stokes' braking experiment in the soft hydrogel is found to give an estimation of the shear modulus. This was achieved by submerging a high density steel ball inside the gel which balances its excess weight with the accumulated strain induced elastic force. The shear modulus (μ) of the gel was also determined from its resonant shear mode in a confined geometry comprising a thin gel slab sandwiched between two glass plates. From the independently determined values of the surface tension and the shear modulus of the gel, a large elasto-capillary length ($\gamma / \mu \approx 1.8$ mm) ascertains the role of elasto-capillarity in the adhesion induced pattern formation with such gels. Experimental results are in accord with a linear stability analysis that predicts that the rescaled wavelength $\lambda(\mu H / \gamma)^{0.27}$ is linear with H. Some new results related to the role of the finite dilation of a material in interfacial pattern formation that may have non-trivial consequences in the adhesive delamination of very thin and/or soft elastic films via self-generated cracks were also found.

(689e) Dynamic of Flooded Adhesion in the Presence of Draining Channels

Dhong, C.,
Wang, Y.,
Gupta, R., Johns Hopkins University
Frechette, J., Johns Hopkins University
Originally presented on: 11/7/2013 14:00:00 - 14:20:00

We will discuss our efforts to understand the mechanisms behind the adhesion of the tree frog under wet conditions. Central to the adhesion and locomotion of tree frogs is their structured toe pads, which consist of hexagonal arrays of epithelial cells separated by a network of interconnected channels. It has been proposed that the channels facilitate the drainage of excess fluid trapped between the toe pads and the contacting surface, and thus reduce the hydrodynamic repulsion during approach. We present direct force measurements of the normal and peeling hydrodynamic interactions during the drainage of fluid from the gap between a structured and a smooth surface using surface force apparatus. We investigate two limits: when the interacting surfaces are driven toward each other at constant velocity and constant force. We compare our results to predictions from Reynolds' theory and discuss how the deviations are related to the surface features. We find agreement between our experiments and the scaling analysis for Reynolds theory in the limit where fluid can drain through the structure.

(689f) Characterization of Bacterial Spore Surfaces Using Nanomechanical Mapping

McCullough, E. J., Virginia Commonwealth University
Haithcock, J. M.,
Ehrhardt, C.,
Yadavalli, V. K., Virginia Commonwealth University
Originally presented on: 11/7/2013 14:20:00 - 14:40:00

Several biochemical signatures have been developed that can differentiate bacterial pathogens (and, potentially, the laboratory of origin). However, most require significant amounts of sample ($>1x10^9$ cells), which prohibits their use in analyzing biocrime evidence. New systems are needed for the detection and characterization of samples with traces of cells. To address this need, this study developed a means of characterizing bacterial spore surfaces for determining forensically-relevant aspects of an organism's growth environment at the single cell level. Specifically, the morphology of the cell surface was investigated using nanomechanical mapping to obtain mechanical "signatures" to distinguish different strains and species. Cultures of *Bacillus cereus* spores, a structural and chemical analog for *Bacillus anthracis* and different strains of *B. cereus* (14579 and T-strain) were grown in sporulating culture medium and analyzed using Atomic Force Microscopy (AFM) based nanoindentation. An image recognition based algorithm was used to classify the maps obtained. Results showed that the topography of the spore could be compared by considering the nanomechanical "signatures" and morphologies of the different spores. Since forensic evidence often contains trace quantities of material, the use of AFM in identifying phenotypic differences of *Bacillus* spores batch cultured in various production conditions can be a novel technique for the attribution of pathogens.

(689g) Membrane Device for Naturally Aiding Child Delivery in Resource-Limited Settings

Kamboj, L., Tulane University
Pesika, N., Tulane University
Sergey, S.,
Originally presented on: 11/7/2013 14:40:00 - 15:00:00

Approximately one third of pregnancies are delivered by one out of three possible operative methods: vacuum extraction, forceps operation, or caesarean section. Using these traditional devices or alternative methods, the risk of injuring to the mother and the fetus is elevated tremendously in the wake of poor training. Here, we propose to develop a polymer-based membrane, which will provide ultra-low friction thus facilitating child delivery but also have the ability to provide high friction when needed for child extraction. Specifically, the friction properties between polydimethylsiloxane (PDMS) and a borosilicate surface were studied using different lubricating media. A low friction coefficient of approximately 0.009 was obtained when using an aqueous 0.367 M sodium dodecyl sulfate solution compared to a relatively high friction coefficient of 2.17 when using deionized water. We will discuss the design of the delivery membrane and how the anisotropic friction properties have been integrated into the device. Such a device has the potential to reduce the risk of injury during childbirth.

(694a) Homogeneous Nucleation of Ice in Droplets and Free-Standing Films of Supercooled Water

Haji-Akbari, A., Princeton University
Debenedetti, P. G., Princeton University
Originally presented on: 11/7/2013 12:30:00 - 12:52:00

The problem of homogeneous ice nucleation is of immense importance to fields as diverse as biology and chemistry, to geology and atmospheric sciences. Of particular interest is the formation of ice in confined geometries. One such example is the nucleation of ice in droplets and free-standing thin films, systems that are especially important in geological and atmospheric sciences. In this work, we use molecular simulations to compute the rates and unravel the mechanism of ice formation in droplets and thin films of supercooled water. The water model used in this study is the monoatomic potential proposed by Molinero et al. [1]. This model has no explicit hydrogen atoms, and the local tetrahedrality of water molecules that arises from the formation of hydrogen bonds is enforced by including a three-body interaction term in the inter-molecular potential. The lack of electrostatic interactions in the Molinero water leads to faster dynamics and increased computational efficiency.

Since the formation of critical nuclei within a supercooled liquid is a rare event, it cannot be properly sampled from a single long Molecular Dynamics (MD) trajectory of the system even at relatively large supercoolings. This makes conventional MD unsuitable for studying nucleation. We therefore combine conventional molecular dynamics with Forward-Flux Sampling (FFS) which is an advanced sampling technique developed for computing the rates of rare events such as protein folding, evaporation and crystallization [2]. We calculate crystallization rates for temperatures 220-235 K, which are then used for computing activation energies and critical nucleus sizes. We also compare our findings with nucleation rates in the bulk to understand how geometric confinement affects the crystallization process. Our findings will help us understand how a free interface affects ice formation in water, which will be an invaluable knowledge for both statistical physicists and atmospheric scientists.

1- Molinero V, Moore EB, *J. Phys. Chem. B* **113(13)**: 4008-16 (**2008**).

2- Allen RJ, Frenkel D, ten Wolde PR, *J. Chem. Phys.* **124**: 194111 (**2006**).

(694b) Growth of Tetrahydrofuran (THF) Clathrate Hydrates: A Molecular Dynamics Study

Wu, J. Y., National Taiwan University
Lin, S. T., National Taiwan University
Originally presented on: 11/7/2013 12:52:00 - 13:14:00

The growth of tetrahydrofuran (THF) clathrate hydrates and methane-THF clathrate hydrate were calculated by Molecular dynamic (MD) simulations. The melting points of THF hydrate determined from MD simulations were 277.5K at 1atm and 270K at 1000atm, in good agreement with the experimental data (277.6K at 1atm and 274K at 1000atm). The melting point of THF hydrate was found to correlate strongly with the solubility of THF in water. The solubility of THF in water may decrease from completely miscible to less than 0.01 (mol THF/mol solution) at 298K and 1atm by changing the interaction parameters between THF and water. In the latter case, the melting point of THF clathrate hydrate decreased to less than 260K at 1atm. The occupancy of THF in the $5^{12}6^4$ cages strongly stabilized the structure II hydrate. In the growth process, the all $5^{12}6^4$ cages are filled with THF (occupancy is 1). The promoting effect of THF on CH_4 clathrate hydrates was successfully simulated. With the adding THF in the system, the melting temperature of CH_4 clathrate hydrate increased about 50 K.

(694c) Molecular Dynamics Simulations Of Vapor Bubble Growth and Detachment On A Heating Surface- Investigation Of The Contact LINE Dynamics

Bao, J., City College of New York ,City University of New York
Koplik, J., City College of the City University of New York
Rumschitzki, D. S., The City College of The City University of New York
Originally presented on: 11/7/2013 13:14:00 - 13:36:00

Based on our previous simulations of conductively-driven quasi-static vapor bubble growth in an axisymmetric, cylindrical cell comprised of solid and liquid phases of finite thicknesses with fluid motion and heat transfer obeying small Reynolds, Peclet, Capillary and Bond numbers, we found that the appearance and the motion of the three-phase contact line (CL) is a critical element, central to physics needed to explain the large heat enhancement in nucleate boiling. However, our simulations also suggested that different ad hoc models for the poorly-understood motion of the CL with phase change occurring at the CL as the vapor bubble grows do have a significant effect on vapor bubble deformation and detachment when the Bond number is no longer small. Therefore, we have used molecular dynamics (MD) to simulate a nano-scale version of our three-phase system.

Our MD simulation domain of interest is a cuboid, which is composed of the fluid region sandwiched between two parallel solid walls. The fluid phase is made of argon atoms interacting via a 12-6 Lennard-Jones (LJ) potential. The atoms of the two solid walls are tethered via springs to their lattice positions and also interact with both the fluid and each other via LJ potentials. We determine the critical interaction parameters between the solid and fluid atoms in LJ form by considering the balance between the wettability of the solid surface, which is key to observe nucleation and the heat transfer rate through the solid to the fluid. Periodic boundary conditions are applied at the 4 vertical boundaries of the domain. After achieving thermal equilibrium of all the phases in the simulation domain (solid walls and liquid argon) at the uniform reduced temperature of T=0.75 (the fluid saturation temperature), we start to expend the top wall gradually at constant reduced temperature T=0.75 to some specified pressure at which no vapor bubble appears via cavitation. Next, we instantaneously increase the temperature at the bottom layer of the bottom wall while keeping the original lower temperature at the top wall, both maintained by thermostating. This maintains a temperature gradient through the fluid phase. Meanwhile, the pressure at the top wall (which matches that in the bulk fluid) is kept fixed by allowing the top surface to freely slide up and down.

Without applying a uniform body force (that in some way might mimic gravity) to the fluid atoms, we adjust the control pressure and the heating temperature and find the region of the pressure vs temperature plot where a vapor bubble nucleates and eventually grows to cover the entire solid surface, where a bubble nucleates but does not grow to cover the entire solid surface and where no bubble nucleates in a reasonable amount of time. By then artificially applying a uniform body force that mimics a relevant Bond number, we then successfully nucleate and grow a vapor bubble to a moderate size and observe bubble necking and detachment. We find that,, despite the existence of temperature slip between solid and liquid, the vapor bubble volume still grows as $t^{3/2}$, t being time, as in our earlier conduction-only continuum calculations. This scaling is also in agreement with numerous experimental observations. Furthermore, our temperature profiles show that the vapor bubble surface at the three-phase CL appears to be at the liquid saturation temperature, which confirms the assumptions made in our continuum calculations' boundary condition at liquid-vapor interface and where it contacts the solid surface. By further increasing this artificial uniform body force (i.e., Bond number) and then keeping it constant, we find the body force indeed deforms the bubble and causes it to detach from the surface. This provides critical information as to how system parameters affect the size of the vapor residue left on the surface and on whether cooler fluid flowing downward to replace the rising detached bubble succeeds in quenching this residue. During this process, we reset/raise the temperature at the top wall to avoid vapor condensation at the top of vapor bubble as the bubble grows, detaches and rises into regions of lower temperature, without changing the magnitude of the uniform body force. These simulations result in, among other things, the time evolution of the CL motion. The radius of CL initially expands with the growth of the bubble at low Bond number. Then, as the bubble grows to a size where the body force becomes significant, the CL contracts sharply but continuously as the bubble deforms until detachment. Finally, we incorporate this physical CL motion into our continuum calculation to simulate the vapor bubble growth until detachment in a continuum model.

(694d) Molecular Simulation of Homogeneous Crystal Nucleation and Growth in Polyethylene Melts

Yi, P., Massachusetts Institute of Technology
Rutledge, G. C., Massachusetts Institute of Technology
Originally presented on: 11/7/2013 13:36:00 - 13:58:00

The crystallization of polymers, like that of small molecules, can be described by a two-stage process of homogeneous nucleation and crystal growth. Yet it is also different from that of small molecules in having chain folding phenomenon and the formation of lamellae structure due to the interplay between long, flexible chain molecules. Chain folding and the lamellar crystalline morphology is essential to the structure and properties of the majority of commodity and high performance plastics, but the understand of their development on the microscopic level remains incomplete.

To study this process, using a realistic united-atom force field, molecular dynamics simulations were performed at about 30% supercooling from the melts of *n*-pentacontahectane (C150) and a linear polyethylene (C1000), both of which are long enough to exhibit the chain folding that is characteristic of polymer crystallization. The nucleation rate was calculated and the critical nuclei were identified using a mean first-passage time analysis. The nucleation rate was found to be insensitive to the chain length in this range of molecular weight. The critical nucleus contains about 150 carbons on average and is significantly smaller than the radius of gyration of the chains, at this supercooling. Using the survival probability method, the critical nucleus size was determined as a function of temperature. A cylinder model was used to characterize the shape of the crystal nuclei and to calculate the interfacial free energies. The interfacial free energies were found to be temperature-dependent. The nucleation free energy barrier and nucleation rate as functions of temperature were also calculated and compare favorably with experiments. At the end of the crystal growth stage, stable crystal lamella forms. A chain segment analysis was performed to characterize the topology of the lamella-melt interface in terms of loops, bridges and tails. The length distribution of loops is broad, supporting the "switchboard model" for the crystals formed at deep supercooling.

References: [1] P. Yi, C.R. Locker and G.C. Rutledge, Macromolecules 2013 (in press).

(694e) Pore-Scale Study of the Effect of Precipitate On Dissolution Using Lattice Boltzmann Method

Kang, Q., Los Alamos National Laboratory
Chen, L., Xi'an Jiaotong University
Originally presented on: 11/7/2013 13:58:00 - 14:20:00

A mesoscopic model based on the lattice Boltzmann method (LBM) is developed to investigate the effects of secondary precipitate on the dissolution of primary solid phase. The model can predict coupled multiple physicochemical processes including fluid flow, mass transport, chemical reaction, dissolution, precipitation consisting of nucleation and crystal growth, as well as dynamical evolutions of pore geometries. Effects of dissolution and precipitation reaction kinetics, molar volumes of primary and secondary solid phases, initial powder size and surface roughness of the primary solid phase, as well as nucleation and crystal growth mechanisms on the dissolution and precipitation processes are investigated in terms of rate and amount of dissolution and precipitation. Several types of dissolution and precipitation processes are identified based on the morphology and structures of the precipitates and the extent the precipitates affect the dissolution of the primary solid phase. Simulation results are also compared with existing experimental results. Depending on the conditions, the precipitates are found to have trivial, favorable or detrimental effects on the dissolution of the primary phase.

(694f) Prediction of Crystal Habits of Urea Using Molecular Dynamics Simulations and Comparision to Experimental Results

Vetter, T., ETH Zurich
Salvalalglio, M., ETH Zurich
Mazzotti, M., ETH Zurich
Parrinello, M., ETH Zurich
Originally presented on: 11/7/2013 14:20:00 - 14:42:00

Crystals grown at moderate and low levels of supersaturation typically exhibit a faceted shape that reflects the ordered arrangement of molecules within the crystal lattice and the environment in which the crystal was grown. In solution crystallization, solvents and additives (or impurities) can significantly change crystal shape by selectivley discouraging (or promoting) the growth of individual crystal faces. While these effects manifest in habit changes on the scale of whole crystals, their underlying reason is the interaction between solvent and additive molecules with the crystal on a molecular scale (Salvalaglio et al., 2012).

Investigating crystal growth at the molecular level is therefore of great importance in order to enhance our current understanding of crystallization processes. Moving towards this goal is a multi-scale (and multi-disciplinary) challenge that can greatly benefit from the interplay between theory, simulations and experimental techniques. In this work we therefore combine advanced molecular simulation methods (Laio and Parrinello, 2002, Barducci et al., 2008, Bonomi et al., 2009) with a theoretical crystal growth model that describes rough crystal growth and growth by the birth and spread (2D nucleation) mechanism (Ohara and Reid, 1973) on a molecular scale. Using molecular simulations that explicitly include solute, solvent and additive molecules allows us to obtain the physical parameters in our theoretical crystal growth model, which allows calculating relative growth rates of different crystal faces. The relative growth rates can then be used to predict crystal habits that originate from the simulated conditions.

Here we examine the paradigmatic case of urea, which grows from vapor exposing three well defined crystal faces: {001}, {110} and {111} (Docherty et al., 1993). Experimental evidence (Vetter, 2012) highlighted that the solvent and additives in solution caused a wide variety of crystal habits that all consisted of combinations (or subsets) of these faces. In order to rationalize this finding we have developed a representation tool, the Shape Diagram (SD), in which all possible steady state morphologies of urea crystals for different growth rate ratios can be assessed.

The rational display of the accessible crystal habits allowed comparing experimental results with computational predictions of the surprisingly wide ensemble of urea growth shapes obtained from different solvents and additives. The experimentally found and computationally predicted crystal habits were found to be in good agreement, thus demonstrating that the developed simulation methodology and theoretical model are able to predict crystal habits from a variety of supersaturations, solvents and additives (Salvalaglio et al.,

2013).

References:

A. Barducci, G. Bussi, and M. Parrinello (2008). Well-Tempered Metadynamics: A Smoothly Converging and Tunable Free-Energy Method. Phys. Rev. Lett., 100:020603

M. Bonomi, A. Barducci, and M. Parrinello (2009). Reconstructing the equilibrium Boltzmann distribution from well-tempered metadynamics. J. Comput. Chem., 30:1615

K. Chow, H. H. Y. Tong, S. Lum, and A. H. L. Chow (2008). Engineering of pharmaceutical materials: An industrial perspective. J. Pharm. Sci., 97:2855-2877.

R. Docherty, K. J. Roberts, V. Saunders, S. Black and R. Davey (1993). Theoretical analysis of the polar morphology and absolute polarity of crystalline urea. Faraday Discuss., 95:11-25.

M. Ohara and R. C. Reid (1973). Modeling crystal growth rates from solution, Prentice Hall Inc., Englewood Cliffs.

A. Laio and M. Parrinello (2002). Escaping free-energy minima. Proc. Natl. Acad. Sci., 99:12562{12566, 2002.

M. Salvalaglio, T. Vetter, M. Mazzotti, and M. Parrinello (2012). Uncovering Molecular Details of Urea Crystal Growth in the Presence of Additives. J. Am. Chem. Soc., 134:17221-17233.

M. Salvalaglio, T. Vetter, M. Mazzotti, and M. Parrinello (2013). Controlling and Predicting Crystal Shapes: The Case of Urea. Submitted to Nature Materials.

T. Vetter (2012). PhD thesis, ETH Zurich, 2012. (http://e-collection.library.ethz.ch/view/eth:6639)

(694g) Predicting Dispersions in Crystallization Process

Singh, M. R., Purdue University
Ramkrishna, D., Purdue University
Originally presented on: 11/7/2013 14:42:00 - 15:04:00

Crystallization is one of the most commonly used and relatively well explored techniques in industries. Yet there are beleaguering issues towards fundamental understanding of this process. One of the curious observations is the appearance of unavoidable dispersions in nucleation and growth rates of crystalline materials grown under homogenous bulk conditions. These dispersions can contribute significantly towards polydispersity in crystalline products. The classical models such as Gibbs critical nucleation theory and Burton-Cabrera-Frank growth models primarily relate nucleation and growth to supersaturation. Therefore, the dispersions in nucleation and growth rates can be related to the local fluctuations in supersaturation. These local fluctuations are driven by the Brownian motion of solute molecules which affects their local concentration and temperature. This article presents a stochastic model to describe fluctuations in supersaturation and thereby dispersions in crystalline materials. The stochastic model is derived from the framework of density and temperature fluctuations. The resulting dispersions in size and shape of nuclei are obtained by multi-dimensional maximization of Gibbs free energy in Morphology Domain of Potassium Acid Phthalate. The model predictions are validated with the experimental measurements of growth rate dispersions in Potassium Acid Phthalate crystals. The proposed framework provides the first rational and quantitative basis for observed dispersions in crystallization processes.

(709a) Computational Study of the Proton Affinity, 1 H NMR Chemical Shifts, and Thermophysical Properties of Imidazole Derivatives

Liu, H., University of Alabama
Bara, J. E., University of Alabama
Turner, C. H., University of Alabama
Originally presented on: 11/7/2013 15:15:00 - 15:36:00

There has been increasing interest in developing new materials for CO_2 capture. One of the approaches is to synthesize polymer membrane using imidazole derivatives as the monomer. These imidazole derivatives are able to react with aqueous CO_2 to form bicarbonate anions, thus enabling facilitated transport of CO_2 through the membrane. In order to better understand the chemical and thermophysical properties of imidazole derivatives, we have employed quantum mechanics and molecular dynamics methods to calculate the proton affinities, ^{1}H NMR chemical shifts, and absorption properties of a variety of imidazole derivatives. From our simulations, a quantitative characterization of the effect of various exocyclic substituents on the proton affinity of imidazole was achieved. In particular, we find that electron-donating groups are able to increase the proton affinity of imidazole. In addition, the calculated ^{1}H NMR chemical shift and thermophysical properties are in excellent with experiment. These results will be helpful for future experimental design of new materials to be used for CO_2 capture.

(709b) Separation Selectivity of Various Gases in the Ionic Liquid 1-Ethyl-3 -Methylimidazolium Tris(pentafluoroethyl)Trifluorophosphate

Althuluth, M., The Petroleum Institute
Peters, C. J., Petroleum Institute
Berrouk, A. S., Petroleum Institute
Kroon, M. C., Eindhoven University of Technology
Originally presented on: 11/7/2013 15:36:00 - 15:57:00

The separation of carbon dioxide (CO_2) from methane (CH_4) is an important process in many industrial areas such as natural gas processing and biogas purification [1]. Natural gas also contains significant amounts of ethane, some propane, butane, and other higher hydrocarbons [2]. In natural gas treating, loss of heavy hydrocarbons is a concern. It is desirable to recover these compounds due to practical problems.

There are many methods available for the removal of acid gases from gas streams. The most commonly used are chemical solvents, physical solvents, membranes and cryogenic fractionation [3]. Physical solvents tend to be favoured over chemical solvents when the concentration of acid gases is very high. However, if the concentration of heavy hydrocarbons is high, a physical solvent may not be the best option due to higher co-absorption of hydrocarbons. An acceptable solvent should have a high capacity for acid gas and a low capacity for hydrocarbons [4].

Therefore, ILs have been proposed to be used as a solvent for CO_2 capture, because of their advantageous properties over conventional ones [5]. We found that the ionic liquid (IL) 1-ethyl-3 methylimidazoliumtris(pentafluoroethyl) trifluorophosphate ([emim][FAP]) shows the highest carbon dioxide (CO_2) solubility of all ILs studied so far and shows higher selectivities for CO_2/CH_4 separations than any other IL, indicating the promising potential of using this IL for the separation of CO_2 from natural gas. Also, the solubilities of C_2H_6, C_3H_8 and C_4H_{10} in ([emim][FAP]) were measured and compared to the CO_2 solubility in the same IL.

The separation ratio between CO_2 and hydrocarbons decreases as the hydrocarbon chain becomes longer. The selectivity increase accordingly to the order: $S_{CO2}/_{C4H10} < S_{CO2}/_{C3H8} < S_{CO2}/_{C2H6} < S_{CO2}/_{CH4}$. A maximum selectivity is achieved at lower temperatures. Therefore, the CO_2 removal from a natural gas stream is recommended to be performed at low temperatures in order to achieve the best separation.

The obtained solubility data are modelled with the Peng-Robinson equation of state combined with quadratic mixing rule. The calculated data have been found to be in a good agreement with the experimental results.

Refrences:

[1] M. Arruebo, J. Coronas, M. MenÃ©ndez, J. Santamaría, Separation and Purification Technology 25 (2001) 275-286.

[1] R.W. Baker, K. Lokhandwala, Industrial & Engineering Chemistry Research 47 (2008) 2109-2121.

[3] G. Aldana, R. Arai, D. G. Elliot, An Evaluation of Sources of CO2 for EOR in Venezuela, Proceedings of the Sixty-Third Annual Gas Processors Association Convention, New Orleans, Louisiana, 1984.

[4] N. Korens, D. R. Simbeck, D. J. Wilhelm, Process Screening Analysis of Alternative Gas Treating and Sulfur Removal for Gasification, Prepared for U.S. Department of Energy by SFA Pacific, Inc., Revised Final Report, December 2002.

[5] E.D. Bates, R.D. Mayton, I. Ntai, J.H. Davis, Journal of the American Chemical Society 124 (2002) 926-927.

(709c) Modelling and Measuring Hydrate Promotion for CO2 Capture

Herslund, P. J.,
Thomsen, K., Technical University of Denmark
Abildskov, J., Technical University of Denmark
von Solms, N., Technical University of Denmark (DTU)
Galfr , A.,
Brantuas, P., Ecole Nationale Sup rieure des Mines de St-Etienne
Kwaterski, M.,
Herri, J. M.,
Originally presented on: 11/7/2013 15:57:00 - 16:18:00

A novel separation concept is investigated here, whereby a precipitated, or slurry, system, enables CO_2 capture by hydrate formation – enhanced by thermodynamic promoters. We have demonstrated that a model for the hydrate phase, combined with an equation of state for associating systems – the Cubic Plus Association (CPA) equation of state enables an accurate determination of mixed hydrate dissociation pressures as well as hydrate composition.

We have subsequently presented experimental work measuring promoted gas hydrate formation with tetrahydrofuran (THF) and cyclopentane (CP), both known low-pressure/high temperature sII hydrate formers. However when applied as a mixed promoter system the hydrate formation pressures are lowered by up to 20% when compared to the pure CP promoter system, which is thermodynamically the more efficient of the two pure promoter systems.

Finally, in order to model this promoted system the complex phase behaviour between water and THF needs to be understood. This system is successfully modeled using CPA, including the complex LLE observed at high pressures. This complex system is best modeled by specifically accounting for the solvation of THF (i.e. that THF forms hydrogen bonds with water but is not self-associating), which is also the correct physical picture of what occurs.

(709d) Experimental Characterization Of Multiply-Substituted Alkylimidazoles As Solvents and Polymeric Membranes For CO2 Capture

Bara, J. E., University of Alabama
Shannon, M. S., University of Alabama
Moon, J., University of Alabama
Reclusado, K., Villanova University
Originally presented on: 11/7/2013 16:18:00 - 16:39:00

Previous work has characterized specific physical properties and CO_2 solubility of 1-*n*-alkylimidazole solvents for use in pre- and post-combustion CO_2 capture during gas treating. This research project explores the CO_2 solubility and physical properties of multiply-branched alkylimidazole solvents to determine the effects of functional group position(s) on various chemical and physical attributes. Using this information, additional correlations between ring substitution and physical properties can be derived to optimize the characteristics best suited for larger-scale pilot testing in industrial applications. Synthesis and purification of these compounds were followed by tests to determine density, viscosity, and vapor pressure as functions of temperature and ring substitution, and these properties were shown to be comparable to previously-tested 1-*n*-alkylimidazoles. The solvents in general were found to have low viscosities (<15 cP) and negligible vapor pressures (<0.5 mbar). CO_2 solubility experiments also showed that multiply-branched alkylimidazoles have similar if not improved absorption capacities relative to ionic liquids and 1-*n*-alkylimidazoles. In future work, polymeric membranes can be synthesized from these compounds to determine CO_2 permeability and selectivity as a function of ring substitution. The ultimate goal of this research is to find an imidazole-based solvent with low volatility and high CO_2 absorption capacity or an imidazole-based membrane structure with high CO_2 permeability and selectivity that can be easily manufactured for application in large-scale industry for the reduction of atmospheric CO_2 emissions.

(709e) Tuning the Physical and Chemical Properties of Novel CO2 Capture Solvents Based On Hybrid Nanomaterials

Petit, C., Imperial College London
Park, A. H. A., Columbia University
Originally presented on: 11/7/2013 16:39:00 - 17:00:00

Nanoparticle Organic Hybrid Materials (NOHMs) represent a novel class of liquid-like materials, whose potential applications include CO_2 capture. Their negligible vapor pressure and high thermal stability have indeed prompted the research community to investigate these nanomaterials as alternative carbon capture media to the common amine-based solvents. NOHMs are synthesized by grafting a nanosized inorganic core with a polymeric canopy. Prior studies have shed light on the mechanisms of CO_2 capture in NOHMs as well as their capture capacity and selectivity. In this study, the investigation of CO_2 capture using NOHMs is continued to investigate the effects of NOHMs' chemistry (*e.g.*, effect of amines functionalities) and structure (*e.g.*, effect of core size) on tuning the kinetic (*e.g.*, CO_2 diffusion), thermodynamic (*e.g.*, heat capacity), absorptive (*e.g.*, selectivity compared to H_2O and SO_2) and physical (*e.g.*, thermal stability) properties of NOHMs in terms of CO_2 capture. A range of analytical techniques are employed including various spectroscopical techniques (*i.e.*, ATR FT-IR, NMR, Raman), thermogravimetry and rheometry.

(709f) The Study of Using Novel Scrubbing Agent Mg(OH)2 Solution to Capture CO2 From Flue Gases

Cheng, L., University of Cincinnati
Li, T., University of cincinnati
Keener, T. C., University of Cincinnati
Lee, J. Y., University of Cincinnati
Originally presented on: 11/7/2013 17:00:00 - 17:21:00

Abstract

A novel CO_2 capture method by using environment-friendly chemical magnesium hydroxide solution has been tested and evaluated in response to minimizing the CO_2 emissions from coal-fired power plant in the post-combustion control area. The goal of this study is to conduct the technical feasibility studies in the area of physical &chemical characteristic of $Mg(OH)_2$, chemistry of the system, lean solution performance in absorption, capacity of regeneration in desorption. It is found that >90% removal efficiency can be achieved in the bubble column absorber; a C/Mg ratio of 9 is achievable at the condition of 52 ℃absorption and 75 ℃ desorption; a fresh feeding rate of 0.008 L/sec is optimized condition in the continuous test.

(709g) Electrochemical Strategies for Reduction of Greenhouse Gas Emissions

Stern, M. C., Massachusetts Institute of Technology
Herzog, H., Massachusetts Institute of Technology
Hatton, T. A., Massachusetts Institute of Technology
Eltayeb, A.,

Originally presented on: 11/7/2013 17:21:00 - 17:42:00

Anthropogenic carbon dioxide (CO_2) in the Earth's atmosphere has been cited as a primary cause of global climate change and threatens global public health and welfare. Carbon Capture and Sequestration (CCS) is an effective and important part of CO_2 emission abatement strategies, with the major CCS efforts to date focusing on the removal of directly from large-scale carbon emitters and storing it in secure geologic reservoirs. Thermal-swing operations using aqueous base scrubbing followed by stripping at elevated temperature have been the chemical sorption processes most investigated over the past two decades for CO_2 capture. Considerable quantities of steam and heat are required to release the CO_2 after capture at low temperatures, and substantial parasitic energy losses result from the need to use excess steam and heat in order to meet the kinetic requirements of the process.

Electrochemically-mediated amine regeneration is a new post-combustion capture technology with the potential to exploit the excellent removal efficiencies of thermal amine scrubbers while reducing parasitic energy losses and capital costs. The improvements result from the use of an electrochemical stripping cycle, in lieu of the traditional thermal swing, to facilitate CO_2 desorption and amine regeneration; metal cations generated at an anode react with the amines, displacing the CO_2, which is then flashed off, and the amines are regenerated by subsequent reduction of the metal cations in a cathode cell. The advantages of such a process include higher CO_2 desorption pressures, smaller absorbers, and lower energy demands. Several example chemistries using different polyamines and copper are presented. Experimental results indicate an open-circuit efficiency of 54% (15 kJ/mole CO_2) is achievable at the tested conditions and models predict that 69% efficiency is possible at higher temperatures and pressures. A bench scale system produced 1.6 mL/min CO_2 while operating at 0.4 volts and 42% Faradaic efficiency; this corresponds to a work of less than 100 kJ/mole.

(728a) Tunable Macroporous Photonic Crystals Enabled By Shape Memory Polymers

Fang, Y., University of Florida
Jiang, P., University of Florida
Originally presented on: 11/7/2013 15:20:00 - 15:40:00

Shape memory polymers (SMPs) have been extensively studied for applications ranging from biomedical devices (e.g., stents) to smart surfaces. However, most of the currently available SMPs are heat or light sensitive. Here we demonstrate a new pressure-sensitive SMP that enables the rapid recovery of the memorized shapes by applying a very small pressure. By combining this new type of SMP with self-assembled colloidal crystals created by a scalable doctor blade coating technology, we have developed tunable macroporous photonic crystals showing large optical stop band shifts triggered by a small pressure change. The optical responses of the pressure-sensitive photonic crystals are simulated by using a scalar-wave approximation (SWA) model and the theoretical results match well with the experimental data. We have also demonstrated that nanoporous smart coatings can be templated by using self-assembled colloidal nanoparticles as sacrificial scaffolds. The resulting SMP coatings exhibit tunable anti-glare properties and the experimental reflection measurements agree with the numerical simulations using a finite-difference time-domain (FDTD) model.

(728b) Electrostatically Assisted Assembling of Silver-Dielectric Core/Shell Nanoparticle Thin Film Capacitor With Uniform Metal Nanoparticle Distribution and Controlled Spacing

Li, X., The City College of New York
Niitsoo, O., City College of New York
Couzis, A., The City College of New York
Originally presented on: 11/7/2013 15:40:00 - 16:00:00

New dielectric materials are sought after for keeping up with device miniaturization and for emerging applications (e.g. energy conservation and storage). Previous reports give great promise to composite dielectric materials for achieving high dielectric constant (*k*). Conductive filler/polymer composite material, identified as conductor-insulator percolative system, is one of these promising method to achieve high-*k*. We present a novel approach to preparation of metal-dielectric composites where metal volume fractions significantly higher than in simple metal particle-polymer mixtures can be achieved. The approach is based on using metal-dielectric core-shell particles, where each individual metal particle is coated with a uniform shell of a dielectric or insulating material. The shell prevents electrical contact between two neighboring metal cores. Changing shell thickness enables to control the separation distance between metal cores and also the metal volume fraction for the same size of metal particles. Additionally, volume fraction of metal core can be varied with changing metal core size.

In our system, silica shell serving as insulating interlayers could cause potential barriers so as to reduce the tunnel current between neighboring Ag cores, endowing the core/shell nanocomposites with stable and relatively high dielectric constants (*k*) and low dielectric loss (*D*). We fabricated and characterized Ag@SiO_2 core-shell particle thin film based parallel plate capacitors. To the best of our knowledge, there are no reports on thickness dependence of *k* and *D* in metal-dielectric composites. For most dielectrics, *D* decreases with thickness. Our thin film composites (1-3 um in thickness) thus show promising *D* values. The dielectric loss tangent of our thin film composites filling with Ag@SiO_2 were below 0.1, which meets the requirement for embedded decoupling capacitors. These results suggest that the conductive metal core of silver effectively enhances the dielectric constant while maintaining the low dielectric loss of the metal-dielectric composites. In addition, detailed dielectric property measurements revealed that the dielectric properties of the thin film Ag-dielectric composites were related to the incorporation and volume fraction of Ag nanoparitcles in the composites film.

(728c) Controlling the Assembly of Semiconducting Nanoparticles With Conjugated Polymers for Solar Cell Applications

Chao, K. P., Rice University
Gullapalli, S., Rice University
Verduzco, R., Rice University
Wong, M. S., Rice University
Biswal, S. L., Rice University
Originally presented on: 11/7/2013 16:00:00 - 16:20:00

Hybrid organic-inorganic solar cells have emerged as a newly-developed branch in the third generation of solar cell technology. It is well-known that the morphology inside the active layer of the solar cell is the crucial factor to the performance. In this study, we propose a fabrication procedure to control the morphology of the active layer, building up the solar cell with a "bottom-up" method. We focus on controlling the arrangement of the inorganic nanocrystal, CdSe tetrapods (TPs), when the conjugated polymer, poly[2-methoxy-5-(2-ethylhexyloxy)-1,4-phenylenevinylene] (MEH-PPV), serves as the templating agent at the air/water interface by utilizing the Langmuir-Blodgett (LB) technique. All the experiments are monitored by Brewster angle microscopy (BAM). The thin film of CdSe TPs/MEH-PPV are transferred by Langmuir-Blodgett or Langmuir-Schaefer (LS) technique and characterized by the atomic force microscopy (AFM) and the transmission electronic microscopy (TEM). The preliminary results show that solar cells made via the LB method has three times higher power conversion efficiency (PCE) than the one made from the conventional spin-coating methods. The higher PCE from the LB method benefits from the percolating network of packed CdSe TPs, which provides a pathway for electrons to travel through the active layer to the cathode. This study highlights the importance in controlling the structure of the inorganic material within an organic matrix for solar cells.

(728d) Assembling Colloidal Clusters From Spherical Codes

Phillips, C. L., University of Michigan
Jankowski, E., University of Colorado at Boulder
Marval, M.,
Glotzer, S. C., University of Michigan
Originally presented on: 11/7/2013 16:20:00 - 16:40:00

Anisotropic building blocks assembled from colloidal particles are attractive building blocks for self-assembled materials because their complex interactions can be exploited to drive self-assembly. In this work we consider the thermodynamically driven self-assembly of terminal clusters of particles[1]. This assembly process forms self-limiting anisotropic clusters with well-defined structures. We predict that clusters related to spherical codes, a mathematical sequence of points, can be synthesized via self assembly. Using Brownian dynamics and free energy calculations we predict the expected cluster sizes and shapes as a function of temperature and ratio of particle diameter. Structural and dynamical analysis of these tiny systems reveal rich and sometimes surprising properties. For example, the equilibrium structure of a 5-cluster, a square pyramid arrangement is preferred over a more symmetric structure. Our results suggest a promising way to assemble anisotropic building blocks from constituent colloidal spheres.

[1] C.L. Phillips, E. Jankowski, M. Marval and S.C. Glotzer, "Self-assembled clusters of spheres related to spherical codes," Phys. Rev. E, 86 (4) 1124 (2012)

(728e) External Field Directed Co-Assembly of Rigid and Ultra-Flexible Permanent Chains

Bharti, B., North Carolina State University
Fameau, A. L.,
Findenegg, G. H., Technical University Berlin
Velev, O. D., North Carolina State University
Originally presented on: 11/7/2013 16:40:00 - 17:00:00

Assemblies of colloidal particles are key components in the development of microscopically and nanoscale patterned or ordered materials. The most popular and widely used technique for tailoring linear structures is the application of an external field (electric or magnetic) to the colloidal dispersion. Application of external AC-electric or magnetic field causes a rapid arrangement of colloidal particles into linear chains by dielectrophoresis (DEP) or magnetophoresis (MAP). DEP (or MAP) induced structures resulting from similarly charged microparticles are not permanent, because of their surface charge repulsion, and last only as long as the field is present. We present two novel approaches to use DEP and MAP induced chaining for the formation of permanently linked structures. For achieving DEP assembly of rigid and permanent chain structures of dielectric particles, we used a bi-particle system, i.e. a dispersion of two types of colloidal particles having opposite surface charges. By replacing similarly charged micro-particles by oppositely charged binary mixtures, we introduce a short-range strongly attractive potential. DEP is then used to drag the particles near the surface contact point, where the strong attractive electrostatic potential binds them together and the structures formed survive even when the electric field is switched off. The results of the field-driven process depend on the size ratio as well as the number ratio of the two interacting species. We derived assembly rules for the resulting structures and a statistical model that is in excellent agreement with the experimental results. The oppositely charged biparticles yield permanent but stiff chains. In order to fabricate chains with soft and flexible connectors we used molecularly self-assembling fatty acid as a new type of binding agent. We fabricated chains of FeO nanoparticles connected via *soft linkers* using an initial burst of magnetic field. The composite chains subjected to 1D low frequency alternating magnetic field (0.1 Hz) showed an instantaneous response by reorienting themselves in the field following the minimum energy constraints. In addition, we showed that in a modulated alternating field, chains can reversibly wind/unwind into rings, hooks and other type structures. To the best of our knowledge, this is the first study where ultra-flexible magnetic/soft matter responsive chains have been shown to form and respond in a robust and reversible manner. Such structures could possess unusual magnetic, electric and rheological properties, which can be used in new functional soft materials.

(728f) Thermodynamics and Kinetics of Electric Field Mediated Colloidal Crystal Assembly

Edwards, T. D.,
Beltran-Villegas, D. J.,
Yang, Y., Johns Hopkins University
Bevan, M. A., Johns Hopkins University
Originally presented on: 11/7/2013 17:00:00 - 17:20:00

The ability of nano- and micro- scale colloidal components to autonomously and reversibly assemble into ordered configurations can potentially serve as a basis to manufacture materials with exotic optical properties. However, a limited understanding of the thermodynamics and kinetics of colloidal assembly has limited the informed design of robust processes that achieve sufficiently low defect densities. To address these issues, in this work we compare measurements and models of system size dependent thermodynamic and kinetic effects in the assembly of quasi-two dimensional colloidal crystals within interfacial quadrupole electrodes. Perturbation theory and Monte Carlo simulations are used to make thermodynamic predictions of the voltages required to obtain all particles within a crystalline phase for different system sizes. These results show good agreement with video microscopy measurements of colloidal crystals containing 75-300 particles. Colloidal assembly kinetic trajectories for different systems sizes are quantified from microscopy experiments using order parameters (reaction coordinates) for local and global hexagonal ordering. These trajectories are fit with a Smoluchowski model to extract configuration dependent free energy and diffusivity landscapes. The resulting landscapes capture system size dependent non-equilibrium thermodynamics and diffusivities for different configurations as particles rearrange from fluid to polycrystalline and single crystal states. These results demonstrate the importance of system size effects to the formation and annealing of polycrystalline states in colloidal crystallization.

(728g) The Effect of the Surface Roughness On Floating Hydrophobic, Micron-Sized Particles At An Oil/Water Interface

Dani, A.,
Maldarelli, C., The City College of New York
Originally presented on: 11/7/2013 17:20:00 - 17:40:00

A spherical particle placed at an oil/water interface (oil above water),which is more dense than either fluid phases, deflects the interface downward as gravity acts to push the particle through the surface. In spite of its weight, the particle can remain suspended on the interface if the interfacial tension or capillary force created by the interfacial deflection is large enough to support the buoyant weight of the particle. The maximum particle size which can be balanced by the capillary force is a function of the interfacial tension and the contact angle at the particle surface (defined as measured through the water phase). The larger the contact angle (or the greater the surface hydrophobicity), the greater is the maximum particle size,which can be supported. Atomically smooth surfaces are characterized by contact angles which are determined by the surface chemistry. However, recent research has shown that the contact angle of water on hydrophobic textured surfaces in air (as measured through the water phase) can be increased significantly relative to the corresponding value on a smooth surface with the same hydrophobic chemistry as that of the texture. This superhydrophobic effect can change the smooth surface contact angle on a hydrophobic surface from 100 degrees to 160 degrees or more, and is due to the pinning of the interface onto the texture as the water does not saturate into the texture (Cassie-Baxter wetting). In the case of a water droplet placed on a hydrophobic,textured surface immersed in oil, this same effect can result in an increase in the contact angle of the water droplet in the oil. This raises the possibility that the size of hydrophobic textured particles,which can be supported at an oil/water interface, is significantly larger than smooth particles with the same surface chemistry because of the increase in the contact angle.

In this investigation, we demonstrate this effect by using silica particles that have been textured by the binding of hydrophilic nanoparticles. In the case of both the smooth and nanoparticle textured surfaces, the surfaces are made hydrophobic by silanization. Experi- ments on planar substrates of smooth silica surfaces and those with the nanoparticles, after silanization, demonstrate the increase in contact angle of water drops placed on the sur- face. We also verify that larger surface textured spherical particles can be supported at the oil/water interface relative to smooth particles with the same (silanated) surface chemistry. We also examine the effect of the interfacial tension of the oil/water interface on the relative difference between the maximum particle size supported by smooth and microtextured par- ticles. The tension is varied by introducing a surfactant into the continuous oil phase. We show that even with the reduction in the tension due to the surfactant adsorption onto the interface, textured particles of considerable size can be supported at the oil/water interface due to the increase in contact angle at the textured surface.

(730a) Double Layer Forces in the Presence of Charge Heterogeneities

Pick, C., Johns Hopkins University
Frechette, J., Johns Hopkins University
Originally presented on: 11/7/2013 15:20:00 - 15:35:00

The study of particle deposition at solid-liquid interfaces has been the subject of numerous investigations due to its great importance in a large number of technological applications. The transport and fate of colloidal particles in underground reservoirs, and its impact in the facilitated spreading of contaminants, makes particle capture and reentrainment a crucial issue in environmental applications. In all cases, deposition is a complex process that includes competing contributions from different electrostatic effects, dispersion forces, Brownian motion and hydrodynamic interactions. Surface charge heterogeneities in particular, provide a means of precisely tailoring the interactions between particles and surfaces.

Here we are investigating the role of charge heterogeneities on double layer forces using direct force measurements in the Surface Force Apparatus. The surface heterogeneities consist of an ordered array of positive patches patterned onto mica via chemical vapor deposition. The force profile between the patterned surface and bare mica are directly measured in the surface force apparatus. The force profiles are compared to the force profiles between two bare (negatively charged) mica surfaces and two positive surfaces. Comparisons are made between the measured force profiles and DLVO theory.

(730b) Microrheology of Bacterial Films At Oil-Water Interfaces

Vaccari, L.,
Allan, D. B., Johns Hopkins University
Sheng, J.,
Leheny, R. L., Johns Hopkins University
Stebe, K. J., University of Pennsylvania
Originally presented on: 11/7/2013 15:35:00 - 15:50:00

Bacterial films play key roles in emerging petroleum technologies such as enhanced oil recovery and oil remediation. In these applications, bacteria assemble at interfaces between the oil and water to form dense films that can strongly affect the interfacial properties of the oil, including its mechanical behavior. Despite this significance, limited previous research has sought to characterize systematically the mechanical properties of these films. The microstructure of the films, which can include not only dense packings of bacteria but macromolecular surfactants secreted by the bacteria and the remains of dead bacteria, resemble from a physical perspective a disordered colloidal suspension in a polymer solution. In this work, we employ passive and active microrheology and more traditional active rheometry to characterize the shear response of bacteria films during formation. We measure a dramatic evolution in film rheology, studied both from a fundamental perspective, as a quasi-two-dimensional, active, glassy system, and in the context of the application of the films to petroleum technologies.

(730c) Tween 80 and Aerosol-OT Adsorption Behavior At An OIL-Aqueous Interface

Reichert, M. D., Carnegie Mellon University
Anna, S. L., Carnegie Mellon University
Walker, L. M., Carnegie Mellon University

Originally presented on: 11/7/2013 15:50:00 - 16:05:00

Dispersants contain a number of different surface active components. We have characterized the impact of two primary components on the adsorption of each other at oil-water interfaces. Pre-adsorbed layers of Tween-80 are found to inhibit adsorption of AOT molecules at the interface. The mechanism appears to be due both to reduction of available interfacial area for AOT molecules, and generation of charge in the layers of nonionic Tween-80 that causes electrostatic repulsion for the negatively charged AOT. Once adsorbed into a mixed layer at the interface, the AOT and Tween-80 molecules interact at the interface, as determined by the nature of the dilatational viscosity and elasticity of the interfaces. A mixed transport and surface reaction model was developed and used to fit adsorption rate constants for AOT to the Tween coated interfaces. Overall, these experiments and analysis serve to describe important interactions that might be expected to play a role in the transport properties of dispersant mixtures to oil-aqueous interfaces. The experimental techniques and analysis approach used in this study are applicable to any binary mixture of surfactant systems where one component adsorbs irreversibly, and also for elucidating the effects of charged surfactant layers. This information is critical for effective dispersant design, as well as for emulsion systems containing multiple surfactant components.

(730d) Volume-Dependent Adhesion Behaviors Between Clathrate Hydrates and Water

Lee, J. W., The City College of New York
Cha, M.,
Couzis, A., The City College of New York
Originally presented on: 11/7/2013 16:05:00 - 16:20:00

Clathrate hydrates are crystalline compounds in which guest molecules such as methane, ethane, carbon dioxide, cyclopentane (CP), etc. are entrapped in host water cages formed by the hydrogen-bonded network of water molecules. During the extraction process of oil and gas, the existence of water and gas can lead to the formation of hydrate inside the pipeline. Hydrate particles can be aggregated/agglomerated down the pipeline. The continuous growth of hydrate particles can lead to a shutdown of the production due to the pipeline blockage. Therefore, the interfacial dynamic behavior needs to be identified to understand the adhesion between hydrate particles and water droplets in the pipeline.

Here, we investigated the volume-dependent adhesion behaviors between cyclopentane (CP) hydrate and solution droplet in a CP/n-decane oil mixture. Three to five times of precise contact force measurements are obtained from 100 to 600 μL of solution droplet by using the z-directional microbalance. In both pure water and surfactant solution system, there is a transition volume over certain level of solution droplet. In addition, we also visually observed the abnormal variations of capillary bridge between CP hydrate and solution droplet through the optical microscope.

(730e) Generation of Biomimicked Leaves for Crude Oil Bioadhesion Studies

Cremaldi, J., Tulane University
Cutting, D., Tulane University
Pesika, N., Tulane University
Originally presented on: 11/7/2013 16:20:00 - 16:35:00

This research focuses on the study of the surface effects of crude oil on costal plant life. With the recent oil spill in the Gulf of Mexico, the study of crude oil's bioadhesion to coastal plants is a topic of high interest. However, living plant tissue makes it difficult to perform testing as it begins to dehydrate immediately on being severed from the main body. This dehydration causes the surface topography to change and introduces time dependency into tests. For this reason, biomimicked leaf samples were created out of polyurethane using a two stage molding process after which the epicuticular wax taken from the original biological sample was redeposited on the artificial surface. The adaxial surface of *Spartina Antiflora*, a common coastal plant, was utilized for these tests. Optical micrographs, SEM, and contact angle analysis then show that the surface chemistry and structure display a very high resolution to the original leaf. Utilizing these biomimicks, pendant drop testing provided a mechanism to explain the oil's adverse effects on coastal plant life which caused suffocation and eventual death. Last, microfluidic devices were utilized to simulate dispersed oil in a controlled manner in order to validate this mechanism.

(730f) Evaluation of Mucilage From the Common Cactus As a Natural Dispersant for Oil Spill Clean-Up Operations

Guo, F., University of South Florida
Peng, T.,
Smith, D. M.,
Stebbins, D. M. L., University of South Florida
Celis, M.,
Alcantar, N., University of South Florida
Originally presented on: 11/7/2013 16:35:00 - 16:50:00

The purpose of this work is to assess the *Opuntia ficus-indica* (*Ofi*) cactus plant-based mucilage as a natural oil dispersant. We have evaluated the biodegradation and efficiency of cactus mucilage to disperse crude-oil in sweet and salty environments. We have also determined that this material is non-toxic to cold blooded animals because the median lethal dose (LD_{50}) of cactus mucilage was found to be large (i.e., above 1000 mg/kg). Cactus mucilage is natural biomaterial that has up to 55 sugars, mainly arabinose, galactose, rhamnose, xylose, glucose, and uronic acids. The extraction process is relatively simple, and yields two types of mucilage: the non-gelling extract (NE) and the gelling extract (GE). The characterization of dispersed oil-mucilage droplets in oil/water emulsions was evaluated by transmission electron microscopy (TEM), optical microcopy and UV-VIS spectroscopy. Mucilage aids the dispersion and stabilization of oil droplets by lowering the surface tension and maintaining the interfacial forces differential close to zero. Biochemical oxygen demand (BOD) was also assessed in seawater samples with crude oil (0.5 to 50 g/L) using concentrations of mucilage ranging from 0.5 to 2000 mg/L. Studies have also shown that mucilage increased biodegradation of the crude oil in seawater samples by adding a readily available source of carbohydrates to the system. The toxicity of the mucilage extracts were performed by a standard test using *Daphnia magna* organisms at room temperature. Daphnia colonies were exposed to both NE and GE mucilage extracts in concentrations ranging from 0 to 5000 mg/L. After 24 hours, the number of living organisms in each the test group was determined. It was observed that mucilage did not influence *Daphnia magna* colonies at all. Cactus mucilage is readily available through sustainable agriculture, it is inexpensive to extract, and has exhibited a long shelf live. Therefore, cactus mucilage can be an alternative technology to mitigate the damage that oil may cause to the aquatic ecosystem and minimize undesired effects associated with the use of synthetic dispersants in oil spills.

(730g) Green Leaf Volatiles and Reactive Oxygen Species On Atmospheric Air/Water Interfaces

Liyana-Arachchi, T. P., Louisiana State University
Hansel, A. K.,
Stevens, C.,
Ehrenhauser, F. S., Louisiana State University
Valsaraj, K. T., Louisiana State University
Hung, F. R., Louisiana State University
Originally presented on: 11/7/2013 16:50:00 - 17:05:00

Atmospheric aerosols play an important role in climate, air quality and fate of pollutants under atmospheric conditions. However, many knowledge gaps still remain for secondary organic aerosols (SOAs), which are particulate matter formed by compounds that originated from chemical reactions of organics in the atmosphere. SOAs are an important factor in the well-known smog-fog-smog cycle, where fog forms by condensation of water on submicron atmospheric particles, which then uptake organics and reactive oxygen species (ROSs, e.g., ozone and radicals such as •OH). These compounds undergo chemical reactions within the fog droplets, yielding products that contribute to the formation of more SOAs. Many recent efforts have concentrated on characterizing the organic fraction in samples of fog water. A significant fraction of these organics are emitted by microbes, bacteria and plants. Some of these compounds include green leaf volatiles (GLVs), which is a group of very reactive oxygenated hydrocarbons emitted in large quantities by plants, especially when they are injured or under stress conditions (e.g., grass cutting, animal grazing, local weather changes). These GLVs can react with ROSs in a fog environment, but the mechanisms and kinetics of these reactions depend on whether the reactions take place within the bulk of the water droplets, in the air/water interface or in gas phase. Here we investigated the adsorption of GLVs and ROSs on atmospheric air/water interfaces, as well as the interactions between these species, using potential of mean force (PMF) calculations and classical molecular dynamics (MD) simulations. Our results indicate that both GLVs and ROSs have a strong thermodynamic preference to remain at the air/water interface, and thus chemical reactions between atmospheric GLVs and ROSs are more likely to take place at the interface, rather than in the water phase or in the gas phase. Relevant interfacial properties (free energies, density profiles, orientations, dynamics and number of contacts between GLVs and ROSs) are investigated and discussed in this study.

(730h) Atmospheric Oxidation of Benzene: Effect of Temperature and Ionic Strength in Both the Bulk Aqueous Phase and At the Air-Water Interface

Heath, A. A., Louisiana Statue University
Ehrenhauser, F. S., Louisiana State University
Valsaraj, K. T., Louisiana State University
Originally presented on: 11/7/2013 17:05:00 - 17:20:00

Atmospheric water droplets, such as fog and clouds, can differ broadly in both ionic strength and pH. In the atmosphere, benzene, a common aromatic pollutant resulting from both stationary (chemical and petrochemical industry emissions) and mobile sources (automobile exhaust), reacts with hydroxyl radicals to form two main products, phenol and biphenyl. Although aromatic compounds are generally very stable, benzene is known to react easily with hydroxyl radicals in both the gas phase ($k_{OH} = 4.3\times10^4$ $\mu M^{-1}\cdot min^{-1}$)[1] and the aqueous phase ($k_{OH} = 4.7\times10^5$ $\mu M^{-1}\cdot min^{-1}$)[2].

In the bulk aqueous phase, higher temperatures and an increase of ionic strength have been shown to enhance the degradation of both benzene and phenol as well as the formation of biphenyl. However, reactions in atmospheric droplets are not solely governed by the bulk phase; instead, it is the reactions at the air-water interface that predominately dominate the reactivity of the droplet. The surface is also known to have an affinity for both aromatic compounds and hydroxyl radicals[3-6], which could influence compound fate given the large specific surface area of fog and clouds[7]. Moreover, the surface can have a different ionic strength than the bulk aqueous phase, which in turn can affect the overall reaction rate. Thus, in this study, a thin film reactor is used to evaluate how both temperature and ionic strength affects the degradation of benzene at the air-water interface. In particular, this study focuses on the photo-generation of hydroxyl radicals by UVB irradiation of hydrogen peroxide, which has relevance for aqueous atmospheric systems. These results from the interface are compared to the previous results found within the bulk aqueous phase.

1. B.J. Finlayson-Pitts, J. James N. Pitts, *Chemistry of the Upper and Lower Atmosphere: Theory, Experiments, and Applications*. Elsevier Science, 1999.

2. G.V. Buxton, C.L. Greenstock, W.P. Helman, A.B. Ross, Critical Review of rate constants for reactions of hydrated electrons, hydrogen atoms and hydroxyl radicals in aqueous solution, *J. Phys. Chem. Ref. Data* 17 (1988) 513-886.

3. J. Chen, F.S. Ehrenhauser, K.T. Valsaraj, M.J. Wornat, Uptake and UV-Photooxidation of Gas-Phase PAHs on the Surface of Atmospheric Water Films. 1. Naphthalene, *J. Phys. Chem. A* 110 (2006) 9161-9168.

4. M. Roeselova, J. Vieceli, L.X. Dang, B.C. Garrett, D.J. Tobias, Hydroxyl Radical at the Air-Water

Interface, *J. Am. Chem. Soc.* 126 (2004) 16308-16309.

5. C.D. Wick, L.X. Dang, Computational Observation of Enhanced Solvation of the Hydroxyl Radical with Increased NaCl Concentration, *J. Phys. Chem. B* 110 (2006) 8917-8920.

6. R. Vacha, P. Jungwirth, J. Chen, K. Valsaraj, Adsorption of polycyclic aromatic hydrocarbons at the air-water interface: Molecular dynamics simulations and experimental atmospheric observations, *Phys. Chem. Chem. Phys.* 8 (2006) 4461-4467.

7. K.T. Valsaraj, *Elements of Environmental Engineering: Thermodynamics and Kinetics.* Taylor and Francis Group, LLC, Boca Raton, 2009.

(730i) Aggregation Characteristics of Radioactive Particles in the Atmosphere

Kim, Y. H., Georgia Institute of Technology
Yiacoumi, S., Georgia Institute of Technology
Lee, I., University of Tennessee at Knoxville
McFarlane, J., Oak Ridge National Laboratory
Tsouris, C., Oak Ridge National Laboratory
Originally presented on: 11/7/2013 17:20:00 - 17:35:00

Understanding the microphysics of radioactive components in air is important for the accurate prediction of radioactivity transport in the environment. Aggregation and deposition rates of radioactive particles depend mainly on surface forces. Radioactivity can modify the surface charge of the transport media (e.g., aerosol particles); however, radioactivity-induced charging and its effects on radioactivity transport are currently not considered in predictive atmospheric transport models. In this study, the mechanisms of radioactivity-induced charging and its influence on aggregation kinetics of radioactive particles are investigated. Radioactivity-induced charging has been investigated by using scanning surface potential microscopy. Radioactive particles can be charged through self-charging and diffusion charging mechanisms, and the radioactivity-induced surface charge can be heterogeneously distributed due to the coupled effects of the charging mechanisms. Modeling based on particle collision probabilities has been carried out to analyze the interaction between radioactive particles (e.g., ^{137}Cs and $Cs^{131}I$ particles; ^{137}Cs and nonradioactive particles), modified by the heterogeneous charge distribution. Simulation results have indicated that radioactivity-induced charging hinders aggregation of highly radioactive particles, but facilitates aggregation between radioactive and nonradioactive particles. The influence of environmental parameters on radioactivity-induced charging and aggregation of radioactive particles are investigated.

(736a) Local Compositions and Structures of Platinum Species in Heterogeneous Catalysts

Jones, L. C., University of California
Chmelka, B. F., University of California
Originally presented on: 11/7/2013 15:15:00 - 15:35:00

The preparations of Pt-containing heterogeneous catalysts generally involve several distinct chemical treatments (e.g., oxidation, reduction, sulfidation) that are collectively influential on the macroscopic adsorption and reaction properties of the final catalyst products. Despite the technological importance of Pt-containing catalysts and compounds, distinguishing the effects of these treatments on local and surface Pt environments is still a challenging task. This is principally due to the inherent disorder present within many heterogeneous Pt-containing materials, which present challenges to elucidating Pt moieties with standard diffraction or spectroscopy methods. Nuclear magnetic resonance (NMR) spectroscopy is a sensitive probe of local bonding environments and can in principle distinguish locally ordered and disordered molecular species. However, solid-state ^{195}Pt NMR spectra can span very large radiofrequency bandwidths (~MHz), which can provide subtle information on the molecular environments of Pt species, but have been impractical to excite and detect with standard techniques. Nevertheless, recent methodological advancements now allow such large spectral widths to be managed enabling detailed characterization of Pt species that have till now been infeasible to measure. Results will be presented for bulk, supported nanoparticle, and molecular Pt catalysts whose local ^{195}Pt environments are readily distinguished, including influences of the supports and of oxidation and sulfidation treatments. These insights are correlated with macroscopic reactions properties, which will be discussed for several examples.

(736b) Selective Hydrogenation Using Supported Metal Catalysts Modified With Alkanethiol Self-Assembled Monolayers

Kahsar, R., University of Colorado
Schwartz, D. K., University of Colorado Boulder
Medlin, W., University of Colorado
Originally presented on: 11/7/2013 15:35:00 - 15:55:00

Of particular interest and difficulty in chemical processes is the chemoselective conversion of molecules containing more than one reactive functional group. To bias a catalyst to produce the less energetically favorable product, we deposit alkanethiol self-assembled monolayers (SAMs) on the surface of supported metal catalysts. These SAMs form a well-ordered structure on the metal catalyst surface, altering the surface chemistry through steric and electronic effects thereby influencing the selectivity of a reaction.

The development of novel alkanethiol SAM coated selective catalysts has wide and promising applications to the field of catalysis. Here we display the usefulness of selectively hydrogenating molecules with multiple functional groups such as epoxybutene, polyunsaturated fatty acids, and α,β-unsaturated aldehydes.

Our group has previously used SAMs to modify supported metal hydrogenation catalysts such as Pd/Al2O3 for the selective hydrogenation of 1-epoxy-3-butene (EpB) [1,2]. Reaction of the epoxide ring is more preferred on an uncoated catalyst but alkanethiol SAM coated catalysts were found to increase the selectivity to the desired 1-epoxybutane from around <20% to >90% under equivalent conditions in the gas phase and similarly in the liquid phase where solvent-SAM interactions were shown to influence the selectivity of the reaction.

Liquid phase studies are now being employed to study larger, less-volatile reactant molecules which have increased ability to interact with the tail groups of the alkanethiol SAMs. The selective hydrogenation of 18-carbon polyunsaturated fatty acids to monounsaturated products was studied over alkanethiol coated Pd/Al2O3 and showed an increased selectivity to monounsaturated fatty acids as compared to the uncoated Pd/Al2O3 catalyst reaction case. Additionally, thiol tail groups have been used to specifically direct the reaction pathway of cinnamaldehyde, an α,β-unsaturated aldehyde, over a Pt/Al2O3 catalyst. Selective hydrogenation of cinnamaldehyde was performed using phenylated SAMs where interaction between the phenyl ring of the reactant (the large terminal end of the cinnamaldehyde molecule) and the SAM was used to direct hydrogenation of either the double bond or the aldehyde over a Pt/Al2O3 catalyst. Control of the near surface environment of a catalyst system, as studied here, provides a unique, interesting, but not yet fully understood method for controlling the reaction environment.

References

1. Marshall, S., O'Brien, M., Oetter, B., & Corpuz, A. (2010). Controlled selectivity for palladium catalysts using self-assembled monolayers. Nature Materials, 9

2. Kahsar, K. R., Schwartz, D. K., & Medlin, J. W. (2012). Liquid- and vapor-phase hydrogenation of 1-epoxy-3-butene using self-assembled monolayer coated palladium and platinum catalysts. Applied Catalysis A: General 445-446

(736c) Investigation of Metal-Metal Oxide Support Interface

Wang, Z., Northwestern University
Shen, J., Northwestern University
Mashayekhi, N. A., Northwestern University
Wu, Y. Y., Northwestern University
Kung, M. C., Northwestern University
Kung, H. H., Northwestern University
Originally presented on: 11/7/2013 15:55:00 - 16:15:00

Although the interface between a support and metal nanoparticles is an integral component of all supported metal catalysts, there is little information on their properties. While direct measurement of these properties is challenging, we found it possible to gather relevant information using a model system in which an oxide ensemble of tunable composition that can be structured to mimic the interface is deposited on the metal particle. As an example, a system that mimics Au metal-support oxide interface is synthesized and examined for selective oxidation of propane, in which the proximity and the nuclearity of the support oxide ensemble can be varied. The synthesis of such catalytic system can be achieved in two ways. In the first method, a poly-methylhydrosiloxane containing both Au precursor and metal oxo species of interest (for example Ti) are used to generate a composite. Following reduction of Au and removal of organics, Au next to Ti oxo clusters of low nuclearity is generated. Another approach is to synthesize short oligomers of siloxane containing the metal-oxo species of interest in the center. The two ends of the oligomers contain functionalities that will adsorb on Au nanoparticles. The reactivities of these two types of catalysts in CO and propane oxidation are compared to gain better understanding of the active site structure.

(736d) Selective Hydrogenation With Ag-Doped Pt Nanoparticle Catalysts

Gordon, M., University of California, Santa Barbara
Jones, L., University of California, Santa Barbara
Originally presented on: 11/7/2013 16:15:00 - 16:35:00

Alloying has been a successful approach to modify the reactivity of metal nanoparticle surfaces; for example, dopant atoms on the nanoparticle surface can increase turnover, improve selectivity, enhance resistance to deactivation, and promote multiple reaction functionality. However, the reactivity of alloyed NP catalysts often evolves during reaction, particularly at high temperatures, where atoms can easily diffuse and alter surface composition and structure. Dynamic rearrangements, coupled with adsorption of strongly bound intermediates, can make "stable" surface compositions which promote high reactivity difficult to achieve over a large range of operating conditions. As such, gaining insight into improving the activity and selectivity of alloy NP catalysts is a challenging task.

In this work, Ag-doped Pt nanoparticles (NP) were tested for continuous-flow C_2H_2 hydrogenation reactivity and CO adsorption–desorption from 100 to 300 °C to investigate how submonolayer levels of an insoluble metal dopant distribute on steps and terraces of a NP surface and modify reactivity [1,2]. Ag incorporation onto Pt step and terrace sites, and resultant temperature-dependent desorption behavior, were probed using *in situ* diffuse reflectance infrared Fourier transform spectroscopy (DRIFTS) with CO. At low coverage, Ag appeared to occupy Pt step sites; for higher coverage, Ag occupied both step and terrace sites, which was observed to lower the CO desorption temperature. The catalytic influence of sub-monolayer Ag incorporation into Pt NP surfaces was also probed *in-situ* during batch and continuous-flow C_2H_2 hydrogenation. NP catalysts were found to be highly active and selective for C_2H_2 to C_2H_4 hydrogenation. Ag-doped Pt NP catalysts with ~0.5 ML Ag showed increased selectivity with temperature up to 200 °C, in contrast to commercial hydrogenation catalysts (e.g., Pt, Pd, PdAg) that are known to have activity-selectivity tradeoffs. Highly correlated reactivity and DRIFTS data highlight the importance of step site blocking and desorption from Pt during selective C_2H_2 hydrogenation. Furthermore, measurements over multiple reaction cycles from 100 to 300 to 100 °C demonstrate how immiscible metal dopants can preferentially modify NP surface compositions and promote active, selective, and stable catalysts over large temperature operating windows.

[1] L.C. Jones, Z. Buras, M.J. Gordon, *J. Phys. Chem. C* **116**, 12982 (2012).

[2] L.C. Jones, M.J. Gordon, *J. Phys. Chem. C* **116**, 23472 (2012).

(736e) Does Surface Roughness Matter: A Case Study With VOx / SBA-15 in the Partial Oxidation of Methanol and Propane

Smith, M. A., Villanova University
Zoelle, A., Villanova University
Hamilton, N. G., Fritz Haber Institute of MPG
Amakawa, K., Fritz Haber Institute of MPG
Nielsen, P. K.,
Trunschke, A.,
Originally presented on: 11/7/2013 16:35:00 - 16:55:00

SBA-15 is a template-synthesized mesoporous silicate that has found extensive use as a model support for studies of supported catalysis.[1, 2] Thorough structural analyses clearly describe the dual micropore-mesopore structure with a broad distribution of micropore sizes.[3] In contrast to other oxides such as titania or ceria, silicas such as SBA-15 have long been considered a relatively inert support. However, we find the effect of microporosity or surface roughness in SBA-15 has an underappreciated effect on catalyst performance. Specifically, samples of VO_x-SBA-15 where the support surface roughness was systematically varied were characterized using UV-vis and Raman spectroscopy, then tested in the catalytic partial oxidation of methanol to formaldehyde, and propane to propene. Results show that supports with smoother surfaces permit the development of more polymeric vanadia species at the same surface density loading. Such smoother-surface catalysts result in a lower selectivity of methanol to formaldehyde, yet conversely show a higher selectivity of propane to propene. This result is significant with respect to our understanding the role of vanadium in in partial oxidation catalysts, and illustrates the importance of considering differences in support surface morphology in analyzing catalytic behavior.

References

[1] V. Dufaud, M. E. Davis, J. Am. Chem. Soc. 125 (2003) 9403-9413.

[2] R. K. Zeidan, S. J. Hwang, M. E. Davis, Angew. Chem.-Int. Edit. 45 (2006) 6332-6335.

[3] M. Kruk, M. Jaroniec, R. Ryoo, J. M. Kim, Chem. Mat. 11 (1999) 2568-2572.

(736f) Molecular-Level Structure-Function Relations in Gold Nanoparticle Catalysis

Nigra, M. M., University of California Berkeley
Katz, A., University of California, Berkeley
Originally presented on: 11/7/2013 16:55:00 - 17:15:00

Coordinatively unsaturated corner and edge atoms have been hypothesized to have the highest activity of sites responsible for many catalytic reactions on a metal surface. Recent studies have demonstrated the validity of this hypothesis in varied reaction systems. However, the separate quantification of different types of coordinatively unsaturated sites, and elucidation of their individual catalytic rates has remained a largely unresolved challenge when dealing with catalysis on metal surfaces. Such a quantification and understanding would permit catalyst optimization based on a detailed understanding of how the degree of coordinative unsaturation affects catalysis rate. In a heterogeneous gold nanoparticle system consisting of gold nanoparticles on a metal oxide support, the catalytic contributions of undercoordinated sites such as corner and edge atoms are investigated in a model reaction system using organic ligands bound to the gold nanoparticle surface. It is demonstrated that the most undercoordinated sites, the corner atoms, are the most active atoms by over an order of magnitude as compared to uncoordinated edge atom sites, whereas terrace sites are completely inactive. Results provide a unique perspective into coordinative unsaturation requirements for catalysis.

(736g) Support Acid-Base Properties As a Tool for Directing Selectivity in the Au-Pt Catalyzed Base-Free Glycerol Oxidation

Prati, L., University of Milan
Villa, A.,
Campisi, S.,
Vindigni, F., Universita di Torino
Manzoli, M., Universita di Torino
Dimitratos, N., Cardiff Catalysis Institute
Originally presented on: 11/7/2013 17:15:00 - 17:35:00

Glycerol is a promising chemical building block for the synthesis of fine chemicals from renewable sources [1]. In particular glycerol oxidation shows a complex reaction pathway that can lead to a large number of valuable organic compounds (glyceric acid, tartronic acid, dihydroxyacetone, etc.). Supported noble metal nanoparticles have been used as catalysts for this reaction when molecular oxygen is employed as the oxidant [2]. Depending on the reaction conditions and the noble metal employed, the reaction pathway can be directed either to the oxidation of the primary or the secondary hydroxyl group. In particular Au NPs reveal peculiar properties in terms of both activity and selectivity. However only recently it was found that the catalytic oxidation of glycerol can be carried out without the use of a basic environment, being it a great advantage from an industrial point of view. In this view it has been shown that basic supports greatly enhanced the activity of AuPt alloyed catalysts. In this work AuPt NPs based catalysts were tested in the base-free glycerol aqueous phase oxidation, using supports with different acid/base properties. Experimental results showed that the support surface properties not only influenced the catalyst activity but also greatly affected the selectivity to C3 compounds. Recent studies [3] [4] showed that a careful choice of the support of AuPt alloyed nanoparticle lead to effective catalytic systems in terms of both activity and selectivity, even in the absence of a basic environment. In particular these works opened the way to a new-generation of bifunctional gold-based catalysts where the acid/base properties of the support play a crucial role, affecting activity and selectivity of the reaction. Herein we supported AuPtNPs on a series of supports with different acid-base character in order to clarify the role of the surface properties of the support on catalyst activity/selectivity in the base-free glycerol oxidation. AuPt nanoparticles were immobilized on, MCM-41, SiO_2, H-Mordenite, NiO and MgO and tested in the base-free glycerol oxidation in a batch reactor (0.3 mol L^{-1} aqueous glycerol solution, glycerol/metal molar ratio 500, pO_2 3 atm, 16 h). Using acidic support (MCM-41, SiO_2, H-Mordenite) a high selectivity to glyceraldehyde/glyceric acid (88-90%) has been obtained. Conversely, using basic supports (NiO, MgO), a low selectivity to glyceric acid was observed (35%) as a consequence of the successive oxidation to tartronic acid/ mesoxalic acids but also to the high occurrence of C-C bond oxidative cleavage. Moreover glyceraldehyde was not detected. In situ IR studies with probe molecules will be presented.

References

[1] A. Corma, S. Iborra, and A. Velty, *Chem. Rev.*, 107 (2007) 2411-2502

[2] T. Mallat and A. Baiker, *Chem. Rev.*, 104 (2004) 3037-3058.

[3] A. Villa, G.M. Veith and L. Prati, *Angew. Chem. Int. Ed.*, 49 (2010) 4499-4502

[4] G. L. Brett, Q. He, C. Hammond, P. J. Miedziak, N. Dimitratos, M. Sankar, A. A. Herzing, M. Conte, J. A. Lopez-Sanchez, C. J. Kiely, D. W. Knight, S. H. Taylor, and G. J. Hutchings et al., *Angew. Chem. Int. Ed.*, 50 (2011) 10136-10139

(742a) Catalytic Nanoparticle Adhesion to Carbon Substrates in Energy Systems

He, Q., University of Tennessee, Knoxville
Keffer, D. J., University of Tennessee, Knoxville
Joy, D. C., University of Tennessee, Knoxville
Originally presented on: 11/7/2013 15:20:00 - 15:40:00

Carbon supported platinum (Pt/C) catalysts are of vital importance to today's fine chemical industry. However, both Pt and its carbon support surface undergo oxidation during operation. There is a lack of information on how surface oxidation affects the durability of Pt/C catalysts. In this study, we report the impact of oxidation of Pt/C on the binding energy and nanoparticle adhesion force. Classic molecular dynamics simulations are performed on systems contain PtO nanoparticles of 4 nm and graphite surfaces oxidized with either epoxy or hydroxyl groups at several oxidation extents (10%, 25% and 50%). Appropriate for service in polymer electrolyte membrane fuel cells (PEMFC), the effect of a thin Nafion film (1 nm thick) at different hydration levels ($\lambda = 3, 6, 9$ and 15) is also included in the system to study the impact of oxidation on the interface structure of the electrolyte and the electrode. This study shows that the adhesion of both the Nafion film and the PtO nanoparticle is drastically affected by the type and degree of oxidation on the carbon surface. The oxidation with hydroxyl groups on the graphite surface enhanced the binding energy between the polymer electrolyte and the carbon electrode, while oxidation with the epoxy group beyond a certain amount caused delamination of the film. The adhesion of the PtO nanoparticle was similarly enhanced by the hydroxylated surface and diminished by the epoxidized surface. The binding energies and adhesive forces are reported for all systems studied.

(742b) Modeling and Simulation of the Electrosorption Process in Heterogeneous Porous Media

Gabitto, J. F., PVAMU
Tsouris, C., Oak Ridge National Laboratory
Originally presented on: 11/7/2013 15:40:00 - 16:00:00

Porous electrodes are widely used in electrochemical applications such as capacity deionization, high-density supercapacitors, and blue energy storage. Electrodes that present a combination of macro- and micropores are sought because of the high area available for ion storage. Mesoporous carbon materials have received attention recently due to their high surface area and pore size that is significantly larger than the size of hydrated ions. The volume averaging method has been employed to simulate the charging-discharging processes in porous materials that present a bimodal pore size distribution. The pore size distribution consists of macroporosity outside the particles through which the ions are transported and mesoporosity inside the particles, where the electrical double layers form. Ion adsorption occurs in the mesopores, where the solution exchanges ions with a charged, electrical double-layer (EDL) at the solid-liquid interface. To describe the charging process, we have proposed an EDL model that includes a Stern layer of constant geometrical thickness and a variable electrolyte dielectric constant calculated by the Booth equation. The potential and ionic concentration profiles inside the diffuse layer were computed by solving numerically the Poisson-Boltzmann equation for cylindrical coordinates. The thermodynamic equilibrium assumption has been used to derive one-equation models for salt concentration and electrostatic potential distribution. Theoretical calculations of the transport parameters, effective diffusivity, and effective mobility have been carried out. Comparison between experimental data and theoretical calculations is performed to validate the proposed model.

(742c) Quantum Statistical Mechanics of Charge Transfer Processes At Electrode-Electrolyte Interfaces

Gupta, C., Stanford University
Howe, R. T., Stanford University
Originally presented on: 11/7/2013 16:00:00 - 16:20:00

Charge transfer at an electrochemical solid-liquid interface is a complex physical phenomenon characterized by an interdependence of microscopic electronic and chemical processes. The inherent coupling of chemistry with the nanoscale charge transport dynamics at electrochemical interfaces yields a unique physical platform capable of electrical transduction of chemical information at the molecular level. However, this "information-rich" phenomenon has usually been described experimentally by measurements of observables like the current flowing across the interface, the voltage drop across the interface and/or the charge accumulated at the interface. The traditional measurements and their subsequent analyses present a macroscopic, ensemble-averaged description of the transfer process wherein the microscopic information is often obfuscated in the background signal that is often discarded. Therefore, sensing platforms utilizing the charge transfer process as an information transduction mechanism tend to be limited in the scope of analytes that can be investigated. Typically, electrochemical sensors are deployed for the detection of redox-active species that are capable of direct electronic interaction with metallic or semi-conducting electrodes through available valence and/or d-shell electrons within the useful electronic energy ranges, which is usually within +-0.7eV relative to an electrolytic reference for an aqueous system. We propose here a design methodology for electrochemical systems wherein the engineered charge transfer process enables transduction of chemical information, like inter-molecular bond vibration frequency, about non-redox-active analyte species. The proposed design strategy leverages a quantum statistical model of the charge transfer process at the electrochemical interface developed by the authors, where the critical parameters affecting the information transduction mechanism are identified and quantified. We will discuss the implications of the critical parameters on the engineering of the physical sensor front-end, as well as on the engineering of the instrumentation necessary for the information acquisition. Preliminary experimental data will also be presented that will highlight the capability to distinguish structural isomers and single-atom-mass-isotope substituents of the detected analytes from an all-electronic signal generated by the sensor front-end.

(742d) The Leveling of Liquid Coatings in the Presence of Substrate or Interfacial Charge

Ramkrishnan, A., University of Minnesota
Kumar, S., University of Minnesota
Originally presented on: 11/7/2013 16:20:00 - 16:40:00

Electrostatic charges that accumulate on substrates and at liquid-air interfaces in various coating processes can drive liquid flows that lead to defects. To better understand this phenomenon, we model the leveling of thin liquid films subject to electrohydrodynamic forces. We consider cases of homogeneous and heterogeneous substrate charge distributions and contamination of the film surface by free charge. The liquid is assumed to be Newtonian, both conductive and non-conductive materials are considered, and lubrication theory is employed. Linear stability analysis and nonlinear simulations reveal different leveling criteria for small- and large-amplitude perturbations to the film surface. Heterogeneous charge distributions on the substrate are found to lead to steady curved interface shapes. Using asymptotic methods, we develop analytical expressions to predict these shapes, and consequently, the magnitude of coating defects. We also employ transient nonlinear simulations and a scaling analysis to track the leveling of disturbances created by contamination of the film surface by free charge. The results of our study enable us to develop heuristics for determining the conditions under which coatings subject to electrohydrodynamic forces will level.

(742e) Molecular Dynamics Study On the Capacitive Performance of Double-Cationic Ionic Liquid Electrolytes

Li, S., Vanderbilt University
Feng, G., Vanderbilt University
Cummings, P. T., Oak Ridge National Laboratory
Originally presented on: 11/7/2013 16:40:00 - 17:00:00

Single-cationic ionic liquids have been extensively studied due to their versatility and outstanding properties. However, the study on double-cationic ionic liquids is rarely reported. In this work, the capacitive performance of double-cationic ionic liquids with different chain lengths was investigated using classical molecular dynamics simulation (MD) for comparison with the single-cationic ionic liquids. Different electrical double layer (EDL) structures formed by double- and single-cationic ionic liquids near an onion-like carbon electrode were observed; the orientation of the cation at the interface was analyzed. The number of the cumulative ions was estimated for both double- and single-cationic ionic liquids. The capacitive performance was evaluated by computing the differential capacitance as a function of electrical potential applied. The effects of the linkage chain length as well as temperature on the capacitance-electrical potential curve were studied. This work rationalizes the different capacitive performances of double- and single-cationic ionic liquids as electrolytes in onion-like carbon-based supercapacitors.

(742f) Three-Dimensional Simulations of Charged Droplets in Electric Fields and Shear

Shardt, O., University of Alberta

Mitra, S. K., University of Alberta

Derksen, J., University of Alberta

Originally presented on: 11/7/2013 17:00:00 - 17:20:00

Fluid interfaces are often charged and these charges play an important role in determining the behaviour of droplets in electric fields and during collisions in flow. We have previously coupled solvers for binary-liquid flow and electrostatic potential as described by the linearized Poisson-Boltzmann equation (Debye-Hückel equation). With this two-dimensional hybrid simulation code, we simulated electroosmotic flow in charged channels, the electrophoresis of a charged droplet, and a collision of two charged droplets. A sample flow field around a drop during electrophoresis is shown in the figure below.

We present an extension of the previous work to three dimensions. To simulate the flow of two immiscible liquids, we use the free-energy lattice Boltzmann method. We employ an iterative method to solve the Debye-Hückel equation. Both the flow and electrostatic potential solvers were implemented on graphics processing units to achieve short simulation times. We present the results of several benchmarks of electroosmotic flow between charged surfaces. A comparison of the computed flow and an analytical solution for the flow between two charged parallel plates in a uniform electric field is shown below. We also simulate the electrophoresis of a charged droplet and collisions of two charged droplets. We compare the results of the two- and three-dimensional simulations as well as the collisions of charged and uncharged droplets. These simulations provide insight into the role of surface charge on the coalescence of charged drops in emulsions.

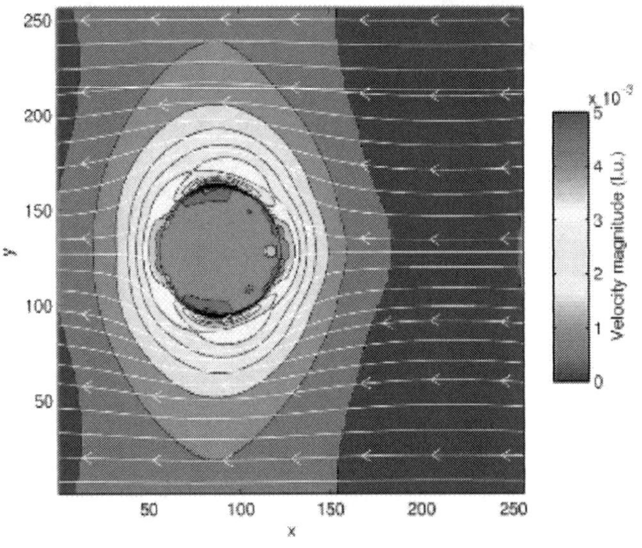

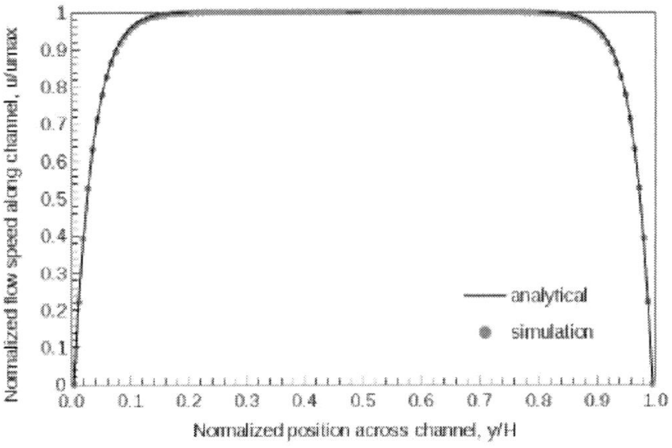
Normalized flow speed along channel, u/umax
Normalized position across channel, y/H
analytical
simulation
0
0.1
0.2
0.3
0.4
0.5
0.6
0.7
0.8
0.9
1
0.0
1.0

(742g) Electrohydrodynamic Instabilities in Thin Viscoelastic Films: AC and DC Fields

Espin, L., University of Minnesota
Corbett, A.,
Kumar, S., University of Minnesota
Originally presented on: 11/7/2013 17:20:00 - 17:40:00

Electrohydrodynamic instabilities in thin liquid films are a promising route for the self-assembly of well-defined topographical features on the surfaces of materials. Here, we study the effect of viscoelasticity on these instabilities under the influence of AC and DC electric fields. Viscoelasticity is incorporated via a Jeffreys model and both perfect and leaky dielectric materials are considered. In the case of DC fields, asymptotic methods are employed to shed light on the nature of a singularity that arises when solvent viscosity is neglected (i.e., the Maxwell-fluid limit). In the case of AC fields, we apply a numerical procedure based on Floquet theory to determine the maximum growth rate and corresponding wavenumber as a function of the oscillation amplitude and frequency. Elasticity is found to increase both the maximum growth rate and the corresponding wavenumber, with the effects being the most pronounced when the oscillation period is comparable to the fluid relaxation time.

(743a) Towards Engineering Advanced Vanadium Flow Battery

Li, L., UniEnergy Technologies, LLC

Originally presented on: 11/7/2013 15:15:00 - 15:37:00

Flow batteries were specifically invented by electric utilities in the 1960's to store large amounts of electricity in a safe and practical way. The complexities and latent dangers of bulk electricity storage in confined spaces have become more evident in recent years with a series of thermal runaway events at multi-MW battery installations. It is precisely in these areas that the inherent control stability, charge uniformity, thermal consistency and flexible form factor of flow batteries is most compelling.

At small scale, flow batteries may appear complex compared with "a few lead-acid cells in series". However, when extended life and significant scale is required, all batteries rapidly require a complex array of components to actively manage and maintain performance. In the case of flow batteries, these components are invariably simpler, cheaper, more reliable, and easier to control. Consequently, a thoughtful flow-battery system design can yield a significantly more compelling product than is possible with conventional static battery technologies.

This paper discusses system level and cost-effective engineering of a compact, resilient and serviceable bulk electricity storage system. The paper further introduces the UniEnergy Technologies (UET) containerized product.

(743b) Development and Experimental Validation of An Automotive System Model for Slurry-Based Chemical Hydrogen Storage Materials

Brooks, K., Pacific Northwest National Laboratory
Pires, R.,
Westman, M.,
Holladay, J.,
Originally presented on: 11/7/2013 15:37:00 - 15:59:00

This paper presents system models for the slurry-based chemical hydrogen storage materials being explored for fuel cell vehicles. As a part of the Hydrogen Storage Engineering Center of Excellence program sponsored by US Department of Energy, PNNL has developed and modeled systems for both exothermic and endothermic storage materials using Simulink® software. The exothermic and endothermic materials modeled were ammonia borane and alane, respectively. These models track the chemical constituents through all the components of the system including the reactor, phase separator, heat exchanger, and product and ballast tanks. To validate the models, experiments with both simulants and actual materials were performed on scaled version of key components. The model was then used with drive cycle simulations for an 80 kWe fuel cell vehicle to determine if the system can meet the DOE targets. Additionally, a sensitivity analysis was performed to provide an operating envelope for other yet-to-be-developed chemical hydrogen storage materials.

(743c) Deployment of An in-Membrane Micro Fuel Cell for Portable Applications

Omosebi, A., New Jersey Institute of Technology
Besser, R. S., Stevens Institute of Technology
Originally presented on: 11/7/2013 15:59:00 - 16:21:00

Advances in nano/microfabrication technology has led to the deployment of smaller, faster, and more efficient portable, wireless, and autonomous devices. Today's devices have significantly improved multitasking and computing capability that leads to increased energy requirements. These devices currently rely on secondary batteries like the Ni-Cd, Li-ion, and Li-polymers as a power source, but the need for constant replacement and recharging, interrupted operation, disposal issues, and increasing power requirements have prompted the development of alternative power sources like polymer electrolyte fuel cells as replacements. Fuel cells are high energy power sources that produce DC electricity directly from the stored chemical potential in their fuel and oxidant input. A conventional Nafion-based polymer electrolyte membrane (PEM) fuel cell is a sandwich-like stack of separate layers that facilitate electronic, protonic, and fluid transport. In the conventional micro PEM fuel cell, the auxiliary layers that serve as flow field, current collectors, and reactant distributors contribute significantly to the size and weight of the device leading to bulkier devices. New planar type micro fuels cell architectures have recently emerged where proton conduction is in-plane versus the traditional through-plane orientation. These planar micro fuel cells can facilitate the deployment of thinner and reduced form factor power sources. Distinct from previously fabricated micro fuel cells by other researchers, in this work we present a novel micro fuel cell design where the micro-flow channels for fuel and oxidant input and the membrane electrode assembly (MEA) are fabricated entirely within a Nafion membrane. Nafion is the most widely used membrane in fuel cells that utilize a polymer electrolyte due to its superior properties. The fabrication process leverages electron beam lithography and dry etching for Nafion patterning and this new fuel cell design can result in the use of fewer materials for the deployment of a low-weight, high energy density power source.

(743d) Experimental Study of Solid Fuel Agglomeration and De-Agglomeration Behaviors in the Reactions With Metal Oxide Particles

Sun, Z., The Ohio State University
Ding, Y.,
Fan, L. S., The Ohio State University
Originally presented on: 11/7/2013 16:21:00 - 16:43:00

The present work studies the solid fuel agglomeration and de-agglomeration behaviors in the context of chemical looping combustion. A Thermal Gravimetric Analysis (TGA) apparatus is used in this study. PRB coal, Pittsburgh #8 coal, Alabama metallurgic coke, and biomass are used as solid fuels, whereas pure Fe_2O_3 and supported Fe_2O_3 are used as oxygen carrier particles, respectively. In each test, four types of samples are tested: 1) solid fuel powder only; 2) solid fuel powder + 30 wt% oxygen carrier powder; 3) solid fuel powder + 1-2 mm oxygen carrier balls; 4) solid fuel powder + 30 wt% oxygen carrier powder + 1-2 mm oxygen carrier balls. At reducing environment, each sample is heated to 1000 °C at 50 °C/min, maintained for 10 minutes, and then cooled down to the room temperature. It is found that PRB coal powder and Alabama coke powder form no agglomeration, whereas the biomass powder forms a loose and soft agglomeration only when heated alone. Pittsburgh #8 coal powder forms agglomeration when heated alone or with 1-2 mm oxygen carrier "balls". However, when mixed with the oxygen carrier "powder", no agglomeration forms for Pittsburgh #8 coal powder. To prevent the agglomeration of Pittsburgh #8 coal, two anti-agglomeration approaches are tested: 1) low-temperature devolatization (below caking temperature), and 2) high-temperature devolatization (above caking temperature) and subsequent pulverization. Both approaches are found to be able to effectively prevent/mitigate agglomeration of Pittsburgh #8 coal.

(743e) Development of Fe-Cu Oxygen Carriers for Solid Fuel Gasification and Hydrogen Production

Galinsky, N., North Carolina State University
He, F., North Carolina State University
Li, F., North Carolina State University
Originally presented on: 11/7/2013 16:43:00 - 17:05:00

Carbon dioxide emission from fossil energy conversion is a major contributor to the greenhouse effect. Therefore, development of processes to effectively convert fossil fuels with smaller carbon footprints is highly desirable. Chemical looping schemes, which include chemical looping combustion (CLC) and chemical looping gasification (CLG), have shown promise to convert carbonaceous fuels while capturing CO_2 efficiently. CLG, in particular, uses a solid metal oxide based oxygen carrier in a redox process to convert carbonaceous fuels into CO_2, hydrogen, and heat in separate reactors, avoiding the energy intensive product separation steps.

Currently, Fe based materials are oxygen carriers of choice for CLG processes since reduced iron oxides can effectively convert steam into hydrogen. The limitation of using Fe-based oxygen carriers is that they are relatively slow when converting carbonaceous fuels, especially those in solid forms. In order to address this challenge, we propose to incorporate a secondary metal oxide with oxygen uncoupling properties, e.g. CuO, to overcome the slow reactivity of Fe-based materials. A recent simulation study has shown that the addition of a small amount of CuO to a Fe_2O_3 oxygen carrier (9:1 molar ratio between Fe and Cu) can increase the efficiency of the process, resulting in higher H_2 production rates.

In the current study, a small amount of CuO based oxygen carrier is incorporated to a batch of Fe based material. TGA and fluidized bed experiments are conducted for chemical looping conversion of coal and methane. Significant increase in reactivity with coal char is observed with 5 w.t. % CuO addition. After the reaction, Fe-Cu spinel structure is not observed under XRD.

(743f) Synthesis and Redox Pathways of MgAl2O4-Stabilized, Fe-Cu-Based Oxygen Carriers for Chemical Looping Water Splitting

Imtiaz, Q., ETH Zurich
Yuzbasi, N. S., ETH Zurich
Broda, M., ETH Zurich
Kierzkowska, A., ETH Zurich
Abdala, P.,
van Beek, W.,
Muller, C. R., ETH Zurich
Originally presented on: 11/7/2013 17:05:00 - 17:27:00

Hydrogen may become an important energy carrier in the future. The current technologies for the large scale production of H_2, *i.e.* steam reforming of natural gas and steam gasification of coal, are very energy intensive owing to the need for various separation steps to obtain pure H_2.[1] However, for H_2 to become a major energy carrier it has to be produced in an efficient and sustainable way. Chemical looping water-splitting (CLWS) is an emerging technology that can inherently produce a pure stream of H_2, thus eliminating the need for energy intensive purification steps.[2] In CLWS, lattice oxygen of an oxygen carrier, typically a transition metal oxide, is used to combust a hydro-carbonaceous fuel. In the subsequent step, the reduced oxygen carrier is re-oxidized with steam to produce a pure stream of H_2. Iron oxide is an attractive material for the CLWS process owing to its wide availability, low cost, minimal environmental impact and high equilibrium partial pressure of H_2.[3] Nevertheless, its reactivity and cyclic redox stability are two major concerns with regards to the CLWS process. Pure Fe_2O_3 that is completely reduced to Fe deactivates after a few cycles, probably due to sintering.[4] To circumvent this rapid deactivation it has been proposed to stabilize iron oxide with a high Tammann temperature support, such as Al_2O_3, ZrO_2, TiO_2, SiO_2 or $MgAl_2O_4$. However, it was found that supported Fe_2O_3 has often a high tendency for carbon deposition. An approach to improve the reactivity of iron oxide with CH_4 and higher hydrocarbons is the addition of more reactive transition metal oxides.[5]

Here, we synthesized $MgAl_2O_4$-stabilized, Fe-Cu-based oxygen carriers containing 70 wt. % Fe_2O_3 for CLWS using CH_4 as a fuel. The reactivity, H_2 yield and carbon deposition characteristics of the oxygen carriers were compared to pure Fe_2O_3 and $CuFe_2O_4$ and $MgAl_2O_4$-stabilized Fe_2O_3. The oxygen carriers were characterized using powder X-ray diffraction (XRD), CH_4 and H_2 temperature programmed reduction,

N_2 adsorption measurements, scanning electron microscopy, transmission electron microscopy, Raman spectroscopy and X-ray absorption spectroscopy (XAS). The redox and carbon deposition characteristics of the synthesized oxygen carriers were determined at 900 Â°C using both a thermo-gravimetric analyzer and a packed bed reactor.

Pure $CuFe_2O_4$ was found to have a low surface area, compact morphology, slow kinetics and high tolerance towards carbon formation. On the other hand $MgAl_2O_4$-supported Fe_2O_3 possessed a 'grainy' surface texture, high surface area, high reactivity towards CH_4 and a high activity for carbon deposition. The surface morphology of $MgAl_2O_4$-supported, Fe-Cu-based oxygen carriers were found to be an intermediate of $MgAl_2O_4$-supported Fe_2O_3 and $CuFe_2O_4$. Thermo-gravimetric and pack bed measurements revealed that the oxidation kinetics and H_2 yield of $MgAl_2O_4$-stabilized, Fe-Cu-based oxygen carriers were, respectively, substantially faster and higher than for $CuFe_2O_4$. For $MgAl_2O_4$-stabilized, Fe-Cu-based oxygen carriers *in-situ* XRD and XAS measurements revealed that Fe_2O_3 was stabilized by both a spinel phase ($MgAl_2O_4$) and a delafossite phase ($CuFeO_2$). Additionally, $MgAl_2O_4$-stabilized, Fe-Cu-based oxygen carriers effectively inhibited the decomposition of CH_4 and the formation of iron carbide. Scanning electron micrographs and XAS measurements indicated that the reduced quantity of carbon deposition on the bimetallic oxygen carriers is due to an alteration of the surface morphology, impeding the Fe-based activation of CH_4.

References

[1] A. T-Raissi and D. L. Block, Power and Energy Magazine, IEEE **2004**, 2, 40.

[2] A. Thursfield, A. Murugan, R. Franca and I. S. Metcalfe, Energy Environ. Sci. **2012**, 5, 7421.

[3] R. D. Solunke and G. T. Veser, Ind. Eng. Chem. Res. **2010**, 49, 11037.

[4] C. D. Bohn, C. R. M?ller, J. P. Cleeton, A. N. Hayhurst, J. F. Davidson, S. A. Scott and J. S. Dennis, Ind. Eng. Chem. Res. **2008**, 47, 7623.

[5] S. Wang, G. Wang, F. Jiang, M. Luo and H. Li, Energy Environ. Sci. **2010**, 3, 1353.

(743g) Study On Performance of the Novel Steam Generation System Using Water-Zeolite Pair

Nakaso, K., Kyushu University

Nakashima, K., Kyushu University

Tanaka, Y., Kyushu University

Iwama, Y., Kyushu University

Fukai, J., Kyushu University

Originally presented on: 11/7/2013 17:27:00 - 17:49:00

The efforts for energy conservation are requested because of the limitation of fossil fuel and the environmental petrochemical and steel industries are especially the largest energy-consuming manufacturing industries. While with the range of 60-90°C is released from these industrial processes, steam is demanded and generally provide fuel. The system to generate high-temperature steam from waste heat is therefore required for the effective util novel steam generation system using a direct heat exchange system with adsorbent-water pair is proposed. Con adsorbent results in evaporation of excess water due to the release of adsorption heat from adsorbent. Because any heat exchangers, not only higher packing density of adsorbent particles in the reactor but also lower heat ca expected. In our previous study, generation of steam under cyclic operation of steam generation and regenerati experimentally confirmed using zeolite as adsorbent. Superheated steam at 0.1-0.4MPa was generated using wa 130°C with lab-scale apparatus [1].

The purpose of this study is to investigate performance of the system using water-zeolite pair. For discussing pe mass of the steam generated in the cyclic steam generation and regeneration processes is particularly studied. A the system satisfied with the energy balance in the steam generation process [1]. This implies that mass of stean based on the energy balance between the initial and final states of the steam generation process. While the fina process is independent from the regeneration process, the initial state of the steam generation process strongly temperature distributions and distributions of water content in the zeolite after the regeneration process. Theref process is carried out first. Performance of the system is then discussed based on the estimation of regeneration

A mathematical model considering transport phenomena during regeneration process is developed. As the gove equations of mass and energy for gas and solid are considered. Ergun's equation is used instead of the momentu gas. The rate of desorption in the model is determined based on the measurement data from thermal analysis. T assumed as constant-rate of drying. For the validation of the model, numerical results are compared to experim laboratory-scale apparatus. As a result, numerical results reasonably agree with the experimental ones with vari temperatures.

Performance of the system is then discussed in terms of mass of steam generated in the cyclic operation under Because the regeneration process is considered as the energy storage process in the system, mass of steam gene regeneration time and reaches the saturated value. The typical result is shown in Fig1(a). When the flow rate is generated at 140°C (0.36MPa) increases with the regeneration time and reaches the saturated value of 0.13 kg- result, appropriate regeneration time seems 600-1000 s. To discuss more clearly, the mean steam generation rat as the mass of steam generated during the cycle operation per operation time. For this case, the mean steam ger maximum value of 0.53×10^{-4} kg-steam/(kg-zeolite·s) at the regeneration time of 690 s (see Fig1(b)). According generation rate is used for the following discussion.

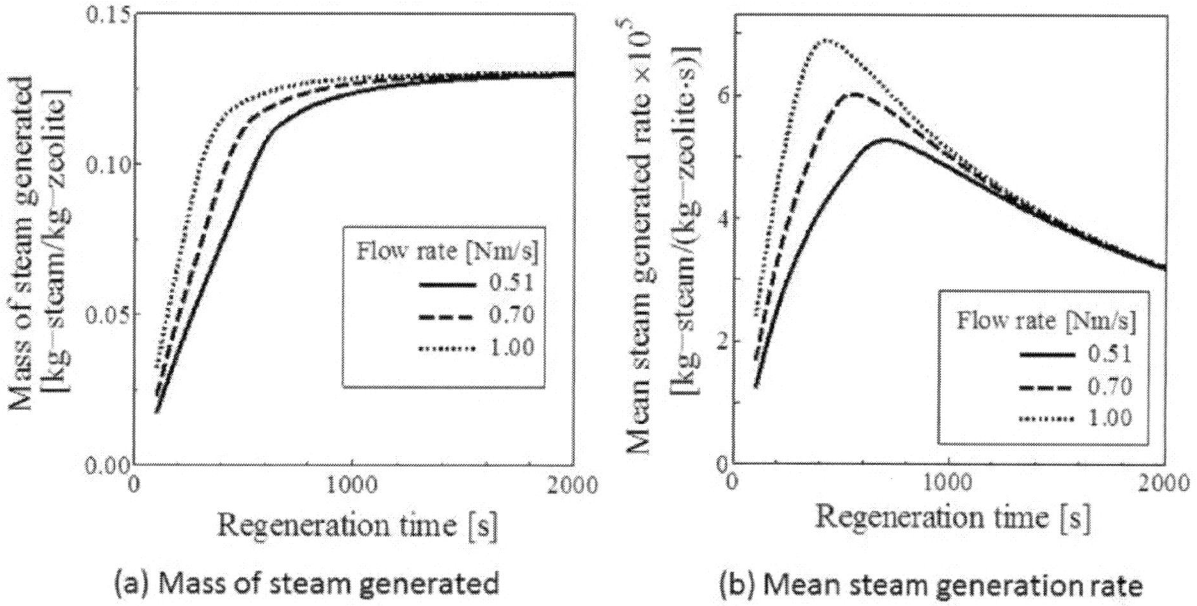

Fig.1 Effect of regeneration time on the mass of steam generated at 140°C in cyclic operation (Inlet gas temper
of 60 °C for regeneration)

Effect of the regeneration condition on the maximum mean steam generation rate is studied. The regeneration c
and dew point of the gas are studied. The typical result is shown in Fig2. The maximum mean steam generation
dew point of inlet gas increases above -20°C. When gas temperature is 200°C, the effect of dew point on the m
rate is not so significant. When the gas temperature is 140°C, dew point of gas should become low to generate s
it was found that not only dry gas but also humid gas can be applied to generate steam when gas temperature is
helpful in terms of effective utilization of waste heat such as exhaust gas.

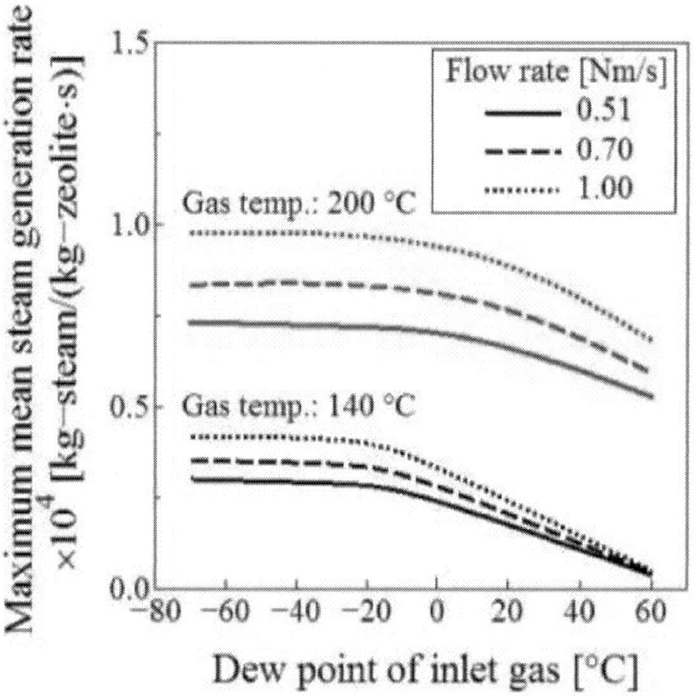

Fig.2 Effect of dew point of the inlet gas for the regeneration process on the maximum mean steam generation r
140°C)

Acknowledgements

This work was partially supported by the NEDO project under Research and Development Program for Innovat
Technology, and also by the Japan Society for the Promotion of Science (JSPS) Grant-in-Aids for Scientific Res

Reference

[1] B. Xue, Y. Iwama, Y. Tanaka, K. Nakashima, A. T. Wijayanta, K. Nakaso, J. Fukai, "Cyclic steam generatio
adsorption heat pump using low-grade waste heat", Experimental Thermal and Fluid Science, Vol. 46, 54–63 (2

(750a) Hydrogen Production From Methanol and Formic Acid Reactions Over Atomically-Dispersed Precious Metal (Au and Pt) Catalysts

Yi, N., Tufts University
Saltsburg, H., Tufts University
Flytzani-Stephanopoulos, M., Tufts University
Originally presented on: 11/7/2013 15:15:00 - 15:35:00

Steam reforming of methanol (SRM), which produces the highest amount of hydrogen among all hydrogen-producing reactions using methanol as reactant, is one of the significant catalytic processes to produce hydrogen in different scales [1]. But methanol is only one option among possible liquid fuel carriers for a hydrogen-based fuel cell. As is the case with introducing any novel industrial application in competitive markets, other possible alternatives, such as formic acid, must be fully considered [2]. From the fundamental viewpoint, it is interesting to compare these two liquids in terms of the pathways involved and the selectivity of different catalysts to hydrogen. Moreover, if precious metals are to be used as catalysts, it is highly desirable to explore efficient approaches that decrease the amount of precious metals used while maintaining high activity at close-to- ambient temperatures for electrocatalytic applications. Here we investigated the potential of two candidate catalysts: dispersed sub-nm gold on ceria and sodium-promoted platinum on silica for catalytic hydrogen production through steam reforming of methanol and decomposition of formic acid reactions[3,4,5].

1. Palo, R. M., Dagle, R. A., Holladay J. D. *Chem. Rev.*107, 3992 (2007).

2. Yu, X., Pickup, P. G. *J. Power Sources*182, 124 (2008).

3. Yi, N., Si, R., Saltsburg, H., Flytzani-Stephanopoulos, M. *Appl.Catal. B* 109, 87 (2010).

4. Yi, N., Si, R., Saltsburg, H., Flytzani-Stephanopoulos, M. *Energy Environ. Sci.* 3, 831 (2010).

5. Yi, N., Saltsburg, H., Flytzani-Stephanopoulos, M. *ChemSusChem* 6, 816 (2013).

(750b) Structure Sensitivity in Pt Nanoparticle Catalysts for Hydrogenation of 1,3-Butadiene: In Situ Study of Reaction Intermediates Using SFG Vibrational Spectroscopy

Michalak, W. D., Lawrence Berkeley National Laboratory
Krier, J., University of California
Komvopoulos, K., University of California
Somorjai, G. A., University of California, Berkeley
Originally presented on: 11/7/2013 15:35:00 - 15:55:00

The product selectivity during 1,3-butadiene hydrogenation on monodisperse, colloidally synthesized, Pt nanoparticles was studied under reaction conditions with kinetic measurements and *in situ* sum frequency generation (SFG) vibrational spectroscopy. Four products are formed at 75 °C: 1-butene, cis-2-butene, trans-2-butene, and n-butane. Ensembles of Pt nanoparticles with average diameters of 0.9 and 1.8 nm exhibit a 30% and 20% increase in the full hydrogenation products, respectively, as compared to Pt nanoparticles with average diameters of 4.6 and 6.7 nm. Methyl and methylene vibrational stretches of reaction intermediates observed under working conditions using SFG were used to correlate the stable reaction intermediates with the product distribution. Kinetic and SFG results correlate with previous DFT predictions for two parallel reaction pathways of 1,3-butadiene hydrogenation. Hydrogenation of 1,3-butadiene can initiate with H-addition at internal or terminal carbons leading to the formation of 1-buten-4-yl radical (metallocycle) and 2-buten-1-yl radical intermediates, respectively. Small (0.9 and 1.8 nm) nanoparticles exhibited vibrational resonances originating from both intermediates, while the large (4.6 and 6.7 nm) particles exhibited vibrational resonances originating predominately from the 2-buten-1-yl radical. This suggests each reaction pathway competes for partial and full hydrogenation and the nanoparticle size affects the kinetic preference for the two pathways. The reaction pathway through the metallocycle intermediate on the small nanoparticles is likely due to the presence of low-coordinated sites.

(750c) Design of Catalysts At the Subnanometer to Nanometer Scale: Tuning Performance Via Size, Composition, Doping, Support and Assembly

Vajda, S., Argonne National Laboratory

Originally presented on: 11/7/2013 15:55:00 - 16:15:00

The elucidation of the size/composition/shape/structure and function relationship, the effect of support along with the determination and control of the nature of the catalytic particles under reaction conditions are instrumental for addressing fundamental aspects of catalysis on the way to the design of new classes of catalytic materials. Highly uniform particles on technologically relevant supports are prerequisites for such studies.

Our experimental studies are based on 1) fabrication of technologically relevant supports, 2) physical size- and composition selected cluster deposition with atomic precision control, 3) *ex situ* and *in situ* microscopies and 4) *in situ* synchrotron X-ray characterization of cluster size/shape and oxidation state under realistic working conditions, combined with mass spectroscopy analysis of the reaction products. The experimental studies are complemented with DFT calculations. Using model size-selected clusters, this contribution will outline the applicability of the outlined approach on example of selective C-H bond activation in the oxidative and non-oxidative dehydrogenation of cyclohexane and cyclohexene. Select examples will be presented on coupling studies of model and practical catalysts.

Oxidative and non-oxidative dehydrogenation of cyclohexane on subnanometer Co_4O_x and $Co_{27}O_x$ clusters and nanometer size Co_3O_4 and $(Au)Co_3O_4$particles. In the absence of oxygen, the clusters became active around 300 °C, with the larger clusters being about 2x as active as the smaller ones. The clusters were active already around 100 °C – with somewhat different selectivities for the two cluster sizes. The performance of the subnanometer Co_nO_x clusters will be compared with that of various size nanometer Co_3O_4 particles prepared by wet-chemical methods - bare as well as Au-doped. Notably, the activity of cobalt clusters under oxidative conditions makes them as potential candidates to replace Pt-based dehydrogenation catalysts, while the spectrum of investigated cobalt oxide clusters is also offering a tunable selectivity and activity with particle size, doping and oxidation state.

Dehydrogenation of cyclohexene on subnanometer $Co_{27}O_x$clusters and their nanoscale assemblies: Increased activity through assembly under reaction conditions and the effect of the support and reaction conditions on catalyst performance. 27-atom cobalt clusters were deposited on TiO_2, ZnO, Al_2O_3, MgO and ultrananocrystalline diamond (UNCD). While the activity and selectivity of TiO_2, ZnO and Al_2O_3 supported clusters were comparable, MgO-supported clusters were 2 to 3 times more active under oxygen rich conditions. *In situ* data reveal that a formation of a fluxional ~2-3 nm structure of a $Co_xMg_yO_z$ composition is responsible for the dramatic increase in activity. Under oxygen lean conditions, UNCD-supported clusters possessed the highest activity as well as highest toward selectivity towards dehydrogetanted C6 products. Along with the dehydrogenation of cyclohexane, these results indicate extraordinary propensities of sub-nm Co_nO_x clusters in low-temperature activation of O_2.

References

1. "Oxidative Dehydrogenation of Cyclohexane on Cobalt Oxide (Co_3O_4) NanoparticlesThe Effect of Particle Size on Activity and Selectivity", by E. C. Tyo, C. Yin, M. Di Vece, Q. Qian, S. Lee, B. Lee, S. Seifert, R. E. Winans, R. Si, B. Ricks, S. Goergen, M. Rutter, B. Zugic, M. Flytzani-Stephanopoulos, Z. Wang, R. E. Palmer, M. Neurock, and S. Vajda, *ACS Catal.* 2, p. 2409–2423 (2012)

2. "Support-Dependent Performance of Size-Selected Subnanometer Cobalt Cluster-Based Catalysts in the Dehydrogenation of Cyclohexene", S. Lee, M. Di Vece, B. Lee, S. Seifert, R. E. Winans and S. Vajda, *Chem. Cat. Chem.* 4, p. 1632-1637 (2012)

3. "Oxidative Dehydrogenation of Cyclohexene on Size Selected Subnanometer Cobalt Clusters: Improved Catalytic Performance via Evolution of Cluster-Assembled Nanostructures", S. Lee, M. Di Vece, B. Lee, S. Seifert, R. E. Winans and S. Vajda, *Phys. Chem. Chem. Phys.*, **14**, p. 9336 - 9342 (2012)

4. L. Curtiss, J. Greeley, M. Flytzani-Stephanopoulos, G. Haller, S. Lee, M. Neurock, R. Palmer, L. Pfefferle, Q. Qian, R. Schlögl, D. Teschner, E. Tyo, S. Vajda, R. Winans et al, papers in preparation

5. "Subnanometre Platinum Clusters as Highly Active and Selective Catalysts for the Oxidative Dehydrogenation of Propane", S. Vajda, M. J. Pellin, J. P. Greeley, C. L. Marshall, L. A. Curtiss, G. A. Ballentine, J. W. Elam, S. Catillon-Mucherie, P. C. Redfern, F. Mehmood and P. Zapol, *Nat. Mater.* 8, 213-216 (2009)

(750d) Morphology Dynamics of Precious Metal Catalysts for Use in Steam Reformation of Oxygenated Fuels

Crowley, S., The City College of New York
Castaldi, M. J., City College of New York
Originally presented on: 11/7/2013 16:15:00 - 16:35:00

The advancement in liquid fuel production from biological sources will result in oxygenated fuel compositions. While oxygen is beneficial for combustion purposes, its presence within the chemical makeup of the fuel is detrimental to the fuel's energy density. In understanding how oxygenated fuel may be reformed to remove the oxygen functional groups, we have been exploring ethanol reforming as a test case. Recently, initial results show for the first time that the oxidation state of catalytic Rh nanoparticles changes dynamically under reaction conditions with ethanol and that the extent of Rh oxidation appears to control catalyst activity and the rate of deactivation. Catalyst deactivation is one of the major challenges faced in fuel reforming and there exists a need not only to determine the primary causes, but model them in order to take preventative measures. Through the use of common analytical techniques such as CO chemisorption and X-ray Diffraction (XRD), the surface morphology of the catalysts has been examined. Each of three formulations was calcined at temperatures ranging from 550°C to 950°C. In addition, X-ray Diffraction (XRD) has been performed in order to detect any conformational changes in the crystal structure of the catalysts as a function of calcination temperature. The extent of sintering of the precious metal catalysts as well as any changes in oxidation state, catalyst poisons adsorbed to the catalyst, and burrowing of the precious metals into the support are suspected to result in lower catalytic activity. This presentation will correlate and explain the active metal distribution and catalytic activity as well as the time dependence of the distribution during the reaction. In this way, we can create a complete profile of the molecular composition of the catalyst surface, which could lead to better overall performance. It is anticipated that these results will provide an important basis for DFT and mechanistic modeling to further elucidate the underpinnings of the observed performance.

(750e) Morphological, Electronic, and Catalytic Properties of Graphene-Supported Pt Nanoclusters

Fampiou, I., University of Massachusetts Amherst
Ramasubramaniam, A., University of Massachusetts Amherst
Originally presented on: 11/7/2013 16:35:00 - 16:55:00

Graphene has received considerable recent attention as a support material for fuel-cell electrodes due to its high surface area and excellent electrical conductivity. Experiments indicate that graphene-supported Pt nanoparticles exhibit greater stability and better electrocatalytic activity as compared to their counterparts on traditional carbon supports, or even bulk Pt electrodes.

In this study we use a combination of empirical-potential molecular-dynamics calculations and first-principles density functional theory calculations to investigate the structural, electronic, and catalytic properties of Pt nanoclusters on graphene supports. We present detailed thermodynamic and electronic structure analyses of Pt clusters on both defective and defect-free graphene substrates and elucidate the role of support point defects in altering the electronic structure of the supported clusters. Defects in graphene supports are found to enhance the metal-support interaction thereby increasing stability towards catalyst sintering. The strong Pt-C interactions are also accompanied by significant charge redistribution between the cluster and the substrate, which alters the catalytic properties of the cluster.

As a model reaction, we investigate CO adsorption and oxidation on graphene-supported Pt clusters. We show that clusters bound to defects in graphene could potentially exhibit increased tolerance toward CO poisoning. We examine the activity of the graphene-supported Pt clusters for the CO oxidation reaction and obtain estimates for CO-oxidation kinetics. Reaction kinetics on supported Pt nanoclusters are comparable with bulk Pt and, on average, faster than on unsupported clusters.

Overall, our results suggest promising avenues for controlling the dispersion and catalytic activity of Pt nanoclusters on graphene supports via defect engineering.

(750f) Nanoparticle SAMs As Molecular Rulers: Two Cu(I) Catalytic Centers Are Required for 'Click' Catalysis

Walker, D. A., Northwestern University
Browne, K. P., Northwestern University
Grzybowski, B., Northwestern University
Originally presented on: 11/7/2013 16:55:00 - 17:15:00

The 'click' reaction has become an exceedingly popular coupling reaction in organic chemistry, but the mechanism by which the reaction operates remains unknown. It has been suggested that two Cu(I) catalytic centers are required for the mechanism, but to date there is no experimental evidence for this assumption. Here we employ on particle mixed self-assembled monolayers (mSAMs) to control the relative spacing between catalytic centers. This technique allows us to (i) experimentally prove that two catalytic centers are necessary for the 'click' reaction to occur and (ii) measure the critical separation at which a catalytically active pair of Cu(I) ions become inactive. This technique, beyond providing useful insights into the mechanism of the 'click' reaction, develops a platform for which other catalytic diads and triads can be investigated. A mixture of experimental and theoretical results will be resented.

(750g) Ni-Based Catalysts Synthesized By Atomic Layer Deposition for Dry Reforming of Methane

Gould, T. D., University of Colorado at Boulder
Lubers, A. M., University of Colorado at Boulder
Medlin, J. W., University of Colorado Boulder
Weimer, A. W., University of Colorado at Boulder
Falconer, J. L., University of Colorado
Originally presented on: 11/7/2013 17:15:00 - 17:35:00

Supported Ni catalysts with particle diameters of approximately 2.5 nm were synthesized by atomic layer deposition (ALD). These catalysts were evaluated for dry reforming of methane activity and stability. The Ni ALD catalysts exhibited a higher activity per gram of metal than catalysts prepared by incipient wetness, and performed slightly better than Ni catalysts in the literature that were promoted with Mg or Pt. Additionally, the Ni ALD catalysts exhibited higher long-term activity than other catalysts for up to 72 h at 600°C. When the Ni ALD catalysts were modified with Pt ALD, their activity increased, sintering decreased, and resistance to coking was enhanced. Modifying the incipient wetness Ni catalysts with 1 cycle of Pt ALD increased their activity per gram of metal by more than a factor of 2. Adding 1 cycle of Pt ALD to the Ni ALD catalysts doubled the activity per gram of metal of the Ni ALD catalyst, resulting in the highest activity per gram of metal of all the samples investigated. Additionally, modifying the Ni ALD catalyst with Pt ALD decreased the loss of activity due to sintering. The Ni ALD catalysts deactivated by ~50% of their initial activity due to sintering within the first 6 h, but the Pt ALD-modified catalysts only deactivated by 2% over 6 h at 600°C. The nature of bimetallic interaction was probed via CO-TPD and temperature-programmed reduction. Additionally, the structure of the as-prepared samples and heat treated samples were analyzed by HRTEM.

(753a) High Throughput Study of Alloy Surface Oxidation Across Composition Space: AlxFeyNi1-x-Y

Gellman, A. J., Carnegie Mellon University
Miller, J. B., Carnegie Mellon University
Payne, M., Carnegie Mellon University
Originally presented on: 11/7/2013 15:20:00 - 15:40:00

Comprehensive study of the interfacial phenomena of alloy materials, $A_xB_yC_{1-x-y}$, across composition space is hampered by the fact that the study must span a multidimensional and continuous composition space, $x = 0 \rightarrow 1$ and $y = 0 \rightarrow x$. Traditional methods of study require the preparation and characterization of large numbers of samples of single composition. This can be circumvented by high throughput methods that use Composition Spread Alloy Films (CSAFs) as platforms or libraries of alloy composition. CSAFs are thin alloy films deposited onto substrates in such a way that there is a gradient in their composition across the surface. As such they can contain all possible compositions of a ternary alloy, $A_xB_yC_{1-x-y}$ with $x = 0 \rightarrow 1$ and $y = 0 \rightarrow x$. Spatially resolved methods for characterizing the CSAF can then be used to determine bulk composition, surface composition, phase, electronic structure, etc. as functions of position on the substrate, i.e. as functions of alloy composition, with arbitrarily high composition resolution. We demonstrate the utility of the CSAF methodology for study of alloy interface properties through a study of the surface oxidation of $Al_xFe_yNi_{1-x-y}$ superalloys used as turbine materials for electrical power generation. The critical problem is to understand the role of Al in prevention of alloy oxidation by formation of a passivating layer of Al_2O_3 at the alloy surface during exposure to air. Using a $Al_xFe_yNi_{1-x-y}$ and a combination of energy dispersive x-ray spectroscopy and depth profiling using x-ray photoelectron spectroscopy, we have been able to map the Al concentration boundary in $Al_xFe_yNi_{1-x-y}$ above which the alloy is passivated. Along the AlNi binary, the passivation is caused by the formation of a surface layer of Al_2O_3. Along the AlFe binary, one can discern two regions of behavior, one in which the passivating layer is at the surface and the other in which the passivating layer is formed in the subsurface region beneath a layer of Fe_2O_3. Mapping these oxidation resistance boundaries across alloy composition space would be impossible without the high throughput CSAF methodology.

(753b) Growth of Gold Nanoshells On Colloidal Silica Characterized By in Situ Second-Harmonic Generation

Sauerbeck, C., University of Erlangen-Nuremberg
Schurer, B.,
Haderlein, M.,
Braunschweig, B., University of Erlangen-Nuremberg
Peukert, W., University of Erlangen-Nuremberg
Klupp Taylor, R. N., University of Erlangen-Nuremberg
Originally presented on: 11/7/2013 15:40:00 - 16:00:00

Metal nanoparticles and metal composite nanoparticles can have unique optical properties due to localized surface plasmon resonances and are therefore of particular interest in various applications. The resonance frequency of such particles is a function of their composition and their geometry. Gold nanoshells are comprised of a dielectric core and a thin gold shell. The optical properties of such particles can be readily tuned over a wide range by changing the ratio of the core diameter to the shell thickness.

In order to impart better control over the optical properties of gold nanoshells a detailed understanding of the shell formation is of great importance. For that purpose it is necessary to monitor changes in surface composition and morphology *in situ* and in real time. However, most experimental methods lack the desired surface sensitivity or cannot be applied *in situ* under the required experimental conditions. For that reason we have applied second-harmonic generation (SHG) spectroscopy as a powerful surface sensitive technique which shows great potential to probe surfaces of nanoparticles under relevant experimental conditions[1-3].

In our contribution, we present results from synchronized SHG and UV/Vis/NIR spectroscopies that were obtained on-line during the seeded growth of gold shells on silica core particles. UV/Vis/NIR spectroscopy allows the monitoring of the coverage of gold and to estimate whether the shells are closed or not, while SHG shows a remarkable sensitivity to changes in the surface morphology which has not been accessible by other *in situ*methods so far. Notably, at intermediate conditions where a still incomplete shell exists on the core particle during shell growth, we observe first a steep increase in the second-harmonic intensity that is followed by a subsequent rapid decrease in SHG intensity. As the SHG signal is a function of the fundamental electric field to the fourth power, we propose that the observed maximum in the temporal evolution of the SHG intensity is caused by a local field enhancement due to crevices that are being formed in between individual gold islands. As shell growth proceeds, the surface coverage of these crevices decreases and closed shells are formed which results in a substantially decreased SHG signal. This hypothesis is corroborated by additional angle-resolved SHG experiments and ex-situ electron microscopy of gold shells with different gold coverages and surface morphologies.

[1] Schürer, B., Wunderlich, S., Sauerbeck, C., Peschel, U., Peukert, W.,Phys. Rev. B, 82 (2010) 241404

[2] Schürer, B., Elser, M. J.; Sternig, A., Peukert, W., Diwald, O., J. Phys. Chem. C, 115 (2011) 12381–12387

[3] Schürer, B., Hoffmann, M., Wunderlich, S., Harnau, L., Peschel, U., Ballauff, M., Peukert, W., J. Phys. Chem. C, 115 (2011) 18302–18309

(753c) Correlating Monolayer Molecular Structure With Interfacial Viscoelasticity

Choi, S. Q., University of Minnesota
Fellows, C., University of California, Santa Barbara
Cao, K., University of Chicago
Kim, K., Univerisity of California, Santa Barbara
Squires, T. M., University of California, Santa Barbara
Lee, K. Y. C., The University of Chicago
Zasadzinski, J. A., University of Minnesota
Originally presented on: 11/7/2013 16:00:00 - 16:20:00

The surface viscosity of dipalmitoylphosphatidylcholine (DPPC) monolayers decreases by two orders of magnitude on addition of 3.7 mol% cholesterol, followed by a sharp increase in the monolayer elasticity above 3.7 mol%. We correlate these intriguing rheological properties with changes in the molecular organization as revealed by Grazing Incidence X-ray Diffraction (GIXD). Adding cholesterol at constant surface pressure decreases the tilt of the DPPC lattice in the same way as increasing the surface pressure at constant cholesterol. The correlation length of the DPPC lattice decreases with cholesterol fraction, suggesting that the increase in defect density leads to the decrease in surface viscosity. Above 4 mol%, the DPPC lattice parameter and correlation length saturates, showing that the observed increase in monolayer elasticity comes about from a percolated, emulsion-like cholesterol network.

Cholesterol is hypothesized to alter the fluidity, permeability and phase transitions of monolayers, bilayers, and cell membranes, but the mechanisms by which this happens are still obscure. At different concentrations, cholesterol organizes phospholipid molecules into "liquid-ordered" phases; while at other concentrations, cholesterol inhibits phospholipid crystallization into solid or liquid condensed phases. Despite their ubiquity in mammalian cell membranes, there is little quantitative information on the surface viscosity and elasticity of lipid/cholesterol monolayers, and how the viscosity and elasticity relate to interactions between cholesterol and phospholipids. For example, nanometer scale variations in membrane fluidity are hypothesized to be essential to the dynamics and function of membrane-associated proteins in self-organizing "rafts." Monolayer fluidity also plays an important, but as yet unappreciated role in the spreading and surface tension lowering properties of lung surfactants. Lung surfactant must reliably and reproducibly reduce the surface tension at the air-water interface of the alveoli to near zero to minimize the work of breathing. This requires fast spreading of the surfactant over the alveolar air-liquid interfaces during inspiration as the lung area rapidly increases. During exhalation, surfactant must resist Marangoni flow due to the surface tension differences between the deep lung and the bronchi. A lack of surfactant in premature infants causes potentially fatal neonatal Respiratory Distress Syndrome, which is currently treated with animal-derived replacement surfactants. However, the optimal concentration of many of the lipid and protein species, especially cholesterol is unknown. Even the presence of cholesterol in replacement lung surfactants is controversial as blood and cell debris extracted with surfactant from animal lungs complicate the cholesterol analysis. FDA-approved clinical surfactants Survanta and Curosurf have all cholesterol removed, while FDA-approved Infasurf has 4-5 wt% cholesterol. One common factor between all replacement surfactants is that DPPC is the dominant (50 – 80 mol%) phospholipid.

Here we show that the rheological properties of mixed DPPC/Chol monolayers at low cholesterol ($\leq$ 3.7 mol%) arise from changes in the DPPC lattice, but the increased elasticity at higher cholesterol mole fractions is the result of the mesoscopic evolution of the nanodomain morphology. GIXD shows that increasing the Chol fraction at constant surface pressure, or increasing the surface pressure at constant Chol fraction decreases the tilt in the same way. DPPC tilts to accommodate the mismatch between the larger projected area of the phosphocholine headgroup relative to the alkane tailgroup at low surface pressure. Palmitic acid (PA, used in Survanta) and hexadecanol (HD, used in the replacement surfactant, Exosurf) also decrease the molecular tilt at a given surface pressure. Cholesterol, like PA and HD, has a relatively larger tailgroup and a smaller headgroup, which mitigates the area mismatch, leading to the reduction in tilt. However, unlike PA or HD, the correlation length of the DPPC lattice decreases from ~ 10 nm to ~ 1 nm as Chol is increased from 0 to 7 mol%. This decreased correlation length is consistent with the cholesterol sterol ring disrupting the packing of the tailgroup lattice of DPPC. The correlation length can be thought of as the distance between lattice defects; hence the reduction in lattice correlations with increasing cholesterol fraction suggest an increased defect density, which corresponds to an increase in the "free area" available to the molecules to enhance diffusion and hence, reduce the surface viscosity. The reduction in lattice correlation and tilt saturates at ~ 4 mol% Chol, at the same concentration we see the change from a viscous to elastic monolayer (Fig. 4). However, the reduction in correlations are not consistent with an increase in the elasticity of the monolayer, but rather suggests the opposite. However, at mole fractions > 3.7%, the nanodomains percolate to form a bicontinuous morphology and the DPPC domain size decreases. In analogy to 3-D surfactant-stabilized emulsions, even small stresses deform the emulsion network structure, leading to an elastic modulus, $G' \sim \lambda R$, in which λ, the line tension, is the two-dimension equivalent of the surface tension in 3-D emulsions. R is a characteristic length scale of the DPPC domains, which decreases with increasing Chol content as the nanodomains form a percolating network.

(753d) AC Impedance Spectroscopy and Equivalent Circuit-Based Modeling As a Means to Probe the Micro and Nanostructure of Charged Colloidal Systems

Riley, M. W.,
Lane, S. D.,
Originally presented on: 11/7/2013 16:20:00 - 16:40:00

Reducing costs and increasing power densities of fuel cell and battery technologies requires improved material utilization. We are studying ionically conductive layered nanoclays (smectites and layered double hydroxides) as possible building blocks for fuel cell and battery structures. The goal of this technique is to locate the ionic conductor in close-proximity to the fuel cell catalyst or battery electrode material by tailoring the structure at the nano-level. Self-assembly characteristics of these charged, colloidal systems can then also be employed to produce structures with appropriate transport characteristics (ionic, electronic, and in the case of fuel cells, fuel and air).

The orientation, self-assembly, and aggregation characteristics of colloidal particles is often probed using rheological or light scattering techniques. Standard rheological techniques do not generally afford a long time window for study on an individual sample if solvent evaporation is a concern. Light scattering techniques (static and/or dynamic) allow for sealed samples to prevent solvent evaporation, but are often limited to dilute and/or transparent dispersions. AC impedance measurements can provide a study of these properties while allowing for sealed samples for long periods of time as well as over a wide range of colloid concentrations.

We have been using AC impedance measurements to characterize the conductivity of our colloidal gels, and have begun to realize a side benefit of also providing insight into the colloidal micro and nanostructure as well, particularly when coupled with relatively simple equivalent circuit-based models. This technique has been used to investigate the micro and nanostructure of nanclay-based gels, with varying clay cation-exchange capacity and primary particle size, as well as varying salt concentrations. We have begun to study the impact of organic solvents and water soluble polymers and have been designing a modified system to provide insight into the self-assembly behavior of these materials. The general technique will be presented, as well as past and current results.

(753e) Interfacial Rheological Responses of Langmuir Monolayers to Magnetic Probes: Exploring Non-Newtonian Behavior

Fellows, C., University of California, Santa Barbara
Zell, Z. A., University of California, Santa Barbara
Zasadzinski, J. A., University of Minnesota
Squires, T. M., University of California, Santa Barbara
Originally presented on: 11/7/2013 16:40:00 - 17:00:00

Various methods have been employed to measure the interfacial rheological properties of surfactant monolayers, but it is common to find disagreement in the literature in the values obtained by these techniques. Many of these measurements cannot fully account for effects such as history-dependence, yield stress, and viscoelastic recovery which are known to exist in some monolayers. We use electromagnets to drive microfabricated, micron-scale magnetic probes of prescribed shape, embedded within insoluble surfactant monolayers at the air-water interface. Small amplitude oscillatory shear experiments provide a linear response, recovering purely shear moduli from the angular displacement of circular probes under applies magnetic torques. We describe a series of experiments to investigate the different rheological responses to shear, dilation, and extension that arise under different forcings with different probes. During each of these experiments, the response of the condensed domains to the magnetic probe is simultaneously visualized with fluorescence microscopy. Thus we correlate the measured response of the monolayer under different driving forces/torques with the observed microstructural deformations, rearrangements, and behavior more generally.

(753f) Local Deformation and Contact Phenomena of Microscopic Objects Near Surfaces: A View From Below

Contreras-Naranjo, J. C., Texas A&M University
Ugaz, V. M., Texas A&M University
Originally presented on: 11/7/2013 17:00:00 - 17:20:00

Interference-based techniques have demonstrated great spatial (nanometer-scale) and temporal (microsecond-scale) resolution capabilities in the study of objects near surfaces. In addition to this, reflection interference contrast microscopy (RICM) offers a unique non-invasive "view from below" perspective that provides valuable information about contact phenomena, if any, and the topography of microscopic objects in close proximity to a transparent substrate. Well-known RICM capabilities, such as contact area and shape determination, have been extensively used in adhesion studies of particles, cells and lipid/polymer vesicles, although a significant drawback has been the complexity of the image analysis required for accurate results.

In this work, we introduce an innovative analysis of RICM interferograms that allows fast reconstruction of an arbitrarily shaped convex object with unprecedented nanometer-scale resolution. In addition, RICM's "view-from-below" is used to observe, for the first time, femtoliter-scale capillary condensation dynamics underneath micron-sized particles. Therefore, our implementation of RICM provides a high-resolution tool that opens new possibilities to realistically capture statistic/probabilistic behavior of dynamic interactions between ensembles of microscopic objects and surfaces.

(753g) A New Device for Measuring 3D Friction Forces and Applying Electrochemical Potential

Kristiansen, K., University of California, Santa Barbara
Banquy, X., University of California, Santa Barbara
Charrault, E., University of Montreal
Zeng, H., University of Alberta
Giasson, S., University of Montreal
Valtiner, M., Max-Planck-Institut fÃ¼r Eisenforschung GmbH
Israelachvili, J. N., University of California, Santa Barbara
Originally presented on: 11/7/2013 17:20:00 - 17:40:00

Current measurements of frictional forces are usually done parallel to the sliding direction. However, when the distances between the moving surfaces and the dimensions of the lubricant approach the nanometer scale, the molecular structure and surface asperities can induce significant frictional forces that are not parallel to the sliding direction. We have developed a new sensor-actuator for the Surface Forces Apparatus which can measure forces and move two surfaces relative to each other in all 3 orthogonal directions with force resolution in the μN range, and distance control in nm range in the normal direction and μm range in the plane of the surfaces. In addition, we can add an electrochemical attachment in order to control the surface potential of one of the surfaces. I will present how "off-axis" friction forces behave when shearing two atomically smooth crystalline (lattice) surfaces separated by nanometer thick layers of hexadecane, and how these friction forces are affected by parameters like relative orientation of the surfaces, temperature, and sliding history and velocity. In the end I will discuss how an applied electrochemical potential may change the friction forces in an aqueous system.

(774a) A Novel Fluidized Bed Hydrator for Steam Reactivation of Calcium-Based CO2 Sorbent

Wang, A.,
Wang, D., The Ohio State University
Deshpande, N., The Ohio State University
Phalak, N., The Ohio State University
Wang, W., The Ohio State University
Fan, L. S., The Ohio State University
Originally presented on: 11/8/2013 8:30:00 - 8:55:00

Calcium-based solid sorbents has been widely applied for the removal of acid gases, such as sulfur dioxide (SO_2), sulfur trioxide (SO_3), and hydrochloric acid (HCl), from coal-fired flue gas. Amid possible carbon dioxide (CO_2) regulations, scientists have explored ways to extend this technology for CO_2 removal especially from sizable point sources at low cost. Here at the Ohio State University (OSU), researchers have developed a 3-step CaO-based carbonation-calcination reaction (CCR) process that operates in the range of 450 – 950 °C for the simultaneous removal of both CO_2 and SO_2 from the post-combustion flue gas. The decay in sorbent reactivity for CO_2 capture over multiple reaction cycles poses a significant challenge to large-scale cyclic carbonation-calcination process. Thermal sintering at elevated temperatures and the subsequent loss in pore volume both contribute to decreasing CO_2 capture capacity. Intermediate hydration has been suggested to be one of the most promising approaches to overcome this limitation by regenerating the particle pore volume and surface area. However, most existing hydration studies have been performed at laboratory scale using thermogravimetric techniques, often investigating conditions not amenable to the 3-step CCR process. In addition, a reactor design specifically suited to high temperature steam hydration is warranted. Experimental data from such a reactor can offer insights to the mass transfer limitations associated with the gas-solid reaction of the extremely cohesive calcium hydroxide ($Ca(OH)_2$) particles under fluidized bed conditions. For the first time, this study reports the design of a bench-scale high-temperature steam hydrator for reactivation of CaO sorbent. The hydrator, consisting of a fluidized-bed reactor with additional internals such as an agitator and a cooling coil, was evaluated using cold-flow tests. Based on the results obtained, a high temperature reactor unit was constructed. The high-temperature steam hydration tests were performed at isothermal conditions in the range of 300-500 °C and 0.6-1.0 atm steam partial pressure. In particular, at an average temperature of 473 °C and P_{H2O} = 1 atm, the 30 min hydration conversion exceeded 70%. Sorbent reactivity toward CO_2 was substantially recovered by steam hydration and the correlation follows a linear trend. Lastly, the cooling coil design demonstrated a proof-of-concept for extracting the heat from this highly exothermic high temperature (>450 °C) reaction. This further expands the application of such reactor design beyond CO_2 capture, for other applications such as chemical heat pumps.

(774b) Dynamic Modeling of a Solid-Sorbent CO2 Capture System

Modekurti, S., West Virginia University
Bhattacharyya, D., West Virginia University
Zitney, S. E., National Energy Technology Laboratory
Miller, D. C., U.S. Department Of Energy, National Energy Technology Laboratory
Originally presented on: 11/8/2013 8:55:00 - 9:20:00

A one-dimensional (1D), two-phase, non-isothermal, dynamic model of a solid sorbent-based CO_2 capture system has been developed in Aspen Custom Modeler (ACM). The system consists of a two-stage bubbling fluidized bed (BFB) adsorber-reactor for CO_2 capture, a single stage moving bed regenerator for the sorbent regeneration, a multi-stage integral-gear centrifugal compressor system for CO_2compression, a pneumatic transport system for solids transport between the adsorber and regenerator, and the balance of the plant. In addition, a heat recovery system has been modeled for the regenerator which includes pre- and post-heat exchangers using steam/boiler feed water as the heating/cooling medium. The solid sorbent is NETL 32D developed at National Energy Technology Laboratory (NETL). The system of partial differential algebraic equations (PDAEs) has been generated using conservation equations, hydrodynamics, and reaction kinetics along with the appropriate boundary and initial conditions.

Flue gas from a supercritical pulverized coal (SCPC) plant enters at the bottom of the lower stage adsorber and regenerated sorbent enters the upper stage adsorber. Embedded heat exchangers have been considered for cooling of the bed using water as the cooling medium. The CO_2-rich sorbent is pneumatically transported to the regenerator through a pre-heater. An embedded heat exchanger in the regenerator maintains the unit at near-isothermal conditions. The regenerated sorbent passes through a low pressure steam generator that helps to recover heat for improving the overall efficiency. The medium selected for transporting the solid sorbent between the adsorber and regenerator is the one that minimizes the auxiliary power requirement without violating the overall target for CO_2 capture. The CO_2 rich gas stream from the regenerator is compressed to pipeline pressure by a multi-stage integral-gear compressor. A glycol absorption system is used to satisfy the pipeline specification for water in the CO_2stream. For off-design operation of the compressors, surge detection and control algorithms are developed.

The PDAE system with the appropriate boundary conditions is solved in ACM using the well-known method of lines which involves discretization of the spatial domain and dynamic integration of the time domain. The transient responses of various process variables such as CO_2 capture, flue gas and sorbent temperatures, and loading of the sorbent are studied by simulating typical disturbances such as change in the flue gas temperature, flowrate, and composition. A lower level control system is designed to maintain the overall CO_2 capture at the desired value in the face of such disturbances.

(774c) Comparative Environmental Life Cycle Assessment of CO2 Capture Systems Based On Ionic Liquid and Monoethanolamine Solvents

Farahipour, R.,
Karunanithi, A. T., University of Colorado Denver
Originally presented on: 11/8/2013 9:20:00 - 9:45:00

Carbon capture, storage and utilization (CCUS) is increasingly being viewed as a global climate mitigation option. Central to a viable CCUS system is the development of cost-effective methods for removal of CO_2 from flue-gas streams of power plants and other stationary CO_2 sources. In addition to economic viability it is important to consider the life cycle impacts of CCUS methods in order to evaluate the net benefits. Moreover, in addition to climate mitigation we would need to consider other issues such as increased life cycle impacts on human and/or ecosystem health. Amin-based (e.g., monoethanolamine) absorption systems for CO_2 capture is a fairly mature technology. However, they can be very energy intensive and less attractive from an environmental stand point due to emissions of the used volatile solvent components. In recent years, because of their low volatility, ionic liquids show significant promise as a CO_2 absorption solvent for post-combustion CO_2 capture. In this work, we present process simulation results of post-combustion CO_2 capture with monoethanolamine (MEA) and ionic liquid bmim [PF6]. The heat and material balances derived from these simulations are utilized to construct the life cycle inventory for the capture process. In addition data related to other upstream processes such as solvent (MEA and ionic liquid) production and downstream processes such as CO_2 compression, transportation, storage and/or utilization is used to perform a life cycle assessment with an aim to compare the environmental impacts of absorption systems based on these two different solvents.

(774d) Whole-Chain CCS System Modelling: Enabling Technology to Help Accelerate Commercialisation and Manage Technology Risk

Matzopoulos, M., Process Systems Enterprise Ltd.
Ramos, A., Process Systems Enterprise, Inc.
Originally presented on: 11/8/2013 9:45:00 - 10:10:00

The commercial implementation of CCS still faces significant challenges. Many of these arise from the fact that the whole chain – and, eventually, whole CO_2 transportation network – needs to be considered as a single system in order to make design and operation decisions that satisfactorily address the commercial imperatives and risk requirements of the various stakeholders along the chain. Even a quick analysis shows that design and operating decisions at the power plant can have a significant effect on storage providers at the other end of the chain, and vice versa. A systems modelling approach is essential, but there are currently no tools that can satisfactorily provide this capability over the whole CCS chain.

In order to address this, a £3m project was commissioned and co-funded by the ETI and project participants, who comprise E.ON, EDF, Rolls-Royce, Petrofac (via its subsidiary CO2DeepStore), Process Systems Enterprise (PSE) and E4tech. The project is aimed at delivering a robust, fully integrated tool-kit that can be used by CCS stakeholders across the whole CCS chain that will be released as a commercially-supported software product at the end of the project.

Because of the requirement for many stakeholders to be able to model areas of the chain beyond their own specific processes (for example, a key requirement for power generators is the ability to investigate the effects on their operation of amine and compression systems attached to their plants), the tool-kit will provide process models for the whole chain. This will also allow it to be used by groups who require a whole-chain view, such as engineering companies or government departments who need to quantify policy decisions.

The tool-kit includes models for conventional generation (pulverised coal and combined cycle gas turbine), new generation (gasification and oxyfuel), solvent-based carbon capture, compression, transmission and injection. The individual process models are mostly 'medium-to-high fidelity': multicomponent streams; equilibrium methods for vapour-liquid separation; comprehensive compressor models allowing multi-stage, multi-section compressors with manufacturers' curves; and distributed pipeline models for construction of pipeline networks that can take elevation into account. The capture models use rate-based techniques for accurate quantification of chemical and physical capture in order to quantify energy penalties accurately and allow meaningful analysis of transient operations.

Using the system modelling tool-kit it will be possible to look at single areas such as amine plants in detail; investigate partial-chain operations – for example, power generation, capture and compression; or analyse interactions across the whole chain and eventually the whole CO2 transmission network with multiple sources and multiple sinks.

The tool-kit also caters for interfacing of external software packages such as E.ON's PROATES, in order to simplify interoperability analysis while allowing companies to preserve existing workflows where necessary. In addition an advanced custom modelling capability means that it is easy to add models of new processes and combine them with existing flowsheeting components.

This presentation describes the status of developments to date in whole chain CCS system analysis and plans to roll the tool-kit software out to the wider 'CCS community' following the current testing phase.

(774e) Optimization of CO2 Capture, Utilization and Sequestration (CCUS) Supply Chain Networks

Hasan, M. M. F., Princeton University
Boukouvala, F., Princeton University
Floudas, C. A., Princeton University
Originally presented on: 11/8/2013 10:10:00 - 10:35:00

CO_2 capture, utilization and storage (CCUS) is an enabling technology toward reducing CO_2 emissions from stationary sources which include power plants, cement production plants, iron and steel plants, refineries, petrochemicals and gas processing plants. More than 60% of the total CO_2 emissions in the United States are attributed to stationary sources [1]. CCUS connects CO_2 sources and utilization/storage sites through three sequential operations: capture, compression, and transportation. CO_2 is first captured and compressed at the source plant, and then transported via pipeline to a site where it is utilized or injected for sequestration for geological storage.

It is critical to integrate capture, compression, transportation, utilization and sequestration in an optimal fashion. However, standalone development of CCUS components have made the design of large-scale CCUS networks a complex and challenging task. In this work, we present a mixed-integer optimization (MILP) model for the synthesis of CCUS supply chain networks for nationwide and regional CO_2 management. We design the CCUS supply chain network by considering simultaneous selection of source plants, capture technologies, capture materials, and locations of utilization and sequestration sites for the first time. We consider real geographic locations and reliable storage estimates from the available database. Furthermore, our cost model includes the costs of capture, compression, transportation and injection for sequestration, and also revenues from CO_2 utilization. The capture and compression costs are derived using advanced modeling, simulation and optimization techniques for a wide range of flue gas compositions and flow rates [2-3], which are crucial to incorporate stationary sources of different types in a large CCUS network. We apply our model to determine the CCUS supply chain network with minimum net expenditure for a given CO_2 reduction level.

Our model selects the optimal nationwide CCUS supply chain network from 3317 stationary sources, 491 saline formations, 1837 unmineable coal areas, and 161 utilization locations (oil and gas reservoirs). We also allow four leading capture technologies (absorption, membrane, pressure swing adsorption (PSA) and vacuum swing adsorption (VSA)), and 9 top materials (2 solvents, 4 zeolites [4-5], and 3 membranes) to choose from for CO_2 capture. Results suggest that it is possible to capture 50% of the current CO_2 emissions from the stationary sources in the United States at a cost less than $30 for each ton of CO_2. Results for the CCUS in South Central United States, and Texas are also favorable. We demonstrate that it is crucial to consider both material and technology selections as integral parts of a large-scale CCUS network design.

References:

[1]. DOE/NETL. Carbon Sequestration Atlas of the United States and Canada, 2012.
[2]. Hasan, M. M. F.; Baliban, R. C.; Elia, J. A.; Floudas, C. A. modeling, simulation, and optimization of postcombustion CO_2 capture for variable feed concentration and flow rate. 1. Chemical absorption and membrane processes. Ind. Eng. Chem. Res. 2012a, 51, 15642-15664.
[3]. Hasan, M. M. F.; Baliban, R. C.; Elia, J. A.; Floudas, C. A. modeling, simulation, and optimization of postcombustion CO_2 capture for variable feed concentration and flow rate. 2. Pressure swing adsorption and vacuum swing adsorption processes. Ind. Eng. Chem. Res. 2012b, 51,15665-15682.
[4]. Hasan, M. M. F.; First, E. L.; Floudas, C. A. Cost-effective CO_2 capture based on in silico screening of zeolites and process optimization. Submitted for publication, 2013.
[5]. U.S. Provisional Patent Application #61/761,436 and #61/765,284.

(775a) A Reactive Interatomic Potential for the Molecular Dynamics Simulations of Saturated Hydrocarbons Using the Modified Embedded-Atom Method

Nouranian, S., Mississippi State University
Baskes, M. I., University of California, San Diego
Tschopp, M. A., Army Research Laboratory
Gwaltney, S. R., Mississippi State University
Horstemeyer, M. F., Mississippi State University
Originally presented on: 11/8/2013 8:30:00 - 8:48:00

In this work, an interatomic potential for saturated hydrocarbons was developed using the modified embedded-atom method (MEAM), a semi-empirical N-body potential based on density functional theory and pair potentials. The potential was parameterized by fitting to a large experimental and first-principles (FP) database consisting of 1) bond distances, bond angles, and atomization energies at 0 K of a homologous series of alkanes and their select isomers from methane to *n*-octane, 2) the potential energy curves of H_2, CH, and C_2 diatomics, 3) the potential energy curves of hydrogen, methane, ethane, and propane dimers, i.e., $(H_2)_2$, $(CH_4)_2$, $(C_2H_6)_2$, and $(C_3H_8)_2$, respectively, and 5) pressure-volume-temperature (PVT) data of a dense high-pressure methane system with the density of 0.5534 g/cc. The atomization energies and geometries of a range of linear alkanes, cycloalkanes, and free radicals calculated from the MEAM potential were compared to those calculated by other commonly used potentials for hydrocarbons, i.e., second-generation reactive empirical bond order (REBO) and reactive force field (ReaxFF). MEAM reproduced the experimental and/or FP data with great accuracy, comparable to or better than REBO or ReaxFF. The experimental PVT data for a relatively large series of methane, ethane, propane, and butane systems with different densities were predicted reasonably well by MEAM.

Although the MEAM formalism has been applied to atomic systems with predominantly metallic bonding in the past, the current work demonstrates the promising extension of the MEAM potential to covalently bonded molecular systems, specifically saturated hydrocarbons and saturated hydrocarbon-based polymers. The MEAM potential has already been parameterized for a large number of metallic unary, binary, ternary, carbide, nitride, and hydride systems, and extending it to saturated hydrocarbons provides a reliable and transferable potential for atomistic/molecular studies of complex material phenomena involving hydrocarbon-metal or polymer-metal interfaces, polymer-metal nanocomposites, fracture and failure in hydrocarbon-based polymers, etc. The latter is especially true since MEAM is a reactive potential that allows for dynamic bond formation and bond breaking during simulation. However, it is important to note that the current parameterization does not accurately reproduce the energetics and structures of unsaturated hydrocarbons and, therefore, should not be applied to such systems.

(775b) A Reactive Force Field for Manganese-Rich Cathodes in Electrochemical Systems

Lastoskie, C. M., University of Michigan
Reddivari, S.,
Originally presented on: 11/8/2013 8:48:00 - 9:06:00

Lithium ion batteries are now the most widely used rechargeable systems. Owing to their high voltage and high charge density, they have replaced conventional alkaline and nickel based batteries for portable devices. They are now also being incorporated into high power applications like hybrid electric vehicles (HEV). Among the various lithium metal oxides that have been used, or proposed for use, as cathode active materials, lithium manganese oxide ($LiMn_2O_4$) is the primary cathode material being exploited for HEV applications, on account of its low cost, low toxicity, and simple preparation process. There are, however, well documented problems with $LiMn_2O_4$, the most critical of which is capacity fade due to active material dissolution. The $LiMn_2O_4$ spinel has an equal proportion of manganese ions in two oxidation states, Mn^{3+} and Mn^{4+}, yielding a mean manganese oxidation state of +3.5. Conventional wisdom holds that trivalent manganese ions are unstable to acid attack (e.g. from hydrofluoric acid formed by electrolyte reactions), and are lost from the cathode surface into the electrolyte following charge disproportionation through the Hunter reaction 2 Mn^{3+} <--> Mn^{4+}_{solid} + $Mn^{2+}_{solution}$. Dissolution of divalent manganese into the electrolyte reduces the effective amount of functional cathode material and also decreases the Li content of the anode, and hence the overall cell capacity, on account of manganese deposition at the anode-electrolyte interface. Therefore, the key to improving the performance of manganese-rich lithium oxide cathode materials is to better understand the Mn dissolution reactions occurring at cathode-electrolyte interface, so as to devise strategies (e.g. coatings, dopants, electrolyte compositions) that curb dissolution and increase the spinel electrode stability.

Force fields based on bond order and bond energy have been developed to simulate bond formation and dissociation, and hence accurately describe reactive systems. The Reactive Force Field (ReaxFF) is one such force field developed to bridge the gap between quantum chemical and empirical force field based computations. ReaxFF has been developed for a wide range of systems including hydrocarbon oxidation, pure metallic systems, catalytic oxidation of metal oxides, and evolution of the anodic solid-electrolyte interface (SEI) in Li-ion batteries. In this paper, we report on our progress to extend ReaxFF so as to incorporate mangenese and fluorine for intended applications in ReaxFF molecular dynamics (MD) simulations of lithium ion battery cells. To capture the chemistries of manganese and fluorine, a DFT training set has been developed consisting of approximately one thousand structures and computed energetics related to bond dissociation, angle and dihedral distortions, and reactions with the other elements in the lithium cell (C, H, O, Li) whose force fields are known and transferable to the present ReaxFF application. Preliminary results are presented for ReaxFF MD simulations of the cathode and the cathode SEI layer spatial composition.

(775c) Application of Bond-Valence Theory for Developing Efficient Interatomic Potential for Oxides

Liu, S., University of Pennsylvania
Grinberg, I., University of Pennsylvania
Rappe, A. M., University of Pennsylvania
Originally presented on: 11/8/2013 9:06:00 - 9:24:00

First-principles density functional theory (DFT) calculations have played an important role in enhancing microscopic understanding of intrinsic effects (e.g., compositions and structures) on material properties of oxides.[1-2] For elucidating the changes (dynamics) in material behaviors with extrinsic effects, such as temperature, strain, and electric field, large system sizes and long simulations are necessary. Conventional first-principles methods for large systems and long-time simulations are limited due to their intense computational costs. There is therefore still a strong need to develop accurate and efficient atomistic potential that could reproduce the full dynamical behaviors of metal oxides. [3-9] The development of general atomistic potentials for oxides has proven difficult due to the complex nature of various metal-oxygen bonds.[10]

In this work, we developed a bond-valence model based on the principles of bond-valence and bond-valence vector conservation. [11-12] The relationship between the bond-valence model and the bond-order potential is derived analytically in the framework of a tight-binding model, demonstrating that our model is formally equivalent to the bond-order potential, but is dramatically more efficient computationally. A new energy term, bond-valence vector energy, is introduced into the atomistic model. The force fields for $PbTiO_3$ and $BiFeO_3$ are parameterized respectively. [13] The new model potential can be applied to both canonical ensemble (*NVT*) and isobaric-isothermal ensemble (*NPT*) molecular dynamics (MD) simulations. Our model potential can reproduce the experimental phase transition in *NVT* MD simulations and also exhibit the experimental sequence of temperature-driven and pressure-driven phase transitions in *NPT* simulations. We expect that our bond-valence model can be applied to a broad range of inorganic materials.

References

[1] W. Zhong, R.D. King-Smith, and D. Vanderbilt, Phys Rev Lett 72, 3618 (1994).

[2] I. Grinberg and A.M. Rappe, Phys Rev B 70, 220101 (2004).

[3] M. Sepliarsky and R.E. Cohen, AIP Conf. Proc. 626, 36 (2002).

[4] T. Shimada, K. Wakahara, Y. Umeno, and T. Kitamura, J. Phys.: Condens. Matter, 20, 325225 (2008).

[5] M. Sepliarsky, A. Asthagiri, S.R. Phillpot, M.G. Stachiotti, and R.L. Migoni, Curr. Opin. Solid State Mater. Sci. 9, 107 (2005).

[6] I. Grinberg, V.R. Cooper and A.M. Rappe, Nature 419, 909 (2002).

[7] W. Zhong, D. Vanderbilt, and K.M. Rabe, Phys. Rev. B 52, 6301 (1995).

[8] U.V. Waghmare and K.M. Rabe, Phys. Rev. B 55, 6161 (1997).

[9] Y.-H. Shin, V.R. Cooper, I. Grinberg and A.M. Rappe, Phys. Rev. B 71, 054104 (2005).

[10] S.R. Phillpot, S.B. Sinnott, and A. Asthagiri, Annu. Rev. Mater. Res. 37, 239 (2007).

[11] I.D. Brown, Chem. Rev. 109, 6858 (2009).

[12] M.A. Harvey, S. Baggio, and R. Baggio, Acta Crystallogr. B62, 1038, (2006).

[13] S. Liu, I. Grinberg, and A. M. Rappe, J. Phys. Cond. Matt. 25, 102202 (2013)

(775d) Reconsidering Dispersion Potentials: Using Reduced Cutoffs in Mesh-Based Ewald Solvers Can Eliminate the Performance Gap Versus Truncation

Isele-Holder, R. E., RWTH Aachen University
Mitchell, W., Loyola University
Hammond, J.,
Ismail, A. E., RWTH Aachen University
Originally presented on: 11/8/2013 9:24:00 - 9:42:00

Long-range dispersion interactions play a major role in determining the physical quantities in simulations of inhomogeneous systems. However, because of the perceived computational cost of long-range solvers, their use in molecular dynamics packages has been discouraged. Here, we demonstrate that reducing the cutoff radius for local interactions in a recently implemented particle-particle particle-mesh (PPPM) method for dispersion [Isele-Holder et al., *J. Chem. Phys.*, **137**, 174107 (2012)] can, in many cases, actually be faster than simulations where dispersion interactions are simply truncated. In addition, because all long-range dispersion interactions are incorporated, this approach avoids the physical inaccuracies that arise from truncating the potential. Simulations using PPPM for dispersion can thus provide results that are both more accurate and more efficient than simulations that truncate dispersion interactions.

We provide a set of parameters for the dispersion PPPM method using either *i***k** or analytic differentiation that we recommend for future use, and demonstrate increased simulation efficiency by using the long-range dispersion solver in a series of performance tests on massively parallel computers using several different architectures.

(775e) The Importance of Intra- and Inter-Molecular Interactions in the Development of Coarse-Grained Models for Chain Fluids Using SAFT-γ Mie in Molecular Simulations

Rahman, S., Imperial College

Braga, C., Imperial College London

Lobanova, O., Imperial College London

Raptis, V., Imperial College London

Galindo, A., Imperial College London, Centre for Process Systems Engineering

Jackson, G., Imperial College London, Centre for Process Systems Engineering

Muller, E. A., Imperial College London

Originally presented on: 11/8/2013 9:42:00 - 10:00:00

Introduction

In common coarse-graining (CG) approaches one starts from an atomistic, bottom-up, description of molecules and proceed to integrate degrees of freedom, simplifying the force field in order to speed up the computer simulation of complex molecules and mixtures. [1-4] By contrast, the CG intermolecular potentials can also be obtained with a top-down approach by reproducing the macroscopic configurational properties of the system with a molecular-based equation of state. The latter approach is considered here, in particular, the use of the SAFT-γ Mie version of the Statistical Associating Fluid Theory (SAFT) which is based on interactions of the Mie form and has successfully been applied to simple organic molecules [5-7]. In our current work we explore the role of inter- and intra-molecular interactions in SAFT-γ Mie force fields of the linear alkane homologous series.

Alkane Study

Despite its success in correctly reproducing the thermodynamic properties of complex fluids, SAFT based approaches do not take into account the structure and flexibility of molecular systems explicitly, apart from the assumption that chains of *m* segments are formed [8]. We have addressed this issue by combining the SAFT-γ Mie top-down approach with a parameterization of the CG intramolecular (bonding) potentials on the basis of atomistic simulations.

As prototypical molecules we first perform the analysis for long chain *n*-alkanes. In order to obtain a CG model, three consecutive backbone carbon atoms and their bonded hydrogen's are all grouped together to form a "super-atom", or CG moiety. The inter-molecular parameterization of the CG beads with the SAFT-γ Mie methodology then leads to two possibilities, depending on the recognition (or not) of the distinct identity of the terminal end CH_3 group. In the first model (homonuclear) the beads representing the end and intervening CG alkyl groups are taken as identical; in the second model (heteronuclear) a distinction is made between the interactions of the two types of CG beads. (Fig 1)

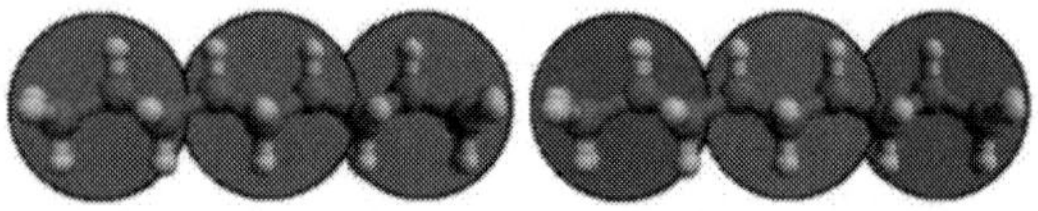

Figure 1 Schematic of the CG strategy for *n*-nonane, lumping three backbone carbon atoms into single beads. For the homonuclear model (left) all beads are treated on an equal footing, and for the heteronuclear model (right) the $CH_2CH_2CH_3$ beads that contain the CH_3 end group and the $CH_2CH_2CH_2$ beads that make up the intervening groups are distinguished.

To address the importance of chain flexibility, we have performed single component molecular dynamics simulations of an equivalent united-atom (UA) model using the NERD force field [9]. From the set of atomistic trajectories the CG intermolecular force field was computed using the potential of mean force of the corresponding CG degrees of freedom, e.g., CG bond distances and angles. The CG bond stretching and bond angle bending interactions were then superimposed to the CG intermolecular parameters. The full model includes both inter- and intra-molecular contributions.

An assessment of the different contributions to the full intermolecular potential is made by comparing the results obtained from the molecular dynamics simulations of fluid phase equilibria as well as the interfacial, transport and structural properties. An example of how the structural properties of nonane are affected when using different intra- and inter-molecular parameters can be seen in Figure 2. The g(r) results obtained for the homonuclear model was the same as the heteronuclear, so has not been included to keep the figure simple.

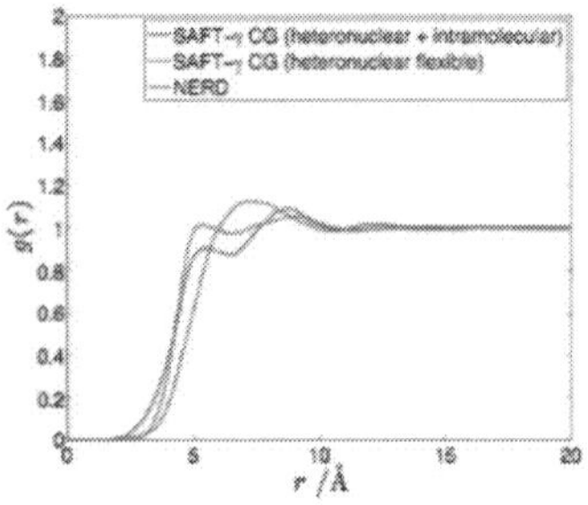

Figure 2: Centre-to-centre radial distribution function for *n*-nonane comparing the heteronuclear SAFT-γ Mie CG descriptions (cf. Figure 1) with completely flexible (pearl-necklace) model or the restrained bonding angle model.

3. Conclusion

The inclusion of intra- molecular interactions CG models allows for an improved description of thermodynamic and structural properties without a detriment to the speed of the calculations. An additional advantage of the heteronuclear description is that they provide fully transferable CG models for linear alkanes.

References

[1] K. A. Maerzke and J. I. Siepmann, *J. Phys. Chem. B*, **115**, 3452 (2011).

[2] S. J. Marrink, H. J. Risselada, S. Yefimov, D. P. Tieleman, and A. H. de Vries, *J. Phys. Chem. B*, **111**, 7812 (2007).

[3] S. O. Nielsen, C. F. Lopez, G. Srinivas, and M. L. Klein, *J. Chem. Phys.*, **119**, 7043 (2003).

[4] W. Shinoda, R. DeVane, and M. L. Klein, *Mol. Simul.* **33**, 27 (2007).

[5] C. Avendano, T. Lafitte, A. Galindo, C. S. Adjiman, G. Jackson and E. A. Muller, *J. Phys. Chem. B*, **115**, 11154 (2011).

[6] T. Lafitte, C. Avendano, V. Papaioannou, A. Galindo, C. S. Adjiman ,G. Jackson and E. A. Muller, *Mol. Phys*, **110**, 1189 (2012).

[7] C. Avendano, T. Lafitte, C. S. Adjiman, A. Galindo, E. A. Muller and G. Jackson, *J. Phys. Chem. B*, **117**, 2717 (2013).

[8] E.A. Muller and K.E. Gubbins, *Mol. Phys.*, 80, 957 (1993).

[9] S.K.Nath, B.J. Banaszak & J.J. de Pablo, *J. Chem. Phys*, **114**, 3612 (2001)

(775f) Development of a Coarse-Grained Force Field for Aqueous Non-Ionic Surfactant Systems Using a Molecular Based Equation of State

Lobanova, O., Imperial College London

Herdes, C., Universidade de Évora

Jackson, G., Imperial College London

Muller, E. A., Imperial College London

Originally presented on: 11/8/2013 10:00:00 - 10:18:00

Development of a Coarse-Grained Force Field for aqueous non-ionic surfactant systems using a molecular based equation of state

Olga Lobanova, Carmelo Herdes, George Jackson, Erich A. Müller

Department of Chemical Engineering, Imperial College London, U.K.

Keywords: force field, coarse-graining, equation of state, non-ionic surfactants

The understanding of phase behavior and thermodynamic properties of amphiphilic systems is an essential component in the design of complex fluid mixtures commonplace, such as health care products, pharmaceutical industry, and environmental protection. However, the sheer complexity and diversity of these systems, the large number of possible components and process conditions makes the experimental study an overwhelming task. While modern simulation techniques allow an insight into phase behavior of complex fluid systems, the appropriate choice of the modeling technique is crucial, as some properties are not accessible due to limited resolution and reproducibility. The use of coarse-grained molecular simulation, as opposed to a fully detailed atomistic approach, overcomes some of the limitations of spanning the time and length scales required to study the kinetics and self-assembly of complex systems. The problem of parameterization of the force field is a key factor since the accuracy and robustness of the force field determines the reliability of the molecular simulation results.

Monoalkyl ethers of poly(ethylene) glycols, (CiEj), are particularly interesting for their wide range of applications, particularly due to their rich mesophase phase behavior. Therefore, those systems have been widely studied both experimentally [1] and by molecular simulation [2]. These surfactants exhibit self-assembly in aqueous solution, as do their more complex biological analogues, with the added simplicity of being non-ionic. Most coarse-grained (CG) simulations are based on intermolecular potentials, which are effectively integrated version of more fine-resolution atomistic-level models [3,4].

In opposition to this, in this work we develop a CG model where the force field parameters are obtained from a top-down approach, by effectively averaging a large number of macroscopic thermodynamic states with the aid of a fluid equation of state. We use the Statistical Associating Fluid

Theory (SAFT-γ Mie EoS) to obtain the potential parameters for coarse-grained moieties [5,6,7]. The potential parameters for these CG "super-atoms" are estimated from thermophysical properties (density, vapour pressure, surface tension, etc.) of simple organic molecules, e.g., alkanes and ethers that share chemical commonalities with the segments of the amphiphile. The cross-interaction parameters are obtained from the properties of the corresponding mixtures.

Molecular Dynamics simulations in an explicit solvent are carried out at different surfactant concentrations to investigate the thermodynamic, interfacial and structural properties of the systems. For a low surfactant concentration, we observe self-assembly into micellar aggregates (Fig. 1), as well as the interchange of the single surfactant with the micelle, the fusion and the separation of two micelles and other equilibrium and dynamic signatures of micellar systems. Using a cluster analysis algorithm, we are able to identify individual micelles and calculate relevant structural properties, such as aggregation number, aggregate size distribution and shape, radius of gyration and density profiles. For higher surfactant concentrations, the formation of hexagonal and lamellar phases is observed. Further validation of the CG model has been undertaken in respect to water-air tension, surface excess properties, nematic order parameter, critical micelle concentration and aggregate properties of amphiphiles of different chain length.

It is very gratifying to find that the proposed force field is transferable for prediction of various properties of these mixtures and also for mixtures of the components of related homologues. In this manner, the SAFT- γ Mie Eos establishes a direct robust link to macroscopic properties and enables the estimation of the CG force field parameters in a very fast and efficient way. The force field parameters obtained from the equation of state, can be used as a direct input in molecular simulation to explore different properties of the mixture that are usually not accessible via an equation of state, such as structural and interfacial properties.

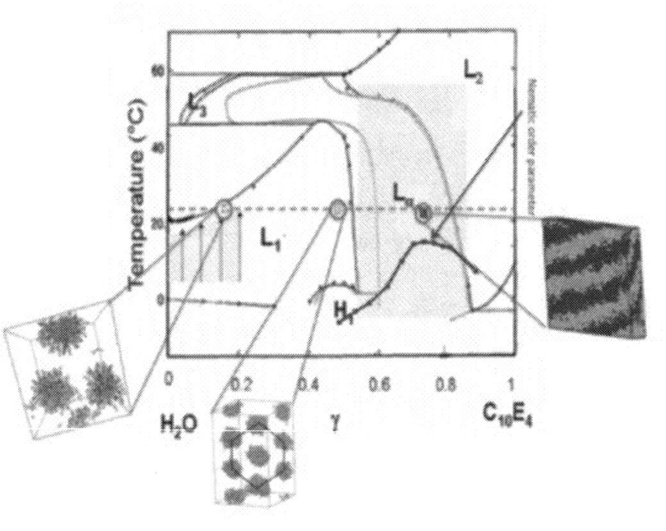

Fig. 1: Experimental phase diagram for C10E4 surfactant in water (taken from Fournial *et al.* [8]) and snapshots corresponding to the simulation results at 298 K at 5 wt%, 50 wt% and 70 wt% surfactant in water. The blue line denotes the nematic order parameter calculated for the surfactant concentration range between 50 and 90 wt%.

References

[1] Shearman, G.C. ; Tyler, A.I.I. ; Brooks, N.J.; Templer, R.H.; Ces, O.; Law R.V. ; Seddon, J.M.; *Liq. Cryst.,* 37, 679 (2010)

[2] Jusufi, A.; Sanders S.; Klein M.L.; Panagiotopoulos A.Z.; *J. Phys. Chem. B,* 115, 990 (2011)

[3] Muller-Plathe, F.; Reith, D.; Putz M.; *J. Comp. Chem.,* 24, 1624 (2003)

[4] Voth G. A.; Izvekov.S.; *J. Phys. Chem. B,* 109, 2469 (2005).

[5] Avendano, C.; Lafitte, T.; Galindo, A.; Adjiman, C. S.; Jackson, G.; Muller, E. A.; *J. Phys. Chem. B*, 115, 11154 (2011)

[6] Lafitte, T.; Avendano, C.; Papagioannou, V.; Galindo, A.; Adjiman, C. S.; Jackson, G.; Muller, E. A.; *Mol. Phys,* 110, 1189 (2012)

[7] Avendano, C.; Lafitte, T.; Adjiman, C. S.; Galindo, A.; Muller, E. A.; Jackson, G.; *J. Phys. Chem. B*, 117, 2717 (2013)

[8] Fournial, A.-G.; Zhu,Y.; Molinier, V.; Vermeersch, G.; Aubry, J.-M.; Azaroual, N; *Langmuir,* 23, 11443 (2007)

(775g) Extension of Mie Potentials to Branched Alkanes, Alkenes, and Ethers

Potoff, J. J., Wayne State University

Originally presented on: 11/8/2013 10:18:00 - 10:36:00

Transferable united-atom force fields, based on n - 6 Lennard-Jones potentials, are presented for branched alkanes, alkenes and ethers. It is shown that by varying the repulsive exponent the range of the potential can be altered, leading to improved predictions of vapor pressures while also reproducing saturated liquid densities to high accuracy. Histogram-reweighting Monte Carlo simulations in the grand canonical ensemble are used to determine the vapor liquid coexistence curves, vapor pressures, heats of vaporization, and critical points for a variety of pentane and hexane isomers, 1-alkenes (ethane to octane), 2-alkenes (cis-2-butene, trans-2-butene), butadiene, dimethyl ether and diethyl ether. These data are combined with quantum chemical calculations to illustrate the effect of bonding environment on the transferability of potential parameters. For all molecules studied, saturated liquid densities and vapor pressures are reproduced to within 1% and 5% of experiment, respectively.

(775h) Charmm-Compatible Lipid Parameters for Ceramides and United Atom Chains

Lee, S.,
Tran, A., University of Maryland
Allsopp, M.,
Klauda, J., University of Maryland
Originally presented on: 11/8/2013 10:36:00 - 10:54:00

Biological membranes form a barrier to protect the cell from its environment and consist of a wide variety of lipids. In molecular simulations, accurate force field (FF) parameters are crucial for representing biological membranes. Although all-atom FFs, such as CHARMM36 (C36), can now accurately represent a variety of lipid chains (saturated, unsaturated, and branched) and head groups (PC, PE, PG, and PS), parameters for ceramide lipids are lacking. These lipids are common in biology (neuronal and ocular lens membranes) and are involved in lipid domain formation with cholesterol. Quantum mechanical calculations are used to develop torsional profiles and partial charges of the head group. Molecular dynamics (MD) simulations of sphingomyelin bilayers (16:0 and 18:0 chains) are used to test new parameters in comparison with available deuterium order parameters (S_{CD}s) and x-ray diffraction experiments. Simulations with this all-atom force field more accurately agree with experimental S_{CD}s and density data and indicate that GROMOS and OPLS force fields may result in lateral packing that is too dense.

The development of a CHARMM-compatible united atom (UA) chain model will also be presented, referred to here as C36-UA. This is an extension of previous work (HÃ©nin et al. *JPC*, **112**: 7008 (2008)) to the C36 FF. MD simulations of pure and dimyristoylphosphatidylcholine/cholesterol bilayers demonstrate a UA-chain model that is accurate and compatible with the CHARMM FF. MD simulations of micelle formation of short-chain lipids and single-chained detergents are used to demonstrate the ability of C36-UA to self-assemble lipids. MD simulations with C36-UA allow for timescales not approachable with all-atom models without a loss of accuracy, which may have applications to lipid organization and studies with membrane-associated proteins.

(778a) Hydrodesulfurization On Re Metal and Sulfide Catalysts: Effects of Catalyst Structure and Metal-Sulfur Binding Energy On Reactivity

Yik, E., University of California at Berkeley
Wang, H.,
Iglesia, E., University of California, Berkeley
Originally presented on: 11/8/2013 8:30:00 - 8:50:00

Hydrodesulfurization on Re Metal and Sulfide Catalysts:

Effects of Catalyst Structure and Metal-Sulfur Binding Energy on Reactivity

Edwin Yik[a], Huamin Wang[b], and Enrique Iglesia[a,*]

[a]Department of Chemical and Biomolecular Engineering, University of California, Berkeley, CA 94720

[b] Pacific Northwest National Laboratory, Energy and Efficiency Division, Richland, WA 99352

*iglesia@berkeley.edu

Recent mechanistic studies of thiophene hydrodesulfurization (HDS) [1,2] on silica-supported Ru and Pt metal clusters, which avoid the structural anisotropy of layered metal sulfides encountered in commercial Co(Ni)Mo(W) sulfides [3], and on the effects of their cluster size have shown that HDS turnover rates are lower on surfaces with higher S-metal binding energies, due to a lower availability of vacant sites required for reaction at a given sulfur chemical potential, which is set by varying H_2S:H_2 ratios [4]. We extend those studies to silica-supported Re metal clusters and layered Re sulfides (ReS_x) and observe that, despite differences in the catalyst identity and structure, HDS of thiophene occurs via elementary steps that are similar to those prevalent on Ru and Pt.Â Therefore, the differences in turnover rates on these materials may be attributed to the inherent properties of the catalysts rather than to a change in the reaction mechanism. Reactivity decreases from Pt to Ru to Re metal clusters, consistent with an increase in the S-metal bond strength [5] and consequent decrease in the density of empty sites as a result of the higher S-coverage.

HDS turnover rates on ReS_x/SiO_2 samples were roughly 100 times higher than those on Re/SiO_2, even when the former were normalized by the total number of Re atoms (because transmission electron microscopy (TEM) images showed well-dispersed lamellar features) and the latter by its surface atoms. Both samples were synthesized from different treatments of a common ReO_x/SiO_2 precursor and remained kinetically stable during the course of the experiments (1-3 MPa of H_2, 1-10 kPa of thiophene, and 0.2-3.0 kPa of H_2S) over a period of more than 100 h. X-ray absorption (XAS), TEM, and S-contents (obtained through temperature programmed reduction (TPR) in H_2) suggested that Re metal retained its metallic bulk during HDS and exhibited an amount of sulfur corresponding to near-monolayer surface coverages after HDS. TEM images also showed that Re clusters did not form lamellar ReS_x structures and that ReS_x samples did not form isotropic Re clusters during HDS, further indicating an absence of interconversion between the two phases despite their exposure to identical experimental conditions. The marked differences in reactivity on surfaces of ReS_x and on those of S-covered Re metal likely reflect weaker S-Re bond strengths in the former [4], which lead to a concomitant increase in S-vacancy surface concentrations. In addition, the structural buckling that may exist in ReS_x could allow higher accessibility to the vacant sites on which the reaction may proceed. Thus, the inhibition of turnover rates as a result of increasing H_2S:H_2 ratios is more pronounced on Re/SiO_2 than on ReS_x/SiO_2.

References

1. Wang, H., and Iglesia, E. *J. Catal.* 270, 245 (2010).

2. Wang, H., and Iglesia, E. *ChemCatChem.* 3, 1166 (2011).

3. TopsÃ¸e, H., Clausen, B.S., and Massoth F.E., in ?Catalysis Science and Technology? (J.R. Anderson and M. Boudart, eds.) Vol. 11. Springer-Verlag, New York, 1996.

4. Bartholomew, C.H., Agrawal, P.K., and Katzer, J.R. *Adv. in Catalysis.* 31, 135 (1982).

5. Alfonso, D.R. *J. Phys. Chem. C.* 115, 17077 (2011).

(778b) Determination of the Transition Between Supported Pdo and Pd Under Various Time-Temperature Trajectories

Schwank, J. W., University of Michigan
McCabe, R. W., Ford Motor Company
Chen, X., University of Michigan
Cheng, Y., Ford Motor Company
Fisher, G. B., University of Michigan
Originally presented on: 11/8/2013 8:50:00 - 9:10:00

Palladium plays an important role as oxidation catalyst in automotive three-way catalyst applications and in catalysed combustion processes. It is still an open question in the literature whether Pd or PdO represent the active sites for oxidation reactions. It is also important to better understand how the reactive environments, i.e. reducing versus oxidizing conditions, affect the state of palladium and its degree of dispersion. Here, a systematic study is carried out to determine the time-temperature trajectories for the transition of supported PdO into metallic Pd and vice versa. In addition, the influence of the support and the Pd particle size on the transition are discussed with special emphasis on its relationship to the activity in various oxidation reactions. The catalysts are extensively characterized by hydrogen and carbon monoxide chemisorption, temperature programmed reduction and oxidation, ex-situ and in-situ XRD, and XPS. As probe reactions, the oxidation of CO, methane, and propylene are used.

(778c) Selective Hydrogenation of 1,3-Butadiene On Pd-Ni Bimetallic Catalyst: From Model Surfaces to Supported Catalysts

Hou, R., Tsinghua University
Wang, T., Tsinghua University
Yu, W., University of Delaware
Porosoff, M. D., University of Delaware
Chen, J. G., Columbia University
Originally presented on: 11/8/2013 9:10:00 - 9:30:00

The selective hydrogenation of 1,3-butadiene serves as a means to purify the butene stream generated in the cracking of naphtha or gas oil. In order to successfully polymerize butene, 1,3-butadiene must be removed below 10ppm, because even small amounts of 1,3-butadiene are sufficient to poison the polymerization catalysts.

In order to identify the selective hydrogenation catalysts, 1,3-butadiene was studied on single crystal Ni/Pd(111) bimetallic surfaces, utilizing density functional theory (DFT) modeling and temperature-programmed desorption (TPD) in ultra high vacuum system. DFT results revealed lower activation barriers on the PdNiPd(111) surface than on Pd(111). Consistent with DFT predictions, TPD results revealed that the PdNiPd(111) surface was more active for hydrogenation of 1,3-butadiene to butene than Pd(111).

Surface science results were verified in a batch reactor on supported catalysts. Both the bimetallic (0.91w.t.%Pd 1.51%w.t.Ni) and monometallic (0.91w.t.%Pd, 1.51%w.t.Ni) catalysts were prepared by incipient wetness impregnation. SiO_2, g-Al_2O_3, CeO_2 were applied as supports for the study of support effect. Extended X-Ray absorption fine structure (EXAFS) and transmission electron microscopy (TEM) confirmed the formation of uniform, bimetallic particles. The PdNi bimetallic catalysts showed higher activity than the monometallic catalysts. Among the supports studied in this work, SiO_2 supported bimetallic catalyst was found to exhibit significant enhancement in hydrogenation activity than the monometallic Pd and Ni catalysts.

(778d) Catalytic Consequences of Reactive Intermediates During Partial Oxidation of Methanol On Pt Clusters

Tu, W., University of Toronto

Chin, Y. H., University of Toronto

Originally presented on: 11/8/2013 9:30:00 - 9:50:00

Catalytic Consequences of Reactive Intermediates during Partial Oxidation of Methanol on Pt Clusters

Weifeng Tu[1,2] and Ya-Huei (Cathy) Chin[1,*]

[1] Department of Chemical Engineering and Applied Chemistry, University of Toronto, Toronto, Canada

[2] College of Power Engineering, Chongqing University, Chongqing, China

(*) corresponding author: cathy.chin@utoronto.ca

Keywords: methanol oxidation, Pt, partial oxidation, alcohol, formaldehyde, chemisorbed oxygen, oxygen binding strengths

Reactions of alkanol and oxygen over transition metal or metal-oxide catalysts lead to value added products of aldehydes and esters, syngas, or combustion products (CO_2, H_2O) at rates and selectivities that vary largely with the operating oxidant-to-alkanol ratios, because these ratios prescribe the chemical state of catalytic sites and the identity and coverages of reactive intermediates on Pt cluster surfaces during steady-state catalysis. The effects of surface coverages of the diverse intermediates and their relative abundances on supported Pt clusters are interpreted here in terms of oxygen and carbon virtual pressures, the rigorous surrogates of their respective chemical potentials, and connected to the rates and selectivities of methanol partial oxidation based on rate measurements at differential conditions and elimination of transport artifacts by extensive dilution within the intra-pellet and catalytic bed.

CH_3OH-O_2 catalytic turnovers (O_2/CH_3OH=0.1-40; 353-473 K) on Pt clusters (1.5-10 nm) form CH_2O, CH_3OCHO, CO_2, and H_2O. The catalytic reactions do not form CO and H_2 at detectable limits under all conditions before the complete depletion of O_2 because O_2 reactions with CO and H_2 are much faster than with CH_3OH. The primary and secondary nature of the CH_2O, CH_3OCHO, and CO_2 product and the relative rates of their reactions with oxygen relative to those of CH_3OH were probed from CH_3OH-O_2, CH_2O-O_2, and CH_3OCHO-O_2 reactions at different oxidant-to-reductant ratios. The rates of CH_3OCHO oxidation were unimportant relative to the CH_3OH oxidation turnovers at Pt cluster surfaces occupied by chemisorbed oxygen (O*) and oxygenated intermediates (e.g. CH_3O*) to varying extents. In contrast, CH_2O oxidation rates became increasingly important relative to CH_3OH oxidation rates as Pt cluster surfaces became populated with chemisorbed oxygen atoms. These results suggest that chemisorbed oxygen atoms are effective for oxidizing CH_2O and CH_3OH and become more reactive towards CH_2O than CH_3OH as they are more weakly bound at high surface coverages.

The individual rates of CH_2O, CH_3OCHO, and CO_2 formation from CH_3OH-O_2 reaction acquire diverse kinetic dependencies with O_2 and CH_3OH pressures that can be categorized in three kinetic regimes according to the operating O_2-to-CH_3OH ratio, because this ratio influences the oxygen and carbon chemical potential at Pt cluster surfaces and thus, determines the relative abundances of the various reactive species of chemisorbed oxygen, unoccupied metal site, and methanol derived intermediates on Pt cluster surfaces. The identity and coverages of the reactive intermediates define the kinetically relevant steps during methanol partial oxidation and, in turn, lead to complex kinetic dependencies for the formation of CH_2O, CH_3OCHO, and CO_2. At high O_2/CH_3OH ratios (> 7.5; 373 K), Pt cluster surfaces are saturated with reactive oxygen species; thus, CH_2O, CH_3OCHO, and CO_2 formation rates do not vary with O_2 pressure but increase with increasing CH_3OH pressure. As the O_2/CH_3OH ratio decreases to intermediate values between 0.5 and 7.5 (373 K), kinetic dependencies for all products undergo a drastic transition: HCHO and CO_2 formation rates are no longer proportional to CH_3OH but instead acquire a linear dependence on O_2 pressure and the CH_3OCHO formation rates increase more than linearly with O_2 pressure. Such kinetic transitions reflect a concomitant surface transition from O* saturation to uncovered of reactive intermediates, and from catalytic involvements of O* in CH_3OH oxidation to * in O_2 activation as the kinetically relevant step. As the O_2/CH_3OH ratio decreases further to below 0.5 (373 K), rates for CH_2O and CO_2 formation decrease with increasing CH_3OH pressure but the rate for CH_3OCHO is unaffected by CH_3OH pressure. All of the species formation rates acquire a much larger dependence on O_2 pressure with apparent reaction orders exceeding 1.5. These kinetic dependencies on CH_3OH and O_2 pressures are consistent with CH_3OH activation at vacancies on Pt cluster surfaces that are saturated with CH_3OH derived intermediates, prevalent only at the very low O_2/CH_3OH ratio.

The complex kinetic phenomena reported here are prescribed strictly by the operating O_2-to-CH_3OH ratio because the ratio determines the oxygen and carbon virtual pressures at the Pt cluster surfaces and in turn, the chemical states and identity and relative abundances of the surface intermediates during steady-state catalysis. Similar relations between the oxygen virtual pressure and the catalytic rate coefficients have been reported for alkane [1-2] and NO oxidation [3]. These relations appear to be ubiquitous for oxidation reaction on metal or metal-oxide clusters and govern the reaction path selectivities. <>Acknowledgements

We acknowledge China Scholarship Council for a Joint-Supervising Ph.D. program for Weifeng Tu and Natural Sciences Council of Canada (NSERC) and Canadian Foundation for Innovation (CFI) for their financial supports.

References

[1] Y-H. Chin, C. Buda, M. Neurock, and E. Iglesia, *J. Am. Chem. Soc.* **133** (2011) 15958.

[2] M. Garcia-Dieguez, Y-H. Chin, and E. Iglesia, *J. Catal.* **285** (2012) 260.

[3] B.M. Weiss and E. Iglesia, *J. Phys. Chem. C* **113** (2009) 13331.

(778e) Ni@Pt and Co@Pt Overlayer Catalysts for Aqueous Phase Reforming

Holles, J., University of Wyoming
Skoglund, M., University of Wyoming
Morris, A.,
Originally presented on: 11/8/2013 9:50:00 - 10:10:00

Ni@Pt and Co@Pt Overlayer Catalysts for Aqueous Phase Reforming

Joseph H. Holles, Michael D. Skoglund, and Allen Morris

Department of Chemical and Petroleum Engineering, University of Wyoming

Creating new and sustainable fuels to meet our world's ever growing energy demands is a national challenge. Aqueous phase reforming (APR) has emerged as a process capable of turning renewable feedstocks into hydrogen gas or alkanes. While most previous APR studies have used feedstocks that are derived from biomass (methanol, ethylene glycol, glycerol, sorbitol, and glucose), further studies using a direct biomass feed would be beneficial. Lactose, a sugar waste product of the cheese and dairy industry, is one possible candidate.

Past APR studies have identified bimetallic Pt/Ni and Pt/Co as the best catalysts for production of hydrogen through APR. Literature computational work has also predicted that a Pt overlayer on Ni or Co will decrease the heat of adsorption of the H_2 and CO products. Thus, we have synthesized a series of Ni@Pt and Co@Pt pseudomorphic overlayer type bimetallic catalysts which aim to increase the activity of the catalyst while reducing the necessary loading of costly platinum.

The directed deposition technique was used to selectively deposit the overlayer (platinum) metal on the base metal without it depositing onto the support surface. Hydrogen chemisorption and ethylene hydrogenation were used as characterization tools to identify catalysts for subsequent APR studies. Pt and Ni or Co X-ray Absorption Spectroscopy (XAS) was performed on all catalysts at beamline 10-ID at Argonne National Laboratory to determine the structure associated with the overlayer metal. APR studies were performed using 3 w/o lactose in water solutions at varying temperatures to determine activity and selectivity to the desired hydrogen product.

As predicted computationally, the heat of adsorption for hydrogen decreased for the overlayer compared to Pt. The weaker hydrogen adsorption for both overlayer catalysts when compared to monometallic Pt should aid in increasing APR reaction rates because the hydrogen product will not be as tightly bound to the metal surface; thus freeing up active sites for adsorption of reactants. The deposition of the Pt overlayer resulted in catalysts that were more active than the parent catalyst of pure Ni or Co, as expected. In addition, when compared to pure Pt, both the Ni@Pt and Co@Pt catalyst showed decreased activity. These results agree with computationally predicted d-band shifts from the literature that would cause weaker hydrogen adsorption on the metal surface, decreased surface coverage, and ultimately reduced activity for ethylene hydrogenation when compared to platinum metal alone. Reaction order studies indicate H_2 surface coverage consistent with this conclusion. Thus, using the chemisorption and reactivity descriptors, we have identified several promising candidates for subsequent APR studies.

XAS studies of the Pt overlayer were conducted to determine the structure of this overlayer metal. If the Pt is in an overlayer structure as desired, the interatomic distance to Ni is expected to be measurably shorter than a Pt-Pt distance. In addition, since the Pt is on the surface, the coordination number should be lower than the bulk Ni or Co. XAS data has been obtained and analysis is currently underway and the results will be discussed.

Finally, APR studies of the Ni@Pt and Co@Pt are currently being conducted. Preliminary data has shown the catalysts capable of converting all of the lactose. Additional studies will determine the reactivity and selectivity for all overlayer catalysts as well as Co, Ni, and Pt only catalysts and Pt/Ni and Pt/Co co-impregnated (non-overlayer) catalysts. All of the APR reactivity data will be correlated with the XAS, hydrogen heat of adsorption, and ethylene hydrogenation results.

(778f) Kinetics Studies On the Hydrodeoxygenation of Propionic Acid Over Supported Monometallic and Bimetallic Catalysts

Lugo Jose, Y.,
Williams, C., University of South Carolina
Monnier, J. R., University of South Carolina
Originally presented on: 11/8/2013 10:10:00 - 10:30:00

The unceasing demand of energy and the need to develop a more efficient process for the conversion of biomass to bio-fuels, has prompted a move towards novel catalytic techniques. We report a catalytic gas-phase hydrodeoxygenation (HDO) of carboxylic acid (i.e. Propionic Acid), to produce hydrocarbons over two different Pd catalysts supported on carbon and silica, and a series of bimetallic systems of Ag-Pd. The reaction studies were carried out in a single-pass plug flow reactor, where PAc is co-fed with hydrogen at temperatures ranges of (200-400ºC) and atmospheric pressure. Pd/C showed high activity and selectivity towards ethane, indicating that the reaction favored the C-OH bond cleavage based on decarbonylation (DCN) and decarboxylation (DCX) reactions. Similarly, Pd/SiO_2, showed to be selective towards ethane and propionaldehyde, indicating DCN and hydrogenation reactions. The TOF values decrease in the following the order: Pd/SiO_2 >Pd/C, suggesting that the particle size and/or support plays an important role in the activity of the catalyst. A detailed kinetic study for the HDO of PAc showed that both Pd/C and Pd/SiO_2, exhibited an apparent reaction order for PAc and H_2 of half and zero order, respectively. The corresponding activation energies of Pd/C was higher (17 ± 0.6 kcal/mole) compared to Pd/SiO_2 (12 ± 0.7 kcal/mole), due to a higher catalytic activity. A series of bimetallic systems were examined, to explore the structure and activity for the HDO of PAc. The addition of Ag, has a significant ensemble size effect for Pd/C and Pd/SiO_2 catalysts. A correlation between the structural properties and the catalytic performance at different coverage of Ag was established. The catalysts were characterized utilizing Fourier transform infrared spectroscopy (FTIR), Transmission Electron Microscopy (TEM), Temperature Programmed Oxidation and Reduction TPO/TPR, Atomic Absorption and H_2 titration of oxygen pre-covered catalyst.

(778g) Deoxygenation of Alkanols, Alkanals and Phenolic Monomers On Metal and Solid Acid Catalysts

Gurbuz, E. I., University of Wisconsin-Madison
Iglesia, E., University of California, Berkeley
Originally presented on: 11/8/2013 10:30:00 - 10:50:00

Selective removal of excess oxygen from biomass-derived molecules is a formidable challenge for the efficient formation of fuels and chemicals. This study demonstrates how the rates of different deoxygenation pathways, such as decarbonylation, C-O hydrogenolysis and dehydration are controlled by the identity of the metal function and, in some cases, by the synergistic presence of acid sites within molecular diffusion distance. Specifically, we examine reactions of alkanols, alkanals and phenolic monomers.

Alkanols and alkanals react predominantly via decarbonylation, forming CO on monofunctional catalysts based on Group VIII metals (e.g. Ru, Pt and Ir with similar cluster sizes of 0.6-0.8 nm, 473-523 K, 1-3 MPa H_2) at turnover rates that are highest on Ru and lowest on Ir. Alkanol-alkanal equilibration is not complete and turnover rates for C-O hydrogenolysis (to form alkanes and H_2O) are ~10 times smaller than for decarbonylation on these catalysts. Methane is also formed through the methanation reaction of CO with H_2, generating H_2O. Chemisorbed CO (CO*) and hydrogen (H*) cover cluster surfaces during reaction, causing the inhibition of decarbonylation by the CO product and lowering turnover rates as conversion increases or CO is added to reactants. CO* hydrogenation on Ru decreases these inhibition effects by converting CO to CH_4, albeit with concomitant loss of H_2 and formation of low-value products.

Decarbonylation rates and selectivities are much lower on monofunctional Cu catalysts than on Ru, Pt, and Ir catalysts and C-O hydrogenolysis rates are ~10-fold higher than decarbonylation rates on Cu. Alkanol-alkanal equilibration is reached at all residence times on Cu catalysts and the effects of H_2 and 1-butanol pressures on C-O hydrogenolysis rate indicate that the most abundant surface intermediate (MASI) is an alcohol-derived species. C-O hydrogenolysis, which can be carried out selectively over Cu/SiO_2, is preferable to decarbonylation as a means of deoxygenation for the production of fuels, because the carbon length of the starting molecule is preserved, and excess depletion of hydrogen through methanation of the decarbonylation product, CO, is avoided.

C-O cleavage reactions of alkanols and alkanals can be observed on surfaces of γ-Al_2O_3 powders when an ad-mixed metal catalyst (Cu/SiO_2) is used to provide hydrogenation-dehydrogenation sites. 1-butanol can be converted to butane through the dehydration/hydrogenation pathway with a 5-fold increase in turnover rates per Cu sites, when γ-Al_2O_3 is mixed with Cu/SiO_2 (at a mass ratio of γ-Al_2O_3 to Cu/SiO_2 equal to 0.05).

Bimolecular dehydration, resulting in the formation of dibutyl ether, becomes a competitive pathway over Lewis acid sites with a 60% selectivity (40% selectivity is towards butane and butenes) at 523 K and 1 MPa H_2. Kinetic studies show that the surface of γ-Al_2O_3 is covered by alkoxide, alkene and water species, and alkenes can act as intermediates for the formation of the ether product on Lewis acid sites. Increasing the hydrogen pressure to 2 MPa decreases the ether selectivity to 40%, demonstrating that selectivity towards the monomolecular and bimolecular dehydration products can be controlled by changing the alkene surface coverage through H_2 pressure or the ratio of Cu to Lewis acid sites. Dehydration turnover rates of alkanols measured on zeolites (e.g. H-FAU, H-BEA) show that the selectivity of bimolecular and monomolecular dehydration products are influenced by the size of microporous voids of these materials through solvation and diffusion effects. Metal-acid site proximity was changed by changing the ratio of metal to acid sites in zeolites containing encapsulated metal (Cu) clusters. Metal sites in close proximity to acid sites result in increased monomolecular dehydration selectivities because alkene intermediates are hydrogenated prior to secondary bimolecular reactions in the acid domain.

Bifunctional (metal along with acid sites) catalytic hydrodeoxygenation can also be utilized to convert lignin-derived phenolic monomers to cyclic and acyclic alkanes. Transition metals, such as Pt and Ru can carry out the hydrodeoxygenation of phenol, anisole and guaiacol when physically mixed with solid acids, γ-Al_2O_3 and zeolite H-BEA. Kinetic studies show that the degree of unsaturation in the cyclic products can be controlled by the H_2 pressure and while metal-acid site proximity does not have significant effects on deoxygenation turnover rates, it can alter the product selectivities when secondary acid catalyzed reactions take place at rates comparable to diffusion rates in the acid domains.

(786a) Exceptional Activity for Methane Oxidation With Catalysts Prepared By Modular Assembly of Subunits

Cargnello, M., University of Pennsylvania
Originally presented on: 11/8/2013 8:30:00 - 8:50:00

Exceptional activity for methane oxidation with catalysts prepared by modular assembly of subunits

M. Cargnello,[1,2] J. J. Delgado JaÃ©n,[3] J. C. HernÃ¡ndez Garrido,[3] K. Bakhmutsky,[4] T. Montini,[1] J. J. Calvino Gamez,[3] R. J. Gorte,[4] P. Fornasiero[1]

[1] *Department of Chemical and Pharmaceutical Sciences, ICCOM-CNR, Consortium INSTM, University of Trieste, via L. Giorgieri 1, 34127 Trieste, Italy*

Â [2] Department of Chemistry, University of Pennsylvania, 231 S. 34th Street, Philadelphia, PA 19104, USA

[3] *Departamento de Ciencia de los Materiales e IngenierÃa MetalÃºrgica y QuÃmica InorgÃ¡nica, Facultad de Ciencias, Universidad de CÃ¡diz, Campus RÃo San Pedro, 11510 Puerto Real, CÃ¡diz (Spain)*

[3] *Department of Chemical and Biomolecular Engineering, University of Pennsylvania, 311A Towne Building, 220 S. 33rdÂ Street, Philadelphia, PA 19104, USA*

There is a critical need for improved methane-oxidation catalysts to both reduce emissions of methane, a green-house gas that will soon be regulated, and significantly enhance the efficiency of power generation in gas turbines. Presently available, emissions-control catalysts are notoriously ineffective at reducing CH_4 in exhaust streams. High-temperature combustion also results in the emission of toxic nitrogen oxides (NO_x) and CO. Combustion of CH_4 promoted by heterogeneous catalysts could utilize the available energy of methane at lower temperatures, increasing efficiency and limiting emissions by drastically reducing the required temperatures. Materials for this application must also be catalytically and mechanically stable at high reaction and flame temperatures. PdO_x is recognized as one of the best catalysts for catalytic CH_4 oxidation. Unfortunately, Pd-based catalysts tend to deactivate, both due to the loss of active surface by sintering and due to the transformation into the less active but thermodynamically favored metallic Pd phase at temperatures above 600 Â°C. Although alumina is a commonly employed support, both experimental and theoretical studies reveal that ceria (CeO_2) can improve the catalytic activity of supported Pd due to the stabilization of PdO_x species. Materials that could simultaneously enhance the performance of Pd-based catalysts at low temperatures and limit deactivation at elevated temperatures would greatly improve the viability of catalytic CH_4 combustion processes.

In this contribution, we report on a novel, hierarchical design of core-shell type catalysts that is inspired by the concepts of supramolecular chemistry. Two active building blocks, Pd and CeO_2, are prepared separately, then self-assemble and organize in solution to form supramolecular core-shell structures held together by metal ion-ligand coordination chemistry. We exploit the pre-organization of the functionalized Pd@CeO_2 core-shell structures to disperse *single units* onto a modified, catalytically inert alumina carrier. Transmission Electron Microscopy (TEM) investigations demonstrate that it is indeed possible to deposit single structures where the metal-promoter interaction is maintained even after severe thermal treatments at temperatures up to 850 Â°C. The special configuration of the hierarchical catalyst gives rise to exceptionally high and stable performance for the catalytic combustion of methane with reduced amounts of Pd and ceria. The particular geometry appears to stabilize the active phase of the catalyst, not only preventing agglomeration of palladium oxide particles during the catalytic reaction but also preventing the PdO_x from being transformed to Pd at its usual transition temperature.

(786b) Shape Dependence of Oxygen Reduction Activity On Ag Nanoparticles

Van Cleve, T., University of Michigan
Linic, S., University of Michigan
Originally presented on: 11/8/2013 8:50:00 - 9:10:00

To mitigate negative economic and environmental consequences arising from our dependence on petroleum, there is a definite need to develop alternative energy systems such as low temperature fuel cells. Low temperature fuel cells generate electrical energy and water by reacting hydrogen and oxygen gases. It has been well established that the majority of the energy losses originates from the oxygen reduction reaction at the cathode. Currently, fuel cells are limited by large kinetic overpotential, limited stability with cycling, and high material cost of platinum electrodes. With recent advances in anion exchange membranes, cheaper materials have become available as potential cathodes in low temperature alkaline fuel cells.

In particular, we are investigating Ag nanoparticles as alternative electrocatalysts for oxygen reduction. Different particle geometries have different distributions of surface facets, which have distinct chemical and catalytic properties. Specifically, we are interested in examining differences in activity between nanoparticles terminated with differing proportions of 100 and 111 facets. In this work, silver nanoparticles of several geometries are synthesized using several colloidal routes prior to being deposited onto a Vulcan XC72R support. Kinetic current and electrochemical surface area measurements are obtained using linear sweep and cyclic voltammetry with a rotating disc electrode in a three electrode electrochemical cell. Silver nanospheres are shown to have slightly higher mass and specific kinetic currents compared to nanocubes. Our results are consistent with previous work on silver single crystal electrodes, which show the Ag 111 surface is more active than Ag 100.[1]

REFERENCES

1. B.B. Bliznanc, P.N. Ross, and N.M. Markovic, J Phys Chem B 110 (2006) 4735.

(786c) Photo-Enhanced Reactive Desulfurization Via Gold-ZnO Nanoparticles

Behl, M., University of Illinois at Urbana-Champaign
Jain, P. K., University of Illinois at Urbana-Champaign
Originally presented on: 11/8/2013 9:10:00 - 9:30:00

Increasingly stringent regulations on the sulfur-content of fuels require continuous development of improved sulfur-removal technology. Technologies involving high temperature adsorptive removal of organo-sulfur impurities from raw fuel sources are widely practiced. Adsorption-based technologies that rely on metal oxides often require higher temperature to accelerate the chemical conversion, which often results in sorbent volatilization, chemical reduction, loss of sorbent surface area and physical attrition. It would be highly desirable to devise a low-temperature *catalytic adsorbent*, which can provide accelerated reaction together with higher capacity of metal oxides.

Here, we show that it is possible to use visible light to achieve room temperature reactive desulfurization by using gold-zinc oxide nano-heterostructures as adsorbents. In preliminary studies, we found that, when exposed to green light Au-loaded ZnO nanoparticles achieve room temperature catalytic decomposition of model organo-sulfur compounds (thioacetamide), which is accompanied by reactive transformation of zinc oxide to zinc sulfide. Without light, the system had to be heated to 373 K to cause the desired conversion. Photo-excited hot-electrons from gold are likely responsible for this photocatalytic process. In order to shed light on this potential mechanism, we are investigating the performance of the heterostructures with different excitation wavelengths, electron/hole scavengers, and other metals. Analysis is being carried out via X-ray diffraction, mass-spectrometery, and analytical electron microscopy. We are particularly interested in probing the possible plasmonic origin of the photocatalytic effect and synergy between the metal and oxide.

(786d) Synthesis and Characterization of Metal Carbide and Nitride Nanoparticle Catalysts for Bio-Oil Deoxygenation

Ruddy, D. A., National Renewable Energy Laboratory
Wang, J., National Renewable Energy Laboratory
Schaidle, J., The University of Michigan
Blackburn, J., NREL
Hensley, J., National Renewable Energy Laboratory
Originally presented on: 11/8/2013 9:30:00 - 9:50:00

Pyrolysis bio-oil is a promising intermediate for biomass-to-fuels processes because this liquid oil holds the potential to supply diesel and jet fuels, in addition to many valuable chemical intermediates. Elemental analysis of bio-oil reveals that despite its dark brown viscous appearance, it is compositionally different from petroleum. Bio-oils contain relatively little S and N, but have a high O content, typically near 40 wt% and as high as 50 wt%. This oxygen is present in the 400 compounds across a variety of functionalities, including acids, aldehydes, esters, alcohols, ethers, ketones, phenolics, sugars, and furans. Due to these stark chemical differences from petroleum, pyrolysis oil cannot be directly used as a transportation fuel nor can it be "dropped-in" to existing petroleum refinery processes. The oil must first be upgraded, however, efficient deoxygenation has proven to be especially challenging. Therefore, deoxygenation remains a critical obstacle to enabling transportation fuel production from pyrolysis oil at the industrial scale, and there is a continuing need for new catalytic processes and chemistries. This presentation will focus on our recent advancements in the syntheses of transition metal carbides and nitrides for use as deoxygenation catalysts. Both high-temperature tube-furnace synthetic methods (>600 ^{o}C) and lower temperature solution-phase methods (<400 ^{o}C) have been employed to prepare nanoparticle carbide and nitride materials with diameters ranging from 5 – 100 nm. Notably, size control was achieved in the high-temperature synthesis of Mo_2N by varying the gas flow rate, and average particle sizes of 5 or 50 nm can be selected. The structural characterization of Mo, W, and V (and mixed-metal) carbide/nitride will be presented, including in-situ experiments to observe compositional changes under flowing reactive gases (e.g., nitride-to-carbide). The carbide/nitride active phases were dispersed on silica for characterization relevant to deoxygenation conditions. The acidity of the catalysts was probed using FTIR spectroscopy with pyridine adsorption to differentiate and quantify Lewis acid sites and Bronsted acid sites on the catalysts. Finally, hydrogen activation is an essential parameter for deoxygenation catalysis, and the hydrogen activation chemistry of these catalysts will be presented and compared.

(786e) Cu3(BTC)2 Metal-Organic-Framework As a Solid Acid Catalyst for the Production of Biodiesel

Nambo, A., University of Louisville
Carreon, M. A., University of Louisville
Originally presented on: 11/8/2013 9:50:00 - 10:10:00

Biodiesel is a fuel derived from the transesterification of fats and oils that has similar properties to that of diesel produced from crude oil and can be used directly to run existing diesel engines or as a mixture with crude oil diesel. The main advantages of using biodiesel is that it is biodegradable, can be used without modifying existing engines, and produces less harmful gas emissions such as sulfur oxide. [1]. The transesterification reaction can be catalyzed by acid or base solids, the advantages of using solid acid catalyst are: (1) they are insensitive to free fatty acids (FFA) content, (2) esterification and transesterification occurs simultaneously, (3) eliminate the washing step of biodiesel, (4) easy separation of the catalyst from the reaction medium, resulting in lower product contamination level, (5) easy regeneration and recycling of catalyst and (6) reduce corrosion problem, even with the presence of acid species. Currently, biodiesel research is focused on exploring new and sustainable solid acid catalysts for transesterification reaction.[2]

In recent years, metal-organic frameworks (MOFs) have received considerable attention as potentially valuable gas storage, sensing, separation and catalytic materials. They are composed of an organic linker, for example a dicarboxylic acid and a transition metal complex or cluster, both units being building blocks that are designed to assemble a three-dimensional open framework. In the present work a highly active and selective Lewis acid catalyst [4], copper-benzentricarboxi metal-organic-framework Cu-MOF was prepared via hydrothermal treatment and used as catalyst for the transesterification of tryglicerides to produce biodiesel.

(786f) Hydrolysis of Glucans Grafted On Inorganic Oxides: Role of OH-Defect Sites

Katz, A., University of California, Berkeley
Gazit, O., University of California, Berkeley
Originally presented on: 11/8/2013 10:10:00 - 10:30:00

We describe the first example of OH-defect-site catalysis in water, with the hydrolysis of grafted glucans on silica and alumina. Catalytic comparisons between these two systems demonstrate that the key feature that controls hydrolysis activity is OH-defect site density on the surface, both during glucan anchoring as well as during hydrolysis catalysis. The latter enforces a high local density of defect sites, which act as an array to activate glycosidic oxygens that are constrained to be near the surface for hydrolysis catalysis. This result has significant repercussions on understanding other weak-acid site catalysts for glucan hydrolysis such as unfunctionalized carbon.

(786g) A Highly Selective Sn-Doped PMO Catalyst for the Isomerization of Glucose to Fructose

Crisci, A., University of Wisconsin-Madison

Alamillo, R., University of Wisconsin-Madison

Coller, D.,

Gallo, A., University of California

Scott, S. L., University of California, Santa Barbara

Dumesic, J. A., University of Wisconsin-Madison

Originally presented on: 11/8/2013 10:30:00 - 10:50:00

Title: A Highly Selective Sn-Doped PMO Catalyst for the Isomerization of Glucose to Fructose

Anthony J. Crisci*, Ricardo Alamillo, Daniel H. Coller, Alessandro Gallo, Susannah L. Scott, and James A. Dumesic

The potential platform chemical, 5-hydroxymethylfurfural (HMF), can be produced selectively from fructose by acid-catalyzed dehydration. Fructose itself is produced by isomerization of glucose. Some efforts to convert glucose directly to HMF using a combination of a high-boiling solvent (e.g., DMSO, ionic liquids) and a metal chloride catalyst (e.g., $CrCl_2$, $PrCl_3$, $NdCl_3$, $DyCl_3$, $YbCl_3$) have resulted in high HMF yields, but product isolation remains a challenge, adding to potential production costs. To produce HMF efficiently, the catalyst(s) must isomerize glucose to fructose, then dehydrate the fructose to HMF, all in a reaction medium from which the products and unconverted reactants are easily separated.

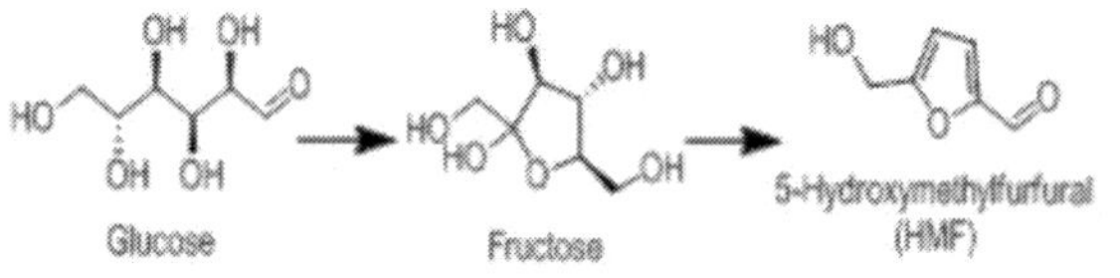

Figure 1. Reaction pathway for the direct production of HMF from glucose by first, glucose isomerization to fructose, followed by fructose dehydration to HMF.

Heterogeneous Lewis acid catalysts may provide an efficient pathway to HMF. These catalysts can be easily coupled with additional catalysts for use in a single pot reaction or in a packed bed reactor for the continuous production of HMF from glucose (to ease catalyst separation). A small number of solid Lewis acid catalysts have shown activity for glucose isomerization in an aqueous solution. Recently, Sn-β zeolite was reported to isomerize glucose to similar fructose yields as the enzyme used industrially. Sn-β zeolite has also been applied in tandem with HCl (dehydration catalyst) for the direct conversion of glucose to HMF. However, the process required a biphasic solvent system to achieve a moderate selectivity (6% single phase and 72% bi-phasic).

In addition to being difficult and time consuming to synthesize, post-synthesis functionalization of crystalline Sn-β zeolite with organosilanes (e.g., MPTMS, APTMS) is limited to silanols generated by surface defects. Similar to Sn-β, Lewis acidity can be incorporated into ordered mesoporous silicas through the addition of metal chlorides during the sol-gel synthesis. These Sn-modified (PMOs) are quick to synthesize (2 days), easily modified with additional functionalities, and potentially scalable. A series of phenylene-bridged periodic mesoporous organosilicas (PMOs) were synthesized and doped with various amounts of Sn (from ca. 0.5 to 6 wt%). A Sn-PMO (0.5 wt%) containing alkylsulfonic acid groups was also functionalized. The materials were extensively characterized (elemental analysis. XRD, TEM/STEM, N_2 physisorption, solid-state NMR, IR, EXAFS and TGA) and glucose isomerization reaction studies were conducted. Reaction studies were also performed using Sn-β zeolite, and a Sn-SBA-15 for comparison.

Each reaction was ran for 2 hours in a 2 wt% glucose solution in a mono-phasic low boiling 4:1 THF:water (w:w) solvent at 413 K. Both the Sn:glucose molar ratio (1:100) and the total catalyst mass (diluted with Sn-free PMO) were held constant. PMO with isolated Sn sites (0.4 wt%) converted 28 % of the glucose to fructose with a selectivity of 47 %. However, as the Sn-loading increased, both glucose conversion and selectivity to fructose decreased, compared to the 0.4 wt% Sn-PMO. Sn-β zeolite under these conditions fully converted glucose, but did not yield any fructose and possessed a low selectivity to HMF (7%). The reaction time was reduced to compare results at the same conversion levels to that of the Sn-PMOs. Between 80 and 90% glucose conversion, the selectivity to fructose is 30-50%. Once glucose is completely consumed, the desired products (fructose and HMF) react and the corresponding selectivity drops significantly and rapidly. In contrast, the 0.4 % Sn-PMO maintains its selectivity to desired products around 40% even at higher conversions (> 85%). Sn-PMOs are capable of selectively isomerizing glucose to fructose in THF/Water without rapidly consuming glucose. The bifunctional PMO (with Lewis and Brønsted acid sites) at 30 % conversion had a selectivity to fructose and HMF of 27 and 14 %, respectively. With double the reaction time, the HMF selectivity improved to 37 % while fructose selectivity decreased to 4 % at 52 % conversion. HMF selectivity is expected to improve with the optimization of the Lewis: Brønsted acid ratio and the reaction conditions. Sn-PMOs are capable of selectively isomerizing glucose to fructose in THF/Water without rapidly consuming glucose. Through the incorporation of alkylsulfonic acid sites, reasonable selectivity to HMF from glucose is achieved without the use of costly and/or separation unfriendly solvents.

(797a) Invited Talk: Thermodynamics of Nanoparticle-Polymer Melt Hybrids

Kumar, S., Columbia University

Originally presented on: 11/8/2013 8:30:00 - 9:00:00

While the miscibility and spatial dispersion of nanoparticles (NPs) in a polymer melt critically affects the properties of the resulting nanocomposite, little simulation work exists on understanding this critical issue. We examine a recent experimentally motivated conjecture by Mackay that miscibility between NPs and polymer melts is achieved when the chains are larger than the nanoparticles using isothermal-isobaric ensemble simulations. In the case of systems with attractive interactions we show that polymer chain length only plays a secondary role, probably because the entropic, depletion-induced inter-NP attractions only occur over length scales comparable to the correlation length in the melt, namely the segment size. We therefore consider athermal systems and show indeed that there is a fascinating role played by polymer chain length relative to the NP size in determining "miscibility" and NP self-assembly. The consequences of these results on experimental results will be the final piece of this story that we shall discuss.

(797b) Thermodynamics of Block Copolymer Micellization in Ionic Liquid Solvents

Hoarfrost, M. L., University of California, Berkeley
Lodge, T. P., University of Minnesota

Originally presented on: 11/8/2013 9:00:00 - 9:20:00

Block copolymers can be designed to self-assemble in solution into micelles with complex shapes and functionality, which are useful for a variety of applications in many industries. However, our understanding of the thermodynamics of block copolymer micellization has been limited by the long experimental time scales for micelle equilibration, which stem from the large energy barrier for extracting long solvophobic copolymer segments from micelle cores into solution. This limitation is exacerbated by the limited temperature ranges of water and volatile organic solvents traditionally used in micelle solutions. In this work, we study the phase behavior of block copolymer micelles in ionic liquid solvents. The negligible volatility of ionic liquids allows for thermal annealing and provides an opportunity to study micelle phase behavior over a much larger temperature window than previously possible. Furthermore, the solvent quality can be precisely tuned by mixing ionic liquids. We first characterize the solvent quality of ionic liquid mixtures for the solvophobic component of a block copolymer using light and neutron scattering. We then show how micelle phase boundaries (i.e. critical micelle temperature and concentration) depend on solvent quality and block copolymer degree of polymerization.

(797c) Phase Behavior of Tapered Diblock Copolymers From Self-Consistent Field Theory

Brown, J. R.,
Sides, S. W., Tech-X Research Corporation
Hall, L. M., Sandia National Laboratories
Originally presented on: 11/8/2013 9:20:00 - 9:40:00

It is well known that polymers composed of two connected segments/blocks of chemically different monomers A and B can microphase separate into various ordered structures. The fraction of A monomers and the quantity χN together determine whether phase separation is preferred and the details of the resulting morphology, where N is the polymer length and χ is the Flory parameter related to how unfavorable the AB interactions are. Thus, it's inherently difficult to independently tune material properties such as the bulk modulus (with increases with N) and the microphase separated state (related to χN). This is unfortunate specifically in the search for high modulus materials with a bicontinuous network structure such as the double gyroid phase, which is generally less favorable at high χN. Such materials are of interest in applications such as solid battery electrolytes where penetrant transport occurs through one continuous phase while the other phase provides mechanical strength.

An attractive way to get around this problem is to use tapered block copolymers. These are composed of a pure A block, a linear gradient "block" whose composition varies from fully A to fully B (tapered) or B to A (inverse-tapered), and a pure B block. The resulting copolymer then contains a new parameter (the fraction of gradient/tapered block) that can be used to tune the microphase separated state independent of N. To determine how this strategy can best be employed, we compute phase diagrams of tapered and inverse-tapered diblock copolymers via self-consistent field theory. Their composition is approximated using a multi-block model in which the tapered region consists of alternating A and B blocks of appropriate lengths to approximate the gradient. Phase diagrams were produced for varying sizes of the tapered region, and show a shift of the ordered phases to higher χN for larger tapered regions (and an even greater shift for inverse-tapered systems). Interestingly, the typically narrow gyroid region widens for direct tapered systems, by about a factor of two at high χN.

(797d) Understanding Fluctuation/Correlation Effects in Diblock Copolymer Melts With a Density-Functional Theory

Zong, J., Colorado State University
Yang, D., Colorado State University
Wang, Q., Colorado State University
Originally presented on: 11/8/2013 9:40:00 - 10:00:00

With the polymer self-consistent field theory well developed and widely applied to many polymeric systems, it is both natural and timely to develop advanced theories that can quantitatively account for the fluctuations/correlations neglected by this mean-field theory. Here we formulate a density-functional theory for a model diblock copolymer system of discrete Gaussian chains interacting with soft, finite-range repulsions as commonly used in dissipative-particle dynamics simulations. Direct comparisons among the self-consistent field theory, the density-functional theory, and the fast off-lattice Monte Carlo simulations[1], all based on exactly the same model system and thus without any parameter-fitting, then unambiguously and quantitatively reveal the fluctuation/correlation effects on both the disordered and ordered phases, as well as the order-disorder transition, of diblock copolymer melts.

[1] Q. Wang and Y. Yin, J. Chem. Phys. 130, 104903 (2009).

(797e) Free Energy of Block Copolymer Systems Via Thermodynamic Integration of a Mesoscale Block-Copolymer Model

Peters, A., Georgia Institute of Technology
Lawson, R., Georgia Institute of Technology
Nation, B., Georgia Institute of Technology
Ludovice, P. J., Georgia Institute of Technology
Henderson, C., Georgia Institute of Technology
Originally presented on: 11/8/2013 10:00:00 - 10:20:00

Directed self-assembly (DSA) of block copolymers (BCPs) is a promising technique for producing sub-30 nm pitch regular patterns, and the development of these DSA techniques could benefit greatly from computer simulation of such methods. Unlike other models, molecular dynamics (MD) combined with realistic potentials for polymer behavior can potentially provide more accurate simulations of the inherent polymer behavior, dynamics, and equilibrium states without a need to guess modes of molecular movement and without oversimplifying interatomic interactions.

Through the use of thermodynamic integration, this model allows us to calculate free energy differences between various states, provided a reversible path can be envisioned between them. Using this technique, various equilibrium and metastable states can be compared. Defect free energies and subsequently defect densities can be calculated as functions of polymer and underlayer properties. Defectivity is the primary hurdle for implementation of BCP-DSA into semiconductor fabrication and defect free energies are critical to guide defectivity reduction techniques and patterning techniques. Free energy maps as a function of system properties like pitch, and polymer parameters can also be calculated. These calculations can provide insight into ideal and practical best case scenarios for both polymer properties and system guiding properties, whereas vast number of experiments would have to be run to achieve the same conclusions.

(797f) Understanding Retention of Block Copolymers in Size Exclusion Chromatography and Liquid Chromatography At the Critical Condition

Struk, K.,
Schure, M. R.,
Siepmann, J. I., University of Minnesota
Originally presented on: 11/8/2013 10:20:00 - 10:40:00

Chromatographic separations involve a delicate interplay of enthalpic and entropic factors arising from interactions of the analyte molecules with the stationary and mobile phases. Particle-based simulations allow one to probe chromatographic systems on the molecular level and to provide detailed information on structural characteristics and retention mechanisms. This talk will highlight recent applications of configurational-bias Monte Carlo simulations in the Gibbs ensemble to understand the retention mechanisms of block copolymers in size exclusion chromatography (SEC) and in liquid chromatography at the critical condition (LCCC) for one of the blocks. Copolymers with up to three block types and up to five blocks are investigated. It is found that sequence and architecture play a major role in determining retention order of copolymers with identical composition. Structural analysis of the simulation trajectories provides and explanation for the retention order.

(797g) Thermodynamic Analysis of the Solubility of Gases and Vapors in Swelling Glassy Polymers: A Predictive Modeling Approach

Minelli, M., Alma Mater Studiorum - Universit di Bologna
Doghieri, F., Universit di Bologna
Originally presented on: 11/8/2013 10:40:00 - 11:00:00

The determination of the solubility of gases and vapors in polymeric systems is of remarkable interest for many different applications, ranging from the fabrication of gas separation membranes to the development of barrier materials for packaging or sensors.

Equation of state approaches have been developed in the past following different routes, which provide reliable means for the calculation of the solubility of gases and vapors in rubbers or polymer melts. On the other hand, no comparable tools are available for the case of glassy polymers. The general approach called Non Equilibrium Thermodynamics of Glassy Polymers (NET-GP) coupled to Lattice Fluid (LF) equation of state was shown to be able to predict the solubility of gases and vapors in glassy polymers at various pressures based on pure component parameters. The corresponding NELF model relies on the use of polymer mass density as an internal state variable for the description of the out-of-equilibrium conditions for the penetrant-polymer system. The model application requires some information on the volumetric behavior of the polymer in sorption conditions, which might be a serious limitation to its predictive ability.

In the present work, a simple and effective rheological model has been considered in order to evaluate *a priori* the swelling induced by the penetrating species. The main assumption is in considering that the polymer relaxation occurs involving two characteristic times on fairly different scales. If temperature is appreciably below the glass transition of the polymer, only the short-time relaxation phenomena are involved during sorption within the usual experimental times.

This model allows the estimation of the polymer swelling at increasing penetrant activity and for the determination of the complete sorption isotherm by the NELF model, in a purely predictive way.

This tool was satisfactorily applied for many sorption isotherms of swelling gases (such as CO_2 or C_2H_4) and organic vapors (xylene, toluene, or hydrocarbons) in a variety of glassy polymers such as polystyrene, polycarbonate, polysulfone or amorphous Teflon. The model was also able to capture complex behaviors such as those given by non linear trends of dilation or when a change in concavity of the sorption isotherm is observed due to a penetrant-induced glass transitions of the polymer. The experimental dilation isotherms in sorption conditions, for those systems for which they were available, confirmed the reliability of the present model.

Furthermore, the model was employed to describe the solubility behavior of CO_2 in glassy polymers exposed to different pretreatment. The effect of thermal annealing, or the exposure of the polymer to a swelling agent, on the polymer sorptive capacity has been also analyzed. The present model was thus able to describe such differences simply by considering different initial polymer density, which is indeed assumed in the model as the only order parameter.

In conclusion, the NELF model, endowed with a rheological tool for the swelling prediction, is able to describe the solubility of gases and vapors in glassy polymer systems predictively. The modeling results, as obtained from a variety of different polymer/penetrant systems in different conditions, are compared with the experimental data retrieved from the literature showing the remarkable accuracy of this approach.

(808a) CO2 Capture By Gas Hydrate Formation: Combined Effect of Thermodynamic and Kinetic Promoters

Daraboina, N., Denmark Technical University
von Solms, N., Technical University of Denmark (DTU)
Originally presented on: 11/8/2013 12:30:00 - 12:55:00

Carbon capture by gas hydrate crystallization is a promising technology currently being considered in the battle to mitigate climate change. However, the requirement of both high pressure and slow formation rates are serious limitations which need to be resolved prior to implementing this technology at demonstration scale. Various additives (e.g.: THF, TBAB, cyclopentane and others) are known to promote hydrate formation thermodynamically by reducing the hydrate formation pressure. Furthermore there are also kinetic promoters such as surfactants which can increase the hydrate crystallization rate based on the concentration. The appropriate combination of thermodynamic and kinetic promoters has the potential to move the technology towards practical application.. In this work, the hydrate-based process for carbon dioxide capture was tested in a rocking cell (RC-5; Systemtechnik, Germany) apparatus. Various combinations of surfactant (kinetic promoters: SDS and CTAB) and THF (thermodynamic promoter) in liquid water were assessed in order to find the optimum operating conditions.

(808b) CO2/Brine/Rock Interactions Under CO2 Sequestration Conditions

Soong, Y., National Energy Technology Laboratory (NETL), Office of Research and Development, Department of Energy

Howard, B. H., National Energy Technology Laboratory

Hedges, S. W., U.S. Department of Energy

McIntyre, D., National Energy Technology Laboratory

Warzinski, R. P., U.S. Department of Energy, National Energy Technology Laboratory

Irdi, G.,

Haljasmaa, I.,

Crandall, D., National Energy Technology Laboratory (NETL), Offi¬ce of Research and Development, Department of Energy

Originally presented on: 11/8/2013 12:55:00 - 13:20:00

CO_2/brine/rock interactions under CO_2 sequestration conditions

Y. Soong*, B. H. Howard, S. W. Hedges, D. McIntyre, R. Warzinski, G. Irdi, I. Haljasmaa[1], and D. Crandall[1]

U.S. DOE, National Energy Technology Laboratory,

P.O. Box 10940, Pittsburgh, PA 15236

[1] URS Co.

It has indicated by Intergovernmental Panel on Climate Change Study that injection of CO_2 into confined geological formations, given their potentially massive carbon storage capacity and widespread geographic distribution, represents one of the most promising options for mitigating anthropogenically-induced global climate change. The Mount Simon sandstone in the Midwest region of the U.S is recognized as a promising candidate host reservoir for carbon sequestration. The Mount Simon is a deep saline aquifer and is a primary reservoir target for large scale carbon dioxide injection tests due to its proximity to CO_2 sources, favorable depth, thickness, permeability, porosity, and the presence of the overlying Eau Claire Formation as a seal. The Midwest Geological Sequestration Consortium (MGSC) has selected the Mount Simon as the reservoir at its major demonstration project. While the Mount Simon Sandstone has ideal reservoir characteristics in some areas, there are significant variations in porosity, permeability, and mineralogy.

Numerous studies have been conducted to investigate changes in host rock properties when exposed to CO_2. Luquot and Gouze conducted CO_2-enriched fluid flow through a carbonate core sample under in situ sequestration conditions (T = 100 °C and P = 12 MPa). Under these conditions, a decrease in permeability and porosity is seen that is linked to precipitation of Mg-rich calcite. Gouze and Luquot also conducted an X-ray microtomography characterization study of porosity, permeability and reactive surface changes during the dissolution of pure calcite in a brine-CO_2 mixture. It showed that the increase in permeability is due to the decrease of the tortuosity for homogeneous dissolution, whereas it is due to the combination of tortuosity decrease and hydraulic radius increase for heterogeneous dissolution. Jove Colon et al., studied the permeability, porosity and reactive surface area evolution during dissolution of nonfractured, clay-free Fontainebleau sandstone cores using a flow-through percolation reactor. The experiments were performed at 80°C with 0.1 M NaOH solution. The permeability was found to vary depending on where the dissolution occurred: in the pore void or the pore throat. Liu et al., conducted a simulation study of CO_2 sequestration in the Mount Simon sandstone via TOUGHREACT. They indicated that a strongly acidified zone (pH 3-5) forms in the areas affected by the injected CO_2 and consequently causes extensive secondary mineral precipitation (calcite, magnesite, ankerite, alunite and anhydrite) and dissolution of feldspars. Thus, the injection of CO_2 into a deep saline aquifer affects porosity since it can cause both mineral dissolution and precipitation in the formation. To predict if there will be any changes to the reservoir during sequestration, it is critical to understand the behavior of the authigenic minerals that line pore spaces when they are in contact with the CO_2 saturated brine.

An experimental study was conducted to assess the impacts of injected CO_2. Core samples from the Mount Simon formation were placed in CO_2 saturated brines at sequestration temperature of 85 °C and pressure of 23.8 MPa in the presence of synthetic Illinois Basin brine for a period of six months in a set of 1.3 liter high-pressure vessels (17-4PH-1150 stainless steel, 10.2 cm I.D. x 16.5 cm depth, manufactured by Thar Technologies, Inc.). One sandstone sample with a porosity of 7.9% was obtained at depth of 1,769.7 m from a well located in Vermillion County, IN. The other sample with a porosity of 1.4% was collected at depth of 2634.2 m from a well located in Knox County, IN. The major mineral composition for the first sample is 78% quartz, 15% Feldspar and with less than 2% microcline and trace minerals. The primary mineral composition of the second sample obtained from Knox County is 70% quartz, 22% Feldspar, and around 4% illite/muscovite. Micro CT, CT, XRD, SEM, petrography, porosity, and permeability analyses were conducted prior to the six month exposure. Upon completion of the experiment, the analyses were repeated.

The permeability of the Vermillion sandstone after the six month exposure to CO_2 and brine was 0.85 mD compared to 1.6 mD for the fresh sample. On the other hand, the permeability of Knox sandstone was 69 nD compared to 50 nD for the fresh sample. The results suggest that the exposure altered the pore characteristics and thus the permeability of the sample. Micro CT analysis as well as SEM backscattered images were used to probe the Vermillion and Knox Co. sandstone before and after the six month exposure to the CO_2/brine environment. It can be seen that mineral dissolution and mineral precipitation occurred after the samples were exposed to CO_2/brine for six months. Mineral precipitation occurred in the Vermillion sample pores and cracks blocking flow and resulting within the observed 50% decrease in permeability. For the case of Knox sandstone, the mineral dissolution occurred in pore though the precipitation was observed in the exterior of the sample. This might result in increasing in permeability. In addition, from the analyses of the reacted brines within the reactor, it indicated the possibly of Feldspar dissolution.

When CO_2 is injected into a saline formation, acidic conditions form in the brine in the nearby portion of the reservoir thus reducing the pH of the brine in the vicinity of the injection system. Minor minerals in the sandstone, such as K-feldspar, calcite, kaolinite, microcline etc., that are stable in the native environment but not stable at lower pH can dissolve. Eventually, the increase of cation concentration produced by mineral dissolution can supersaturate the fluids with respect to other minerals. If this occurs, precipitation of new minerals can result in a change of porosity and permeability which could affect both CO_2 storage capacity and fluid movement. The issue is further complicated by the evolution of pore geometry due to these geochemical reactions that may affect the overall hydrologic properties of the host rock and therefore the performance of the unit as an effective reservoir.

The decreasing permeability found in Vermillion sandstone was probably due to Feldspar dissolution, migration and precipitation altering the sandstone pore/crack structure. For the case of sandstone from Knox county, the permeability increased by 38%. It might be due to the Feldspar dissolution in the pore and mineral precipitation at exterior surface of the sandstone. Investigations are continuing to more clearly elucidate the mechanisms and geochemistry of this system, especially at the pore structure scale.

(808c) Novel CO2 Capture and Carbonation Process With Waste Concrete

Shuto, D.,
Yamasaki, A., Seikei University
Hongo, T., Seikei University
Iizuka, A., Tohoku University
Originally presented on: 11/8/2013 13:20:00 - 13:45:00

A novel CO_2 capture and carbonation process is developed. In this process, wastes containing calcium, such as waste cement particles or steel slug, are used as a calcium source for carbonation reaction with CO_2 emitted from point sources. The process is composed of several steps. CO_2 in the flue gas is captured through absorption process with alkaline solution such as sodium hydroxide to form sodium carbonate and sodium bicarbonate. The calcium content in the waste cement particles is extracted with acid such as nitric acid or acetic acid to form solution containing calcium. Mixing of these two solutions leads to the precipitation of calcium carbonate: CO_2 fixation via carbonation reaction. Electrodialysis is applied for the regeneration of acid and alkaline from the remained solution after filtration of the precipitant. Thus, the net input of the process is CO_2 in the flue gas and waste cement particles, and the output is calcium carbonate. The process feasibility was examined through laboratory -scale experimental studies under various operation conditions. Based on the experimental results, the process was optimized and evaluated in terms of the power consumption and cost for the CO_2 recovery, and compared with conventional processes. Net CO_2 sequestration can be achieved by the process under certain conditions.

(808d) Direct Capture and Conversion of CO2 From Flue Gas Via Ex-Situ Mineral Carbonation Using Chemical and Biological Catalysts

Swanson, E. J., Columbia University
Brady, P. V., Sandia National Laboratories
Park, A. H. A., Columbia University
Originally presented on: 11/8/2013 13:45:00 - 14:10:00

Converting carbon dioxide from flue gas directly into a solid magnesium carbonate presents a compelling option for the mitigation of power plant emissions. The carbonate product is environmentally benign and thermodynamically stable, thus creating a safe and reliable alternative to geological storage. However, given the low partial pressure of carbon dioxide in flue gas and the low reactivity of magnesium bearing minerals, achieving the kinetics required to make such a process feasible is challenging. In addition, magnesium carbonates formed at low temperature tend to incorporate significant waters of hydration, which increases transportation requirements and limits utilization of by-products. Each of these issues was addressed in the design of a multistep process for the conversion of magnesium silicates to carbonates as a carbon capture and storage option. The mineral dissolution reaction is accelerated using a mixture of organic chelating agents to produce a concentrated magnesium solution. This acceleration appears to occur through a stabilization effect on the pores formed as the silicate dissolves. The concentrated magnesium solution produced during the dissolution of the silicate is used to drive the precipitation of carbonates directly from flue gas. Through strict control over the chemistry of a simultaneous absorption and precipitation process, the phase of magnesium carbonate formed can be controlled. In addition, the mass transfer rate of carbon dioxide from flue gas into the precipitating liquor is accelerated using biological catalysis, which is shown to provide significant savings in process equipment size and operating parameters. Finally, the process as a whole is demonstrated to have a negative overall carbon balance through life cycle assessment.

(808e) A Green Approach to Mitigate CO2 Using Mixed Microbial Culture

Raghuvanshi, S., Birla Institute of Technology & Science, Pilani
Gupta, S., Birla Institute of Technology & Science, Pilani
Mishra, S.,
Originally presented on: 11/8/2013 14:10:00 - 14:35:00

Carbon dioxide is one of the GHG's (Green House Gases) & responsible for Global warming.. As of July 2011, CO_2 in the earth's atmosphere is at a concentration of 392 ppm by volume which is 35% higher as compared to the mid-1800s. Carbon dioxide, which is estimated to contribute 50-60% of the Global warming in the last century, is majorly released during the combustion of fossil fuel. The CO_2 concentration in earth atmosphere can be reduced by biological, chemical and technological techniques available through either mitigating or sequestering emissions. In the recent years, biological route has been identified as one of the most viable options available for the mitigation of CO_2. There are six carbon fixing pathway's. Out of which, the Calvin cycle (Calvin-Benson-Bassham) is used by photoautotroph's to fix CO_2. There are microbes which are ubiquitous in nature and thrive in extreme environment classified as chemoautotrops. They also classified as Extremophiles. The requirement of industries is to develop a economic, efficient and effective system which not only fix CO_2 but also capable to withstand in the severe industrial environment. Mixed culture is proved to be a superior candidate, as the consortium of different engineered species develop a symbiotic relationship which helps each other not only to survive in harsh conditions but also curb down CO_2. The another possibilities are simultaneous waste water treatment, production of Methane, and bio-hydrogen production. The present work deals in identification of microbes present in sludge of waste water treatment plant having CO_2 fixation capabilities. It also gives the insight of the relation which exists between natural selected mixed culture and to understand the capabilities of CO_2 fixation. The batch study has been conducted to measure the capabilities of mixed culture to mitigate CO_2 for five different V/V % concentration of CO_2 4, 12, 16, 22 and 25. It was observed that CO_2 is completed consumed by microorganisms in 4 days. Mitigation rate for different concentration of CO_2, bio-mass growth rate and resulting product has been analyzed on per hour basis and per day basis. Mixed culture is found to be superior candidate than photoautotrophic organisms for CO_2 mitigation. They show a promising future to develop a low cost, compact size closed system for *in-situ* CO_2 mitigation.

AUTHOR INDEX

AUTHOR INDEX

AUTHOR INDEX

AUTHOR INDEX

AUTHOR INDEX

AUTHOR INDEX

AUTHOR INDEX

AUTHOR INDEX

AUTHOR INDEX

AUTHOR INDEX

AUTHOR INDEX

AUTHOR INDEX

AUTHOR INDEX

AUTHOR INDEX

AUTHOR INDEX

AUTHOR INDEX

AUTHOR INDEX

AUTHOR INDEX

AUTHOR INDEX

AUTHOR INDEX

AUTHOR INDEX

AUTHOR INDEX

AUTHOR INDEX

AUTHOR INDEX